41st Riso International Symposium on Materials Science 2020

Materials and Design for Next Generation Wind Turbine Blades

Held online due to COVID-19

IOP Conference Series: Materials Science and Engineering
Volume 942

Roskilde, Denmark
7 - 10 September 2020

ISBN: 978-1-7138-3195-2
ISSN: 1757-8981

Printed from e-media with permission by:

Curran Associates, Inc.
57 Morehouse Lane
Red Hook, NY 12571

CURRAN ASSOCIATES INC. proceedings.com

Some format issues inherent in the e-media version may also appear in this print version.

Institute of Physics
Dirac House, Temple Back
Bristol BS1 6BE UK

Phone: 44 1 17 929 7481
Fax: 44 1 17 920 0979

techtracking@iop.org

Additional copies of this publication are available from:

Curran Associates, Inc.
57 Morehouse Lane
Red Hook, NY 12571 USA
Phone: 845-758-0400
Fax: 845-758-2633
Email: curran@proceedings.com
Web: www.proceedings.com

TABLE OF CONTENTS

INVITED PAPERS

CONTRIBUTED PAPERS

IOP Conf. Series: Materials Science and Engineering **942** (2020) 011001 doi:10.1088/1757-899X/942/1/011001

4 September 2020

Preface for the 41st Risø Symposium Proceedings, IOP publication

PREFACE

Proceedings of the 41st Risø International Symposium on Materials Science: *Materials and Design for Next Generation Wind Turbine Blades*

L P Mikkelsen, B Madsen, S Fæster

The 41st Risø Symposium aims to address new requirements for materials and design for next generation wind turbine blades. Over the last 40 years, wind turbine blades have grown an order of magnitude. Today, the longest blades are exceeding 100 meters and weighing 50 tons. Because of this growth in blade size, the cost of wind energy can now compete with fossil-based energy sources on market terms. Together with a wish in society for a zero-emission future, the wind energy sector is foreseen to further expand. This will require new solutions and approaches for next generation wind turbine blades, such as improved mechanical performance and structural reliability, coupled to more sustainable solutions including end-of-life considerations.

These new requirements for next generation wind turbine blades will be addressed at the symposium by covering the themes of manufacturing, mechanical properties and performance, experimental characterization and testing, polymer matrices and coatings, and sustainability, life-time extension and end-of-life.

The Proceedings contain 5 invited papers, and 40 contributed papers. The 41st Risø International Symposium is organised by the Division of Materials and Components, Department of Wind Energy, Technical University of Denmark (DTU), at the Risø Campus. We would like to thank all those at DTU Wind Energy who assisted in the preparation of the Symposium. We appreciate additionally the support from the international advisory committee consisting of: Kristofer Gamstedt, SE; Roberts Joffe, SE; Hiroyuki Kawada, JP; Marino Quaresimin, IT; Michael Sutcliffe, UK; Ramesh Talreja, US; Michael Wisnom, UK.

We gratefully acknowledge financial support from the following foundations: Fabrikant Mads Clausens Fond, Knud Højgaards Fond, Otto Mønsteds Fond and Civilingeniør Frederik Leth Christiansens Almennyttige Fond.

The Risø International Symposium on Materials Science has been organized on-site Risø Campus for the last 40 years. The Covid-19 and travel restrictions, forced us for the first time to organize this symposium as a virtual meeting. More than 100 participants have registered for the symposium, participating from more than 17 different countries. The Virtual meeting run over 4 days, 7-10th of September, organized as 15-25 minutes Zoom Webinar presentations followed by parallel Zoom breakout rooms for each of the presenting papers. During the symposium, we will ensure to maintain the Risoe Symposium as a high-quality conference, well organized and with high-level scientific presentations based on peer-review papers, and still with the possibility for the participants to discuss and interact with each other, but now using conference online tools.

IOP Conf. Series: Materials Science and Engineering **942** (2020) 011002 doi:10.1088/1757-899X/942/1/011002

Peer review declaration

All papers published in this volume of IOP Conference Series: Materials Science and Engineering have been peer reviewed through processes administered by the Editors. Reviews were conducted by expert referees to the professional and scientific standards expected of a proceedings journal published by IOP Publishing.

- **Type of peer review:**
  - **Single-blind of reviewer 1: the handling editor selected the reviewer and the identities was kept hidden from the authors.**
  - **Open of reviewer 2: the second reviewer was mainly taken care of the format of the paper and was known to the authors.**

- **Conference submission management system:**
  - **Symposium email (symp41@windenergy.dtu.dk, and an excel-sheet organized by the handling editors**

- **Number of submissions received:**
  - **45**

- **Number of submissions sent for review:**
  - **45**

- **Number of submissions accepted:**
  - **45: 30: revision needed, 8: revision optional, 7: Accepted as is**

- **Acceptance Rate (Number of Submissions Accepted / Number of Submissions Received X 100): 100%**

- **Average number of reviews per paper: 2**

- **Total number of reviewers involved: 17**

- **Any additional info on review process**

- **Contact person for queries: Lars P. Mikkelsen: lapm@dtu.dk**

IOP Conf. Series: Materials Science and Engineering **942** (2020) 012001 doi:10.1088/1757-899X/942/1/012001

A Critical Review of Damage and Failure of Composite Wind Turbine Blade Structures

Xiao Chen* **and Martin A. Eder**

Department of Wind Energy, Technical University of Denmark, Frederiksborgvej 399, 4000 Roskilde, Denmark

*xiac@dtu.dk

Abstract. Advancing beyond the historic feasibility limits deemed by designers, wind turbine rotor blades have eventually surpassed the 100m milestone, making them reside among the largest single components in the world made of fiber composite materials. In the development of wind energy, aerodynamics has been an essential field of research, just as it will still be, the recent two decades witnesses structures becoming more critical due to the ever-increasing size of the rotor blades. The fact is simple: for wind turbines to operate continuously and cost-efficiently, rotor blades must maintain their structural integrity and reliability. This study reviews recent advances in the field of the structural integrity of large scale composite wind turbine blades. A particular focus is placed on damage and failure in such structures from the triad of field observations, laboratory experiments and numerical modeling. This study also identifies the established knowledge, the latest achievements, the topical research fields and the current challenges for future research and development.

1. Introduction

Wind turbine blades are surprisingly simple structures from a construction point of view, c.f. Figure 1. They are either manufactured by adhesively joining a few structural sub-components, i.e., aerodynamic shells, spar caps and shear webs, or by virtue of an infusion process as a single component in one shot. Four primary materials are commonly used to build wind turbine blades: fiber fabrics (glass, carbon), resin (epoxy, polyester), sandwich core material (e.g. PVC foam, Balsa wood), and adhesives (e.g. epoxy, polyurethane). The blades are manufactured by a vacuum-assisted resin infusion process, unavoidably introducing defects of different types and sizes distributed at different locations. Before the blade is cleared to leave the factory, nondestructive testing (NDT) is carried out to detect major defects located in critical regions such as spar caps and trailing edge bondlines, which are typical for the prevailing production process. During their designated 20 to 25 years of operation in the field, the blades are expected to withstand high cycle fatigue and environmental degradation. Already born with various manufacturing defects and imperfections, (typically in the form of small air bubbles any ply waviness in the laminates) some of these remain latent while others act as damage initiators potentially leading to catastrophic structural failure.

The manufacturing-induced defects are only one of many other additional culprits responsible for the damage and failure of blades: operational loads, environmental effects (e.g. erosion, thermal cycles, lightning), structural design, material properties, manufacturing processes and their interplay determine the destiny of a blade in terms of how, why and possibly when it fails c.f. Figure 2. These circumstances indeed pose a highly complex and multidimensional problem whose severity is soaring with the ever-increasing demand for longer blades that compel a multi-disciplinary approach in order to fulfill the reliability requirements. The considerable research effort of the past two decades has provided deep insights and a broad understanding of damage mechanisms and fracture processes in composite materials and adhesive bondlines.

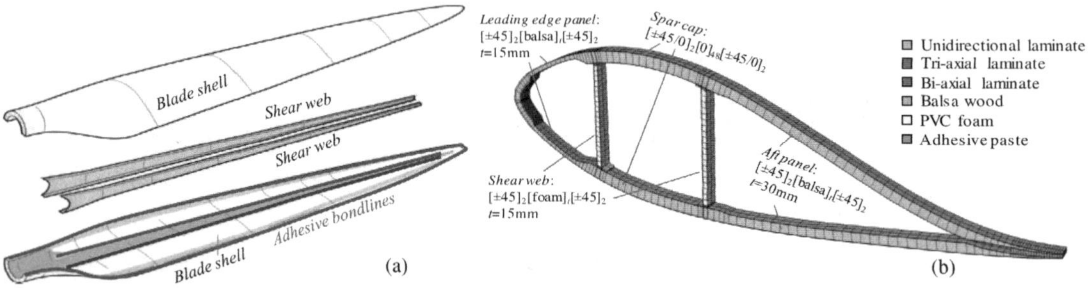

Figure 1. A typical structural construction of wind turbine blade. (a)A sketch showing a blade is assembled by adhesively joining four parts, adapted from [1]. (b) A cross section of a commercially available large utility wind turbine rotor blade [2]. Four adhesive bondlines are used to join two shear webs with a pair of spar caps. Two adhesive bondlines are used to join two lift generating shells at the leading edge (left end) and the trailing edge (right end). Two spar caps comprise a considerable amount of unidirectional laminates responsible for carrying bending moment in the flapwise direction (i.e. bending around the minor axis). Sandwich constructions are used to build the complex geometry of the aerodynamic shells that need to be lightweight and simultaneously resistant towards buckling.

Building on this bedrock on a microscopic material level, a variety of advanced experimental and numerical analysis methods evolved with the aim to more closely approach realistic test scenarios and to predict failure more accurately on an intermediate and large scale. Despite the enhanced methodology, the fierce time-to-market requirements put a strong emphasis on the development of more efficient experimental and numerical methods in terms of full-scale testing time and computation time. Both, the improved fidelity as well as the gain in efficiency facilitate the development of longer blades with a high strength-to-mass utilization ratio, which eventually leads to more cost-efficient blades for the wind energy industry.

This study is dedicated to reviewing the most recent research advances in the structural integrity of large scale composite wind turbine blades. A particular focus is placed on damage and failure of such structures from the perspective of field observation, laboratory experiments and numerical modeling. Also identified in this study are the state-of-the-art knowledge, the latest achievements, the current understanding, the on-going topics and the remaining challenges associated with this research field for future work. An overview of the research status is shown in Figure 3 and consequently discussed in detail in the following sections.

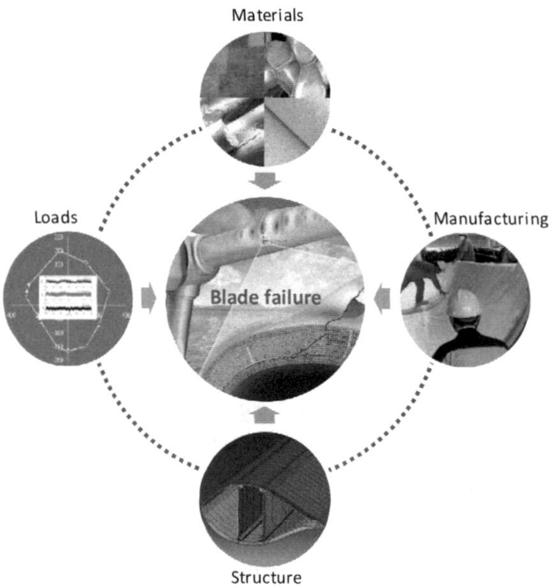

Figure 2. Interactive aspects affecting rotor blade failure [3]. The interplay among material properties, manufacturing processes, structural designs and operational loads (and environmental conditions) are essential to be considered in their collectivity if the structural integrity of composite rotor blades should live up to the designated service life of 20 to 25 years.

IOP Conf. Series: Materials Science and Engineering **942** (2020) 012001 doi:10.1088/1757-899X/942/1/012001

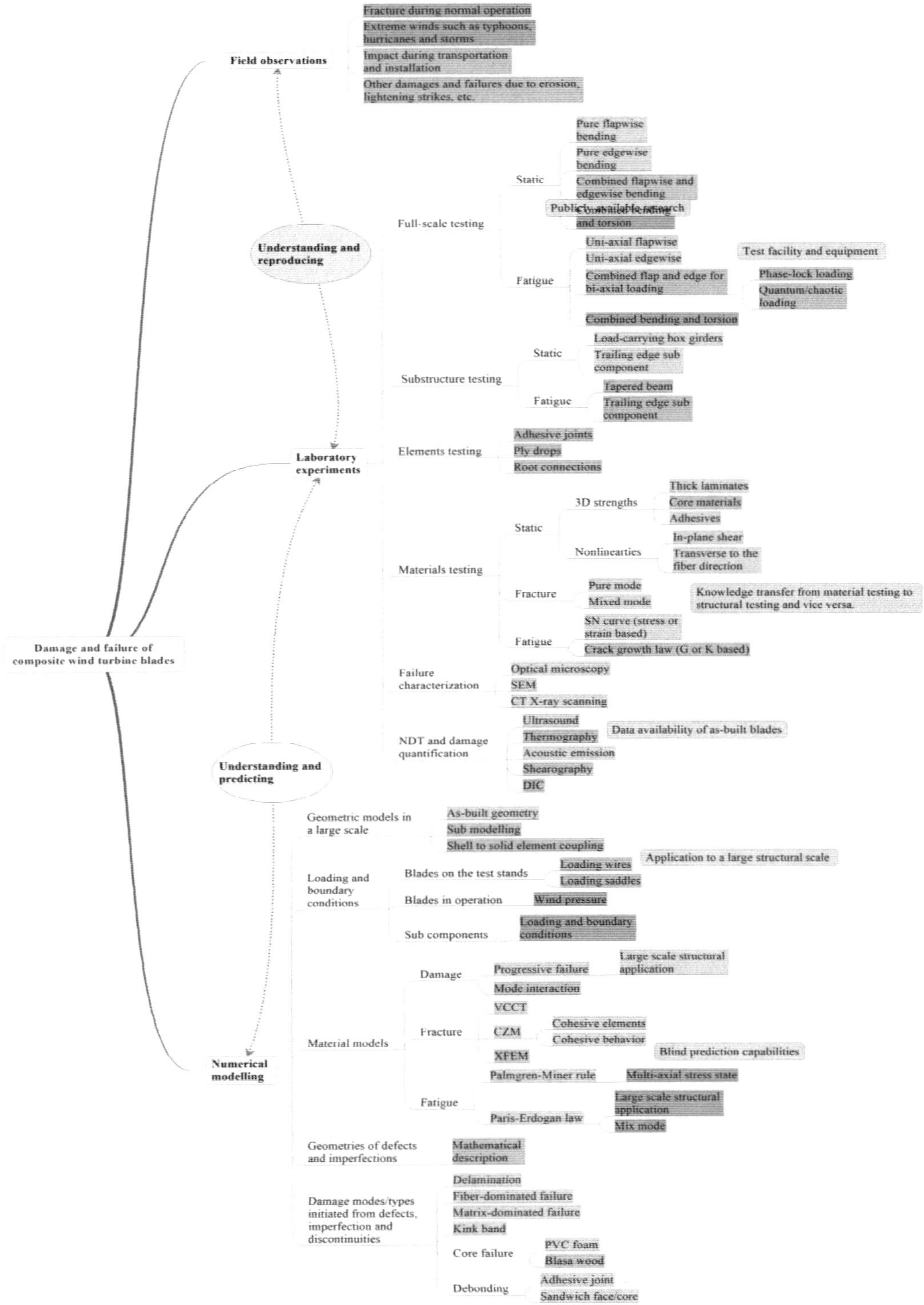

Figure 3. An overview of research status in damage and failure of composite wind turbine blades. Color codes: green – well studied with considerable understanding achieved; orange – ongoing studied with a few recent research publications; red – remaining challenges with a few or limited research publications; blue – less studied with a few or limited research publications; yellow – key issues to be addressed; Note that the color categorization may be debatable as continuous and interlinked research usually blurs boundaries between different categories.

2. Field observations

It is reported that with an estimated 700 000 blades in operation globally, there are, on average, 3800 incidents of blade failure each year [4]. Based on this report [4], blades are susceptible to a number of different failures caused by one or a combination of the following incidents:

- Failure of control system to detect vibration, imbalance or insufficient power
- Failure at root connection leading to blade throw (entire blade becomes separated from the hub at the metal to metal root joint)
- Extreme load buckling
- Lightning damages, including subsurface effect and moisture ingression thereafter
- Manufacturing defects leading to debonding (deterioration of the bonding agents used at the interface between the various structural elements of a blade)
- Blade overspeed striking the tower
- Environmental events, including natural perils, outside design envelops
- Incorrect design for fatigue loads
- Crane impact during scheduled maintenance or on-site repair
- Poor manufacturing quality control/assurance leading to delamination (air traps between the piles of a blade or poor infusion of resin in a given area that causes poor or no bonding)
- Nacelle fire spreading to blades
- Human error (making an unauthorized adjustment during construction/maintenance)

These failures are rarely published in academic literature, mainly because relevant companies are reluctant to disclose such issues and technical data is hardly available due to commercial confidentiality. This inaccessibility of course deprives others from the opportunity to improve blade design and manufacturing processes in order to avoid similar blade failure from happening in the future. Nevertheless, a few available studies [5-9] are shedding light on the structural failure of rotor blades under extreme wind conditions such as typhoons, Figure 4. These incentives provide useful information on the different root causes of blade failure and can be used to infer recommendations for possible improvements in blade design and control strategies to reduce the risk of failure. This valuable knowledge contributes to the successful development of wind energy in regions prone to the impact of typhoons and hurricanes.

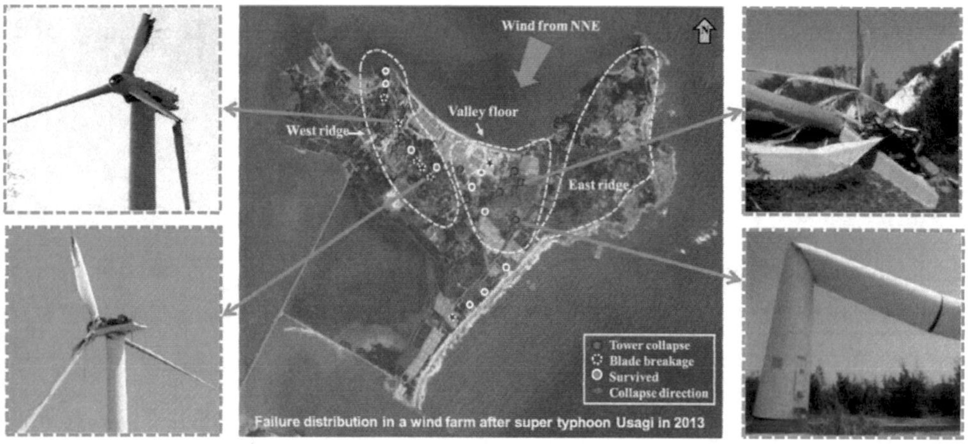

Figure 4. A wind farm located on a coastal island is impacted by a typhoon in 2013 [5]. In total, 35 out of 75 22.9-meter long blades are totally fractured in the wind farm. The fracture location is from 6 to 11 m from the blade root. The estimated wind speed during the typhoon is higher than the design survival wind speed of the turbine.

In surprisingly frequent cases rotor blades fail during normal operation when deemed structurally safe due to a presumably large safety margin. A recently conducted comprehensive study [3] scrutinized the reasons for rotor blade failure during normal operation in a wind farm. Another notable study published in [10], investigates fatigue damage, i.e., laminate cracks, of operational wind turbine blades by examining the slices of the damaged area with an abrupt thickness transition pf laminates at the blade root. Field investigations concerning damages inflicted by erosion and lightning strikes are scarce and mostly limited to experimental work under laboratory conditions [11-14].

The bottleneck concerning the availability of field observation data is the restricted accessibility due to proprietary rights preventing the dissemination to the public domain. The quality of a field failure investigation is a strong function of the availability of high-quality data and reliable information that needs to be collected, verified and analyzed before conclusions can be drawn on a rational basis. In particular material data, manufacturing processes, structural design details and loads represent crucially important elements giving credence to the root cause analysis, but are closely guarded secrets with highly restricted accessibility.

3. Experimental investigation

Experimental investigation on damage and failure of the blades can be categorized into full-scale testing, substructure/subcomponent testing, elements testing and materials testing. Complimentary utilization of material and failure characterization (e.g. scanning electron microscopy, X-ray tomography), nondestructive testing (e.g. ultrasound, thermal imaging) and experimental campaigns at different length scales provide indispensable information on a holistic level necessary to advance the understanding of failure mechanisms and their emergence at increasing length scales under controlled conditions (load magnitudes, loading frequencies, moisture, temperature, etc.).

3.1. Full-scale testing

The damage and failure modes of full-scale blades under monotonic ultimate loading conditions, in particular under pure flapwise (i.e. bending around the weak axis) and edgewise bending (i.e. bending around the strong axis), have been studied extensively over the past two decades. Full-scale structural tests enable an in-depth apprehension of the structural response of composite blades to quasi-static loads. Such tests can be used to validate the blade strength and serve the purpose to probe for necessary modifications in order to improve the structural performance and/or to reduce the blade weight. Prominent studies have been carried out by [15-26] and - despite the variance in structural design and material properties - came to a few common conclusions, notably:

- The governing failure phenomenon is primarily buckling driven whereby structural (geometric) nonlinearity plays an important role in the failure process.
- The ultimate failure is typically characterized by a sequence of multiple failure modes, in which three-dimensional stresses/strain states may play an important role.

These conclusions agree well with the general behavior of an imperfection sensitive thin-walled shell-type structure that can undergo a tip deformation of 25% of its span. The three-dimensional stress state stems from 1) thick laminates and sandwich panels; 2) material and geometric discontinuities; 3) the doubly curved lift generating shells (airfoil shape curvature in-plane and the taper curvature along the blade axis direction) which are prone to fail in a cascade of buckling modes.

In reality, however, the blades are subject to highly complex combined loading conditions which include the simultaneous action of flapwise-, edgewise bending and/or torsion. A full-scale blade test under the combined flapwise and edgewise bending is reported in [27], where the buckling of trailing edge panels subject to critical levels of compressive stress initiated the failure of the blade. The full-scale blade collapse tests under combined flapwise bending and torsion bending are pioneered by [28] are shown in Figure 5 and were later continued by [29]. In [28], failure was caused by crushing of the spar cap laminates under compressive bending stress leading to delamination that is driven by local buckling. Note that all publicly available studies use blades made of glass fiber composites. It is practically unavoidable though, to employ (hybrid) carbon fiber composites in order to compete in the global blade upscaling battle. The failure behavior of such large-scale carbon fiber composite blades is to date not publically available although a previous experimental study [30] has been carried out on three small scale (9 m) blade, showing the failure of two blades due to panel buckling near the max chord and good buckling resistance of the blade using a flat back airfoil feature.

Figure 5. Setup of a full-scale blade test under combined flapwise bending and torsion [28]. The desired state of combined bending and torsion is achieved by shifting the loading lines (force action lines) in such a way that they do not pass the shear centers of blade sections. The step-wise static load is applied to collapse the blade.

In the same vein as static tests, fatigue tests of full-scale blades are traditionally conducted under pure flapwise and pure edgewise bending, respectively. This testing mode is commonly referred to as uniaxial fatigue testing. Some notable studies [2, 31-34] report on the damage progress inspected during the fatigue test. It can be inferred from literature, that the observed damage is highly blade-specific and catastrophic structural failure did not occur during these tests. A categorization and consequential generalization of the observed damage mechanisms is strongly exacerbated by the limited number of publicly available test results reported in the literature. The most popular approach for uniaxial testing is the equivalent force approach, in which the first natural mode of the blade is excited and additional tuning masses are mounted on the blade in order to fit a prescribed equivalent bending moment distribution at maximum amplitude. Operating blades are exposed to a vast number of cycles in the order of $\sim 10^{10}$ during their lifespan of 20 - 25 years. For monetary reasons, blades are typically tested to approx. 2×10^6 cycles where the amplitude is artificially amplified with the intent to induce a damage level equivalent to $\sim 10^{10}$ cycles. It is noteworthy to mention that the validity of the uniaxial testing approach is questionable for two principal reasons: First, an operating blade is exposed to multiaxial loading conditions and operates at an entire frequency spectrum. Second, it is to date not scientifically corroborated that the equivalent force approach indeed induces the same damage levels and failure modes at the same location after 2×10^6 cycles the same blade would exhibit under operating conditions after $\sim 10^{10}$ cycles.

The key limiting factor associated with full-scale blade testing is the exorbitantly high cost, primarily caused by the testing time of up to six months (e.g., for a 90 m long blade) and the energy consumption due to aerodynamic damping effects. The full-scale blade testing in the industry was mainly concerned in developing methods able to reduce the cost of uniaxial testing using aerodynamic drag breakers (fairing) and by developing excitation methods in which the exciter mass is relocated from the blade to the ground. A method for significantly reducing the testing time was proposed by [35] which utilizes a multi-frequency excitation approach allowing to accumulate the desired cycle count in a shorter period of time. Moreover, the multi-frequency approach is a closer approximation to the operational loading conditions. Another limiting factor inhibiting the availability of publically available fatigue damage data pertains to current blade certification procedures driving the incentive to demonstrate that the blade does not show any damage e.g. after 2×10^6 cycles. While the pursuit of such an approach is intelligible from an economical perspective, it evenhandedly foils the fundamental idea of testing which relies on the observation of damage for the sake of knowledge gain.

More recently, research focused on the development of multi-axial fatigue testing methods in which the blade is simultaneously excited in two different (e.g. orthogonal) directions. This approach represents a major advancement in fatigue testing technology. The reports on the damage of the full-scale blades under multi-axial fatigue are scarce although the multi-axial fatigue tests themselves are reported frequently [36-38] and demonstrated such as shown in Figure 6. The development of improved full-scale blade testing methods is a very active field with an international research effort. A substantial investment in test facilities, equipment, instrumentation and time sets the bar rather high for most research organizations to enter this field if without close collaboration with industry.

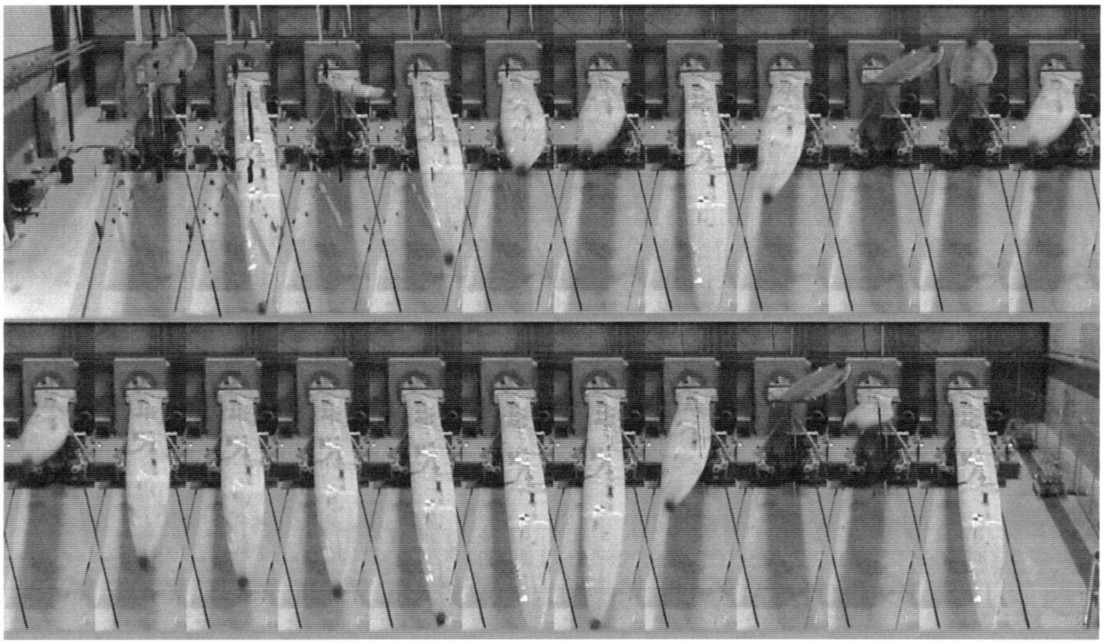

Figure 6. A 14.3 meter long blade under biaxial fatigue that is excited by a ground-based backlash constrained dual-axis exciter developed by R&D A/S within the BLATIGUE project (Fast and Efficient Fatigue Test of Large Wind Turbine Blades, 64016-0023) funded by the Danish Energy Technology Development and Demonstration Program (EUDP). A clown's nose is mounted at the tip of the blade to visualize its trajectory under bi-axial fatigue. The trajectory of the tip over 1 second of period is superposed to the figure.

3.2. Substructure and element testing

Considering the dominating burden of the cost involved in full-scale blade testing, it appears natural to develop a more cost-efficient trade-off. Equally important to consider is the distinct level of scattering inherent to fatigue test results. It is, therefore, desirable to establish statistically significant datasets by testing a sufficiently large number of identical specimens that permits the application of statistical methods. In this light, the high cost involved in full-scale tests alas prevents the acquisition of the required sample size. Testing a part of the blades – a subcomponent – is a cost-effective alternative solution to experimentally evaluate the structural integrity of the potentially critical regions. Subcomponent testing or a.k.a. substructure testing can be conducted at a fraction of the time and cost of a full-scale test. Consequently, a larger sample size can be tested endowing the data with lower uncertainty levels. Subcomponents lend themselves to experimentally investigate a wide range of different design details, e.g., load-carrying beams [39-43], buckling-prone trailing edge segments [44-46], and elements such as ply drops and adhesive connections in various parts of the blade to name a few. Subcomponent tests provide valuable insights into local structural behavior that are difficult and/or costly to investigate via full-scale blade testing and hardly possible to gain on a smaller scale by virtue of e.g. coupon tests.

The biggest challenge of developing meaningful subcomponent test methods is to emulate the force and displacement boundary conditions along the free body cut of the very same component as-integrated in the blade. A well-known hurdle in the development of subcomponents is failure initiation in the load application points and the restraints. It is natural that in these particular regions stress concentrations arise (e.g. high stiffness gradients due to local reinforcements) which precipitate failure in regions remote from the gauge area. Moreover, the stress state in the boundaries and local force application points (e.g. three-point- or four-point bending tests) is typically largely at a variance of the stress state present in the real blade. The critical question which arises pertains to the value such subcomponent test data provides if a failure occurs outside the designated region at unrepresentative load levels in an unrepresentative failure mode. When it comes to fatigue, the challenge may be worsened as the effects of both loading and boundary conditions on the premature damage are more significant compared to the static loading. The first step towards meaningful subcomponent testing was done by [43]. Later it was shown that topology optimization can be used to develop a subcomponent geometry that satisfies the failure location and facilitates the investigation of a specific failure mode [47]. One further application of this approach is shown in Figure 7 where the hourglass-shaped subcomponent was developed to test the web-to-spar-cap adhesive bondline under fatigue loading. By utilizing a finite element optimization-based approach, it was possible to successfully investigate the desired shear failure mode inside the desired gauge region [48]. The other possible solution would be to reinforce the boundaries during (not after) the infusion/manufacturing of the specimens.

Figure 7. A tapered beam with adhesive joint interfaces [48]. The adhesive failure initiates in the gauge area remote from boundary conditions and load application points.

Testing the so-called elements is one additional way to evaluate structural details of composite blades. The difference between the element and substructures and the subcomponents is that the elements are smaller in size, allowing a larger number of specimens to be tested more cost-effectively. The element test usually focuses on rather local structural details with two or more failure modes although its distinction from materials testing is not yet uniquely defined in the current research community. A typical element test is shown in [49] where thick adhesive joints used in wind turbine blades are tested under static and fatigue loads.

3.3. Materials testing

This section only focuses on materials testing that is directly relevant to damage, failure and fracture of composite wind turbine blade structures. Four primary materials commonly used to build wind turbine blades include fiber fabrics, resin, cores, and adhesives. Their ultimate strength, fracture toughness and fatigue resistance are key material properties relevant to the structural integrity of the composite blades. A few notable studies have been published. They include a) three-dimensional static strength parameters of a thick glass/epoxy unidirectional laminate [50, 51] where stress-strain curves are also reported; b) bi-axial static strength [52] and fatigue strength [53] of thick adhesives using tubular specimens; c) composite fatigue test database SNL/MSU/DOE [54], FACT [55], and OptiDATA [56] with the SNL/MSU/DOE database recently updated and remain publicly available. A short review of these databases can be found in [57]; d) material strength of PVC foams under multiaxial static loading [58, 59] and fatigue [60]. A short review [61] on sandwich materials for wind turbine blades identified a few challenges, some of which, such as NDT methods, are still valid nowadays.

The availability of publically accessible databases is highly celebrated and constitutes an endeavor that should under all circumstances be carried forward. However, it is noteworthy to mention that both resin systems and fiber fabrics underwent soaring developments in the past decade, outdating large parts of existing material data available in these databases. State-of-the-art composite materials used in modern wind turbine rotor blades exhibit distinctly elevated fatigue strengths as compared to previous systems. Another aspect concerns the size effect (especially the thickness effect) of fiber composite materials under fatigue loading. Additional experimental research is required to shed more light on the size effect of composite materials under high cycle loading conditions. Other topics that can only be briefly be touched upon in this work concern the influence of fiber architecture and curing induced residual stress on fatigue performance as well as the effect of the load cycle sequence dependency [62].

In the context of wind turbine blade damage and failure, the future trend of material testing will be not only continuing advanced tests in the on-going topics such as multiaxial strength, high cycle fatigue beyond tension and material degradation due to environmental effects but also will focus on establishing knowledge to close the gap between failure mechanisms observed in a material scale and those found in a structural scale; discontinuities between different materials, as well as within one material; the new failure modes due to the application of new materials such as hybrid and bio-materials in large wind turbine blades.

3.4. Damage detection and failure characterization

When damages and failures occur in the field or the laboratories, they should be detected, identified and preferably quantified to help understand the failure phenomena. Damage detection and failure characterization are a huge topic themselves. Regarding the application in wind turbine blades, ultrasound, thermography and acoustic emission are among the most widely used NDT methods, while shearography is less common but promising. A recent work by the lead author uses thermography to detect fatigue damage during the full-scale blade test, see Figure 8. Note that each NDT method has its advantages and shortcomings, the choice depends on the purposes of the testing. A trend is to enrich test results thus enhance the detectability and accuracy by either combining different methods in the test campaign or utilizing digital techniques such as machining learning and computer vision.

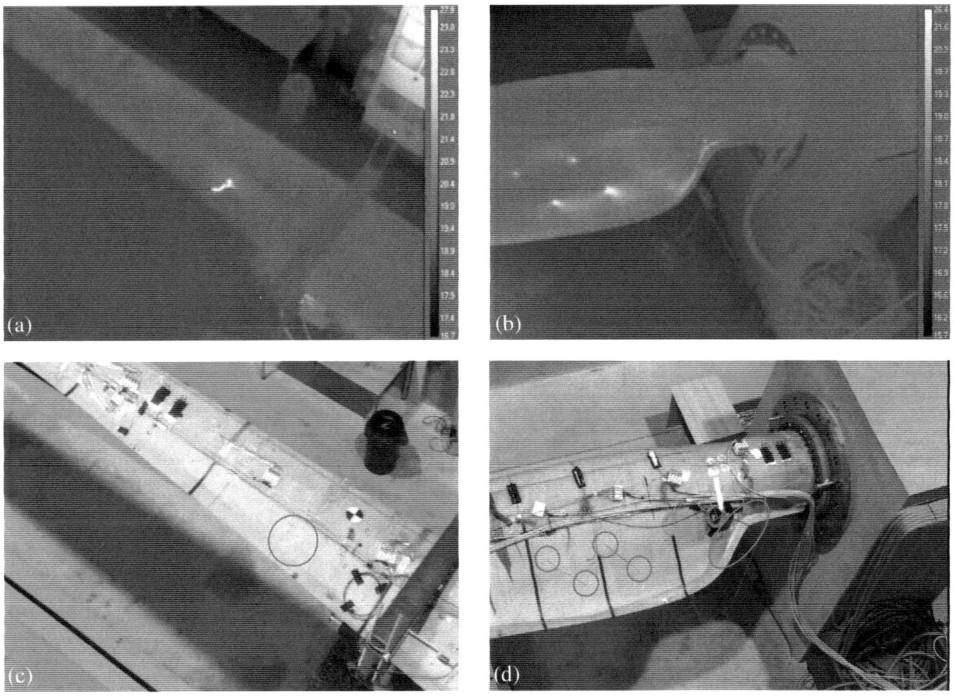

Figure 8. Remotely detecting fatigue damage of a 14.3 meter long blade under bi-axial fatigue using infrared thermography. (a) thermal footprints of progressive surface cracks in the sandwich panel with a pre-cut in the trailing edge. (b) thermal footprints of multiple damage sites in the max-chord sandwich panels and in the root transition region. (c) the regular RGB image of the progressive surface cracks in the sandwich panel marked with a circle. (d) the regular RGB image shows surface damages at the tips of two pre-cracks in the sandwich panel marked with four circles and subsurface delamination damage in the root transition region marked with an ellipse. A detailed study to be published in a separate work coined by AquaDa. The study is supported by the DARWIN Project (Drone Application for pioneering Reporting in Wind turbine blade INspections, 6151-00020B) funded by Innovation Fund Denmark.

For failure characterization, the widely used techniques regular cameras, optical microscopy, Scanning Electron Microscopes (SEM), computer tomography (CT) scanning. Notable literature include fractographic analysis of composite materials [63] is systemically introduced; fatigue damage characterization using micro CT scanning images [64]; post-mortem investigation on two large blades collapsed during normal operation [3] and a subcomponent specimen after a collapse test [65], establishing the relationship between microscale material failure characteristics and macroscale structural failure behavior.

Driven by the field application aiming for the cost-effective operation and maintenance (O&M) in the wind energy industry, the future trend in this area will be more automated damage detection and characterization to evaluate damage criticality of the structures under concern. The data-driven approach will be gaining more and more popularity simply due to the overall trend of digitalization in the industry [66].

4. Numerical modeling

This section only focuses on numerical modeling of damage and failure in full-scale blades and their substructures and subcomponents. Modeling damage and failure of composite wind turbine blades on a large scale is challenging and it is worthwhile to briefly elaborate on the gravity of this statement. It is abundantly clear that damage and failure are distinctly local phenomena which for the sake of argument can be said to occur on a characteristic length scale range of 10^{-3}-10^{-6} m for all practical purposes in the fiber composites world. Now unlike e.g. plasticity phenomena, the characteristic length e.g. of a fracture process zone does unfortunately not scale with the characteristic macroscopic length – which for a modern wind turbine blade is in the order of $\sim 10^2$ m. This illustrates the dilemma of numerical damage and failure analysis of large scale blades and the ordeal to bridge an overwhelming length scale gap of several orders of magnitude whilst simultaneously struggling to maintain reasonable computation times – even on modern supercomputers. Literature shows that high fidelity modeling approaches such e.g. as cohesive zone models are predominantly applied on manageable macroscopic length scales such as coupons where the scale bridging dilemma is significantly less apparent. Therefore, numerical modeling of test elements and coupons are not discussed here to maintain this section at a manageable level.

Naturally, research has developed methods to close this gap from both sides. On the macroscopic side, e.g. sub-modeling techniques are deployed providing the opportunity to limit the high mesh discretization levels to comparatively small areas of the blade. Sub-modelling requires prior ad-hoc knowledge about the location and size of the damage zone rendering this approach increasingly inapplicable to the simulation of operating wind turbine blades with multisite damage distributions. That is to say, resolving the blade with different sub-models like a jigsaw compromises the computational advantage. On the microscopic side, e.g. continuum damage models utilize a phenomenological approach to smear over microscopic length scales enabling the adoption of comparatively coarse mesh discretization levels. In other words, instead of modeling the discrete fracture process, the material softening effect caused by the fracture process is implicitly considered by modifying the local element constitutive matrix as a function of the prevailing damage index. Continuum damage approaches represent indeed a computationally efficient alternative. However, their inability to accurately resolve the physical crack, makes them less compatible with full-scale experimental observations and inspection based repair frameworks. Other approaches such as the celebrated extended finite element family are better suited for this purpose but despite significant software developments still suffer to a considerable degree from the wide applicability to anisotropic materials, numerical instability (i.e. convergence issues) and crack-path mesh-size/ element-type dependency.

While the aforementioned circumstances can be reasonably dealt with under monotonic loading conditions, the situation faces a game changer when it comes to high cycle fatigue analysis of full-scale blades. In the latter case the burden of spatial discretization is aggravated by the temporal discretization of a highly time-dependent process (note that load cycles can be considered as a pseudo-time parameter). It needs to be borne in mind that blades under operating conditions are subject to dynamic multiaxial loading conditions contained in an entire frequency spectrum. These variable amplitude (quasi-stochastic) loads containing vast cycle numbers are typically non-proportional and inherent nonlinearities (see below) let the superposition principle break down. The latter was often used as a lifeline inasmuch it enabled the analysis of complex loading situations by applying linear combinations of finite element model runs with unit forces. This approach does not hold anymore if large scale blades are concerned.

In a nutshell, the main focus of future research on numerical modeling strategies will need to focus on not only further improvements of their fidelity but also, maybe more importantly, drastic improvements of computational efficiency capable of dealing with the challenges outlined above.

It should be noted that any numerical model is essentially a mathematical representation of a physical process on a specific characteristic length scale underlying certain simplifications and assumptions. Depending on the objectives of the simulation, different modeling techniques can be used. Nevertheless, to perform reliable numerical simulations on damage and failure of large-scale blades, a few points need to be addressed:

(1) Geometric models that are detailed enough at least in the regions of interest where damage and failure may occur. In this context, 'detailed' does not only refer to mesh refinement but also more importantly thorough consideration of the damages to be simulated. For example, finer shell element meshes alone are not able to capture debonding and delamination that require through-thickness stresses/strains. Imagine the modeling effort if the detailed geometric modeling pertains to a predominant portion of the blade and even through its wall thickness. Typical examples for such cases are the adhesive bondlines running along the entire span of the blade with varying geometry. Adhesive web foot joints feature local stress concentration zones and singularities in corners caused by the adhesive flow front and the substrate [67]. Figure 1 only shows one slice of a blade that is modeled with solid elements and with different elements representing different materials, e.g., the outside skin of the sandwich laminate, the PVC foam core, the inside skin of sandwich laminate, through the wall thickness. Obviously, there is a balance between the level to which details are resolved and the time spent on the modeling i.e. the modeling effort. Automated modeling tools for research purposes are built in-house, e.g., Figure 9. Some software is commercially available for general industrial applications such as [68, 69] and they mainly focus on completing the entire aerodynamic-aeroelasticity-structure design loop for a blade and a wind turbine, respectively. One open-source software is NuMAD [70] developed by Sandia National Laboratories as a general blade modeling tool.

(2) Publicly available data of the blade details that are needed to model blades is another challenge in this area. Although some blade models such as the DTU 10 MW wind turbine blade [71], the NREL 5 MW wind turbine blade [72], and the Sandia 100 m wind turbine blade [73] are publicly available, they are artificially created and do not necessarily reflect the actual blade details used in the industry. Due to the proprietary data, existing publications are only able to partly disclose the information when the commercial blades are modeled, making the reproduction of simulation results difficult or even impossible. A set of publicly available data of real modern blades are highly valuable to the entire research community.

(3) Loading and boundary conditions. For the simulation of full-scale blade tests, it is sometimes important to model the loading wires in order to consider the change of load directions during static loading. Moreover, it may be also necessary to model loading saddles [74] if they may affect the failure of the blades. It is challenging to model proper loading and boundary conditions of substructure and subcomponent tests when the connections between specimens and loading points affect the failure behavior [44, 45, 75-78].

(4) Structural nonlinearities. The consideration of geometric nonlinearity (i.e. large deformation theory) is straightforward in commercially available finite element software packages. Material nonlinearity is usually considered when the specific damage types are under concern. The importance of contact nonlinearity has not received a great deal of attention but proves to have a considerable influence on the failure sequences [44, 76] and needs to be assessed. A unified modeling approach that

considers all three types of nonlinearities, Figure 10, will be a valuable advancement in this field. A step towards this approach is presented in [79].

(5) Highly sophisticated material models have been developed to predict the damage and failure of composites, core materials, and adhesives used in wind turbine blades. The current challenge is to apply these models to the damage prediction of large-scale blade structures. The widely used modeling technique is continuum damage mechanics for progressive damage analysis of the blades or their subcomponents. The fracture mechanics based approach is also used but primarily limited to small regions, e.g., trailing edge bondline and spar cap. The future trend will be a combination of two approaches in one single simulation to capture different damage modes and their interaction.

(6) Fatigue simulation of rotor blades is mainly based on phenomenological models and usually the distribution of equivalent damages over the blades is calculated as damage indexes. Arguably the most popular method foresees the computation of the damage index according to the Palmgreen-Miner rule and to superimposing it as a scalar field on the blade model. Despite the convenience of using such an – arrantly speaking – intelligent post-processing scheme, the biggest drawback manifests itself in the negligence of the damage evolution since all predictions are based on the initial undamaged configuration. Strictly speaking, such approaches are only valid to predict fatigue damage initiation which can be a useful tool to screen blades for potential hot-spots e.g. for mitigation at an early design phase. For a more faithful prediction of fatigue crack growth in a large-scale blade requires overcoming the challenges outlined at the beginning of this section by radically new approaches and a shift of paradigm. A recent work [80], coined as FASTIGUE, is a first attempt to thrust forward and presents an efficient discrete fatigue crack growth simulation approach for the large-scale blade application. It is done by outsourcing the computationally demanding fracture analysis into a pre-processing module whereas the crack-growth simulation is conducted subsequently and independently in a separate fatigue analysis module, see Figure 11.

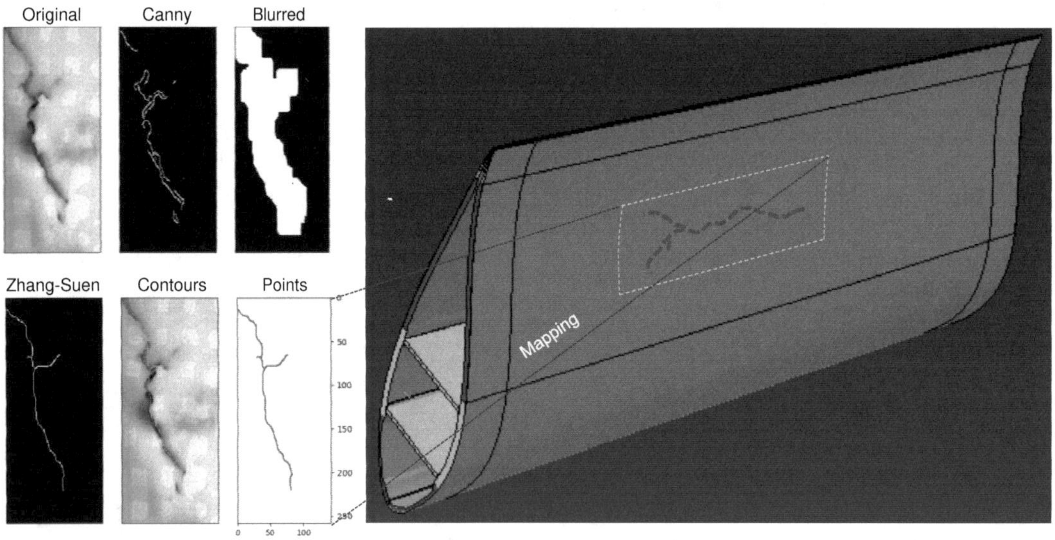

Figure 9. Automated damage identification and modelling (AUDIN) tool developed by X. Chen and S. Semenov. The tool first digitalizes complex crack geometries based on images from regular RGB cameras, thermography and other NDT methods, and then automatically imports the information into the FE model through mapping and scaling. The structural integrity of the damaged blade can be analyzed online via cloud-computing. The evaluation results can be sent back to inspectors in near real time who may work in the field, facilitating fast and reliable decision making for cost-effective O&M strategies. The work is supported by the DARWIN Project (Drone Application for pioneering Reporting in Wind turbine blade INspections, 6151-00020B) funded by Innovation Fund Denmark and the RELIABLADE project (Improving Blade Reliability through Application of Digital Twins over Entire Life Cycle, 64018–0068) funded by the Energy Technology Development and Demonstration Program (EUDP) of Denmark.

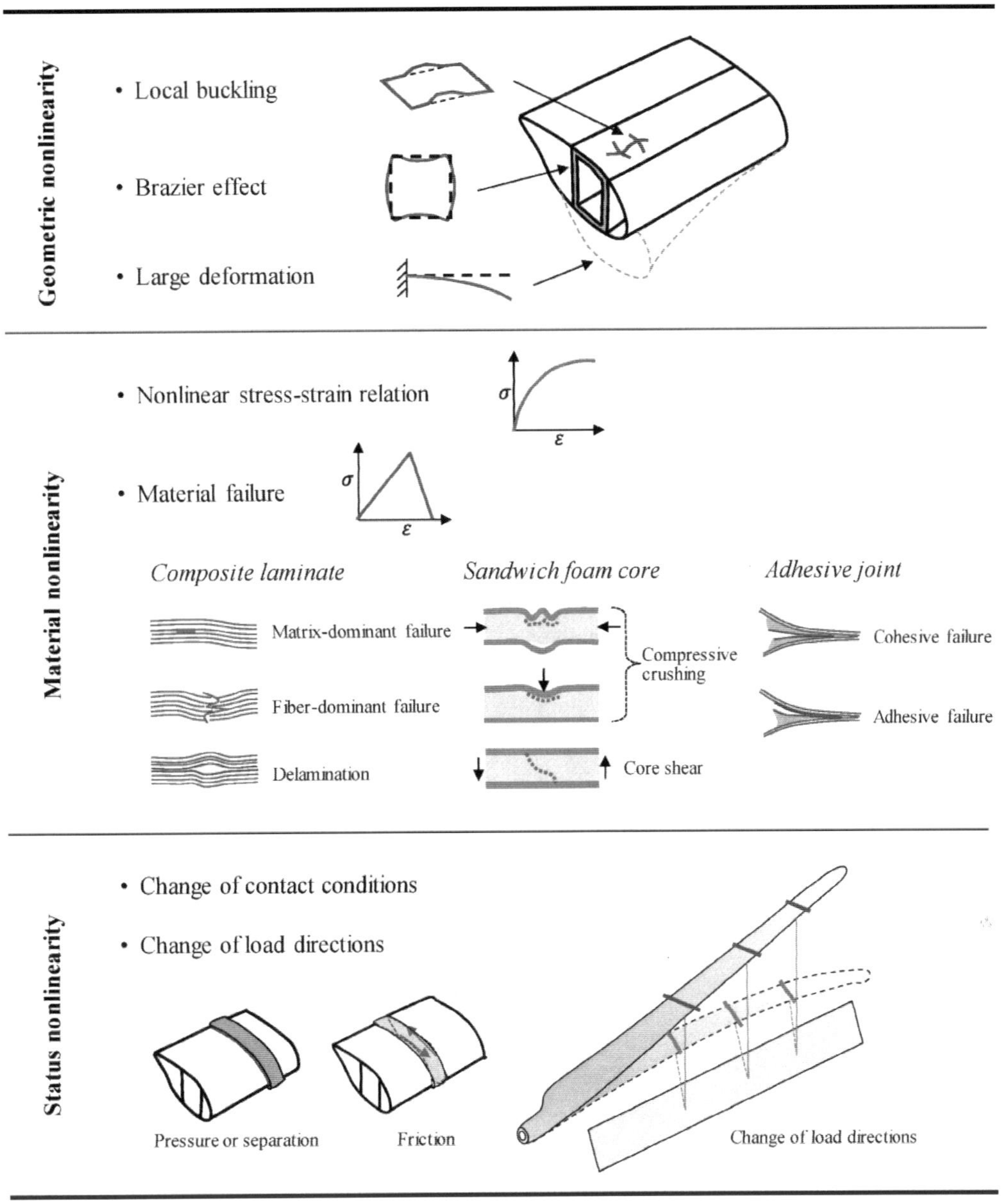

Figure 10. Examples of three types of structural nonlinearities that might be experienced by composite wind turbine blades [79]. Depending on the purpose of the simulation, relevant nonlinearities have to be modeled to capture damage and failure of the blades.

(7) Numerical modeling of damage and failure has been mainly focused on understanding failure behavior and reproducing experimental observations of the as-designed blades in their pristine state without pre-existing damage. The future trends are towards focusing on scenarios in which it is accepted that blades are born with major defects and imperfections in their as-built state and to predict and trace the damage evolution under service conditions. How these defects, imperfections and damages behave in the large-scale blades and their influence on the overall structural integrity are essential and of practical importance for the life-cycle performance evaluation of wind turbine blades. Instead of considering the faith of individual blades, future numerical modeling strategies need to be fit for the integration of all the blades of a wind turbine into interconnected digital twin systems.

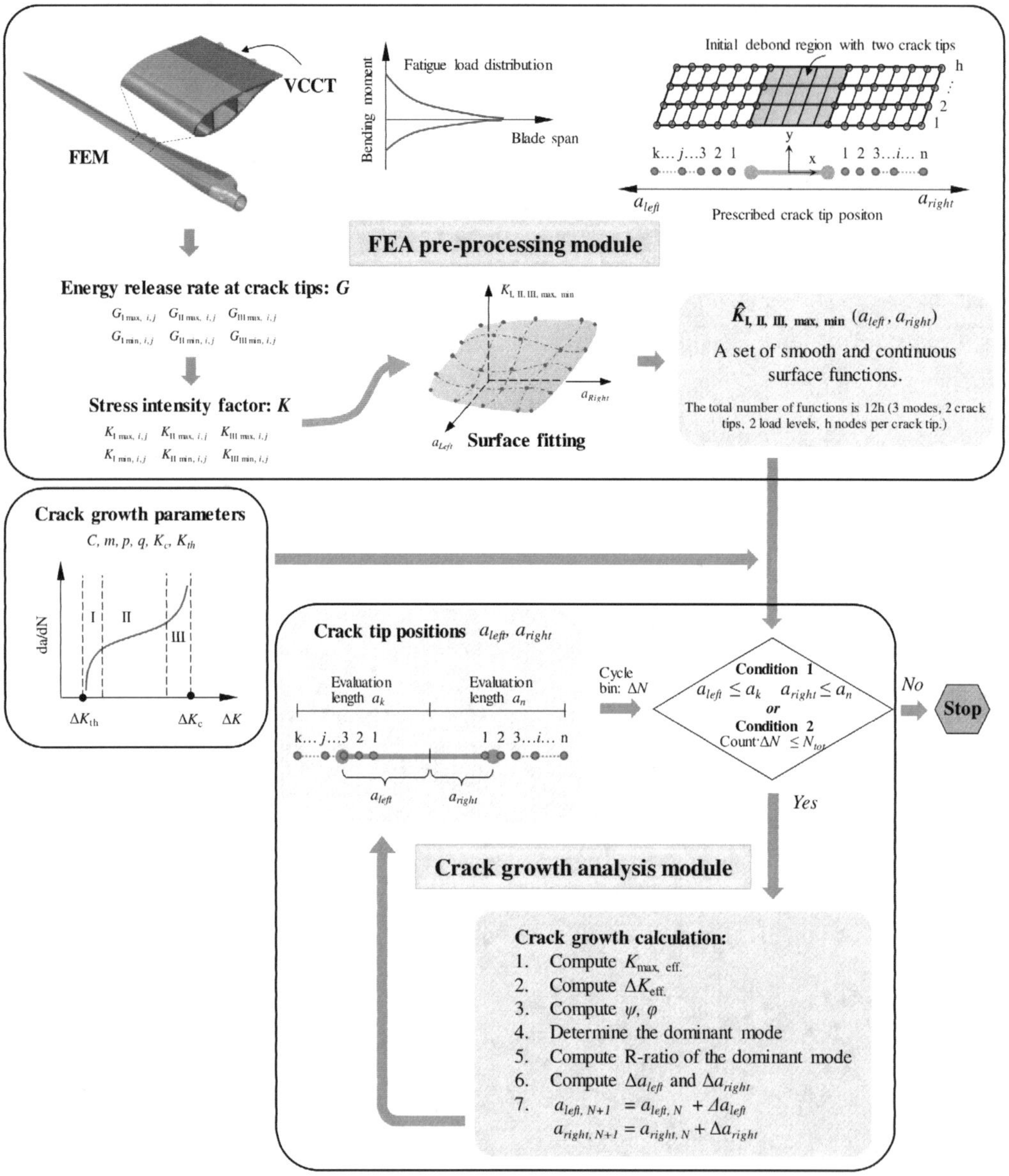

Figure 11. The flow chart of the FASTIGUE [80] approach comprising an FEA pre-processing module (top) and a crack growth analysis module (bottom); the grey arrows indicating the computational flow show that a feedback between the FEA preprocessing module and the crack growth analysis module is avoided, resolving heavy computational demands of fatigue crack simulation for composite wind turbine blades.

5. Concluding remarks

The recent two decades of research show significant advances in the understanding of damage and failure of composite rotor blades for multi-megawatt wind turbines, particularly in the case of ultimate loading situations. Important macroscale failure modes in blades observed in the large-scale experiments have been categorized and entrenched in the underlying theory. Numerical methods capable of predicting these structural damage and failure modes have been developed and successfully vindicated. The remaining challenges associated with the damage and failure prediction in large utility wind turbine rotor blades are identified as follows:

(1) Damage and failure under complex static loads: Structural failure mechanisms under combined complex (multiaxial) loading, in conjunction with failure phenomena associated with the advent of rather novel material systems, e.g., carbon/glass hybrid fiber fabrics and bio-materials whose utilization become increasingly popular. A missing keystone for quasi-static ultimate loading situations is a general/unified numerical modeling approach capable of predicting key failure modes in large scale structures. The success of these modeling strategies available in the literature should be measured and ranked objectively using blind tests during which experimental results are only disclosed in retrospect.

(2) Efficient (and maybe radically novel) methods to predict fatigue damage in large structures: The supreme discipline of predicting high cycle fatigue damage in composite rotor blades for multi-megawatt wind turbines is still a very active field of research and will remain so for decades to come. Conventional stress/strain-based methods hinging on damage accumulation, such as using the linear Palmgren-Miner rule, are well established for historic reasons and owing to their simplicity. However, as blades grow in size, structural behavior and failure modes gain complexity and become less intuitive. In fact, the apparent speed of the design evolution combined with the relentless time-to-market demands increase the risk of experience-based designs. Large scale fatigue tests and fatigue failure cases observed in the field provide ample evidence for the need of deploying more sophisticated fatigue prediction tools. Especially local damage modes such as delamination and debonding mostly inherent to the regions of geometric and material transitions/discontinuities (e.g. adhesive bondlines, ply drops) require more advanced modeling formulations on the basis of fracture mechanics and continuum damage mechanics.

It needs to be conceded that such methods are partially readily available albeit, do not enjoy great popularity for application on large scale because of two reasons: First, computational efficiency is compromised by high discretization demands and poor numerical stability rendering them impractical for high cycle fatigue analysis in practice. Secondly, many advanced fatigue prediction models require a comprehensive set of material parameters that might simply not be available for the desired material system. Even worse, testing methods to obtain these parameters might be inexistent. Large as it is, nowadays philosopher's stone is to integrate available material models into large scale systems such as digital twins which will be only achievable with research effort dedicated to a significant improvement of their computational efficiency. The recently developed FASTIGUE [80] approach shows one promising application of fracture mechanics based fatigue prediction to large scale structures.

(3) Interfacing the boundary between experiments and simulations: The development of improved numerical models goes hand in hand with the advancement of reliable experimental methodologies more accurately emulating prevailing stress/strain states, reproducing relevant failure modes and considering more realistic loading scenarios. More sophisticated fatigue test methodologies at full-scale, substructure, subcomponent and element levels have to be devised with a particular emphasis on reproducing damage and failure of the blades in operation and proper documentation to benefit reproducibility.

In contrast to other fields such as e.g. the automotive industry or earthquake engineering, it comes at a surprise that well established test methodologies on intermediate sub-component level for blades are scarce. This emphasises a strong research demand for novel sub-component testing techniques in the wind turbine blade industry. In particular an approach known as 'hybrid testing' craves for application in experimental blade research. Hybrid testing refers to a subcomponent of a structure that is loaded by an arrangement of different actuators in such a way that forces and displacements are closely emulated. An array of sensors enables real time feedback communication between the subcomponents response with a numerical full-scale model that is run in parallel. In this way, the cyclic global response is provided by the full-scale model and the force output consequently applied to the subcomponent. In turn, the stiffness degradation of the subcomponent due to fatigue damage is fed back into the numerical model in terms of the displacement. Such a versatile hybrid-testing approach offers a striking time and cost efficient alternative to full-scale tests. In the future, driven by digitalization, the historic sharp boundaries between experiment and simulation will strongly interlink and the two paradigms will become indistinguishable. To take it a step further, the real world wind turbines will become an integrated part of the digital world simply because the control of such complex systems and the decision making process will not be possible without the aid of machine learning. These research efforts will continue to be the trend in the foreseeable future.

(4) Operational damage and failure in the real world: The importance of public accessibility of real-world failure incidents for the greater good of research cannot be overemphasized. Previous experience has shown that the formation of consortia between universities and key players of the industry successfully facilitates broader dissemination of research data with the shared incentive to make existing design guidelines less conservative (i.e. reduction of manufacturing cost) through public research. The research on the structural integrity of rotor blades under extreme winds such as typhoons, hurricanes and tornadoes will be still conducted by researchers outside of calm Europe. With the increasing development of coastal and offshore wind farms in typhoon/hurricane-prone regions, the necessity to boost the research in this area is still urgent for the industry.

As the wind energy industry moves towards digitalization, the research activities in the topic structural integrity of composite blades in recent years have involved more and more industry 4.0 technologies such as big data, internet of things, cloud computing and drone technology [81]. These technologies not only facilitate the development of more reliable and accurate testing and modeling of the rotor blades but also may change the way the structural performance is evaluated through the entire life cycle of the blades using digital twins [82]. This digital trend will be undoubtedly long-lasting and profound for the entire wind energy industry and beyond.

Author contributions

X. Chen conceived this study with inputs from M.A. Eder. X. Chen wrote the draft of this manuscript. M.A. Eder contributed significantly to modifications and revision of the draft.

Acknowledgments

The authors would like to acknowledge funding from the DARWIN Project (Drone Application for pioneering Reporting in Wind turbine blade INspections, 6151-00020B) funded by Innovation Fund Denmark; and the RELIABLADE project (Improving Blade Reliability through Application of Digital Twins over Entire Life Cycle, 64018-0068) and the BLATIGUE project (Fast and Efficient Fatigue Test of Large Wind Turbine Blades, 64016-0023) funded by the Energy Technology Development and Demonstration Program (EUDP) of Denmark.

References

[1] Sørensen B.F. 2017. Introduction to scaling issues of damage and fracture in wind turbine blades, Wind Energy Denmark, October 2-3, Herning, Denmark.

[2] Chen X. 2019. Experimental observation of fatigue degradation in a composite wind turbine blade. *Compos. Struct.* **212**:547-551.

[3] Chen X. 2018. Fracture of wind turbine blades in operation-Part I: A comprehensive forensic investigation. *Wind Energy*. **21**(11):1046-1063.

[4] GCube Insurance Services, Inc. GCube report: breaking blades: global trends in wind turbine downtime events, http://www.gcube-insurance.com/, accessed on May 28, 2020.

[5] Chen X, Li C, Xu J. 2015. Failure investigation on a coastal wind farm damaged by super typhoon: A forensic engineering study. *J. Wind Eng. Ind. Aerodyn.* **147**:132-142.

[6] Chen X, Li C, Tang J. 2016. Structural integrity of wind turbines impacted by tropical cyclones: A case study from China. *J. Phys. Conf. Ser.* Oct 3;753.

[7] Chen X, Xu JZ. 2016. Structural failure analysis of wind turbines impacted by super typhoon Usagi. *Eng. Fail. Anal.* **60**:391-404.

[8] J.S. Chou, C.K. Chiu, I.K. Huang, K.N. Chi, 2013. Failure analysis of wind turbine blade under critical wind loads, Eng. Fail. Anal., **27**: 99-118.

[9] T. Ishihara, A. Yamaguchi, K. Takahara, T. Mekaru, S. Matsuura, An analysis of damaged wind turbines by typhoon Maemi in 2003, The Sixth Asia-Pacific Conference on Wind Engineering (APCWE-VI), Seoul, Korea, September 12-14 2005.

[10] J.C.Marín, A.Barroso, F.París, J.Cañas, 2009. Study of fatigue damage in wind turbine blades, *Eng. Fail. Anal.* **16**: 656-668.
[11] Yan, J, Wang, G, Ma, Y, et al. 2019. Electrical and thermal performance of different core materials applied in wind turbine blades under lightning strikes. *Wind Energy*. **22**: 1603-1621.
[12] Branko M. Radičević, Milan S. Savić, Søren Find Madsen, Ion Badea, 2012. Impact of wind turbine blade rotation on the lightning strike incidence – A theoretical and experimental study using a reduced-size model, *Energy*, **45**: 644-654.
[13] Shizhong Zhang, Kim Dam-Johansen, Sten Nørkjær, Pablo L. Bernad, Søren Kiil, 2015. Erosion of wind turbine blade coatings – Design and analysis of jet-based laboratory equipment for performance evaluation, *Prog. Org. Coat.* **78**: 103-115.
[14] Bartolomé, L, Teuwen, J. 2019. Prospective challenges in the experimentation of the rain erosion on the leading edge of wind turbine blades. *Wind Energy*. **22**: 140-151.
[15] Jorgensen E.R., Borum K.K., McGugan M., Thomsen C.L., Jensen F.M., Debel C.P., Sorensen B.T., 2004. Full Scale Testing of Wind Turbine Blade to Failure-flapwise Loading, Risø -R-1392 (EN), Risø National Laboratory: Roskilde, Denmark, June 2004.
[16] B.F. Sørensen, K. Branner, H. Stang, H.M. Jensen, E. Lund, T.K. Jacobsen and K. M. Halling, 2005. Improved Design of Large Wind Turbine Blades of Fibre Composites (Phase 2) – Summary Report," Risø-R-1526(EN), Risø National Laboratory, Roskilde, Denmark, August 2005.
[17] Jensen F.M., Falzon B.G., Ankersen J., Stang H. 2006. Structural testing and numerical simulation of 34 m composite wind turbine blade, *Compos. Struct.* **76**: 52-61.
[18] Jensen FM, Weaver PM, Cecchini LS, Stang H, Nielsen RF. 2012. The Brazier effect in wind turbine blades and its influence on design. *Wind Energy*. **15**: 319-333.
[19] Overgaard LCT, Lund E, Thomsen OT. 2010. Structural collapse of a wind turbine blade—part A: static test and equivalent single layered models. *Composites, Part A*. **41**: 257–270.
[20] Yang JS, Peng CY, Xiao JY, Zeng JC, Xing SL, Jin JT, Deng H. 2013. Structural investigation of composite wind turbine blade considering structural collapse in full-scale static tests. *Compos. Struct.* **97**: 15–29.
[21] Lee HG, Park JS. 2016. Static test until structural collapse after fatigue testing of a full-scale wind turbine blade. *Compos. Struct.* **136**: 251–257.
[22] Chen X, Zhao X, Xu J. 2017. Revisiting the structural collapse of a 52.3 m composite wind turbine blade in a full-scale bending test: Structural collapse of a 52.3 m composite wind turbine blade. *Wind Energy*. **20**: 1111-1127.
[23] Chen X, Zhao W, Zhao XL, Xu JZ. 2014. Failure test and finite element simulation of a large wind turbine composite blade under static loading. *Energies*. **7**: 2274-2297.
[24] Chen X, Qin ZW, Yang K, Zhao XL, Xu JZ. 2015. Numerical analysis and experimental investigation of wind turbine blades with innovative features: Structural response and characteristics. *Sci. China Technol. Sci.* **58,** 1–8. https://doi.org/10.1007/s11431-014-5741-8
[25] Chen X, Zhao W, Zhao XL, Xu JZ. 2014. Preliminary failure investigation of a 52.3 m glass/epoxy composite wind turbine blade. *Eng. Fail. Anal.* **44**:345-350. https://doi.org/10.1016/j.engfailanal.2014.05.024
[26] Lei an Zhang, Yanzhen Guo, Liangfeng Yu, Xiuting Wei, Weisheng Liu, Xuemei Huang, 2019. Structural collapse characteristics of a 48.8 m wind turbine blade under ultimate bending loading, *Eng. Fail. Anal.* 10.1016/j.engfailanal.2019.104150, 106.
[27] Haselbach, P. U., Eder, M. A., and Belloni, F. 2016. A comprehensive investigation of trailing edge damage in a wind turbine rotor blade. *Wind Energy*, **19**: 1871– 1888. doi: 10.1002/we.1956.
[28] Chen X. 2017. Experimental investigation on structural collapse of a large composite wind turbine blade under combined bending and torsion. *Compos. Struct.* **160**:435-445. https://doi.org/10.1016/j.compstruct.2016.10.086

[29] Lei an Zhang, Yanzhen Guo, Jinghua Wang, Xuemei Huang, Xiuting Wei, Weisheng Liu, 2019. Structural failure test of a 52.5 m wind turbine blade under combined loading, *Eng. Fail. Anal.* 10.1016/j.engfailanal.2019.04.069, 103, (286-293).

[30] Joshua Paquette, J van Dam, Scott Hughes, 2007. Structural Testing of 9 m Carbon Fiber Wind Turbine Research Blades, 45th AIAA Aerospace Sciences Meeting and Exhibit, 08 January 2007 - 11 January 2007, Reno, Nevada, USA, https://doi.org/10.2514/6.2007-816.

[31] Joshua Paquette, J van Dam, Scott Hughes, Jay Johnson, 2008. Fatigue Testing of 9 m Carbon Fiber Wind Turbine Research Blades, Structural Testing of 9 m Carbon Fiber Wind Turbine Research Blades, 46th AIAA Aerospace Sciences Meeting and Exhibit, 07 January 2008 - 10 January 2008, Reno, Nevada, USA, https://doi.org/10.2514/6.2008-1350.

[32] Jeppe B. Jørgensen, Bent F. Sørensen, Casper Kildegaard, 2018. Tunneling cracks in full scale wind turbine blade joints, *Eng. Fract. Mech.*, **189**: 361-376.

[33] Othman Al-Khudairi, Homayoun Hadavinia, Christian Little, Gavin Gillmore, Peter Greaves, Kirsten Dyer, 2017. Full-Scale Fatigue Testing of a Wind Turbine Blade in Flapwise Direction and Examining the Effect of Crack Propagation on the Blade Performance, *Materials*, **10**, 1152; doi:10.3390/ma10101152

[34] Hak Gu Lee, Min Gyu Kang, Ji sang Park, 2015. Fatigue failure of a composite wind turbine blade at its root end, *Compos. Struct.*, **133**: 878-885.

[35] Eder, M. A., Belloni, F., Tesauro, A., & Hanis, T. 2017. A multi-frequency fatigue testing method for wind turbine rotor blades. *J. Sound Vib.*, **388**: 123–140.

[36] Darris White, 2004. New Method for Dual-Axis Fatigue Testing of Large Wind Turbine Blades Using Resonance Excitation and Spectral Loading, National Renewable Energy Laboratory, April 2004.

[37] Nathan Post, Falko Bürkner, 2019. Fatigue Test Design: Scenarios for Biaxial Fatigue Testing of a 60-Meter Wind Turbine Blade, National Renewable Energy Laboratory, Technical Report NREL/TP-5000-65227 Revised August 2019.

[38] Melcher, D., Bätge, M., and Neßlinger, S. 2020. A novel rotor blade fatigue test setup with elliptical biaxial resonant excitation, *Wind Energ. Sci.*, **5**: 675–684, https://doi.org/10.5194/wes-5-675-2020, 2020.

[39] Tang J, Chen X. 2018. Experimental investigation on ultimate strength and failure response of composite box beams used in wind turbine blades. *Compos. Struct.* **15**:198:19-34. https://doi.org/10.1016/j.compstruct.2018.05.042

[40] B. Cole, A. Miyase, T.P. Yu, K.H. Lo, S.S. Wang, 2015. Failure modes and strength of composite box beam structures, Proceeding of the American Society for Composites 30th Annual Technical Conference..

[41] Qin, Z., Wang, J.H., Yang, K., Yu, G.G., Xu, Y., Xu, J.Z. 2018. Design and nonlinear structural responses of multi-bolted joint composite box-beam for sectional wind turbine blades. *Compos. Struct.* **206**, 801–813.

[42] Sayer, F., Antoniou, A., Wingerde, A. 2012. Investigation of structural bond lines in wind turbine blades by sub-component tests. *Int. J. Adhes. Adhes.* **37**: 129–135.

[43] Zarouchas, D.S., Makris, A.A., Sayer, F., Van Hemelrijck, D., Van Wingerde, A.M. 2012. Investigations on the mechanical behavior of a wind rotor blade subcomponent. *Compos. Part B-Eng.* **43**: 647-654.

[44] Chen X, Berring P, Madsen SH, Branner K, Semenov S. 2019. Understanding progressive failure mechanisms of a wind turbine blade trailing edge section through subcomponent tests and nonlinear FE analysis. *Compos. Struct.* **214**:422-438. https://doi.org/10.1016/j.compstruct.2019.02.024

[45] F. Lahuerta, N. Koorn, D. Smissaert, 2018. Wind turbine blade trailing edge failure assessment with sub-component test on static and fatigue load conditions, *Compos Struct*, **204**: 755-766.

[46] Antoniou A, Rosemeier M, Pueyo C, Makris A, Roukis G, Lahuerta F, et al. 2017. Testing report on blade subcomponents. Work Package 7.1: Efficient blade structure Deliverable

number 7.1.2, Tech. rep., IRPWind, Grant agreement no 609795; 2017.

[47] Belloni F., Eder M.A., and Cherrier B., 2018. An Improved Sub-component Fatigue Testing Method for Material Characterization, *Exp. Tech.*, **42**:533–550.

[48] M. A. Eder, S. Semenov, M. Sala, 2020. Multiaxial Stress Based High Cycle Fatigue Model for Adhesive Joint Interfaces, H. Okada and S. N. Atluri (eds.), Computational and Experimental Simulations in Engineering, Mechanisms and Machine Science 75, Springer Nature Switzerland AG.

[49] Daniel D. Samborsky, Aaron T. Sears and John F. Mandell, 2009. Static and Fatigue Testing of Thick Adhesive Joints for Wind Turbine Blades, 2009 ASME Wind Energy Symposium.

[50] Daniel D. Samborsky, John F. Mandell and Pancasatya Agastra, 2012. 3-D Static Elastic Constants and Strength Properties of a Glass/Epoxy Unidirectional Laminate, Department of Chemical and Biological Engineering, Montana State University, Bozeman, MT, 2012.

[51] H. Kotik and J. Perez Ipiña, 2018. 3D Quasi-Static Strengths and Elastic Constants of Glass Fiber Reinforced Polyester Composite Extracted From a Wind Turbine Blade, *J. Test. Eval.* **46**: 55-66.

[52] Dimitrios Zarouchas, Rogier Nijssen, 2016. Mechanical behaviour of thick structural adhesives in wind turbine blades under multi-axial loading, *J. Adhes. Sci. Technol.*, **30**:13, 1413-1429, DOI: 10.1080/01694243.2016.1146392

[53] Garbiñe Fernandez, Dirk Vandepitte, Hodei Usabiaga, Stijn Debruyne, 2018. Static and cyclic strength properties of brittle adhesives with porosity, Int. J. Fatigue. **117**: 340–351

[54] SNL/MSU/DOE 2019 Composite Material Fatigue Database, version 29.0, https://www.montana.edu/composites/ accessed on May 28, 2019.

[55] De Smet, B. J., & Bach, P. W. 1994. DATABASE FACT, fatigue of composites for wind turbines. 3d IEA symposium on wind turbine fatigue. The Netherlands: ECN Petten.

[56] Nijssen, R., OptiDAT database reference document, 2006.

[57] K.A.M. Vallons, 2015. Databases for fatigue analysis in composite materials, Fatigue of Textile Composites, Woodhead Publishing Series in Composites Science and Engineering, Pages 75-82

[58] E.EGdoutos, I.M Daniel, K.-A Wang, 2002. Failure of cellular foams under multiaxial loading, *Composites, Part A.* **33**: 163-176.

[59] Isaac M. Daniel, Jeong-Min Cho, 2010. Characterization of Polymeric Foams under Multi-Axial Static and Dynamic Loading, Proceedings of the SEM Annual Conference, June 7-10, 2010, Indianapolis, Indiana USA.

[60] Zenkert, Dan, Andrey Shipsha, and Magnus Burman. 2006. Fatigue of Closed Cell Foams. *J. Sandwich Struct. Mater.* **8**: 517–38. doi:10.1177/1099636206065886.

[61] Thomsen, Ole Thybo. 2009. Sandwich Materials for Wind Turbine Blades — Present and Future. *J. Sandwich Struct. Mater.* 7–26. doi:10.1177/1099636208099710.

[62] Brøndsted, P., Lilholt, H. and Lystrup, A., 2005. Composite materials for wind power turbine blades, *Annu. Rev. Mater. Res.* **35**:505–38.

[63] E.S. Greenhalgh, 2009. Failure Analysis and Fractography of Polymer Composites. Woodhead Publishing in Materials; 2009 978-1-84569-217-9.

[64] Kristine M. Jespersen, Jens A. Glud, Jens Zangenberg, Atsushi Hosoi, Hiroyuki Kawada, Lars P. Mikkelsen, 2018. Ex-situ X-ray computed tomography, tension clamp and in-situ transilluminated white light imaging data of non-crimp fabric based fibre composite under fatigue loading, *Data in Brief*, **21**.

[65] Chen X. 2020. Fractographic analysis of sandwich panels in a composite wind turbine blade using optical microscopy and X-ray computed tomography. *Eng. Fail. Anal.*. 2020;111. 104475. https://doi.org/10.1016/j.engfailanal.2020.104475.

[66] Aabo, C.; Scharling Holm, J.; Jensen, P.; Andersson, M.; Lulu Neilsen, E. 2018. MEGAVIND Wind Power Annual Research and Innovation Agenda; MEGAVIND: Copenhagen, Denmark, 2018.

[67] Eder, M.A., Sarhadi, A., 2020. A semi-analytical correction method for bi-material Mode-III V-notches, *Theor. Appl. Fract. Mech.*, **106**.

[68] WMC Laboratories, FOCUS6, https://www.lmwindpower.com/en/products-and-services/technology-centers/wmc-the-netherlands, accessed on May 28, 2020.

[69] DNV-GL, Bladed, https://www.dnvgl.com/services/wind-turbine-design-software-bladed-3775, accessed on May, 28, 2020.

[70] Jonathan C. Berg and Brian R. Resor, 2012. Numerical Manufacturing And Design Tool (NuMAD v2.0) for Wind Turbine Blades: User's Guide, Sandia National Laboratories, 2012.

[71] Bak, Christian; Zahle, Frederik; Bitsche, Robert; Kim, Taeseong; Yde, Anders; Henriksen, Lars Christian; Hansen, Morten Hartvig; Blasques, José Pedro Albergaria Amaral; Gaunaa, Mac; Natarajan, Anand, 2017. The DTU 10-MW Reference Wind Turbine v9_1, 2017.

[72] Jonkman, J., S. Butterfield, W. Musial, and G. Scott, "Definition of a 5-MW Reference Wind Turbine for Offshore System Development," NREL/TP-500-38060, Golden, CO: National Renewable Energy Laboratory, February 2009.

[73] D. Todd Griffith and Thomas D. Ashwill, The Sandia 100-meter All-glass Baseline Wind Turbine Blade: SNL100-00, Sandia National Laboratories, 2011.

[74] Chen X. Collapse of a 47-meter composite blade under combined bending and torsion in a full-scale static test. In 35th Wind Energy Symposium. 2017. (35th Wind Energy Symposium, 2017). https://doi.org/10.2514/6.2017-1166

[75] Yu, T.P., Miyase, A., Lo, K.H., Wang, S.S. Composite box-beam failure modes and strength: 3D modeling and analysis and comparison with experimental results. In: Proceeding of the American Society for Composites 31th Annual Technical Conference, 2016.

[76] Chen X, Haselbach PU, Branner K, Madsen SH. 2019. Effects of different material failures and surface contact on structural response of trailing edge sections in composite wind turbine blades. Compos. Struct. **226**. 111306. https://doi.org/10.1016/j.compstruct.2019.111306

[77] Rosemeier, M, Antoniou, A, Chen, X, Lahuerta, F, Berring, P, Branner, K. 2019. Trailing edge subcomponent testing for wind turbine blades–Part A: Comparison of concepts. *Wind Energy*. **22**: 487– 498. https://doi.org/10.1002/we.2301

[78] Tang J, Chen X, Yang K. 2019. Evaluating Structural Failure of Load-Carrying Composite Box Beams with Different Geometries and Load Conditions. *Appl. Compos. Mater.* **26**(4):1151-1161. https://doi.org/10.1007/s10443-019-09776-4

[79] Chen X, Tang J, Yang K. 2019. Modeling multiple failures of composite box beams used in wind turbine blades. Compos. Struct. **217**:130-142. https://doi.org/10.1016/j.compstruct.2019.03.018.

[80] Eder MA, Chen X. 2020. FASTIGUE: A computationally efficient approach for simulating discrete fatigue crack growth in large-scale structures. *Eng. Fract. Mech.* **233**. 107075. https://doi.org/10.1016/j.engfracmech.2020.107075

[81] Shihavuddin ASM, Chen X, Fedorov V, Christensen AN, Riis NAB, Branner K et al. 2019. Wind Turbine Surface Damage Detection by Deep Learning Aided Drone Inspection Analysis. *Energies*. **12**(4). 676. https://doi.org/10.3390/en12040676

[82] ReliaBlade, https://www.reliablade.com/, accessed on May 28, 2020.

Upgrading and reuse of glass fibre recycled from end-of-life composites

J L Thomason[*], L Yang and K Pender

Department of Mechanical and Aerospace Engineering, University of Strathclyde, 75 Montrose Street, Glasgow G1 1XJ, Scotland

[*]E-mail: james.thomason@strath.ac.uk

Abstract. The value of recycled glass fibres is significantly reduced due to the loss of fibre strength and surface functionality that occurs during recycling. Results are presented from the ReCoVeR project on the regeneration of the strength of thermally conditioned glass fibres. Thermal recycling of end-of-life glass fibre reinforced composites or composite manufacturing waste delivers fibres with virtually no residual strength or value. Composites produced from such fibres also have extremely poor mechanical performance. Data is presented showing that a short hot alkali treatment of glass fibres which have been heat treated at typical composite recycling temperatures can more than triple their strength and restore their ability to act as an effective reinforcement in second life composite materials. Glass fibre recovered from fluidised bed recycling of composite materials exhibited much greater levels of mechanical abrasion damage. However, the strength of these fibres could also be increased to levels required in composite reinforcement by longer or more aggressive alkali treatment. The implications of these results for real materials reuse of recycled glass fibres as replacement for pristine reinforcement fibres are discussed.

1. Introduction

The disposal of composite manufacturing waste and end-of-life composite products in an environmentally friendly manner is one of the most important challenges currently facing the industrial and academic composites community. Glass fibre was the reinforcement of choice in more than 90% of all these composites and more than six million tons of reinforcement grade E-glass fibre was used in composite production in 2019. This will have potentially been associated with the production of up to one million tons of glass fibre manufacturing waste [1] most of which is landfilled. Approximately 70 % of reinforcement glass fibre is used to manufacture thermoset based glass reinforced plastics (GRP) which also produces approximately 15 % manufacturing waste. Such GRP materials (both end-of-life and manufacturing waste) are difficult to recycle in an efficient manner and have historically also been disposed of in landfills. Such landfilling is rapidly becoming untenable due to legislative and landfill cost developments. The perspectives on this issue have been recently highlighted due to the accelerating growth in the use of such composite materials in transportation and wind energy sectors [2-4]. A number of processes are available for recycling such composites and thermal recycling is probably the most technologically advanced [2,3]. However, nearly all options deliver recycled glass fibres (RGF) that suffer from a lack of cost competitiveness with pristine first-pass materials.

A critical technical challenge in the development of GRP recycling technology is the 80-90 % drop in the performance (and value) of RGF in comparison to its original state [5-9]. Recent studies have confirmed that the room temperature glass fibre strength can be drastically reduced by exposure to temperatures in the 300°C-600 °C temperature range [5,6] typical of the many different potential GRP recycling processes (see Figure 1). Similar behaviour has also been observed in silica and basalt reinforcement fibres [6]. Consequently, RGF have a very poor performance to cost ratio, and in most cases are unsuitable for reprocessing and reuse as a valuable reinforcement for composites. A breakthrough in this field could enable RGF to compete with pristine materials in high volume discontinuous fibre reinforced composite applications which would have major technological, societal, economic and environmental impacts. Yang et al. [7] recently demonstrated that the reinforcement potential of RGF could be significantly improved by a post treatment with hydrogen fluoride (HF) to remove the "damaged" fibre surface layer. However, given the very aggressive nature of HF and the associated safety issues, it seems unlikely that a cost-effective regeneration process could be based on HF treatment. We have been researching less aggressive chemical routes to achieve similar results. The literature generally refers to the effects of acid or alkali treatments in terms of their deleterious effects on glass fibres and the resultant lowering of composite strength. However, this body of work is almost universally predicated on maintaining the performance of strong fibres. Little if any work has been reported on the use of acid or alkali treatments to change the strength of very weak fibres. However, it is well known that a silica network can be dissolved by the use of hot alkali solutions [1].

The ReCoVeR project (**Re**generated **Co**mposite **V**alu**e** **R**einforcement) focuses on enabling cost-effective recycling of glass fibre thermosetting composites. In particular ReCoVeR investigates the possibility of regenerating the strength of recycled glass fibres to enable their reuse as a composite reinforcement replacing the use of pristine fibre materials. In this paper we report some recent research results on the use of hot alkali solutions and subsequent silane sizing to modify the surface of thermally recycled glass fibres and the resultant effects on glass fibre and composite strength.

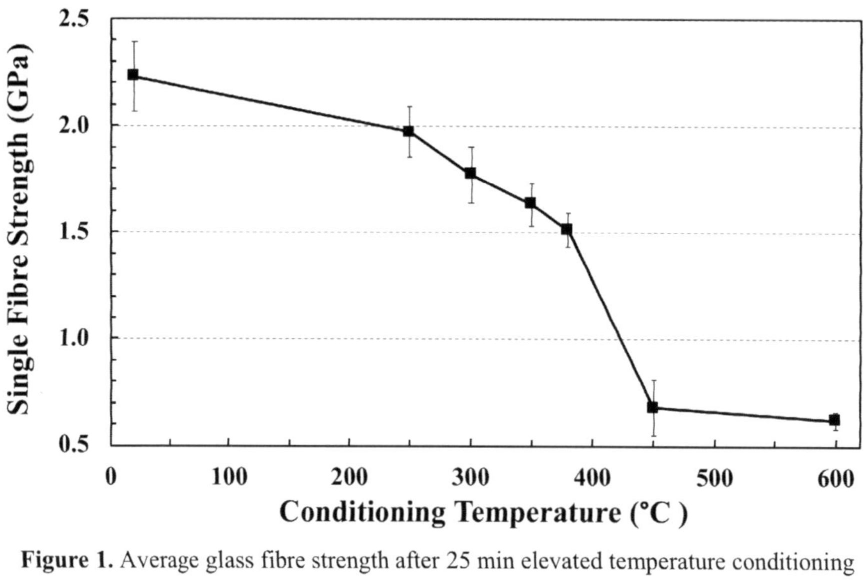

Figure 1. Average glass fibre strength after 25 min elevated temperature conditioning

2. Experimental materials and methods

2.1. Materials

Thermal and chemical treatments were carried out on aminosilane coated (OC-APS) boron-free E-glass fibres supplied by Owens Corning [5]. Glass mat thermoplastic (GMT) composites were prepared using PPG 8069 chopped glass fibres, a product designed for easy dispersion in aqueous media for the production of wet-laid non-wovens, supplied by PPG Fibreglass (now part of NEG) combined with DA3/60 chopped homopolymer polypropylene (PP) fibres purchased in the form of Goonvean. Injection moulded composites were prepared using DS 2200-13P chopped glass fibres provided by 3B Fibreglass company, SABIC® PP 579 S polypropylene (PP) pellets with 1 % added Polybond 3200 maleic anhydride-grafted polypropylene pellets. Glass fibre reinforced epoxy (GF-EP) composites for recycling were produced in-house by infusing an E-glass fibres tri-axial 3-ply mat supplied by Hexcel Reinforcements with PRIME 27 Resin and PRIME 20LV extra slow hardener supplied by Gurit. After ashing in a furnace, it was established that the resulting GF-EP had a fibre weight fraction of 60%. The cured GF-EP feedstock was cut to 50 × 15 mm so the recycled fibres were long enough to be tested with a 20 mm gauge length.

2.2. *Fibre treatments*

Fibre bundles were arranged in a steel rig for thermal conditioning, which was carried out in air. A Carbolite furnace was used to treat the fibres at a range of temperatures up to 600 °C for 25 minutes. The rig was then extracted from the furnace and left to cool at room temperature, before the heat-treated fibre bundles were removed for further chemical treatments. These fibres are defined are glass fibres recycled thermally (rGFt). More details of the heat conditioning procedure are given in [5]. NaOH solutions were prepared according to the desired molarity and heated to 95 °C before treating the fibre bundles for the required times. After the fibres had been treated in alkaline solution, they were rinsed in 37 % HCl and deionised water for 1 minute respectively. The purpose of this rinsing procedure was to ensure the effective removal of residual deposits which developed on the fibre surface as a result of interaction with alkaline solution [1]. The SEM micrograph in Figure 2a shows a typical example of these deposits on an NaOH treated heat conditioned fibre. Where required, a 1 vol% γ-aminopropyltriethoxysilane (APS) aqueous solution was used to regenerate the surface functionality of the heat treated and alkali treated fibres. The fibres were immersed in the APS solution for 15 minutes. Samples for fibre testing were dried, however samples for GMT wet production were processed directly into the composites. Figure 2b shows the fibre surface of a heat conditioned fibre after NaOH treatment and silane treatment. This fibre exhibits the virtually featureless surface which is typically observed with new glass fibres and confirms the removal of the deposits seen in Figure 2a which can be very detrimental to the efficacy of the silane treatment [1].

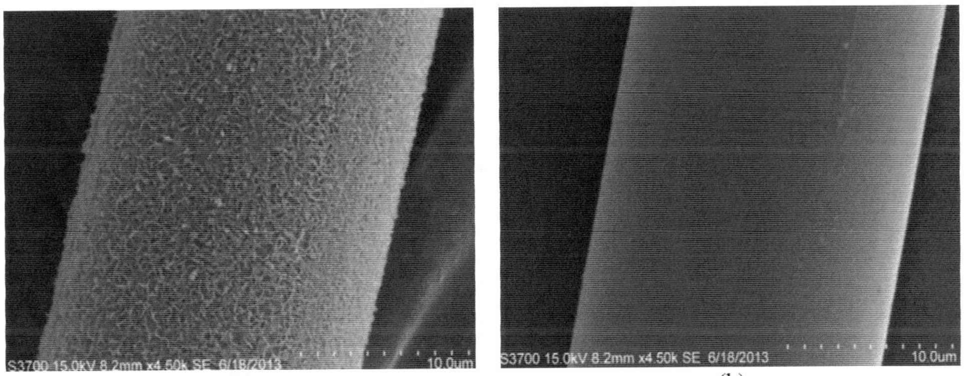

(a) (b)

Figure 2. SEM images of rGFt after a) NaOH treatment, b) NaOH + HCl rinse + APS coating.

2.3. Recycling glass fibre reinforced epoxy composites using fluidised bed

The GF-EP composites were recycled using an in-house built fluidised bed recycling process [10]. The thermosetting polymer matrix was thermally decomposed within the fluidised bed reactor, which comprises silica sand fluidised with air pre-heated using an electric heater. The composite feedstock was fed into the reactor through an access pipe, located above the expanded bed. The fluidising air was supplied to the reactor via a centrifugal fan and was sufficient to entrain the liberated glass fibres from the bed and into the reactor freeboard. An induced draft fan was used to draw the glass fibres and polymer volatiles from the freeboard and to keep the reactor under a slight negative pressure. The glass fibres were recovered from the gas stream using a cyclone separator. The bed temperature, static bed height and fluidisation velocity remained constant throughout this work at 500 °C, 150 mm and 1.5 m/s respectively. These fibres are defined as glass fibres recycled from composites (rGFc).

2.4. Single fibre tensile testing

Single fibre tensile testing was performed following the standard ASTM C1557-03 at a gauge length of 20 mm. Fibre diameters were measured using an optical microscope before testing for tensile strength using a Testometric tensile testing machine at ambient environment. The load cell was 5 N with an applied strain rate of 1.5 %/min. At least 30 fibres of each sample were tested. Error bars associated with the strength measurements represent 95 % confidence limits [1,5].

2.5. GF-PP Composite Preparation and Testing

The influence of the properties of glass fibres after thermal recycling and chemical regeneration on composite performance was further investigated by the preparation of 30 % in-plane randomly oriented glass fibre reinforced polypropylene laminates (GMT) prepared by a wet-laid papermaking process and compression moulding as previously described [1]. A water jet cutter was used to cut dog bone shaped tensile test specimens. Tensile tests were performed following ISO 527 using an Instron 5969 universal testing machine equipped with a 50 kN load cell and a crosshead speed of 1 mm/min. For injection moulded composites a Betol BC25 extruder was first used to compound PP pellets with 30% by weight glass fibres. The extruded material was drawn through a water bath and cut into pellets using a rotary cutter. An Arburg 170-90/200 injection moulding machine was used to produce dog-bone shaped tensile test specimens according to ASTM 638. The processing temperatures were set between 170°C and 230°C. All samples were conditioned at room temperature for three weeks before mechanical testing. An Instron 596 testing machine was used to perform the tensile tests. Sample strain was recorded with a video extensometer. A constant head displacement rate of 1 mm/min was used to test the composites. The PP reference sample was tested with a head displacement rate of 1mm/min up to 3 % strain. Then the head displacement rate was increased to 5 mm/min. Unnotched Charpy impact test specimens according to ISO 179-1 were cut from injection moulded tensile bars. A Tinius Olsen Impact 503 impact tester with a 25 J hammer was used to perform the impact tests.

3. Results

3.1. Fibre Strength after Thermal and Chemical Treatment

The results for the average single fibre tensile strength of the OC-APS fibres conditioned at 450, 500 and 600 °C and then treated in hot NaOH and APS are summarised in Figure 3 [1]. It can be seen that thermal conditioning caused a considerable reduction in fibre strength, with a loss of over 85% of the original strength. After a 10 minute treatment in hot NaOH solution the fibre strength initially increases significantly. It was also found that with prolonged treatment times and higher alkali concentrations the fibre strength improvement was significantly reduced [8]. It is generally accepted that the measured strength of glass fibres is dependent on their brittle nature and the state of the fibre surface region. Fundamental research on the thermally induced strength loss of glass fibre indicates that the phenomenon may have a complex nature but that it appears to be caused mainly by changes on the fibre surface. We have also recently demonstrated that the reinforcement potential of rGF can be significantly improved by a post treatment with hydrogen fluoride (HF) to remove this "damaged" surface layer [7].

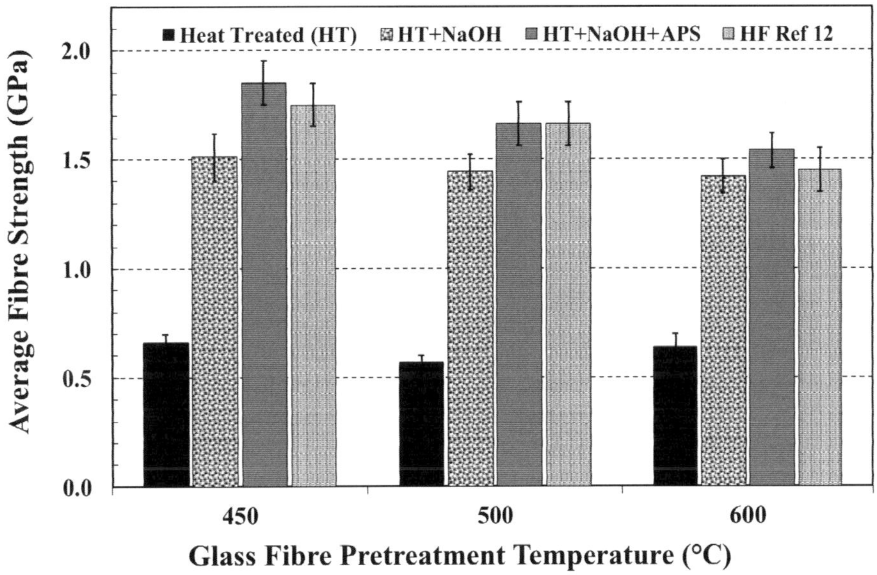

Figure 3. Influence of recycling temperature and ReCoVeR treatments on single fibre strength.

The results of a further investigation of fibre strength regeneration are also shown in Figure 3. OC-APS fibres were thermally conditioned at three different temperatures and subjected to either hot NaOH alone or hot NaOH followed by application of a simple APS sizing. The results obtained by Yang et al on the same fibre, thermally conditioned in an identical manner but then treated using HF etching, are also included in Figure 3 for comparison [7]. The large drop in average fibre strength (from an untreated reference value of 2.3 GPa) was observed after HT alone at all three temperatures. It was also noted that, after thermal conditioning, the fibres were extremely brittle and difficult to handle without breakage. Hence thermal recycling not only results in a very large drop in the glass fibre performance but also makes rGF virtually impossible to process in any standard composite production equipment. The results for the fibre strength recovery achieved with the hot NaOH treatment are impressive, with the strength regeneration for all three HT temperatures reaching approximately 1.5 GPa from the initial value of the HT fibre strength of (less than) 0.6 GPa. The addition of a simple APS sizing after the hot NaOH treatment leads to a further significant (at the 95 % confidence level) increase in fibre strength regeneration to well above 1.5 GPa. We consider this value of 1.5 GPa to be an important target for strength regeneration of recycled fibre in order to compete, on performance, with commercial first-pass glass fibre products [9].

From the above results, it can be seen that significant increases of fibre strength were obtained through hot alkali regeneration treatments, achieving greater than a tripling of fibre strength over the thermally treated fibre values. Consequently, hot alkali treatments enable regeneration of fibre strength to a level which makes reusing these fibres as a composite reinforcement a viable technical option.

3.2. Performance of Regenerated Fibres in GF-PP Composites

Figure 4 presents the results for the tensile strength of the GF-PP GMTs produced using thermally conditioned (HT 25 min at 500°C) and chemically treated fibres (NaOH and APS) [1]. The sample containing the HT fibres exhibits a large drop in tensile strength due to the low strength of the fibres

IOP Conf. Series: Materials Science and Engineering **942** (2020) 012002 doi:10.1088/1757-899X/942/1/012002

and the low level of interfacial adhesion in this composite. GMT tensile strength recovers by 44 % when the HT fibres are given a hot NaOH treatment. Addition of silane coupling agent after the NaOH treatment results in a 74 % recovery of the GMT tensile strength. To the best of our knowledge the result shown here is one of the first published demonstrations of a chemical treatment, other than HF etching, that significantly increases the strength of thermally recycled glass fibres. Moreover, the results on GMT laminates clearly show that this regeneration of glass fibre strength can be translated directly into composite strength. There are many non-technical factors which will play a role in the eventual commercial profitability of any GRP recycling process. However, maximizing the performance regeneration of rGF will, enable replacement of pristine fibre products from a performance viewpoint, maximize the value of such a recycled fibre product, and increase the economic attraction of composite recycling. The development of a non-HF based glass fibre strength regeneration treatment is an exciting development in the progress towards a cost-effective GRP recycling technology.

Figure 5 summarises the data on the tensile strength and unnotched Charpy impact performance of the injection moulded, glass reinforced polypropylene composites [11]. These composites were prepared with glass fibres that had been thermally conditioned for 25 min, across a range of temperatures. It can be seen that the tensile strength of these composites is also significantly reduced by the glass fibre thermal conditioning prior to composite manufacture. The glass fibre heat-treatment at 300°C caused a tensile strength reduction of more than 40%. A glass fibre heat treatment at higher temperatures caused a drop of the tensile strength of greater than 50%. The tensile strength of unreinforced PP was measured as 35.8 MPa. Consequently, the thermal conditioning of the glass fibres above 300°C essentially removed 100% of the reinforcement effect of the fibres in terms of increasing tensile strength of the composite in comparison with the tensile strength of the PP polymer alone. Similarly, the thermal conditioning of the glass fibres resulted in a loss of more than 50% of the unnotched impact performance of the composite. These results clearly reveal the potential poor reinforcement performance of any glass fibres obtained by thermal recycling of composites. The stiffness of the GF-PP composites was not seriously comprised by the thermal conditioning but the large drop in strength means that there is unlikely to be any significant commercial interest in large-scale application of rGF unless the properties of the rGF can be cost-effectively regenerated.

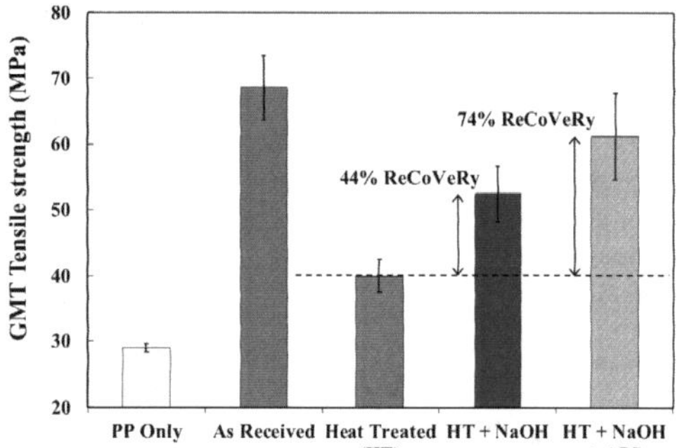

Figure 4. Influence of ReCoVeR fibre treatments on GF-PP GMT tensile strength

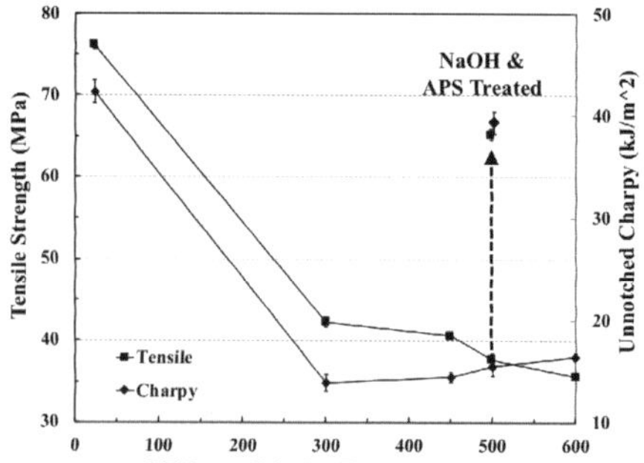

Figure 5. Influence of ReCoVeR treatments on injection moulded GF-PP performance

Injection moulded glass fibre reinforced PP composites were produced where the heat-treated fibres were also subjected to NaOH regeneration treatment and APS coating prior to extrusion and moulding. Data on the tensile strength of these composites is also shown in Figure 5. It can be seen that 71% of the composite strength loss due to thermal preconditioning of the glass fibres at 500°C was recovered by use of this treatment. The unnotched impact strength recovery of these composites was even more impressive with 87% of the heat treatment loss in composite impact strength performance recovered. This level of composite performance recovery exceeds the 50-70% regeneration of the properties of GF-epoxy composites reported by Yang et al [7] when using HF treatment to regenerate the strength of heat-treated glass fibres. The application of a full commercial PP compatible sizing [12,13] to the ReCoVeR treated fibres may well increase the final composite performance further.

3.3. Tensile strength of fibres recycled from composites

Thomason has recently reviewed the available literature on the thermally induced strength loss in glass fibres and concluded that the mechanism of strength loss in thermally recycled glass fibre involves a combination of thermal and mechanical damage. Clearly the potential for greater levels of mechanical damage are present when considering potential large-scale industrial recycling processes [6,14]. Figure 6 shows the single fibre tensile strength for the same glass fibres after only thermal conditioning rGFt compared to fully recycled rGFc from fluidised bed processed composites, both following treatment in the NaOH solutions. The NaOH treatment time was kept constant at 10 min and solution concentrations of 1.5, 3 and 5 mol/L were used. The strength of glass fibre after thermal conditioning increases with NaOH concentration. Higher concentrations may continue to improve the strength of these fibres since a plateau does not appear to have been reached in the range of concentrations studied. However, these NaOH treatments showed little effect on the strength of rGFc, with a relatively consistent single fibre strength of 0.5 - 0.6 GPa in all cases. It has been hypothesised that the reaction of silica in the glass fibre with hydroxide ions from the NaOH solution blunts and/or removes the severe surface flaws that manifest during exposure to elevated temperatures, and thus increases the tensile strength of the fibre [1,8]. Given their significantly lower strength, it seems likely that there are more severe flaws on the surface of the rGFc compared to the rGFt.

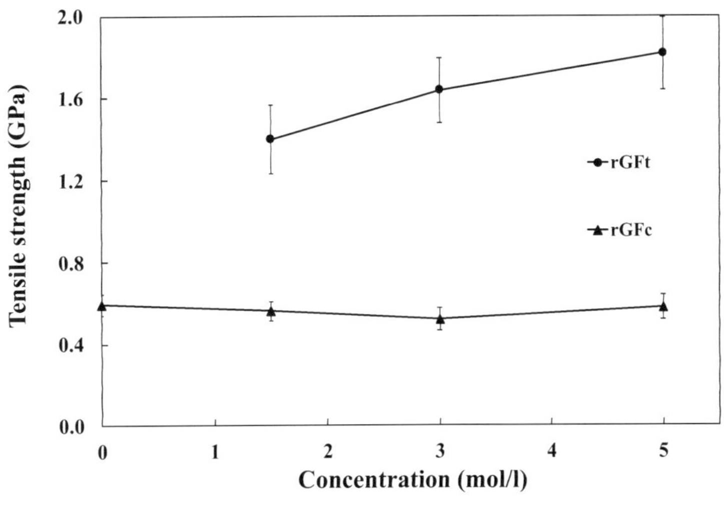

Figure 6. Tensile strength of rGFt and rGFc after NaOH treatment at different concentrations

The surface of fibres recycled in the fluidised bed was further studied under SEM, as seen in Figure 7. The presence of apparent damage was found on the surface of rGFc in the form of scratches and depressions. No such features were observed on the surface of glass fibres simply thermally conditioned within a furnace at the same temperature, suggesting the damage is a result of mechanical attrition during recycling. Given the high hardness of silica sand, abrasion from this sand in the fluidised bed may be a likely cause. Pickering et al. [3], Fenwick [15] and Kennerley [16,17] reported the properties of glass fibres recycled using a fluidised bed, however, no such features were discussed.

IOP Conf. Series: Materials Science and Engineering **942** (2020) 012002 doi:10.1088/1757-899X/942/1/012002

The depth of the damage cannot be determined using SEM; however the topography of the fibre surface could be further explored using atomic force microscopy.

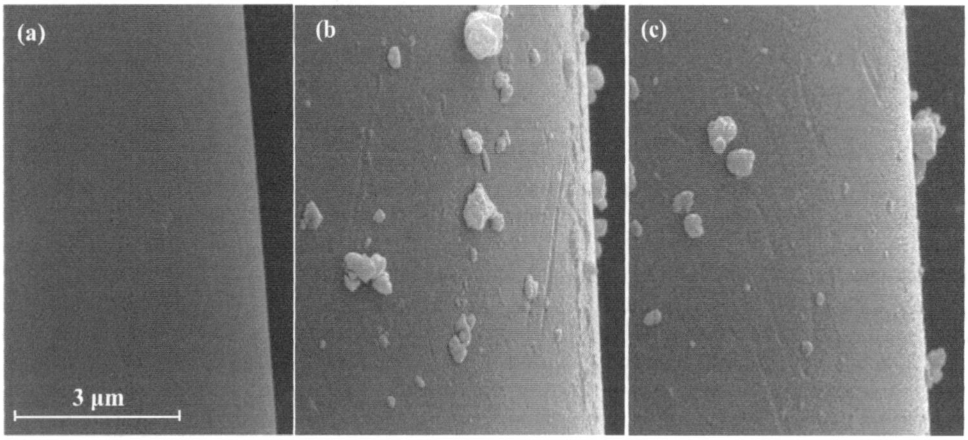

Figure 7. SEM images comparing surface features of (a) rGFt, (b) rGFc (c) rGFc

Given the lack in strength increase under the conditions used in Figure 6, rGFc were treated with NaOH for extended times. Figure 8 gives the strength and relative diameter of rGFc after treatment with a 5 mol/L solution of NaOH for times up to 180 min. There is a clear increase in tensile strength with treatment time when treating for longer than 20 min. The fibre strength appears to begin to plateau at 1.25 GPa after treating for 180 min, suggesting only minimal additional strength gain can be achieved by extending treatment duration. The reduction in rGFc diameter with treatment time is more pronounced when extending the range of NaOH treatment times. Similar to the tensile strength data, significant effect of the NaOH on rGFc diameter was not observed for treatments shorter than 60 min.

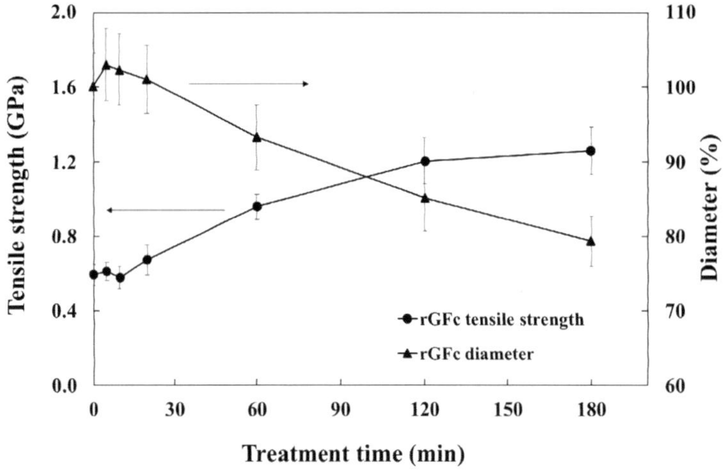

Figure 8. Tensile strength and relative diameter of rGFc treated with NaOH at 5 mol/L

It was anticipated that removing additional damaged surface material could lead to re-strengthening of recycled fibres; therefore, additional concentrations of NaOH solutions were applied to rGFc at extended treatment times. Figure 9 gives the tensile strength of rGFc then treated with NaOH at the various concentrations and times. All concentrations investigated, except 1.5 mol/L NaOH solution, increased the strength of rGFc. In agreement with the rGFt data in Figure 6, the effectiveness of NaOH on rGFc strength tends to increase with NaOH concentration at these longer treatment times. rGFc treated with 1.5 mol/L NaOH solution may exhibit an increase in strength if the treatment times were extended further [8,18]. It appears there is a maximum strength increase range of 1.2-1.4 GPa for 7 and 10 mol/L treatments, where extending the treatment time does not improve the fibre strength. This plateau occurs more rapidly when higher concentrations are used. Figure 9 shows the strength of fibres treated at 7 and 10 mol/L appear to plateau after treatments lasting 120 and 60 min respectively. The 5 mol/L treatment also appears to be reaching the plateau level and may reach its maximum level for about 4 hr treatment time. No plateau in regenerated fibre strength is observed for 3 mol/L treatments which approximate a linear increase in strength within the treatment times studied. Extended treatments at 3 and 5 mol/L are required to determine if/when the fibre strength plateaus when using these concentrations. The maximum possible strength increase may in fact be the same for all concentrations, with lower concentrations requiring longer treatments to converge.

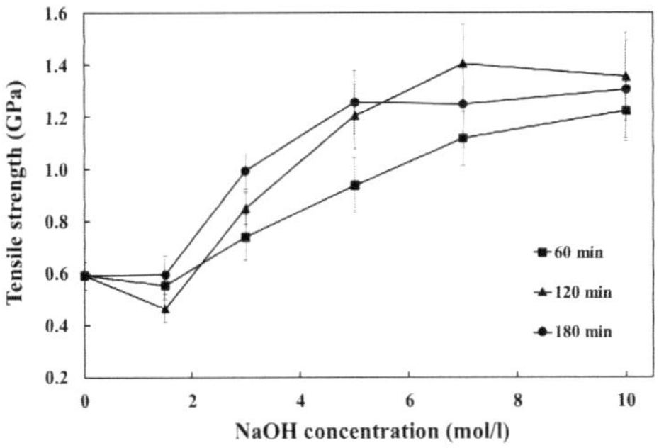

Figure 9. Tensile strength of rGFc treated with NaOH at different concentrations and times

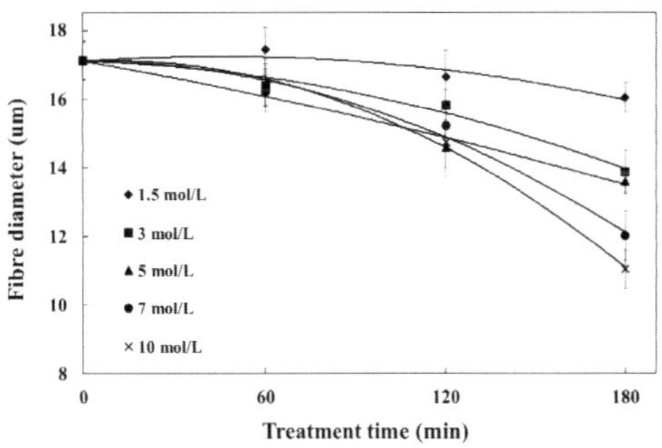

Figure 10. Diameter of rGFc treated with NaOH at different concentrations and times

Figure 10 shows the rGFc diameter after NaOH treatment at the various times and concentrations used in Figure 9. The trend lines represent second order polynomials with a minimum R^2 value of 0.91. rGFc treated with 1.5 mol/L solution exhibit no significant change in diameter until treated for 180 min. All treatments which exhibit an increase in fibre strength in Figure 9, also show a significant reduction in fibre diameter. Both fibre diameter reduction and rate of reduction tend to increase with treatment time as shown in Figure 10. Yang et al. observed an inverse relation between rate of fibre reduction and treatment time when etching glass fibre with HF [7]. This was attributed to the build-up of insoluble residue particles on the fibre surface masking the fibre surface and impeded the etching process over time. However, this does not seem to occur when using NaOH within the treatment times studied. The observations made in Figure 10 are in line with Bashir et al. who examined the kinetics of dissolution of glass fibres in NaOH solution, finding that a radial diffusion model does not describe the reaction well [18].

Under the conditions investigated, the maximum rGFc tensile strength attained was approximately 1.4 GPa after treating for 120 min at 7 mol/L. This is comparable to the strength level obtained for the heat treated fibres shown in Figure 3 which could be further processed to give the excellent composite performance discussed in Figures 4 and 5. It was previously demonstrated by Thomason and Kalinka that the actual average strength of fibres in a commercial 4 mm chopped glass product used in extrusion and injection moulding was as low as 1.5 GPa at a gauge length of 2 mm [9]. It is well established that there is an inverse relation between fibre tensile strength and gauge length, therefore, the regenerated strength of 1.4 GPa at 20 mm gauge length found in this work may provide a fibre strength distribution that is higher than that obtained in typical commercial discontinuous glass fibre products. It seems likely that injection moulded composite applications requiring short glass fibres will be a better starting point for the use of recycled fibres than those requiring strength at longer lengths, such as the 10-20 mm fibres used in long fibre thermoplastic processes as these will have a higher probability of still having a critical flaw on their larger surface.

4. Conclusions

Glass fibres lose 80 % or more of their strength when exposed to temperatures used in GRP thermal recycling processes making them unsuitable for reuse as a composite reinforcement. The results of single fibre tensile testing clearly show that more than 50% of that strength loss can be recovered by a short treatment of the fibres using hot alkali solution. This strength recovery can be maintained and improved upon by a further application of a silane sizing which also acts to recover the compatibility of the recycled fibre surface with a polymer matrix. The regeneration of the performance of thermally treated glass fibres by these simple chemical methods resulted in GF-PP GMT and injection moulded composites which exhibited a 70-90 % property recovery compared to composites based on thermally recycled fibres which had received no further treatment. It is proposed that further optimization of these chemical treatments can lead to further improvements in performance enabling then to be used to replace glass fibre products used in many discontinuous glass-fibre reinforced composite applications.

However, real-life recycling of glass fibres from composites exposes fibres to both thermal and mechanically abrasive conditions. Consequently, the treatment time and/or NaOH solution concentration required for strength regeneration was found to be significantly higher for glass fibres recycled from composites in a fluidised bed, compared to fibres thermally conditioned in a furnace at the same temperature. This was attributed to fibres sustaining a greater degree of damage during the recycling processes, therefore, requiring a greater degree of surface etching to remove/sufficiently modify the surface flaws. It was found that treating such fibres from fluidised bed composite recycling in 7 mol/L NaOH solution for 2 hr yielded approximately a 130% increase in fibre tensile strength.

References

[1] Thomason JL, Nagel U, Yang L and Sáez ER 2016 *Compos. Pt. A* **87** 220-7
[2] Job S 2013 *Reinf. Plast.* **57** 19-23
[3] Pickering SJ 2006 *Compos. Pt. A* **37** 1206-15
[4] Oliveux G, Dandy LO and Leeke GA 2015 *Prog. Mater. Sci.* **72** 61-99
[5] Thomason JL, Yang L and Meier, R 2014 *Compos. Pt. A* **61** 201–8
[6] Thomason JL, Jenkins P and Yang L 2016 Fibres 4 18
[7] Yang L, Sáez ER, Nagel U and Thomason JL 2015 *Compos. Pt. A* **72** 167–74
[8] Bashir ST, Yang L, Anderson R, Tang PL, Liggat JJ and Thomason JL 2017 *Compos. Pt. A* **102** 76-89
[9] Thomason JL and Kalinka G 2001 *Compos. Pt. A* **32** 85-90
[10] Pender K and Yang L 2019 *Polym. Compos.* **40** 3510-19
[11] Nagel U, Yang L, Kao CC and Thomason JL 2018 *Polym. Compos.* **39** 1032-40
[12] Thomason JL 2019 *Compos. Pt. A* **121** 56-63
[13] Thomason JL 2015 Glass Fibre Sizing: A Review of Size Formulation Patents (Blurb Books) ISBN: 798-0-9573814-3-8
[14] Jenkins PG, Yang L, Liggat JJ and Thomason JL 2015 *J. Mater. Sci.* **50** 1050-57
[15] Fenwick N 1996 *Recycling of Composite Materials Using Fluidised Bed Processes* (PhD Thesis, University of Nottingham)
[16] Kennerley JR, Fenwick NJ, Pickering SJ and Rudd CD 1998 *Compos. Pt. A* **29** 839-45
[17] Kennerley J 1998 *Recycling Fibres Recovered from Composite Materials Using a Fluidised Bed Process* (PhD Thesis, University of Nottingham)
[18] Bashir ST, Yang L, Liggat JJ and Thomason JL *2017 J. Mater. Sci.* **53** 1710-22

IOP Conf. Series: Materials Science and Engineering **942** (2020) 012003 doi:10.1088/1757-899X/942/1/012003

Numerical and experimental analyses in composites processing: impregnation, heat transfer, resin cure and residual stresses

Filip Bo Salling[1*], Michael Sandberg[1], Jon Spangenberg[1] and Jesper H Hattel[1*]

[1]Department of Mechanical Engineering, Technical University of Denmark (DTU), 2800 Kgs. Lyngby, Denmark.

Corresponding authors: *fsras@mek.dtu.dk and *jhat@mek.dtu.dk

Abstract. This is a literature review of the published scientific results by the authors on numerical modelling and experimental characterization of composites processing with a focus on the resin injection pultrusion (RIP) process. Differential scanning calorimetry is used to obtain cure kinetic parameters for an epoxy resin system. The cure kinetic parameters are used for thermo-chemical simulations of the RIP process which are subsequently validated by temperature measurements. Scanning Electron Microscopy (SEM) and X-ray Computed Tomography (XCT) are used to characterize the microstructure of a glass fiber reinforced polyurethane profile. SEM was used to obtain the variation in cross-sectional fiber volume fraction and subsequently relating this variation to permeability. XCT was used to obtain individual fiber inclination and relating these inclinations to the observed reduction in longitudinal stiffness as proposed in literature by Herbert Krenchel. Considering multi-physics modelling of the RIP process two new state-of-the-art approaches are summarized in this work: i) a coupled flow-thermo-chemical model necessary for thick-section profiles, and ii) a steady state 3D-Eulerian approach for thermo-chemical-mechanical simulations showing a significant increase in computational performance compared with the traditional Lagrangian approach. Finally a study by Mortensen et al., on how to minimize residual stresses by minimizing the gelation temperature is reviewed.

1. Introduction

Composite materials, in the form of Fiber Reinforced Polymers (FRP), have in recent years been increasingly used for various structural applications, e.g. in the construction-, automotive-, and wind energy sectors. The demand for composite materials is expected to grow from USD 90.6 billion in 2019 to USD 131.6 billion by 2024, hence an average yearly growth rate of 7.7% [1]. The market growth is a result of a rising demand for high performance materials, i.e. light weight structures with high strength. Although the manufacturing cost of FRPs are high compared to metals they often require less post-processing, i.e. machining and fastening operations.

The use of FRPs in load bearing structural applications requires a high degree of control over process induced residual stresses and shape distortions. This is of utmost importance since internal residual stresses affect the in-service load conditions, and shape distortions might lead to not meeting geometrical specifications [2]. The inherent non-uniform distribution of residual stresses results in deformation, which are represented as matrix cracking, delamination, spring-in or warpage. On production sites most problems are solved using a trial-and-error approach which is very time consuming and labor intense, hence expensive. This approach is sought to be replaced by numerical analysis of the manufacturing process. In order to do so it is necessary to consider the fact that the residual stresses arising from the manufacturing process are a result of the interaction between the diverse multi-physics involved, i.e.: impregnation flow, heat transfer, polymerization or crystallization and solid mechanics. On top of that comes constitutive material behavior, tool-part interaction, micromechanics and laminate theory. The inherent multi-physics nature of the manufacturing process necessitates experimental analysis, i.e. materials characterization, e.g.: Differential Scanning Calorimetry (DSC) is used to obtain cure kinetics [3]; Light Optical Microscopy (LOM), Scanning Electron Microscopy (SEM) and X-ray Computed Tomography (X-ray CT) are used to characterize microstructure in 2D and 3D [4-6]; and special resin injection setups are necessary in order to characterize permeability and compaction behavior of the fiber-bed [7].

Hence, in order to minimize or even avoid the expensive trial-and-error based approach it is necessary to establish state-of-the-art numerical process models including input from advanced experimental materials analyses.

FRPs are manufactured using either a stationary- or continuous process. Vacuum Assisted Resin Transfer Moulding (VARTM) and manufacturing in an autoclave are classical examples of stationary FRP manufacturing processes while Pultrusion is a well-known continuous process.

The scope of the current work is to summarize state-of-the-art work on numerical and experimental analysis of composite manufacturing processes with a focus on thermoset resin systems and the continuous Resin Injection Pultrusion (RIP) process. However, most of the work is generic and can easily be extended to thermoplastics and composites manufacturing processes in general.

Finally, recommendations are given on how to minimize residual stresses by controlling the gelation temperature.

2. Numerical Analyses in Composites Processing

Manufacturing of high quality FRPs is directly related to the multi-physics involved during production, i.e. impregnation flow, heat transfer, chemical cure or polymerization, constitutive material behavior, and solid mechanics. The multi-physics involved is inherently connected but for the purpose of clarity presented separately in the following subsections.

2.1. Impregnation-Flow Modelling

Resin impregnation is the process step in FRP manufacturing where the fluid resin flows into the fibrous porous medium. In this paper, the governing equations for a saturated approach are presented as the following results are obtained by applying such methodology. This approach is characterized by a sharp flow front that splits the saturated and unsaturated domains. The saturated pressure-driven flow is modelled by Darcy's law [8]:

$$\mathbf{q} = \mathbf{u}\theta_v = -\frac{\mathbf{K}}{\mu}\nabla p + \theta_v \mathbf{U} \tag{1}$$

where $\mathbf{q}$ is the flux (i.e., the superficial resin velocity), $\mathbf{u}$ is the resin velocity, θ_v is the initial volume fraction of void in the fibrous porous medium, $\mathbf{K}$ is the permeability tensor, μ is the viscosity of the resin, and ∇p is the pressure gradient. Furthermore, $\mathbf{U}$ is the bulk velocity of the porous medium, which is zero in the case of stationary processes (e.g. most conventional resin transfer moulding processes), but non-zero in the case of continuous processes (e.g. pultrusion where $\mathbf{U}$ equals the pulling speed).

When modelling resin impregnation, the resin and fibres are typically assumed incompressible and thus the resin flow obeys the mass conservation (continuity) equation:

$$\nabla \cdot \mathbf{q} = 0 \tag{2}$$

Combining eqs. (1) and (2), yields the final equation for the resin pressure:

$$\nabla \cdot \left(-\frac{\mathbf{K}}{\mu}\nabla p + \theta_v \mathbf{U}\right) = 0 \tag{3}$$

The equation is solved with a numerical scheme such as the Finite Difference Method (FDM), Finite Volume Method (FVM), or Finite Element Method (FEM), in order to find the pressure field. Based on the pressure, the resin velocity is found by usage of Eqn. (1). Subsequently, the velocity is used to update the flow front of the resin via a free-surface algorithm [9-11].

The permeability of the porous medium is a very important parameter for the impregnation flow. The permeability is depending on the geometrical shape of the porous medium as well as the volume fraction of the fibres [12]. Researchers have estimated the permeability of porous media through experimental work [5,13] semi-analytical solutions and empirical relations [14], as well as simulations [15]. The most common equation for estimating the longitudinal permeability was presented by Kozeny and Carman [16,17]:

$$K_L = \frac{R_f^2(1-\theta_f)^3}{4k\theta_f^2} \tag{4}$$

where R_f is the fibre radius, θ_f is the volume fraction of the fibres, and k is the empirical Kozeny constant. In the case of unidirectional fibres, Gebart [14] has derived an analytical solution for the longitudinal and transverse permeability:

$$K_L = \frac{8R_f^2}{c}\frac{(1-\theta_f)^3}{\theta_f{}^2} \tag{5}$$

$$K_T = C_1 R_f^2\left(\sqrt{\frac{\theta_{f,max}}{\theta_f}} - 1\right)^{5/2} \tag{6}$$

where c and C_1 are parameters that depend on the fibre arrangement and $\theta_{f,max}$ is the maximum obtainable volume fraction of the fibres for the given arrangement.

2.2. Thermo-Chemical Modelling

Transient heat transfer is an essential phenomenon because thermal gradients and thermally activated chemical gradients, i.e. inhomogeneous vitrification or solidification, are driving forces for the build-up of residual stresses. The transient heat transfer is governed by the same main principle independent of resin type and manufacturing process. The transient temperate profiles are predicted numerically by solving the energy equation given in eq. (7).

$$\frac{\partial(\rho C_p T)}{\partial t} + \nabla \cdot (\rho C_p T\boldsymbol{u}) - \nabla \cdot (k\nabla T) = s \tag{7}$$

where T is the temperature, ρ is the density, C_p is the specific heat capacity, $\boldsymbol{k}$ is the thermal conductivity tensor, $\boldsymbol{u}$ is the velocity vector and ∇ is the divergence operator. It should be noted that lumped material properties are assumed when modelling the composite material. The source term s in eq. (7) is used to capture the internal heat generation related to chemorheological behavior, i.e. polymerization of thermoset resin systems and crystallization of thermoplastics resin systems. The source term representing the internal heat generation can be expressed as in [3,18]:

$$s = (1 - V_f)\rho_r H_{tr} R_r \tag{8}$$

where H_{tr} is the total heat of reaction, ρ_r is the resin density, V_f is the fiber volume fraction and R_r is the rate of cure or crystallization.

For thermoset resin systems the reaction of cure, i.e. R_r in eq. (8), is typically described using a cure kinetic equation with an Arrhenius-type temperature dependency. An example is the well-known autocatalytic cure kinetic model expressed as in [19]:

$$R_r(\alpha,T) = A_0 \exp\left(\frac{-E_\alpha}{RT}\right) \alpha^m (1-\alpha)^n \quad (9)$$

where α is the degree of cure defined as the ratio of the heat generated at a given time to the total heat of reaction which can by expressed as:

$$\alpha = \frac{H(t)}{H_{tr}} \quad (10)$$

In eq. (9) A_0 is a pre-exponential factor, E_α is the activation energy and m, n are reaction orders, all of which can be obtained by DSC experiments [3], which is the subject of subsection 3.1

For thermoplastic resin systems the chemorheological phenomenon taking place is crystallization. The crystallization transformation takes place between the melting temperature and glass transition temperature. For further introduction on modelling of composites manufacturing using thermoplastics confer [2].

The energy equation given in eq. (7) is typically solved using either the finite element method (FEM) [20] or a control volume based finite difference method [3,4,18].
In the case of a stationary composite manufacturing process, e.g. VARTM, the velocity in the energy eq. (7) would be $\boldsymbol{u} = \boldsymbol{0}$. Hence, the energy equation is simplified to the well-known heat conduction equation, which is readily solved as in the work by Bogetti et al. [21]. However, in the case of a continuous composite manufacturing process, e.g. pultrusion, a velocity component in the pulling direction should be included and the advective nature of the pultrusion process should also be accounted for when updating the degree of cure (thermosets) [18,22]. The latter can be expressed in the following way:

$$\frac{\partial \alpha}{\partial t} = R_r - u_1 \frac{\partial \alpha}{\partial x_1} \quad (11)$$

where u_1 is the velocity components in the pulling direction.

2.3. Constitutive Material Modelling

The stiffness of a thermoset resin matrix material is inherently related to the degree of cure. It is of great importance to capture this phenomenon in order to predict process induced residual stresses and distortions. A model for the cure dependent instantaneous resin modulus (E_r) was proposed by Bogetti et al. in [21] and expressed as:

$$E_r = (1-\alpha_{mod})E_r^0 + \alpha_{mod}E_r^\infty + \gamma\alpha_{mod}(1-\alpha_{mod})(E_r^\infty - E_r^0) \quad (12)$$

and [2]:

$$\alpha_{\mathrm{mod}} = \frac{\alpha - \alpha_{mod}^{gel}}{\alpha_{mod}^{diff} - \alpha_{mod}^{gel}} \quad (13)$$

where E_r^0 and E_r^∞ are the elastic modulus of the fully- uncured and cured resin, respectively. γ has a value between -1 and 1 and is used to capture the competing mechanism of chemical hardening and stress relaxation [23].

Eq. (12) was modified to incorporate the temperature dependency, i.e. thermal softening [24,25]. This model is known as the cure hardening instantaneous linear elastic (CHILE) approach, expressed as:

$$E_r = \begin{cases} E_r^0 & T^* \le T_{C1} \\ E_r^0 + \frac{T^*-T_{C1}}{T_{C2}-T_{C1}}(E_r^\infty - E_r^0) \quad for & T_{C1} < T^* < T_{C2} \\ E_r^\infty & T_{C2} \le T^* \end{cases} \quad (14)$$

where T^* is the difference between the instantaneous glass transition temperature and resin temperature, i.e. $T^* = T_g - T$, where the glass transition temperature is defined as a function of the degree of cure,

$$T_g = T_g^0 + \alpha_{T_g} \cdot \alpha \tag{15}$$

where T_g^0 is the glass transition temperature of the essentially uncured resin material and α_{T_g} is a constant.

Two other constitutive modelling approached, commonly used in literature, are the viscoelastic model and the path dependent model. The viscoelastic constitutive models are used to predict residual stresses by accounting for stress relaxation during polymerization, e.g. curing [2]. The viscoelastic models are used in either differential or integral form, where the latter is extensively used in literature [26-28]. The path dependent constitutive model is a simplified version of the viscoelastic model, i.e. the material behavior is considered path dependent as opposed to rate dependent [2]. The path dependent model was proposed in [29,30]

2.4. Thermo-Chemical-Mechanical Modelling

Based on an incremental approach where $\dot{()}$ denotes rate-changes or discrete increments, the constitutive relation between the stress, $\dot{\sigma}_{ij}$, and strains rates, $\dot{\epsilon}_{ij}$, are denoted by:

$$\dot{\sigma}_{ij} = L_{ijkl}(\dot{\epsilon}_{kl} - \dot{\epsilon}_{kl}{}^{pr}) \tag{16}$$

Where indices i, j, k, l take values between 1 and 3 and repeated indices imply summation. In eq. (16), components of the instantaneous stiffness tensor, L_{ijkl}, are based on the instantaneous modulus in eq. (14) together with a suitable micromechanics model (for example, the self-consistent field mechanics model by Bogetti and Gillespie [21]). The total strain rate, $\dot{\epsilon}_{ij}$, is derived from the displacement field using, for example, small-strain theory. In [31-33], the process-induced strain rate, $\dot{\epsilon}_{ij}{}^{pr}$, was approximated as an additive decomposition of the process-induced thermal and chemical strain rates, respectively $\dot{\epsilon}_{ij}{}^{pr} = \dot{\epsilon}_{ij}{}^{th} + \dot{\epsilon}_{ij}{}^{ch}$.

In [31-33], the process-induced thermal strain rate was estimated as the product of the thermal expansion coefficient, $\alpha_{ij}{}^{th}$, and the temperature rate. The chemical shrinkage rate was estimated as the effective shrinkage contraction coefficient, $\alpha_{ij}{}^{ch}$, and the volumetric shrinkage of the resin system, V_{sh}. In summary, these relations are:

$$\dot{\epsilon}_{ij}{}^{th} = \alpha_{ij}{}^{th}\dot{T}, \qquad \dot{\epsilon}_{ij}{}^{ch} = \alpha_{ij}{}^{ch}\dot{\epsilon}_r{}^{ch} \ \left(\text{with } \dot{\epsilon}_r{}^{ch} = \sqrt[3]{1 + V_{sh}\dot{\alpha}} - 1\right) \tag{17}$$

In most conventional solid mechanics problems, the strain and deformation states are typically associated with a point that follows the deformation or flow of the material. This approach has led to the current state-of-the-art 2D and 3D methods for the RIP process by Baran et al. [32-33], where a cross section or profile part travels with the pulling speed, u_1, through the thermal- and cure-history of the profile, see fig. 1. In [31], Sandberg et al. discussed that the approaches by Baran et al. [32-33] required many intermediate time-steps to reach the stationary solution. For the 3D approach, in particular, this was undesirable as it led to long computation times. Instead, an Eulerian approach is proposed where the rate-changes of strain and deformation are associated with the spatial variation along a streamline [31]:

$$\dot{()} = \frac{\partial \dot{()}}{\partial t} + u_1 \frac{\partial \dot{()}_{ij}}{\partial x_1} \tag{18}$$

where $\partial\dot{()}/\partial t = 0$ at steady state and u_1 is the pulling speed (fig. 1). In this new approach, stress and strain increments can be solved everywhere in one step, which greatly reduces computation times (see also subsection 4.3).

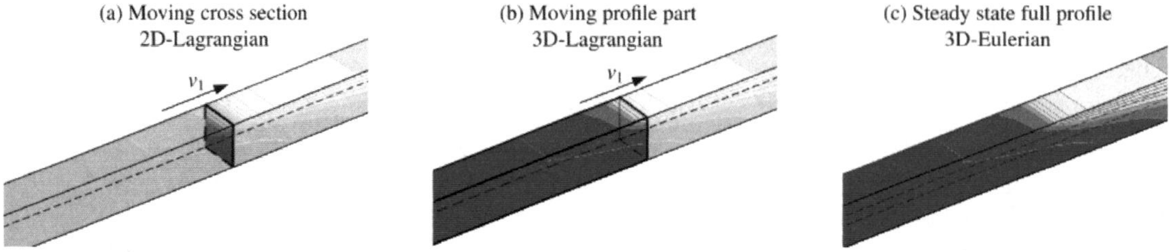

Figure 1. (a-b) Show the traditional modelling approach by Baran et al. [31,32], which are based on the Lagrangian framework. (c) Illustrates the new state-of-the-art approach by the authors considering the full profile in one step [33].

3. Experimental Analyses in Composites Processing

Experimental materials characterization is a necessity in order to obtain the needed input parameters for the numerical models to properly represent the real physical manufacturing process. In the case of FRPs, examples of process dependent material properties and material parameters are: viscosity, elastic modulus of the resin and in the case of a thermoset resin, the cure kinetic parameters (cf. eq. (9)).

In the following subsections examples of experimental analyses of FRPs are presented, i.e. chemorheology, microstructure investigations and permeability characterization.

3.1. Chemorheology

In the following the focus will be on processing of FRPs with a thermoset resin system, e.g. epoxies, polyesters, and polyurethanes. The state of the thermoset resin is changing during the manufacturing process, i.e. the polymer changes from a liquid to a solid, through a rubbery state, due to cross-linking of the polymer chains. This cross-linking phenomenon is referred to as curing of the resin and is captured in the numerical models through the reaction of cure, repeated here for clarity (cf. eq. (9)).

$$R_r(\alpha, T) = \frac{d\alpha}{dt} = A_0 \exp\left(\frac{-E_\alpha}{RT}\right) \alpha^m (1-\alpha)^n$$

The kinetic parameters in such models are generally obtained from DSC experiments. Fig. 2 shows a typical heat flow curve obtained from DSC. Recommendations for how to collect thermal analysis data are given in [34].

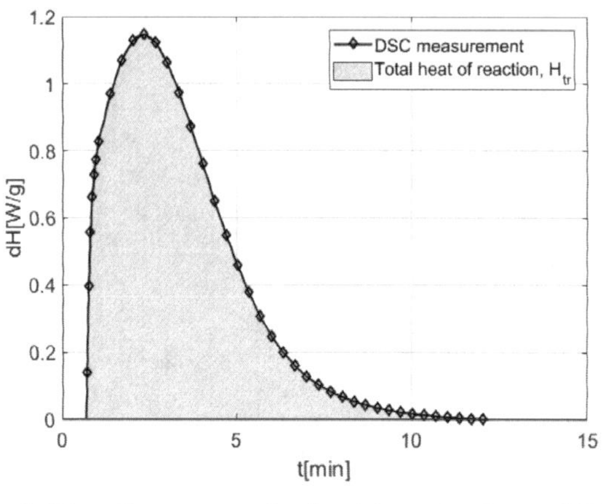

Figure 2. Heat flow curve obtained from isothermal DSC scan.

The total heat of reaction, i.e. H_{tr}, is readily found as the area between the heat flow curve and the baseline. The baseline is often assumed to be a straight line and in the current case it is simply the first axis, cf. fig. 2. Recommendations for how to perform kinetic computations are given in [35]. The kinetic parameters for the cure kinetic model presented in the equation above was obtained using a least square non-linear regression analysis in [3]. The results are summarized in table 1. together with another example of kinetic parameters also obtained from DSC experiments.

Table 1. Kinetic parameters obtained from DSC experiments.

	$A_0[s^{-1}]$	$E_\alpha\left[\frac{\text{kj}}{\text{mol}}\right]$	m	n
Polyester [18]	7.5581e9	82.727	0.63	1.85
Epoxy [3]	6.125e6	67.857	0.45	1.03

Examples of isothermal and dynamic DSC results and the corresponding cure kinetic model predictions of an epoxy resin system (cf. table 1) are depicted in figs. 3 and 4, respectively.

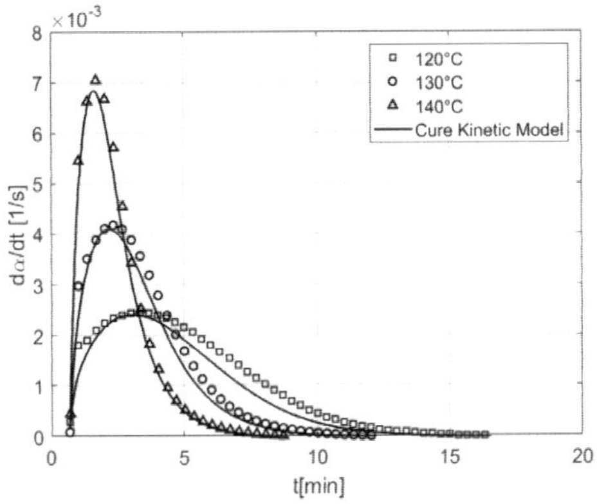

Figure 3. Isothemral DSC results and the corresponding cure kinetic model predictions.

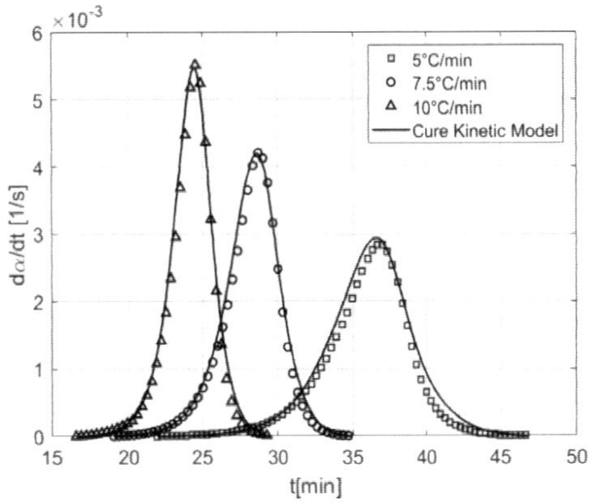

Figure 4. Dynamic DSC results and the corresponding cure kinetic model predictions

An overall good fit is observed between the measured DSC results and the fitted cure kinetic model, cf. figs. 3 and 4. Hence, the fitted cure kinetic model could readily be used to capture the exothermic heat generation in the numerical model through the reaction of cure, hence source term (cf. eqs. (8) and (9)). This is the subject of subsection 4.1.

The rheological behavior of the resin system, directly affects the instantaneous viscosity. A temperature and degree of cure dependent viscosity can be modelled as in [19]:

$$\eta(\alpha, T) = \eta_\infty \exp\left(\frac{\Delta E_\eta}{RT} + K\alpha\right) \tag{19}$$

where ΔE_η is the viscous activation energy, η_∞ is the initial viscosity, K is a constant, R is the universal gas constant and T the absolute temperature. Using a rheometer to measure the viscosity as a function of time and temperature and subsequently, using a least squares non-linear regression analysis, the constant in eq. (19) is readily found [19].

3.2. Microstructure Investigation

The quality of a FRP composite product is directly linked to the microstructure features of the composite part, e.g. fiber diameter, fiber volume fraction and fiber configuration. The quality of a FRP is inherently also a product of the manufacturing process, which is affected by the microstructure and vice versa, e.g.: i) porosities due to insufficient wet-out of the fiber-bed, ii) varying permeability due to fiber agglomeration; and iii) a decreased longitudinal elastic modulus due to fiber inclination.

In the following scanning electron microscopy (SEM) is used to assess the influence of varying fiber volume fraction on transverse permeability and X-ray computed tomography (XCT) is used to estimate the fiber inclination of textured continuous fiber yarns.

The material of interest in the present study is a glass fiber reinforced polyurethane (GF-PUR) rod produced by RIP. The glass fibers are continuous textured yarns. A reference case is used to pinpoint the influence of texturing of the glass fiber yarns. The reference case is a continuous unidirectional (UD) glass fiber reinforced polyester rod also produced by RIP.

3.2.1. Scanning Electron Microscopy (SEM). SEM is used to investigate the 2D microstructure of a material. The GF-PUR rod investigated has a diameter of Ø5mm and the microstructural feature of interest is the variation in cross-sectional fiber volume fraction as well as the fiber diameter distribution. The fiber volume fraction and fiber diameter are of specific interest because they dictate the permeability of the initial fiber-bed [5]. As mentioned in subsection 2.1 the permeability is typically assumed to follow the Kozeny-Carman equation modified by Gebart [14,16,17], cf. eqs. (5) and (6), here repeated for clarity:

$$K_{\mathrm{L}} = \frac{8R_f^2}{c}\frac{\left(1-V_f\right)^3}{{V_f}^2}$$

$$K_{\mathrm{T}} = C_1 R_f^2 \left(\sqrt{\frac{V_{f,max}}{V_f}} - 1\right)^{5/2}$$

where c and C_1 are fiber packing constants, assuming either hexagonal or quadratic fiber packing and R_f is the fiber radius.

Examples of SEM micrographs obtained for the GF-PUR rod and the reference case are depicted in fig. 5. It should be noticed how the textured GF-PUR rod has a highly non-uniform fiber distribution compared to the reference case of having solely UD fibers.

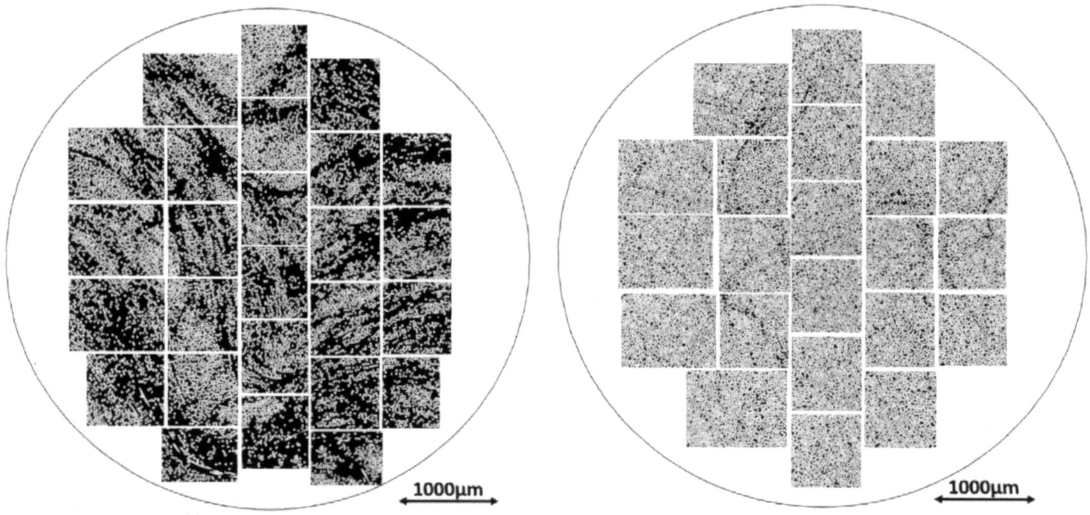

Figure 5. SEM micrographs of a glass fiber reinforced polyurethane rod produced using textured fiber yarns (left) and a reference case produced with solely unidirectional fibers (right).

The high contrast obtained from SEM, due to the high density difference of the constituent materials, ease the subsequent image analysis making a simple binerization analysis sufficient to evaluate the local fiber volume fraction. The resulting fiber volume fraction and the corresponding transverse permeability are depicted as a function of cross-sectional position in figs. 6 and 7, respectively. It should be noticed how fig. 6 shows a great resemblance with the original SEM micrographs (cf. fig. 5 left).

The transverse permeability estimates, depicted in fig. 7, are calculated using eq. (6) with a fiber diameter of $D_f = 23.2\mu m$ [6] assuming a hexagonal fiber packing [14]. The calculated permeability values range from approximately [2e-13 – 2e-9], hence a difference of a factor 1e4. However, in order to obtain a distinguishable contrast in fig. 7 the colour scheme is cropped at a maximum value of 4e-11. Hence it is not possible to distinguish between permeability values above 4e-11 (all the bright yellow areas in fig. 7).

The high cross-sectional variation in transverse permeability would very likely have a significant influence on the impregnation flow behaviour during manufacturing. Hence, this could be accounted for in flow-impregnation simulations using the local permeability values presented in fig. 7.

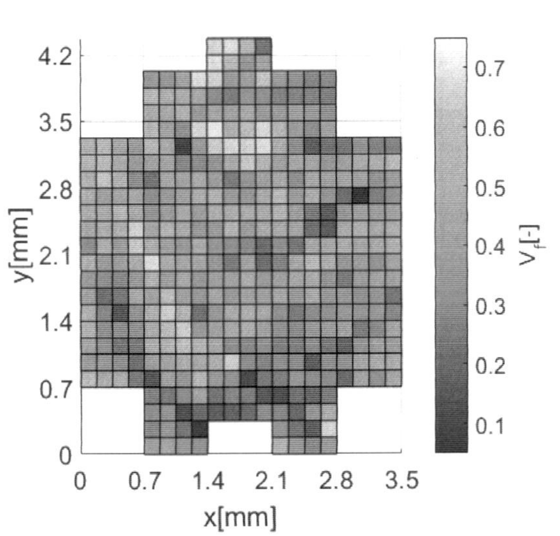

Figure 6. Fiber volume fraction distribution.

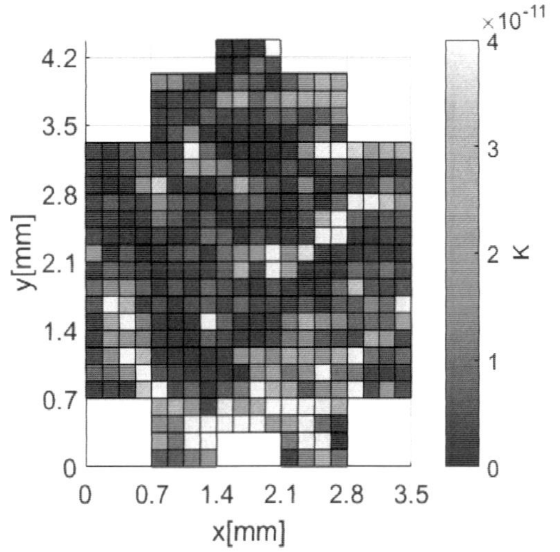

Figure 7. Transverse permeability, Eq. (6)

3.2.2. X-ray Computed Tomography (XCT). XCT is used to investigate the 3-dimensional microstructure of the GF-PUR composite rod described in the previous subsection. FRPs are known for their inherent anisotropic nature. One of the most well-known anisotropic properties is the elastic modulus (E). In the simple case of a unidirectional FRP the rule of mixture (ROM) is typically used to predict the idealised lumped elastic modulus of the composite. The ROM equations, predicting the longitudinal and transverse elastic modulus are expressed as

$$E_{cL} = V_f E_f + (1 - V_f) E_m \tag{20}$$

$$E_{cT} = \left(\frac{V_f}{E_f} + \frac{(1-V_f)}{E_m}\right)^{-1} \tag{21}$$

Eq. (20) and (21) are known as the upper- and lower-bound elastic modulus, respectively. As mentioned the ROM is applicable for idealised cases assuming solely UD fiber reinforcement. However, there are many FRP composite products where this assumption does not comply. Take for example the GF-PUR case with textured glass fiber yarns (cf. subsection 3.2.1). Some of the 2D micrographs depicted in fig. 5. (left) clearly show how the textured fiber yarns result in a FRP with non-unidirectional fiber reinforcement. Hence, the ROM, eq. (20), would overestimate the longitudinal modulus. The GF-PUR composite is the subject of ongoing research by the authors, but recent experimental results regarding the constituent- and lumped material properties are summarised in table 2.

Table 2. Preliminary experimental results on fiber volume fraction and fiber modulus together with the corresponding ROM prediction and measured longitudinal elastic modulus.

	$V_f[-]$ microscopy	E_f[GPa] single fiber test	E_m[GPa]* -	E_{cL}[GPa] ROM	E_{cL}[GPa] tensile test
Glass Fiber/Polyurethane	0.42	76.3	3	33.8	31.4

*material data from supplier

From table 2. it is clear how the actual measured elastic modulus (tensile tests) is somewhat lower than the ROM prediction, i.e. a decrease of 7%. This effect on the elastic modulus was sought accounted for by Krenchel [36] using a modified ROM with an efficiency factor η_0 which is a function of the fiber inclination θ. The relation is expressed as,

$$E_{cL}^{*} = \eta_0 V_f E_f + (1 - V_f) E_m \tag{22}$$

$$\eta_0 = \sum_n a_n \cos^4(\theta_n) \tag{23}$$

where a_n represents the fraction of fibers with an angle θ_n. This approach was applied in [37] but on a macro scale level assuming bundles of fibers with the same orientation. It was concluded that a more precise knowledge of individual fiber orientation is needed. This problem could be solved using XCT. Fig. 8 shows examples of 3D micrographs, obtained from XCT, of the GF-PUR sample with textured fiber yarns and the reference case with UD fibers in a polyester resin.

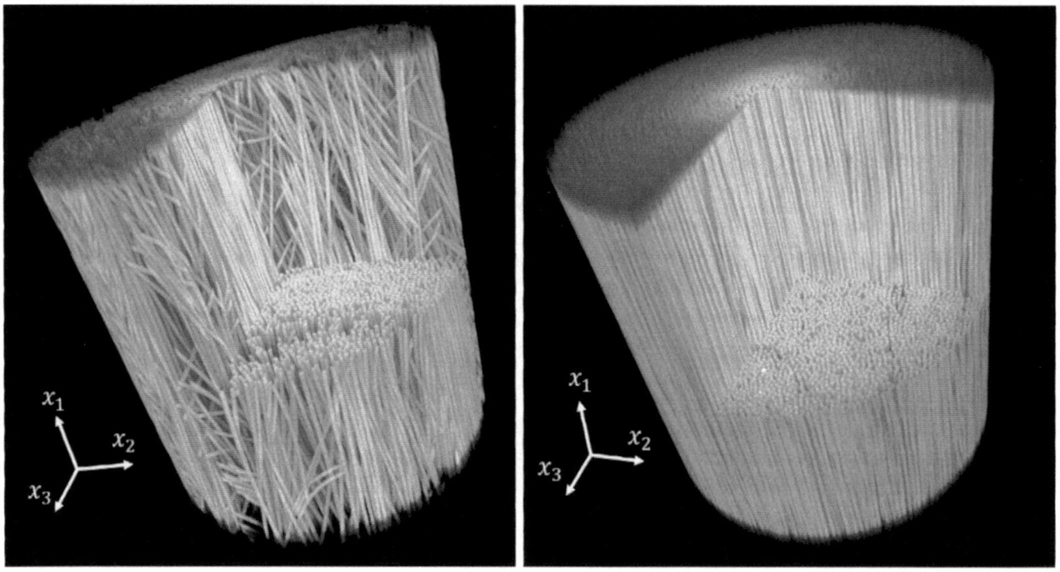

Figure 8. 3D microstructure from XCT: textured GF-PUR (left) and UD GF-polyester (right).

From a comparison of the two composite rods in fig. 8. it becomes crystal clear how the texturing of GF reinforcement introduces a large variation in fiber inclination compared to the reference case with solely UD fibers. In order to characterize the individual fiber inclination advanced state-of-the-art segmentation is needed. This problem is currently solved using an interactive segmentation tool [38]. Subsequently the fiber inclinations are estimated using singular value decomposition for 3D line fitting. The preliminary results indicate a *Krenchel* efficiency factor of $\eta_0 = 0.935$ (cf. eq. (23), i.e. a 6.5% decrease in elastic modulus). Hence, the XCT characterization of the fiber angles can to a great extent account for the observed decrease in elastic modulus (cf. table 2.).

3.3. Experimental characterisation of permeability

Several empirical relations exist to quantify the permeability of technical fabrics, mats, or stacks of rovings. For example, as we discussed in sec. 2.1, a bank of aligned fibers resembles the permeability of unidirectional rovings, which is commonly described by the models developed by Gebart as well as Kozeny and Carman [14,16,17] (see eqs. (4-6)). For many other fiber-architectures, however, there exist no empirical relations, and therefore, the permeability must be determined experimentally.

Experimental characterization of permeability is normally conducted in a controlled injection test. Such an experiment is performed by injecting a test fluid (e.g. a silicone oil) into the fiber material at an overpressure or by utilizing vacuum. The experiments are normally conducted in special molds where the cavity thickness and the mold walls control the fiber volume fraction and the topology of the flow front, see e.g. [39-42]. By monitoring how the flow front of the saturated zone develops, often supplemented with local pressure-readings, the permeability can be estimated using closed-form solutions of Darcy's law (see [8] and eqs. (1-3)). Since this type of characterization is conducted while the saturated zone still develops, this particular methodology is used to determine the unsaturated permeability.

In pultrusion processes, rovings of fibers can directly enter the profile layup. Instead of utilizing weaving techniques or stitching multiple layers to a non-crimp technical fabric, texturization can be applied to individual rovings to achieve the desired mechanical and impregnation properties of the profile. When applying texturization, the arrangement of fibers no longer follows e.g. an idealized fiber architecture such as a bank of aligned fibers (see e.g. the fiber architecture illustrated in fig. 8). Therefore, experimental permeability characterization of rovings is essential to model and predict the impregnation-flow in pultrusion processes.

While most experimental approaches utilize radial and linear injection tests in custom molds specifically designed for testing fabrics and mats, Sandberg et al. [7] found that rovings can be prepared into a ply-like sample. In summary, rovings were drawn back and forth between two perforated plates where a small amount of a powder soluble (FILCO 661MG020, Coim) was applied to the edges of the samples. Once the samples were prepared using this approach, the rovings could be tested in a permeability cell specifically designed for mats and fabrics (See fig. 9).

The characterization work in [7] showed that the normalized permeability of air-texturized rovings (similar to the type in fig. 8) is much higher than conventional UD-rovings. Specifically, the normalized permeability was three times higher along the fiber direction and 40 times higher transverse to the fiber direction. Accordingly, the degree of anisotropy was approximately one magnitude lower compared to conventional UD-rovings. The results indicate that texturized rovings are well-suited for pultrusion of profiles where resin-impregnation is difficult. This is the case for pultrusion of thick profiles as well as profiles drawn at high pulling speeds.

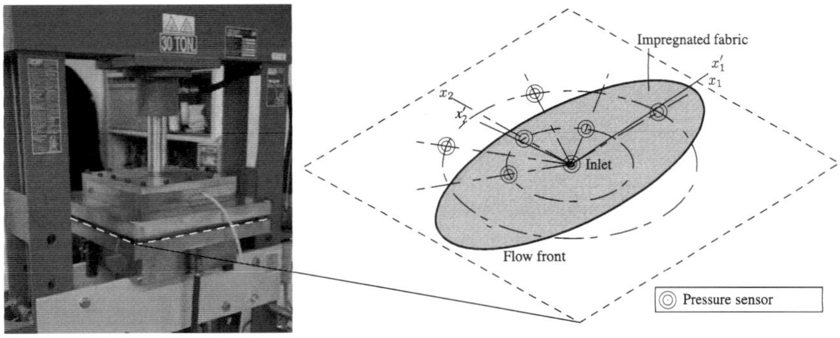

Figure 9. The permeability cell for radial injection tests (incl. schematic of the pressure sensors and the inlet location) used in [7] to characterize the permeability of texturized rovings. Due to anisotropy, the saturated zone is ellipse-shaped, where the principal axes follow the local x_1', x_2'-coordinate system [7].

4. Case Studies

The multi-physics theory from section 2 and the material data obtained from experimental analysis (section 3) are applied for real composite manufacturing processes. In the following, 3 cases of modelling the RIP process will be presented. The first case will be used to evaluate the cure kinetic parameters obtained from DSC (cf. subsection 3.1) by comparison of simulation results and in process temperature measurements; the second case will illustrate the importance of a coupled flow-thermo-chemical modelling approach considering RIP of thick-section profiles; and the third case will illustrate state-of-the-art thermo-chemical-mechanical simulations of the RIP process. Finally a case study about minimizing residual stresses in a laminate plate is presented.

4.1. Thermo-Chemical Simulation with Experimental Cure Kinetics and Temperature Validation

Two of the most significant drivers for build-up of residual stresses during manufacturing of FRPs are the thermal gradients and chemical gradients, i.e. the local variation in degree of cure. The most significant gradients are inherently observed while the thermoset resin cures. Hence, it is of utmost importance to ensure that a process simulation captures the exothermic heat generation and resulting temperature field during curing. The cure kinetic parameters obtained from DSC (cf. subsection 3.1) have been used for a process simulation of a thin carbon fiber reinforced epoxy plate produced by RIP [3]. The numerical degree of cure predictions are depicted in fig. 10. It should be noticed how the curing is taking place along the second quarter of the die length, i.e. $x = [0.2 - 0.5]$. Hence, it is along this length of the pultrusion die the exothermic reaction is taking place. The corresponding numerical temperature predictions and temperature measurements are illustrated in fig. 11. Evaluating the temperature predictions against the measurements along the second quarter of the pultrusion die (cf. fig. 11.) it is clear how the cure kinetic model obtained from DSC adequately describes the exothermic heat generation, hence temperature field. Hence, it is concluded that the cure kinetic parameters obtained from DSC adequately describes the exothermic heat generation due to curing.

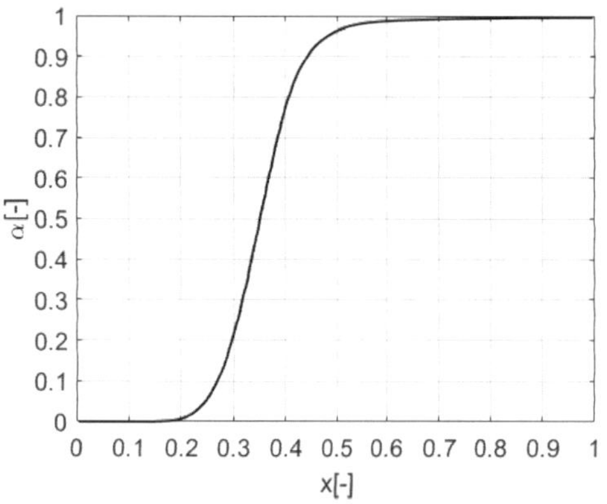

Figure 10. Degree of cure prediction from simulation along the pultrusion die length.

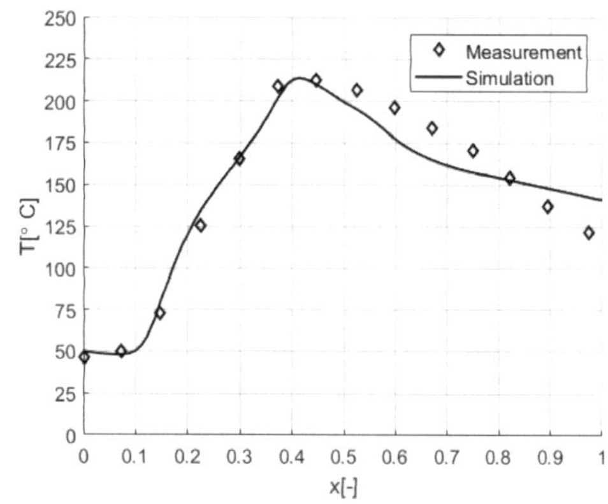

Figure 11. Temperature- simulation and measurement along the pultrusion die length [3].

4.2. Flow-Thermo-Chemical Simulation and Experimental Temperature Validation in a Resin-Injection-Pultrusion Process

In RIP processes, the velocity components of the resin and fibers diverge. This is particularly the case near the resin inlet, where the in-feed of resin takes place. Depending on the process conditions, resin cure and resin flow can take place simultaneously. When this is the case, it is no longer valid to consider heat-transfer and cure as an isolated event as we did in subsection 4.1. Therefore, a coupled flow-thermo-chemical analysis is needed to model the resin flow, heat-transfer, and cure in RIP processes.

In [43], a coupled flow-thermo-chemical model for the simulation of RIP was developed. Compared to the theory discussed in sec. 2.2, the convective term is now split in two since the resin and fibers' velocity components diverge. Consequently, eq. (7) now reads:

$$\nabla \cdot \left(\left(1 - V_f\right)\rho_r C_{pr} \boldsymbol{u_r} T + V_f \rho_f C_{pf} \boldsymbol{u_f} T \right) - \nabla \cdot (k \nabla T) = s \tag{24}$$

where subset $(\,)_{\boldsymbol{r}}$ refers to the resin and $(\,)_{\boldsymbol{f}}$ refers to the fibers. In eq. (24), the transient term was omitted since we only seek stationary solutions. When the resin flow is considered, the real resin velocity enters the species equation for the degree of cure (eq. (11)):

$$\nabla \cdot (\boldsymbol{u_r} \alpha) = R_r \tag{25}$$

In eq. (25), we again omitted transient terms since we only seek stationary solutions.

In RIP, the resin flow front split the saturated and unsaturated domains. To model this free surface flow, a flow front tracker is needed [44-46]. While fully Eulerian approaches such as the level-set and control-volume finite element methods have previously been used to model RIP processes [44,47], Sandberg et al. [43] utilized an Arbitrary Lagrangian-Eulerian approach (ALE, cf. [48]). Since this approach adapts the mesh to the resin flow front, it was found very suitable to model a sharp transition between the unsaturated and saturated fibers in RIP processes [43].

The case study in [43] considered the manufacture of a thick glass fiber profile saturated with a thermoset resin. In the simulation model (fig. 12), inputs from a material characterization study (curing kinetics, chemorheology, and permeability) entered, and the results were compared to temperature measurements from an industrial pultrusion line (fig. 13).

The stationary solutions of pressure, temperature, and degree of cure of the case study are depicted in fig. 12. As the fiber material was pulled into the die from the left, the resin was injected at an overpressure. The pressure was highest near the inlet and fell to atmospheric pressure towards the die exit and the flow front. This established the pressure gradient that feed-in fresh resin to the profile and kept the flow front at rest downstream behind the resin inlet. Since the pultrusion die was actively heated, the profile's temperature rose from the edge of the profile. At the same time, the fiber material entered the die at an elevated temperature. This, together with thermal insulation, caused a rapid temperature-rise in the center of the profile. The build-up of temperature and degree of cure occurred faster in the center of the profile than at the edges of the profile. The temperature measurements from the industrial pultrusion line confirmed this behavior and validated the simulation model. Although the simulation model slightly underestimated the temperature rise in the center of the profile, a good correlation between the simulations and the experiment results are found.

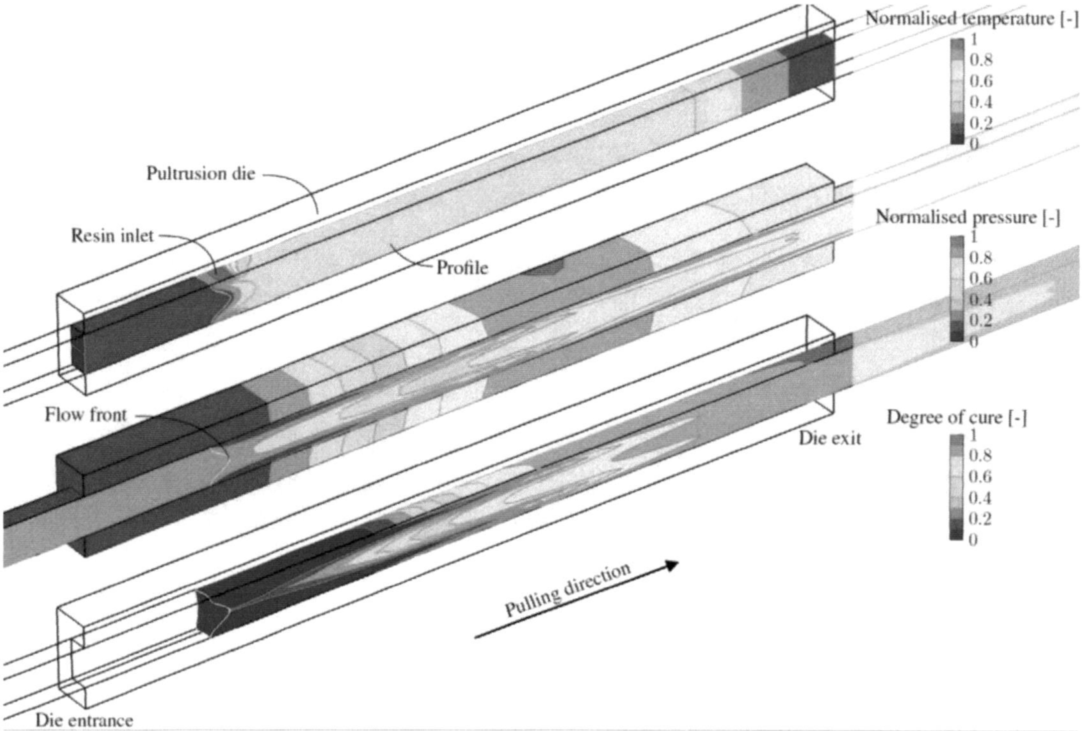

Figure 12. Example of stationary solutions of pressure, temperature, and degree of cure obtained in a RIP process [43].

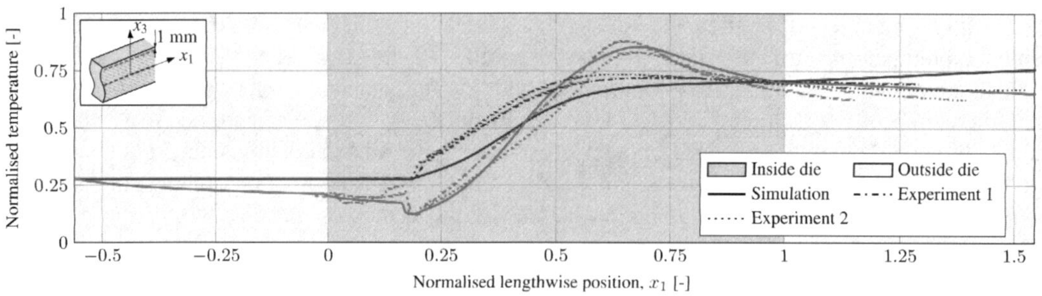

Figure 13. Stationary temperature profiles measured at an industrial pultrusion line and simulation results. Red lines indicate the measured or simulated temperature near the edge of the profile, while temperature at the center of the profile is highlighted with at blue line [43].

4.3. Thermo-Chemical-Mechanical Simulation of Thermoset Pultrusion Processes

As earlier stated, thermal expansion and chemical shrinkage, which are inherent, conditions in composite manufacturing processes introduce stress and strain in the composite profile. As the composite cools down the resin system cures and vitrify often leaving a residual stress state.

In [31,32], the authors considered pultrusion of a 25.4 mm thick square glass fiber profile saturated with an Epoxy resin system. Figure 14 illustrates the achieved transversal stress and strain components. We refer to [32,33] for a presentation of the process-induced deformation and the longitudinal stress and strain components. In figure 14, results were generated using the original framework by Baran et al. [31,32] as well as the steady-state approach by Sandberg et al. [33]. As shown in the figure, the steady-state approach achieved the same results as the traditional Lagrangian approaches by Baran et al. [31,32], which validates the new framework and implementation.

In table 3, computation times for the different approaches are listed. The original 2D framework by Baran et al. [31] is superior in computation times, which means if a 2D-reduction of the problem at hand is acceptable, the original framework in [31] remains attractive. For the two 3D approaches, the computation time for the approach by Sandberg et al. [33] lasted only 98 s, which is only approximately 5% of the computation time for Baran et al. [32]. Hence, this improvement marks a 20 times increase in computational performance for the 3D case.

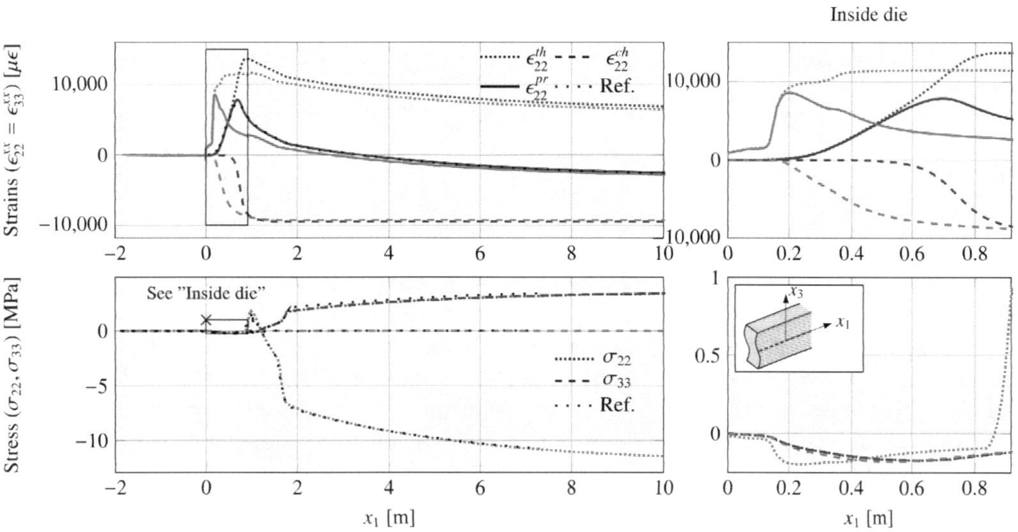

Figure 14. Transversal stress and strain components at the center and edge of the profile. The dashed lines (Ref.) show results from Baran et al. [31] to validate the numerical framework.

Table 3. Simulation times for reaching steady state for the three modelling approaches.

2D-Lagrangian [31]	3D-Lagrangian [32]	3D-Eulerian [33]
45 s	1920 s	98 s

4.4. Optimization – Minimizing Residual Stresses

Apart from simulation with the purpose of describing and analyzing composite manufacturing processes as discussed earlier, the presented modelling methods can also be taken one step further and used as the engine in an automatic optimization calculation. Here, a certain objective of e.g. increasing production output or improving load carrying capacity is combined with adjusting certain so-called design variables which can actually be controlled during production. One such very interesting example is the work given in [49,50]. In this work, it is firstly documented that the process induced residual stresses from manufacturing, not surprisingly, affect the subsequent load carrying capacity of produced laminate plates in a negative way. It is moreover shown that the magnitude of the residual stresses correlate quite strongly to the temperature of resin gelation via the process induced strain, i.e. thermal strain + chemical shrinkage strain. In essence, reducing the gelation temperature correspondingly reduces the effective thermal strain, which drives the thermal stress part of the final residual stresses. The chemical strain remains unchanged. This reduction of the gelation temperature is achieved by the authors via a two-step procedure.

In figs. 15 and 16, schematics of the classical one-step curing profile and the proposed two-step profile are presented, respectively. The major advantage of introducing the latter is that it reverses the sign of the thermal strain of the second heating step right after gelation (i.e. A2 in figure 16) such that the overall induced thermal strain becomes considerably smaller, effectively being driven by the temperature difference between the lowered gelation temperature and the final temperature.

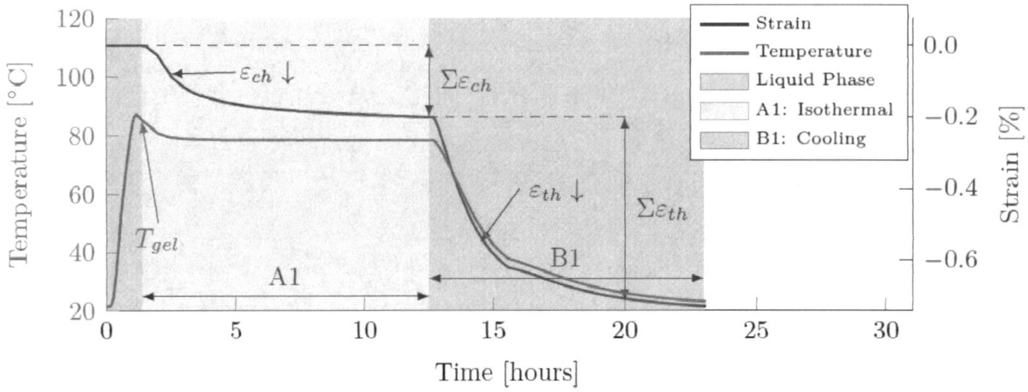

Figure 15. Strain and temperature development over time for a one stage cure with 80°C of curing temperature for 12 hours [50].

So instead of making a full-blown 3D thermo-mechanical FE analysis and using the reduction of the calculated residual stresses as objective, the authors of [50] make a very elegant and simple solution in which they use the minimization of the gelation temperature as the objective, because they know that this will also reduce residual stress levels. The advantage of this is that they can suggest an optimization scheme, which is very fast and easy to implement in an industrial setting. However, it comes with a price tag, i.e. that the curing time increases, hence decreases productivity. Note, that the specific value of acceptable production time is to be defined by the user as input and subsequently it is treated as a constraint in the optimization algorithm.

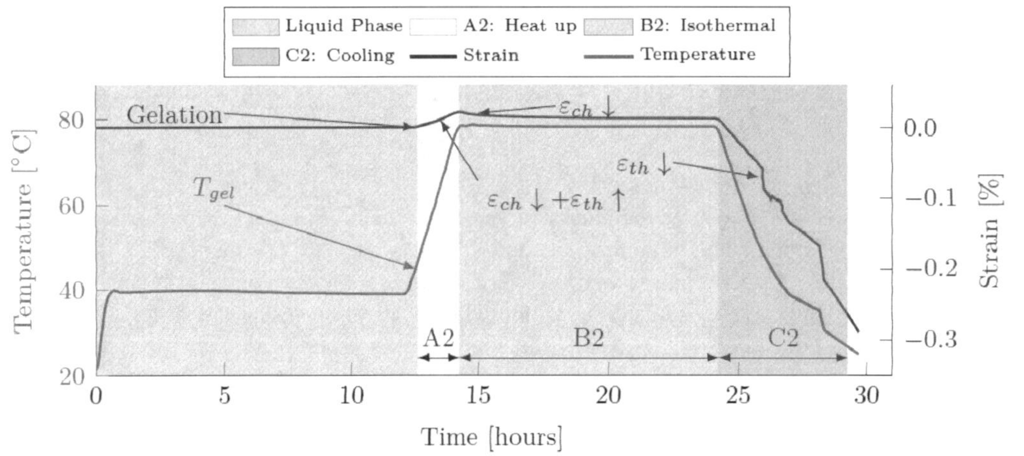

Figure 16. Strain and temperature development over time for a two stage cure with 40°C for 12 hours followed by a 80°C for 12 hours [50].

In fig. 17, the result of accepting increasing production times of 8, 16, 24 hours are shown [50] for a prescribed final degree of cure of 0.99 for a laminate plate. The longer times, the lower gelation temperature, the lower process induced strain and ultimately the lower residual stresses.

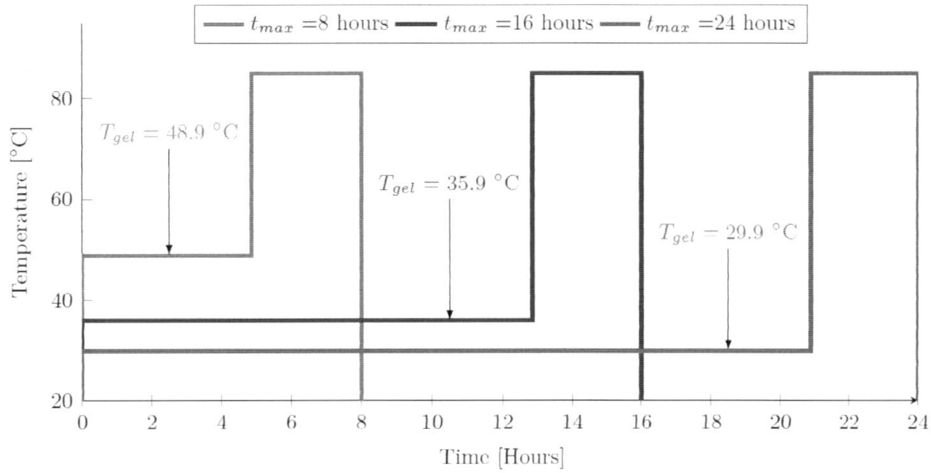

Figure 17. Reduction of resin gelation temperature with increasing curing time [50].

5. Conclusion

In this paper, the main mechanisms generating residual stresses in FRP composites during manufacturing were reviewed. The review includes state-of-the-art modelling approaches of the inherent multi-physics nature of composites manufacturing, i.e. impregnation flow, heat transfer and cure, and the associated mechanical response, e.g. build-up of residual stresses.

The multi-physics nature of the manufacturing of FRPs necessitates advanced materials analyses to obtain input parameters for the process simulations. DSC experiments were successfully used to obtain cure kinetic parameters of an epoxy resin (cf. table 1 and figs. 3 and 4). The resulting cure kinetic parameters were applied for thermo-chemical simulations capturing heat generation in a pultrusion process. The temperature predictions were validated by comparing with temperature measurements from an industrial pultrusion line. The temperature predictions showed a great resemblance with the temperature measurements throughout the part of the pultrusion die where the resin cures but also some divergence in the latter part of the pultrusion die (cf. fig. 10 and 11).

Advanced material analysis studies were presented, including i) SEM investigations of cross-sectional fiber distribution and the corresponding variation in permeability following the well-known Kozeny-Carman relation [14-16] (cf. fig. 6 and 7); ii) XCT was used to characterize 3-dimensional fiber agglomeration and relating fiber inclination to reduction in longitudinal stiffness (cf. fig. 8); and iii) results from advanced permeability characterization of fiber-beds was presented showing a significant increase in transverse permeability due to texturing of fiber rovings, hence improving impregnation capabilities (cf. sec. 3.3).

A state-of-the-art study were presented showing the necessity of a coupled flow-thermo-chemical modelling strategy for thick sections composites produced by the RIP process. This modelling strategy was validated by comparing the stationary temperature solution with corresponding experimental results showing a good correlation (cf. fig. 13).

Another state-of-the-art approach was presented considering thermo-chemical-mechanical simulations of the RIP process. This approach encompasses a steady state solver using a 3D-Eulerian framework [33] and it was verified by comparing with the traditional Lagrangian approach [31,32] (cf. fig. 14) and showed a 20 times increase in computational performance (cf. table 3) when obtaining a 3D steady state solution.

Finally an optimization study, minimizing residual stresses, was presented based on the work by Mortensen et al. [49,50]. This study introduces a simple and elegant method to reduce residual stresses using a two-step procedure to minimize the gelation temperature (cf. fig. 15 and 16). The idea is to reverse the sign of the thermal strain in the second step, hence minimizing the overall thermal strain and as a result of that also the thermal part of the overall residual stress. Such an optimization approach has the advantage of being fast and easy to implement in an industrial setup but at the expense of increased production time. Hence, the optimal solution is found as an acceptable tradeoff between the residual stress level and productivity.

References

[1] Markets and Markets, Lakshmi Narayanan, Dec. 2019, Composites Market by Fiber Type (Glass Fiber Composites, Carbon Fiber Composites, Natural Fiber Composites), Resin Type (Thermoset Composites, Thermoplastic Composites), manufacturing process, End-use Industri and Region – Global Forecast to 2024. https://www.marketsandmarkets.com/Market-Reports/composite-market-200051282.html, Visited 02-07-2020

[2] Baran I, Cina K, Ersoy N, Akkerman R and Hattel J H 2016. A review on the mechanical modeling of composite manufacturing processes. *Arch. of Comp. Meth. in Eng.,* 1-31.

[3] Rasmussen F S, Sonne M R, Larsen M, Spangenberg J, Lilleheden L T and Hattel J H 2017. 2D numerical modelling of the resin injection pultrusion process including experimental resin kinetics and temperature validation. *Proc. 21st Int. Conf. on Compos. Mater.* (Xi'an, China).

[4] Rasmussen F S, Klingaa C G, Larsen M, Sonne M R, Spangenberg J and Hattel J H 2018. Modelling the effect of non-uniform fiber distribution on the uring behaviour in resin injection pultrusion. *Proc. 18th European Conf. on Compos. Mater.* (Athens, Greece)

[5] Rasmussen F S, Sonne M R, Larsen M, Mikkelsen L P and Hattel J H 2018. A characterization study relating cross-sectional distribution og fiber volume fraction and permeability. *Proc. 22nd Int. Conf. on Compos. Mater.* (Melbourn, Australia).

[6] Rasmussen F S, Emerson M J, Sonne M R, Hattel J H, Mikkelsen L P and Dahl V A 2019. Fiber segmentation from 3D X-ray computed tomography of composites with continuous and textured glass fiber yarns. *Proc. of 2019 Int. Conf. on Tomography of Mater. & Struct.* (Cairns, Australia).

[7] Sandberg M, Kabachi A, Volk M, Salling F B, Ermanni P, Hattel J H and Spangenberg J 2020. Permeability and compaction behaviour of air-texturised glass fiber rovings: A characterisation study. *Jour. of Compos. Mater..* Accepted/in press - 2020.

[8] Darcy H 1856. Les fontaines publiques de la ville de Dijon, Victor Dalmont, Libraire des Corps imperiaux des ponts et chauss´ees et des mines.

[9] Voorakaranam S, Joseph B and Kardos J L 1999. Modeling and control of an injection pultrusion process. *Jour. of Compos. Mater.* **33**:1173–204.

[10] Jeswani A and Roux J 2007. Manufacturing modeling of three-dimensional resin injection pultrusion process control parameters for polyester/glass rovings composites. *Jour. of Manufacturing Sci. and Eng.* **129**:143–56.

[11] Sandberg M, Hattel J H and Spangenberg J 2018. Numerical modelling and optimisation of fibre wet-out in resin-injection pultrusion processes. *Proc. 18th European Conf. on Compos. Mater.* (Athens, Greece).

[12] Jiang J, Su Y, Zhou L, Guo Q, Xu C, Deng G, Chen X, Yao X and Fang L 2017. Effect of nesting on the permeability of multilayer unidirectional fabrics. *App. Compos. Mater.* **24**:625–42.

[13] Schell J, Siegrist M and Ermanni P 2007. Experimental determination of the transversal and longitudinal fibre bundle permeability. App. Compos. Materials **14**:117–28.

[14] Gebart B R 1992. Permeability of unidirectional reinforcements for RTM. *Jour.og Compos. Mater.* **26**:1100-33.

[15] Orlik J, Pietsch K, Fassbender A, Sivak O and Steiner K 2018. Simulation and experimental validation of spacer fabrics based on their structure and yarn's properties. *Appl. Compos. Mater.* **25**:709–24.

[16] Kozeny J 1927. Uber kapillare leitung der wasser in boden. *Royal Academy of Sci., Vienna, Proc. Class I* **136:**271–306.

[17] Carman P C 1937. Fluid flow through granular beds. *Transactions, Institution of chem. engineers, London.* **15**:150-66.
[18] Baran I, Hattel J H and Tutum C C 2013. Thermo-chemical modelling strategies for the pultrusion process. *App. comp. mater.* **20**:1247-63.
[19] Baran I, Akkerman R and Hattel J H 2014. Material characterization of polyester resin system for the pultrusion process. *Comp part B: Eng.* **64**:194-201.
[20] Liu X L, Grouch I G and Lam Y C 2000. Simulation of heat transfer and cure in pultrusion with a general-purpose finite element package. *Comp. Sci. and Tech.* **60**:857-64.
[21] Bogetti T A and Gillespie J W 1992. Process-induced stress and deformation in thick-section thermoset composite laminates. *Jour. of Comp. Mater.* **26**(5):626-60.
[22] Sandberg M, Yuksel O, Comminal R B, Sonne M R, Jabbari M, Larsen M, Salling F B, Baran I, Spangenberg J and Hattel J H 2020. *Numerical modeling of the mechanics of pultrusion.* In Silberschmidt V V (Ed.), *Mechanics of mater. in modern manufacturing methods and processing techniques*, Elsevier.
[23] Dillman S H and Seferis J C Kinetic viscoelasticity for the dynamic mechanical properties of polymer systems. *Jour. Of Macromolecular Sci.-Chemistry,* **24** (1987).
[24] Johnston A. *An integrated model of the development of process-induced deformation in autoclave processing of composite structures*, Ph.D. thesis, The University of British Columbia, Vancouver (1997)
[25] Johnston A, Vaziri R and Poursartip A A 2001. plane strain model for process-induced deformation of laminated composites structures. *Jour. Of Comp. Mater.* **35**:1435-70.
[26] Kim Y K and White SR. Stress relaxation behavior of 3501-6 epoxy resin during cure. *Polym. Eng. Sci.,* **36(23)**:2852-62.
[27] Kim Y K, White SR. Viscoelastic analysis of processing induced residual stresses in thick composite laminates. *Mech. Compos. Struc.*, **4(4)**:361-87.
[28] Ding A, Li S, Wang J, Zu L. A three-dimensional thermos-viscoelastic analysis of process-induced residual stress in composite laminates. *Compos. Struc.*, **129**:60-9.
[29] Svanberg J M, Holmberg J A. An experimental investigation on mechanisms for manufacturing induced shape distortions in homogeneous and balanced laminates. *Compos. A*, **32**:827-38.
[30] Svanberg J M, Holmberg J A.prediction of shape distortions. Part II. Experimental validation and analysis of boundary conditions. *Compos. A*, **35**:723-34.
[31] Sandberg M, Yuksel O, Baran I, Hattel J H, Spangenberg J, Steady-state modelling and analysis of process-induced stress and deformation in thermoset pultrusion processes, (in preperation).
[32] Baran I, Tutum C C, Nielsen M W and Hattel J H 2013. Process induced residual stresses and distortions in pultrusion. *Compos. Part B: Eng.* **51:**148-61.
[33] Baran I, Hattel J H, Akkerman R and Tutum C C 2015. Mechanical modelling of pultrusion process: 2D and 3D numerical approaches. *App. Compos. Mater.* **22**(1):99-118.
[34] Vyazovkin S, Chrisafis K, Lorenzo M L D, Koga N, Pijolat M, Roduit B, Sbirrazzuoli N and Suñol J J 2014. ICTAS kinetic committee recommendations for collecting experimental thermal analysis data for kinetic computations. *Thermochimica Acta.* **590**:1-23.
[35] Vyazovkin S, Burnham A K, Criado J M, Pérez-Maqueda L A, Popescu C and Sbirrazzuoli N 2011. ICTAC kinetics committee recommendations for performing kinetic computations on thermal analysis data. *Thermochimica Acta.* **520**:1-19.
[36] Krenchel H 1964. *Fiber reinforcement: Theoretical and practical investigations of the elasticity and strength of fiber-reinforced materials.* Akademisk forlag, Danmarks tekniske højskole.
[37] Rosenthal J 1992. A model for determining fiber reinforcement efficiencies and fiber orientation in polymer composites. *Polymer Comp.* **13**:462-6.
[38] Emerson M J, Jespersen K M, Dahl A B, Conradsen K and Mikkelsen, L P. 2017. Individual fiber segmentation from 3D x-ray computed tomography for characterising the fibre orientation in unidirectional composite materials. *Comp. part A: App. Sci. and Manuf.* **97**:83-92.

[39] Arbter R, Beraud J, Binetruy C et al. 2011. Experimental determination of the permeability of textiles: A benchmark exercise. *Compos. Part A: App. Sci. and Manufacturing,* **42(9)**: 1157–68.

[40] Vernet N, Ruiz E, Advani S G et al. 2014. Experimental determination of the permeability of engineering textiles: Benchmark II. *Compos. Part A: App. Sci. and Manufacturing.* **61**:172–84.

[41] May D, Aktas A, Advani S G et al. In-plane permeability characterization of engineering textiles based on radial flow experiments: A benchmark exercise. *Composites Part A: Applied Science and Manufacturing* 2019; **12**: 100–14.

[42] Schell J S, Siegrist M and Ermanni P 2007. Experimental determination of the transversal and longitudinal fibre bundle permeability. *Applied Composite Mater.* **14(2)**:117–28.

[43] Sandberg M, Yuksel O, Baran I, Hattel J H and Spangenberg J 2019. Numerical and experimental analysis of resin-flow, heat-transfer, and cure in a resin-injection pultrusion process, (submitted).

[44] Sandberg M, Rasmussen F S, Hattel J H and Spangenberg J 2019. Simulation of resin-impregnation, heat- transfer and cure in a resin-injection pultrusion process, *AIP Conf. Proc.* **2113** 20022.

[45] Sandberg M, Hattel J H and Spangenberg J 2019. Simulation of liquid composite moulding using a finite volume scheme and the level-set method. *Int. Jour. Multiphas. Flow.* **118**:183–92.

[46] Comminal R, Spangenberg J and Hattel J H 2015. Cellwise conservative unsplit advection for the volume of fluid method. *Jour. of comp. phy.* **283**:582-608.

[47] Simacek P and Advani S 2015. Simulating tape resin infiltration during thermoset pultrusion process. *Compos. Part A.* **72**:115–26.

[48] Donea J, Huerta A, Ponthot J-P and Rodríguez-Ferran A 2014. Arbitrary Lagrangian-Eulerian Methods. *Encyclopedia of Comp. Mech.*

[49] Mortensen U A. Process parameters and fatigue properties of high modulus composites (*Ph.D. thesis*). DTU Wind Energy, 2019. 196 p.

[50] Mortensen U A, Andersen T L and Mikkelsen L P, 2018. pragmatic approach to cure profile enhancement for improved fatigue performance of thermoset matrix composites. *Proc. Of the 39th Risø Inter. Symp. on Mater. Sci.* - IOP Conf. Series: Materials Science and Engineering, **388**.

IOP Conf. Series: Materials Science and Engineering **942** (2020) 012004 doi:10.1088/1757-899X/942/1/012004

Microscale testing and modelling for damage tolerant composite materials and structures

Bent F Sørensen

The Technical University of Denmark, Department of Wind Energy, Section of Composites Mechanics and Structures, Frederiksborgvej 399, DK-4000 Roskilde, Denmark

E-mail: bsqr@dtu.dk

Abstract. There is a need for damage tolerant composite material for very large engineering structures such as wind turbine rotor blades. The development of improved composite materials with higher damage tolerance will be most efficiently guided by models of the relevant damage modes. Two damage modes are considered in the present paper: Fatigue damage due to in-plane tensile stresses in the fibre direction and cyclic delamination crack growth from a ply drop. For both failure modes the damage evolution is considered at macro- and microscale. Micromechanical models are used to illustrate how changes in the mechanical properties of fibre, matrix and the fibre/matrix interface can lead to an increased damage tolerance material. Also, micromechanical testing methods for characterizing the relevant micromechanical parameters are discussed.

1. Introduction

Composites based on long aligned high-stiffness fibres in a polymer matrix are widely used in large load-carrying structures where low weight and long fatigue life are of major importance. Examples of use are wind turbine rotor blades, which are to be in service for 20-25 years, while being subjected to varying cyclic loads (aerodynamic and gravitational) and varying environmental exposure [1, 2].

However, structures made of composite materials can fail by a number of complicated failure modes, since they typical possess weak interfaces at multiple length scales. Damage is a multiscale and progressive phenomenon: Damage can start as one failure mode at the microsale and then develop into other failure modes at the macro- and structural scales and prediction of progressive damage growth is an ongoing research field [3, 4]. Typically, damage prediction has been accomplished for a limited number of interacting failure modes.

Therefore, many engineering structures built using composites are designed quite conservatively against damage. Or more precisely, they are designed using criteria that does not allow damage growth. The manufacturing process then aims at making "perfect" structures, i.e. without defects that can lead to damages [1, 2]. Structures are inspected after manufacturing and defects are identified. Some defects are deemed insignificant, while other leads to repairs. Severe manufacturing defects can lead to structures being discarded.

The traditional approach of obtaining structures with a safe life by enforcing strict quality control and discard parts with manufacturing defects is an expensive strategy for components made of very large parts since any discarded part is expensive simply due to the cost of materials used.

The present paper considers a way to go beyond the "no growth" approach for damage. Instead, the philosophy will be to assume that damage is present, but the materials and structures should be designed such that damage growth is stable (i.e., propagates slowly or even stops growing) and the damage growth is predictable.

In aerospace, damage tolerance concepts (allowing damage initiation and grow) combined with regular inspection interval are established for structures made of metals but not yet for composites [5]. The present paper considers how damage tolerant materials behaviour can be obtained by control of material properties at the microscale, i.e., by controlling the mechanical properties of the fibre, matrix and the fibre/matrix interface. Then, it is essential to have micromechanical models that can predict which microscale properties that give the optimal macroscale damage tolerance. It is likewise essential to have microscale experimental testing methods to establish what the microscale mechanical properties are. Combined, micromechanical modelling and microscale testing should be efficient tools in the development of future damage tolerant materials.

In the present paper we will discuss two important damage types for composite structures: In-plane fatigue and delamination crack growth from ply-drops. Both damage types are of great relevance for the load-carrying laminate (main spar) in wind turbine rotor blades. The focus will be on the development of composites materials with enhanced damage tolerance to fatigue, i.e., damage growth under cyclic loads.

The paper is organised as follows. In Section 2, concepts of damage tolerance will be given. Next, in Section 3, so-called "fatigue damage mechanisms maps" are introduced to visualize damage evolution. Examples of micromechanical models for the prediction of in-plane fatigue damage are presented in Section 4. Section 5 discusses delamination from ply-drops, Section 6 considers microscale experiments and Section 7 gives a brief perspective for the future.

2. Concepts of damage tolerance

2.1. Definition of damage tolerance

In the present paper, damage tolerance relates to a structures ability to undergo distributed damage and crack growth. The first mode of damage must not lead to imminent failure. Instead, damage should grow in a stable manner while being detectable and predicable, such that damage can be found well before it becomes critical. There should be plenty of time to react, e.g. stop using the component, make inspection and decide if it is necessary make repairs to the damaged structure. A high damage tolerance indicates that the damage or crack growth occurs stably over a large physical size (e.g., a long crack length) before the damage growth becomes unstable, or that the component undergoes a very high number of load cycles before unstable crack growth occurs.

Let us clarify what the statement above means. Under static loading, "stable manner" means that damage growth must arrest under fixed loading and a higher applied load is required to develop further damage. Under cyclic loading, fatigue damage should advance slowly, i.e., over many load cycles.

"Damage being detectable" means that the the presence of damage should result in changes in the materials stiffness locally (e.g. being detectable by strain measurements by strain gauges, digital image correlation and/or embedded sensors such as fibres with Fibre Bragg Grating) or other types of sensors, e.g. acoustic emission sensors. Preferably, damage should be identified by visual inspection, but this is often not possible for composite structures. We should be able to identify that damage is developing, what kind of damage it is, its physical location within a structure as well as its size and depth by the use of various non-destructive techniques (NDT).

"Predictable" means that the damage state and the load-carrying capability can be predicted for static loading and the damage growth rate and remaining number of cycles to failure can be predicted for cyclic loading. The predictions should be made using reliable models and the

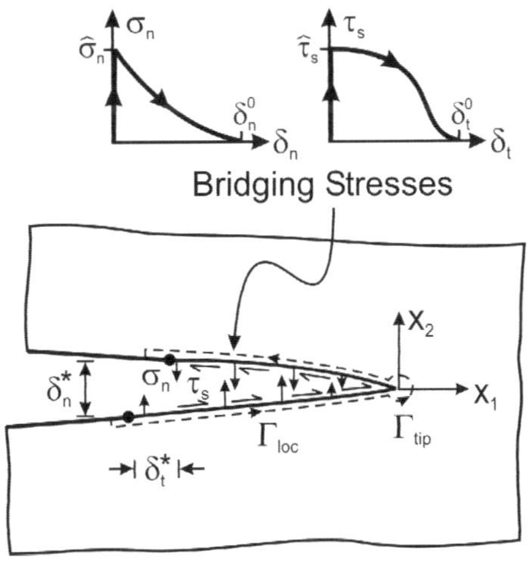

Figure 1. Mixed mode cohesive laws and integral path, Γ_{loc} for the J integral (dashed line).

relevant mechanical materials properties can be determined by independent experiments.

When the connection between microscale and macroscale has been obtained, micromechanics modelling then opens the possibility to tailor-make composites by microstructral optimization. As mentioned, we aim to make "damage controlled" materials. By this we mean to control the material properties such that the progressive damage evolution behaves in a pre-determined, damage tolerant manner.

The tailor-making of composites links micromechanics to materials processing: Modifications of the processing steps or processing conditions are needed for making more damage tolerant composite materials. The "optimized" processing condition that give the desired macroscale mechanical properties can then be obtained in iterations, where each iteration consists of (a) materials or components testing, (b) identification of damage types, and (c) micromechanic modelling to find the best microscale parameters. Then follows iterations of (d) microscale experiments to characterize the current microscale mechanical properties, and (e) change processing conditions in a direction towards the "optimal" processing conditions.

It should be noted that it is possible to design the geometry of structures such that they become more damage tolerant. Damage tolerance design is a combined material-structure property. This will be detailed in the next section.

2.2. Damage tolerance as a combined materials-structure property

As an illustration on how damage tolerance depends on both materials properties and structural properties (shape and size), we consider the growth of a crack that experiences crack bridging. Crack bridging can occur by fibres that connect the crack faces and create tractions that act against the opening of the crack. In mechanics terms, the crack bridging can be described by so-called cohesive laws, where the local bridging tractions are taken to be a function of the local crack opening displacements. For plane problems, a crack can have two tractions, a normal traction, σ_n and a shear traction, τ_s, see Fig. 1. In the general case they will both be functions of the local normal opening displacement δ_n and the local tangential opening displacement, δ_t,

$$\sigma_n = \sigma_n(\delta_n, \delta_t) \quad and \quad \tau_s = \tau_s(\delta_n, \delta_t). \tag{1}$$

The bridging tractions contribute to the fracture resistance of the material. This can be understood by an application of the path-independent J integral [6] locally around the crack tip and bridging zone. The result becomes [7]

$$J_{loc} = \int_0^{\delta_n^*} \sigma_n(\delta_n, \delta_t) d\delta_n + \int_0^{\delta_t^*} \tau_s(\delta_n, \delta_t) d\delta_t + J_{tip}, \tag{2}$$

where δ_n^* and δ_t^* are the end-opening and end-sliding of the bridging zone, respectively and J_{tip} is the J integral evaluated around the crack tip. When J_{tip} equals the crack tip fracture energy (denoted J_0), J_{loc} becomes equal to the fracture resistance, J_R. The fracture resistance J_R thus increases with increasing end-opening and end-sliding. Such rising fracture resistance is called R-curve behaviour. J_{loc} can be considered as being a function of end-opening and end-sliding of the bridging zone, $J_{loc} = J_{loc}(\delta_n^*, \delta_t^*)$.

In case the bridging tractions are derived from a potential function $\Phi(\delta_n, \delta_t)$

$$\sigma_n(\delta_n, \delta_t) = \frac{\partial \Phi(\delta_n, \delta_t)}{\partial \delta_n} \quad and \quad \tau_s(\delta_n, \delta_t) = \frac{\partial \Phi(\delta_n, \delta_t)}{\partial \delta_t}, \tag{3}$$

and with $J_{tip} = J_0$, eq. (2) becomes

$$J_R(\delta_n^*, \delta_t^*) = \Phi(\delta_n^*, \delta_t^*) + J_0. \tag{4}$$

Now we are ready to consider the stability of growth of cracks with bridging. For materials with R-curve behaviour, a criterion for stable crack growth is [8]

$$J_{ext} = J_R \quad and \quad \frac{dJ_{ext}}{d\ell} \le \frac{dJ_R}{d\ell}, \tag{5}$$

where J_{ext} is the J integral evaluated around the external boundaries (sometimes called the "applied energy release rate" in linear elastic fracture mechanics) of the specimen, and ℓ is the crack length. Eq. (5) expresses that stable crack growth happens if the fracture resistance, J_{loc} (the right hand side), increases more rapid than the applied energy release rate (the left hand side). By eq. (4)

$$\frac{dJ_{ext}}{d\ell} \le \frac{d\Phi(\delta_n^*, \delta_t^*)}{d\ell}. \tag{6}$$

From the chain rule we have

$$\frac{d\Phi(\delta_n^*, \delta_t^*)}{d\ell} = \frac{\partial \Phi(\delta_n^*, \delta_t^*)}{\partial \delta_n^*} \frac{\partial \delta_n^*}{\partial \ell} + \frac{\partial \Phi(\delta_n^*, \delta_t^*)}{\partial \delta_t^*} \frac{\partial \delta_t^*}{\partial \ell}. \tag{7}$$

or by eq. (3)

$$\frac{d\Phi(\delta_n^*, \delta_t^*)}{d\ell} = \sigma_n^* \frac{\partial \delta_n^*}{\partial \ell} + \tau_s^* \frac{\partial \delta_t^*}{\partial \ell}, \tag{8}$$

where σ_n^* and τ_s^* are the normal traction and shear traction at the end of the bridging zone, $\sigma_n^* = \sigma_n(\delta_n^*, \delta_t^*)$ and $\tau_s^* = \tau_s(\delta_n^*, \delta_t^*)$. Inserting eq. (8) into eq. (6) leads to the following criterion for stable crack growth:

$$\frac{dJ_{ext}}{d\ell} \le \sigma_n^* \frac{\partial \delta_n^*}{\partial \ell} + \tau_s^* \frac{\partial \delta_t^*}{\partial \ell}. \tag{9}$$

Equation (9) expresses that crack growth will be stable if the right hand side of the equation exceeds the left hand side.

In eq. (9), the terms σ_n^* and τ_s^* obviously depend on the cohesive law and are thus material properties of the fracture process zone. The term $\partial \delta_n^* / \partial \ell$ expresses how the normal end-opening increases as the crack length increases. Likewise, $\partial \delta_t^* / \partial \ell$ expresses how the end-sliding (tangential opening displacement) increases as the crack length increases. Both these two terms depends on the specimen geometry and stiffness and are thus structural properties. In conclusion, the criterion for stable crack growth, eq. (9), depends on both material and structural parameters.

3. Fatigue damage mechanisms maps

Since damages often evolve sequentially in composite structures, it is possible to understand each damage mechanism within progressive damage evolution one by one. If each damage mechanism is understood (and modelled), damage controlled materials may be realized by considering how the evolution of each damage mechanism can be slowed down or completely stopped.

In the following we will introduce a graphical way of representing the interaction of various fatigue damage mechanisms with the purpose of identifying which individual fatigue damage models are required and how they should interact. We will call this representation "fatigue damage mechanisms maps" and will illustrate this concept by two examples: in-plane fatigue in a unidirectional laminate and delamination fatigue crack growth from a ply-drop.

We contemplate a situation where a composite specimen is subjected to cyclic tension-tension loading between a fixed maximum strain value, $\bar{\varepsilon}_{max}$, and a minimum strain value, $\bar{\varepsilon}_{min}$. A fatigue damage mechanisms map is a representation with the number of load cycles, N, on the horizontal axis and the applied maximum cyclic strain level, $\bar{\varepsilon}_{max}$, on the vertical axis, similar to S-N graphs. However, while S-N graphs only contain the relationship between the maximum applied strain (or maximum applied stress) and the associated number of cycles to failure, N_f, a fatigue damage mechanisms map shows the most important fatigue damage mechanisms from virgin specimen to complete specimen failure. The central idea is that a fatigue damage mechanisms map should be constructed from experimental observation of microscale damage state. In order to make consistent model predictions, a model should be formulated for each damage mechanism and these models can then be linked together. This illustration is to some extend inspired by the concepts of "Fatigue Life Diagram" introduced by Talreja [9] and the damage scenario illustration of Carraro and Quaresimin [10].

3.1. A fatigue damage mechanisms map for in-plane fatigue

Fig. 2 shows a fatigue damage mechanisms map for in-plane tension-tension fatigue of a uniaxial fibre composite. There are no scales on the axis to illustrate that the concepts are generic. Never the less, Fig. 2 is constructed from literature observation of fatigue damage in unidirectional fibre composites with polymer matrix [11, 12, 13, 14]. For low values of $\bar{\varepsilon}_{max}$, the first mode of damage is isolated fibre failures accompanied by limited fibre/matrix debonding. The location of the fibre failure is "random" (unpredictable) in the sense that it occurs at the position along the fibre that possesses the worst defect. With increasing number of load cycles, a second damage mechanism evolves: an increase in the debond length with increasing number of load cycles. Microscale observations indicates that this situation can be stable (i.e., constant debond length, no further fibre failures [11, 12, 15]), indicating that a fatigue limit might exist [11]. For higher values of $\bar{\varepsilon}_{max}$, a second damage mechanism is activated: Breakage of neighbour fibres as a result of the stress field from the debond crack tip stress field of broken and debonded fibres [11, 12, 13, 14]. This establishes a progressive failure mode with a "damage front" growth rate, da/dN, where a is the size of the damage zone consisting of broken and debonded fibres [16]. When a damage zone reaches a certain size, a third damage mechanism can be activated: Splitting cracks (cracks extend from the edge of the damage zone parallel to the fibre direction) [16, 17] propagate along the fibre direction to other fibre damage zones, resulting in fatigue failure of the specimen. For in-plane fatigue, the main composite properties to predict are: (i) the maximum applied cyclic strain that does not lead to fatigue failure (the fatigue limit), $\bar{\varepsilon}_{fl}$, and when $\bar{\varepsilon}_{max} > \bar{\varepsilon}_{fl}$, (ii) the fatigue damage growth rate da/dN and the (iii) the associated fatigue life N_f. Both da/dN and N_f are expected to depend on $\bar{\varepsilon}_{max}$.

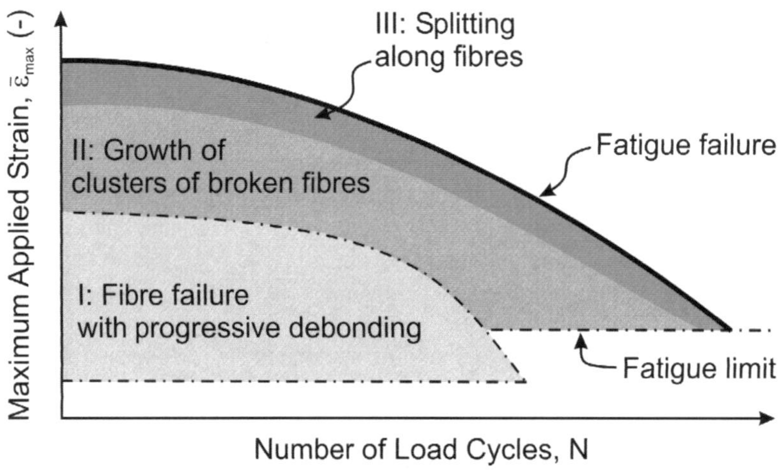

Figure 2. Fatigue damage mechanisms map for in-plane fatigue of unidirectional fibre composites.

3.2. Fatigue damage mechanisms map for delamination from ply-drops

A fatigue damage mechanisms map for delamination cracks originating from two ply-drops is shown in Fig. 3. The first mode of damage is the formation of a so-called tunneling crack at the end of the ply drop, i.e., a crack that is limited in growth in one direction (the thickness direction) but can increase in width direction. The initiation is "random" in the sense that it depends on the worst defect across the width of the specimen. For tunneling cracks it is well-known [18, 19] that once the length of a tunneling crack exceeds a few times the ply thickness, the energy release rate at the crack front attains a steady-state value that is independent of the crack length. For low values of $\bar{\varepsilon}_{max}$, the steady-state energy release rate can be so low (below a threshold) that no cyclic crack propagation can occour. For a higher value of $\bar{\varepsilon}_{max}$, the tunneling crack can propagate at a constant rate, in accordance with a Paris-Erdogan law [20]. For even higher value of $\bar{\varepsilon}_{max}$, the tunneling crack can occur as static crack propagation in the very first load cycle.

The second damage mechanism that can evolve after the formation of a tunneling crack is the initiation and growth of a delamination crack at the tips of the tunneling crack. For low values of $\bar{\varepsilon}_{max}$, the energy release rate of the delamination crack can propagate at a constant rate, in accordance with a Paris-Erdogan law [20]. We now consider the situation of a laminate made of layers with identical thicknesses and several ply-drops. There is experimental evidence that indicates that tunneling cracks and delamination cracks develop at all ply-drops but do not form simultaneously and that they propagate with different growth rates [21].

Thus, models are needed to be able to predict the delamination growth rate $d\ell/dN$ as a function of applied strain levels and the ratio between the ply thickness, h_1, and the thickness h_2 of the underlying layers.

3.3. Microscale parameters

Although these damage modes appears as macroscale damages and cracks, we will analyse them from the perspective of both macroscale and micromechanics in the next section. The basic microscale parameters for fibre, matrix and interface are listed in Table 1 and additional composite parameters are listed in Table 2.

Some comments to the choice of parameters. They represent a balance between accuracy and simplicity, sufficient to describe the most important phenomena while still allow micromechanical

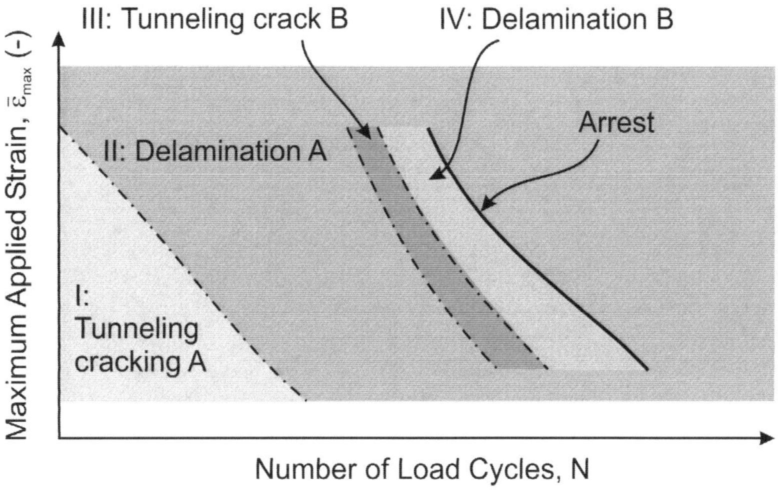

Figure 3. Fatigue damage mechanisms map delamination cracks growing from multiple ply-drops in a unidirectional fibre composites.

Table 1. Microscale parameters for fibres, matrix and interface

E_f	Young's modulus of the fibre
α_f	Thermal expansion coefficient of the fibres
r	Fibre radius
$\sigma_0(L_0)$	Characteristic strength (Weibull model)
m	Weibull modulus (Weibull model)
L_0	Reference length (Weibull model)
E_m	Young's modulus of the matrix
α_m	Thermal expansion coefficient of the matrix
$\mathscr{G}^m_{Ic}$	Mode I fracture energy of matrix
$\mathscr{G}^m_{IIc}$	Mode II fracture energy of matrix
$\mathscr{G}^i_c$	Interface fracture energy (debond energy)
τ_s	Interfacial frictional sliding shear stress

Table 2. Composites and processing parameters

V_f	Fibre volume fraction
T_0	Stress-free temperature
T	Ambient temperature
$\Delta\varepsilon^T$	Misfit strain associated with processing

models to be formulated as relative simple closed-form analytical solutions. For instance we disregard Poisson's effects of fibres and matrix. We take the fibre distribution to be uniform. Assuming the fibres to be distributed in a hexagonal array, the inter-fibre spacing d^* can be calculated as [22]

$$\frac{d^*}{r} = \sqrt{\frac{2\pi}{\sqrt{3}V_f}} - 2, \tag{10}$$

The fibre/matrix interface is particular challenging to model in terms of mechanics. In the present paper we characterise the mechanics of the fibre/matrix interface by two parameters: The interfacial fracture energy, $\mathscr{G}_c^i$, represents the breakage of chemical and physical bonds (covalent bonds, ionic bonds, hydrogen bonds) while a frictional sliding shear stress τ_s represents the stress-transfer between debonded fibres and the surrounding matrix as they slip relative to each other. This discrimination between debonding and friction sliding are considered important in the present paper.

Debonding, being cracking along a bimaterial interface involves a more complicated crack tip stress field than for cracks in homogeneous materials [23]. There is empirical evidence that shows that the interfacial fracture energy depends on the mode mixity ψ [18]. Usually the interfacial fracture energy is lowest for normal crack opening dominated fracture and highest for tangential crack opening displacements. Obviously, a more rigorous description should take $\mathscr{G}_c^i$ as being a function of the mode mixity, $\mathscr{G}_c^i = \mathscr{G}_c^i(\psi)$. However, with the present state of relative simple measurement methods for fibre/matrix interfaces, this is left for the future.

The present formulation of τ_s can be taken to be a combination of friction due to fibre roughness asperities and Coulomb friction [24]

$$\tau_s = \tau_s^0 - \mu\sigma_{rr} \quad for \quad \sigma_{rr} < 0, \tag{11}$$

where τ_s^0 is a friction stress, μ is a friction coefficient and σ_{rr} is the normal stress (taken positive in tension and negative in compression) acting normal to the fibre/matrix interface. In eq. (11) the first term represents friction due to roughness asperities of the fibre surface and the second term in a Coulomb-like friction. Note from eq. (11) that there will be a frictional shear stress due to roughness of the fibre/matrix interfaces even when there is zero normal pressure across the fibre/matrix interface ($\sigma_{rr} = 0$).

The misfit strain $\Delta\varepsilon^T$ is the difference in the in-elastic strains of the fibre and matrix occurring during processing of the composite. The major contribution to $\Delta\varepsilon^T$ is likely to be the differences in thermal expansion coefficients,

$$\Delta\varepsilon^T = (\alpha_f - \alpha_m)(T - T_0), \tag{12}$$

where α_f and α_m are the linear thermal expansion coefficients of fibre and matrix, respectively, T the present temperature and T_0 the stress-free temperature.

4. Models of in-plane fatigue

In this section we review recent models for the prediction of fatigue damage growth and fatigue failure [25]. As indicated in the fatigue damage mechanisms map for in-plane fatigue (Fig. 2), there are three mechanisms that should be modelled independently: First fibre failures (damage initiation), progressive fibre/matrix debonding and growth of clusters of broken fibres, and splitting along fibres (longitudinal splitting). We will ignore the very first mechanisms (isolated fibre failures) and consider only the two latter progressive mechanisms: Growth of the damage front by progressive fibre failure and longitudinal splitting. In the present work, we utilize a recent model by Sørensen *et al.* [25]. In this model, the fibre-to-fibre breakage is attributed to the stress field of the debond crack tip - it induces an extra stress to un-broken neighbour fibres. The progressive increase in the fibre/matrix debond length during cyclic loading is controlled by a decrease in the interfacial friction stress, τ_s originating from abrasion of interfacial asperties due to repeated forward and reverse slip. The decrease in τ_s from its initial (first forward slip) value τ_s^0 to a final value τ_s^c takes place over $\mathcal{N}_c$ load cycles (counting from the number of cycles the fibre broke and experienced it first slip) and is controlled by a shape parameter denoted n.

Then, as the debond crack tip moves along the broken fibre, it exposes a larger length of neighbour fibres to the debond crack tip stress field (the so-called K-field [26]), which then can induce failure in a neighbour fibre. A just-broken fibre will also undergo progressive debonding during the subsequent load cycles (due to the gradual decrease in τ_s). The stresses of the debond crack tip (the K-field) can induce failure in one of its neighbours as it gradually moves up along the fibre. In this manner, microscale damage in the form of progressive fibre failures at a damage front can propagate across the cross section of a specimen (see [25] for details). When the damage zone is sufficiently large, another failure mechanism becomes favourable: Longitudinal splitting. Longitudinal splitting is the growth of mode II like cracks [17]. Longitudinal splitting will be possible when the energy release rate at the tip of the splitting crack is equal to (or larger than) the Mode II fracture energy of the composite. Longitudinal cracks can connect damage zones of clusters of broken fibres, leading to specimen separation.

4.1. Prediction of the damage growth rate, da/dN

The model by Sørensen *et al.* [25] predicts the damage growth rate of a straight damage front. Denote a as the damage size (length), the fatigue damage growth rate is found as

$$\frac{da}{dN} = \sqrt{\frac{\sqrt{3}\pi}{2V_f}\frac{r}{\mathcal{N}^*}}, \tag{13}$$

where $\mathcal{N}^*$, the number of cycles it takes (on average) to cause breakage of the next row of fibres, is found as:

$$\mathcal{N}^* = \mathcal{N}_c\left\{\frac{\tau_s^0}{\tau_s^0-\tau_s^c}\left(1-(1-R)\left\langle\frac{2}{\bar{\varepsilon}_{max}}\left\{2\frac{\tau_s^0}{E_f}\frac{\ell_d^*}{r}+\frac{\bar{\sigma}_i}{E_c}\right\}-(1+R)\right\rangle^{-1}\right)\right\}^{1/n}. \tag{14}$$

In eq. (14), the parameter R is the so-called R-ratio ($R = \bar{\varepsilon}_{min}/\bar{\varepsilon}_{max}$), E_f the Young's modulus of the fibres, r is the fibre radius, ℓ_d^* is the debond length after $\mathcal{N}^*$ cycles, given by

$$\frac{\ell_d^*}{r} = \frac{3ln2}{2}\frac{L_0}{r}\left[\left(\frac{\sigma_f^+ + \hat{\sigma}_f^K}{\sigma_0}\right)^m + 2\left(\frac{\sigma_f^+}{\sigma_0}\right)^m\right]^{-1}, \tag{15}$$

where σ_f^+ is the stress in the fibre far ahead of the debond crack tip at an applied strain level corresponding to the maximum applied strain:

$$\frac{\sigma_f^+(\bar{\varepsilon}_{max})}{E_f} = -\Delta\varepsilon^T\frac{(1-V_f)E_m}{E_c} + \bar{\varepsilon}_{max}, \tag{16}$$

with $\Delta\varepsilon^T$ being the misfit strain.

Also, in eq. (15) σ_0 and m are the characteristic strength and Weibull modulus associated with the gauge length L_0 (parameters of the Weibull model [27]), $\hat{\sigma}_f^K$ is the the stress induced by the K-field of the debond crack tip to the neighbour fibres, given by [22]

$$\hat{\sigma}_f^K = 0.435\frac{E_f}{E_m}\sqrt{\frac{\mathcal{G}_c^i E^*}{d^*}}, \tag{17}$$

with d^*, the fibre spacing, given by (10) and E_m is the Young's modulus of the matrix and E^* is (plane stress)

$$\frac{1}{E^*} = \frac{1}{2}\left(\frac{1}{E_f} + \frac{1}{E_m}\right). \tag{18}$$

Furthermore, in (14), the so-called macroscopic debond initiation stress, $\bar{\sigma}_i$, is given by [22]

$$\frac{\bar{\sigma}_i}{E_c} = \frac{(1-V_f)E_m}{E_c}\Delta\varepsilon^T + 2\sqrt{\frac{(1-V_f)E_m}{E_c}\left(\frac{\mathscr{G}_c^i}{E_f r}\right)}, \tag{19}$$

and the parameter E_c, used in (16) and (19), is the Young's modulus of the composite which is (assuming no porosity) is given by the rule of mixtures

$$E_c = V_f E_f + (1-V_f)E_m. \tag{20}$$

4.2. Number of cycles to fatigue failure, N_f

We now model the final fatigue damage mechanism, the growth of longitudinal splitting cracks. As a first approximation, we assume that the splitting occurs as the growth of a sharp crack tip, so that linear-elastic fracture mechanics is applicable. It is thus implicitly assumed that the crack faces are traction-free (no friction and no fibre bridging). When the length of the splitting cracks is a few times the damage zone size, splitting becomes a steady-state problem where the energy release rate takes a constant value. Then, splitting can occur as static crack growth at the maximum applied strain, $\bar{\varepsilon}_{max}$. This can happen when the steady-state energy release rate equals the mode II fracture energy of the composite, $\mathscr{G}_{IIc}^c$. The critical size for growth of splitting cracks around a hexagonally shaped column is [17]

$$g_s = \frac{4\mathscr{G}_{IIc}^c}{E_c\bar{\varepsilon}_{max}^2}, \tag{21}$$

where g_s is the half of the width of a hexagonal-shaped column forming during splitting (see [17] for more details).

Using fracture area arguments we can express $\mathscr{G}_{IIc}^c$ in terms of microscale parameters as [17],

$$\mathscr{G}_{IIc}^c = \left(1 - \sqrt{\frac{2\sqrt{3}V_f}{\pi}}\right)\mathscr{G}_{IIc}^m + \frac{\pi}{2}\sqrt{\frac{2\sqrt{3}V_f}{\pi}}\mathscr{G}_c^i, \tag{22}$$

where $\mathscr{G}_{IIc}^m$ is the Mode II fracture energy of the matrix.

In case the fatigue damage growth rate, da/dN, is determined, the number of cycles to failure can be estimated as

$$N_f = \frac{g_s}{da/dN}. \tag{23}$$

This reasoning is based on the assumption that the damage zone extends normal to its damage front at a constant rate, da/dN, from failure of the first fibres to the critical cluster size. Inserting g_s from (21) and da/dN from (13) into (23) gives

$$N_f = \frac{4\mathscr{G}_{IIc}^c}{E_c r\bar{\varepsilon}_{max}^2}\sqrt{\frac{2V_f}{\sqrt{3}\pi}}\mathcal{N}^*, \tag{24}$$

where $\mathcal{N}^*$ is given by (14).

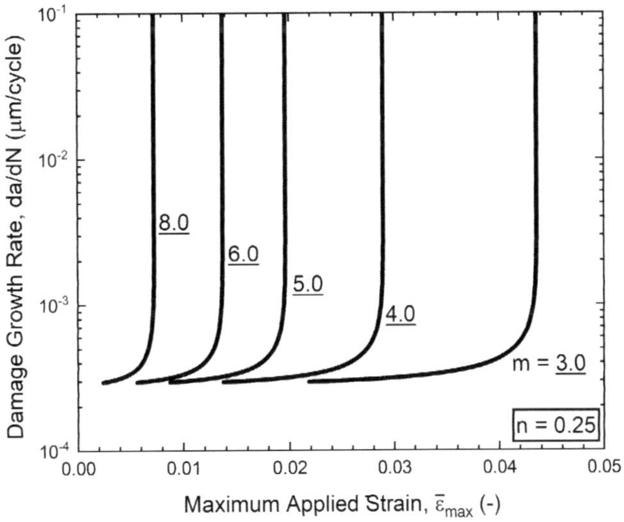

Figure 4. Predicted damage growth rates for in-plane fatigue of unidirectional fibre composites - effect of Weibull modulus, m.

As an example, we show the effect of varying the variation in the fibre strength, by varying the Weibull modulus, m. Fig. 4 shows the damage growth rate da/dN, calculated from eqs. (13) and (14) as a function of the maximum applied strain, $\bar{\varepsilon}_{max}$. The curves have similar form: They all start at a low strain value giving the same minimum damage growth rate, increasing with increasing $\bar{\varepsilon}_{max}$, reaching a maximum value of $\bar{\varepsilon}_{max}$ where the damage growth rate increases asymptotically (corresponding to infinite growth rate, resulting in failure in the very first load cycle). Increasing m (corresponding to a narrower fibre strength variation) increases the damage growth quite significantly or equivalently decreases the strain range at which fatigue occurs (between the two strain values, $\bar{\varepsilon}_{max}$ and $\bar{\varepsilon}_{min}$) and narrows the range between the strain values of slowest and first cycle failure.

Fig. 5 shows predicted S-N-curves for various values of m, calculated using eq. (24): The maximum applied strain is shown as a function of the number of cycles to failure, N_f. The curves have a horizontal part that asymptotically (for $N_f \to 1$) approaches the predicted static strength. For lower values of $\bar{\varepsilon}_{max}$, the predicted S-N-curves turn downwards in a near-linear form. An increase in m moves the curves down (lower strain value for the same fatigue life, N_f) and correspondingly lower fatigue limit.

5. Delamination from ply-drops

5.1. Delamination growth under static, monotonic loading for brittle interface

A simple model of delamination of a symmetric tri-layer (a laminate consisting of three layers) is developed in Appendix A. The two surface layers have a Young's modulus, E_1 and a thickness, h_1, and the middle layer as a Young's modulus, E_2 and thickness h_2. An interface crack exist along both interfaces.

We consider first the case without bridging tractions ($\tau_s = 0$). Only the first term in eq. (45) remains. Then, debonding will happen at a constant applied stress value, given by

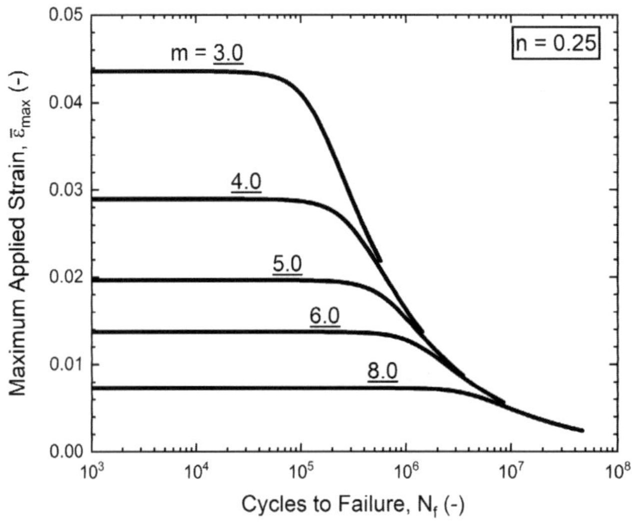

Figure 5. Predicted S-N-curves for in-plane fatigue of unidirectional fibre composites - effect of Weibull modulus, m.

$$\frac{\bar{\sigma}}{E_2} = \sqrt{\frac{2\,(1+2\Sigma\eta)}{\Sigma\eta}\frac{\mathscr{G}^i_{IIc}}{E_2 h_2}} = \sqrt{\frac{2\,(E_2h_2+2E_1h_1)}{E_1h_1}\frac{\mathscr{G}^i_{IIc}}{E_2 h_2}}, \tag{25}$$

irrespective of the debond length ℓ_d. Likewise, the J integral result, eq. (44), becomes

$$J_{tip} = \frac{\Sigma\eta}{2\,(1+2\Sigma\eta)}\frac{\bar{\sigma}^2 h_2}{E_2} = \frac{E_1h_1}{2\,(E_2h_2+2E_1h_1)}\frac{\bar{\sigma}^2 h_2}{E_2}. \tag{26}$$

Fig. 6 shows the normalized J integral result (energy release rate) as a function of the thickness ratio and stiffness ratio. The normalisation is done such that the properties (stiffness and thickness) of the middle layer is held constant and properties of the surface layers are varied. It is seen in Fig. 6 that the energy release rate increases with increasing thickness and stiffness of the surface layers. For softer surface layers ($E_1 < E_2$), the energy release rate increases almost linearly with the thickness of the surface layer - a linear relation applies for thickness ration up to about 0.3. When the surface and middle layers have identical stiffness, the relationship is more non-linear; the linear relation only applies for small thickness ratios (say, less than 0.1). For stiffer surface layers, the relationship is non-linear.

As another example, consider a laminate with two ply-drops as shown in Fig. 7a). All plies have the same thickness, t. There are pre-existing delamination at both ply-drops. We wish to use the model to predict which of the two delaminations, at the thin end (position A) or the thick end (position B) will propagate first under monotonically increasing load. We use (26) for the two cases. For simplicity, we take $E_1 = E_2$ so that $\Sigma = 1$. For Case A (Fig. 7b)), $h_1 = t$ and $h_2 = 2t$ so that $\eta = 1/2$. The stress in the thin end of the delaminated laminate, $\bar{\sigma}_A$ is equal to $\bar{\sigma}_0$. Then (26) gives $J_{tip} = \bar{\sigma}_0^2 t/4E_2$. For Case B (Fig. 7c)), $h_1 = t$ and $h_2 = 4t$ so that $\eta = 1/4$. Moreover, the stress in the middle layer, $\bar{\sigma}_B = \bar{\sigma}_0/2$. Then (26) gives $J_{tip} = \bar{\sigma}_0^2 t/12E_2$. It is clear that J_{tip} is lowest for Case B. Conversely, it can be expected that under a monotonically increasing load, the delamination at position A will start propagating at a lower value of $\bar{\sigma}_0$

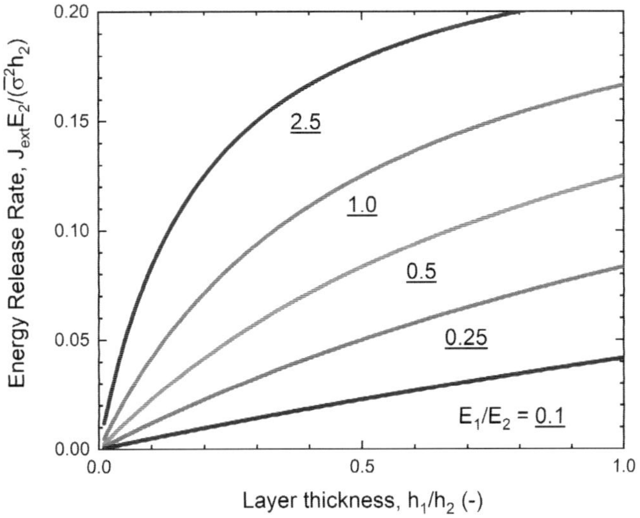

Figure 6. Normalized J integral for the traction-free case shown as a function of ply thickness ratio for several stiffness mismatch.

than required for growth of the delamination at position B. The J_{tip} value is independent of crack length, so the crack will propagate quasi-stable at a constant value of $\bar{\sigma}_0$. As the crack tip of delamination A propagates past the tip of delamination B, then (as shown in Fig. 7d) effectively $h_1 = 2t$ and $h_2 = 2t$, so that $\eta = 1$. With $\bar{\sigma}_A = \bar{\sigma}_0$ we then get $J_{tip} = \bar{\sigma}_0^2 t/3E_2$. This value of J_{tip} is higher than J_{tip} was when the delamination was shorter (Case A). Conversely, as the delamination crack tip A reaches the crack tip of delamination B at under quasi-static growth, delamination crack A will then propagate unstably along the remainder of the specimen. In contrast, delamination B will not be able to propagate under static loading (if the fracture energy of the two interfaces are the same) since its value of J_{tip} is lower than that of delamination A.

5.2. Prediction of crack growth rate, $d\ell/dN$ under cyclic loading for brittle interface

Now we consider cyclic loading applied to the ply-drop specimen shown in Fig. 7. The applied stress is assumed to alternate between a fixed maximum applied tensile stress, $\bar{\sigma}_{max}$, and a fixed minimum tensile stress, $\bar{\sigma}_{min}$. Also, we assume that the cyclic crack advancement of the delamination crack tip follows a Paris-Erdogan law [20, 26]:

$$\frac{d\ell}{dN} = C\left(\Delta K - (\Delta K)_{th}\right)^{\tilde{m}}, \tag{27}$$

where ΔK is the stress intensity range, $(\Delta K)_{th}$ is a threshold value of the stress intensity factor range and C and $\tilde{m}$ are constants.

At the microscale, the crack advancement of crack tip can be considered as consisting of a debond crack growth along the fibre/matrix interface and cyclic crack growth in the matrix. The local crack front of the debond crack and the matrix crack front will adjust themselves (and their local stress intensity ranges) such that they both advance in the same rate, equal to the macroscale delamination crack growth rate, $d\ell/dN$.

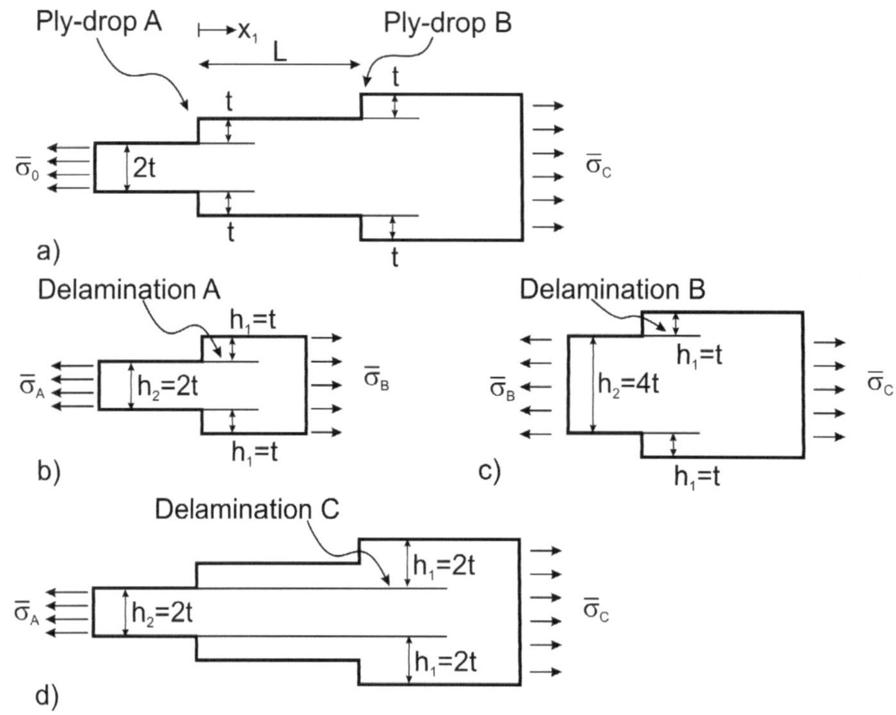

Figure 7. Example of symmetric laminate with two ply-drops with delaminations. The overall problem (a) can be analysed as three "elementary" cases, Case A (b), Case B (c) and Case C (d).

Returning to the macroscale, let us consider the specimen with two ply-drops, Fig. 7. For simplicity, assume that $(\Delta K)_{th} = 0$ and take $\bar{\sigma}_{min} = 0$ (so that the minimum J integral value, J_{min}, is zero) and (again) $E_1 = E_2$. Then, with the Irwin relationship [8, 26] we have

$$\Delta K = \sqrt{J_{max} E_2}, \tag{28}$$

so that we find from eq. (27) that the ratio between the crack growth rate of delamination crack tip B $(d\ell/dN)|_B$ to the crack growth rate of delamination crack tip A $(d\ell/dN)|_A$ becomes: $(d\ell/dN)|_B/(d\ell/dN)|_A = 1/(3)^{\tilde{m}/2}$. As an illustrative example, with $\tilde{m} = 2.5$, we get $(d\ell/dN)|_B/(d\ell/dN)|_A \approx 0.25$. Thus, during cyclic loading, both delamination crack tips A and B propagate during cyclic loading, but delamination B propagates significantly slower than delamination A. Then the crack tip of delamination A can catch up and propagate past the crack tip of delamination B. Then the situation becomes Case C. We find $(d\ell/dN)|_C/(d\ell/dN)|_A = (2/\sqrt{3})^{\tilde{m}}$. Again, as an example, with $\tilde{m} = 2.5$, we find $(d\ell/dN)|_C/(d\ell/dN)|_A \approx 1.4$. Thus, once the delamination crack tip A caches delamination crack tip B, its growth rate increases. From the perspectives of avoiding a faster delamination crack growth, a criterion could be to ensure that crack tip A does not extend to crack tip B. We progress to set up such a criterion.

Consider the situation that tunneling cracks have formed across the full specimen width at both ply-drop A and ply-drop B in the very first load cycle. Assume further that both delamination cracks A and B propagate under the subsequent load cycles with crack growth rates $(d\ell/dN)|_A$ and $(d\ell/dN)|_B$, respectively. With the distance between the ply-drops given by L, the critical number of load cycles for crack tip A to catch crack tip B can be estimated as:

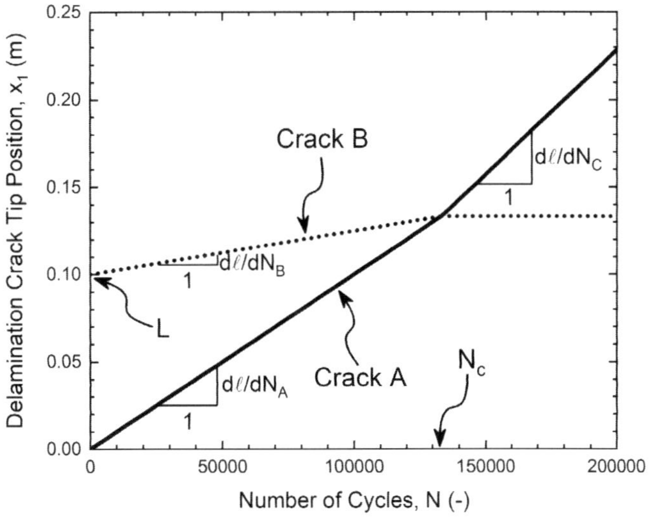

Figure 8. Crack length of two delamination cracks starting from different ply-drops (separated a distance L) in a symmetric laminate as a function of number of load cycles.

$$N_c = \frac{L}{(d\ell/dN)|_A - (d\ell/dN)|_B}. \tag{29}$$

Fig. 8 shows predicted crack growth using (26), (27) and (28). Crack A starts from a ply-drop at the thinnest part of the laminate and crack B starts from the next ply-drop, a distance L away from ply-drop A. Both cracks grow with constant growth rates (crack A grows faster than crack B, $(d\ell/dN)|_A > (d\ell/dN)|_B$) until crack A has caught up crack B after N_c cycles. Thereafter, crack A continues to grow with a faster crack growth rate ($(d\ell/dN)|_C$) while crack B stops growing, since it is now unloaded. These predictions are qualitatively in agreement with experimental findings [21].

It is clear from these considerations that delaminations from ply-drops do not possess a damage tolerant behaviour in case the interface between plies are brittle and crack growth is controlled by a Paris-Erdogan law.

Now we turn our attention to bridging fibres which can bring in damage tolerant behaviour.

5.3. Increasing damage tolerance by crack bridging - microscale aspects

In some cases, unidirectional fibre composites can experience crack bridging by single fibres or ligaments of multiple fibres that connect crack faces and create tractions at the fracture surfaces. As the crack openings these tractions perform work that contributes to the fracture resistance, as described in Section 2.2. This creates damage tolerance, as will be shown in the following.

For a delamination crack that initiate from a ply-drop, the crack openings are dominated by tangential crack opening displacements. Rectangular ligaments (width b and height h) being peeled-off under pure tangential crack opening displacement creates a shear traction [28]

$$\tau_s = \left[\frac{2\mathcal{G}_{IIc}^c}{E_c h}\right]^{1/2} \tilde{\eta} b h E_c, \tag{30}$$

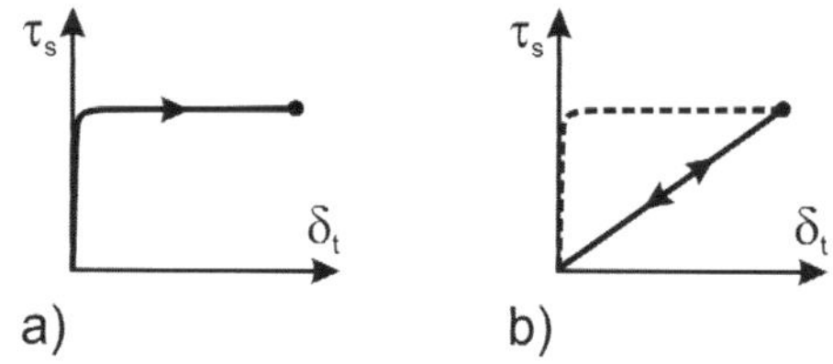

Figure 9. Schematics of the shear traction of bridging ligaments and bridging fibres as a function of tangential opening displacements, under monotonic opening (a) and cyclic opening and closure (b).

where $\mathscr{G}^c_{IIc}$ is the Mode II fracture energy of the composite, and $\tilde{\eta}$ is the number of bridging ligaments per unit cracked area. For bridging by single fibres a similar relation is found:

$$\tau_s = \left[\frac{2\mathscr{G}^i_c}{E_f r}\right]^{1/2} \tilde{\eta}\pi r^2 E_f, \tag{31}$$

where $\mathscr{G}^i_c$ is the fracture energy of the fibre/matrix interface. Note that for both cases, the shear traction is independent of the tangential opening displacement, but increases proportionally with the square root of the fracture energy. Thus, for a crack undergoing pure tangential opening displacements, bridging ligaments or bridging fibres can create a constant shear traction along the bridging zone (failure of ligaments and fibres is disregarded for simplicity). This relationship is shown schematically in Fig. 9a). Under cyclic opening/closure the ligaments are expected to unload linear-elastically, with a linear relationship between opening and traction as indicated in Fig. 9b).

5.4. Delamination growth under static, monotonic loading under large-scale bridging

We take up the model of delamination of a symmetric tri-layer presented in Appendix A. In this model, the bridging shear tractions are represented in terms of a constant shear stress τ_s. Equation (45) gives the relationship between debond length ℓ_d and the applied stress, $\bar{\sigma}$. The applied stress required to start the debond crack is identical to that of a non-bridged crack, eq. (25) and then increases linearly with debond length. This implies that under monotonically increasing loading, the debond crack growth is stable: Although debond cracking has started, a higher value of the applied stress is needed to grow the crack further; the delamination crack growth is stable. This is a damage tolerant behaviour.

5.5. Prediction of crack growth rate, $d\ell/dN$ under cyclic loading under large-scale bridging

We now consider the ply-drop specimen subjected to cyclic stress between a maximum tensile stress, $\bar{\sigma}_{max}$ and minimum applied tensile stress, $\bar{\sigma}_{min}$. In the following, we take for simplicity $\bar{\sigma}_{min} = 0$ (so that the minimum J integral value, J_{min} at $\bar{\sigma}_{min}$, is zero) and $(\Delta K)_{th} = 0$.

Replacing the value of the applied stress, $\bar{\sigma}$, with the maximum value of the applied stress, $\bar{\sigma}_{max}$, during cyclic loading, eq. (45) can also be useful in connection with cyclic loading:

$$\frac{\bar{\sigma}_{max}}{E_2} = \frac{(1+2\Sigma\eta)}{\Sigma\eta}\frac{\tau_s}{E_2}\frac{\ell_d}{h_2} + \sqrt{\frac{2(1+2\Sigma\eta)}{\Sigma\eta}\frac{J_{max}}{E_2 h_2}}, \tag{32}$$

where we have used J_{max} (the value of J_{tip} at the maximum applied stress) instead of $\mathscr{G}^i_{IIc}$. Now (32) can be understood as the relationship between J_{max} and ℓ_d at $\bar{\sigma}_{max}$. With fixed, $\bar{\sigma}_{max}$, an increase in ℓ_d during cyclic loading leads to a decrease in J_{max}. Thus, during a cyclic experiment, J_{max} decreases with increasing ℓ_d. The decrease in J_{max} decreases ΔK (according

IOP Conf. Series: Materials Science and Engineering **942** (2020) 012004 doi:10.1088/1757-899X/942/1/012004

to eq. (28)) and thus a decreasing crack growth rate from the Paris-Erdogan law, eq. (27). Thus, with sufficiently long debond crack length, $J_{max} = 0$ so that $\Delta K = 0$ and thus the cyclic crack growth will stop (crack arrest).

Setting $J_{max} = 0$ in (32) we can find the associated debond length as

$$\frac{\ell_d^a}{h_2} = \frac{\Sigma\eta}{(1+2\Sigma\eta)} \frac{\bar{\sigma}_{max}}{\tau_s}. \tag{33}$$

where superscript "a" has been used to indicate crack arrest.

A criterion for ensuring crack arrest under cyclic loading will then be:

$$\ell_d^a < L, \tag{34}$$

where L is the distance between the ply-drops. Combined, eqs. (34) and (33) can be used to design the maximum applied stress $\bar{\sigma}_{max}$ and the spacing between ply-drops L and laminate dimensions such that delamination cracks originating from ply-drops will arrest, in the case that the delamination occurs with fibre bridging. From eq. (33) it is apparent that a critical parameter in such a design approach is the maximum cyclic shear traction τ_s acting along the delaminated interface.

Crack bridging thus provides the possibilities to have delaminations from ply-drops propagate in a stable manner under increasing applied load and propagate slowly and potentially arrest. Thus, crack bridging is a promising mechanism to incorporate in materials to create damage tolerance behaviour.

6. Microscale experiments

The spirit of the present paper is to consider composites from the microscale, acknowledging that the macroscale behavior of composites is controlled by the microscale parameters. An advantage of such multiscale approach is that the effects of actual physical processing parameters can be understood by micromechanical modelling. Linking to the microscale should thus be an advantage primarily from the perspective of the development of new composite materials with improved properties, such as enhanced damage tolerance.

However, this pushes the mechanical characterisation down to the microscale. In order to be useful in materials development, mechanical characterisation of microscale parameters should become a key activity of future materials development.

The models used in the present paper are relative simple and have analytical solutions. Obviously, they have less accuracy than more advanced computational models that require numerical solutions (e.g. by the finite element method). However, analytical models provide clear insight into the effects of model parameters, something that can be more difficult or time consuming to obtain from numerical models. Also, the simple models balance well with the present state of relative simple experimental methods used for microscale characterisations.

Within the last few years, X-ray tomography has emerged as a very useful tool to study fatigue damage in composites at the microscale. This provides a lot of input and clarification to microscale damage evolution [16]. More such studies are expected in the coming years.

6.1. Characterisation of fibre properties

The major microscale parameters for fibres are stiffness and strength properties, see Table 1. These are determined from single-fibre tensile testing experiments. This used to be tedious and difficult work. But within recent years, automatized fibre testing equipment, which enables fast and reproducible tensile tests of fibres has been commercialized. This enables the tensile testing of larger batches of fibres increasing the relability of the parameters for fibre strength variation, the characteristic strength $\sigma_0(L_0)$ and the Weibull modulus, m.

6.2. Characterisation of matrix properties

Stiffness and Mode I fracture energy of matrix materials should be relative easy to obtain from standard materials testing (tensile testing and fracture mechanics testing). The Mode II fracture energy of matrix, $\mathscr{G}_{IIc}^m$, can be more challenging but can be measured using sandwich specimens consisting of a thin resin layer between stiff beams [29].

6.3. Characterisation of fibre/matrix interface properties

Experimentally, the most challenging area is the characterisation of the mechanics of the fibre/matrix interface. Microscale experiments are required for this purpose. As mentioned, debonding is essentially cracking along a bimaterials interface. For some problems like in-plane fatigue, the fibre/matrix debonding takes place due to a tensile load in the fibre direction and fibre/matrix debonding is predominantly "Mode II" (debonding is truly a mixed mode cracking problem [23]) and is influenced by frictional sliding along the interface.

There is a relative large number of experimental techniques for the charcterisation of the fibre/matrix interface under "Mode II" debonding. Examples are fibre pull-out, single fibre fragmentation test, microbond testing and fibre push-out [30, 31]. The analysis of the experiments are sometimes rather simple, so that only one parameter (Mode II debond energy or friction sliding shear stress) but not both are determined. In other cases, complicated numerical models are used to determine debond energy $\mathscr{G}_c^i$ and a friction shear stress τ_s (or in connection with assumed Coulomb friction, a friction coefficient, μ). However, the parameter derermination is far from simple, and there are examples of different "advanced" models that - applied to the same experimental data - have given debond fracture energy values that differ by a factor of ten [32, 33, 34].

There are relative few experimental studies of the cyclic behaviour of fibre/matrix interfaces, experiments where the debond crack length has been recorded as a function of the number of cycles of forward and reverse slip [12, 15]. Such studies, which can be conducted under light microscopy, are essential for the formulation of the evolution laws for τ_s as function of number of slip cycles etc. This deserver much more research in the future.

In contrast to in-plane fatigue, fibre/matrix debonding in connection with crack bridging of cracks parallel to the fibre direction (delamination and splitting) usually involves fibre debonding without friction. Also, such "peel"-like debonding ocours under mixed mode (involving both normal and shear stresses at the debond crack tip) [35]. Other types of single fibre peel-off experiments could be useful, both under static and cyclic loading.

7. Perspectives

7.1. Measurements of mechanical material properties

Currently, the macroscale is the length scale where the majority of materials testing takes place. There are a large number of test specimens (so-called "coupons") established to characterize quite a number of engineering strength and failure parameters that are considered necessary to describe the fairly large number of complicated failure modes that exist for composite materials and structures. For some of these, there are several test specimen geometries and test standards. A full materials characterisation at the macroscale can thus be quite costly and involving. Still, all these macroscale strength properties must be controlled by the microscale properties of the fibres, the matrix and the fibre/matrix interface. It is thus anticipated that in a long term perspective, the mechanical behaviour at the macroscale can be predicted from micromechanical models with microscale properties of fibre, matrix and the fibre/matrix interface as the "basic" material properties; materials characterisation then moves to the microscale.

7.2. NDT detection of fatigue damage mechanisms

NDT methods can play an important role in the detection of fatigue damage at various length scales. The use of fatigue damage mechanism maps could be useful in linking mechanisms and sensor signals. For instance, sensor signals e.g. from acoustic emission sensors, can be recorded and characterised for individual damage mechanism at the microscale. These signals can then help identify the damage evolution in coupon specimens at the macroscale and elements at the substructural scale. This could help uncovering invisible or barely visible damages and thus help the establishment of generic description of damage evolution in terms on fatigue damage mechanism maps at a coarser length scale.

7.3. Multiscale modelling and multiscale testing

The multiscale nature of the damage evolution in composite materials and composite structures has lead to a multiscale approach to modelling of damage evolution. Clearly, some sort of averaging and loss of details are needed when transferring material laws from a small length scale to a coarser length scale. It remains a non-trivial challenge to establish rigorous correct connections between models at various length scales.

In parallel, as new materials testing methods will be established at the microscale, the testing at the macroscale by coupons and elements could shift from the measurement of "strength" properties and load-carrying capacity to check if the actual damage evolution is similar to the predicted damage evolution. This may require other, new generic test specimens designed for the study of damage evolution.

8. Conclusions

The development of very large composites structures such as wind turbine blades requires materials with high damage tolerance, such that manufacturing defects are of less importance. The use of models of fatigue damage evolution holds the potential to lead the development of future composite materials with enhanced damage tolerance. It is however still a challenge to measure the relevant mechanical properties at the microscale.

Acknowledgements

This research was partly supported by the project DACOMAT (Damage Controlled Composite Materials), funded by the European Union's Horizon 2020 research and innovation programme under GA No. 761072. Useful discussion with Daan J H Cederløf, Jens K Jørgensen and Stergios Goutianos are acknowledged.

References

[1] Nijssen RPL and Brøndsted, P 2013 *Advances in wind turbine blade design and materials* (Ed. Povl Brøndsted and Rogier P. L. Nijssen). (Oxford: Woodhead Publishing) pp. 175-209
[2] Vasilopoulos, AP., *Advances in wind turbine blade design and materials* (Ed. Povl Brøndsted and Rogier P. L. Nijssen). (Oxford: Woodhead Publishing) pp. 251-97
[3] Sevenois RDB and Van Paepegem W. 2015 *Applied Mechanics Review* **67** 020802-1-11
[4] Iarve EV, Hoos K, Braginsky M, Zhou E and Mollenhauer DE 2017 *J. Comp. Mater.* **51** 2143-61
[5] U.S. Department of Transportation, Federal Aviation Administration, 'Composite aircraft structure', Advisory Circular AC No: 20-107B, Change 1, 24 August 2010
[6] Rice JR 1968 *Journal of Applied Mechanics* **35** 379-86
[7] Sørensen, BF and Kirkegaard P 2006 *Engineering Fracture Mechanics*, **73**, 2642-61
[8] Broek D 1986 *Elementary engineering fracture mechanics*, 4th edition, (Dordrecht and Boston: Mertinus Nijhoff Publishers)
[9] Talreja R 1981 *Proc. R. Soc. Lond. A* **378** 461-75
[10] Carraro P and Quaresimin M 2018 *Procedia Engineering* **213** 17–24
[11] Gamstedt EK and Talreja R 1999 *Journal of Materials Science* **34** 2235-46
[12] Goutianos S and Peijs T 2010 *Advanced Composites Letters* **10** 21-32

[13] van den Heuvel PWJ, Peijs T and Young RJ 1997 *Composites Part A* **27** 855-859.
[14] van den Heuvel P.WJ. Peijs T and Young RJ 1998 *Composite Science and Technology* **58** 933-44
[15] Pupurs A Goutianos S Brøndsted P and Varna J 2013 *Composites Part A* **44** 86-94
[16] Jespersen KM and Mikkelsen LP 2017 *Composites Science and Technology* **153** 261-72
[17] Sørensen BF 2019 *Composites Science and Technology* **172** 172-81
[18] Hutchinson JW and Suo Z 1992, *Advanced in Applied Mechanics*, **29** (Ed. JW Hutchinson and TY Wu) (Boston: Academic Press Inc) pp. 63-191
[19] Nakamura T and Kamath SM 1992 *Mechanics of Materials* **13**, 67-77
[20] Paris P and Erdogan F 1963 *Journal of Basic Engineering, Series D* **85** 528-34
[21] Goutianos S, 2019 private communication
[22] Sørensen, BF and Goutianos S 2019 *Mechanics of Materials* **131** 169-87
[23] Rice JR 1988 *Journal of Applied Mechanics* **55** 98-103
[24] Mackin TJ, Warren PD and Evans AG 1992 *Acta Metallurgica et Materialia* **40** 1251–1257
[25] Sørensen BF, Goutianos S, Mikkelsen LP and Søren Fæster 2020 *Composite Science and Technology* submitted
[26] Kanninen MF and Popelar CH 1985 *Advanced fracture mechanics*, (New York: Oxford University Press)
[27] Weibull W 1939 *Ingeniörsvetenskapsakademiens Handlingar* **151** (Stockholm: Generalstabens Litografiska Anstalts Förlag)
[28] Sørensen BF, Gamstedt EK, Østergaard RC and Goutianos S 2008 *Mechanics of Materials* **40** 220-34
[29] Chai H 1993 *International Journal of Fracture* **60** 311-26
[30] Herrera-Franco and Drzal LT 1992 *Composites* **23** 1-27
[31] Moosburger-Will J, Greisel M, Schulz M, Löffler M, Mueller WM and Horn S 2020 *Composites Interfaces* **27** 227-47
[32] Kim BW and Nairn JA 2002 *Journal of Materials Science*, **37** 3965-72
[33] Graciani E, Mantic V, Paris F and Varna J 2009 *Composites Science and Technology* **69** 2514-20
[34] Sørensen, BF 2017 *Mechanics of Materials* **104** 38-48
[35] Cederløf DJH and Sørensen BF 2020 *Proceedings of the 41st Risø International Symposium on Materials Science*
[36] Budiansky B, Hutchinson JW and Evans AG 1986 *Journal of the Mechanics and Physics of Solids* **34** 167-78

Appendix A - J integral analysis of delamination from ply-drop of a tri-layer

8.1. Problem description

The problem to be analysed is shown in Fig. 10. It is a symmetric, three-layer specimen consisting of two surface layers enclosing a central layer, all with the width, B. All layers are taken to be made of linear-elastic materials. The surface layers are made of material #1 that has a Young's modulus E_1 and a thickness h_1, and the Young's modulus of the central layer (material #2) is E_2 and its thickness is h_2. Pre-existing debond cracks of identical lengths, ℓ_d, exist from the left hand side. The Mode II interfacial fracture energy of the debond crack tip is denoted $\mathscr{G}^i_{IIc}$. It is here assumed that the second Dundurs' parameter is zero, so that the complex stress intensity factor decouples into traditional Mode I and Mode II stress intensity factors, so the pure Mode I and Mode II readily exists for the interface crack [18]. A shear stress, τ_s, is present along the debonded interface; it can represent friction or crack bridging by fibres. A Cartesian coordinate system, x_i, is placed such that its origin is located at the tip of the upper delamination crack (see Fig. 10). A tensile stress, $\bar{\sigma}$, is applied to the left hand end of the central layer. The problem will be analysed under plane stress condition (but can easily be extended to a plane strain condition). We aim to establish a relationship between $\bar{\sigma}$ and ℓ_d during delamination crack growth by the use of the J integral [6].

8.2. Upstream and downstream stresses

Stresses far ahead of the crack tip ($x_1/(h_1+h_2) >> 1$) are called "upstream stresses" and stresses in the debonded layers left of the crack tip ($-\ell_d < x_1 < 0$) are called "downstream stresses". The upstream stress (in the x_1-direction, i.e., stress component, σ_{11}) far ahead of the debond crack tip in the surface layer, is denoted σ_1^+ (subscript 1 indicates layer number, superscript + indicates upstream), and the upstream stress in the centre layer, σ_2^+, can be found by assuming

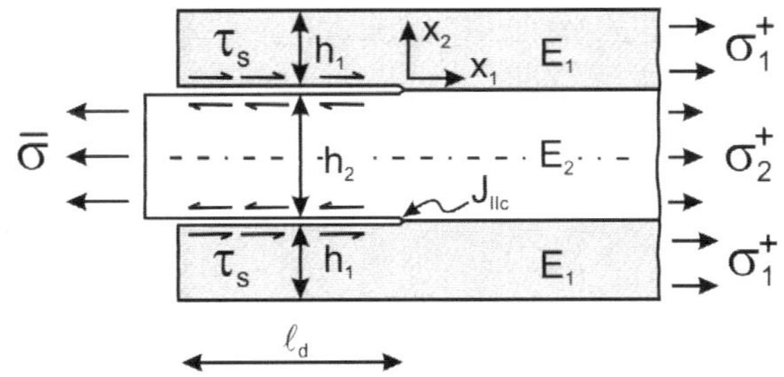

Figure 10. The analysed problem: A tri-layer specimen undergoing delamination with a shear stress τ_s in the delaminated zone.

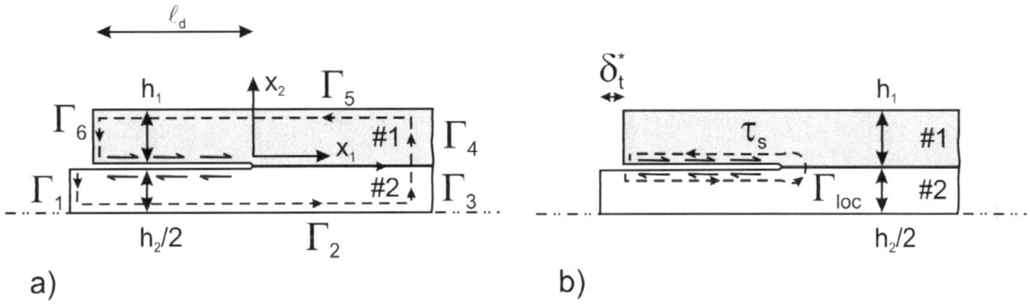

Figure 11. The integration paths for the J integral around the external boundaries of the upper half of the problem (a) and locally around the fracture process zone (b).

the same axial strain in all the layers and by elementary force balance

$$\sigma_1^+ = \bar{\sigma}\frac{\Sigma}{1+2\Sigma\eta} \quad and \quad \sigma_2^+ = \bar{\sigma}\frac{1}{1+2\Sigma\eta}, \tag{35}$$

where the non-dimensional parameters Σ and η are given by

$$\Sigma = \frac{E_1}{E_2} \quad and \quad \eta = \frac{h_1}{h_2}. \tag{36}$$

Downstream stresses at the debonded zone ($-\ell_d < x_1 < 0$) are modelled by a shear-lag approximation as

$$\sigma_1^- = \tau_s\frac{x_1+\ell_d}{h_1} \quad and \quad \sigma_2^- = \bar{\sigma} - 2\tau_s\frac{x_1+\ell_d}{h_2} \quad for \quad -\ell_d \leq x_1 \leq 0. \tag{37}$$

8.3. J integral analysis along external boundaries, Γ_{ext}

Due to symmetry, we only need to analyse the upper half of the specimen ($x_2 > 0$). The problem to be analysed is sketched in Fig. 11. First, we calculate the J integral evaluation along the external boundaries. The integration path is divided into 6 paths as indicated in Fig. 11.

For Γ_1, $\sigma_{11} = \bar{\sigma}$, the rest of the stress components $\sigma_{ij} = 0$, the normal vector, $n_j = (-1; 0; 0)$, the integration path increment $dS = -dx_2$. Furthermore, $u_{1,1} = \varepsilon_{11}$ (here, u_1 and ε_{11} are the displacement and the normal strain in the x_1-direction respectively, and comma denotes partial derivation). Using Hooke's law for material #2 ($\varepsilon_{11} = \sigma_{11}/E_2$) we obtain

$$J_1 = \int_0^{-h_2/2}\left(-\frac{\bar{\sigma}^2}{2E_2}\right)dx_2 = \frac{\bar{\sigma}^2 h_2}{4E_2}. \tag{38}$$

IOP Conf. Series: Materials Science and Engineering **942** (2020) 012004 doi:10.1088/1757-899X/942/1/012004

For Γ_2, $dx_2 = 0$, and due to symmetry $\sigma_{12} = 0$ and $u_{2,1} = 0$ (where u_2 is the displacement in the x_2-direction) and $n_j = (0; -1; 0)$. As a result, $J_2 = 0$.

For path Γ_3 and Γ_4 we find

$$J_3 = -\frac{\bar{\sigma}^2 h_2}{4E_2 (1+2\Sigma\eta)^2} \quad and \quad J_4 = -\frac{\bar{\sigma}^2 h_2}{4E_2} \frac{\Sigma\eta}{(1+2\Sigma\eta)^2}. \tag{39}$$

For Γ_5 and Γ_6 we find $J_5 = J_6 = 0$. Collecting all J integral terms (J_1 to J_6) gives

$$J_{ext} = \frac{\bar{\sigma}^2 h_2}{2E_2} \frac{\Sigma\eta}{(1+2\Sigma\eta)}. \tag{40}$$

8.4. J integral analysis locally around delamination, Γ_{loc}

An application of the J integral locally around the slip zone and the Mode II crack tip gives [7]

$$J_{loc} = \int_0^{\delta_t^*} \tau_s d\delta_t + J_{tip} = \tau_s \delta_t^* + J_{tip}, \tag{41}$$

where δ_t^* is the end-sliding (displacement difference between upper and lower crack faces at $x_1 = -\ell_d$) and J_{tip} is the J integral evaluated around the crack tip.

Displacements can be obtained by integration of the strains ε_{11} (from $x_1 = 0$ to $x_1 = -\ell_d$). The downstream strains in material #1 and #2 can be obtained from the stresses σ_1^- and σ_2^- (eq. (37)) using Hooke's law. Defining integrations constants such that there is zero displacement difference at the crack tip ($x_1 = 0$) leads to

$$\delta_t^* = \frac{\bar{\sigma}}{E_2}\ell_d - \frac{\tau_s \ell_d^2}{2E_1 h_1}(1+2\Sigma\eta). \tag{42}$$

Next, inserting δ_t^* from eq. (42) into eq. (41) and using eq. (36) leads to:

$$J_{loc} = \frac{\bar{\sigma}\tau_s}{E_2}\ell_d - \frac{\tau_s^2 \ell_d^2}{2E_2 h_2}\frac{(1+2\Sigma\eta)}{\Sigma\eta} + J_{tip}. \tag{43}$$

8.5. Path-independence of the J integral

Due to path-independence of the J integral [6], $J_{ext} = J_{loc}$. Setting eq. (40) equal to eq. (43) gives

$$J_{tip} = \frac{\Sigma\eta}{2(1+2\Sigma\eta)}\frac{\bar{\sigma}^2 h_2}{E_2} - \frac{\bar{\sigma}\tau_s \ell_d}{E_2} + \frac{(1+2\Sigma\eta)}{2\Sigma\eta}\frac{\tau_s^2 \ell_d^2}{E_2 h_2}. \tag{44}$$

During cracking, J_{tip} must be equal to the Mode II interfacial fracture energy, $\mathscr{G}_{IIc}^i$. Setting $J_{tip} = \mathscr{G}_{IIc}^i$ in eq. (44) leads to the desired relationship between the applied stress and debond length during delamination cracking:

$$\frac{\bar{\sigma}}{E_2} = \frac{(1+2\Sigma\eta)}{\Sigma\eta}\frac{\tau_s}{E_2}\frac{\ell_d}{h_2} + \sqrt{\frac{2(1+2\Sigma\eta)}{\Sigma\eta}\frac{\mathscr{G}_{IIc}^i}{E_2 h_2}}. \tag{45}$$

For $\ell_d = 0$ (or $\tau_s = 0$), eq. (44) reduces to eq. (26) and eq. (45) reduces to eq. (25).

It was checked that the same results can also be obtained using the potential energy loss approach of Budiansky *et al.* [36] for large-scale frictional sliding problems (using that approach, τ_s was taken to be a frictional sliding shear stress).

State-of-the-art models for mechanical performance of carbon-glass hybrid composites in wind turbine blades

Yentl Swolfs, Babak Fazlali, Arsen Melnikov, Francisco Mesquita, Vincent Feyen, Christian Breite, Larissa Gorbatikh and Stepan V Lomov

Department of Materials Engineering, KU Leuven, Kasteelpark Arenberg 44 box 2450, 3001 Leuven, Belgium

yentl.swolfs@kuleuven.be

Abstract. Wind turbine blades are a key growth market for fibre-hybrid composites, as they offer the potential for higher turbine efficiency at a reduced cost. This paper therefore overviews the developments of KU Leuven models for mechanical properties relevant to wind turbine blades, including fibre-hybridisation, size scaling effects, flexure, transverse cracking, stress relaxation and fatigue. The strengths and limitations of the models will be highlighted with a particular focus on relevance for wind turbine blades. While significant steps forward have been made, the various developments still need to be incorporated in one all-encompassing model, and further work is needed on gathering the required input data and detailed experimental validation studies.

1. Introduction

Over the last decade the interest in fibre-hybrid composites from academia as well as industry continuously increased [1,2]. New applications are appearing in many different markets, such as automotive and sporting goods [1]. One of the major growth markets, however, is wind turbine blades. The blades are often not only the most expensive component of a turbine [3,4], but also the limiting factor for the lifetime of the turbine [3]. It is therefore important to thoroughly understand how damage evolves over their lifetime, and have simulation tools at hand to help designing the blades taking into account potential degradation.

Improving the stiffness of wind turbine blades is vital to increase the overall turbine efficiency. One of the key methods for improving efficiency is through the use of longer blades. This is particularly important for off-shore turbines [5,6]. Nearly all commercial wind turbine blades are made almost entirely of glass fibre composites [4], which are not viable for turbine blades longer than 80 m. One approach would be to switch completely to carbon fibre composites [7,8], which would lead to lighter blades with improved fatigue performance and lower blade tip deflection. The cost of such all-carbon fibre composite blades was, however, found to be prohibitive. In the meantime, the industry has settled on using carbon/glass hybrid composites as the best compromise [9]. The use of such hybrid composites is particularly helpful in the spar flanges, where the additional carbon fibres can have the largest positive impact. They also have the added benefit of enabling passive actuation, as their higher anisotropy allows the blades to twist away when the wind gets too strong [10,11].

Introducing fibre-hybrid composites into wind turbine blades creates additional complexity. Firstly, the failure mechanisms of fibre-hybrid composites are not as well understood as for non-hybrid composites [1,2]. This has also hampered the development of simulation tools that are industry-friendly, but still capture the relevant failure mechanisms. Secondly, the smaller diameter of carbon fibres may make impregnation more difficult. Thirdly, some non-destructive testing techniques may not work anymore. Ultrasonic waves may, for example, be reflected at the carbon/glass interface, which would hamper a reliable ultrasonic C-scan [12].

Delving deeper into the more complex failure behaviour, it should be noted that synergetic effects are commonly observed in fibre-hybrid composites [1,2,13]. These are often referred to as hybrid effects. Many of these synergetic effects only become significant when the fibres are finely dispersed [14-19]. Such fine dispersion is rarely achieved in the fibre-hybrid composites relevant to wind industry. This, however, depends on how narrowly the term hybrid effect is defined. In its broadest sense, a hybrid effect is a deviation from the relevant rule-of-mixtures [1,2,13]. The tensile stress-strain behaviour of a fibre-hybrid composite, for example, commonly deviates from the linear rule-of-mixtures even if the degree of dispersion is low [20,21].

The majority of the published fibre-hybrid research focused on the tensile response [17-24]. Here the development of failure is well understood, and synergetic effects can be predicted with reasonable accuracy [19,25,26]. However, the required predictive models tend to be time-consuming and hence their application is limited to small specimens sizes of a few thousands of fibres of a few millimetres in length [27]. Recently, model predictions have been validated in detail against synchrotron computed tomography experiments [27-29]. These experiments enable tracking of microscopic failure of individual fibres (fibre breaks) as well as the development of clusters. It was concluded that current models are not yet able to capture the real development of clusters. A major contributor to the mismatch is the difficulty in reliably measuring the model input parameters [30,31]. In addition, it also seems that there are still micromechanisms at work that even the most advanced models miss [32]. To increase the applicability of strength prediction models to the load cases in wind turbine blades, the following features require particular attention:

- Size scaling effect: It is unclear how well models capture size effects, and whether the computational limitations may be prohibitive to upscale current approaches.
- Out-of-plane loading, such as flexure: Most models assume uniaxial tension, and cannot capture the effect of bending loads.
- Multidirectionality of laminates: The models that are sufficiently advanced to capture fibre break development assume only load-carrying 0° oriented plies or bundles are present.
- Time-dependent features: Only a few models are able to capture time-dependent effects, such as stress relaxation/creep [33] and fatigue [34].

The size scaling effect is the decrease in strength with an increase in specimen size. Its existence is assumed for composites, but has not been thoroughly investigated for unidirectional tensile strength. Since wind turbine blades are much larger than standard coupons, a solid understanding of size scaling of unidirectional composites is vital to predict component failure. Okabe and Takeda [35] performed tensile tests over a wide range of volumes, reporting a reduction of about 10%. Unfortunately, their study misses key details on which dimensions were changed and whether the specimens failed in the gauge section or near the grips. Wisnom et al. [36] measured a reduction of 14% when the specimen size was scaled up in all three dimensions by a factor of 8. They reported gauge length failures for their smallest specimens, but admitted they could not deduce failure initiation for larger specimen sizes. This study deliberately scaled up all specimen dimensions, meaning that the length and cross-sectional scaling effects could not be distinguished. Given its practical importance, there have been several modelling studies to capture the size scaling effect [35,37-39]. Two of these studies did report a good match with the experimental data [35,39]. None of these models analysed how the predicted size scaling effect was affected by the boundary conditions.

Fibre break models typically assume a uniform strain or stress being applied over the entire volume. So far, published models have not attempted to predict the flexural response. This is likely due to the common assumption that the compressive strength is lower than the tensile strength [40]. While this seems true for composites with significant fibre misalignment or with high modulus carbon fibres, it cannot be assumed in general. Czél et al. [41] showed that the compressive failure strain was higher than the tensile failure strain when the experiments are very carefully performed. It is therefore likely that the common idea of compressive strength being lower than tensile strength is due to the difficulties in reliably measuring compressive strength. If true, this would mean that flexural failure is likely governed by the tensile side. Fibre break models including flexural strain gradients would then become valuable tools for the prediction of bending strength.

Fatigue performance is a vital design driver for wind turbine blades [5,8,42-45]. The stiffness loss occurring during fatigue damage development can be important, as a large blade tip deflection could lead to collision with the tower. Luckily, this stiffness degradation is limited to a few percent for composites with primarily continuous 0° fibres that are well bonded [46]. Many fatigue models in the literature are essentially curve-fitting approaches that yield limited to no insight into the damage mechanisms. Mechanistic models therefore have greater value, as they can offer new understanding of the mechanism and have greater predictive capabilities [47,48]. In recent years, mechanistic models for unidirectional composite fatigue have been developed [34,44,49,50]. These models still have significant limitations:

- Alves and Pimenta [34] capture only a single damage mechanism, which is fibre-matrix debond growth. They assumed a Paris-Erdogan law for the growth [51], which implies that the debond keeps growing until failure.
- Dai and Mishnaevsky [44] developed an advanced finite element (FE)-based model, but still captured only a single fatigue damage mechanism, which is fibre-matrix debond growth. They also assume a Paris-Erdogan law [51].
- Sørensen and Goutianos [49,50] did take a limit on the debond length into account, but only predicted the critical state rather than how that critical state was achieved.

All these models ignore matrix creep and stress relaxation, matrix cracking/fatigue, fibre fatigue and other mechanisms that could potentially contribute.

This paper will provide an overview of developments at KU Leuven towards an all-encompassing model to predict failure modes that are governed by fibre break development, with a particular focus towards aspects relevant in wind turbine blades. After describing the baseline model, the more recent extensions will be briefly explained. The potential of the model will be demonstrated by presenting six different features relevant to wind turbine blades: fibre-hybridisation, size scaling, flexure, transverse cracking, stress relaxation and fatigue. As well as identifying the strengths of the current model, its limitations are reported. Future improvements needed to apply the model to the material design of wind turbine blades are highlighted.

2. Materials and methods

2.1. FE models for stress redistribution around a fibre break

FE models are used to create a library of stress concentrations around individual fibre breaks, which are then used by the strength model described in the next section. These models have been extensively described in earlier papers [52-54]. First, two-dimensional random fibre packings are generated using the basic algorithm of Melro et al. [55]. These packings are then extruded in the fibre direction to create a 3D model of parallel fibres (see Figure 1). Symmetry boundary conditions are applied to the entire top surface, apart from the broken fibre and a tiny matrix ring around the broken fibre. Displacement boundary conditions are applied on the bottom to load the bundle to an axial strain of 2%. The model size is chosen large enough, so that adding additional fibres or extending the length of the fibres does not change the calculated local stress concentrations.

The carbon fibres are transversely isotropic whereas S-glass fibres are isotropic, with the engineering constants reported in Table 1. The matrix properties that were used for the different models were adapted to the investigated scenario with the elastoplastic yielding law described by Okabe et al. [56] as default behaviour.

Table 1: Engineering constants of the fibre in the FE model.

Type	E_{11} (GPa)	E_{22}=E_{23} (GPa)	ν_{12}=ν_{13}=ν_{23} (-)	G_{12}=G_{13} (GPa)	G_{23} (GPa)
Carbon	230	15	0.25	13.7	6
S-glass	88	88	0.22	36.1	36.1

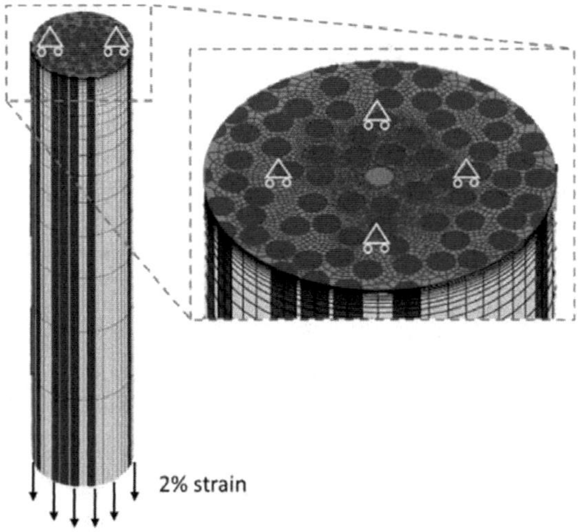

Figure 1: Example 3D FE model for calculating stress redistributions around a single fibre break in a non-hybrid composite with indication of the applied boundary conditions.

The longitudinal stress of each fibre at a given axial location in the FE model is extracted by averaging the stress in the elements with the same axial position over the fibre cross-section. Dividing this average stress by the average far-field stress, the stress concentration factor (SCF) values are obtained. This was typically repeated for five different packing realisations of the model shown in Figure 1. Initially, the resulting SCF profiles in the broken and intact fibres were transferred through the use of trend lines [15,29,53], but later this was improved towards the use of trend surfaces [57]. For carbon/glass fibre-hybrid composites, two sets of models were built to capture the stress field around broken fibres of each used type [15].

2.2. Longitudinal tensile strength model

The basic strength model was developed between 2010 and 2015, and has been described in a range of publications [15,27,29,31,53]. A simplified version was also developed to help in analysing the effect of dispersion for fibre-hybrid composites [14]. The model was validated in detail using synchrotron computed tomography [29] and compared against two other state-of-the-art strength models [27]. Only the key points and assumptions will be described here, as some of the details have been improved over the years.

Figure 2 displays the flow chart of the strength model. It uses the same generator to create random fibre packings [55] as the FE model. Fibres are then extruded along the 0° direction typically over a length of a few millimetre. Each fibre is split up into elements that are typically one fibre radius long. The unimodal Weibull distribution was scaled down to the element length, and then a randomly generated strength value within that distribution was assigned to each element. During loading, the model gradually increases the strain in steps. Whenever the element stress exceeds the element strength, that element is considered broken. The model then checks for final failure, and if that is not the case yet, it will try to find additional fibre break clusters. Then, the stress concentrations from the FE models are used to update stress concentrations assuming that the breaks do not interact. Interactions between breaks in the same cluster is then accounted for using the enhanced superposition principle described in Swolfs et al. [53]. This process repeats itself until an unstable growth of a cluster is detected within one strain increment, at which point the model is stopped.

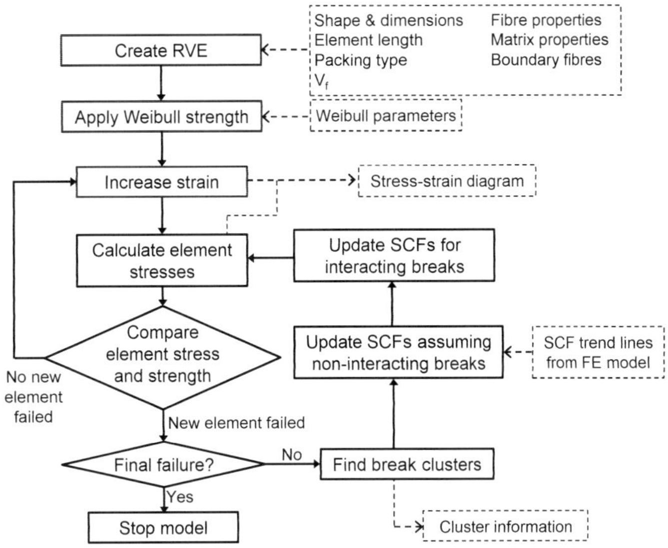

Figure 2: Flow chart of the longitudinal tensile failure model (reprinted from Swolfs et al. [53], with permission from Elsevier).

2.3. Extensions

The modelling framework is intrinsically able to handle fibre-hybridisation, although it was initially limited to predicting the initial failure strain of the carbon fibres or carbon fibre plies. Recent extensions described in Mesquita et al. [57] also enabled it to predict ply fragmentation in interlayer hybrid composites, taking into account interaction of ply fractures both within the same ply and in different plies. This required developing FE models on the ply-level so that the SCF values due to ply fractures could be incorporated in the longitudinal tensile failure model.

Another recent development was the extension of the model towards non-uniform strain fields. A first approach was to model flexural behaviour, as the global strain distribution can be estimated from simple beam bending equations. This enables the model to predict flexural behaviour on the condition that the failure is governed by the tensile rather than compressive stresses. A second approach was to incorporate strain fields in 0° plies in the vicinity of transverse cracks. The necessary fields were either obtained from analytical equations [58] or from FE models [59].

The baseline model was extended to include viscoplastic deformation of the matrix. Detailed measurements of this viscoplastic behaviour were performed for several epoxy matrices, and an advanced user subroutine was developed to incorporate this behaviour into the FE model. This subroutine captures the time as well as the strain dependency of the stress concentrations. The strength model was then extended to capture time rather than strain-based evolution of fibre breaks and clusters at a given strain rate.

A final extension was to include fatigue damage development by using four key assumptions:

- An initial debond length is present around every fibre break.
- The fibre-matrix debond grows according a Paris-Erdogan law, as measured by Pupurs et al. [60].
- The debond growth stops at a certain length, analogous to the approach of Sørensen and Goutianos [49].
- The fibre fatigues by assuming a strength degradation that is linear in a logarithmic scale, and reaches 40% degradation after 1 million cycles.

After an initial static run of the model, the model uses the concept of fatigue cycle jumps to predict how the debond length grows and how this affects other fibre elements. Using this extension, the model can predict the entire S-N diagram.

3. Results

3.1. Fibre-hybridisation

The most important parameter in hybridisation is arguably the degree of dispersion. It has a direct effect on the hybrid effect, which is the delay in the initial failure strain of the brittle ply in a hybrid composite relative to the failure strain in a composite with only brittle plies. Here, we will limit ourselves to the simple scenario of carbon plies sandwiched in between glass plies.

In a collaboration with the University of Bristol, we showed that our model predicts the correct trend: for thinner and hence better dispersed carbon fibre plies in a carbon/glass hybrid composite, the failure strain of the carbon plies increases (see Figure 3). However, a critical reflection shows that this effect is likely not relevant for wind turbine blades, as they tend to use heavy tow yarns to minimise cost. Blade materials therefore likely do not have sufficient dispersion to yield a measurable hybrid effect.

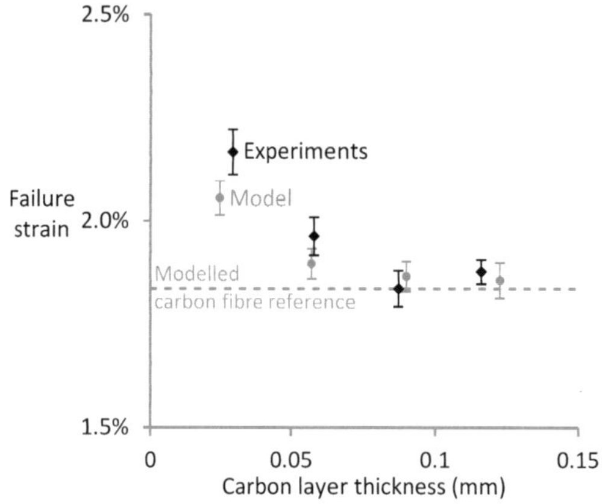

Figure 3: Failure strain of the carbon fibre ply in a carbon/glass hybrid composite (reprinted from Wisnom et al. [19], with permission from Elsevier).

Initially, the model was stopped when the first critical cluster started to propagate. In a fibre-hybrid composite, this, however, does not necessarily coincide with the final failure. The model was therefore extended to predict the four potential scenarios after the initial failure of a carbon ply:

- Failure of the glass fibre ply;
- Delamination of the entire carbon fibre ply, followed by failure of the glass fibre plies;
- Fragmentation without delamination;
- Fragmentation with small delaminations around each fragment.

Figure 4 shows the predicted stress-strain diagram for a carbon/glass/carbon interlayer hybrid composite, alongside experimental data and a simpler analytical model. This particular material system fragments with very limited delaminations. The analytical model predicts a horizontal plateau when the material starts fragmenting, which is not in line with the slope seen in the experiments. This is because the analytical model assumes a uniform strength of the carbon fibre ply. The fibre break model does not assume this uniform strength, but rather inherently predicts its stochastic nature. Therefore, it correctly approximates the observed slope, but ply fragmentation starts too early. The ultimate failure strain is significantly overpredicted relative to the experiments. The too early fragmentation and too large ultimate failure strain are primarily attributed to uncertainty in the Weibull distribution for the carbon and glass fibres, respectively.

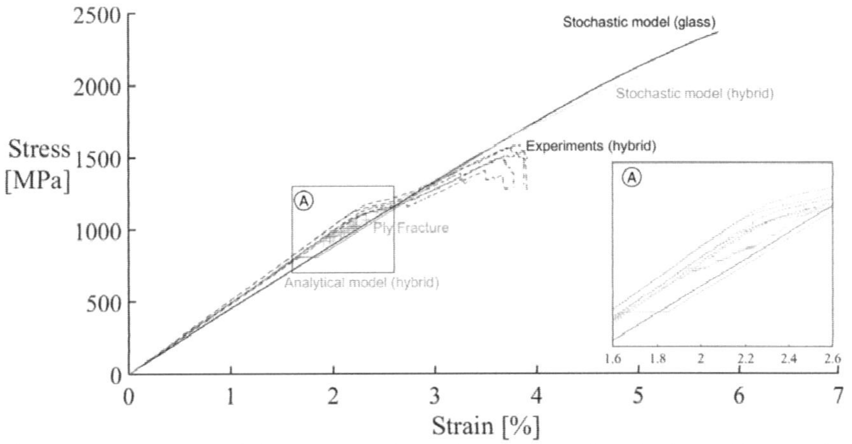

Figure 4: Representative stress-strain behaviour predicted by the fibre break model for glass/carbon/glass hybrid composite, including experimental data from Czél et al. [23] and analytical predictions from Jalalvand et al. [61]. Each "+" marker represents a ply fracture (reprinted from Mesquita et al. [57], with permission from Elsevier).

The model does not only capture the complex stress-strain behaviour of fibre-hybrid composites, but also the underlying mechanisms of fibre break and cluster development and ply fragmentation. This implies the model can be used to guide the design of fibre-hybrid structures that are primarily subjected to tensile loading. At the moment, it is limited to intrayarn and interlayer hybrid configurations. A similar strategy as described in section 2.3 and used for Figure 4 could be used to extend it towards intralayer or bundled hybrid configurations. Such configurations have been much less studied in the literature, but are more relevant for pultruded spar caps.

3.2. Size scaling effects

Wind turbine blades are very large compared most coupons tested in lab conditions. Given the difficulties in testing very large coupons, models are essential to better understand how tensile strength and failure strain scales with size. Boundary effects are a vital feature in size scaling, but are often overlooked in other studies [37-39,56,62]. This is particularly relevant for scaling of the width and thickness.

Figure 5 plots the predicted failure strain for a 10 mm long model with two different boundary conditions applied. In the first case, the model boundary was left free, whereas in the second case, a ring of boundary fibres was added. These boundary fibres can carry stress concentrations, but possess infinite strength. This boundary condition eliminates the preferential cluster development at the perimeter of the model that occurs in the first case. Figure 5 shows a small difference in the predicted failure strain for both cases, but more importantly a large difference in how the failure strain evolves with the number of fibres. Without including boundary fibres, two competing mechanisms occur. On the one hand, an increased total number of fibres increases the likelihood of a combination of weak fibre elements located within each other's vicinity, which will lead to a decrease in failure strain. On the other hand, more fibres make the preferential cluster development at the perimeter less important, which leads to an increased failure strain. The interaction of these two mechanisms results in a near absence of size scaling in the failure strain predictions. For the models with boundary fibres, the size scaling is much more pronounced, because the second mechanism is eliminated.

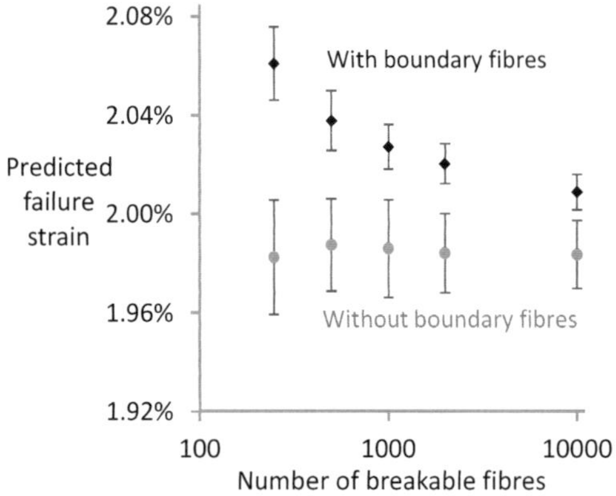

Figure 5: The predicted failure strain as a function of the number of breakable fibres, with and without boundary fibres. The length was kept constant at 10 mm (reprinted from Swolfs et al. [31], with permission from Elsevier).

The case without boundary fibres is likely the more representative case for comparing to experimental size scaling studies. It seems to indicate that size scaling is negligibly small. The study should however be extended to even larger specimen sizes to confirm this observation. For longer specimens or specimens with more than 10000 fibres, it is possible that the boundary effects will stop being relevant and more significant size scaling starts appearing. This would also seem to bring the model more in line with the experimental reports mentioned in the introduction. It should however be mentioned that those reports tended to scale all dimensions. In unpublished studies, we have observed that the length scaling in our model is much more significant than the cross-sectional scaling.

3.3. Flexure

With a strain field that varies linearly along the length (X) and thickness (Y) direction, the model can predict flexural strength. For the presented case study, a span length of 28 mm and specimen thickness of 1 mm were considered for estimating the strain gradient, but only the central 1 mm of the length was modelled. Figure 6 shows how this stress varies in the model, and how the fibre breaks (red dots) are concentrated in the regions with higher stress levels. In each model, the location of the largest cluster at the end of the simulation can be tracked. For the simulation run visualised in Figure 6, this happened to be very close to the middle (X≈0.01mm). Figure 7 plots how this location varied for 100 simulations. This illustrates that a 1 mm long model is long enough to capture a span length of 28 mm, as the failure is unlikely to occur more than 0.5 mm away from the middle. This model is fully set up to include multiple fibre types needed to model carbon/glass hybrid composites.

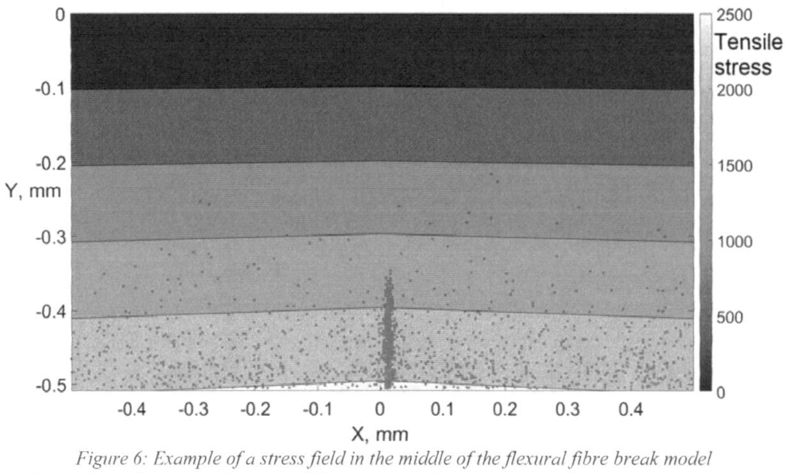

Figure 6: Example of a stress field in the middle of the flexural fibre break model with red dots indicating the location of fibre breaks at failure (reprinted from Melnikov et al. [63]).

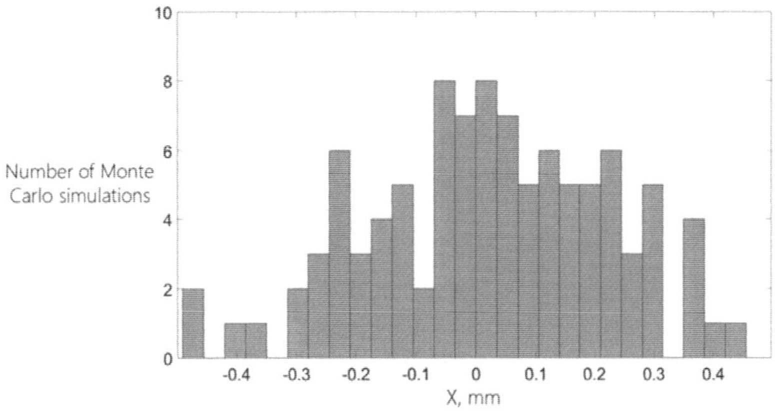

Figure 7: Histogram of the position of final largest cluster with x=0 being in the middle of the span length (reprinted from Melnikov et al. [63]).

3.4. Transverse cracking

Similarly to the flexural model in the previous section, transverse cracking in a cross-ply laminate introduces non-uniform strain fields in the 0° plies. These longitudinal strain fields were captured either using analytical equations [58] or finite element calculations, but as Figure 8 shows, both options lead to very similar predictions. Three different cross-ply laminates with varying 90° ply thicknesses were modelled and compared to a UD composite. The predicted reduction in tensile strength is up to 17.1% relative to the UD composite. Interestingly, the experimental and modelling evidence of Noda et al. [64] showed no such reduction.

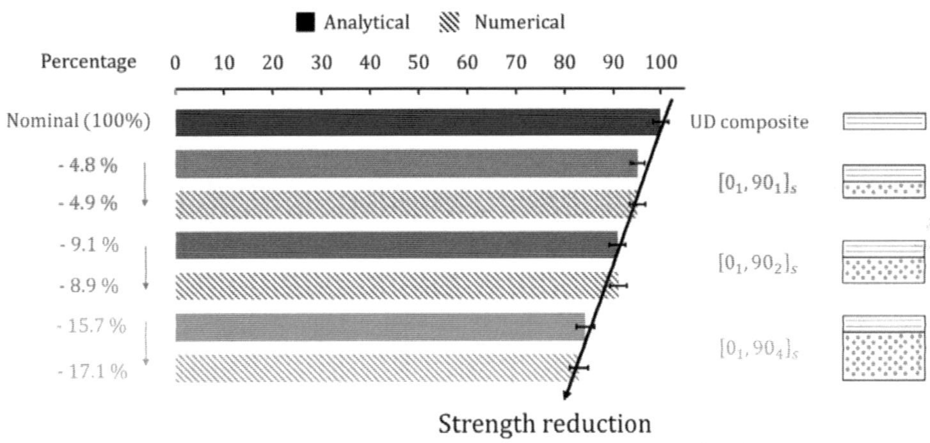

Figure 8: Reduction in tensile strength in cross-ply laminates relative to a UD composite where the strain fields were either based on analytical equations or on FE models (reproduced from Melnikov et al. [59]).

Wind turbine blade materials often contain a certain fraction of off-axis bundles or plies to stabilise the fabrics during manufacturing. Such bundles are known to reduce the fatigue strength of the 0° plies or bundles [65,66]. The modelling evidence here is line with computed tomography results [65,66], showing that fibre breaks develop preferentially at the tip of the off-axis cracks.

3.5. Stress relaxation

The stress relaxation model was initially loaded up statically and then held at a constant displacement for $3\cdot10^6$ s (about 1 month). Figure 9 shows the outcome of an initial study on stress relaxation. The matrix stress relaxation causes the stress concentrations around fibre breaks to spread out more over time, which can lead to additional fibre break and cluster development (see Figure 9). A later study extending the simulated time frame to 10^9 s (about 30 years). Models that were run at 94% of the mean static failure strain revealed no composite failure over time. For 20 models that were run at 98% of the mean static failure strain, however, 4 failed during the initial load-up due to the scatter in composite strength, 11 failed during the relaxation, and only 5 survived the entire modelled period.

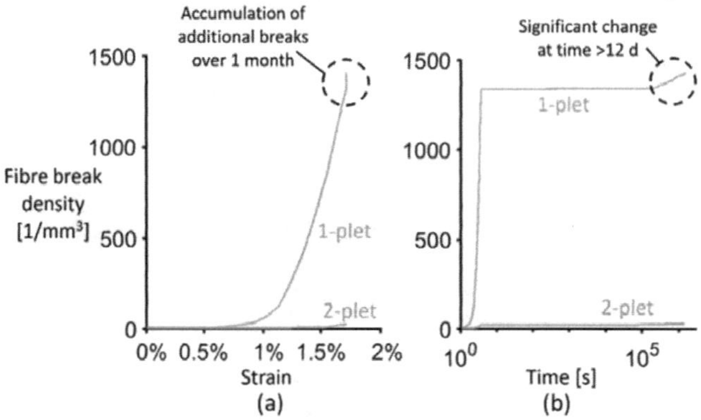

Figure 9: Outcome of the stress relaxation model in terms of fibre break density evolution over $3 \cdot 10^6$ s: (a) as a function of applied strain and (b) as a function of time (reproduced from Breite et al. [67]).

3.6. *Fatigue*

Now that we have included fibre-matrix debond growth as a function of the number of cycles, the model can predict the entire S-N diagram of a UD composite. Figure 10 plots this diagram with and without the inclusion of fibre fatigue. Without the inclusion of fibre fatigue, the predicted fatigue performance is significantly better than the experimentally measured fatigue performance. By including fibre fatigue, the predictions are more or less in line with the experimental data. However, it should be noted that percentage of fibre fatigue was arbitrarily chosen and that such strong fibre fatigue is likely not realistic for carbon fibres. Some authors even indicate that carbon fibres do not fatigue [68], although this study has not been convincingly replicated yet. Glass fibres are known to fatigue, but there are ongoing discussions on whether this really is fatigue or just stress corrosion. Such corrosion is likely much less significant when the fibres are surrounded by matrix [69]. Another reason for this discrepancy could be in the difficulty to perform reliable fatigue tests on UD composites. Such composites are known to be very sensitive to gripping effects, which can be even worse in fatigue.

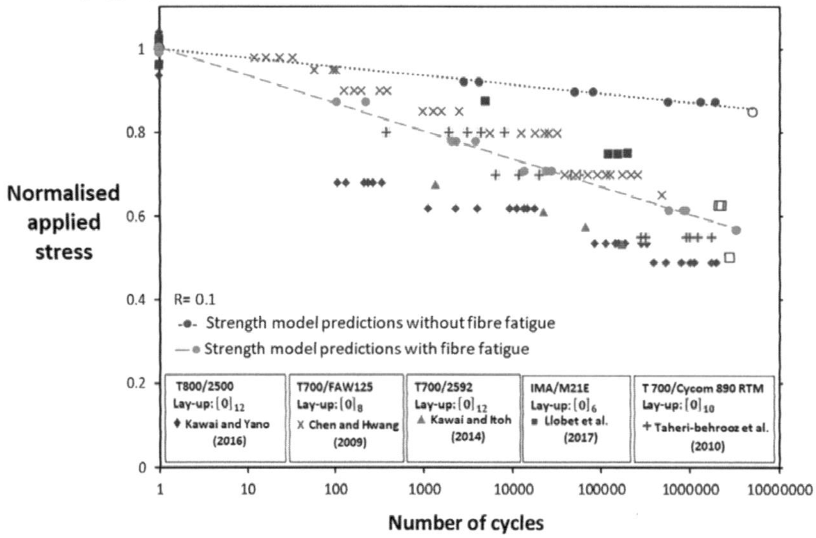

Figure 10: Predicted S-N diagrams for unidirectional carbon fibre composites without and with the inclusion of fibre fatigue, and compared against experimental data sets for unidirectional carbon fibre composites [70-74].

Unfortunately, the fibre-matrix debond growth data used as input to create Figure 10 was based on glass fibre/epoxy measurements. Similar data for carbon fibres is not available in the literature.

The model can be readily adapted for fibre-hybrid composites, provided that the necessary input data are available. This development is currently ongoing.

4. Conclusion and outlook

Fibre break models are powerful tools that can predict a much broader range of properties than just longitudinal tensile strength of unidirectional composites. This paper showed a range of extensions to increase the practical relevance and usefulness of the strength model developed at KU Leuven. A key issue in all of the presented extensions is their experimental validation. While the model seems to capture the key damage mechanisms, it does not capture all of them and hence extrapolations beyond the validated range of materials and behaviours can be unjustified.

In the future, we should focus on better integrating the various features. For example, the flexural and transverse cracking studies have so far limited themselves to quasi-static failure of non-hybrid composites, but can easily be extended to include fatigue, stress relaxation/creep and fibre-hybridisation. Two extensions in particular would be very useful in the modelling of wind turbine blades. The first extension would be to include compressive failure mechanisms. This would enable modelling quasi-static flexure in more detail, and it would open the door towards modelling flexural fatigue. This is the key behaviour driving the design of wind turbine blades. Secondly, the model can be incorporated in a multiscale modelling strategy so that much larger components can be simulated. A proper understanding of size scaling effects in the base model is vital, as they will need to be transferred to the larger scale.

Acknowledgements
This research was performed within the framework of the FiBreMoD and HyFiSyn projects and has received funding from the European Union's Horizon 2020 research and innovation programme under the Marie Skłodowska-Curie grant agreements No 722626 and No 765881. SVL holds the Toray Chair at KU Leuven, the support of which is gratefully acknowledged.

References

[1] Swolfs Y, Verpoest I and Gorbatikh L Recent advances in fibre-hybrid composites: materials selection, opportunities and applications 2019 *Int. Mater. Rev.* **64** 181-215.
[2] Swolfs Y, Gorbatikh L and Verpoest I Fibre hybridisation in polymer composites: a review 2014 Compos. Part A: Appl. S. 67 181-200.
[3] Zhou HF, Dou HY, Qin LZ, Chen Y, Ni YQ and Ko JM A review of full-scale structural testing of wind turbine blades 2014 Renew. Sust. Energ. Rev. 33 177-87.
[4] Mohamed MH and Wetzel KK 3D Woven Carbon/Glass Hybrid Spar Cap for Wind Turbine Rotor Blade 2006 Journal of Solar Energy Engineering 128 562-73.
[5] Mikkelsen LP A simplified model predicting the weight of the load carrying beam in a wind turbine blade 2016 IOP Conf. Ser.: Mat. Sci. 139 012038.
[6] Joosse P, van Delft D, Kensche C, Hahn F, Jacobsen T and van den Berg R. Economic use of carbon fibres in large wind turbine blades? 2000 ASME Wind Energy Symposium: American Institute of Aeronautics and Astronautics; 2000. p.
[7] Griffin DA and Zuteck MD Scaling of Composite Wind Turbine Blades for Rotors of 80 to 120 Meter Diameter 2001 J. Sol Energ. Eng. 123 310-8.
[8] Sutherland HJ A summary of the fatigue properties of wind turbine materials 2000 Wind Energy 3 1-34.
[9] Swolfs Y Perspective for Fibre-Hybrid Composites in Wind Energy Applications 2017 Materials 10 1281.
[10] Brøndsted P, Lilholt H and Lystrup A Composite materials for wind power turbine blades 2005 Annu. Rev. Mater. Res. 35 505-38.
[11] de Goeij WC, van Tooren MJL and Beukers A Implementation of bending-torsion coupling in the design of a wind-turbine rotor-blade 1999 Appl. Energ. 63 191-207.
[12] Eihusen JA and Peters AR. Characterization of the transverse thermal conductivity of intraply hybrid composite laminates. In: Green J, Howell DD, editors. 31st International SAMPE Technical Conference, Chicago, USA: SAMPE; 1999. p. 211-20.
[13] Kretsis G A review of the tensile, compressive, flexural and shear properties of hybrid fibre-reinforced plastics 1987 Composites 18 13-23.

[14] Swolfs Y, McMeeking RM, Gorbatikh L and Verpoest I The effect of fibre dispersion on initial failure strain and cluster development in unidirectional carbon/glass hybrid composites 2015 Compos. Part A: Appl. S. 69 279-87.
[15] Swolfs Y, Verpoest I and Gorbatikh L Maximising the hybrid effect in unidirectional hybrid composites 2016 Mater. Design 93 39-45.
[16] You YJ, Park YH, Kim HY and Park JS Hybrid effect on tensile properties of FRP rods with various material compositions 2007 Compos. Struct. 80 117-22.
[17] Ren PG, Zhang ZP, Xie L, Ren F, Jin YL, Di YY, et al. Hybrid Effect on Mechanical Properties of M40-T300 Carbon Fiber Reinforced Bisphenol A Dicyanate Ester Composites 2010 Polym. Compos. 31 2129-37.
[18] Pitkethly MJ and Bader MG Failure modes of hybrid composites consisting of carbon fibre bundles dispersed in a glass fibre epoxy resin matrix 1987 J. Phys. D Appl. Phys. 20 315-22.
[19] Wisnom MR, Czel G, Swolfs Y, Jalalvand M, Gorbatikh L and Verpoest I Hybrid effects in thin ply carbon/glass unidirectional laminates: accurate experimental determination and prediction 2016 Compos. Part A: Appl. S. 88 131-9.
[20] Bunsell AR and Harris B Hybrid carbon and glass fibre composites 1974 Composites 5 157-64.
[21] Czél G and Wisnom MR Demonstration of pseudo-ductility in high performance glass/epoxy composites by hybridisation with thin-ply carbon prepreg 2013 Compos. Part A: Appl. S. 52 23-30.
[22] Hayashi T. On the improvement of mechanical properties of composites by hybrid composition. Proc. 8th Intl. Reinforced Plastics Conference, Brighton, UK1972. p. 149-52.
[23] Czél G, Jalalvand M and Wisnom MR Design and characterisation of advanced pseudo-ductile unidirectional thin-ply carbon/epoxy–glass/epoxy hybrid composites 2016 Compos. Struct. 143 362-70.
[24] Tang J, Swolfs Y, Longana ML, Yu H, Wisnom MR, Lomov SV, et al. Hybrid composites of aligned discontinuous carbon fibers and self-reinforced polypropylene under tensile loading 2019 Compos. Part A: Appl. S. 123 97-107.
[25] Guerrero JM, Mayugo JA, Costa J and Turon A A 3D Progressive Failure Model for predicting pseudo-ductility in hybrid unidirectional composite materials under fibre tensile loading 2018 Compos. Part A: Appl. S. 107 579-91.
[26] Tavares RP, Melro AR, Bessa MA, Turon A, Liu WK and Camanho PP Mechanics of hybrid polymer composites: analytical and computational study 2016 Comput. Mech. 57 405-21.
[27] Bunsell A, Gorbatikh L, Morton H, Pimenta S, Sinclair I, Spearing M, et al. Benchmarking of strength models for unidirectional composites under longitudinal tension 2018 Compos. Part A: Appl. S. 111 138-50.
[28] Scott AE, Sinclair I, Spearing SM, Thionnet A and Bunsell AR Damage accumulation in a carbon/epoxy composite: Comparison between a multiscale model and computed tomography experimental results 2012 Compos. Part A: Appl. S.43 1514-22.
[29] Swolfs Y, Morton H, Scott AE, Gorbatikh L, Reed PAS, Sinclair I, et al. Synchrotron radiation computed tomography for experimental validation of a tensile strength model for unidirectional fibre-reinforced composites 2015 Compos. Part A: Appl. S. 77 106-13.
[30] Berger MH and Jeulin D Statistical analysis of the failure stresses of ceramic fibres: Dependence of the Weibull parameters on the gauge length, diameter variation and fluctuation of defect density 2003 Journal of Materials Science 38 2913-23.
[31] Swolfs Y, Verpoest I and Gorbatikh L Issues in strength models for unidirectional fibre-reinforced composites related to Weibull distributions, fibre packings and boundary effects 2015 Compos. Sci. Technol. 114 42-9.
[32] Swolfs Y, Verpoest I and Gorbatikh L A review of input data and modelling assumptions in longitudinal strength models for unidirectional fibre-reinforced composites 2016 Compos. Struct. 150 153-72.
[33] Blassiau S, Thionnet A and Bunsell AR Micromechanisms of load transfer in a unidirectional carbon fibre-reinforced epoxy composite due to fibre failures. Part 2: Influence of viscoelastic and plastic matrices on the mechanisms of load transfer 2006 Compos. Struct. 74 319-31.
[34] Alves M and Pimenta S A computationally-efficient micromechanical model for the fatigue life of unidirectional composites under tension-tension loading 2018 Int. J. Fatigue 116 677-90.
[35] Okabe T and Takeda N Size effect on tensile strength of unidirectional CFRP composites - experiment and simulation 2002 Compos. Sci. Technol. 62 2053-64.

[36] Wisnom MR, Khan B and Hallett SR Size effects in unnotched tensile strength of unidirectional and quasi-isotropic carbon/epoxy composites 2008 Compos. Struct. 84 21-8.
[37] Curtin WA Dimensionality and size effects on the strength of fiber-reinforced composites 2000 Compos. Sci. Technol. 60 543-51.
[38] Curtin WA Size scaling of strength in heterogeneous materials 1998 Phys. Rev. Lett. 80 1445-8.
[39] Pimenta S and Pinho ST Hierarchical scaling law for the strength of composite fibre bundles 2013 J. Mech. Phys. Solids 61 1337-56.
[40] Smith PA. Carbon fiber reinforced plastics - properties. In: Kelly A, Zweben C, Chou TW, editors. Comprehensive composite materials, vol. 2.04, Amsterdam: Elsevier; 2000. p. 107-50.
[41] Czél G, Jalalvand M and Wisnom MR Hybrid specimens eliminating stress concentrations in tensile and compressive testing of unidirectional composites 2016 Compos. Part A: Appl. S. 91 436-47.
[42] Hayman B, Wedel-Heinen J and Brøndsted P Materials Challenges in Present and Future Wind Energy 2011 MRS Bull. 33 343-53.
[43] Joosse PA, van Delft DRV, Kensche C, Soendergaard D, van den Berg RM and Hagg F Cost Effective Large Blade Components by Using Carbon Fibers 2002 J. Sol. Energ. Eng. 124 412-8.
[44] Dai G and Mishnaevsky Jr L Fatigue of hybrid glass/carbon composites: 3D computational studies 2014 Compos. Sci. Technol. 94 71-9.
[45] Vallons K, Adolphs G, Lucas P, Lomov SV and Verpoest I Quasi-UD glass fibre NCF composites for wind energy applications: a review of requirements and existing fatigue data for blade materials 2013 Mech. & Ind. 14 175-89.
[46] Gamstedt EK, Berglund LA and Peijs T Fatigue mechanisms in unidirectional glass-fibre-reinforced polypropylene 1999 Compos. Sci. Technol. 59 759-68.
[47] Sevenois RDB and Van Paepegem W Fatigue Damage Modeling Techniques for Textile Composites: Review and Comparison With Unidirectional Composite Modeling Techniques 2015 Appl. Mech. Rev. 67 020802--12.
[48] Quaresimin M, Carraro PA and Maragoni L Early stage damage in off-axis plies under fatigue loading 2016 Compos. Sci. Technol. 128 147-54.
[49] Sørensen BF and Goutianos S Micromechanical model for prediction of the fatigue limit for unidirectional fibre composites 2019 Mech. Mater. 131 169-87.
[50] Sørensen BF The critical damage state controlling the tension-tension fatigue life of unidirectional fibre composites 2019 Compos. Sci. Technol. 172 172-81.
[51] Paris P and Erdogan F A Critical Analysis of Crack Propagation Laws 1963 J. Basic Eng. 85 528-33.
[52] Swolfs Y, Gorbatikh L, Romanov V, Orlova S, Lomov SV and Verpoest I Stress concentrations in an impregnated fibre bundle with random fibre packing 2013 Compos. Sci. Technol. 74 113-20.
[53] Swolfs Y, McMeeking RM, Verpoest I and Gorbatikh L Matrix cracks around fibre breaks and their effect on stress redistribution and failure development in unidirectional composites 2015 Compos. Sci. Technol. 108 16-22.
[54] Swolfs Y, Gorbatikh L and Verpoest I Stress concentrations in hybrid unidirectional fibre-reinforced composites with random fibre packings 2013 Compos. Sci. Technol. 85 10-6.
[55] Melro AR, Camanho PP and Pinho ST Generation of random distribution of fibres in long-fibre reinforced composites 2008 Compos. Sci. Technol. 68 2092-102.
[56] Okabe T, Ishii K, Nishikawa M and Takeda N Prediction of Tensile Strength of Unidirectional CFRP Composites 2010 Adv. Compos. Mater 19 229-41.
[57] Mesquita F, Swolfs Y, Lomov SV and Gorbatikh L Ply fragmentation in unidirectional hybrid composites linked to stochastic fibre behaviour: A dual-scale model 2019 Compos. Sci. Technol. 181 107702.
[58] Varna J and Berglund L Multiple Transverse Cracking and Stiffness Reduction in Cross-Ply Laminates 1991 J. Compos. Tech. Res. 13 97-106.
[59] Melnikov A, Swolfs Y, Lomov SV and Gorbatikh L. Influence of transverse cracking on strength and fibre break development in cross-ply laminates. 7th ECCOMAS Thematic Conference on the Mechanical Response of Composites, Girona, Spain2019.

[60] Pupurs A, Goutianos S, Brondsted P and Varna J Interface debond crack growth in tension–tension cyclic loading of single fiber polymer composites 2013 Compos. Part A: Appl. S. 44 86-94.
[61] Jalalvand M, Czél G and Wisnom MR Damage analysis of pseudo-ductile thin-ply UD hybrid composites - a new analytical method 2015 Compos. Part A: Appl. S. 69 83-93.
[62] Ibnabdeljalil M and Curtin WA Strength and reliability of fiber-reinforced composites: Localized load-sharing and associated size effects 1997 Int. J. Solids Struct. 34 2649-68.
[63] Melnikov A, Swolfs Y, Mesquita F, Lomov SV and Gorbatikh L. Modelling of fibre break development in unidirectional composites under longitudinal bending. 27th International Workshop on Computational Mechanics of Materials, Leuven, Belgium, 2017.
[64] Noda J, Okabe T, Takeda N and Shimizu M Tensile strength of CFRP cross-ply laminates containing transverse cracks 2006 Adv. Compos. Mater 15 81-93.
[65] Garcea SC, Mavrogordato MN, Scott AE, Sinclair I and Spearing SM Fatigue micromechanism characterisation in carbon fibre reinforced polymers using synchrotron radiation computed tomography 2014 Compos. Sci. Technol. 99 23-30.
[66] Garcea SC, Sinclair I and Spearing SM Fibre failure assessment in carbon fibre reinforced polymers under fatigue loading by synchrotron X-ray computed tomography 2016 Compos. Sci. Technol. 133 157-64.
[67] Breite C, Feyen V, Chevalier J, Pardoen T, Lomov SV, Gorbatikh L, et al. Effect of viscoplasticity of the epoxy matrix on long-term stress redistribution around fibre breaks in a composite subjected to high static tensile load. 7th ECCOMAS Thematic Conference on the Mechanical Response of Composites, Girona, Spain2019.
[68] Bunsell AR and Somer A The tensile and fatigue behavior of carbon fibers 1992 Plast. Rubber Compos. Process. Appl. 18 263-7.
[69] Mandell JF, McGarry FJ, Hsieh AJY and Li CG Tensile fatigue of glass fibers and composites with conventional and surface compressed fibers 1985 Polym. Compos. 6 168-74.
[70] Kawai M and Yano K Anisomorphic constant fatigue life diagrams of constant probability of failure and prediction of P–S–N curves for unidirectional carbon/epoxy laminates 2016 Int. J. Fatigue 83, Part 2 323-34.
[71] Chen H-S and Hwang S-F A fatigue damage model for composite materials 2009 Polym. Compos. 30 301-8.
[72] Kawai M and Itoh N A failure-mode based anisomorphic constant life diagram for a unidirectional carbon/epoxy laminate under off-axis fatigue loading at room temperature 2013 J. Compos. Mater. 48 571-92.
[73] Taheri-Behrooz F, Shokrieh MM and Lessard LB Progressive Fatigue Damage Modeling of Cross-ply Laminates, II: Experimental Evaluation 2010 J. Compos. Mater. 44 1261-77.
[74] Llobet J, Maimí P, Mayugo JA, Essa Y and Martin de la Escalera F A fatigue damage and residual strength model for unidirectional carbon/epoxy composites under on-axis tension-tension loadings 2017 Int. J. Fatigue 103 508-15.

IOP Conf. Series: Materials Science and Engineering **942** (2020) 012006 doi:10.1088/1757-899X/942/1/012006

ReliaBlade Project: A Material's Perspective towards the Digitalization of Wind Turbine Rotor Blades

F Sayer[1,*], A Antoniou[1], S Goutianos[2], I Gebauer[1], K Branner[2] and C Balzani[3]

[1] Fraunhofer Institute for Wind Energy Systems, Am Seedeich 45, 27572 Bremerhaven, Germany

[2] Technical University of Denmark; Frederiksborgvej 399; 4000 Roskilde, Denmark

[3] Leibniz University Hannover – Institute for Wind Energy Systems, Appelstr. 9A, 30167 Hannover, Germany

[*] e-mail: florian.sayer@iwes.fraunhofer.de

Abstract. In many industries, digitalization is expected to have a significant economic potential. The digitalization of wind turbine rotor blades including their materials could contribute to accelerate the development of novel and tailored materials, to improve the blades' reliability, and to make wind energy more cost efficient. However, the digitalization of the blades through their entire life cycle is challenging e.g. due to the dependence of the material properties on the manufacturing process parameters, the complex structural health monitoring and the challenging modelling of blade response under complex loading. In the presented work based on the results of the ReliaBlade project, a theoretical approach is attempted towards describing the blade in-situ structural performance, based on the material properties, the blade manufacturing processes and loading history. In the first phase of the ReliaBlade project, an experimental blade for full-scale testing with three pre-defined internal damage modes is designed. Based on these damage modes the digitalization approach towards increasing the blade structural reliability is exemplarily shown.

1. Introduction

A huge technical and economic potential is foreseen from the digitalization of many structural components (see e.g. [1], [2]). This is also valid for wind turbines. Nevertheless, the specific economic potential from digitalization of wind turbine blades is not explored in the literature.

The objectives of digitalization are typically to contribute to an improvement of the design, manufacturing and operation. There is a strong dependence of the manufacturing process parameters on the material properties. This together with the quality of structural health monitoring systems and data availability, the development of reliable digital blade twins is very challenging and requires collaborative work between different fields of expertise (see e.g. [3]).

The work started in the framework of the ReliaBlade project [4] is aiming at contributing to the formulation of a wind turbine blade digital twin concept. Thus, it is attempted to describe the structural blade performance based on the material properties, the blade manufacturing processes and loading history. Furthermore, application of non-destructive inspections (NDI) and structural health monitoring (SHM) methods are investigated, as these technologies could enhance modeling of the blades at defined times through the lifetime [3]. Within the project, major wind turbine manufacturers (including blade design and manufacturing), relevant industry partners and OEMs contribute with their specific know-how. Figure 1 gives an overview of the ReliaBlade project activities during the different life phases of a blade. Blades of 12.6 and 30 m will be investigated through the design, the manufacturing and the operation phase (represented by full-scale blade testing). Beside innovative technologies, artificial intelligence will support the design, the manufacturing data and the test measurements by optimizing the processes and the data evaluation.

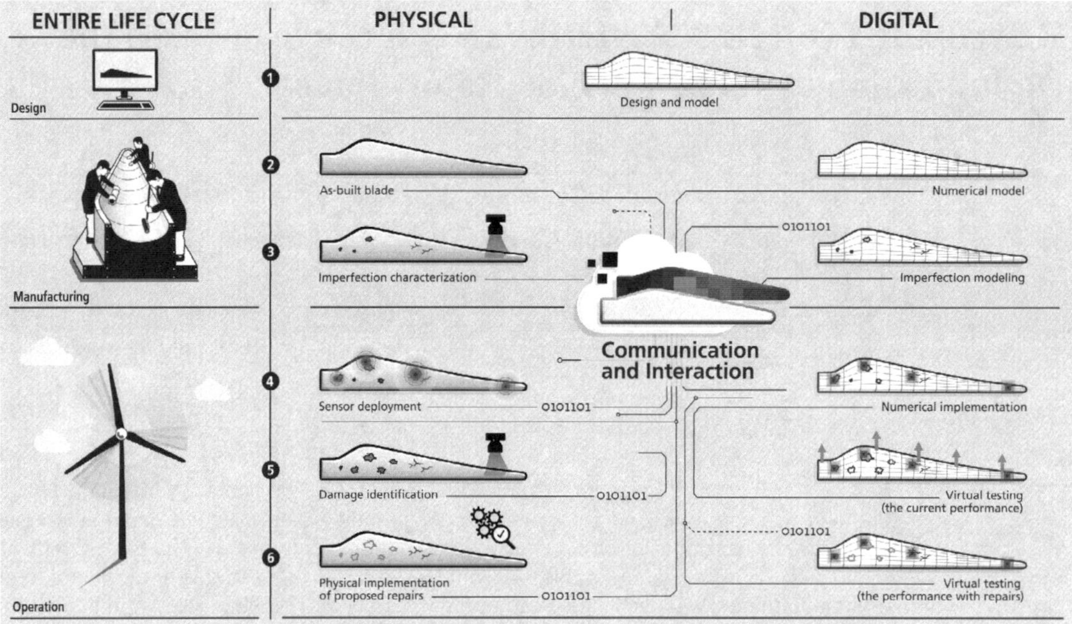

Figure 1: Schematic illustration of activities within the ReliaBlade project [4].

Advanced physically based material models will be one major objective within the ReliaBlade project to support the digitalization approaches and the development of digital blade twins. These are aiming to enhance the understanding of the mechanical material behavior on one hand, thus reducing the property uncertainties, resulting in increased structural reliability as introduced by Toft et al. [5]. On the other hand, they reduce the number of variables and the calculation effort in the digitalization procedures.

Figure 2 illustrates schematically the implemented digital material twin concept. In contrast to traditional design philosophy the material properties shall not be described as a set of properties, but as a function of the manufacturing parameters and the operational conditions. This is essential for composite materials, as the in-situ material properties are a function of the manufacturing and operational/environmental variables, see e.g. [7].

For the manufacturing of wind turbine blades a so-called vacuum bag infusion process is common. In this process dry fibres and core materials are infiltrated with a resin under a vacuum bag and the resin bonds the materials and hardens due to a chemical reaction (so-called curing). Therefore, the infusion and curing processes are of high relevance for the material properties and the physical based modelling.

In the following, an introduction to the process related behaviour of epoxy (EP) systems is given. Based on this description, a typical blade manufacturing process is presented. In a third step the relevance of such measurement based digital material twins will be exemplary investigated for inter-fibre failure in shell laminates.

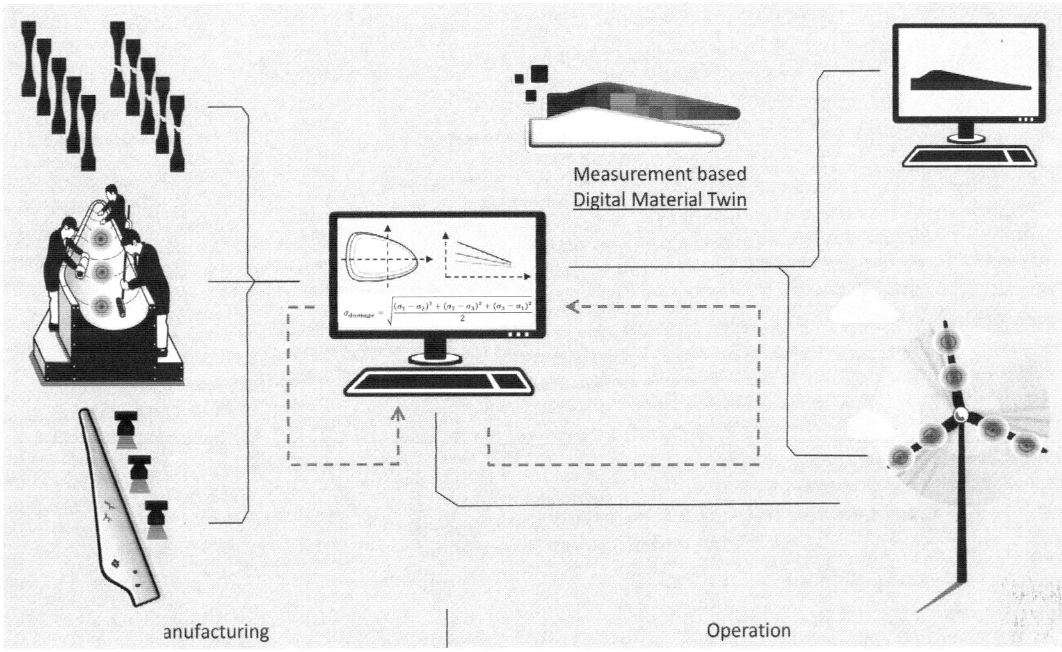

Figure 2: Measurement based digital material twin in ReliaBlade.

2. Introduction to process related behaviour of epoxy systems

Although it is not represented in the traditional stress calculation procedures in the wind industry, it is known that the local resulting stress state and the strength in epoxy based composites or adhesive bond lines are influenced by the manufacturing process ([23], [25]). In this work only process induced effect will be considered. It might be needed, that other effects such as moisture content and/or uptake are taken into account. Although in the stress/strength analysis of rotor blades the thermal and chemical shrinkage effects are covered through safety factors (e.g. according to DNV-GL [6]), it might be that they are not covered sufficiently, e.g. [7] and [10].

Figure 3 gives a schematic and idealized illustration of the thermal and chemical shrinkage during the curing of an EP system by plotting the density over the temperature. Density changes due to a degassing of a solvent or similar are not considered for EP systems as they cure by a polycondensation. An uncured system is heated up from room to curing temperature (A-B). The material expands with the coefficient of thermal expansion of the uncured system and therefore the density reduces. At the curing temperature the chemical shrinkage takes place (B-C). The description between A and C is a simplification as the cross linking starts already at room temperature and the cross linking rate increases with temperature. Therefore a density increase will start already with the temperature increase during ramp up. After point C the thermal shrinkage will take place. As EP systems show two different coefficients of thermal expansion (above and below T_g) this is a bi-linear function. In practice the given description of the curing process is a strong simplification as the described processes will overlap.

This schematic description in Figure 3 does not represent the behavior in an industrial curing process. In Figure 4 the density is plotted together with a possible curing temperature profile over time. The process consists of a ramp up, a holding and a cool down phase. The material will start to expand due to the increasing temperature and at the same time will start to develop cross links. With increasing temperature the cross linking speed will increase. For this reason the start of the holding phase is described by A-B*. During the holding phase the cross linking will develop further and reach, with a small safety margin, the required value before the cool down phase. In an ideal process the level of cure will reach the desired level (B*-C*). In the cool down phase the bi-linear thermal shrinkage will take place (C*-D* and D*-E). The stages A and E have to be interpreted in the same way as A and E in Figure 3.

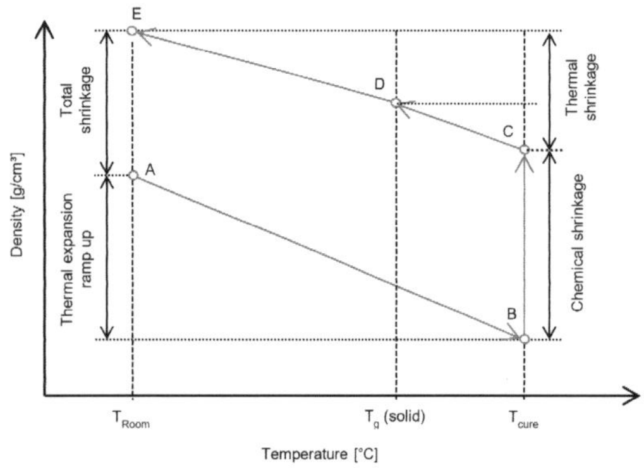

Figure 3: Schematic illustration of thermal and chemical shrinkage during curing of EP system according to Holst [8].

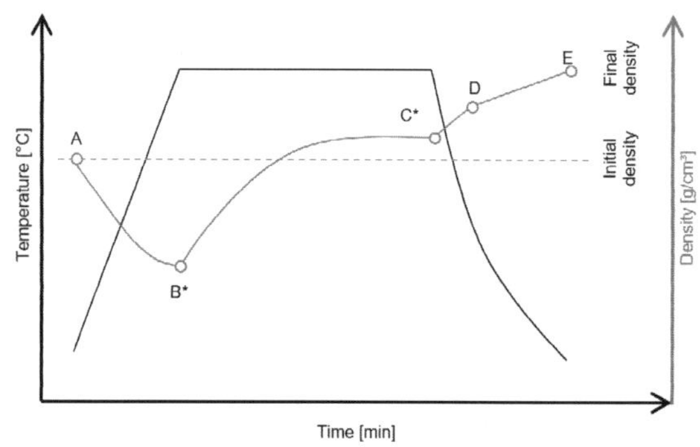

Figure 4: Interpretation of thermal and chemical shrinkage (expressed as density) of EP system for a typical curing.

From an engineering perspective the chemical and thermal shrinkage effects must be mathematically described from the gelation point, which it can be interpreted as the degree of cure at which a polymer network is developed that can transfer mechanical loads (e.g. shear loads). Furthermore shrinkage effects, such as moisture absorption should be modelled if required.

The chemical shrinkage for epoxy systems will lead to a volume change, of which one part will occur in the liquid phase and the second part in the solid phase (see Figure 5). Typical values for the chemical shrinkage are 6 % in volume, of which approx. 30 % is happening in the solid phase (see e.g. Khoun et al. [9], Holst et al. [28] and Nielsen [10]).

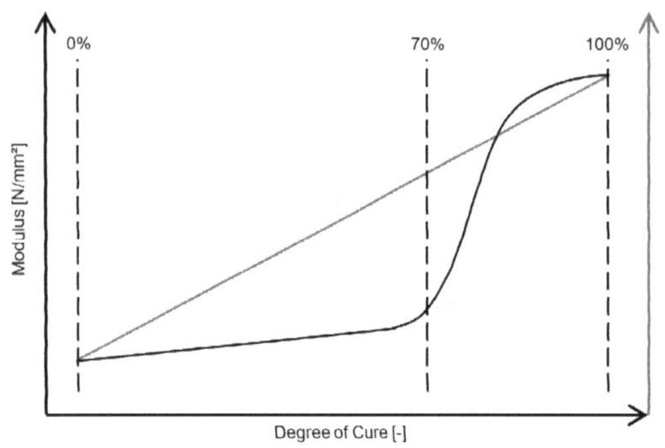

Figure 5: Schematic illustration of curing degree dependency of the elastic modulus and the volume shrinkage for isothermal conditions (see e.g. Khoun et al. [9] and Nielsen [10]).

The thermal shrinkage can be calculated with the coefficients of thermal expansion. As described earlier one value below and above the T_g is required for epoxy systems.

To calculate the resulting stresses from the shrinkage effects in a structural application the "effective modulus" at the strain development must be described. A schematic illustration of the modulus over temperature is given in Figure 6 together with the coefficient of thermal expansion.

An examination on the development of residual stresses in epoxy based systems can be performed for a "1-D" simplification of a sandwich panel exposed to a temperature change. Such a situation can be found in bond lines (e.g. trailing or spar cap to web bond line) as well as between two glass fibres and the surrounding resin (strong idealization).

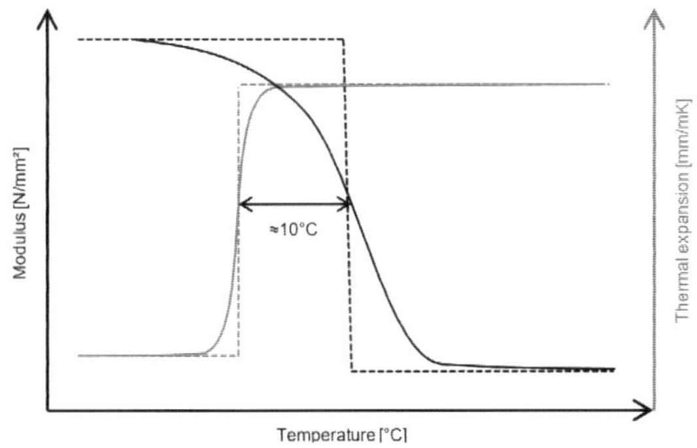

Figure 6: Schematic illustration of temperature dependency of the elastic modulus and the coefficient of thermal expansion (see also Khoun et al. [9] and Nielsen [10]).

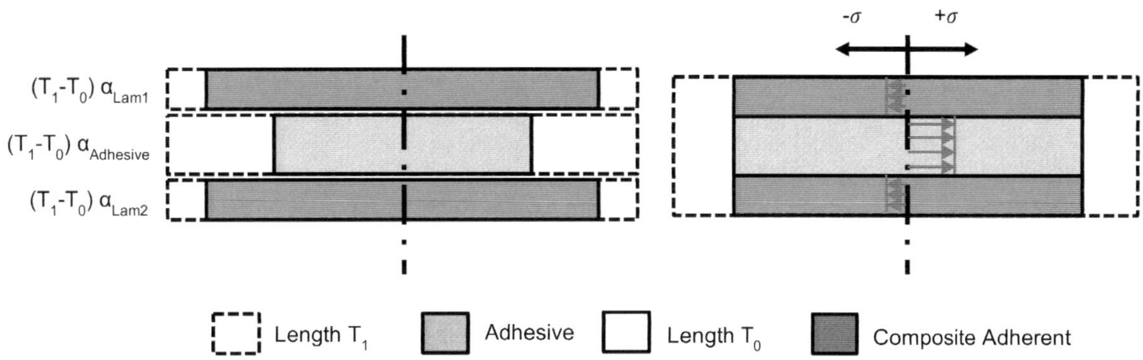

Figure 7: Schematic illustration for the development of thermal residual stresses in a bonded structure.

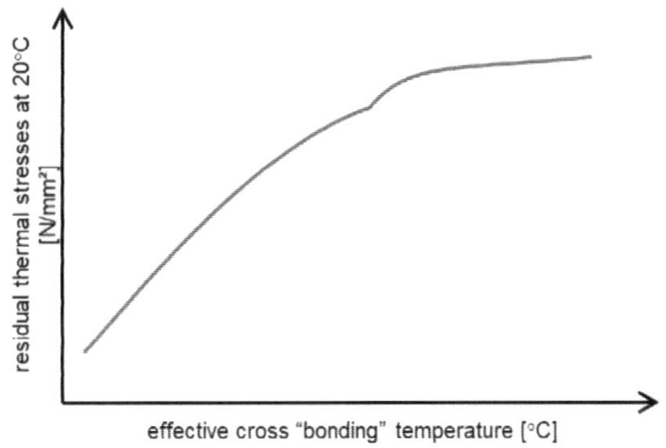

Figure 8: Schematic illustration of resulting residual stresses in epoxy adhesive bond line.

Figure 7 illustrates the potential free thermal shrinkage of adherents and adhesive as well as the total shrinkage and the resulting residual stresses in a bonded configuration. In the left side of the figure the two adherents and the adhesive are shown. The dotted lines indicate the length of these at the bonding temperature T_1. The solid line shows the dimensional changes of these bond line elements after a temperature drop to T_0 assuming a free deformation of the parts without any bonding between the adherents and the adhesive. In the right part of the figure the same temperature change is assumed after a bonding of the parts at T_1 (due to the bonding, the length of all parts will be identical at all temperatures). Due to the different coefficients of thermal expansion residual stresses have to develop. Based on material data, the thermal induced residual stresses can be calculated (see e.g. [10]). Such a model does not consider potential relaxation effects that might reduce the resulting residual stresses. As relaxation effects are more pronounced at high temperatures and epoxy systems consist of a solid cross linked polymer network relaxation effects are often considered to be small below the glass transition temperature (T_g).

3. Blade Manufacturing Process

Details on the manufacturing process of modern WTB are typically not public available. However, a general description of the manufacturing process can be given and such a process is also followed within the ReliaBlade processes. Most WTB are manufactured by vacuum infusion process. Manufacturers typically produce two blade shells which are bonded with at least one additional web to form the WTB. A schematic overview on the manufacturing process of WTBs is shown in Figure 9.

The lay-up in the mold is in most cases a manual process. However, often pre-manufactured spar caps and root inserts are put into the mold with dry materials (the thickness of single glass or carbon fibre plies usually between 0.5 and 2 mm). While the shells are typically made from relative thin biax (up to 2 mm) laminates with a PVC, PET or Balsa core (between 10 mm and 50 mm thick), the spar caps, the transition area and the root section are made from thick laminates. The maximum thickness of spar caps typically varies between 40 and 80 mm and the thickness in the transition area and the root depends on the design and the root connection and is around 30 to 150 mm (with the root having the higher values). Due to the thickness of the composite materials and the typical curing temperatures of 80 °C, exothermal peaks of approx. 120 to 130 °C are observed.

In the bonding process an adhesive is typically applied to the bonding areas of the blade shell with a mixing and dosing machine. After the adhesive application the molds are closed with a hinge system. The adhesive is cured in the mold by the mold heating system (again approx. 80 °C) and exothermal peaks above 100 °C have to be expected.

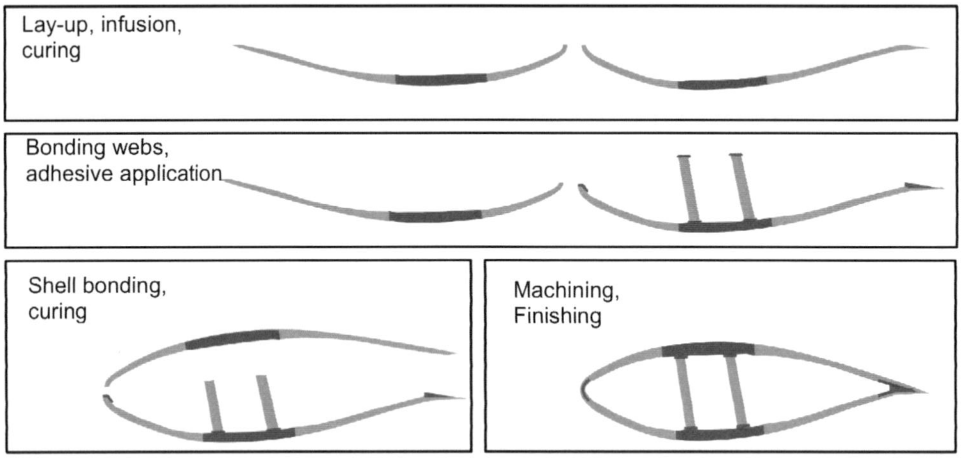

Figure 9: Typical wind turbine blade manufacturing process.

In the manufacturing of the ReliaBlade test blade, the temperatures will be monitored at critical locations with integrated sensing elements like thermocouples. Relevant WTB areas are observed non-invasively by picking up heat signatures with thermal imaging methods. While single temperature measurements are state of the art for process control in industry, instrumentation with both temperature sensors and thermography cameras is atypical. The resin gelation and the development of the curing degree will also be monitored with an advanced network of sensors.

In addition to this, a few local measurements of the fibre orientation will be conducted with a laser scanning system. The major focus is on the continuous improved documentation method of the manufacturing process for each blade which means a proper description of potential imperfections.

An example of thermal imaging is shown in Figure 10. Local effect, e.g. from hose shading or the crimped vacuum foil need to be excluded from the image. This indicates the effort to integrate such methods in an automated monitoring process and digital twins.

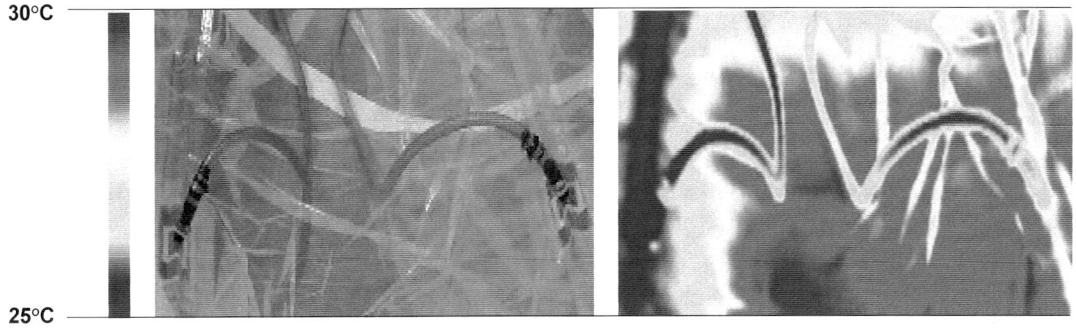

Figure 10: Example of thermal image.

4. Shell laminates and inter-fibre failure as an example for digitalization of material performance

Since wind turbine rotor blades are designed/qualified against first ply failure, i.e. the development of matrix inter-fibre fracture, it is of vital importance to estimate accurately the polymer properties and the stresses that they are bearing. The resin can be seen as an adhesive, which bonds the fibres together. Again residual stresses can develop in the resin and thus the mechanical performance including the inter-fibre fracture resistance, is strongly dependent on the manufacturing process and the applied curing cycle ([12], [13]). According to [12] and [13] the evolution of the curing degree is a function of the applied curing temperature profile, therefore it is a prerequisite for the estimation of the matrix mechanical performance. A number of models published in literature could enhance the estimation of the conversion degree of polymers ([14], [15]).

The evolution of the curing degree is directly impacting the stiffness and the strength of the polymers which can also be influenced from the operational temperature ([7], [13], [18]), see Figure 11 and Figure 12. These are the basic building block input parameters for the estimation of the in-situ laminate stiffnesses, residual stresses and strengths. The observed stiffness peak at curing degrees below one is surprising, however is also described by other authors, see e.g. [7], [17].

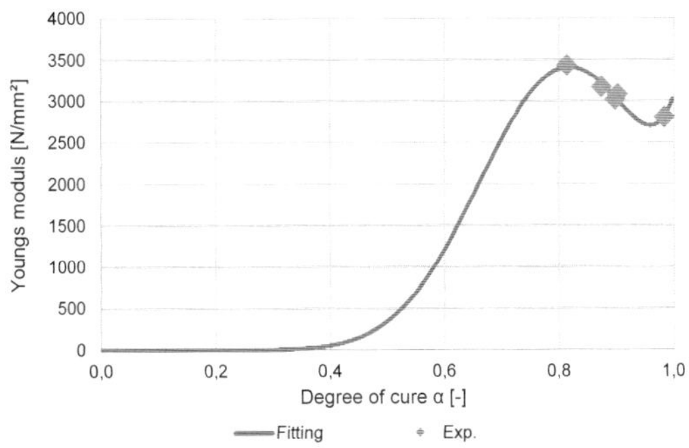

Figure 11: Evolution of matrix stiffness with curing degree [18].

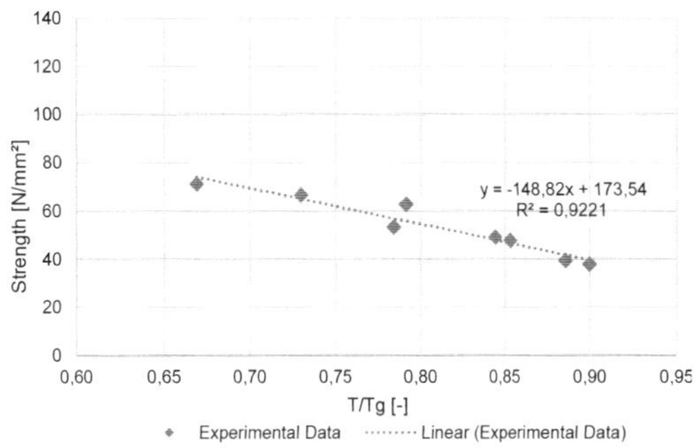

Figure 12: Evolution of strength as a function of operational and glass transition temperature [7].

On the laminate level, the composite properties are depending on parameters with an uncertainty mirroring the manufacturing process. The fibre orientation and the resulting stiffness and strength ([19], [21]), the roving pattern geometry in the laminate cross section (quadratic, hexagonal, etc.), see e.g.

[22], [23] and the Fibre Volume Fraction (FVF) [24] are typical examples. Physically based micromechanical formulations [25] and the application of the classical lamination theory [26] can describe the variability of the aforementioned parameters and thus reduce modelling uncertainty, while increasing the overall reliability [5] at the same time.

Depending on the specific blade manufacturing outcome, each separate digital twin e.g. trailing edge shell or spar cap laminate, can be updated accordingly. For example in the case of varying laminate thickness resulting from the infusion, the corresponding model FVF will have to be adjusted in order to mirror the actual properties. Therefore, the in-situ material properties can be optimally estimated.

For the typical case of infused laminates with Vacuum Assisted Resin Infusion Moulding (VARIM) with an FVF range of 50-60% [24], the model results concerning the in-plane stiffness variation of a unidirectional (UD) glass-epoxy laminate are illustrated in Figure 13, while being compared with experiments [26].

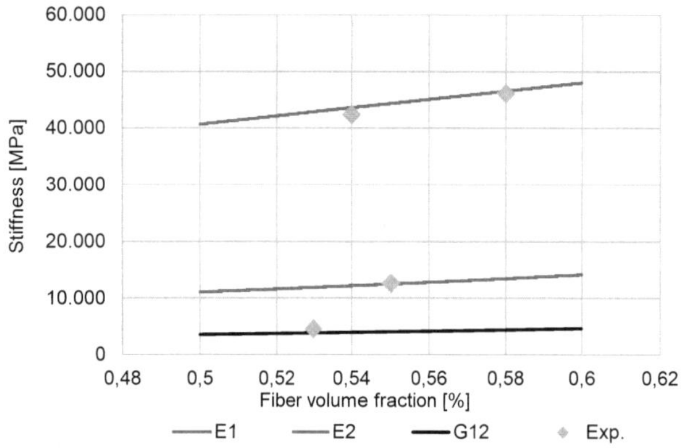

Figure 13: Typical variation of in-plane laminate properties with FVF.

For the ply stress analysis, 3-dimensional (3D) orthotropic stress components are implemented, considering also the residual stresses that are developed in the matrix [25]. Moreover, relaxation effects of the residual stresses during the curing process are taken into account, as described in ([8], [12]). Continuum mechanics is used for the prediction of the static limit stress state, i.e. the prediction of the inter-fibre fracture initiation. A strain energy equivalent stress criteria was applied to describe the behaviour of the isotropic matrix, resulting in an equivalent stress of the 3D matrix element within a lamina [29]. Assuming that the inter-fibre fracture is initiating in the matrix, for a given stress tensor, it can be estimated in terms of the stress effort e_M as defined from Puck [30]. When the ratio of the equivalent stress to the static tensile strength becomes unity, then the crack initiation occurs. The static strength is derived by means of standard material tests.

The same principle of the equivalent matrix stress is also employed for the alternating load calculations of matrix dominated laminates. The stress limit levels are derived with a fitting procedure of uniaxial fatigue experimental data of pure resin material [31]. The applied model results from the convolution of Stüssi [34] and Goodman formulations [35]. To describe the S/N performance with an S-shape formulation, as suggested by Stüssi [36] was used. The Goodman formulation is used to correlate the alternating stress and the allowable cycle number depending on the applied R ratio. Moreover, the implementation of a Weibull distribution for the fatigue strength is enhancing the description of the reliability level.

A validation example for the fatigue inter-fibre fracture model of a UD laminate loaded transverse to the fibres (UD90) is presented in Figure 14. The prediction is correlated to the experimental crack initiation which is detected during a fatigue test. A typical matrix crack is illustrated in Figure 15.

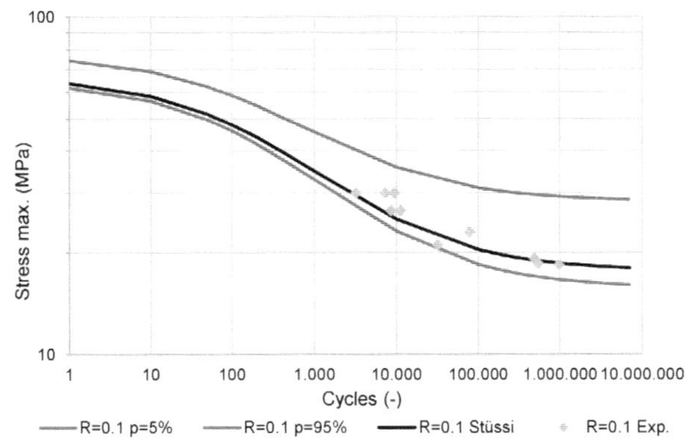

Figure 14: Inter-fibre fracture initiation. Comparison of model with fatigue experimental data of a UD90 laminate.

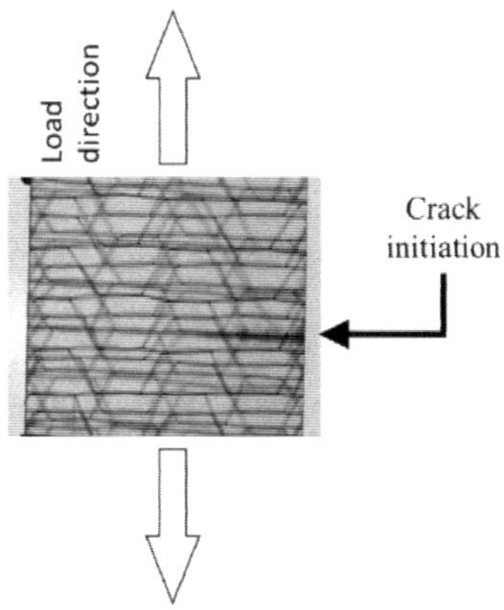

Figure 15: Typical inter-fibre fracture.

This work on the inter-fibre failure damages shows the connection between manufacturing process data and the material modelling results in a more realistic representation of the material performance. The application of such concepts is practically only possible by the use of digitalization, due to the bandwidth of parameter variations in real structures and the resulting amount of data.

It is important to have a look at the measurement technologies needed. Measurement technologies for the temperature distribution in the blade during manufacturing are available. As the blade molds are not closed, mold based sensors and thermography are suitable methods. To determine the local fibre volume content, which would most likely only be conducted locally, ultrasonic inspections or laser / optical based geometry measurements might be applied. With regard to the fibre orientation, existing methods have to be developed further or new technologies need to be applied to cover most of the blade fibre material. However, also statistical methods and simulation procedures might be applied to a larger degree as today.

From a material modelling point of view many details need to be investigated and validated. The variety of materials is large and the loading scenarios manifold. Also the effect of defects should be understood better.

5. Conclusions

In this paper, a future approach for the design and operation of wind turbine blades has been outlined. This approach is based on a physical based material modelling and measurements in the manufacturing and the operation life phase of the blade. The measurement data from the manufacturing can be used to model the local material properties in the blade. Together with measurements from the operational blade phase, the local material strength can be compared to the local blade loading and thus the material usage, the material reserve factor or the remaining life can be calculated more realistically.

With the inter-fibre failure an example was shown and the potential and relevance of material models that consider the manufacturing process parameters was shown. As the material model is connected with measurement data for a specific blade (location) such a concept is interpreted by the authors as a step in the direction of digitalization of wind turbine blades from a material perspective.

Nevertheless, only small and exemplary steps towards a digitalization for structural wind turbine blade applications were presented. Open topics for the inter-fibre failure damage mode remain and other damage modes need also to be evaluated. In addition to that, for the application of such methods, further work is required e.g. the damage models and the in line measurement methods need to be validated, the uncertainties quantified and the certification methods need to be adapted. Also the role of sensor systems in the operation phase for a continuous model updating should be analysed.

Besides the physical based modelling of damage modes, the validation of such models is in focus of the second half of the ReliaBlade project. In a first step, blades up to 30m length will be manufactured under lab conditions and many process parameters will be measured during the process. These data will enable to update the numerical blade models and better predict the blade structural performance. In full scale blade tests, the structural performance of the blades for different damage modes will be determined and compared with the numerical models of the structure. In addition to that the blade will be instrumented with structural health monitoring sensors and the sensor readings will be used to update the digital blade twin.

Acknowledgments
Fraunhofer IWES and Leibniz University Hannover acknowledge the support provided by the German Federal Ministry for Economic Affairs and Energy (BMWi) grant 0324335A and 0324335B and DTU Wind Energy acknowledge the support from the Danish Energy Agency through the EUDP grant 64018-0068. Both grants is within the joint ReliaBlade project.

References

[1] Marr B 7 Amazing Examples of Digital Twin Technology In Practice, https://www.forbes.com/, April 2019
[2] Hightechsoftwarecluster, https://hightechsoftwarecluster.nl/technology/what-is-a-digital-twin-and-what-value-does-it-deliver/, download July 2020
[3] McGugan M, Pereira G, Sørensen BF, Toftegaard H and Branner K Damage tolerance and structural monitoring for wind turbine blades Phil. Trans R. Soc. **A 373**: 20140077, 2015
[4] ReliaBlade project, www.reliablade.com, June 2020
[5] Toft H S Branner K Mishnaevsky L and Sørensen J D Uncertainty modelling and code calibration for composite materials J. Compos. Mater. **47** 14 1729–47
[6] DNV.GL DNVGL-ST-0376 https://rules.dnvgl.com/docs/pdf/DNVGL/ST/2015-12/DNVGL-ST-0376.pdf, download August 2020
[7] Antoniou A, Rosemeier M, Tazefidan A, Krimmer A and Wolken-Möhlmann G Impact of Site-Specific Residual Stress on the Fatigue of Wind-Turbine Blades, presented at AIAA Scitech 2020 Forum and Exposition, Orlando, Florida, DOI:10.5281/zenodo.3600151
[8] Holst M Reaktionsschwindung von Epoxidharz-Systemen, Dissertation, TU Darmstadt, 2001
[9] Khoun L, Centeam T and Hubert P Characterization Methodology of Thermoset Resins for the Processing of Composite Materials – Case Study: CYCOM 890RTM Epoxy Resin, J. Compos. Mater., 2010, pp. 1397-1415
[10] Sayer F Sub-Component testing for structural adhesive joint assessment in wind turbine rotor blades, PhD Dissertation Leibniz University Hannover, 2019
[11] Sørensen B F, Holmes J W, Brøndsted P and Branner K Blade materials, testing methods and structural design. Tong W. (Ed.), Wind Power Generation and Wind Turbine Design, 44, WIT Press, Boston, 2010, pp. 417-460.
[12] Nielsen M 2012 Prediction of process induced shape distortions and residual stresses in large fiber reinforced composite laminates Ph.D. Dissertation Technical University of Denmark
[13] Fiedler B Hochleistungs-Faserverbundwerkstoffe mit Duroplastmatrix, Werkstoffe Struktur Eigenschaften Modellierung 2009 TuTech Innovation GmbH Hamburg Germany
[14] Karkanas P I, Partridge I K and Attwood D 1996 Modelling the cure of a commercial epoxy resin for applications in resin transfer moulding Polym. Int. **41** 2 183–91
[15] Duemichen E Javdanitehran M Erdmann M Trappe V Sturm H Braun U and Ziegmann G 2015 Analyzing the network formation and curing kinetics of epoxy resins by in situ near-infrared measurements with variable heating rates Thermochim. Acta **616** 49-60

[16] Gao J LeBel F Ravalli M Gurnon K Kelley J Rubinsztajn S and Oyerokun F 2015 Prediction of modulus evolution of a silicone composite using DMA for cure kinetic profile development 43rd Annual Conf. of the North American Thermal Analysis Society (NATAS)
[17] Schümann J-P Zur zeit-, temperatur- und umsatzabhängigen Entwicklung polymerphysikalischer Vernetzungsschwindung aminisch vernetzender Epoxide, PhD Thesis Technical University Clausthal, 2018
[18] Tazefidan K, Der Einfluss fertigungsbedingter Eigenspannungen auf das Ermüdungsverhalten von Faserverbundmaterialien Master Thesis, Technical University Berlin, 2020
[19] Halpin J C and Kardos J L 1976 The Halpin Tsai equations. A review Polym. Eng. Sci. **16** 5 344-352
[20] Chamis C C and Sendeckyj G P 1968 Critique on Theories Predicting Thermoelastic Properties of Fibrous Composites J. Compos. Mater. **2** 3 332-358
[21] Emerson M J Jespersen K M Dahl A B Conradsen K and Mikkelsen L P 2017 Individual fibre segmentation from 3D X-ray computed tomography for characterising the fibre orientation in unidirectional composite materials, Compos. Part A Appl. Sci. Manuf. **97** 83-92
[22] Shaffer B W 1964 Stress-Strain Relations of Reinforced Plastics Parallel and Normal to the Internal Filaments AIAA **2** 348-352
[23] Krimmer A 2014 Mikromechanische Modellierung von Fasergelege-Kunststoff-Verbunden auf Basis von Normprüfungen unter Berücksichtigung der in-situ-Eigenschaften der Matrix Ph.D. thesis (TU Berlin). http://opus4.kobv.de/opus4-tuberlin/frontdoor/index/index/docId/4440 Chapter 3 pp 43-53
[24] Mesogitis T S, Skordos A A and Long A C 2014 Uncertainty in the manufacturing of fibrous thermosetting composites: A review Compos. Part A Appl. Sci. Manuf. **57** 67–75
[25] Krimmer A Leifheit, R and Bardenhagen A 2016 Assessment of quasi-static and fatigue performance of uni-directionally fibre reinforced polymers on the basis of matrix effort 6th EASN International Conf. on Innovation in European Aeronautics Research
[26] Antoniou A Mikkelsen L P Goutianos S Bagemiel O Gebauer I Flindt R and Sayer F Influence of the glass non-crimp fabric intrinsic undulation on the stiffness of the composite ply: A micromechanical approach Proc. 41st International Symp. on Materials Science Roskilde Denmark Preprint
[27] Mishnaevsky L Jr Branner K Petersen H N Beauson J McGugan M and Sørensen B F 2017 Materials for Wind Turbine Blades: An Overview Materials **10** 1285-1309
[28] Holst T Material model based analysis of the curing process of a short fiber reinforced epoxy resin adhesive in rotor blades of wind turbines Ph.D. Dissertation Leibniz University of Hannover Preprint
[29] Beltrami E 1885 Sulle Condizioni di Resistenza dei Corpi Elastici Il Nuovo Cimento **18** 1 145–155
[30] Puck A 1996 Festigkeitsanalyse von Faser-Matrix-Laminaten: Modelle für die Praxis Hanser Munich
[31] Barbosa J F Correia J A Freire J R Zhu S P and De Jesus A M 2019 Probabilistic SN fields based on statistical distributions applied to metallic and composite materials: State of the art Adv. Mech. Eng. **11** 8 1-22
[32] Toasa Caiza P D and Ummenhofer T 2018 A probabilistic Stüssi function for modelling the SN curves and its application on specimens made of steel S355J2+ N Int J Fatigue **117** 121–134
[33] Rosemeier M and Antoniou A 2021 Probabilistic Approach for Predicting the Fatigue Strength of Polymers submitted in AIAA Scitech Forum and Exposition Nashville Tennessee Preprint
[34] Stüssi F 1955 Tragwerke aus Aluminium Springer-Verlag doi:10.1007/978-3-642-92661-7
[35] Goodman J 1899 Mechanics Applied to Engineering, Longmans, Green and Co URL https://archive.org/details/mechanicsapplie03goodgoog
[36] Suresh S 1998 Fatigue of materials Cambridge University Press

IOP Conf. Series: Materials Science and Engineering **942** (2020) 012007 doi:10.1088/1757-899X/942/1/012007

Proof of concept: elliptical biaxial rotor blade fatigue test with resonant excitation

David Melcher[1], Heiko Rosemann[1], Bernd Haller[1], Sebastian Neßlinger[2], Enno Petersen[1] and Malo Rosemeier[1]

[1]Department of Rotor Blades, Fraunhofer IWES, Fraunhofer Institute for Wind Energy Systems, Am Seedeich 45, 27572 Bremerhaven, Germany

[2]Nordex Energy GmbH, Langenhorner Chaussee 600, 22419 Hamburg, Germany

E-mail: david.melcher@iwes.fraunhofer.de

Abstract. Rotor blades of wind turbines are subjected to high-cycle fatigue during their lifetime of 20-25 years. To show that a blade design fulfils the normative requirements, the test campaign usually consists of two consecutive cyclic tests, i.e., in the flapwise and lead-lag directions. The fatigue test excites the blade close to its corresponding natural frequency. Combining the two uniaxial tests into one biaxial test excites the blade in both directions simultaneously. Using the same excitation frequency for both directions results in the blade cross-sections describing an elliptical deflection path. To realize such a test while still exciting close to resonance, the natural frequencies for the two directions need to be equalized with the aid of decoupled masses and stiffness elements. This approach reduces the testing time, and induces a more realistic loading which is comparable with field conditions while keeping the energy consumption of the hydraulic actuation low. This work describes the concept of the elliptical biaxial rotor blade fatigue test with resonant excitation using a commercial blade design. To this end, the design model, which uses both a transient and a harmonic simulation, is validated with the experimental results of the test. The simulation model and the experiment agree well with each other in terms of displacements and loads along the blade.

1. Introduction

Rotor blades of wind turbines are subjected to high-cycle fatigue during their expected lifetime of 20-25 years. As proof that a rotor blade design is able to withstand the fatigue loads occurring, experimental validation tests are necessary, as described in IEC 61400-23 [1]. The test campaign usually consists of several static extreme load tests and two consecutive cyclic fatigue tests, i.e., in the flapwise and lead-lag directions. To achieve certification, these tests require the target bending moment distributions to be exceeded within a specific area of interest (AoI) along the blade without excessive overloading, which could damage the blade prematurely. The fatigue tests are time consuming since they require two to seven million load cycles. During the fatigue tests, the blade is excited close to its corresponding natural frequency to reduce the amount of energy which has to be fed in. The associated bending moments over the blade length are realized so as to be damage-equivalent to field loads. This represents a major drawback of these unidirectional tests: The loads introduced into the blade do not necessarily represent the realistic load conditions, as shown by Rosemeier et al. [2].

One approach to realizing more realistic testing is to combine the two unidirectional tests into one so-called biaxial test [3]. Some attempts at a possible realization can be found in the literature [3–7]. Conducting the biaxial fatigue test is much more challenging, however, as the unidirectional frequencies and the phase angle between the two displacement excitations have to be chosen carefully so as to apply the desired combined loading.

Several attempts at performing a biaxial fatigue test with random phase resonant excitation can be found [3, 6, 7]. The concepts here involve summing the damage introduced for the fatigue resistance evaluation using complex computations. A concept with more predictable loads requires the excitation frequencies in both directions to be at a fixed integer ratio, leading to Lissajous motion of the blade tip [4, 5, 8].

Recently, we developed and realized a biaxial test concept with elliptical resonant excitation which has a frequency ratio of 1:1. The concept requires the near equalization of the flapwise and lead-lag natural frequencies using masses and spring elements, as described by Melcher et al. [9]. Decoupled load elements modify the blade properties independently in each direction, as introduced by Post et al. [10].

The approach requires a precise definition of the set-up to match the frequencies and achieve the target bending moments. In practice, this precision can only be provided by an optimization-based computational process. Lee and Park [11] used an algorithm to optimize the overloading by determining the optimal mass distribution, actuator position, and excitation frequencies in unidirectional set-ups. A different optimization algorithm was evaluated by Zhang et al. [12], which was especially used to determine the position of blade tip cut off. Melcher et al. [9] presented an optimization-based tool to determine several test objectives and variables such as positions and values of decoupled masses and stiffnesses as required for an elliptical biaxial fatigue test. Of specific relevance for the elliptical test is the adequate determination of the excitation phase angle. It has a very high impact on the desired bending moments in the combined regions, the so-called off-axis directions, as investigated in our earlier work [13].

This work presents the experimental validation of the elliptical biaxial fatigue test approach with resonant excitation. The aim is to demonstrate the feasibility of the approach, and to quantify the test prediction accuracy by comparison of simulation and experiment, as shown in figure 1.

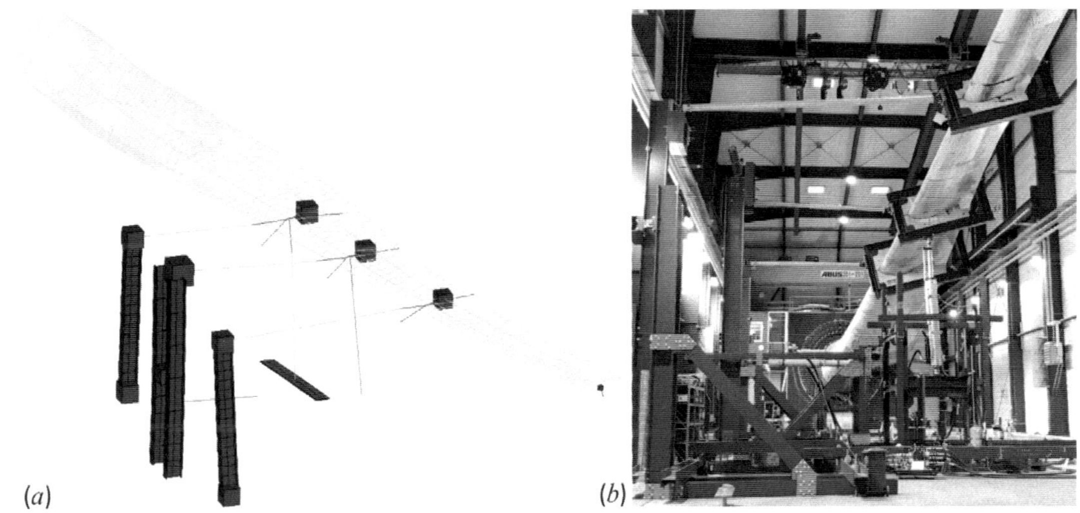

Figure 1. Simulation model (a) and experimental test setup (b) for elliptical biaxial rotor blade fatigue test.

2. Method

To validate the biaxial fatigue test concept, numerical simulations and a corresponding experiment are performed on a state-of-the-art commercial rotor blade, which is between 60 m and 70 m long[1]. To reduce the complexity of this feasibility study, a reduced AoI between 14 % and 37 % of the blade length is chosen. The test setup selected is optimized accordingly. This includes the position and required impact of the spring and mass elements.

2.1. Numerical setup definition

The biaxial test setup is defined using the optimization approach of Melcher et al. [9, 13] which is based on a preloaded harmonic simulation of a parametrized finite element (FE) model implemented in ANSYS APDL 18.1 [14]. In the optimization, the magnitude of the spring stiffnesses and the masses, and their positions, and the excitation frequency and phase angle are used as design parameters to find a feasible fatigue test setup which matches the target bending moment distributions as closely as possible, and equalizes the natural frequencies for the flapwise and lead-lag directions to allow for elliptical biaxial resonant excitation.

The final FE model of the test setup is shown in figure 1(a). The blade shell displayed serves only visualization purposes, as the blade is actually represented by 120 BEAM188-elements in the model using preintegrated composite beam cross-sections. The load-elements are modeled using a combination of MPC184-rigid-beam-elements for rigid connections, MPC184-rigid-link-elements for hinged pushrods, MASS21-elements for masses, BEAM188-elements for mass-levers and spring-beams, and LINK11-elements for actuators.

Once the test setup is defined, another approach using transient simulation evaluates the test setup. The same FE model is used, but unlike the harmonic simulation, this approach considers nonlinear effects. In this simulation, the actuator elements are excited using sinusoidal stroke input signals with frequency, phase angle and amplitudes taken from the preceding harmonic simulation. This transient approach is deemed more precise, but also requires significantly more computational power, and is less stable than the harmonic simulation and therefore unsuitable for use in the optimization.

2.2. Experimental setup

For the fatigue test, the blade root is attached to an inclined test block, tilting the blade pitch axis 7.5° upwards from the horizontal. Moreover, the blade is rotated through 15° about the pitch axis in order for the vertical and horizontal load elements to better align with the actual mode shapes of the blade, which helps to influence the natural frequencies independently. Although this means the load elements are not aligned exactly with the flapwise and lead-lag directions of the blade, they will nevertheless be referred to as such in the following. The load elements attached in the validation setup consist of the following, as shown in figure 2:

- (a) One spring element to increase the flapwise natural frequency of the test setup
- (b) Two decoupled mass elements to decrease the lead-lag natural frequency of the test setup
- (c) Two hydraulic actuators to excite the test setup close to its natural frequencies

To introduce the loads into the test blade, we adapt the loadframe design proven at the Fraunhofer IWES test facility, which consists of a steel frame with wood inserts clamped to the blade. A total of three loadframes are mounted to allow the load elements to be added at 38 %, 48 % and 61 % blade length, as shown in figure 1.

The passive elements (masses and spring) are connected to the loadframes using pretensioned steel pipes (40 kg m^{-1}) as pushrods with swivel bearings at both ends. To minimize the effect

[1] Detailed information on the blade cannot be provided due to confidentiality constraints.

Figure 2. Load elements: spring beam (*a*), decoupled masses (*b*) and hydraulic actuators (*c*).

of the decoupled masses on the flapwise direction, the pushrod's angle of incidence on the blade must change as little as possible. This is achieved by increasing the length of the horizontal pushrods up to 6.7 m. Hence, the blade pitch axis is tilted by 2.5° to the side for optimal utilization of the available hall space.

The first hydraulic actuator, which excites the blade in the flapwise direction, is directly attached to the loadframe at 48 % blade length. The second, smaller hydraulic actuator, which excites the blade in the lead-lag direction, is connected to the same loadframe with a pushrod and a lever to reach higher displacement amplitudes at the blade than the actuator itself is capable of. Also, the actuator has to be decoupled from the flapwise movement, as this is much larger than the lead-lag movement. The lever with added mass of 3.6 t at the top further acts as a third decoupled mass and reduces the lead-lag natural frequency.

The sinusoidal strokes of the two actuators are controlled independently, only the phase angle is fixed. Apart from this synchronization, the actuators are controlled as in a uniaxial fatigue test, by applying their sinusoidal strokes to the rotor blade and observing the resulting loads.

Adding the decoupled masses to the other two loadframes also calls for a vertical lever construction, supported at a pivot in the center, with the required mass divided into two separate steel blocks at the top and bottom of the lever, and the pushrod transferring the load to the blade connected to the top. The total inertia of these constructions acting on the blade corresponds to that of masses of 4.0 t at 38 % blade length and 3.8 t at 61 % blade length.

To provide the required stiffness of 31 kN m^{-1} with a maximum deflection of 300 mm, the spring at 38 % blade length is designed as a beam 6 m in length made of glass-fiber reinforced epoxy.

2.3. Measurement data acquisition

The measurement data are sampled at 50 Hz and collected in a single measurement system.

All load elements are equipped with load cells to measure the force at the pushrod. The hydraulic differential forces of the actuators are also recorded.

To track the displacement of the blade by means of a three-dimensional camera system, two reflectors are attached to each loadframe and one to the tip of the blade, as shown in figure 3(*a*).

To determine the bending moments along the blade, the method described by Lee et al. [15] is used. By measuring the strain at multiple positions along the circumference of selected cross-sections, the flapwise and lead-lag bending moments can be computed on the basis of preceding calibration tests. In the present case, this was used to derive four values for both the flapwise and lead-lag bending moments at every cross-section selected by using different strain gauge combinations. The mean values for each cross-section are used for further computations and are shown in the results below. The off-axis bending moments M_α in the 30°, 60°, 120° and 150° directions are derived from the flapwise moment M_F (0°) and the lead-lag moment M_L (90°) using equation (1).

$$M_\alpha = M_\mathrm{F} \cdot \cos\alpha + M_\mathrm{L} \cdot \sin\alpha \tag{1}$$

3. Results and discussion

The experiment and the corresponding simulations are performed at different load levels up to 90 % target load. A higher load is not applied so as not to damage the specimen unnecessarily. Furthermore, different frequencies and phase angles are tested to evaluate their influence on the fatigue test. These parameter studies show good correlation between experiment and simulation model but exceed the scope of this work.

The results presented in the following use a combination of a frequency of 0.548 Hz and a phase angle of 78° at 90 % load, which results in the best match for the target bending moments.

3.1. Displacement

Figure 3(*a*) shows the position of the numbered points on the blade whose motion was tracked during the experiment. The corresponding measured motion trajectories (red lines) are presented in figure 3(*b*). These data are in the test hall coordinate system, whose origin is below the blade root. Since the pitch axis is tilted, the trajectories are offset correspondingly.

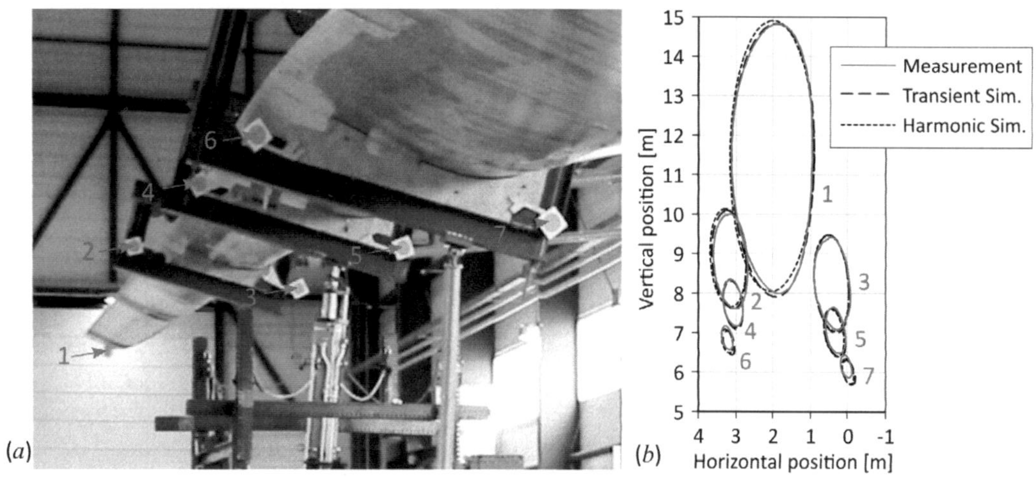

Figure 3. Positions of motion-tracked reflectors using a three dimensional camera system (*a*) and coresponding measured and simulated displacement trajectories (*b*).

The black dotted and dashed lines in figure 3(*b*) represent the results from the harmonic and transient simulation, respectively. The correlation between measurement and transient simulation is good, especially for point 1 at the tip. The mean value of the harmonic simulation deviates slightly from the experiment, since the geometric nonlinear effects are neglected in this simulation. The other points exhibit less displacement, leading to smaller differences between the two simulation methods. Moreover, the shape of their motion in the measurement agrees well with that in the simulation. The offset between measurement and simulation at points 2, 6 and

IOP Conf. Series: Materials Science and Engineering **942** (2020) 012007 doi:10.1088/1757-899X/942/1/012007

7 may be due to measurement errors while finding the undeformed position of the corresponding FE nodes.

3.2. Bending moments

The test to target bending moment ratios are shown in figures 4 and 5. Figure 4 shows the moment envelopes for three exemplary blade cross-sections at 14 %, 29 % and 43 % blade length. The axis scale ratio of the plots conforms to the actual flapwise to lead-lag moment ratio. The positions of the corresponding cross-sections are also indicated by black dashed lines in figure 5, which displays the moment ratio distribution along the blade for the flapwise and lead-lag directions (figure 5(*a*)) and for the off-axis direction (figure 5(*b*)). The legend describes how color and graph type are combined for the individual data plots.

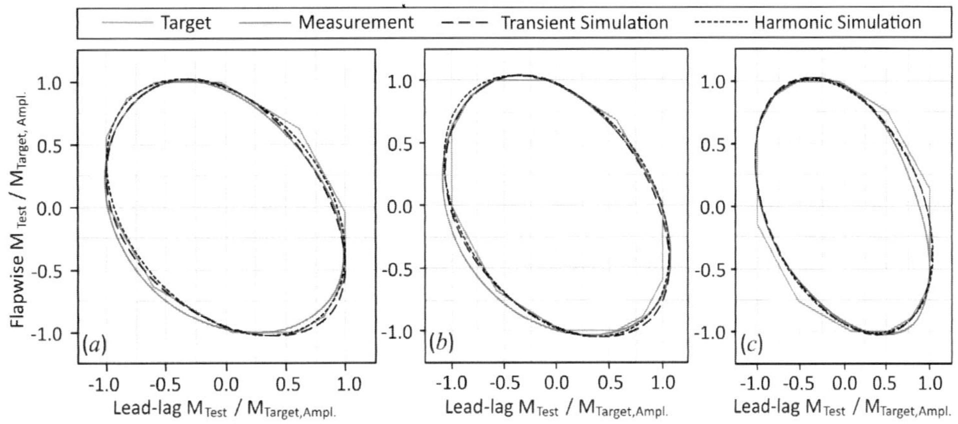

Figure 4. Target, measured and simulated flapwise moment over lead-lag moment envelope of cross-sections at 14 % (*a*), 29 % (*b*) and 43 % (*c*) of blade length.

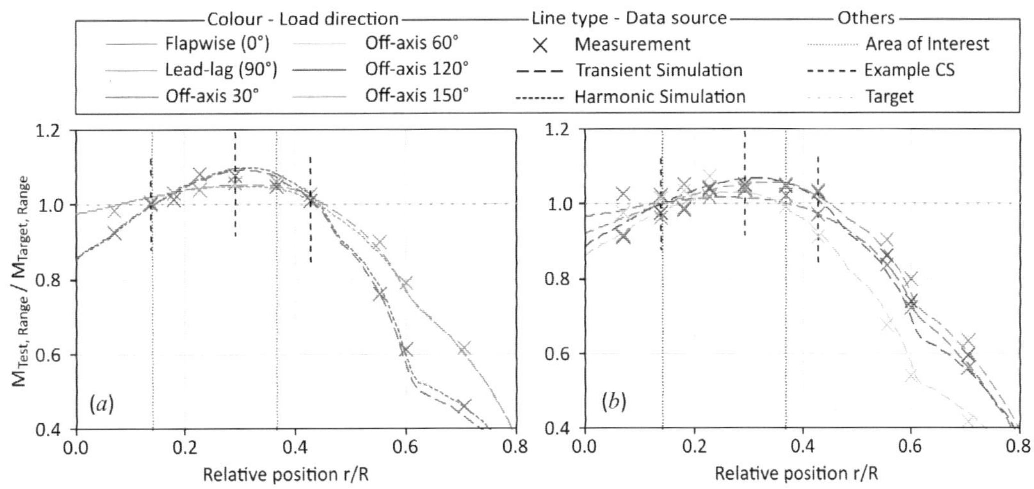

Figure 5. Measured and simulated test to target moment ratio along blade length for flapwise and lead-lag direction (*a*) and for off-axis directions (*b*). Color and graph type combination describe the case.

In general, all simulated load directions show good agreement with the measurements. The lead-lag moments only exhibit noteworthy differences outboard of 60 % blade length. Hence, in figure 5(*b*), the harmonic simulation results are omitted for reasons of clarity.

When only the flapwise and lead-lag moments are considered (figure 5(*a*)), the simulation results are in good agreement with the measurement data. Within the AoI, in particular, the deviation between measurement and simulation is less than 3 %. Off-axis directions deviate by less than 5 % within the AoI. In addition, the deviation between off-axis simulation and measurement data seems to change along the blade. At 14 % blade length (figure 4(*a*)), the simulation amplitude is higher than the measurement amplitude in the 150° direction and lower in the 60° direction, whereas at 43 % blade length (figure 4(*c*)), the opposite is the case. This behavior may be caused by inaccuracies in the beam element properties representing the mass and stiffness distributions of the blade.

Even though the area of interest was chosen to be between 14 % and 37 % blade length, the data for the flapwise and lead-lag directions indicate it could be extended to 43 % blade length. Only the 30° and 60° off-axis moments are slightly below the target values in the extended area.

3.3. External loads
Figure 6 shows the loads which are generated by the different load elements.

The actuator strokes over time are shown in figure 6(*a*). Here, good agreement between simulation and experimental results is observed. Only the simulated amplitude of the flapwise actuator is almost 4 % higher than in the measurement.

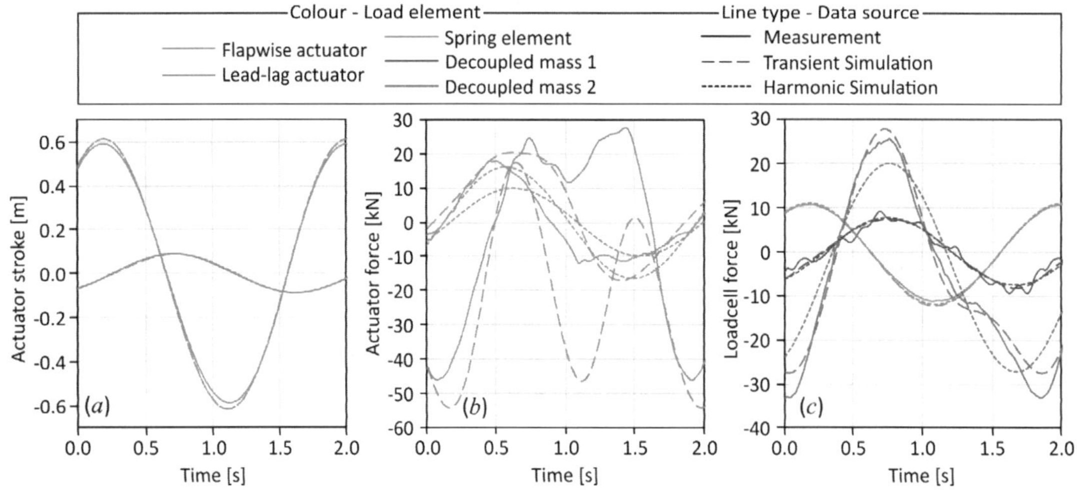

Figure 6. Measured and transient simulated actuator stroke (*a*) actuator forces (*b*) and loadcell forces at spring element and decoupled masses (*c*). Color and graph type combination describe the case.

Figure 6(*b*) shows the actuator forces. These values differ from sinusoidal curves, which are assumed in the harmonic simulation. Also, the deviation between transient simulation and measurement is large, especially for the lead-lag actuator. The cause of this deviation might lie within the hydraulic system and/or the friction in the actuators, neither of which is taken into account in the simulation.

Finally, the forces at the pushrods of the spring element and the decoupled masses are displayed in figure 6(*c*). The force of the spring shows good agreement. The force of the decoupled mass 1, at 38 % blade length, also exhibits good agreement apart from a superimposed signal at higher frequency. This signal might result from an unknown resonance within the load element structure itself, which is not captured by the simulation. The force of decoupled mass 2, at 61 % blade length, reveals a nonlinear behavior which is caused by the changes in the pushrod's angle of inclination over a range of 22°. This behavior is largely reproduced by the transient simulation.

4. Conclusion

Overall, we have shown that a biaxial fatigue test with elliptical resonant excitation can be realized by equalizing both main excitation frequencies in a 1:1 ratio. This test concept is therefore feasible. In addition, the optimization-based test setup definition provided the accuracy required to match the target bending moments. Moreover, the transient simulation was in good agreement with the experimental displacements, bending moments, and forces, with only negligible deviations, and is therefore validated.

However, for demonstration purposes, the AoI considered in this study was reduced to 23 % of the blade length. For certification, a larger AoI is necessary (about 55 %), requiring more load elements. Furthermore, the time-saving potential of the approach presented and insights into the increased structural reliability of the blade need to be investigated.

Acknowledgments

We acknowledge the support within the Future Concept Fatigue Strength of Rotor Blades project granted by the German Federal Ministry for Economic Affairs and Energy (BMWi) (0325939) and the Senator for Economic Affairs, Labour and Europe of the Free Hanseatic City of Bremen within the ERDF programme Bremen 2014-2020 (201/PF_IWES_ZK_Phase I/2017).

References

[1] IEC 2014 *IEC 61400-23 - Wind Turbines Part 23: Full-scale Structural Testing of Rotor Blades* (Geneva, Switzerland: International Electrotechnical Commission)
[2] Rosemeier M, Basters G and Antoniou A *Wind Energy Science* **3** 163–172
[3] White D 2004 New method for dual-axis fatigue testing of large wind turbine blades using resonance excitation and spectral loading Tech. rep. National Renewable Energy Lab., Golden, CO (United States)
[4] White D, Desmond M, Gowharji W, Beckwith J A and Meierjurgen K J 2011 *ASME 2011 International Mechanical Engineering Congress and Exposition* pp 1163–1172
[5] Bürkner F and van Wingerde A 2011 *Proceedings of the 8th International Conference on Structural Dynamics, EURODYN 2011*
[6] Greaves P R, Dominy R G, Ingram G L, Long H and Court R 2012 *Proceedings of the Institution of Mechanical Engineers, Part C: Journal of Mechanical Engineering Science* **226** 1693–1704
[7] Snowberg D, Dana S, Hughes S and Berling P 2014 Implementation of a biaxial resonant fatigue test method on a large wind turbine blade Tech. rep. National Renewable Energy Lab.(NREL), Golden, CO (United States)
[8] Heijdra J, Borst M and Van Delft D 2013 *Advances in Wind Turbine Blade Design and Materials* Woodhead Publishing Series in Energy (Woodhead Publishing, Sawston, Cambridge, UK) pp 432–445 ISBN 978-0-85709-426-1
[9] Melcher D, Bätge M and Neßlinger S 2020 *Wind Energy Science* **5** 675–684
[10] Post N and Bürkner F 2016 Fatigue test design: Scenarios for biaxial fatigue testing of a 60-meter wind turbine blade Tech. rep. National Renewable Energy Laboratory, Golden, CO, (United States)
[11] Lee H G and Park J s 2016 *Wind Energy* **19** 371–380 URL `https://doi.org/10.1002/we.1837`
[12] Zhang J, Shi K and Liao C 2018 *Journal of Renewable and Sustainable Energy* **10** 053303
[13] Melcher D, Petersen E and Neßlinger S 2020 *The Science of making Torque from Wind - submitted* **2020**
[14] ANSYS, Inc. 2017 *ANSYS Mechanical APDL User's guide, Release 18.1*
[15] Lee H G and Lee J 2018 *Composite Structures* **200** 306–312 ISSN 0263-8223

IOP Conf. Series: Materials Science and Engineering **942** (2020) 012008 doi:10.1088/1757-899X/942/1/012008

On the composite design of wood branches leading to improved bending strength

M Hartwig[1] and E K Gamstedt[1,2]

[1] Uppsala University, Division of Applied Mechanics, Department of Materials Science and Engineering, Box 534, SE-751 21 Uppsala, Sweden

[2] kristofer.gamstedt@angstrom.uu.se

Abstract. Wooden branches are designed to carry large bending moments, and so are longer composite structures, e.g. rotor blades for wind turbines. Being a natural fibrous composite material, wood is made from relatively simple biopolymer building blocks. In this preliminary work, we describe the composition and structure of softwood branches, including 3D images from X-ray computed tomography. The main difference in branch structure compared with other wood tissues is the reaction wood formed on the compressive side in e.g. spruce and pine. A simple beam is used to show that maximum bending moment is multiplied several times solely from the reaction wood. This is noteworthy, since chemical composition of the reaction wood does not differ significantly from the rest of the wood. The tissue gradients in the branch resulting from variation in density, microfibrillar angle and cell geometry contribute to the strength improvements. In composite structures, sandwich design is used to improve the load carrying capacity. From a general perspective, also local features as found in wood, such as smooth gradients and controlled cellular structure, could be further explored to improve bending strength in engineered composite materials.

1. Introduction

Structures in trees have developed by evolution to carry load efficiently. In particular, branches show a structure designed to support bending moments. Like manufactured composite components, trees are made of an anisotropic material, that can be optimized for its loading conditions. In this work, we will address a feature in softwood branches which may be of relevance also for composite components carrying significant moment forces, e.g. rotor blades in wind turbines. This feature concerns the formation of reaction wood close to the intersection of the branch to the stem. First, we review some of relevant material features of wooden branches.

Wood itself is composed of a few key components, mainly cellulose, hemicellulose and lignin. However, wooden material is still able to adapt to different environmental conditions and forces acting upon the tree during growth and leading to wide variation in properties, e.g. mechanical properties [1].

One such adaptation is reaction wood. This type of wood is formed to counteract gravity's impact in leaning stems or branches. In broad-leafed trees (angiosperms) the reaction wood is called tension wood, while in conifers (gymnosperms) it is called compression wood. Tension wood is formed on the upper side of the branch or stem, i.e. the tensile side. Compression wood is formed on the lower side of the branch or stem, i.e. the compressive side [2,3]. In figure 1 the formation of compression wood due to growth stresses on the compressive side is shown.

When looking closer at compression wood in branches of gymnosperms, it can be observed on the lower side of the branch that compression wood is formed, while on the upper side opposite wood is formed. Opposite wood is very similar in morphology and properties to normal wood. Compression wood, however, differs substantially from normal wood in its chemical composition and morphology. It contains more lignin and less cellulose, as well as a less hemicellulose (mannan and xylan), but more of the hemicellulose galactan. In figure 2 the hierarchical structure and the angle of the cellulose microfibrils is shown for normal wood and compression wood. The angle of the cellulose microfibrils in the cellular wall is higher in compression wood, especially in the middle layer of the secondary cell wall (S2-layer). The outer layer of the secondary cell wall (S1-layer) is thicker. There is no inner layer of the secondary cell wall in compression wood. The tracheids (wood fibers) have overall a thicker cell wall and are more round in cross sectional shape. The annual rings are wider in compression wood, giving the branch an elliptical rather than circular shape (eccentric growth), as can be seen in figure 1 [2,4–6].

All these factors lead to a change in mechanical properties and better adaption to sustain the load. Hence it is mechanically different to normal wood due to its higher density, increased longitudinal shrinkage, reduced stiffness and reduced tensile strength, but increased compressive strength [3].

This study aims to understand the mechanisms that leads to improved mechanical properties in compression wood of branches, even though it consists of the same building stones as regular wood. This understanding can be used as an inspiration in the future for improving engineered composite materials.

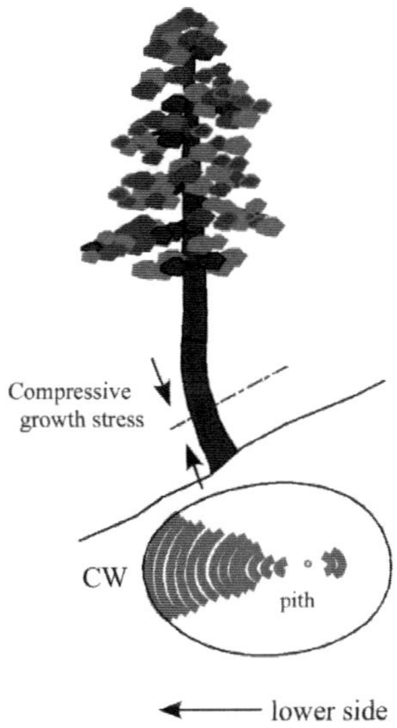

Figure 1. Macroscopic changes of the tree trunk due to stem leaning inducing compressive growth stress and hence formation of compression wood (CW). The figure is taken from [7].

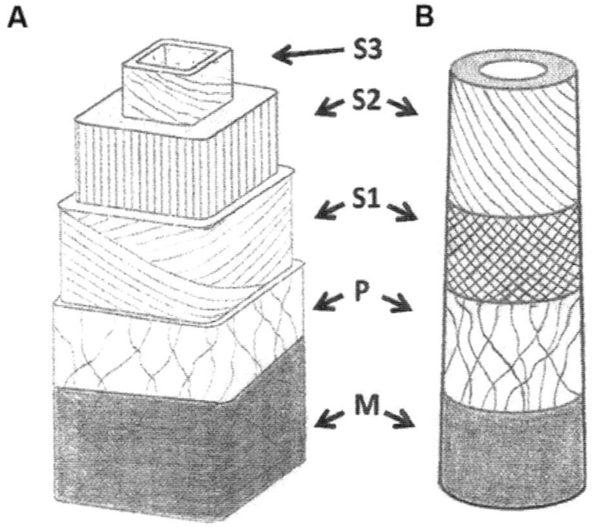

Figure 2. Comparison of tracheid of normal wood (A) and compression wood (B). Differences in shape of the tracheids and the angle of cellulose microfibril in the different layers of the cell wall are shown. M is the middle lamella. P is the primary cell wall. S1 is the outer layer, S2 is the middle layer, and S3 is the inner layer of the secondary cell wall. The figure is taken from [8].

2. Methods

2.1. Computer tomography

Samples for morphological studies were taken from a young Norway Spruce (*Picea abies)* tree, approximately 10 years old. After conditioning the samples at an ambient environment (21°C and 65 %

relative humidity) for at least 2 months, the sample was scanned in a computer tomograph (Bruker Skyscan 112). X-ray computed tomography allows to investigate the anatomy of wood and analyzing wooden structures in three dimensions without needing to dissect the sample [9]. The scan was carried out with a resolution of 6.5 µm. This allows identifying annual year rings and larger structures, but not cellular structures in the scanned wooden sample. Resolution was limited due to size of the sample and the design of the CT-scanner. Computer tomography is based on imaging sample with the help of X-ray showing density differences. After reconstruction of the images from different angles, the sample was three-dimensionally visualized in the software Avizo (ThermoFisher Scientific).

2.2. *Beam model*

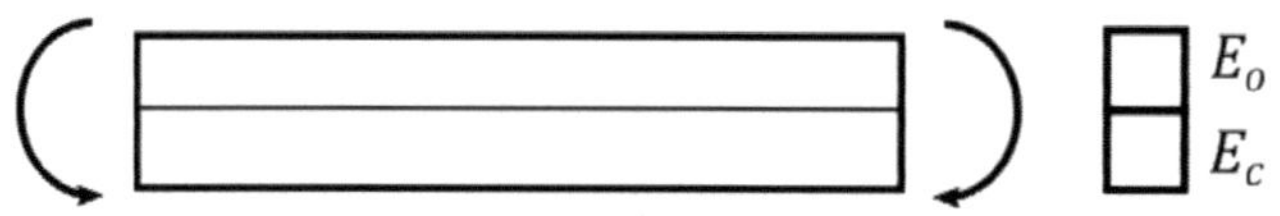

Figure 3. Pure bending of composite beam of two materials with different Young's moduli and strength. E_o is the Young's modulus in opposite wood, while E_c is the Young's modulus in compression wood.

To capture the main deformation mechanism, it is sufficient to consider basic beam theory. Looking at a bending moment in the branch, the branch can be simplified to a composite beam of two materials with different Young's moduli and strength, as shown in figure 3. The height and width of one layer is equal, so as total height of the beam 0.2 m and 0.1 m were chosen. The Young's modulus and strength of tensile loading were taken from [10]. Micro-tensile testing was performed on thin sheets of wood from the compressive side and tensile side of branches of Norway spruce to obtain these properties. However, no ultimate strength was found, so the yield strength at a strain of 1.5 % was used instead as a limit value. The compressive strength of opposite wood is estimated based on results from [11]. It was shown that the compressive strength in is approximately 50 % of the tensile strength in normal wood and clear samples. The compressive strength of compression wood is based on findings since 1904, where several studies showed that the compressive strength in compression wood is 50 % higher than the compressive strength in normal wood [3]. All values used in this model are listed in table 1.

This model is an example of the potential of using materials of different Young's modulus and strength. However, the mechanical performance of tree branches is also depending volume ratio of opposite wood and compression wood, shape of the cross section and further factors.

Table 1. Young's modulus and strength used for calculation of one model of branch with only normal wood and a second model with compression wood.

		normal wood	compression wood
Young's modulus opposite wood in MPa	E_o	3250	3250
Young's modulus compression wood in MPa	E_c	3250	1000
Tensile yield strength in MPa	σ_o	40	40
Compressive strength in MPa	σ_c	20	30

3. Results

3.1. *Computer tomography*

Computer tomographic images of the branch of a young spruce tree, as e.g. in figure 4, show that the annular rings on compression side of the branch are thicker, leading to an eccentric cross section of the branch. However, no difference in density could be measured between opposite wood and compression wood for this particular sample. This might be due to the juvenile state of the tree. Juvenile wood usually has a low specific gravity. Former studies have shown that the mechanical properties are closely related to density of the wood [3].

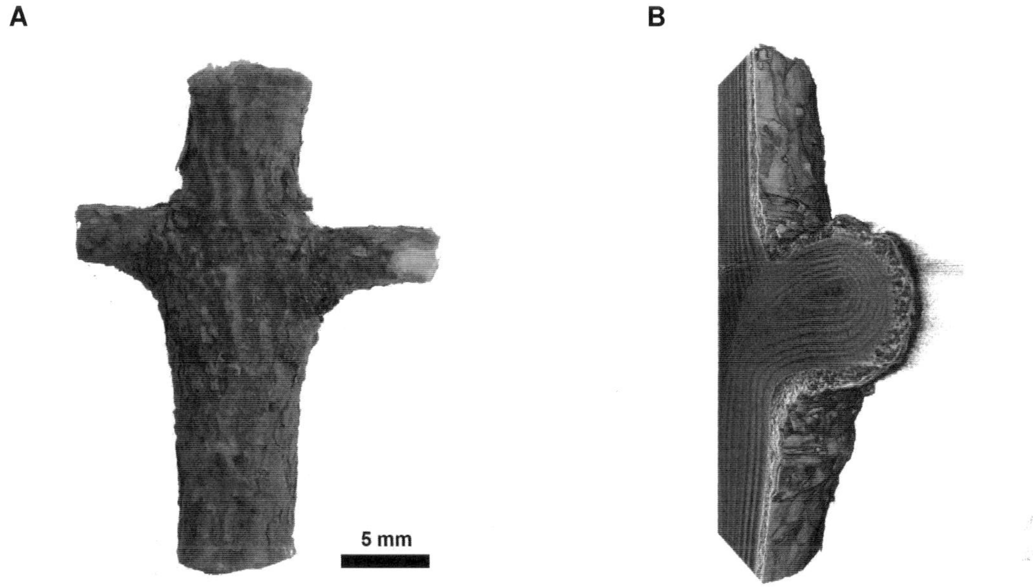

Figure 4. Image of a sample of a stem-branch joint of a young spruce tree (A). Computer tomographic scan of a part of same sample with a resolution of 6.5 µm (B). For the CT-image, black means low density of the material and white means high density.

3.2. Beam model

The ratio n of the Young's modulus in opposite wood E_o and compression wood E_c is

$$n = \frac{E_o}{E_c} \tag{1}$$

The neutral axis Y can be determined to be

$$Y = \left(A\frac{h}{4} + nA\frac{3h}{4}\right)(A + nA)^{-1} \tag{2}$$

The second moment of area I is

$$I = \frac{b}{12}\left(\frac{h}{2}\right)^3 + A\left(Y - \frac{h}{4}\right) + n\frac{b}{12}\left(\frac{h}{2}\right)^3 + nA\left(\frac{3h}{4} - Y\right) \tag{3}$$

The maximal bending moment M_{max} before breaking of opposite side $M_{max,o}$ and compressive side $M_{max,c}$, based on the strength in compression wood σ_c and opposite wood σ_o, is

$$\left|M_{max,o}\right| = \frac{|\sigma_o|\, I}{|0 - Y|} \tag{4}$$

$$\left|M_{max,c}\right| = \frac{|\sigma_c|\, I}{|h - Y|} \tag{5}$$

The two models showed, that the compressive side of the branch will fail first. However, in compressive wood the moment which can be applied before failure is approximately 1.8 times higher than for a branch only consisting of normal wood. The results are shown in table 2. In this model the non-rectangular shape and eccentric growth of the branch was not considered, which could lead to further increase of the maximum moment the branch with compression wood in this model would be able to sustain. Former studies showed that in other tree species, the work needed to the reach the maximum load for static bending is around 2 higher for compression wood than normal wood [3].

Table 2. Maximal bending moment of normal wood and compression wood for compressive side and opposite side.

		normal wood	compression wood
Maximal bending moment of compressive side in Nm	$M_{max,c}$	0.4	0.7
Maximal bending moment of opposite side in Nm	$M_{max,o}$	0.8	1.7

4. Concluding remarks

In slender composite components, a sandwich structure is usually adopted, with composite sheaths and a lightweight foam core. In wooden branches subjected to bending moments, the upper and lower sides are composed of different types of material, where the former is strong in tension and the latter strong in compression. This is not unlike the situation for reinforced concrete beams, which different designs are used for the compressive and tensile sides. In this case, the tensile side is reinforced by steel bars to compensate for the low tensile strength of unreinforced concrete. In softwood branches, a reaction wood is formed on the lower side to accommodate the compressive loads more efficiently.

Despite very similar molecular composition as the normal wood, the compression wood contributes to an astonishing increase in bending strength, estimated to several times the value for a beam that does not contain any compression wood. The main differing features of the compression wood is the more cylindrical shape of the cells and the higher microfibrillar angle, which contributes to its significant compressive properties. The wood material can be regarded as a fibre composite material. In engineered composite structures, a sandwich design is used to carry large moments, typically as two composite faces separated by a foam core. Features found in wood branches, including smoother material gradients, with varying fibre orientation and density, could be explored further for engineered composite structures. Manufacturing costs are of course a limiting factor, but evolving processing techniques could be developed to make composite structure with improved bending properties in a cost efficient way.

References

[1] Mattheck C and Tesari I 2003 The mechanical self-optimisation of trees *Institute of Physics Conference series* **180** 197–206

[2] Donaldson L A and Adya P. Singh 2016 Chapter 6 - Reaction Wood *Secondary Xylem Biology* ed Yoon Soo Kim, Ryo Funada and Adya P. Singh (Academic Press) pp 93–110

[3] Tore E. Timell 1986 *Compression Wood in Gymnosperms* (Berlin Heidelberg: Springer-Verlag)

[4] Gardiner B, Barnett J, Saranpää P and Gril J 2014 *The Biology of Reaction Wood* (Berlin, Heidelberg: Springer Berlin Heidelberg)

[5] Färber J, Lichtenegger H C, Reiterer A, Stanzl-Tschegg S and Fratzl P 2001 Cellulose microfibril angles in a spruce branch and mechanical implications **36** 5087–92

[6] Lichtenegger H, Reiterer A, Stanzl-Tschegg S E and Fratzl P 1999 Variation of Cellulose Microfibril Angles in Softwoods and Hardwoods—A Possible Strategy of Mechanical Optimization *Journal of Structural Biology* **128** 257–69

[7] Gril J, Jullien D, Bardet S and Yamamoto H 2017 Tree growth stress and related problems *J Wood Sci* **63** 411–32

[8] Badel E, Ewers F W, Cochard H and Telewski F W 2015 Acclimation of mechanical and hydraulic functions in trees: impact of the thigmomorphogenetic process *Front. Plant Sci.* **6** 1–12

[9] Brodersen C R 2013 Visualizing wood anatomy in three dimensions with high-resolution X-ray micro-tomography (μCT) – a review – *IAWA J* **34** 408–24

[10] Burgert I and Jungnikl K 2004 Adaptive Growth of Gymnosperm Branches-Ultrastructural and Micromechanical Examinations *J Plant Growth Regul* **23** 76–82

[11] J.M. Dinwoodie 2000 *Timber: Its Nature and Behaviour* (London: CRC Press Inc)

The influence of different glass fiber/epoxy matrix combinations on the durability under severe moisture impact

D Gibhardt*, C Fleschhut and B Fiedler

Hamburg University of Technology, Institute of Polymer and Composites, Denickestraße 15, 21073 Hamburg, Germany

E-mail: dennis.gibhardt@tuhh.de

Abstract. The growing success of fiber reinforced polymers (FRP) as material for the construction of high-performance lightweight structures used under maritime environmental conditions, requires foremost the knowledge about their long term durability. As the supplier market is still growing fast, methods which allow manufacturers to distinguish between suitable and less suitable composites are needed. In this study, the effects of moisture on the mechanical properties of glass fiber reinforced polymers (GFRP) using different glass fiber fabrics are investigated under several ageing and testing conditions. Focusing on the fabrics and introducing an ageing method prior to composite manufacturing, allows to describe the proportions of fiber, matrix, sizing and interphase damage to the composites durability in more detail. Absorption quantity related testing after ageing at temperatures between 8 °C and 50 °C highlights the resulting effects on the tensile strength of mostly unidirectional GFRPs. The strength of composites based on fabrics with high resistance to moisture degradation decreases steady but moderate during absorption. This effect is mainly associated with changes of matrix properties. However, less durable composites show a two-stage behavior. In this case, severe interphase damage and cracking leads to an additional drastic strength decrease when exceeding a defined amount of water absorption.

1. Introduction

The use of FRP in increasingly large, load-bearing structures like rotor blades of wind energy or tidal turbines requires detailed knowledge of their short and long-term properties to ensure safe use. Especially, years of exposure to high humidity and water, e.g. in terms of condensation or rain, can affect the material properties due to absorption related ageing [1]. Although the impact of moisture on mechanical properties of composites has been investigated for a long time, still to date, the interactions have not yet been fully described. In the past, various studies have been published reaching from short term up to 10 or 15-years in- and outdoor conditioning [2–6]. Therefore, it is known, that diffusion of water within composites affects the matrix [7], the fibers [8] and the interphase region [9–11] in different manners. But even for alike GFRPs, e.g. Starkova et al. [6] found recently that the effects of extreme long term ageing might result in entirely different material behavior and durability. In this particular example, the interlaminar shear strength of GFRP bars, which is strongly dependent on interphase properties, either did not decrease at all or decreased significantly. For amine-hardened epoxy resins used as matrix materials, absorption related effects like plasticization and swelling often weaken the polymer and increase its ductility to some extent [7]. In this context, glass fibres are often associated with stress corrosion cracking [8], whereas these are usually protected from direct contact with water in the composite. However, as most often, authors concentrate on one specific material or extreme conditions like full saturation at elevated temperatures, it is hard to find general valid conclusions. Moreover, a comparison of different studies often fails due to different conditioning circumstances. Until today methodologies to investigate the durability of new material combinations within a reasonable amount of time are not standardized. Most recommended accelerated procedures are carried out under high temperature and raise the question of whether this causes additional damage. Therefore, this study aims to improve the understanding of how different glass fiber reinforced polymers perform under harsh environmental conditions by concentrating on accelerated ageing methods in combination with the choice of fiber systems. To avoid the need to carry out time-consuming studies for each possible combination in the future, the objective is to identify generally valid correlations. In particular, this study analyzes the effects of the ageing temperature, duration and amount of absorbed water on the tensile performance in detail.

The presented study is generally based on two different parts. One presents an accelerated ageing method for fiber fabrics that allows a specific manipulation of the composites fiber-matrix interphase. The results enable to identify the impact of a moisture degraded interphase on the composites tensile properties in and transverse to the fiber direction without changing the matrix properties. The second part focuses on conventionally produced and wet-aged unidirectional composites. It aims to enhance the knowledge about correlations between mechanical performance and various ageing or environmental conditions. The key issue in this part is the incremental influence of ageing temperature, duration and amount of absorbed water on tensile properties.

2. Materials and methods

2.1. Design of experiments

As it is difficult to distinguish between the proportions of fiber, matrix and interphase damage after ageing when performing one single mechanical test, the effects have to be separated somehow. The designed ageing method for pure glass fiber fabrics by humidity and temperature provides a possibility to concentrate on interphase effects. It uses the advantage that the humidity has direct contact with the fiber surface and sizing, thus, no diffusion through the matrix is necessary. So, the method provides fast results independently of the sample dimensions. Fabrics where stored at 50 °C and 80 % rel. humidity for 35 days in a climate chamber for ageing prior to resin infusion. A pre-investigation confirmed, that the duration of 5 weeks at 50 °C is sufficient for ageing as longer time was not decreasing the composites interphase performance further. Tensile fiber bundle tests and longitudinal tensile test of pre-aged composites were used to check that there is no essential damage that decreases glass fiber strength itself. Finally, a comparison of transverse tensile tests using the new method and classical accelerated ageing by immersion in 50 °C distilled water allows conclusions about the durability of individual composite systems. The test procedure is schematically presented in figure 1.

2.2. Materials, manufacturing and specimen preparation

All GFRP composites were manufactured using a resin transfer moulding (RTM) process at 50 °C for 16 h and an additional post-curing process at 80 °C according to the manufacturers' recommendation. Three glass fiber fabrics of different manufacturers were chosen for the study. The fabrics are used in wind energy application. Their denotation and most important properties are shown in table 1. While two of them contain backing fibres in each fabric layer, the SE 2020 fabric by Saertex consists only of aligned fibres stitched together by polyester yarn. In case of

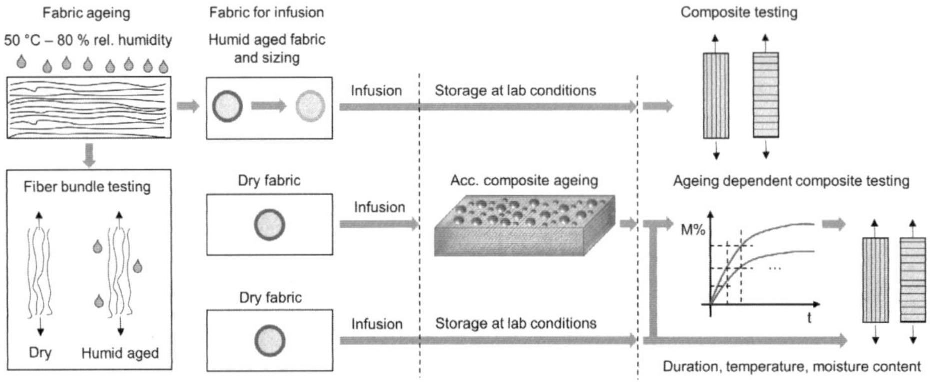

Figure 1. Methodology and test scheme for fabric and composite ageing.

the Hacotech G300U fabric, the backing fibers account for 15 % while the proportion is 10 % for the R&G 600 UD fabric. For reasons of a better comparison, results of the transverse tensile strength are presented in two ways. First, in terms of relative changes with regard to the measured strength at the knee point of the stress-strain diagram and second in absolute strength normalized to the proportion of backing fibers. This is done through micromechanics by the calculation of the stress acting in the 90-layers of the global stress acting on the total specimen cross-section. Furthermore, the thickness of the laminates is either 1 mm, 2 mm or 3 mm. The variations in thickness result from the targeted fiber volume content of 50 %, the fabric weight and the RTM frame availability. Results based on the duration are therefore normalized to the thickness. Focusing on the effects of fiber system selection, the two-component epoxy resin system consisting of Hexion EPIKOTE^{TM} Resin MGS^{TM} RIMR 135 and the amine hardener EPIKURE^{TM} Curing Agent MGS^{TM} RIMH 137 was used as matrix resin for every composite. GFRP specimen were prepared using an automatic saw with water-cooled aluminium oxide blades and subsequently checked for uniform quality. The fiber volume content φ_f was determined by burn of testing.

Table 1. Overview of the E-glass fiber fabrics structure and properties. All fabrics are sized with epoxy compatible sizes.

Supplier	*Fabric*	*Fiber*	*Areal weight*	0°	90°	*Paper denotation*
Saertex	UVE 977	SE 2020	977 g/m^2	99 %	1 %	SE 2020 UD
R&G	600 UD	—	600 g/m^2	90 %	10 %	R&G 600 UD
Hacotech	G300U	PPG 2002	373 g/m^2	85 %	15 %	PPG 2002 UD

2.3. Conditioning

In order to obtain data depending on ageing temperature, duration and water absorption, the chosen GFRP samples were aged immersed in distilled water at 8 °C, 30 °C and 50 °C, the fabrics in an environmental chamber (CTC256 from Memmert) at 50 °C at 80 % rel. humidity. Specimen were removed from conditioning for testing at defined points in time or respectively concerning the absorbed amount of moisture. This results in a crosswise comparable experimental scheme as shown exemplarily in figure 1. The presented results refer to ageing periods of up to 4000 h. Cutting edges were not coated to be able to evaluate the weight increase of single specimens.

2.4. Mechanical testing

Tensile tests in and transverse to the main fiber direction were performed with reference to DIN EN ISO 527-4 standard. Rectangular specimen of 250 mm length, 15 mm width for longitudinal and 25 mm width for transverse tests and their respective thickness were tested using a universal testing machine (Z100 from ZwickRoell) with 100 kN load cell and a mechanical extensometer. Fiber bundle tests were performed with reference to DIN EN 1007-5 standard, using a universal testing machine (Z2.5 from ZwickRoell) with 2.5 kN load cell and strain measurement via contactless video extensometer. The fiber bundles were glued to sandpaper at their gripping edges. This allowed a uniform load introduction.

3. Results and discussion

3.1. Water absorption characteristics

Diffusion characteristics of FRP are mainly determined by the absorption behavior of the matrix and the fiber-matrix interphase region [12–14]. In particular physical and chemical properties such as the polymer networks' polarity are the cause for water attraction and absorption [15]. Because all composites within this study are based on the same matrix resin, the water absorption can be analyzed in relation to the fabric used. Furthermore, the fiber volume, as well as pore and defect contents, have a direct influence on diffusion properties. Therefore, the characteristic water absorption at 50 °C is shown for the composites normalized to the fiber volume content and laminate thickness in comparison to the pure resin in figure 2(a). While the water absorption rate is almost identical, there are clear differences in the maximum content at the apparent saturation level. With about 3.57 m% (PPG), 3.12 m% (SE 2020) and 3.05 m% (R&G) the composites absorb more water in relation to the matrix proportion than the pure epoxy resin with 2.95 m%. The additional water absorption is a consequence of the absorbance due to the interphase areas, stitching yarns, as well as of pores and defects. Especially, for the PPG and the R&G composites extensive fiber-matrix debondings occurred beyond 2.40 m% weight gain and provided space for additional water uptake. In the case of the SE 2020 composite sporadic pores around the stitching yarns were found and can explain the additional uptake. These findings can be seen in section 3.4.

In figure 2(b) the water absorption at three different temperatures is exemplarily represented for the PPG 2002 composite. Elevated temperatures increase the absorption rate as well as the maximum moisture content significantly. During 170 days of ageing at 50 °C and 30 °C saturation was reached at 3.57 m% and 2.91 m% respectively. The further weight increases after reaching apparent saturation are supposed to be the result of arising space due to continued cracking and fiber-matrix separations. At 8 °C the saturation level is expected to be about 2.50 m% whereof already 2.25 m% were reached. So far, no debondings were identified.

3.2. Mechanical properties of dry, pre-aged and saturated composites

In figure 3 the bundle, longitudinal and transverse tensile strength of the fabric bundles and composites in dry condition are compared. Interestingly, only the transverse tensile strength differs significantly. While the SE 2020 and PPG composites have a sufficiently high transverse strength, this is already significantly lower for the RG composite. Thus, it can be seen that the interphase strength depends strongly on the compatibility of sizing and matrix. It should be mentioned that the given values have been mathematically corrected in the case of the fabrics with a cross-weft component in order to ensure better comparability.

Figures 4 and 5 show the effects of the fabric pre-ageing and the accelerated immersed ageing on tensile strength properties. By combining the tests, it can be demonstrated that the sizing of the fabrics degrades due to humid ageing in all cases. But as long as the matrix is dry and unaltered, this does not affect tensile strength in fiber direction. Furthermore, the pre-ageing

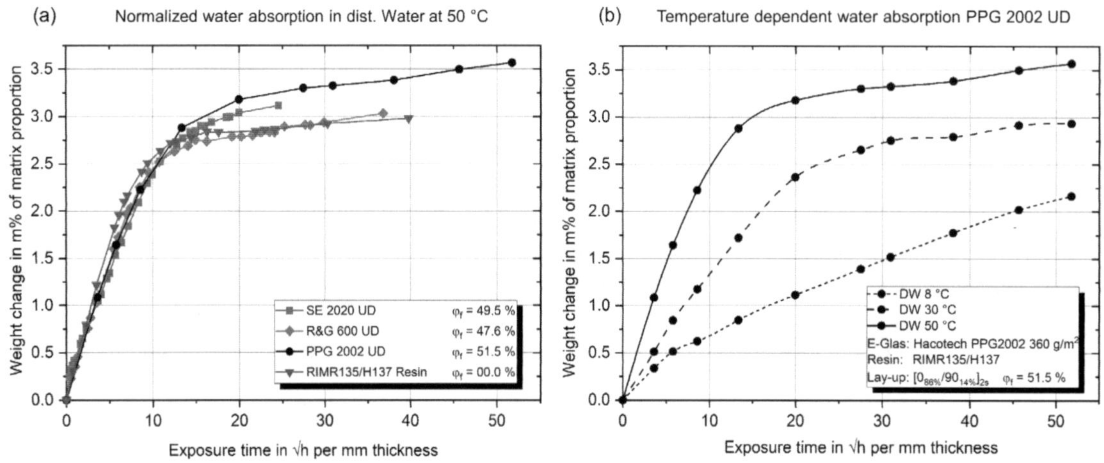

Figure 2. Water absorption characteristics of the composites and pure resin normalized to the fiber volume content immersed at 50 °C (a). Exemplary temperature dependent water absorption of Hacotech PPG 2002 fabric based GFRP immersed at 8 °C, 30 °C and 50 °C (b).

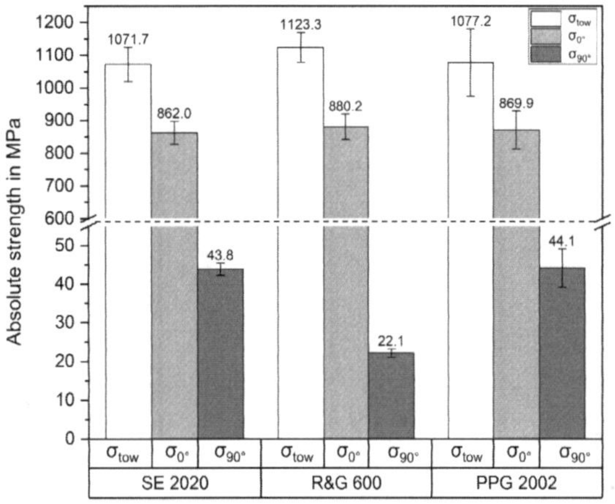

Figure 3. Absolute tow, longitudinal and transverse tensile strength of the studied fabrics and composites in dry reference condition.

indicates differences regarding the durability of the sizings, as the residual transverse strength in relation to the dry strength was found to be between 70 % in the best case and only 44 % in the worst. The wet ageing in comparison decreased the transverse strength just slightly stronger, which is likely mainly a result of a decrease in matrix properties and the additional weakening of the interphase. The differences in wet longitudinal tensile strength are also significant. Again, the SE 2020 composite performs best. Since the effects of the wet matrix should affect all systems equally, the results provide a clear indication that two overlying processes reduce the tensile strength in fiber direction. Classical ageing in a water bath not only causes plasticizing of the matrix, but also degradation within the interphase and perhaps of the fiber strength itself. Therefore, the extensive decrease of both, tensile strength in and transverse to the fiber direction, of the PPG and R&G composites is assumed to be a consequence of the more susceptible fabric sizes.

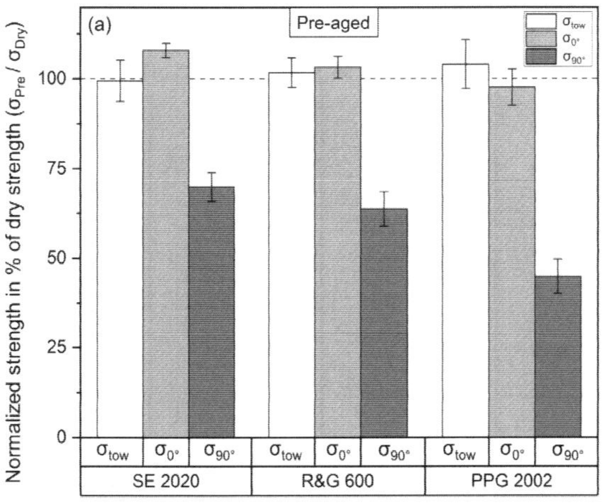

Figure 4. Normalized tow, longitudinal and transverse tensile strength of the studied fabrics and composites in pre-aged condition.

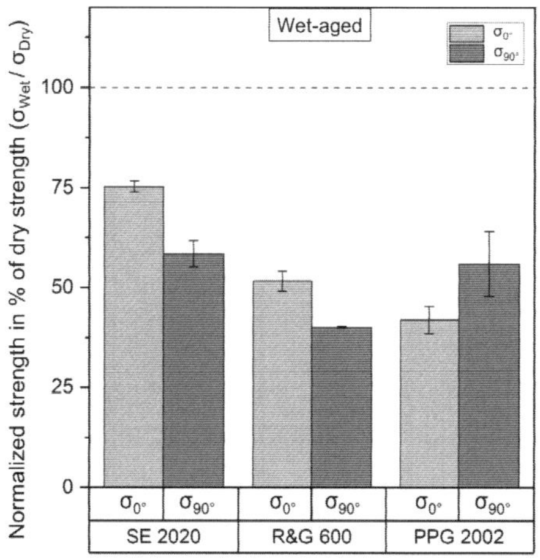

Figure 5. Normalized longitudinal and transverse tensile strength of the studied fabrics and composites in wet-aged condition.

3.3. Absorption and duration related strength properties

To improve understanding of the differences described above, the second part of the study does not only consider dry and saturated samples. In figure 6, the tensile strength in fiber direction is presented in relation to the amount of absorbed water for a total of 215 individual specimens. All specimen are categorised by their laminate type and ageing temperature. It becomes evident that there is a correlation between the increase in water content and strength. This correlation moreover shows a two-stage behavior. Up to a certain weight gain of about 2 m%, a moderate strength decrease by about 50 $\frac{MPa}{m\%}$ results almost independently of the specific composite type and ageing temperature. Above this level, the strength decrease becomes more drastic. Only in this region, significant differences between the single composites can be found. In the case of the PPG 2002 and R&G 600 based specimen, the additional strength decrease proceeds with about 200 $\frac{MPa}{m\%}$. At the same time, it was found by regular visual inspection, that local fiber-matrix debonding and cracking starts to occur within both composites as water absorption continues to increase. Finally, this results in a total strength reduction of about 50 %. The SE 2020 composite on the other hand, shows no extensive debondings or cracking. Consequently, the strength reduction rate remains moderate and leads to a total strength decrease of 25 %.

3.4. Absorption related damage behaviour

Microscopy of polished specimen cross-sections after ageing allowed to uncover the extent of damage which was introduced by water absorption. In figure 7 the damage pattern is shown representative for the R&G FRP. The defects, which can be seen in terms of opaque areas within the transparent laminate even by eye, are found to be massive cracks and delaminations. Typically, these defects arise within the densely packed areas of single fibre bundles and can reach all through a bundle. The details in figure 7 (b) and (c) show also, that fibres within the delaminated areas suffer from direct water contact. Since these kinds of defects have not been found for composites as long as the water absorption was explicitly lower than 2.4 m% extensive absorption should be avoided. Damages like this are not reversible by drying and are expected to decrease the fiber strength.

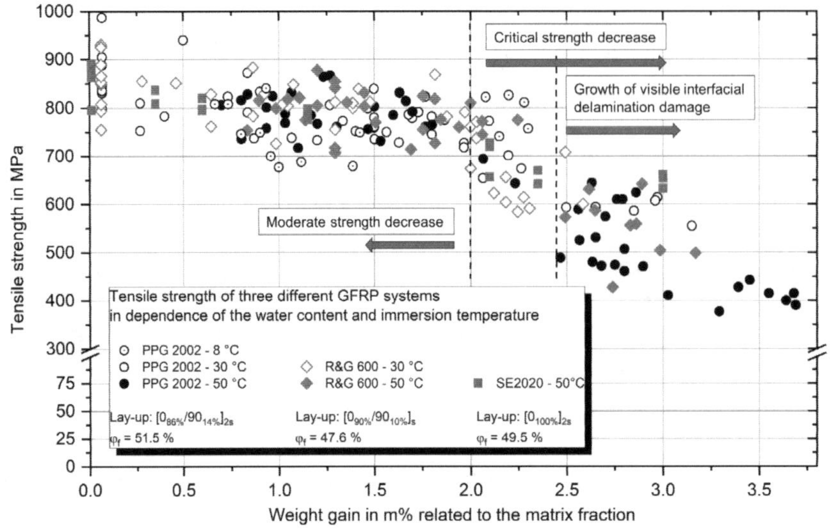

Figure 6. Interrelation between longitudinal tensile strength and amount of absorbed water in terms of matrix related mass increase. All fabrics are manufactured using the same epoxy resin.

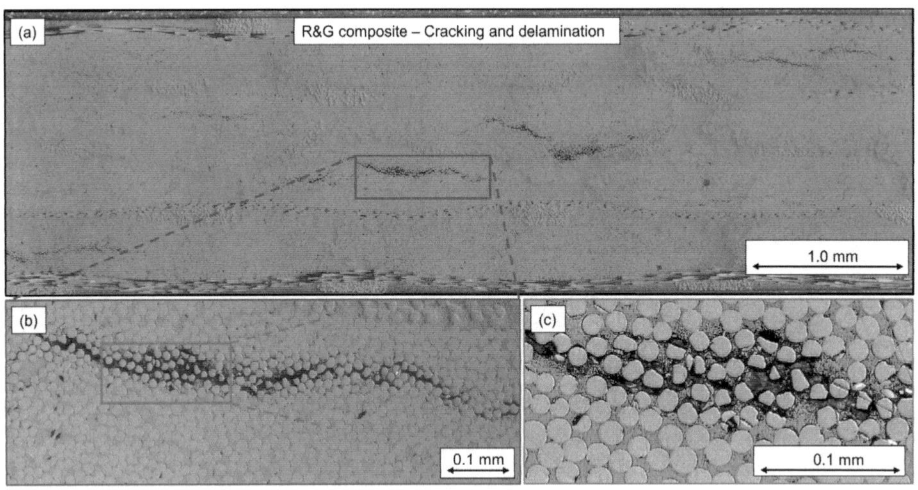

Figure 7. Exemplary microscopy of cracking and delamination damage due to water absorption for R&G composites after 4000 h in 50 °C distilled water.

4. Conlusions

In summary, the main conclusions of the presented study are:

- The introduced method of fabric conditioning allows producing composites of reduced interphase strength. But in contrast to wet-ageing methods, the matrix properties remain unchanged in this case. By combining both methods, ageing effects can now be separated. Thus, it could be demonstrated that the tensile strength in fiber direction does not change at all by weakening the interphase strength only. However, water absorption reduces the tensile strength due to plasticization and swelling stresses within the matrix and interphase. As far as the interphase damage leads to fiber-matrix debondings, the effects on the tensile strength become severe.

- A comparison of the different GFRPs investigated gives reasons to suppose that two superimposing effects characterise the durability of the composites tensile strength under marine conditions. Up to a certain level of water uptake, the strength decrease is mainly determined by the wet matrix resins characteristics. During this stage, the strength decrease is moderate and mostly independent of the specific fabric and sizing used. However, with increasing water absorption composites with sensitive fiber-matrix interphases tend to lose their integrity, which results in a drastic loss of strength. This is the main difference between marine suitable and unsuitable systems.
- The incremental study of unidirectional GFRP composites' tensile strength in fiber direction in terms of duration, temperature and amount of absorbed water led furthermore to three decisive conclusions. First, the amount of absorbed water or humidity is the main characteristic which affects the maximum tensile strength of unidirectional GFRP composites. Secondly, the ageing duration beyond saturation is less crucial, as strength was not further decreased when the apparent saturation level was reached. And thirdly, it has been shown that the only effect of elevated ageing temperatures (below T_g) seems to be, that they allow additional water to be absorbed.
- To find suitable fiber-matrix combination in terms of interfacial strength and to compare the durability of fabrics and their sizes under severe humid conditions the use of transverse tensile tests in combination with the fast fabric ageing method was found to be an efficient process.

Acknowledgments
The authors would like to thank Hexion and Saertex GmbH & Co. KG for providing materials. Furthermore, they thank Devin Meyer and Bentley Schmidt for help with the specimen preparation.

References
[1] Belec L, Nguyen T H, Nguyen D L and Chailan J F 2015 *Compos. Part A Appl. Sci. Manuf.* **68** 235–41 ISSN 1359835X
[2] R P L Nijssen 2006 *Fatigue Life Prediction and Strength Degradation of Wind Turbine Blade Composites* (Wieringswerf / Delft)
[3] Humeau C, Davies P and Jacquemin F 2018 *Compos. Part A Appl. Sci. Manuf.* **107** 94–104 ISSN 1359835X
[4] Rocha I, van der Meer F P, Raijmaekers S, Lahuerta F, Nijssen R, Mikkelsen L P and Sluys L J 2019 *Eur J Mech A Solids* **73** 407–19 ISSN 09977538
[5] Klamer E, Tromp L, de Boer A and Nijssen R 2015 *IABSE Symposium Report* **105** 1–8 ISSN 2221-3783
[6] Starkova O, Aniskevich K and Sevcenko J 2020 *Mater. Today-Proc.* ISSN 22147853
[7] Krauklis A E, Gagani A I and Echtermeyer A T 2018 *Open Eng.* **8** 447–54
[8] Ramirez F A, Carlsson L A and Acha B A 2008 *J. Mater. Sci.* **43** 5230–42 ISSN 0022-2461
[9] Thomason J L 2015 *Glass fibre sizing : a review of size formulation patents* (Glasgow: James L. Thomason) ISBN 9780957381438
[10] Gagani A I, Monsås A B, Krauklis A E and Echtermeyer A T 2019 *Compos. Sci. Technol.* **181** 107703 ISSN 02663538
[11] Gibhardt D, Doblies A, Meyer L and Fiedler B 2019 *Fibers* **7** 55
[12] Rocha I, Raijmaekers S, van der Meer F P, Nijssen R, Fischer H R and Sluys L J 2017 *Compos. Sci. Technol.* **151** 16–24 ISSN 02663538
[13] Gagani A I, Krauklis A E and Echtermeyer A T 2018 *Compos. Part A Appl. Sci. Manuf.* **115** 196–205 ISSN 1359835X
[14] Joliff Y, Rekik W, Belec L and Chailan J F 2014 *Compos. Struct.* **108** 876–85 ISSN 02638223
[15] Li L, Yu Y, Wu Q, Zhan G and Li S 2009 *Corros. Sci.* **51** 3000–06 ISSN 0010938X

IOP Conf. Series: Materials Science and Engineering **942** (2020) 012010 doi:10.1088/1757-899X/942/1/012010

Mechanical properties characterization of the Ichu fibers composites

S Charca and S Candiotti

Department of Mechanical Engineering, Universidad de Ingenieria y Tecnologia – UTEC, Medrano Silva 165, Barranco, Perú
Corresponding author: samuel.charca@upr.edu

Abstract. In the recent years, natural fibers are gaining an outstanding position as a reinforcement material, due to diverse advantages like high specific properties, low cost and low environmental impact. On the other hand, according to the market studies, in the next years the offer of the natural fibers will be in deficit; therefore, exploration of new natural fibers from new resources can be an alternative to fill this gap. Along the Andean region there are diverse natural resources from which cellulosic fibers can be extracted, like the *Stipa obtusa* grass, also commonly known as Ichu. The main purpose of this research was to determine the mechanical properties of unidirectional composite laminates made from Ichu fibers mixed with polyester resin; tensile, compressive and shear properties were evaluated. Results show that the stiffness in the principal direction (E_1 ~ 16 GPa) and the transverse direction (E_2 ~ 4.7 GPa) are over the average values compared to the composites made from the commercial fibers; similar behavior was observed in its shear and compression characteristics. Strength shows values around σ_1 ~ 118 MPa, σ_2~ 13 MPa, τ_{12}~ 46 MPa and τ_{21}~ 18 MPa, which is similar to the values reported in the literature for the commercial natural fiber composites. Based on these results, the Ichu fiber has a great potential as a reinforcement in composites materials.

1. Introduction

During the last decades, environmentally friendly processes and materials are gaining great momentum and demand [1], due to the diverse benefits they when offer compared to current man-made materials [2]; particularly in the area of composite materials, natural fibers have been successful used as a reinforcement due to their low density, excellent specific mechanical properties and low cost; likewise, the processing of this kind of fibers is less hazardous for the environment and processing equipment during their manufacturing [3]. For these reasons, natural fibers are being used in nonstructural parts in automobile and marine industry [4-6]. On the other hand, natural fibers are a combination of cellulose, hemicellulose, lignin, and other extractive components; from these, cellulose is the main component which give the natural fibers their strength and stiffness characteristics [7]. However, raw fibers have low cellulose content and generally, poor bonding characteristics with a polymeric resin matrix, if the purpose is to manufacture composites; as a consequence, in order to increase the cellulose content and improve the bonding characteristics, natural fibers would need to be treated. One of the simplest and most commonly treatment used in the industry is the mercerization; this treatment is a thermochemical process, where an alkaline solution is the commonly chemical used [8-9]. With this treatment, dimensional and thermal stability can be reached; furthermore, an increase in their density and mechanical properties is observed as well, mainly due to the increment in their relative cellulose content [8-11]. Raw natural fibers are naturally hydrophilic with high surface energy, this give to the raw fibers a poor bonding characteristic with synthetic resins (nonpolar matrix); however, after the alkali treatment, substantial improvement in the compatibility to the resin can be observed [12-13].

In recent years, different sources of natural fibers are being studied, with one of those being the fibers extracted from Ichu plant. Ichu is a grass type plant, which can grow above 3000 m.a.s.l. This plant was commonly used as a rope and roofing material; however, with the development of new materials their uses have been reduced. Some recent studies presented an optimal extraction method of the Ichu fibers and their mechanical properties, showing potential to be use as a reinforcement in composites materials [11, 14-15]. Furthermore, this fiber has a particular characteristic: In their natural state, raw fibers are hydrophobic (low surface energy), due to the presence of waxes and other extractives on the fiber surface; after the alkali treatment these components are removed and, while the surface energy increases, the fiber maintained their hydrophobicity [11, 14]. The main purpose of this research is to evaluate the tensile, compressive, shear and bending mechanical properties of unsaturated polyester composites reinforced with extracted Ichu fibers; considering the fiber longitudinal and transverse direction.

2. Materials and methodology

Raw Ichu fibers were harvested from Cusco (Peru) at 3600 m.a.s.l. The fiber extraction process was performed by thermochemical treatment, using a solution of 1.5 M NaOH diluted in distilled water at 70 °C, for a treatment of 45 minutes; with these parameters being established by Mori et al and Tenazoa et al as the optimal, to bring the best physical properties and chemical characteristics of the fibers [14-16], although, based in their last studies, treatment with lower NaOH solution can reach equivalent results, however it requires a prolonged time of treatment [11, 17]. After treatment, the fibers were cleaned and dried, making sure to keep the axis of the fibers in one direction to form a unidirectional matt; and afterwards, were maintained in a desiccator to avoid the moisture intake. To manufacture composites, vacuum assisted resin transfer molding (VARTM) technique was used, and to make economically feasible the use of this fiber as a reinforcement, an unsaturated polyester resin was used (Tekno 285, $\sigma_m \sim 102 \pm 20.11\ MPa, E_m \sim 3.13 \pm 0.17\ GPa, \varepsilon_m \sim 3.3 \pm 0.89\ \%$ tested in three point bending according to ASTM D790 [18]). In order to get a 45 minutes delay before the resin begins to cure, an appropriate mixing proportion of resin, catalyzer (Methyl Ethyl Ketone Peroxide, MEKP) and accelerator (Cobalt Octoate) were set. After composites were made, laminates were post cured in an oven for 12 hours at 60 °C; afterwards, the testing specimens were cut using a water jet cutting system, in dimensions according to ASTM D3039 [19], D3410 [20], D5379 [21] and D7264 [22] standards.

In order to measure the deformation of the samples, a digital image correlation (DIC) technique was used. The deformation process was recorded using a Manta G505 2/3" Monochrome CCD Camera (Allied Vision) with AF Micro-Nikkor 60mm f/2.8D lenses. In order to validate results, strain gauges were glued to one face of the samples, and the other face was painted with white color (basecoat) followed by fine black speckles. Data acquisition was synchronized appropriately in order to get a reliable result. Images obtained were processed using the open source 2D DIC MATLAB Software Ncorr v1.2. Samples were tested using a MTS Landmark system with a 100 kN load cell. According to the standards, tensile, compressive and bending tests were performed at a 0.01 strain/min rate, and for the shear test a crosshead displacement rate of 2 mm/min was used. Finally, in order to determine the fiber volume fraction of the composites, optical microscopy was used on a polished cross-sectional surface of specimens obtained from the laminates.

3. Results and analysis

3.1. Testing results

From the measurement, fiber volume fraction on the composites was determined to be $v_f = 0.36 \pm 0.020$. On the other hand, Figure 1 to Figure 4 shows the typical stress-strain curves observed during the tensile, compressive, shear and bending tests. For tensile test (Figure 1), curves show linear behavior, with brittle failure in the end, for the longitudinal (axial) and transverse direction of the fibers. Contrary to the tensile test, compressive behavior is non-linear (Figure 2), for the longitudinal and transverse direction of the fibers, mainly due to the mechanics of failure, where debonding and micro buckling happens; and afterwards, the overall bucking and development of these processes is progressive. Shear stress-strain curve present a non-linear behavior along the 12 direction (Figure 3(a)), as when the load is transversely applied to the axial direction of fibers, the principal failure mode is the debonding of the fiber/matrix interface near the notch, followed by the transversely deformation of the fibers. On the other hand, if the load is applied along the 21 direction, aligned to the fiber longitudinal axis, is the interface who fails upon reaching their maximum capacity with a sudden failure (Figure 3(b)). Bending stress-strain curve shows a slight non-linearity for the longitudinal fiber direction (Figure 4(a)), with sudden failure at the end, while the transverse stress-strain curve is perfectly linear with sudden failure at the end as well.

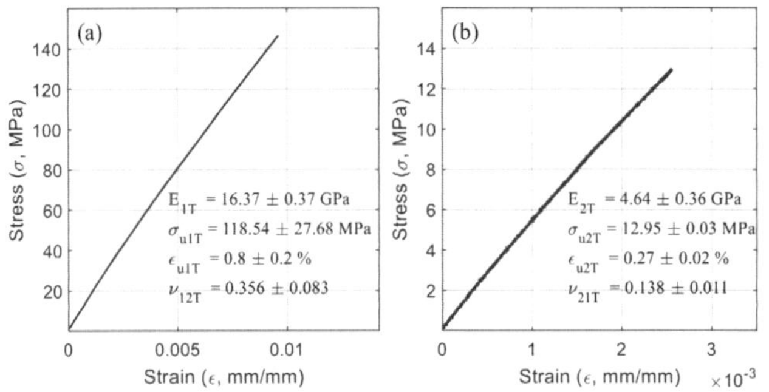

Figure 1. Typical tensile stress-strain curve for the Ichu laminate composites, a) longitudinal direction, b) transverse direction.

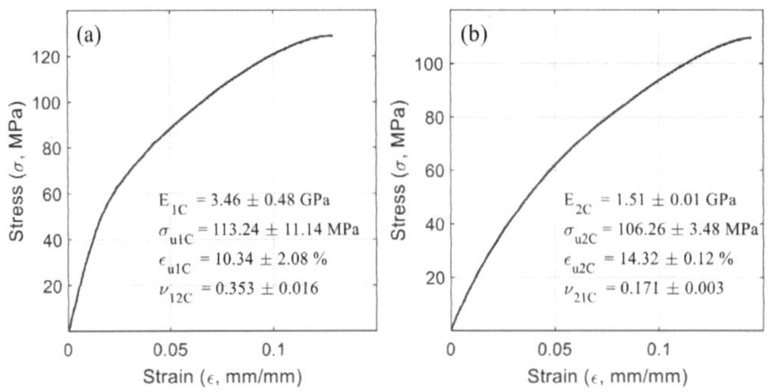

Figure 2. Typical compressive stress-strain curve for the Ichu laminate composites, a) longitudinal direction, b) transverse direction.

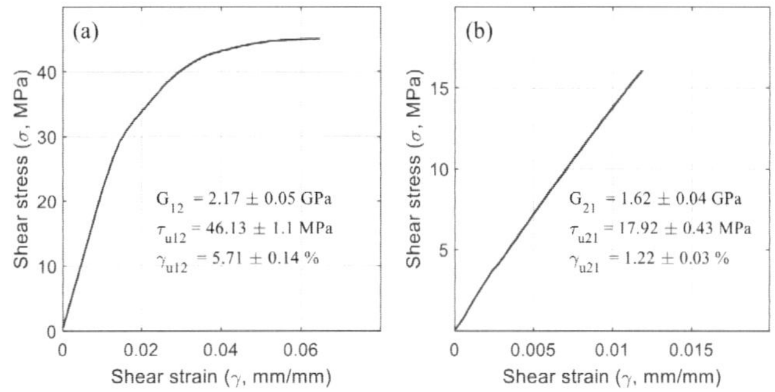

Figure 3. Typical shear stress-strain curve for the Ichu laminate composites, a) load applied transverse to the fibers direction, b) load applied in fibers direction.

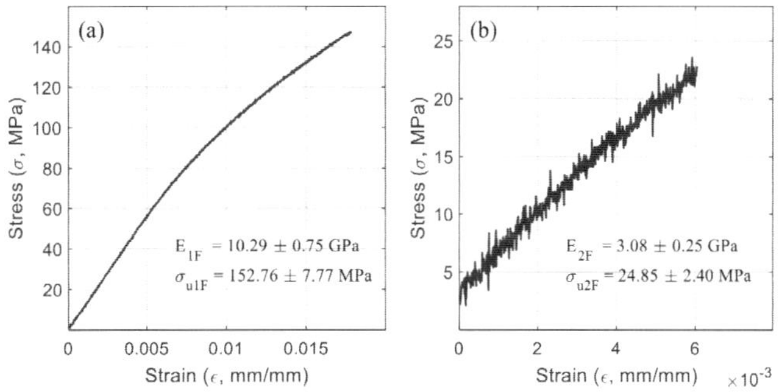

Figure 4. Typical bending stress-strain curve for the Ichu laminate composites, a) longitudinal direction, b) transverse direction.

3.2. Comparative results analysis

Comparative charts (Figure 5 to Figure 8) show the results obtained in this study compared to results found in the literature [23-34]; unfortunately, in the case of transverse testing results there is no data available in the reviewed literature. Furthermore, it is well known that epoxy resin is excellent material to work with natural fibers; however, the main disadvantage is their cost, which is considerable high compared to the unsaturated polyester (UP); so, in order to have a general comparison, composites made with epoxy resin are include as well. Volume fraction (v_f) of the fibers into the composites is reported besides the fiber/matrix used, since it can be variable from literature to literature, due to the natural fibers characteristics and the method used to manufacture composites.

Composites made with Ichu and UP (v_f = 0.36 ±0.020) present tensile strength over 100 MPa (Figure 5), which can be equivalent to the composites made with jute/epoxy (v_f = 0.5), flax/vinyl (v_f = 0.25), jowar/UP (v_f = 0.4), bamboo/UP (v_f = 0.4) and jute/UP (v_f = 0.4); furthermore, their strength is higher than sisal/UP (v_f = 0.4) and kenaf/PLA (v_f = 0.4) composites. For the stiffness, it is evident that modulus is considerable higher than jute/epoxy (v_f = 0.5), flax/vinyl (v_f = 0.25), jovar/UP (v_f = 0.4), sisal/UP (v_f = 0.4), bamboo/UP (v_f = 0.4), jute/UP (v_f = 0.4), and kenaf/PLA (v_f = 0.4) composites; and equivalent to the harakeke/epoxy (v_f = 0.5), sisal/epoxy (v_f = 0.48) and hemp/epoxy (v_f = 0.65) composites.

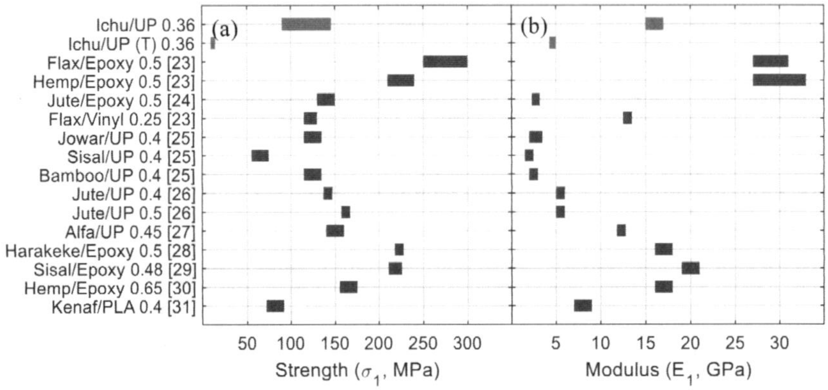

Figure 5. Comparative strength and stiffness for different natural fibers composites under tensile test, a) longitudinal strength and b) longitudinal modulus.

Longitudinal compressive strength and modulus comparisons are shown in the Figure 6. The compressive strength (Figure 6(a)) is quite variable; however, their values are close to the results for Flax/Vinyl (v_f = 0.25), flax/epoxy (v_f = 0.4) and coir/epoxy (v_f = 0.4); nevertheless, lower than the bamboo/epoxy (v_f = 0.4) composites. Modulus in the other hand (Figure 6(b)), is considerable low compared to the flax/vinyl (v_f = 0.25), flax/epoxy (v_f = 0.4) and bamboo/epoxy (v_f = 0.4); however, similar to the values obtained for coir/epoxy (v_f = 0.4). At the bottom of the Figures 6 (a) and (b), Ichu/UP (T) correspond to the transverse direction properties.

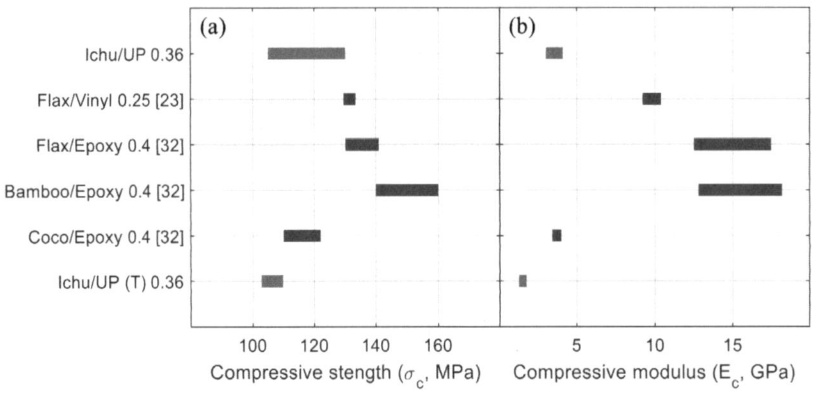

Figure 6. Comparative strength and stiffness for different natural fibers composites under compression test, a) longitudinal strength and b) longitudinal modulus.

Shear strength and shear modulus comparative chart is shown in Figure 7. Unfortunately, the information of shear properties available in the literature for the unidirectional composites reinforced with natural fibers is quite limited; although there are results for strand mat and for woven fibers composites [33]; which are included in the chart as well. From the results, τ_{12} is over 45 MPa; however, τ_{21} are considerable low around 17 MPa. At the bottom of the Figures 7 (a) and (b), Ichu/UP (T) correspond to the transverse direction properties.

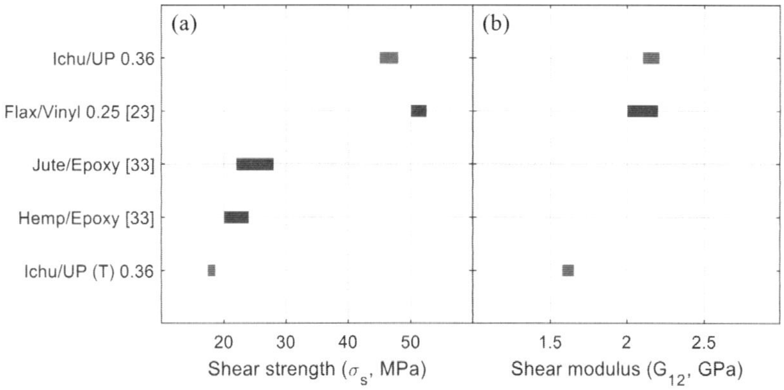

Figure 7. Comparative strength and stiffness for different natural fibers composites under iosipescu shear test, a) shear strength and b) shear modulus.

Bending strength and stiffness comparison chart is shown in the Figure 8. The bending strength of Ichu/UP composites is higher than Jowar/UP, Sisal/UP, Bambu/UP and Kenaf/PLA; however, lower than Flax/Epoxy, Hemp/Epoxy, Jute/Epoxy, Flax/Vinyl. Harakeke/Epoxy and Flax/PP composites, although, in these cases, the fiber v_f is higher than the Ichu/UP composites. On the other hand, bending stiffness of the Ichu composites shows values around 10 GPa, which is higher to the values observed for composites made using unsaturated polyester; however, lower for most of the composites made with epoxy as a matrix. At the bottom of the Figures 8 (a) and (b), Ichu/UP (T) correspond to the transverse direction properties.

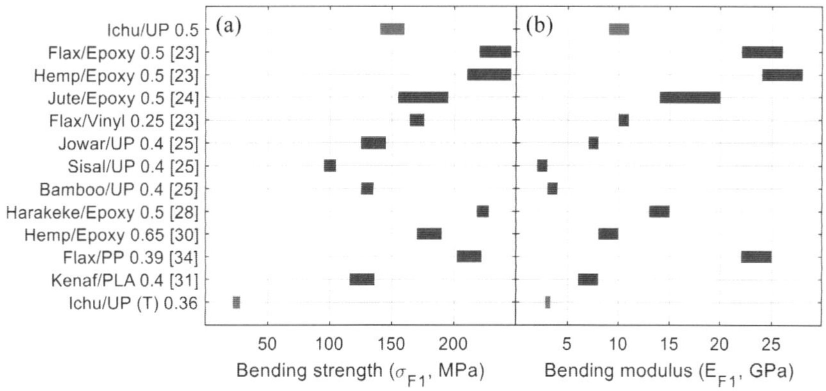

Figure 8. Comparative strength and stiffness for different natural fibers composites under flexural test, a) longitudinal strength and b) longitudinal modulus.

4. Conclusion

Ichu fiber composites were manufactured using the VARTM technique, reaching fiber volume fraction around 0.36; furthermore, the strain stress curves from mechanical testing were built, including: tensile, compressive, shear and flexural tests. The composites show a linear and brittle behavior for tensile and flexural test, like most of the results reported in the literature. On the other hand, comparative analysis of the mechanical properties to the literature, shows that Ichu fiber composites have properties above the average when compared to the composites made with unsaturated polyester resin; and furthermore, their properties can be equivalent and superior to some composites made using epoxy resins.

5. Acknowledgment

The authors wish to thank the Cienciactiva (Perú, contract number 117-2016) and the UTEC Seed fund 2019, for providing the funds to conduct this research.

6. References

[1] Adekomaya O, Jamiru T, Sadiku R and Huan Z 2017 Negative impact from the application of natural fibers. *J. Clean. Prod.* 143 843-846

[2] Joshi SV, Drzal LT, Mohanty AK and Arora S 2004 Are natural fiber composites environmentally superior to glass fiber reinforced composites? *Compos. Part A Appl. Sci. Manuf.* 35(3) 371-376

[3] Pickering KL, Efendy MA and Le TM 2016 A review of recent developments in natural fibre composites and their mechanical performance *Compos. Part A Appl. Sci. Manuf.* 83 98-112

[4] Mohammed L, Ansari MN, Pua G, Jawaid M and Islam MS 2015 A review on natural fiber reinforced polymer composite and its applications *Int. J. Polym. Sci.*

[5] Verma D and Sharma S 2017 Green biocomposites: a prospective utilization in automobile industry *Green Biocomposites (Green Energy and Technology)* ed Jawaid M et al (Springer) pp. 167-191.

[6] Castegnaro S, Gomiero C, Battisti C, Poli M, Basile M, Barucco P, Pizzarello U, Quaresimin M and Lazzaretto A 2017 A bio-composite racing sailboat: Materials selection, design, manufacturing and sailing *Ocean Eng.* 133 142-150

[7] Komuraiah A, Kumar NS and Prasad BD 2014 Chemical composition of natural fibers and its influence on their mechanical properties *Mech. Compos. Mater.* 50(3) 359-376

[8] Boopathi L, Sampath PS and Mylsamy K 2012 Investigation of physical, chemical and mechanical properties of raw and alkali treated Borassus fruit fiber *Compos. Part B Eng.* 43(8) 3044-3052

[9] Das M and Chakraborty D 2008 Evaluation of improvement of physical and mechanical properties of bamboo fibers due to alkali treatment *J. Appl. Polym. Sci.* 107(1) 522-527

[10] Borchani KE, Carrot C and Jaziri M 2015 Untreated and alkali treated fibers from Alfa stem: effect of alkali treatment on structural, morphological and thermal features *Cellulose* 22(3) 1577-1589

[11] Mori S, Charca S, Flores E and Savastano Jr H 2019 Physical and Thermal Properties of Novel Native Andean Natural Fibers *J. Nat. Fibers* 1-17

[12] Doan T-T-L, Brodowsky H and Mäder E 2012 Jute fibre/epoxy composites: Surface properties and interfacial adhesion *Compos. Sci. Technol.* 72(10) 1160-1166

[13] Heng JY, Pearse DF, Thielmann F, Lampke T and Bismarck A 2007 Methods to determine surface energies of natural fibres: a review *Compos. Interfaces* 14(7-9) 581-604

[14] Mori S, Tenazoa C, Candiotti S, Flores E and Charca S 2018 Assessment of Ichu fibers extraction and their use as reinforcement in composite materials *J. Nat. Fibers* 1-16

[15] Tenazoa C, Savastano H, Charca S, Quintana M and Flores E 2019 The Effect of Alkali Treatment on Chemical and Physical Properties of Ichu and Cabuya Fibers *J. Nat. Fibers* 1-14

[16] Mori S, Charca S and Flores E 2018 Ichu: New natural fibers for composites and its extraction methodology *Non-Conventional Materials and Technologies – NOCMAT for the XXI Century* ed Ghavami K and Herrera Franco PJ (Material Research Forum) pp. 764-774.

[17] Candiotti S, Mantari JL, Flores E, Charca S, 2020, Assessment of the mechanical properties of peruvian Stipa Obtusa fibers for their use as reinforcement in composite materials, *Compos. Part A Appl. Sci. Manuf.* 135.

[18] ASTM D790-15 2015 Standard Test Methods for Flexural Properties of Unreinforced and Reinforced Plastics and Electrical Insulating Materials *ASTM International*

[19] ASTM D3039/D3039M-00 2000 Standard Test Method for Tensile Properties of Polymer Matrix Composite Materials *ASTM International*

[20] ASTM D3410/D3410M-03 2003 Standard Test Method for Compressive Properties of Polymer Matrix Composite Materials with Unsupported Gage Section by Shear Loading *ASTM International*
[21] ASTM D5379/D5379M-98 1998 Standard Test Method for Shear Properties of Composite Materials by the V-Notched Beam Method *ASTM International*
[22] ASTM D7264/D7264M-07 2007 Standard Test Method for Flexural Properties of Polymer Matrix Composite Materials *ASTM International*
[23] Crosky A, Soatthiyanon N, Ruys D, Meatherall S and Potter S 2014 Thermoset matrix natural fibre-reinforced composites *Natural Fibre Composites Materials, Processes and Applications* ed Hodzic A and Shanks R (Woodhead Publishing) chapter 9 pp. 233-270
[24] Mishra HK, Dash BN, Tripathy SS and Padhi BN 2000 A study on mechanical performance of jute-epoxy composites *Polym. Plast. Tech. Eng.* 39(1) 187-198
[25] Prasad AR and Rao KM 2011 Mechanical properties of natural fibre reinforced polyester composites: Jowar, sisal and bamboo *Mater. Des.* 32(8-9) 4658-4663
[26] Laranjeira E, De Carvalho LH, De L. Silva SM and d'Almeida JRM 2006 Influence of fiber orientation on the mechanical properties of polyester/jute composites *J. Reinf. Plast. Compos.* 25(12) 1269-1278
[27] Brahim SB and Cheikh RB 2007 Influence of fibre orientation and volume fraction on the tensile properties of unidirectional Alfa-polyester composite *Compos. Sci. Technol.* 67(1) 140-147
[28] Le TM and Pickering KL 2015 The potential of harakeke fibre as reinforcement in polymer matrix composites including modelling of long harakeke fibre composite strength *Compos. Part A Appl.Sci. Manuf.* 76 44-53
[29] Oksman K, Wallström L, Berglund LA and Filho RDT 2002 Morphology and mechanical properties of unidirectional sisal–epoxy composites *J. Appl. Polym. Sci.* 84(13) 2358-2365
[30] Islam MS, Pickering KL and Foreman NJ 2011 Influence of alkali fiber treatment and fiber processing on the mechanical properties of hemp/epoxy composites *J. Appl. Polym. Sci.* 119(6) 3696-3707
[31] Graupner N and Müssig J 2011 A comparison of the mechanical characteristics of kenaf and lyocell fibre reinforced poly (lactic acid) (PLA) and poly (3-hydroxybutyrate) (PHB) composites *Compos. Part A Appl.Sci. Manuf* 42(12) 2010-2019
[32] Van Vuure AW, Baets J, Wouters K and Hendrickx K 2015 Compressive properties of natural fibre composites *Mater. Lett.* 149 138-140
[33] Fajrin J and Sari NH 2018 Shear properties evaluation of natural fibre reinforced epoxy composites using V-notch shear test *MATEC Web Conf.* 195
[34] Van de Velde K and Kiekens P 2003 Effect of material and process parameters on the mechanical properties of unidirectional and multidirectional flax/polypropylene composites *Compos. Struct.* 62(3-4) 443-448

Blade materials selection influence on sustainability: a case study through LCA

G Chiesura [1], H Stecher [2] and J Pagh Jensen [3]

[1] Design Engineer – Blade Design Dept. – Siemens Gamesa Renewable Energy A/S, Assensvej 11, 9220 Aalborg, Denmark, Gabriele.Chiesura@siemensgamesa.com
[2] Blade Materials Engineer – Blade Design Dept. – Siemens Gamesa Renewable Energy A/S, Assensvej 11, 9220 Aalborg, Denmark, Harald.Stecher@siemensgamesa.com
[3] EHS Specialist – Health, Safety & Environment Dept. – Siemens Gamesa Renewable Energy A/S, Assensvej 11, 9220 Aalborg, Denmark, Jonas.Jensen@siemensgamesa.com

Abstract. In this study, the environmental influence that blade materials have on the entire wind turbine lifecycle is considered. In order to quantify and compare the effects that a specific material's choice has with respect to one another, Life-Cycle Assessment (LCA) tools are used. Specific focus is addressed on aspects related to the wind turbine rotor-blade and how that affects the whole turbine. The blade lifecycle is divided into five main stages: (1) raw material sourcing, (2) blade manufacturing, (3) blade installation, (4) Operation and Maintenance (O&M) and (5) End-of-Life (EoL). Materials sourcing is found responsible for the largest contribution in Green House Gas (GHG) emission in the turbine and blade lifecycle. Particularly, a case study on a representative blade production in Siemens Gamesa Renewable Energy (SGRE) is disclosed and compared to a hypothetical blade production where a recyclable resin system is being used. The lifecycle assessment shows a 28% reduced amount of GHG emissions when the recyclable resin system is employed. By employing such resin system most of blade materials could be separated and re-used at the blade EoL, enabling a 90% blade recycling rate. An even greater advantage of achieving blade circularity is therefore envisioned.

1. Introduction

Compared to fossil-fuel power plants, renewable-energy power plants are considerably less impactful in terms of the environmental footprint and the equivalent CO2 released into the environment. During its lifetime, a wind power plant emits less than 1% of the CO2 emitted per kWh by an average power plant using fossil fuels [1]. This in itself represents a strong reason to transition from non-renewable to renewable energy sources. Renewable energy sources are often also considered sustainable energy sources, because their intrinsic ability to replenish (or renew) themselves naturally, allows for an endless supply of energy that can meet the needs of the present demand without compromising the ability of future generations to meet even an higher demand [2]. However, this definition does not consider the energy conversion technology (i.e. the wind turbine generator) that is used to transform the renewable energy source into the final product, being the electricity. For instance, aspects like the consumption of resources used to produce and deploy the wind turbine, the disposal of the wind turbine and its induced environmental impact should also be considered. To evaluate the relative burden of energy systems within the environment, full energy supply chains need to be considered on a lifecycle basis, including all system components, and across all impact categories [3]. To properly account for all these aspects, Life-Cycle Assessment (LCA) tools are used to quantify the footprint that a conversion technology has on the environment [4]. The most used quantity in LCA to define the environmental footprint of a conversion technology is the equivalent CO2 (CO2eq) which is directly related to Green House Gas (GHG) emissions [3]. Additionally, such assessments need to address a wide range of issues, such as land and water use, as well as air, water, and soil pollution, which are often location specific. To have a holistic view on the environmental footprint of a conversion technology, all stages in its lifecycle need to be identified and properly accounted, using a "cradle-to-grave" so to say approach. To fully prove that a new conversion technology has a lower footprint with respect to a more traditional technology (e.g. offshore wind turbine technology vs. coal power plant technology) all implications that come with the new technology should be addressed. This can often be a complicated task and a certain amount of assumptions needs to be made.

When analysing the lifecycle of a wind power plant, raw materials sourcing is considered responsible for most of the total CO2eq emissions [4]. This accounts for the extractions of raw constituents such as minerals, wood and oils, the processing to produce the raw materials (i.e. concrete, copper, steel, glass fibres, resin, lubricants), the transportation of those materials to the production site and the waste production associated with them. These emissions can, to a large extent, be re-gained by recycling and indeed the current practice for wind turbine materials is to recycle as many of the components as possible. To date, up to 85-90% of the turbine components can be recycled [5]. However, not all the turbine components can be recycled. The rotor blade and part of the nacelle canopy are produced out of thermoset based composite materials. To date, no well-established best-practice for recycling of thermoset composites materials exists on industrial scale [6]. This is due on the one hand because no clear legislation is yet in place on circularity for composite applications and on the other hand because the volumes of composite waste coming from different applications are still relatively small and discontinuous, making it difficult to create a solid business around recycling. Most of the composite blade waste is sent to landfill, or in some cases the blades are shredded into pieces and incinerated. In the latter case, an energy recovery from burning the composite organic fraction is obtained and the remaining glass fibres are used as fillers in cement production or other secondary applications [1].

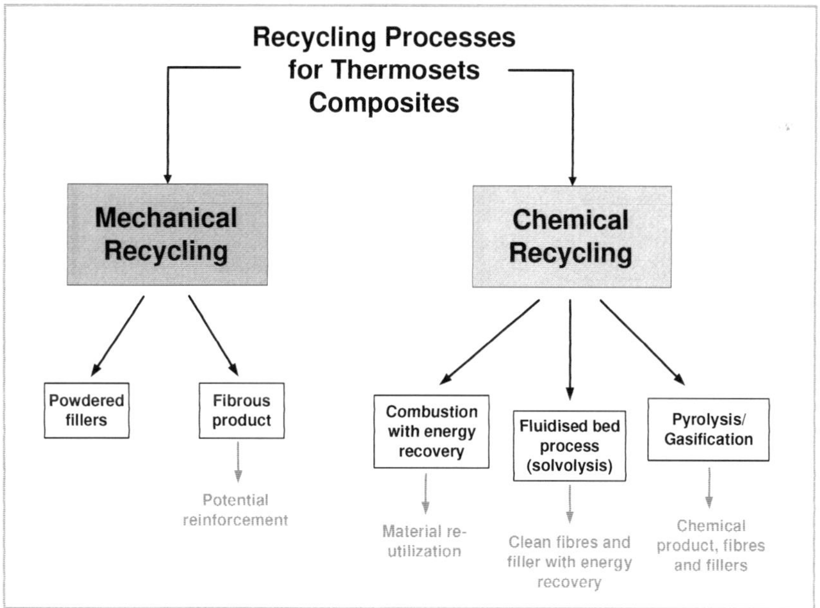

Figure 1. Recycling processes for composites.

Figure 1 shows the currently existing methods for recycling of thermosets composite materials. Two major classes of recycling methods exist: (i) mechanical recycling where composites are shredded into pieces and either used as fillers or short fibres for other applications and (ii) chemical recycling where the fibre fraction is recovered, or an energy recovery is obtained through combustion of the organic composite fraction [7].

The European Composites Industry Association (EuCIA) agreed with the European Commission that co-processing of composite material in cement kilns is compliant with the European Waste Framework Directive 2008/98/EC and is considered both recycling and energy recovery [8]. However, this recycling/recovering method cannot be categorized as re-use and therefore it fails to fulfil the requirements for circularity. Other technological methods exist to recycle thermosets composites. The two major ones that rely on thermal processing of the material are pyrolysis and solvolysis. Pyrolysis is the thermal decomposition of materials at elevated temperatures in an inert atmosphere. It involves a change of chemical composition. The main drawbacks of this process are that it is intensively energy demanding and due to its high temperatures, it can only be used for glass fibres reinforced composites and not for carbon fibres that would undergo combustion. Solvolysis is a chemical depolymerisation that involves using a reagent to decompose the polymer matrix. This process largely depends on the type of polymer that is dissolved and generally it does not involve too high temperature (e.g. 200°C) to deteriorate the fibre fraction in the composite. The recovery process allows for reuse of the fibre reinforcement fraction usually in lower composite applications, whereas the resin fraction is degraded with the solvent and therefore lost. As a drawback, this process creates a certain amount of hazardous waste that still need to be dissolved or disposed.

Some research projects investigated the possibility of using blade sections as roofing for affordable housing but this will only be possible for limited applications and cannot be considered as a valuable option for circularity [9]. Because of the decarbonization and energy transition plans ongoing worldwide, the forecast for blade waste production are expected to increase exponentially in the decades to come to reach in the lowest case scenario a 43 Mt of cumulative blade waste by 2050 [10]. Therefore, blade circularity will be a strategic area of development within the industry in the next decade [5]. It becomes clear to the authors that a technological method to create circularity in the wind energy composite sector is necessary.

2. LCA study on SG 8.0-167 DD turbine lifecycle

To first understand where the most environmental values lies in the whole turbine lifecycle a reference LCA study was performed on one of the SGRE product the 8MW offshore wind turbine with 167m rotor diameter and 81m composite blade length [11]. This study will then be used as baseline when the LCA impact of a new technology needs to be assessed.

2.1. LCA Assumptions

The blade lifecycle is been divided into five main stages: **(i)** raw material sourcing, **(ii)** blade manufacturing, **(iii)** blade installation, **(iv)** Operation and Maintenance (O&M) and **(v)** End-of-Life (EoL) [1]. **(i)** Main types and quantities of materials and energy that had to be extracted and consumed to produce the turbine components and the elements needed to connect the wind power plant to the grid, i.e. substations and connecting cables were considered. **(ii)** For the manufacturing, data from Siemens Gamesa Renewable Energy's (SGRE) own production sites and from main suppliers were collected. Consumption data for manufacturing as well as waste and subsequent treatment are based primarily on annual manufacturing data from European production sites. **(iii)** Transport of materials to the manufacturing site is included in the data. Components, auxiliary resources, and workers are transported to the wind turbine site during the turbine installation. On-site installation includes preparing the site, erecting the turbines and connecting the turbines to the grid. These installation activities result in the consumption of resources and production of waste. Associated data has been collected from actual on-site installations. **(iv)** The SG 8.0-167 DD turbine is designed to a 25 years lifetime. Actual site data, including manpower, materials, and energy required for service and maintenance over the turbine's lifetime were collected. Wake, availability, and electrical losses have been included in the assessment to define a realistic estimate of annual energy production delivered to the grid. **(v)** At the wind power plant's EoL the components will be disassembled and the materials transported and treated according to different waste handling methods. For the turbine components, recycling was assumed to all recyclable material e.g. metals. Recycling leads to the recovery of materials, which subsequently reduces primary material extraction. The rest of the materials are either thermally treated or disposed in landfills. The EoL stage described here represents the current status of waste management options in Northern Europe.

Planning a new wind power plant includes assessing the environmental impact of the installation and operation phases to minimize negative impacts. Often these assessments focus on birds, marine wildlife and visual impacts. How a wind power plant impacts its surroundings varies depending on its location. Being this site-specific, it was decided not to consider it in the LCA assessment.

2.2. Breakdown of CO2eq per turbine lifecycle stage
When breaking down the CO2 emissions at each stage of the turbine lifecycle, it results immediately evident that the raw materials sourcing is responsible for more than 70% of the total CO2eq. This is understandable as most of the emissions are embodied within the materials that constitutes the turbine. The CO2eq/kWh breakdown per turbine lifecycle stage is depicted in Figure 2.

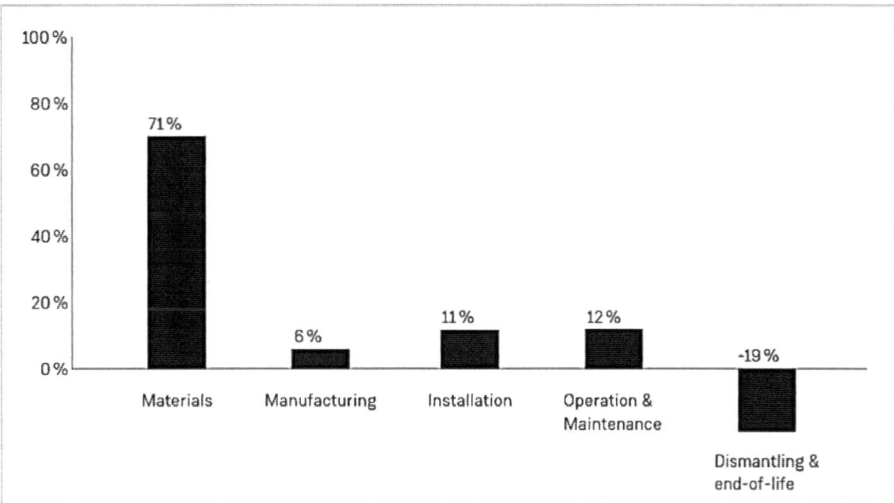

Figure 2. Breakdown of CO2eq/kWh per turbine lifecycle stage. Environmental Product Declaration SG 8.0-167 DD [1].

When a turbine is dismantled, it has not necessarily reached its EoL. Turbines are often replaced by larger turbines, allowing the dismantled turbines to be refurbished and sold for installation elsewhere. When disposing of wind turbines, recycling is usually the preferred solution. This not only prevents the materials from being sent to landfills, but also reduces the need for the extraction of primary materials. Therefore, a negative CO2eq impact is seen at the turbine EoL and dismantling stage in Figure 2. As already discussed in the introduction, metallic materials in the wind power plant components are to a great extent recycled at their EoL. Magnets from the direct drive turbines can be demagnetized, remagnetized and used or reused for new magnet production. Blades may be shredded and incinerated for energy recovery. The residues from fiberglass incineration can be used in other secondary applications e. g. for cement production. This process would not be allowed for carbon fibres composite blades and in both cases this process doesn't allow to reuse the materials. A better recycling strategy for blade materials seems therefore necessary.

2.3. Blade materials influence on turbine LCA
To better understand where the major CO2 contribution on blade materials lies and therefore decide which recycling strategy to favour, a blade materials LCA study limited to the CO2 equivalent emissions was performed. A normalized CO2eq. profile for each of the most common blade materials is presented in Figure 3. The carbon fibres impact profile is very significant compared to the remaining materials, underpinning the need for defining a suitable recycling strategy that would enable its reuse. This becomes even more relevant as the design drivers for larger offshore rotor blades will require employing an increased amount of carbon fibres in the blade bill of materials [12].

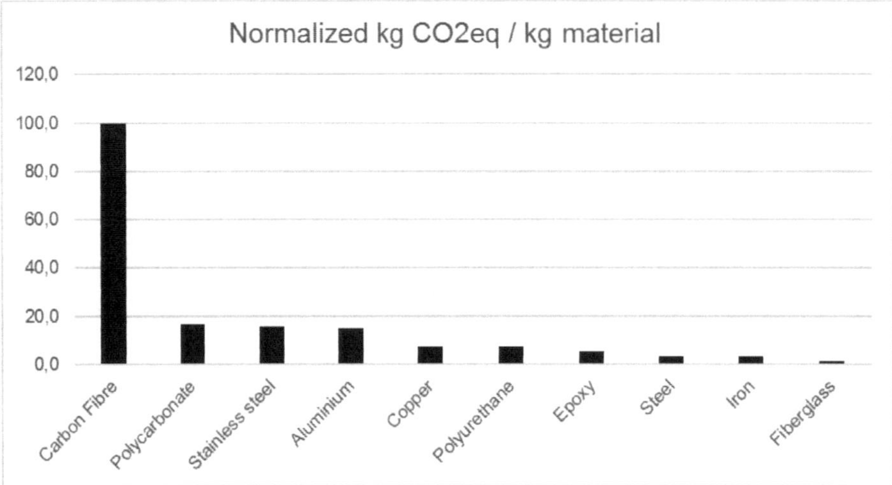

Figure 3. Wind turbine blade materials breakdown for a SG 8.0-167 DD turbine.

Figure 3 alone cannot give the full overview on why not only carbon fibres, but also glass fibres and Epoxy resin materials are important element to consider when defining an appropriate recycling strategy. In fact, the two materials combined still represent more than 80% in weight of the blade bill of materials. Glass fibres make use of large amounts of sand, that is becoming a lack of resources and represents a growing environment damage with future direct impact on landscapes. Rare-earth as trace components are also used in fibre reinforcements. Even though these are small amounts (<1% in glass reinforcement weight), future sourcing demand for the forecasted business growth will result in a considerable amount. Epoxy resin materials are also to be seen in terms of their large share in future offshore rotor blades bills of materials, therefore their recycling/reuse would be the favourable scenario. All combined, it can be concluded that a recycling strategy that enables separation and reuse of carbon/glass reinforcement and resin without needing any major mechanical shredding of the composite material would be preferred. This should be somehow represented by a solvolysis process, where the process chemicals would allow to precipitate the resin material and reuse it for either the same or other composite applications. In addition to this, the process chemicals should also be recovered and reused in a closed loop process.

3. LCA case study on different blade resin system

In this section, the environmental impacts of substituting a conventional Epoxy resin system with a recyclable alternative for large offshore wind turbine blade manufacturing was investigated. It is out of the scope of this paper to discuss in detail the resin system considered, but it is worth to mention that this would enable blade material recycling at the turbine blade EoL. This will lead to a scenario, where the baseline model of landfilling used for the majority of blade LCA studies is no longer the most likely scenario, but rather a recycling of the resin (in the form of a thermoplastic), fibres, core materials and metals is considered. The earlier developed SG 8.0-167 DD turbine LCA assumptions are being compared to the new scenario using the recyclable resin system allowing for recycling of the main parts of the wind turbine blade and a new turbine LCA derived.

3.1. Methodology

Data were collected for each phase of the lifecycle. Primary, data derives directly from SGRE internal data (i.e. material composition), whereas manufacturing, installation and operations and maintenance data is directly taken from previous SGRE internal LCA assessments. As for this study, the main differentiator is at the EoL scenario, the analysis will focus on the impact from going from 90% landfill and 10% incineration to recycling of the main parts of the blade such as resin and fibres. Figure 4 below is to illustrate the interchanges between the ecosphere and technosphere evaluated in this study [13]. For the impact assessment, the software used is SimaPro using the EcoInvent Data using the Impact2002 methodology to evaluate environmental impacts.

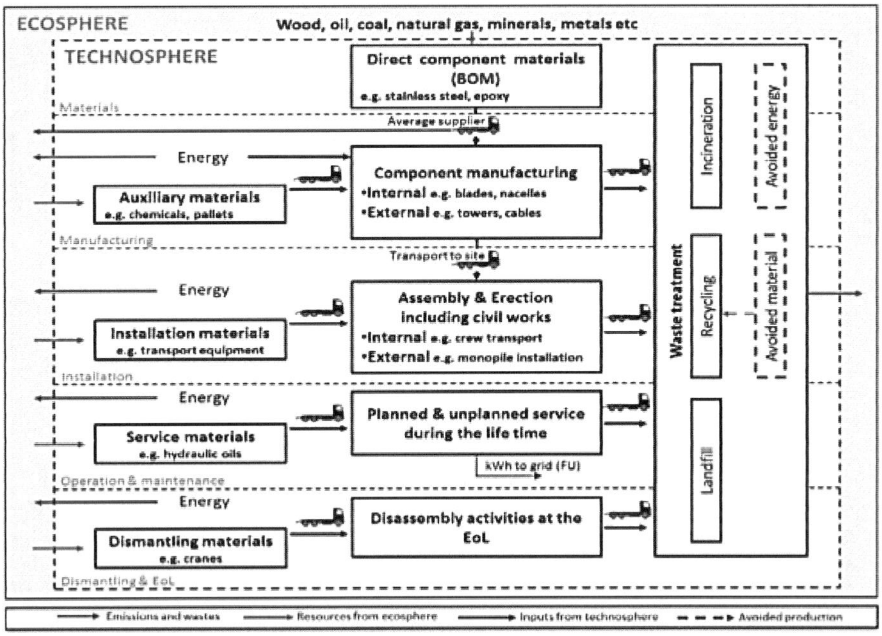

Figure 4. System boundaries – illustrative figure based on Bonou et al. 2016 [13].

Table 1 outlines the main assumptions used for the baseline study of the wind turbine blade and the comparison with its recyclable alternative. From the table, it is quite clear that the difference is within the end-of-service-life phase, where landfilling and incineration are replaced by recycling of the resin system (in the form of thermoplastic, or substitute to PA or PC), the fibres (assuming 90% recycling rate), metals and incineration of the remaining parts of the blade.

Table 1. LCA comparison study assumptions.

	Baseline assumptions	Recyclable alternative
Materials	Net weight per blade ~34t • Fibres ~60% • Resin ~28% • Wood ~8% • Plastics ~2% • Paint ~1%	Main difference is in the hardener, so the resin system is modelled as similar to baseline with the exception that production waste is being recycled instead of incinerated.
Transportation	Modelled from suppliers to Aalborg and from Aalborg to Esbjerg pre-assembly site-	Similar to baseline
Manufacturing	Manufacturing in Aalborg, use of real production data for	Similar to baseline, however, waste can be recycled
Installation	Use of installation data from recent project	Similar to baseline
Operations & Maintenance	Use of O&M data from installed base regarding maintenance need, change of component, etc.	Similar to baseline

End-of-life	Non-recyclable • 90% landfilling, 10% incineration	Recyclable • Resin can be recovered and used as replacement for Polyamide or Polycarbonate thermoplastics • Glass fibres can be recovered, with expected 10% lower properties (due to sizing removal, fibre misalignment etc) and recycled • Carbon fibres can be recovered with expected 10% lower properties (due to sizing removal, fibre misalignment etc) and recycled • Metals can be recycled • Paint, coating and core materials will be incinerated

3.2. CO2eq LCA comparison

When analysing the full lifecycle of a set of wind turbine blades, it is evident that the recyclable solution is favourable from a CO2eq perspective as the overall lifecycle emissions will be reduced by 28% through recycling of resin, fibres and metals and feed this back into the material cycles. Another interesting observation is that from the EcoInvent data, polycarbonate has a CO2eq. profile that is much higher than epoxy resin, so the recycling of the resin in a fairly light process (i.e. acidic solution with 90°C process temperature) will result in a favourable situation as high impact plastic is being substituted by the recycled material.

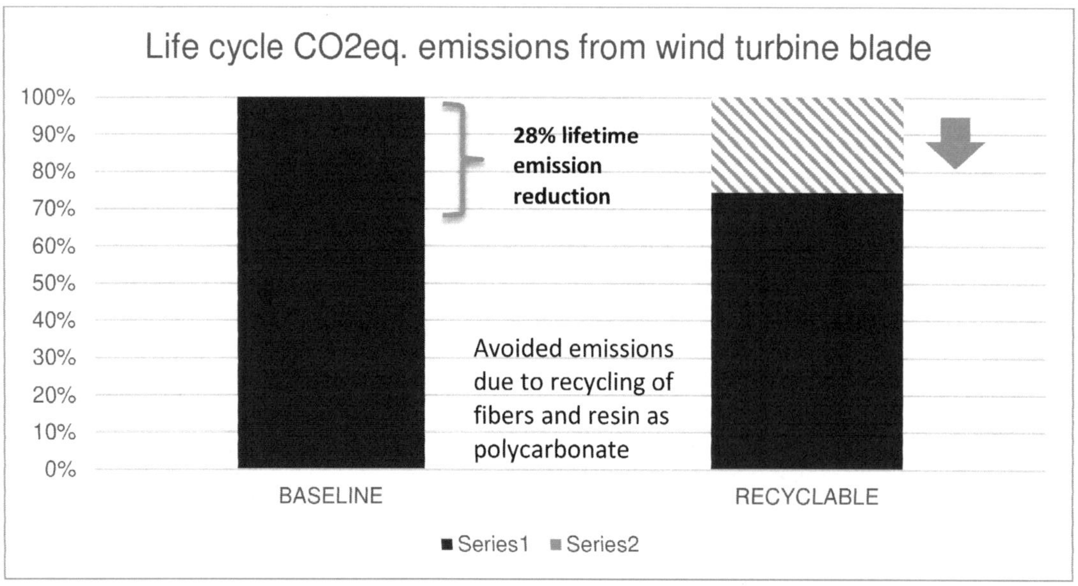

Figure 5. Estimated tons of CO2eq saved per turbine (SG 8.0-167 DD) by using a recyclable resin system as opposed to a conventional epoxy resin system.

4. Conclusions

To summarise, the current study showed that raw materials are responsible for 71% of the CO2eq emission on a representative 8MW offshore wind turbine lifetime and due to the non-recyclable nature of blade materials, their CO2eq emissions impact is remarkable. Therefore, investing efforts in research and innovation to diversify and scale-up composite recycling technologies, to develop new, high-performance materials with enhanced circularity and to design methodologies to enhance circularity and recycling abilities of blades are required. At the same time, existing treatment routes like cement co-processing must be deployed more widely to deal with the current waste streams. Finally, the scientific understanding of the environmental impacts associated with the choice of materials and with the different waste treatment methods should also be improved by extensive use of LCA tools.

An LCA comparative study employing an innovative recyclable resin system has proven to cut on CO2eq emissions by 28% with respect to a conventional epoxy resin system. This under the assumptions that the blade materials can be recycled with a 90% rate. Only CO2eq impact has been calculated using the data available based on assumptions, extrapolation of recent installations as well as using the EcoInvent database for material emission profiles. This study can therefore only be considered a screening study, and a full study must be carried out to validate these preliminary findings. Nonetheless, the indicators show that recyclable resin systems are environmentally superior to the baseline alternative under the assumptions used.

References

[1] Environmental Product Declaration SG 8.0-167 DD 2019 Available online: (PDF) Retrieved 19 May 2020

[2] Renewable Energy & Efficiency Partnership 2004 "Glossary of terms in sustainable energy regulation" Available online: (PDF). Retrieved 19 May 2020

[3] Bruckner T, *et all*, 2014 Energy Systems. In: Climate Change 2014: Mitigation of Climate Change. *Contribution of Working Group III to the Fifth Assessment Report of the Intergovernmental Panel on Climate Change* (Cambridge: United Kingdom) chapter 7 pp 511–597

[4] Martínez E, Sanz F, Pellegrini S, Jiménez E and Blanco J, 2009 Life cycle assessment of a multi-megawatt wind turbine, *Renewable Energy* **34** 3 doi: /10.1016/j.renene.2008.05.020

[5] WindEurope – Cefic – EuCIA 2020 Accelerating Wind Turbine Blade Circularity Available online: (PDF). Retrieved 28 May 2020

[6] Psomopoulos C, Kalkanis K, Kaminaris S, Ioannidis G and Pachos P, 2019 A Review of the Potential for the Recovery of Wind Turbine Blade Waste Materials, *Recycling* **4** 7 doi:10.3390/recycling4010007

[7] Van Oudheusden A, 2019 Recycling of composite materials, Engineering (Delft University of Technology, the Netherlands) pp 3 – 68

[8] European Waste Framework Directive 2008/98/EC Available online: (https). Retrieved 28 May 2020

[9] Bank L, Arias F, Yazdanbakhsh A, Gentry R, Al-Haddad T, Chen F and Morrow R, 2018 Concepts for Reusing Composite Materials from Decommissioned Wind Turbine Blades in Affordable Housing, *Recycling* **3** 3 doi:10.3390/recycling3010003

[10] Liu P, and Barlow C, 2017 Wind turbine blade waste in 2050, *Waste Management* **62** 229-240 doi.org/10.1016/j.wasman.2017.02.007

[11] SG 8.0-167 DD turbine product brochure 2019 Available online (PDF) Retrieved 28 May 2020

[12] Liu P and Barlow C Y 2016 The environmental impact of wind turbine blades *IOP Conference: Materials Science and Engineering* **139** doi:10.1088/1757-899X/139/1/012032

[13] Bonou A, Laurent A, Olsen SI, 2016 Life cycle assessment of onshore and offshore wind energy from theory to application, *Applied Energy* **180** doi: 10.1016/j.apenergy.2016.07.058

IOP Conf. Series: Materials Science and Engineering **942** (2020) 012012 doi:10.1088/1757-899X/942/1/012012

A comparison of surface preparation techniques for wind turbine field repairs

Ariel Lusty*, **Douglas Cairns, David Miller, and Daniel Samborsky**

Department of Mechanical and Industrial Engineering, Montana State University, Bozeman, MT, 59715, USA

*Corresponding author's email: arilusty@gmail.com

Abstract. Wind turbine service lifetimes can exceed 20 years. This extreme operational demand necessitates reliable and consistent field repairs; however, the effects of current wind turbine field repair surface preparation techniques are not well-documented and are thus not well-understood. This leads to reliance on technician experience rather than scientific data for repair procedures, which causes variability in the quality of repairs. Solvent wiping is a common procedure for contaminant removal after damaged material is removed from a wind turbine blade, but it was unknown if solvent wiping reduced the surface energy and consequently the likelihood of a durable repair. The goal of this study was to quantify the effects that various surface preparation techniques have on the overall strength and reliability of wind turbine repairs. Type of reagent-grade solvent, fiber direction, matrix, and adhesive were varied for contact angle, lap shear, Double Cantilever Beam (DCB), and End-Notch Flexure (ENF) testing. The type of solvent used did not affect surface energy, maximum lap shear stress, or modes I and II fracture toughness values. However matrix-adhesive combinations significantly affected maximum lap shear stress and fracture toughness values.

1. Introduction

Fixing structural damage is of utmost priority for owners and operators of wind turbines. Structural damage is typically from manufacturing defects, but may include in-service damage from mechanical fatigue, bird collisions, or lightning strikes. Damage that inhibits a wind turbine from efficiently harvesting energy should be repaired if the repair does not exceed the cost of blade replacement or if the procedure is economically viable for other reasons. The goal of any repair is to return the structure to its original strength, stiffness, shape, and surface finish [1]. The most common and preferred structural wind turbine blade repair method is the scarf repair due to the need to restore strength, avoid load concentration and eccentricities, and maintain aerodynamic efficiency.

Regardless of the repair design, contamination will be present before, during, and after damage removal. A commonly used technique for contaminant removal involves wiping a surface with reagent grade solvents to remove contamination and wiping again with a dry cloth to remove the solvent before it evaporates. It is essential to ensure that the solvent being used is capable of dissolving the contaminant without further damaging the composite structure [2]. If the use of solvents damages the surface of the composite structure, bond strengths and repair durability could be compromised.

In this research, the effects of current solvent wiping techniques on composite bond strengths were investigated using direct adhesion measurements and surface characterization techniques [3]. Direct adhesion measurements are destructive methods that measure the force required to break, tear and delaminate surfaces at the interface. Direct adhesion measurements used in this study were lap shear, DCB, and ENF tests. Understanding failure initiation and propagation through failure surface examination is crucial after direct adhesion measurements, so adhesive failure modes were documented for each direct adhesion measurement. Four adhesive failure modes were considered: interfacial failure, cohesive failure, stock-break failure, or fiber-tear failure. Interfacial failure occurs along the adhesive-substrate interface and is usually due to a lack of chemical bonding. Cohesive failure occurs within the adhesive and is typically the preferred type of failure in a bonded joint because cohesive failure usually corresponds with a higher maximum lap shear stress. Stock-break failure occurs when the substrate is either damaged or weaker than the adhesive bond. Fiber-tear failure occurs when the top layer of the substrate is damaged, resulting in fiber re-exposure.

Vacuum assisted resin transfer molding (VARTM) was used to manufacture glass fiber reinforced polymer (GFRP) composite substrates. Sessile drop contact angle testing was used for surface characterization, and was conducted after composite samples were sanded and wiped with reagent grade solvents including acetone, methyl ethyl ketone (MEK), methyl isobutyl ketone (MIBK), or isopropyl alcohol (ISP), as these are commonly used solvents in composite repairs [1]. Contact angle testing was also conducted on plasma-treated specimens, as plasma treatment is a common method used to increase surface energy [4].

Lap shear, DCB, and ENF testing using two different adhesives was used to compare the effects of wiping substrate surfaces with different solvents on resulting maximum lap shear stress, mode I fracture toughness (G_{IC}), and mode II fracture toughness (G_{IIC}) values. Dillingham demonstrated using DCB testing that cohesive failure corresponded with higher G_{IC} values, and interfacial failure corresponded with lower G_{IC} values [5]. Overall, if a solvent had deleterious effects on bond strength more than another, it would be reflected in differences in contact angles, lap shear strengths, G_{IC} values, G_{IIC} values and adhesive failure modes.

2. Materials

Materials were selected to replicate the surfaces that would be encountered in a wind turbine blade field repair. Wind turbine blade face sheets are often manufactured using both 0° and ±45° glass fibers, with either an epoxy or a vinyl ester matrix. Accordingly, VARTM was used to make $[0]_4$ epoxy, $[45]_6$ epoxy, $[45]_6$ vinyl ester, and $[0]_4$ vinyl ester composites for contact angle and lap shear testing. VARTM was also used to make $[0]_6$ epoxy and $[0]_6$ vinyl ester specimens for DCB and ENF testing. Unidirectional Vectorply E-LT-2900 fabric was used for unidirectional layups, and ±45° PPG D810 fabric was used for $[45]_6$ layups. Hexion Epikote MGS RIMR 135 epoxy was used for the epoxy composites, and Fibreglast 1110 vinyl ester resin was used for the vinyl ester composites. Two adhesives were compared in lap shear, DCB, and ENF testing: Huntsman Araldite 2051 and Loctite EA 9396 Aero. Huntsman Araldite 2051 is a methyl methacrylate adhesive, and Loctite EA 9396 Aero is an epoxy adhesive. Wypall X60 cloths were used for solvent wiping.

3. Experimentation

3.1 Solvent Wipe Contact Angle Tests

Sessile drop contact angle testing can be used to calculate surface energy, which is the work spent in forming a unit area of a solid surface. Contact angle testing involves the placement of a water droplet on a surface and measuring the resultant angle the droplet makes with the surface. Low contact angle values correspond with hydrophilicity, high wettability, and higher surface energy. Alternatively, high contact angles indicate hydrophobicity, low wettability and lower surface energy [6]. Additionally, low surface energy has been shown to result in more interfacial failure modes, and high surface energy will typically result in more cohesive failure modes [5].

Sessile drop contact angle testing was conducted after samples were sanded and wiped with solvents. For contact angle testing, 25 x 50 mm^2 samples were first wet sanded using Federation of European Producers of Abrasives (FEPA) 1000 grit sandpaper, then wiped with a dry cloth to remove water. The water break test was used to ensure mold release was removed from composite surfaces [7]. Next, the samples were wiped with another cloth that had been saturated with either distilled water, ISP, acetone, MIBK, or MEK. Then, samples were wiped with a dry cloth. Six contact angle tests were conducted on the surface of each sample.

Contact angle testing was conducted immediately after solvent wiping. A 1 μL droplet of MilliQ 0.2 μm water was formed at the tip of the needle. Next, the substrate was slowly brought up to the droplet.

Once the droplet contacted the surface, the substrate was lowered. The contact angle decreases linearly with time after a water droplet is placed on a surface. Conclusions from preliminary testing demonstrated that water droplets would stabilize, but not completely evaporate, after approximately 10 seconds. Thus, after the droplet was on the composite surface for 10 seconds, an image was captured using VCA Optima software. Contact angles were also measured using VCA Optima.

3.2 Plasma Contact Angle Tests
Specimens cut from the same composite plates as the solvent wipe contact angle tests were shipped to Enercon Industries Corporation for plasma treatments, followed immediately by contact angle testing. Specimens were sanded with FEPA 1000 grit sandpaper prior to shipment to remove mold release. Two areas on each sample were contact angle tested before and after being treated by two surface treatment types: a blown ion treatment and a flame treatment. The line speed was run at 30.5 cm per minute for both blown ion and flame treatments. The output used for the blown ion treatment was at 0.5 kW, and the output for the flame burner was 200 liters per minute. The distance between the blown ion treatment and the sample was 6.35 mm, and the distance between the flame treatment and the sample was 50.8 mm.

3.3 Lap Shear Tests
In addition to contact angle tests, lap shear tests were used to compare the effects of solvent wipe techniques on the resulting maximum lap shear stress values and lap shear failure modes. It is important to note that lap shear testing does not produce true ultimate stress because the stress distribution in the adhesive is not uniform over the bond area [8]. Additionally, the stiffness of adherends influences peel stress concentrations at the end of the joint in lap shear testing. Increasing the modulus or thickness of adherends will lead to an increase in apparent strength of the bond. Nevertheless, lap shear data does have value from a comparative standpoint and was thus used to check the effectiveness of different surface preparation techniques [9].

Coupons with dimensions 25 x 400 mm^2 were cut from composite plates. Instead of the FEPA 1000 sand grit that was used for contact angle testing, FEPA 320 grit sandpaper was used to create a rougher surface and to promote adhesion for lap shear testing. Solvent wipe techniques identical to those used in contact angle testing were used prior to adhesive bonding. Wire with a 0.33 mm diameter was placed in tab areas as spacers, and weights were used to reduce variation in adhesive thicknesses. The maximum lap shear stress is reported as the failure stress in the adhesive, which is found by dividing the failure load by the bond area. The $[45]_6$ laminates were tested at 13 mm/min load rate to minimize creep, and $[0]_4$ laminates were tested at 1 mm/min load rate [10]. No significant differences in maximum lap shear stress were found between the two load rates. The failure surfaces of each lap shear specimen were visually inspected and assigned failure mode types.

3.4 DCB Tests
Similarly to the lap shear tests, coupons were cut to dimensions 25 x 400 mm^2 and FEPA 320 grit sandpaper was used to hand-sand samples on a flat granite surface plate. Solvent wipe techniques identical to those used in contact angle testing were used prior to adhesive bonding. In addition, the same two adhesives from lap shear testing were used to adhere coupons together. Teflon films were used as starter cracks and were placed at the ends of the specimens. Unlike for lap shear specimen preparation, wire spacers were not used for DCB specimen preparation. Instead, binder clips were used to secure coupons together while adhesives cured. After adhesives were cured, all edges of each specimen were trimmed and polished. Aluminum tabs were adhered to specimens using Plexus MA310 methacrylate adhesive. DCB specimens were tested in tension at a loading rate of 0.3 mm/sec. Data was acquired using HP Logger software. G_{IC} was calculated from [11]. Values of 40270 MPa and 125 MPa were used for E_{11} and E_{22}, respectively, and were found using [12]. P was found using the 5% rule, which involves finding the slope of the force-displacement curve generated in DCB tests, then taking 95% of the slope and using the load where the 95%-of-the-slope line intersects plotted data points. The 5% rule is a common approximation used to account for the nonlinearity in DCB test data [13].

3.5 ENF Tests

Post-tested DCB specimens were used for ENF tests. Cracks from the DCB specimens were manually propagated and remeasured because a longer crack was necessary for ENF testing to avoid local face sheet delamination from the loading nose. Specimens were placed in the loading fixture such that the crack tip was approximately halfway between the supporting roller and the loading roller. G_{IIC} was calculated using an equation from [11]. The compliance term in [11] was not needed in calculations for this study because DCB and ENF tests were conducted separately, rather than with a mixed-mode bending apparatus. The E_{11} value from DCB tests was used. Shear modulus G_{13} (5050 MPa) was found using Iosipescu notched-shear tests and digital image correlation (DIC) [14]. After Iosipescu tests, GOM-ARAMIS software was used to calculate strain. The surface component used was based on a facet size of 25 and a point distance of 17. A spatial average filter of 2 was used. After ENF tests, specimens were manually peeled apart and, similarly to after lap shear tests, adhesive failure modes were identified for DCB and ENF failure surfaces.

4. Results and Discussion

4.1 Solvent Wipe Contact Angle Results

Results for contact angle testing after wiping with solvents are summarized in Table 1. The *unprepared* samples were neither sanded nor solvent wiped, so mold release was still present on the surfaces. The *dry wipe* samples were wet sanded and wiped with dry cloths before contact angle testing.

Table 1. Average contact angle (degrees) for each surface treatment.

Treatment	$[0]_4$ Epoxy	$[45]_6$ Epoxy	$[0]_4$ Vinyl Ester	$[45]_6$ Vinyl Ester
Unprepared	87.7 ± 3.5	73.4 ± 5.0	87.8 ± 9.6	70.6 ± 12.8
Dry Wipe	64.3 ± 5.4	66.8 ± 9.7	67.1 ± 6.8	67.6 ± 6.3
Distilled Water	56.5 ± 5.9	79.0 ± 2.1	59.5 ± 9.9	82.1 ± 5.6
ISP	59.4 ± 5.4	70.2 ± 6.4	65.0 ± 9.1	72.0 ± 6.7
Acetone	54.5 ± 7.2	73.0 ± 2.6	67.3 ± 9.4	68.1 ± 3.7
MIBK	62.3 ± 6.1	58.7 ± 6.4	61.1 ± 5.2	71.8 ± 3.7
MEK	62.4 ± 7.3	65.5 ± 8.5	60.9 ± 8.9	68.3 ± 5.8

The contact angle was neither significantly altered by wiping samples' surfaces with solvents, nor the type of matrix used. However, removing mold release through sanding is necessary to decrease the contact angle, as shown in the drops in contact angle between *unprepared* and solvent wiped samples in the $[0]_4$ epoxy and vinyl ester columns.

4.2 Plasma Treatment Contact Angle Results

Contact angles across all solvent-wiped specimens were high, indicating poor wettability and poor adhesion. To decrease contact angle, samples were plasma treated followed by contact angle testing.

Table 2. Average contact angles and standard deviations on composite surfaces before and after blown ion treatment.

Treatment	$[0]_4$ Epoxy	$[45]_6$ Epoxy	$[0]_4$ Vinyl Ester	$[45]_6$ Vinyl Ester
Initial (before blown ion treatment)	73.5 ± 0.2	71.1 ± 0.7	75.0 ± 3.7	70.4 ± 3.7
After blown ion treatment	30.2 ± 2.2	30.6 ± 2.3	16.2 ± 0.0	22.9 ± 0.8

Across all composite surface types in Table 2, there was a 66% drop between the average contact angles before and after blown ion treatment.

Table 3. Average contact angles and standard deviations on composite surfaces before and after flame treatment.

Treatment	$[0]_4$ Epoxy	$[45]_6$ Epoxy	$[0]_4$ Vinyl Ester	$[45]_6$ Vinyl Ester
Initial (before flame treatment)	70.2 ± 0.1	78.5 ± 2.5	63.5 ± 0.2	77.7 ± 1.3
After flame treatment	25.8 ± 0.4	28.9 ± 0.9	15.9 ± 0.4	18.8 ± 2.5

Across all composite surface types in Table 3, there was a 69% drop between the average contact angles before and after flame treatment. Overall, contact angle dropped significantly between initial and plasma treated composite surfaces for both blown ion and flame treatments, indicating that wettability and adhesion properties were improved.

4.3 Lap Shear Results

The moduli of the $[0]_4$ and $[45]_6$ adherends were 39.4 and 13.3 GPa, respectively. Lap shear test results for the Huntsman Araldite 2051 adhesive are shown in Table 4. Figures 1, 2, and 3 include common fracture surfaces for lap shear tests for each adhesive. Lap shear test results for the Loctite EA 9396 Aero adhesive are shown in Table 6. Five test results were averaged for each tabulated value.

Table 4. Average maximum lap shear stress (MPa) and standard deviations for specimens adhered with a 658 ± 19 mm^2 average bond area of Huntsman Araldite 2051.

Surface Treatment	$[0]_4$ Epoxy	$[45]_6$ Epoxy	$[0]_4$ Vinyl Ester	$[45]_6$ Vinyl Ester
Sanded	14.89 ± 0.77	6.89 ± 0.29	22.61 ± 1.56	7.69 ± 0.63
ISP	14.52 ± 0.55	7.31 ± 0.58	22.17 ± 1.45	6.57 ± 0.93
Acetone	13.67 ± 2.33	6.88 ± 0.78	22.80 ± 0.46	8.48 ± 0.48
MIBK	14.13 ± 1.38	6.99 ± 0.94	22.75 ± 1.65	7.96 ± 0.13
MEK	13.49 ± 2.14	6.83 ± 1.21	23.63 ± 0.86	8.07 ± 0.38

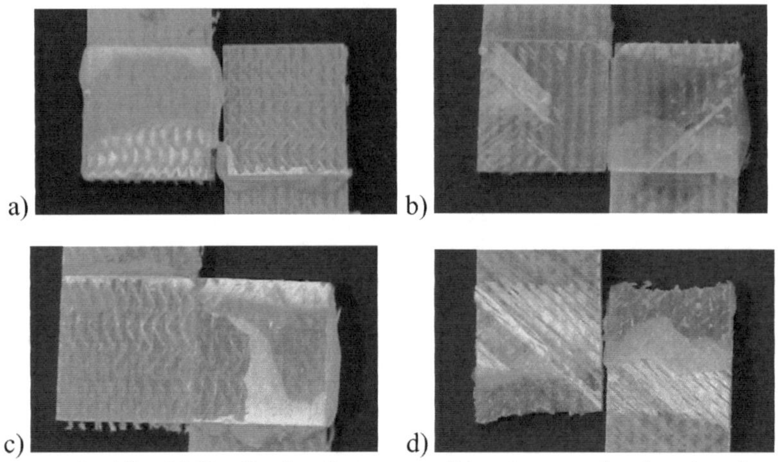

Figure 1. Huntsman Araldite 2051 adhesive fracture surface photographs showing the primary failure modes for each substrate type: a) $[0]_4$ epoxy substrates (ISP wipe) with interfacial failure mode; b) $[45]_6$ epoxy substrates (MIBK wipe) with fiber-tear failure mode; c) $[0]_4$ vinyl ester substrates (ISP wipe) with interfacial failure mode; d) $[45]_6$ vinyl ester substrate (MEK wipe) with fiber-tear failure mode.

In contrast to Figure 1c with ISP wipe, the fracture surfaces in Figure 2 with MEK wipe show more cohesive failure.

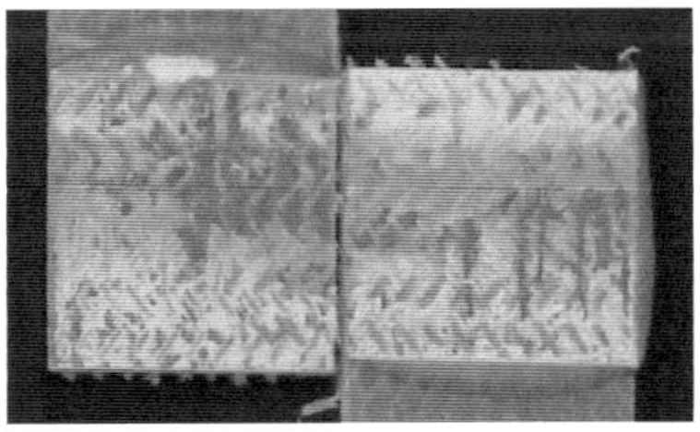

Figure 2. Primary failure mode for $[0]_4$ vinyl ester specimens wiped with MEK.

The $[0]_4$ vinyl ester specimens had higher maximum lap shear stresses than the $[0]_4$ epoxy specimens, while the epoxy and vinyl ester $[45]_6$ specimens all had around the same maximum lap shear stresses. There were no significant differences in maximum lap shear stress values from using different solvents. The $[0]_4$ specimens resulted in mostly interfacial failure, and the $[45]_6$ specimens had mostly fiber-tear failure. Although most $[0]_4$ vinyl ester specimens exhibited interfacial failure modes, those wiped with MEK exhibited the most instances of cohesive failure modes but did not have significantly higher maximum lap shear stresses. The $[0]_4$ specimens resulted in higher maximum lap shear stresses and $[45]_6$ layups resulted in the lowest maximum lap shear stresses. The $[45]_6$ specimens failed at lower ultimate tensile stresses than the $[0]_4$ specimens due to their higher susceptibility to bending in tension. Overall, the layup-matrix combinations had the most differences in resulting maximum lap shear stresses.

Table 5. Average maximum shear stress (MPa) for specimens adhered with a 658 ± 16 mm² average bond area of Loctite EA 9396 Aero Adhesive.

Surface Treatment	$[0]_4$ Epoxy	$[45]_6$ Epoxy	$[0]_4$ Vinyl Ester	$[45]_6$ Vinyl Ester
Sanded	13.84 ± 0.82	8.12 ± 0.47	5.81 ± 0.26	4.62 ± 0.50
ISP	13.05 ± 0.91	8.23 ± 0.30	6.56 ± 0.48	5.67 ± 0.32
Acetone	12.54 ± 0.27	8.16 ± 0.19	5.40 ± 0.28	4.96 ± 0.62
MIBK	13.39 ± 0.72	8.21 ± 0.14	5.63 ± 0.61	4.94 ± 0.48
MEK	12.62 ± 0.60	8.24 ± 0.11	5.81 ± 0.47	4.77 ± 0.34

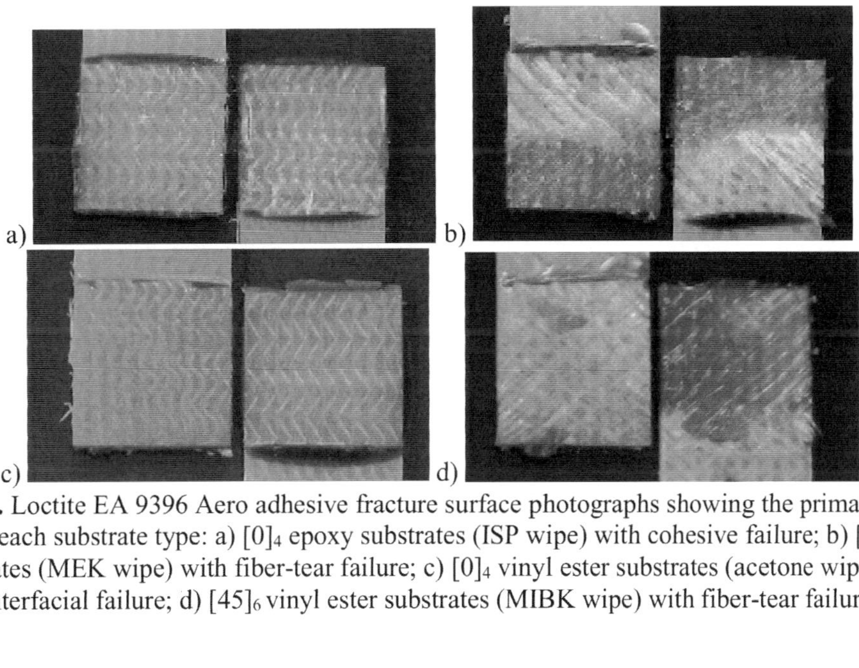

Figure 3. Loctite EA 9396 Aero adhesive fracture surface photographs showing the primary failure modes for each substrate type: a) $[0]_4$ epoxy substrates (ISP wipe) with cohesive failure; b) $[45]_6$ epoxy substrates (MEK wipe) with fiber-tear failure; c) $[0]_4$ vinyl ester substrates (acetone wipe) with interfacial failure; d) $[45]_6$ vinyl ester substrates (MIBK wipe) with fiber-tear failure.

When using the Loctite EA 9396 Aero adhesive, the $[0]_4$ epoxy specimens had the highest maximum lap shear stresses and the vinyl ester specimens had the lowest maximum lap shear stresses. The Loctite EA 9396 Aero epoxy adhesive likely has a more similar molecular structure to the epoxy matrix than the vinyl ester matrix, which could explain why maximum lap shear stresses were higher than with the $[0]_4$ vinyl ester specimens. The $[0]_4$ epoxy specimens resulted in mostly cohesive failure, while the $[45]_6$ epoxy and $[45]_6$ vinyl ester specimens resulted in mostly fiber-tear failure.

4.4 DCB and ENF Test Results

Mode I and mode II fracture toughness values are presented, and DCB and ENF failure modes are discussed. DCB and ENF results for the Huntsman Araldite 2051 adhesive are in Table 6, and DCB and ENF results for the Loctite 9396 Aero adhesive are in Table 7. Five test results were averaged for each tabulated value. Figure 3 is an example of DCB load-displacement curves for specimens wiped with acetone.

Table 6. Modes I and II fracture toughness values for glass fiber reinforced epoxy and vinyl ester substrates adhered with the Huntsman Araldite 2051 adhesive.

	Epoxy Substrates		**Vinyl Ester Substrates**	
Surface Treatment	**G_{IC} (J/m²)**	**G_{IIC} (J/m²)**	**G_{IC} (J/m²)**	**G_{IIC} (J/m²)**
Sanded	390 ± 222	8079 ± 2143	553 ± 222	4468 ± 1821
ISP	605 ± 26	9170 ± 1297	426 ± 417	5425 ± 3581
Acetone	551 ± 203	10235 ± 3664	590 ± 168	2537 ± 1036
MIBK	583 ± 97	6601 ± 2630	867 ± 408	4621 ± 3443
MEK	602 ± 104	7079 ± 1780	702 ± 310	2209 ± 322

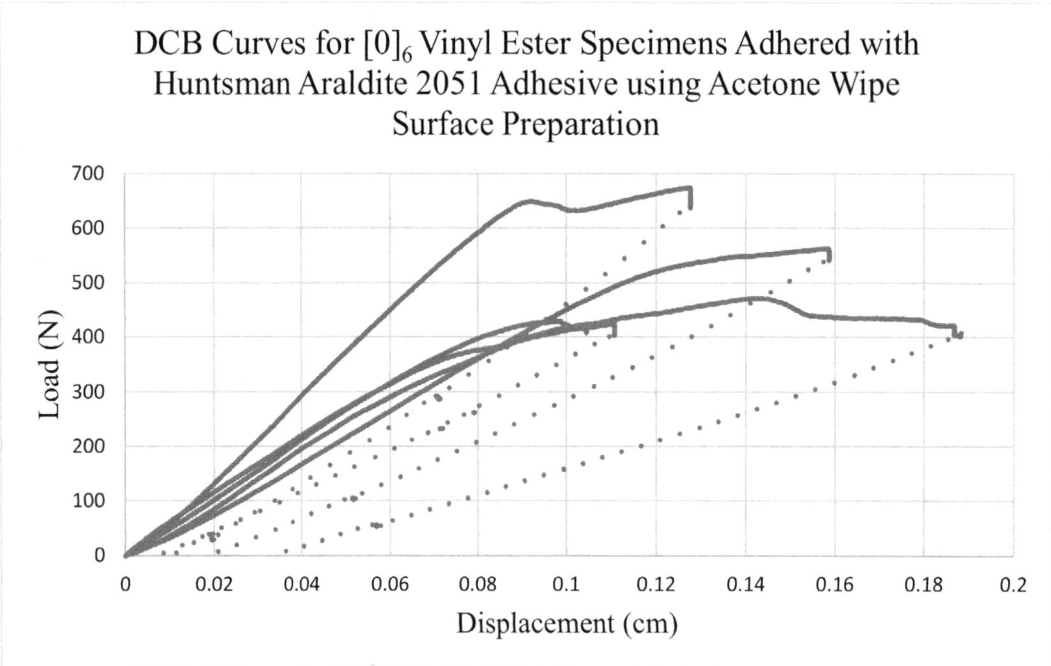

Figure 4. An example of five load-displacement curves from DCB tests.

There were no significant differences in G_{IC} or G_{IIC} values between surface treatment types for neither the epoxy nor the vinyl ester composites adhered together with the Huntsman Araldite 2051 adhesive.

Standard deviations were large for two reasons. First, even when using a flat granite sanding plate, hand sanding can result in surface variations from inevitable uneven pressure applied. Second, variations may also be due to complex stress distributions in adhesive layers. Crack and damage growth do not always occur in a simple and predictable fashion, which was observed in post-test crack length measurements and failure surface observations [15]. DCB tests for epoxy substrates resulted primarily in cohesive failure modes and ENF tests resulted in primarily interfacial failure modes. Both DCB and ENF tests resulted primarily in cohesive failure modes for vinyl ester specimens adhered with Huntsman Araldite 2051 adhesive.

Table 7. Modes I and II fracture toughness values for glass fiber reinforced epoxy coupons adhered together with Loctite EA 9396 Aero adhesive.

Surface Treatment	G_{IC} (J/m²)	G_{IIC} (J/m²)
Sanded	234 ± 72	2772 ± 269
ISP	166 ± 47	2418 ± 523
Acetone	149 ± 42	2930 ± 661
MIBK	157 ± 48	2173 ± 468
MEK	227 ± 83	2943 ± 1115

There were no significant differences in G_{IC} or G_{IIC} values between surface treatment types for epoxy coupons adhered together with Loctite EA 9396 Aero adhesive. Overall, epoxy coupons adhered together with Loctite EA 9396 Aero adhesive resulted in cohesive failure for both DCB and ENF tests. All glass fiber reinforced vinyl ester specimens adhered together with Loctite EA 9396 Aero adhesive, regardless of surface treatments tested, resulted in $G_{IC} = 0$ J/m^2. This result indicates that the Loctite EA 9396 Aero adhesive was incompatible with the vinyl ester composite substrate. ENF results were not collected for vinyl ester specimens adhered together with Loctite EA 9396 Aero adhesive. All G_{IC} values were lower than G_{IIC} values because cracks typically spread preferentially in a direction perpendicular to tensile stresses [16]. Epoxy substrates adhered with the Loctite EA 9396 Aero adhesive resulted in significantly lower G_{IC} and G_{IC} values compared to the Huntsman Araldite 2051. Overall, the vinyl ester substrates adhered with the Huntsman Araldite 2051 adhesive resulted in the highest G_{IC} values, and the epoxy substrates adhered with the Huntsman Araldite 2051 adhesive resulted in the highest G_{IIC} values.

5. Conclusions

Generally, wiping a composite substrate with either distilled water, ISP, Acetone, MIBK, or MEK does not significantly affect the resulting water contact angle, maximum lap shear stress, G_{IC} or G_{IIC} values. Contact angle testing demonstrated the importance of removing mold release from composites prior to bonding, and plasma treatment delivered promising results for decreasing contact angle but may be challenging to implicate into wind turbine field repairs due to already-present challenges with wind turbine blade accessibility. Furthermore, future work will examine the effectiveness of applying sizing to the surface of a composite in increasing resultant repair bond strengths.

Differences observed when testing different lap shear substrates were mostly due to differences in adherend thicknesses and moduli rather than the type of solvent used, which is expected because the stiffness of adherends influences peel stress concentrations at the end of the joint in lap shear testing. Adhesive-matrix combination also had a significant effect on the maximum lap shear stresses. Unidirectional vinyl ester lap shear specimens resulted in more cohesive failure when wiped with MEK than for those wiped with other solvents, but corresponding maximum lap shear stresses were not significantly higher. In addition, failure modes after DCB and ENF testing for MEK-wiped vinyl ester specimens adhered with methyl methacrylate were not visibly different. Even so, the causes for the increases in cohesive failure modes for the MEK-wiped vinyl ester substrates adhered with the methyl methacrylate adhesive merit further investigation.

Across all DCB and ENF tests, G_{IIC} values were consistently higher than G_{IC} values, which is optimal for wind turbine field repairs, particularly scarf repairs, because patches are typically subjected to more shear stresses than peel stresses. Data from DCB and ENF tests may be used in modelling software such as ABAQUS to simulate repairs with the composite-adhesive combinations used in this study. SEM will be used in future work to confirm visually observed lap shear, DCB and ENF failure modes. Similarly to lap shear results, adhesive-matrix combinations had significant effects on G_{IC} and G_{IIC} values and on failure modes. G_{IIC} values for specimens adhered using the Huntsman Araldite 2051 adhesive were twice as high for the epoxy specimens than the vinyl ester specimens. Both epoxy and vinyl ester specimens had similar G_{IC} values when adhered with the Huntsman Araldite 2051 adhesive. Most notably, vinyl ester composite substrates adhered extremely poorly to the Loctite EA 9396 Aero adhesive, with no adhesive-matrix compatibility in mode I loading. Based on maximum lap shear stress values, vinyl ester composite substrates adhered with the Loctite EA 9396 Aero adhesive may have some shear strength, but since $G_{IC} = 0$ J/m^2, the Loctite EA 9396 Aero adhesive should be considered unsuitable for repairs to Fibreglast 1110 vinyl ester composites.

Additional testing using different adhesives and the effects of common contaminants and surface treatments using direct adhesion measurements and surface characterization techniques will be conducted in future work. Procedures will then be developed for the evaluation of solvent effects on composite surfaces, the evaluation of the effectiveness of solvents on the removal of common contaminants, and the evaluation of surface treatment effectiveness on increasing repair bond strength and durability. The development of these procedures is crucial to the advancement of wind turbine blade repair technology and thus the extension of wind turbine blade lifetimes.

References

[1] Dorworth L C 2013 Repairs for Advanced Composite Structures Automotive Composites Conference and Exhibition (Reno: Abaris Training)
[2] Dorworth L C, Gardiner G L and Mellema G M 2015 Essentials of Advanced Composite Fabrication and Repair (Newcastle: Aviation Supplies & Academics)
[3] Awaja F, Gilbert M, Kelly G, Fox B and Pigram P J 2009 Adhesion of Polymers *Prog. Polym. Sci.* vol 34 no 9 pp 948–68 DOI: https://doi.org/10.1016/j.progpolymsci.2009.04.007
[4] Mattox D M 2010 Substrate ('Real') Surfaces and Surface Modification (PVD) Processing Elsevier pp 25–72 DOI: https://doi.org/10.1016/b978-0-8155-2037-5.00002-2
[5] Dillingham R G and Oakley B R 2006 Surface Energy and Adhesion in Composite–Composite Adhesive Bonds *J. Adhes.* vol 82 no 4 pp 407–26 DOI: 10.1080/00218460600683944
[6] Erbil H Y 2006 Surface Chemistry of Solid and Liquid Interfaces (Oxford: Blackwell Publishing Ltd)
[7] Hart-Smith L J, Ochnsner R W and Radecky R L 1990 Surface Preparation of Composites for Adhesive-Bonded Repair *Engineered Materials Handbook: Adhesives and Sealants* (Cleveland: ASM International)
[8] Campbell F C 2004 Manufacturing Processes for Advanced Composites (New York : Elsevier)
[9] Duncan B and Crocker L 2001 Review of Tests for Adhesion Strength (Teddington : National Physics Laboratory Materials Centre)
[10] ASTM International 2014 ASTM D5868-01: Standard Test Method for Lap Shear Adhesion for Fiber Reinforced Plastic (FRP) Bonding DOI: 10.1520/D5868-01R14
[11] Reeder J R and Crews J H 1990 Mixed-Mode Bending Method for Delamination Testing *AIAA J* vol 28 no 7 pp 1270–76 DOI: 10.2514/3.25204
[12] ASTM International 2017 ASTM D3039/D3039M-17 Standard Test Method for Tensile Properties of Polymer Matrix Composite Materials DOI: https://doi.org/10.1520/D3039_D3039M-17
[13] Agastra P 2003 Mixed Mode Delamination of Glass Fiber/Polymer Matrix Composites Montana State University
[14] ASTM International 2020 D7079/D7078M-20 Standard Test Method for Shear Properties of Composite Materials by V-Notched Rail Shear Method DOI: https://doi-org.proxybz.lib.montana.edu/10.1520/D7078_D7078M-20

IOP Conf. Series: Materials Science and Engineering **942** (2020) 012012 doi:10.1088/1757-899X/942/1/012012

[15] Vintilescu I and Spelt J 1998 Mixed Mode I, II, and III Fracture Characterization of Adhesive Joints *J. Compos. Technol. Res.* DOI: 10.1520/ctr10053j
[16] Broek D and Ric J R 1975 Elementary Engineering Fracture Mechanics *J. Appl. Mech.* DOI: 10.1115/1.3423697

Chemical recycling of End-of-Life wind turbine blades by solvolysis/HTL

C Mattsson*, A André, M Juntikka, T Tränkle and R Sott

RISE Research Institutes of Sweden, SE-431 22 Mölndal, Sweden

*Email: cecilia.mattsson@ri.se

Abstract. The focus of this contribution is to highlight the challenges of chemical recycling of End-of-Life glass fiber composite (GFRP) waste from wind turbine blades utilizing solvolysis/HTL (hydrothermal liquefaction) methods based on subcritical water as solvent. A multitude of investigations have been published during the years regarding solvolysis of newly produced composite laminates and known thermoset composition (epoxy, polyester, and vinyl ester). However, a real wind turbine blade is more complex and constitutes of thermosets, thermoplastics, and other materials such as balsa wood. It is a very challenging task to separate these materials from each other within the wind turbine blade structure, so the premise for recycling is a mixed waste stream where little is known about the chemical composition. In the present study, the solvolysis process for GFRPs based on sub/supercritical water at 250-370 °C and 100-170 bar process conditions with catalyst (acid and base) and additives (alcohols and glycols) was studied and optimized. The samples used are representative for End-of-Life wind turbine blades. The aim is therefore to investigate if it is possible to develop a general process that can accept all material constituents in a real wind turbine blade, resulting in recycled glass fibers and a hydrocarbon fraction that can be used as a refinery feedstock.

1. Introduction

Wind power is an important source for production of renewable energy, and an increasing number of wind farms are built around the globe. However, the End-of-Life issues of glass fiber composite (GFRP) wind turbine blades are yet to be solved. Estimates have suggested that 330,000 t/year until 2028 and 418,000 t/year between 2028 and 2040 of composite materials from wind turbine blades will be deposited/incinerated worldwide [1]. Therefore, new innovative recycling processes for GFRP are needed. As the main driving force for recycling GFRP is the environmental impact rather than economic profitability, this question has been neglected until recently.

The focus of this work is to highlight the challenges of chemical recycling of composite waste materials from the wind turbine blades. The general structure of a wind turbine blade is based on both laminate and sandwich constructions in various parts, but the exact chemical constituents are very seldom disclosed. The wind turbine blade is made of glass fiber, thermosets (epoxy, polyester, and vinyl ester), thermoplastics [i.e. polyvinyl chloride (PVC), polyethylene terephthalate (PET) and polyurethane (PU)], balsa wood, adhesives and coatings [2]. This complex material mix leads to a great challenge for material recycling since the different materials are difficult to separate. For an economically viable process for chemical recycling of GFRPs, both major constituents of the blade (i.e. glass fiber and plastic) need to find usage in new future material streams i.e. recycled glass fiber and chemical building blocks.

One strategy is therefore to separate the insoluble GFRPs by utilizing a high temperature process based on sub/supercritical solvents i.e. solvolysis/HTL (hydrothermal liquefaction). At high temperatures and pressure (200-370 °C, 100-250 bar) water and alcohols can act as both solvent and reactant for GFRPs similarly to organic solvents. By hydrolysis reactions the plastic components are degraded to smaller fragments and the glass fibers are released from the composite network [3]. Solvolysis of GFRP with subcritical water results in glass fiber (solid material) and resin degradation products such as monomers, oligomers and polymers which separate into the oil or water phase depending on their polarity (Figure 1). The solvolysis also result in the formation of new repolymerized polymer structures i.e. oligomers or high molecular weight structures (char) [7].

Previous studies related to the use of chemical recycling by solvolysis of epoxy, polyester, and vinyl ester fiber reinforced composites (FRPs) can be found in the literature (reviews [3, 4]; water [5-7]; alcohols [4, 8]; glycols [9-14]; acetone [5]: acetic acid/hydrogen peroxide [15, 16]; wind turbine blade recycling [2]). Many of these studies focus on recycling of the more valuable carbon fiber composites (CFRP) due to higher market value of the recycled carbon fiber product vs. glass fiber. Another important issue is that most of the studies utilize newly produced composite laminates with one known thermoset and no blends of thermoset GFRPs. Therefore, it is difficult to translate the reported solvolysis studies to real samples from wind turbine blades where different optimized thermoset systems are used for distinct parts within the blade.

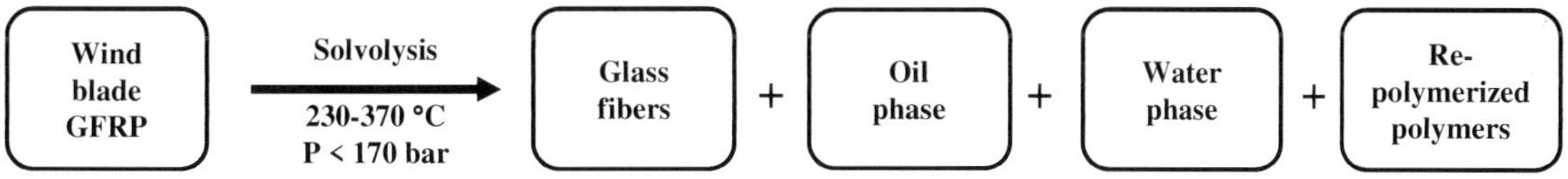

Figure 1. General flow diagram of solvolysis of wind turbine blade GFRP and the resulting product fractions of degradation in subcritical water.

A few studies have utilized material from wind turbine blades for chemical or thermal (pyrolysis) recycling [2, 17-19]. The focus of all studies was on polyester based GFRP. In the GENVIND project solvolysis with subcritical water at 300 °C and 100 bar was used for recycling GFRP from polyester wind blades into a solvolysis oil with high calorific value (40 MJ/kg). The recovered glass fibers were free of the resin and reported to have a tensile strength in the range of 80–90% of virgin fibers [2, 20]. Thermal recycling with a microwave pyrolysis (300-600 °C) on wind turbine blade waste based on polyester GFRP was studied by Åkesson et. al. [19]. Glass fiber with 75% strength of the virgin glass fibers could be recovered along with an oil with calorific value of about 36 MJ/kg. In another project EURECOMP, glass-fiber reinforced polyester was recycled by solvolysis based on hot water and high pressure. The process conditions were optimized at small scale reactor systems and upscaled to a large batch reactor system of 20 L. Solvolysis reaction at 270 °C led to 70 % in mechanical strength of the recovered glass fibers vs. virgin fibers [17].

Critical conclusions from literature available to date on chemical recycling of FRPs by solvolysis:

a. The optimization of reaction time and temperature for retaining highest possible strength of the recovered glass fibers leads to longer reaction times and a higher energy consumption. Increased reaction temperature yields significant reduction of the recovered glass fiber strengths [17].

b. High temperature water has been reported to cause detrimental chemical reactions with glass fiber, reducing their strength. Reports on solvolysis with organic solvents (i.e. alcohols and acetone) have claimed less damage on the recovered glass fibers and increased diffusion of solvent into the FRP due to changed hydrophobicity of the sub/supercritical media [4, 5, 8, 9].

c. Recovered glass fibers need to be washed with organic solvent (acetone) and resized before usage as reinforcement in secondary products [7, 17].

d. Polyester/vinyl thermosets are easier to degrade than epoxy based due to ester linkages can be hydrolyzed/trans esterified [3].

e. Heavily crosslinked matrix networks hinder the diffusion of solvent molecules and are therefore more difficult to degrade by chemical recycling. Sample size also affects the reaction rates i.e. grinded sample needs shorter time than larger chopped pieces [15, 16].

f. Pretreatment with solvent (acetic acid/hydrogen peroxide) for swelling of the FRP can reduce the reaction time by increased diffusion of solvent into the composite structure [21].

The aim is therefore to investigate if it is possible to develop a general solvolysis/HTL process that can accept all material constituents in used wind turbine blades. Therefore, a hit-to-lead approach is adopted for depolymerizing End-of-Life GFRPs in subcritical water (250-370 °C, 100-170 bar) in small-scale equipment (7 mL), which enables screening of various reaction conditions. Experimental data from the solvolysis process of End-of-Life GFRPs and the resulting resin degradation products will be discussed.

2. Materials and experimental section

2.1. Materials

Two different blades with unknown material composition was used: glass fiber composite (GFRP, 1A, 1B, 1D, 2A, 2C and 3A), balsa wood (1C), polyester foam core (2B), rubber (3B) and PVC foam core (3C) (Figure 2 and 3). The different parts of the wind turbine blade were cut in pieces (approx. 10×10 cm) and were grinded to particles <1 mm size. In order to check the performance of the solvolysis equipment, a polyester based GFRP was used as reference sample in solvolysis experiments (PE).

Figure 2. Wind turbine blade 1 sample parts: glass fiber composite (GFRP, 1A, 1B, 1D, 2A and 2C), balsa wood (1C) and polyester foam core (2B) (Enercon).

Figure 3. Wind turbine blade 2 sample parts: glass fiber composite (GFRP, 3A), rubber (3B) and PVC foam core (3C) (blade type LM 29.0 An Bonus 1.3 MW) (Anmet)

2.2. *Solvolysis*
Solvolysis reactions were performed in sealed high-pressure vessels (10 mm inner diameter, stainless steel) with an internal volume of 7 mL, connected to a safety pressure valve (180 bar). In a typical solvolysis experiment 0.3-0.5 g of sample was heated with 5 mL of liquid in sealed pressure vessels to temperatures between 230-370 °C in a tube furnace. After the reaction, the pressure vessels were cooled with water to room temperature and the solvate was rinsed with water, acetone or alcohol and the solid material were separated from the liquids by filtration. The product oil was isolated by extraction with ethyl acetate and acetone, followed by rotary evaporation. Due to the low sample size the oil and the water phases did not separate without the addition of extraction solvent unless at minimum of 5 experiments were combined. The resin degradation yield was calculated by equation (1) using the resin content in the sample before and after solvolysis equation (1).

Equation (1) $$Resin\ degradation\ yield\ (\%) = \frac{W1-W2}{W0} * 100$$

W_1 – *weight of composite before solvolysis*
W_2 - *weight of solid material after solvolysis*
W_0 – *weight of epoxy resin in composite before solvolysis*

Unfortunately, the yield calculated on the weight of solid material before and after solvolysis resulted in a high uncertainty since the sample parts are often deposited in the reaction vessel. Instead, the fraction of epoxy resin in the sample before and after solvolysis, determined by TGA, was used to calculate W0 and W2 based on the sample weight W1. For plastics without glass fibers, the yield was calculated on the weight of solid material before and after solvolysis using the average results of triplicate experiments due to the high result variation.

2.3. *Chemical characterization*
The TGA analysis were performed on a Mettler Toledo Star System. The samples (10 mg) were heated in ceramic pans from 50 °C to 600 °C under N_2 at a heating rate of 10 K/min. Non-homogenous samples were grinded before analysis. The GC-MS analyses were performed on an Agilent 6890N gas chromatograph coupled with an Agilent Model 5975 Inert XL Mass Selective Detector, equipped with a DB-5MS column (30 m × 0.25 mm ID, 0.25 µm film thickness), using helium as carrier gas at 1.0 ml/min and a EI detector of 70 eV. Compounds were identified by NIST 08 library of mass spectra.
The Fourier Transform Infrared (FTIR) spectra were recorded on a NICOLET 6700 apparatus (Thermo Electron Corporation, USA) equipped with a Smart Orbit ATR (Attenuated Total Reflection).

3. Result and discussions
After screening a multitude of different solvolysis reactions with different catalyst and co-solvents such as acids, bases, organic solvents, peroxides, surfactants and catalysts (unpublished results), the most promising solvolysis reaction conditions for degradation of GFRP appears to be ethylene glycol, propylene glycol or water combined with alcohols and catalyzed with base at temperatures of 270-330 °C. The solvolysis efficiency of degrading the organic material (thermoset and thermoplastics) from the GFRP blade parts are presented in Table 1 and 2.

Table 1. Resin degradation efficiency in solvolysis of GFRP wind turbine blade parts (1A, 1B, 1D, 2A and 2C), foam core (2B) and polyester GFRP reference (PE) from different reaction conditions.

Solvolysis experiment	1A (%)	1B (%)	1D (%)	2A (%)	2B (%)	2C (%)	PE (%)
Ethylene glycol, 270 °C, 16h	74	85	50	75	>90[a]	60	>90[a]
Propylene glycol, 270 °C, 16h	52	80	40	85	-	55	90[a]
Water, 1-propanol (20%) and KOH (10%), 330 °C, 4-16h	15-55	10-50	-	10-64	88[a]	-	85[a]
Water, 330 °C, 6h	<10	-	-	<10	-	-	-
Resin content of starting material[b]	70	35	46	46	-	37	20

[a]Yield after 4h; [b]Resin content determined by TGA.

Table 2. Comparison of the solvolysis reaction times and temperatures effects of degradation efficiency in water, 1-propanol (20%) and potassium hydroxide (10%) for GFRP sample 1A.

Experiment	Sample (°C)	Time (h)	Resin degradation (%)
1	270	2	< 10
1	270	16	15
2	330	2	20
2	330	6	40
2	330	16	55

3.1. Solvolysis comparison between water and ethylene/propylene glycol-based systems of GFRPs

Subcritical solvolysis of the epoxy-based samples 1A and 2A in water without additives resulted in poor yields (Table 1). During the optimization of the solvolysis, the use of glycols, 1-propanol and KOH turned out to be the best ways to improve the degradation efficiency. Ethylene glycol and propylene glycol at 270 °C were found to work more efficiently for the solvolysis wind turbine blade 1 (1A, 1B, 1D, 2A and 2C, yield 50-85% resp. 40-85%, Table 1) compared to water based solvolysis (yield 10-55%, Table 1). However, when the recovered glass fiber fraction from solvolysis of 1A in water/1-propanol/KOH for 4 hours was further extracted in acetone for five days in room temperature, the resin was fully removed from the glass fiber. This preliminary result indicates that major parts of thermoset resin in 1A is degraded in water/1-propanol/KOH at 330 °C, forming a solid organic residue on the glass fiber. This suggests that the epoxy thermoset is degraded to a thermoplastic that can be removed from the glass fiber by extraction in organic solvents, and that the epoxy resins are degraded to a higher degree than the yields presented in Table 1 for solvolysis of GRFP in water/1-propanol/KOH at 330 °C. The use of organic solvents such as acetone, alcohols and glycols has a great impact on the sustainability of the solvolysis. Further work is needed to determine the most economical and sustainable solvolysis system for GFRPs.

Comparison of the degradation yields of sample 1A and 1B (70% and 35% resin content respectively) indicates that the resin vs. glass fiber content has minor importance for degradation efficiency.

3.2. Solvolysis and influence of reaction time and temperature of GFRPs
In general, the most effective solvolysis is achieved when the reaction time is prolonged overnight (16 h), while reactions at higher temperatures and shorter reaction time do not result in higher yields. A study on the degradation efficiency of sample 1A with varied temperatures and reaction times in water 1-propanol and base illustrates this relationship (Table 2). The solvolysis performs poorly at 270 °C (yield <10-15%) and gives low yields at 330 °C with short reaction time (20-40%, 2-6 h) and after reaction overnight at 330 °C finally yields 55% removal of the thermoset.

3.3. Solvolysis of polyester- and epoxy-based GFRPs
The system investigated in the present contribution was wind turbine blade 1 (Figure 2) with unknown chemical composition of all ingoing materials. Therefore, a polyester based GFRP was used in the solvolysis for comparison (PE, Table 1). The polyester sample yielded higher degradation yields than the wind turbine blade 1 samples (Table 1). These yields are in good agreement with observations from other studies on polyester GFRPs and illustrate the large difference in resistance to solvolytic degradation between polyester- and epoxy-based composites [3].

3.4. Chemical analysis of solvolysis fractions from GFRPs degradation of wind turbine blade 1
Wind turbine blades' chemical compositions are well-kept trade secrets. Therefore, the chemical contents of the plastics were characterized by FTIR and GC-MS of the starting materials as well as of the solvolysis products. The GFRP thermosets give poor signals and require long scanning times (high glass fiber content up to 65%). However, the GFRP samples from wind turbine blade 1 are identified as epoxy resins according to FTIR signals and bisphenol A detected by GC-MS in the solvolysis extracts. The foam core 2B is identified as polyester both by FTIR and GC-MS of the solvolysis extracts.

Chromatographic analysis of the oil phase fractions with GC-MS generally reveal low content of monomeric organic compounds (generally < 1% of the oil mass) which indicates that the polymers break down into larger oligomeric fragments or hydrophilic compounds that end up in the aqueous phase. Therefore, the GC-MS analysis only reveals a small part of the reaction products. Elemental analysis (CHN) and characterization of ground elements with X-ray fluorescence (XRF) are important tools for identifying starting materials and oil phase composition. For the GFRP samples (Figure 2 and 3), calcium and silicon are the main elements due to glass fiber content, with traces of chlorine (≤0.2%) and iron (≤0.8%), while the hydrocarbon content varies between 30-50%. CHN analysis of samples 1A and 1B showed nitrogen content (2.7 resp. 2.4%) indicating that these GFRPs contain amine cured epoxy thermosets. The corresponding solvolysis oil/wax products showed less than 0.1% of nitrogen, silicon, chlorine and iron. This indicates that the solvolysis can be a future tool for separating unwanted additives such as nitrogen-containing polar compounds and inorganic ions usually found in the wind turbine blades. For further upgrading and use of the produced solvolysis oil into resins, chemicals or fuels, the removal of chlorine and nitrogen atoms is essential. This study indicates that solvolysis have the possibility to act as a recycling method for cleaning mixed plastic waste streams like GFRPs from wind turbine blades before further processing and upgrading in a refinery [22].

3.5. Solvolysis and chemical analysis of thermoplastics and balsa wood parts from wind turbine blades
The wind turbine blades sandwich structures generally contain a foam core composed of other plastics such as PVC, polyester and PU or balsa wood. It seems possible to separate these core materials from the GFRP at an early stage of the solvolysis. Foam core samples 2B and 3C are nearly fully degraded in water/1-propanol/KOH at 330 °C after 4 h. These reaction conditions result in ≤15% degradation of the

epoxy based GFRP samples. The foam core 3C was found to have a chlorine content of >10% according to XRF, indicating PVC as the main component. Preliminary results from the XRF of the resulting oil phase fraction show only 2% chlorine which suggests that the chlorine content can be reduced from polymeric materials by solvolysis. Polymers containing chlorine are problematic in all types of chemical recycling due to formation of strong acids at high temperatures (i.e. hydrochloric acid, HCl) [23]. Therefore, this solvolysis technique seems to be a promising tool for removal of unwanted chlorine atoms prior to use in industrial chemical processes or combustion facilities.

Balsa wood core material was also degraded by solvolysis in water/1-propanol/KOH at 330 °C and yielded a fraction that shows the characteristics of paper pulp. These reaction conditions are similar to those used by the pulp and paper industry in existing plants (Kraft process) to produce cellulose pulp [24, 25]. The degraded balsa wood fraction ends up in the filtrated solid materials after solvolysis and is possible to separate from the glass fibers for small scale experiments.

3.6. Solvolysis experiment in larger scale of wind turbine blade 1

In order to examine the solvolysis efficiency on larger samples, a part of wind turbine blade 1 was sawn to pieces of 5 ×35 ×4 mm (0.5 g), displayed in Figure 4. A two-step process with 16 hours in ethylene glycol at 270 ° C followed by 3 hours in water with 1-propanol (20%) and KOH (10%) at 330 °C yields 75% resin removal. The recovered glass fiber fraction from the first step of the treatment yields a material that can easily be cut into smaller fragments. In a larger process set up this could be a way forward to reduce time and cost for shredding or grinding of the GFRPs which is resource consuming.

Figure 4. Solvolysis of a larger piece GFRP (5×35×4 mm, left) of wind turbine blade 1 in two reaction steps: 1. ethylene glycol (270 °C, 16 h) 2. Water,1-propanol (20%) and KOH (10%) at 330 °C for 3h. Recovered glass fiber fraction from solvolysis to the right in the figure.

4. Conclusions

A chemical recycling method for samples from decommissioned wind turbine blades through solvolysis have been investigated and optimized to be able to separate plastic resins from glass fibers in a resource efficient approach. Most of the sample resins are characterized as epoxy which are difficult to degrade and separate from the glass fiber without the use organic solvents in addition to long reaction times. The best reaction conditions are achieved in a two-step process where the GFRP material is heated to 270 °C in ethylene glycol for 16 hours followed by heating to 330 °C in water/1-propanol/KOH for 3 hours.

Preliminary analysis of the produced oil phase products from wind turbine blade GFRP and foam cores indicate removal of nitrogen-containing hardeners in epoxy and a reduction of the chlorine content from PVC-based foam cores. Producing solvolysis oil with low levels of nitrogen and chlorine is important for further upgrading of the solvolysis oil to useful end products i.e. resins, chemicals or fuels. This study indicates that solvolysis have the possibility to act as a recycling method for cleaning mixed plastic waste streams like GFRPs from wind turbine blades before further processing and upgrading in a refinery.

However, large challenges still persist regarding chemical recycling with solvolysis of mixed waste streams of End-of-Life GFRPs and requires continued method development in order to reduce the consumption of energy and chemicals. One of the major challenges are the long reaction times due to thick and dense materials, and future work will focus on the use of flow reactors and recycled solvents for the optimization of the solvolysis.

Acknowledgments
The authors would like to thank Swedish Energy Agency (Chemical recycling of glass fiber composite from wind turbine blades - Rekovind), 2018-007282) and Vinnova (ReComp – Circular streams from fiberglass composites, 2018-04132) for financial support. Enercon (Germany, https://www.enercon.de) and Anmet (Polen, https://www.anmet.com.pl) are greatly acknowledged for support of wind turbine blade samples.

References

[1] Naqvi S R, Prabhakara H M, Bramer E A, Dierkes W, Akkerman R and Brem G 2018 A critical review on recycling of end-of-life carbon fibre/glass fibre reinforced composites waste using pyrolysis towards a circular economy *Resour Conserv Recycl* **136** 118-29
[2] Jensen J P and Skelton K 2018 Wind turbine blade recycling: Experiences, challenges and possibilities in a circular economy *Renew. Sust. Energ. Rev.* **97** 165-76
[3] Oliveux G, Dandy L O and Leeke G A 2015 Current status of recycling of fibre reinforced polymers: Review of technologies, reuse and resulting properties *Prog. Mater. Sci.* **72** 61-99
[4] Karuppannan Gopalraj S and Kärki T 2020 A review on the recycling of waste carbon fibre/glass fibre-reinforced composites: fibre recovery, properties and life-cycle analysis *SN Appl Sci* **2** 433
[5] Sokoli H U, Beauson J, Simonsen M E, Fraisse A, Brøndsted P and Søgaard E G 2017 Optimized process for recovery of glass- and carbon fibers with retained mechanical properties by means of near- and supercritical fluids *J Supercrit Fluids* **124** 80-9
[6] Oliveux G, Bailleul J-L, Le Gal La Salle E, Lefèvre N and Biotteau G 2013 Recycling of glass fibre reinforced composites using subcritical hydrolysis: Reaction mechanisms and kinetics, influence of the chemical structure of the resin *Polym. Degrad. Stab.* **98** 785-800
[7] Oliveux G, Bailleul J-L and Salle E L G L 2012 Chemical recycling of glass fibre reinforced composites using subcritical water *Compos. Part A Appl. Sci. and Manuf.* **43** 1809-18
[8] Piñero-Hernanz R, García-Serna J, Dodds C, Hyde J, Poliakoff M, Cocero M J, Kingman S, Pickering S and Lester E 2008 Chemical recycling of carbon fibre composites using alcohols under subcritical and supercritical conditions *J Supercrit Fluids* **46** 83-92
[9] Yang P, Zhou Q, Li X-Y, Yang K-K and Wang Y-Z 2014 Chemical recycling of fiber-reinforced epoxy resin using a polyethylene glycol/KOH system *J Reinf. Plast. Compos* **33** 2106-14
[10] Shin S-R, Mai V D and Lee D-S 2019 Chemical Recycling of Used Printed Circuit Board Scraps: Recovery and Utilization of Organic Products. *Processes* **7** 22
[11] El Gersifi K, Destais-Orvoën N, Durand G and Tersac G 2003 Glycolysis of epoxide-amine hardened networks. I. Diglycidylether/aliphatic amines model networks *Polymer* **44** 3795-801
[12] Destais-Orvoën N, Durand G and Tersac G 2004 Glycolysis of epoxide–amine hardened networks II—aminoether model compound *Polymer* **45** 5473-82
[13] El Gersifi K, Durand G and Tersac G 2006 Solvolysis of bisphenol A diglycidyl ether/anhydride model networks *Polym. Degrad. Stab.* **91** 690-702
[14] Yildirir E, Onwudili J A and Williams P T 2014 Recovery of carbon fibres and production of high quality fuel gas from the chemical recycling of carbon fibre reinforced plastic wastes *J Supercrit Fluids* **92** 107-14

[15] Ma Y and Nutt S 2018 Chemical treatment for recycling of amine/epoxy composites at atmospheric pressure *Polym. Degrad. Stab.* **153** 307-17

[16] Ma Y, Kim D and Nutt S R 2017 Chemical treatment for dissolution of amine-cured epoxies at atmospheric pressure *Polym. Degrad. Stab.* **146** 240-9

[17] 2009 - 2012 EURECOMP (Recycling Thermoset Composites of the SST). (https://cordis.europa.eu/project/id/218609/reporting

[18] Pico D, Seide G and Gries T J C E 2014 Thermo chemical processes: potential improvement of the wind blades life cycle **36**

[19] Åkesson D, Foltynowicz Z, Christéen J and Skrifvars M 2012 Microwave pyrolysis as a method of recycling glass fibre from used blades of wind turbines *J Reinf. Plast. Compos.* **31** 1136-42

[20] Sokoli H U, Simonsen M E, Nielsen R P, Arturi K R and Søgaard E G 2016 Conversion of the matrix in glass fiber reinforced composites into a high heating value oil and other valuable feedstocks *Fuel Process. Technol.* **149** 29-39

[21] Wang Y, Cui X, Yang Q, Deng T, Wang Y, Yang Y, Jia S, Qin Z and Hou X 2015 Chemical recycling of unsaturated polyester resin and its composites via selective cleavage of the ester bond *Green Chem.* **17** 4527-32

[22] Wong S L, Ngadi N, Abdullah T A T and Inuwa I M 2015 Current state and future prospects of plastic waste as source of fuel: A review *Renew. Sustain. Energ. Rev.* **50** 1167-80

[23] Al-Salem S M, Antelava A, Constantinou A, Manos G and Dutta A 2017 A review on thermal and catalytic pyrolysis of plastic solid waste (PSW) *Supercrit. Fluid Sci. Technol., J. Environ. Manage.* **197** 177-98

[24] Sasaki M, Fang Z, Fukushima Y, Adschiri T and Arai K 2000 Dissolution and Hydrolysis of Cellulose in Subcritical and Supercritical Water *Ind. Eng. Chem. Res.* **39** 2883-90

[25] Bose S K, Omori S, Kanungo D, Francis R C and Shin N H 2009 Mechanistic Differences Between Kraft and Soda/AQ Pulping. Part 1: Results from Wood Chips and Pulps *J. Wood Chem. Technol.* **29** 214-26

[26] Caplar R and Kulisic P 1973 *Proc. Int. Conf. on Nuclear Physics (Munich)* vol 1 (Amsterdam: North-Holland/American Elsevier) p 517

Recyclable epoxy systems for rotor blades

Pradip K Dubey[1], Satish K Mahanth[2], Amit Dixit[3] and Sirirat Changmongkol[4]

Aditya Birla Chemicals (Thailand) Limited (Advanced Materials)
Maptaphut Industrial Estate, No. 2, I-5 Rd., Maptaput, Muangrayong, Rayong, Thailand

E-mail:[1]pradipkumar.dubey@adityabirla.com,[2]satishkumar.mahanth@adityabirla.com, [3]amit.dixit@adityabirla.com, [4]sirirat.changmongkol@adityabirla.com

Abstract. The global energy demand is largely met by power generated from fossil fuels (Thermal Power plants) and nuclear means (Nuclear Power plants). Even though both these technologies have been used for decades, they have huge environmental impact and severe sustainability issues with depleting fossil fuel and serious concerns about the spent nuclear rods disposal. With growing awareness of impact on the environment and finding a sustainable solution for power generation, wind energy offers a lasting solution that addresses both issues, however it is still not truly sustainable as the primary materials used for manufacturing of rotor blades are non-recyclable. A new revolutionary technological solution has been invented where in, the epoxy thermoset can be recovered as thermoplastic and the reinforcement in composite can be re-used, making it a first of its kind in the thermoset industry segment. The Recyclable Epoxy Systems for rotor blades provide a path breaking solution that enables recycling of epoxy thermoset, recovery and re-use of the reinforcement and the matrix in a fibre reinforced composite. Leveraging from the proprietary Recyclamine® platform technology, series of recyclable epoxy systems are developed for wet lay-up and infusion processes to meet the requirements for both on-shore and off-shore rotor blades, including next generation rotor blades. The systems are characterized by determining process and performance properties and found to provide distinct advantages in comparison to the conventional non-recyclable epoxy systems. The recycling of composites made from recyclable epoxy systems is successfully demonstrated by low energy solvolysis followed by recovery of the reinforcement and epoxy matrix as thermoplastic. The recovered reinforcement and epoxy thermoplastic are re-used to make composite and representative thermoplastic object respectively, deriving value from the waste and closing the loop to make the wind energy industry sustainable for a circular economy.

1. Background

Wind Energy as a renewable source of energy has come a long way in being recognized as a major alternative source to the conventional energy sources based on fossil fuels. The global cumulative installed capacity of wind power generators reached 651GW at the end of 2019 witnessing a 10 % increase over the previous year [1]. The growth and maturation of the wind industry is duly supported by technological advancements in wind turbine rotor blades. The emergence of new – longer, stiffer and aerodynamically advanced rotor blade designs has enabled to harness more energy from the wind.

Designers and rotor blade manufacturers are continuously looking for new performance driven, robust materials to support the latest and future developments.

Incidentally, the industry is also facing compelling issues, such as process waste management and end of life management of aged rotor blades. It is estimated that 85% of the components of a wind turbine can be recycled and re-used except the rotor blades due to the non-recyclable thermoset matrix [2]. In addition it is estimated that 10% of the process waste generated during manufacturing of a rotor blade is non-recyclable. The present methods of waste management (Figure 1) such as landfills, pyrolysis, incineration, and pelletization are either not environment friendly or are energy intensive and fail to derive value. Solvolysis, involving nucleophilic substitution or elimination, is a clean and efficient method, however its use is limited due to non-recyclable nature of the epoxy thermoset.

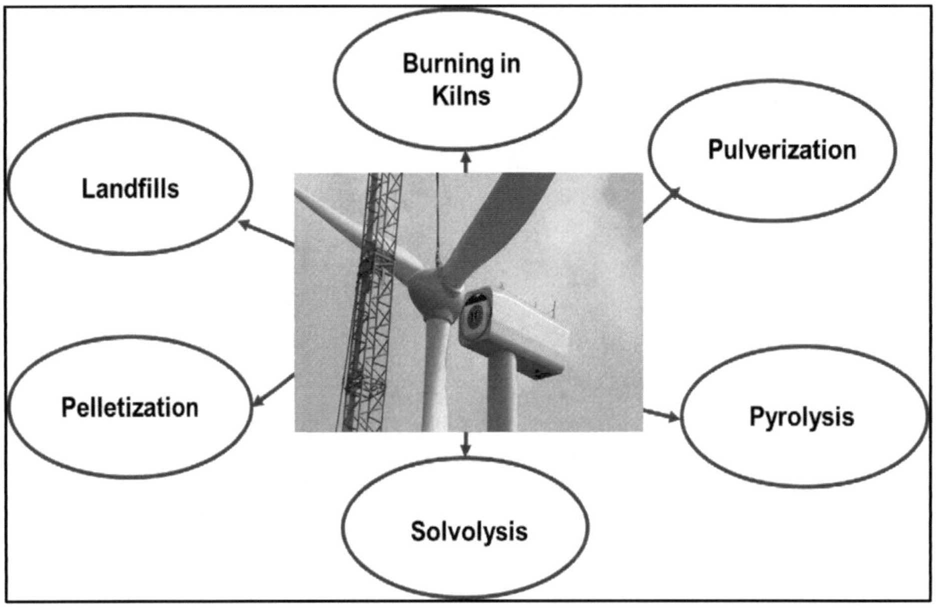

Figure 1: Present Methods for End of Life Waste Management of Rotor Blades

With >4000 aged rotor blades scheduled to be decommissioned annually [2] and lack of full-proof or robust methods to manage waste, the growth and sustenance of wind energy is at stake.

2. Recyclamine® : Concept and Technology

Epoxy resins are preferred as polymer matrix materials for manufacturing of rotor blades on account of their versatility and outstanding process & performance properties. The epoxy resin systems used as matrix for manufacturing of rotor blades comprise of epoxy resin component and curing agent or hardener component which react to cross link and cure into rigid three dimensional infusible network that cannot be re-formed, re-used or re-cycled. With emergence of new rotor blade designs, epoxy resin systems are being continuously developed to align with rotor blade manufacturer and the designer's requirement, however the non-recyclable nature of epoxy thermosets does not address the compelling issue of process waste management and end of life management of rotor blades.

Recyclamine® technology provides an opportunity with path breaking and lasting solution that enables recyclable epoxy thermosets [3][4].The technology is based on specifically engineered recyclamine® curing agents or hardeners which cross link with epoxy resins to provide a network with cleavage points. The matrix comprising of epoxy resin and recyclamine hardeners in polymer composites, can be cleaved by solvolysis under specific conditions, leading to recovery of reinforcement such as carbon fiber, glass fiber, kevlar® and the epoxy matrix as recyclable thermoplastic. The recovered reinforcement and epoxy thermoplastic can be re-used and re-purposed. The concept is depicted pictorially in Figure 2.

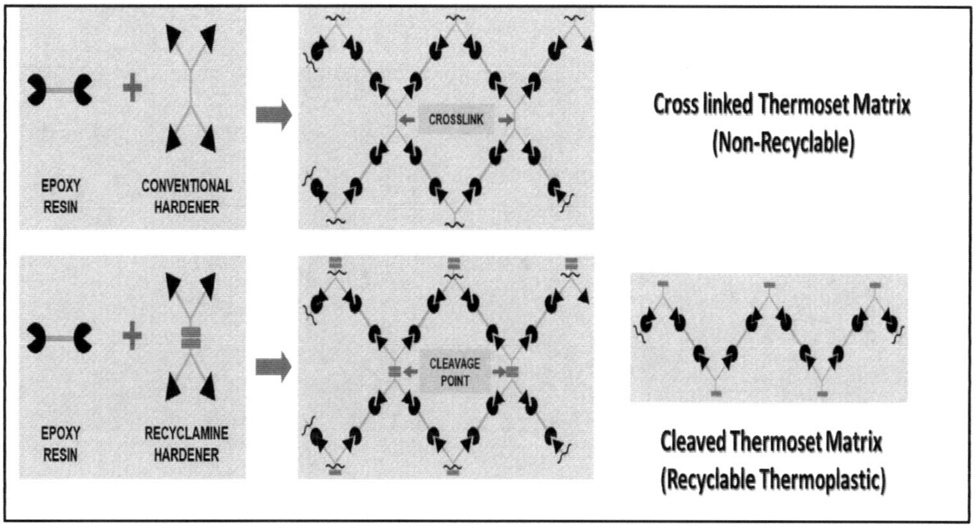

Figure 2: Schematic depicting Curing of Epoxy Resin Systems
(Non-Recyclable v/s Recyclable)

The Recyclamine® technology is a platform chemistry with series and wide spectrum of curing agents providing fast to slow reactivity & latency, low to high temperature resistance, chemical resistance and recyclability (Figure 3).

Platform 1	R101	R102	R103		
Platform 2	R301	R302	R303		
Platform 3	R401	R402	R403		
Platform 4	R501	R502	R503		
Platform 5	R701	R702	R703	R704	R705
Platform 6	R801	R802	R803	R804	R805
Platform 7	R901	R902	R903	R904	R905

Reactivity: Fast → Slow (top to bottom)
Temp. Resistance: Low → High (top to bottom)
Chemical Resistance: Low → High (left to right)
Reactivity: Fast ← Slow (left to right)

Figure 3: The Recyclamine® Curing Agent Platform

3. Recyclable Epoxy Systems from Recyclamine® Technology:

The Recyclamine® technology was leveraged for the development of recyclable epoxy resin systems for rotor blades. Laminating resin systems for wet lay-up process and infusion process were developed by selecting suitable epoxy resins [5] and appropriate Recyclamine® hardeners. Series of recyclable systems with varying reactivity were developed and characterized for process and performance properties (Table 1).

Property	Unit	System A	System B	System C	System D	System E
Physical and Process Properties		Wet lay-up Systems		Infusion Resin Systems		
Mixing ratio	by wt.	100:32	100:22	100:20	100:24	100:29
AHEW (calculated)	gm/ eq	57.3	40.5	33.5	40.5	47.5
Resin viscosity @ 25°C	mPa.s	11,447	12,270	2,606	1,243	1,170
Hardener viscosity @ 25°C	mPa.s	9.11	8.83	4.92	7.29	8.3
Initial mix viscosity @ 25°C	mPa.s	780.4	889.8	283.7	219	194.1
Pot life, 100g mix @ 25°C	minutes	17.5	80.5	82.5	365	>800
T_g, cured @ 80°C/ 4 hours	°C	99.75	89.74	84.17	80.69	86.72
Mechanical Properties						
• Tensile Test (ISO 527-2) [5]						
Tensile Strength	MPa	73.49	65.32	67.32	67.30	76.37
Tensile Strain	%	4.84	4.52	5.80	4.47	4.44
E-Modulus	MPa	3,280	2,958	2,302	2,872	3,104
• Flexural Test (ISO 178) [6]						
Flexural Strength	MPa	131.11	114.26	128.89	108.51	132.12
Flexural Strain	%	6.38	5.20	4.78	6.06	5.90
E-Modulus	MPa	3,388	3,144	4,604	2,804	3,289

Table 1: Physical, Process and Mechanical Properties of Recyclable Epoxy Systems

System A and B were developed for wet lay-up process and designed to provide fast and moderate reactivities whereas Systems C,D and E were developed for infusion process and designed to provide moderate, slow and ultra-slow reactivities. Properties of conventional, non-recyclable epoxy systems were used as benchmark for development of the recyclable systems. The processabilty of the recyclable systems was measured by measuring initial mix viscosity [9] while the reactivity was determined by pot life test. Performance of the systems was evaluated by mechanical properties; tensile [6] and flexural [7] of the neat cured system.

4. Process and Performance Characterization of Recyclable Epoxy Infusion System E

The properties of recyclable epoxy systems were found to be comparable to the conventional wet lay-up and infusion systems used for rotor blades manufacturing. In order to illustrate advantages of recyclable systems, infusion System E is selected as a representative example in this paper. The system E was specifically developed considering the requirements of offshore blades which are large structures typically > 70 meter in length and necessitate slow reacting matrix systems to ensure controlled and defect free infusion of the reinforcement during processing.

The recyclable resin system E was designed as an extension of the proprietary ultra-slow fast cure technology from Aditya Birla Chemicals [8]. The resin component of the system was formulated by selecting epoxy resins that could enable low viscosity and provide ultra-slow reactivity with the hardener component based on appropriate Recyclamine® molecules. The system was characterized for process and performance properties in comparison to conventional slow reacting non-recyclable system used for offshore blades.

4.1. Processing Properties of Conventional and Recyclable System E

The processing properties of the conventional and recyclable system were determined under identical conditions (temperature, humidity, process and test method) and results were compared (Table 2).

Property	Test Standard	Unit	Conventional System	Recyclable System E
Viscosity @ 25°C • Resin • Hardener	ASTM D 2196 [8]	mPa.s	1,464 13.7	1,170 8.3
Mixing ratio	-	by weight	100:33	100:29
Initial mix viscosity @ 25°C	ASTM D 2196 [8]	mPa.s	250.1	194.1
Rise in viscosity @ 30°C (time to reach 1,000 mPa.s)	ISO 3219 [9]	minutes	198.0	356
Pot life 1000 gms. mix @ 25°C (time to reach 60°C)	ASTM D 2471 [10]	minutes	300	555
Glass transition temperature, Tg (cured @ 70°C/ 6 hrs.)	ISO 11357-2 [11]	°C	78.56	77.08
Ultimate Tg (Tg2)		°C	87.91	87.49

Table 2: Processing Properties of Conventional and Recyclable System E

The recyclable epoxy system E provided lower initial mix viscosity and significantly slower rise in viscosity, longer pot life compared to the conventional system. These properties are indicative of improved processability, faster and controlled infusion of the reinforcement leading to defect reduction in rotor blade manufacturing. Importantly the glass transition temperature (Tg) which relates to degree of cross-linking was found to be comparable for the conventional and recyclable system E, which indicates that the slow reactivity of recyclable system will not have an influence on in-mold time thus the cycle time to manufacture rotor blade will be unchanged.

4.2. Performance Properties of Conventional and Recyclable System E

The performance properties of conventional and recyclable system were determined by mechanical properties, fatigue behavior under identical conditions (temperature, humidity, test method and equipment) and results were compared.

Property	Unit	Conventional System	Recyclable System E
• Tensile test (ISO 527-2) **[5]**			
Tensile strength	MPa	67.44	76.37
Elongation at tensile strength	%	4.63	4.44
Ultimate strength	MPa	63.34	70.66
Ultimate elongation	%	8.07	8.15
Tensile modulus	MPa	2,758	3,104
• Flexural test (ISO 178) **[6]**			
Flexural strength	MPa	113.82	132.12
Elongation at flexural strength	%	6.19	5.90
Ultimate strength	MPa	96.03	113.33
Ultimate elongation	%	9.88	9.12
Flexural modulus	MPa	2,882	3,289

Curing Condition : 80°C / 8 hours

Table 3: Comparative Mechanical Properties of Cured Epoxy Matrix

The tensile and flexural properties, strength and stiffness of the cured epoxy matrix for 10 specimens per series were found to be considerably higher for the recyclable system E compared to the conventional system (Table 3). The characterization of mechanical properties for glass reinforced epoxy composite was done by preparing laminates by vacuum assisted resin transfer molding (VARTM) process using uni-directional (1200 gsm) and bi-directional (800 gsm) fabrics with fiber volume fraction, 50-55%. Tensile, shear and fracture toughness (G1c) properties were determined, results indicated higher values for the recyclable system E compared to the conventional system (Table 4).

Property	Unit	Conventional System	Recyclable System E
• Tensile properties of fiber reinforced plastic in 0° direction of UD (4 layers) according to ISO 527-5 **[12]**			
Tensile strength	MPa	815.64	856.48
Fracture strain	%	2.02	2.00
Modulus (Chord 0.05%-0.25%)	MPa	43,507	44,355
• Shear properties of fiber reinforced plastic in ± 45° direction of 2AX according to ISO 14129 **[13]**			
Shear stress at 5% shear strain	MPa	39.11	43.47
Shear modulus	MPa	3,220	3,555
Tensile strength	MPa	124.39	136.20
• GIc of fiber reinforced plastic test according to ASTM D 5528 **[14]**			
G1c (at 1 mm crack, MBT method)	kJ/m^2	0.372	0.387

Curing Condition : 80°C / 8 hours

Table 4: Comparative Mechanical Properties of Glass Reinforced Epoxy Composite

The fatigue behavior of conventional and recyclable system was determined by conducting test for reinforced epoxy laminate made from uni-directional glass fabrics. Test was conducted in strain control mode, at 40% of ultimate strain and ratio R=0.1, frequency = 4 hertz. Change in load with number of cycles was monitored, test was continued until 800,000 cycles. Residual tensile properties; strength, elongation and E-modulus, of specimens were determined and compared with specimens before fatigue (Figure 4).

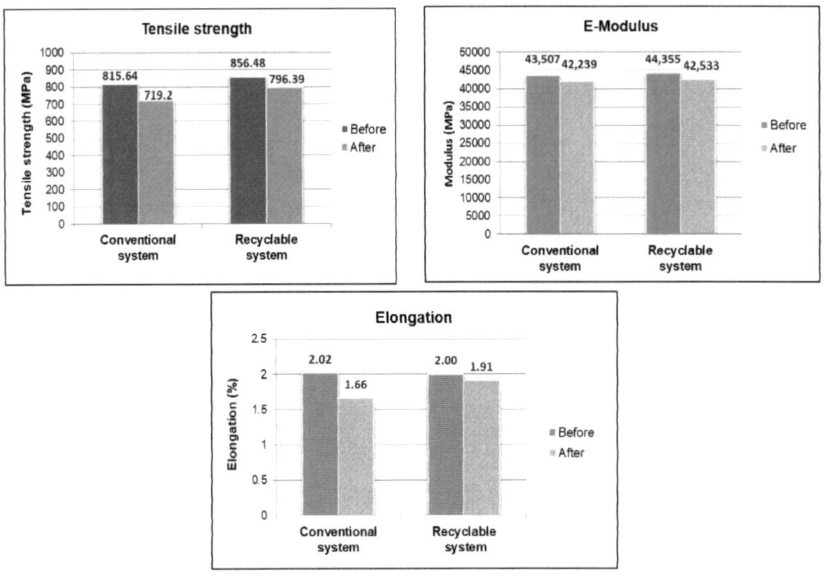

Figure 4: Comparative Tensile Properties, before and after fatigue test

5. Recycling of Glass Reinforced Epoxy Composite and Recovery of Matrix and Reinforcement

The recycling of glass reinforced epoxy composite made from recyclable epoxy system E and recovery of matrix and reinforcement was demonstrated by preparing a laminate with VARTM process using tri-axial fabrics (1200 gsm). The laminate was trimmed and subjected to solvolysis by immersing in recovery solution and conditioned at 80^{0}C for 6 hours to enable cleavage of the recyclable epoxy matrix and its dissolution in the recovery solution. The glass fabric reinforcement of the composite, recovered from the recycling process was rinsed and dried. Recovery solution containing the dissolved epoxy matrix was filtered, neutralized and coagulated to recover the epoxy matrix as thermoplastic (Figure 5).

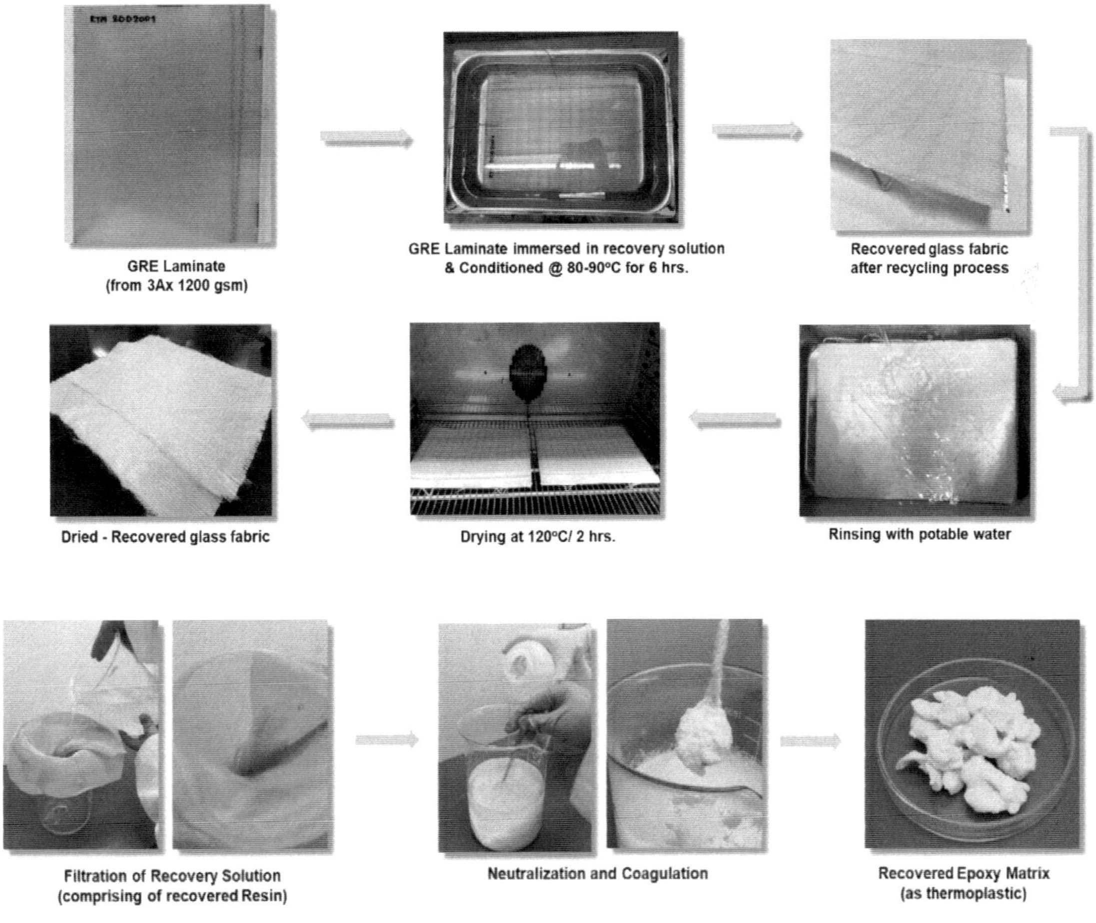

Figure 5: Recycling of Composite and Recovery of Matrix and Reinforcement

6. Re-use of Recovered Reinforcement and Thermoplastic

The tri-axial reinforcement recovered from recycling process was re-used by preparing laminate by VARTM process using recyclable epoxy system E as the matrix. Laminate was also prepared with fresh tri-axial (1200 gsm) reinforcement keeping all parameters identical. Specimens were extracted from both laminates and characterized for tensile properties using Universal testing machine Instron 5569 and clip on extensometer. Results indicated 10% lower tensile strength and modulus for the laminate prepared from recovered reinforcement evidently due to distortion in fibre alignment, however the tensile strain was found to comparable to the laminate prepared from fresh reinforcement (Figure 6).

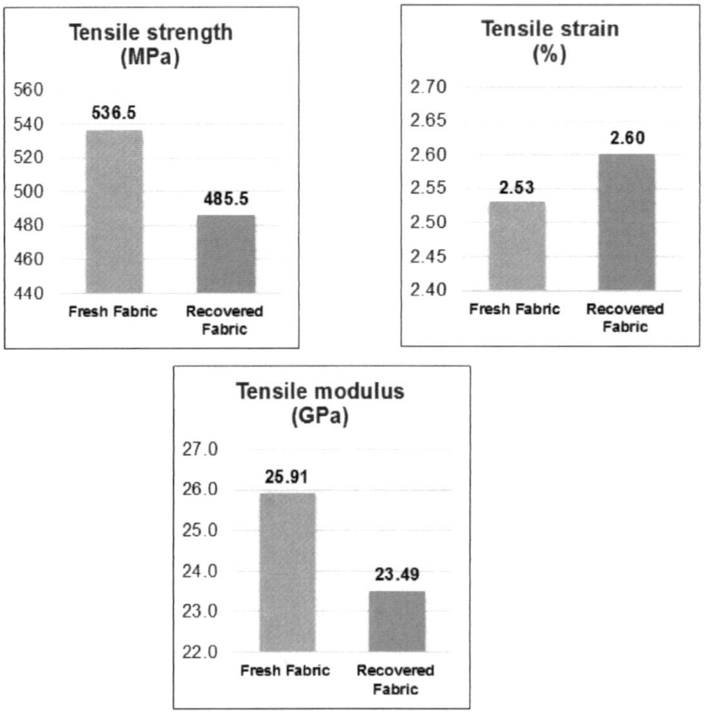

Figure 6: Comparative Tensile Properties of Laminate from Fresh and Recovered Fabrics

The epoxy thermoplastic recovered from recycling process was dried and characterized for physical, thermal and mechanical properties such as melt flow index, glass transition temperature, tensile and flexural properties (Table 5).

Property	Test Standard	Unit	Results
Melt Flow Index (190°C @ 2.16 kg)	ASTM D 1238 [15]	minutes	10
Specific gravity	ASTM D 792 [16]	gm/cm^3	1.19
Glass Transition Temperature (Tg)	ASTM D 3418 [17]	°C	78
Tensile Strength	ASTM D 638 [18]	MPa	51.2
Tensile Modulus		GPa	2.90
Elongation at break		%	30-110
Flexural Strength	ASTM D 790 [19]	MPa	100
Flexural Modulus		GPa	2.7

Table 5: Characterization of Recovered Thermoplastic

The thermoplastic was compounded with 80% polyethylene by weight and injection molded to re-purpose and make foot strap insert used in surfboards (Figure 7).

Figure 7: Foot Strap Insert from Recovered Thermoplastic

7. Summary and Conclusion

The Recyclamine® technology enables recycling of epoxy thermosets and further recovery, re-use of reinforcement, epoxy matrix in fibre reinforced composite. The technology was successfully leveraged to develop series of recyclable epoxy resin systems for rotor blades thereby providing path breaking and lasting solution to the compelling issue of waste management faced by the wind industry.

The recyclable epoxy systems were characterized for process and performance properties and benchmarked with conventional, non-recyclable epoxy systems. The recyclable epoxy infusion system E, developed for longer, aerodynamic offshore blades provided low viscosity and ultra-slow reactivity which is advantageous for improved impregnation and process defect reduction and importantly cured in the same time as the conventional system. Composite laminate prepared from the system was recycled by solvolysis to recover reinforcement and epoxy matrix as thermoplastic. Re-use of the reinforcement and re-purposing of the recovered thermoplastic was demonstrated to close the loop to make wind energy *"Sustainable for a Circular Economy"*.

8. References

[1] Global Wind Energy Council , GWEC Global Wind Report 2019.
[2] Chris Martin, *Wind turbine blades can't be recycled, so they're piling up in landfills*, Bloomberg Green, February 5, 2020
[3] Connora Technologies, US Patent 20190016667A1, *"Salts of Diaminoacetals and Diaminoketals and Their Synthesis and Their Transformations to Diaminoacetals and Diaminoketals"*.2019.
[4] Connora Technologies, US Patent 20190016667A,*"Synthesis of and compositions containing Diaminoacetals and Diaminoacetals."*2019.
[5] Lee,Henry.,Neville,Kris.*Handbook of Epoxy Resins;* New York: McGraw-Hill, 1967.
[6] ISO Standard 527-2, "Plastics – Determination of tensile properties – Part 2: Test conditions for moulding and extrusion plastics".
[7] ISO Standard 178, "Plastics — Determination of flexural properties".
[8] Aditya Birla Chemicals Thailand Limited, WO/2019/053645, *"An Epoxy Resin System for Structural Composites."2019.*
[9] ASTM Standard. D2196-05, "Standard Test Methods for Rheological Properties of Non-Newtonian Materials by Rotational (Brookfield type) Viscometer".

[10] ISO Standard 3219, "Plastics — Polymers/resins in the liquid state or as emulsions or dispersions — Determination of viscosity using a rotational viscometer with defined shear rate".
[11] ASTM Standard D2471-99, "Standard Test Method for Gel Time and Peak Exothermic Temperature of Reacting Thermosetting Resins".
[12] ISO Standard 11357-2, "Plastics — Differential scanning calorimetry (DSC) — Part 2: Determination of glass transition temperature and glass transition step height".
[13] ISO Standard 527-5, "Plastics – Determination of tensile properties –Part 5: Test conditions for unidirectional fibre-reinforced plastic composites".
[14] ISO Standard 14129, "Fibre-reinforced plastic composites — Determination of the in-plane shear stress/shear strain response, including the in-plane shear modulus and strength, by the plus or minus 45 degree tension test method".
[15] ASTM Standard D5528-01, "Standard Test Method for Mode I Interlaminar Fracture Toughness of Unidirectional Fiber-Reinforced Polymer Matrix Composites".
[16] ASTM Standard D1238-13, "Standard Test Method for Melt Flow Rates of Thermoplastics by Extrusion Plastometer".
[17] ASTM Standard D792-13, "Standard Test Methods for Density and Specific Gravity (Relative Density) of Plastics by Displacement".
[18] ASTM Standard D3418-99, "Standard Test Method for Transition Temperatures of Polymers by Differential Scanning Calorimetry.
[19] ASTM Standard D638-14, "Standard Test Method for Tensile Properties of Plastics".
[20] ASTM Standard D790-17, "Standard Test Methods for Flexural Properties of Unreinforced and Reinforced Plastics and Electrical Insulating Materials".

IOP Conf. Series: Materials Science and Engineering **942** (2020) 012015 doi:10.1088/1757-899X/942/1/012015

Re-use of wind turbine blade for construction and infrastructure applications

A André*, J Kullberg, D Nygren, C Mattsson, G Nedev and R Haghani

RISE Research Institutes of Sweden, SE-431 22 Mölndal, Sweden

*Email: alann.andre@ri.se

Abstract. To achieve a more resource-efficient society with a future with reduced carbon dioxide emissions, new technological challenges must be dealt. One way to reach a more sustainable world is to start re-using end-of-life structures and waste and give them a 'Second Life" with a new function in the society. As composite structures are lightweight, strong, stiff and durable materials, there is great potential to re-use decommissioned composite for new resource-efficient solutions in the building and infrastructure sector. The present paper investigates innovative solutions in re-using wind turbine blades as elements in new bicycle and pedestrian bridge designs. Several conceptual bridge designs where wind blades utilized as load bearing elements were developed and studied. The main design requirements for pedestrian bridges were considered and assumptions regarding wind blades quality and their mechanical properties addressed based on interviews with industries working with wind turbine blades repair and recycling. The aim of this paper is to contribute to a sustainable use of fibre reinforced polymer (FRP) waste and at the same time provide a more cost-effective FRP bridges. In a larger perspective, the authors would like to highlight the economically profitable potential of recovering and reusing / re-manufacturing end-of-life glass FRP composites.

1. Introduction

In 2017, 1118 kilotons of glass fibre reinforced polymer composites (GFPRs) were produced in Europe, of which 40 kilotons in the Nordic region, which showed an increase of 10% since 2012. At the present, GFRPs represent 98% of the composite market in the boat and wind turbine sectors [1]. In the wind turbine sector only, it is estimated that the amount of GFRP products reaching their end-of-life (EoL) in the coming decades will be considerable [2]. This is a major concern in terms of sustainability as reuse or recycling of GFRP is rarely considered in the industry today. Current EoL GFRP parts are either incinerated to produce heat (over 100€/ton) or are discarded in landfills with detrimental effect on the environment [3-5].

Wind energy is currently one of the fastest growing renewable energy sources and is expected to increase in the future. The lifetime of a wind turbine, whose blades are mainly made of GFRPs, is approximately 20-25 years, which will lead to a large number of decommissioned turbines in a near future. By 2025, about 36 000 blades in Europe are expected to be dismantled, which corresponds to 240 000 tons of polymer composite waste [6]. Today, there is no sustainable and efficient solution to recycle these structures, as they are complex to recycle [7-8]. If the EoL treatment is not addressed in a sustainable way, wind energy cannot be fully considered as sustainable energy source. Therefore, new solutions to undertake these growing issues must be developed [9]. These challenges are global, and solutions used currently are far from the idea of considering waste as a resource (figure 1)

Figure 1: Landfill of EoL wind turbine blades in USA [10].

The recent EU directive from 2015 (2000/53 / EC) suggests that 95% of the EoL/waste should be reused or recycled. In addition, new strict EU directives against landfill have been legislated. Therefore, new technical solutions for reusing and/or recycling of GFRP composites must be developed. Considering the low cost of new glass fibres (1-2€/kg), re-use of EoL GFRP seems to be a more economically promising and viable alternative.

Simultaneously, an increasing demand on FRP lightweight structures has been observed in transport and infrastructure applications in the last decade [11]. In addition, the market for pedestrian and bicycle bridges is rapidly growing in Europe as policies promoting shift to healthy and sustainable mobility are introduced. FRP bridges are proven to be the one of the best alternatives for infrastructure owners when considered in LCCA (Life-Cycle Cost Analysis) [12]. In remote locations the lightweight FRP bridges are the perfect solution due to little needed maintenance, and in highly populated areas due to their fast installation where they offer significantly reduced traffic disturbances. However, the initial investment cost is often a barrier for selection of this solution [13].

The development of "circular economy-based design of FRP bridges", where decommissioned GFRP parts are used as base production material, will address both the issue of sustainability in dealing with GFRP waste handling, as well as decreasing investment cost for FRP bridges. The possibility of prolonging the service life of GFRPs while at the same time minimizing their impact on the environment is considered highly favorable and industrially relevant. To this end, investigations to develop innovative pedestrian and bicycle bridge concepts where the base material is EoL wind turbine blades has been carried out.

2. State-of-the-art

So far, very few research projects have been investigating this area. The reason probably being that the problem related to managing EoL wind turbine blades was not as obvious as it has become today. A thorough and inspiring study investigating conceptual design in reusing wind turbine blades for a slow traffic bridge was presented in 2018 at Delft University [14]. The conceptual design included connectors, railing, deck, as well as preliminary structural analysis for performance and safety.

Re-Wind project is an ongoing collaborative project between US and Ireland. It investigates the conceptual re-use of FRP wind turbine blades for housing, buildings, infrastructure, landscape and public art applications [15-16]. The results of the study showed a great potential for the construction industry, and the process of re-purposing decommissioned FRP parts such as wind turbine blades is shown to be a very sustainable way of handling FRP parts reaching EoL. However, the authors clearly indicate that further studies should be conducted to clarify aspects such as structural analysis, logistics, detailing, cost and ease of dis- and re-assembly; social accessibility and acceptability, together with quality control were also identified as paramount aspects to investigate.

Wild project is another ongoing project in Aalborg, Denmark, where two decommissioned blades are planned to be used for the construction of a pedestrian bridge, see Figure 3b [17]. The procurement procedure has been conducted by the municipality of Aalborg. This could be the first pedestrian bridge of this type in the world. In both projects, industrial partners are actively participating, showing a clear industrial interest of finding a solution to the management of EoL blades.

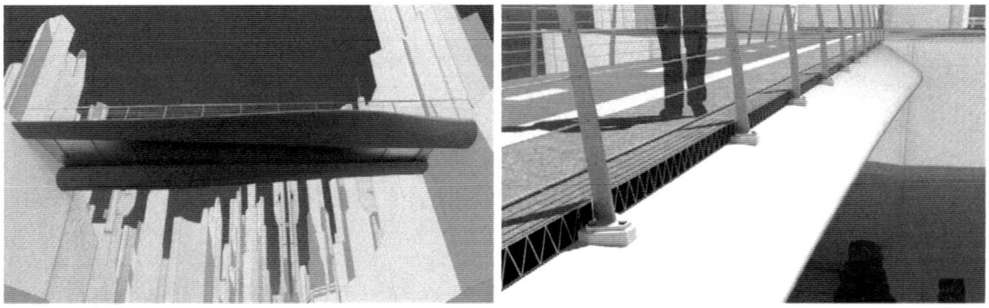

Figure 2: Concept Design of a bridge made of wind blades [14, Stijn Speksnijder].

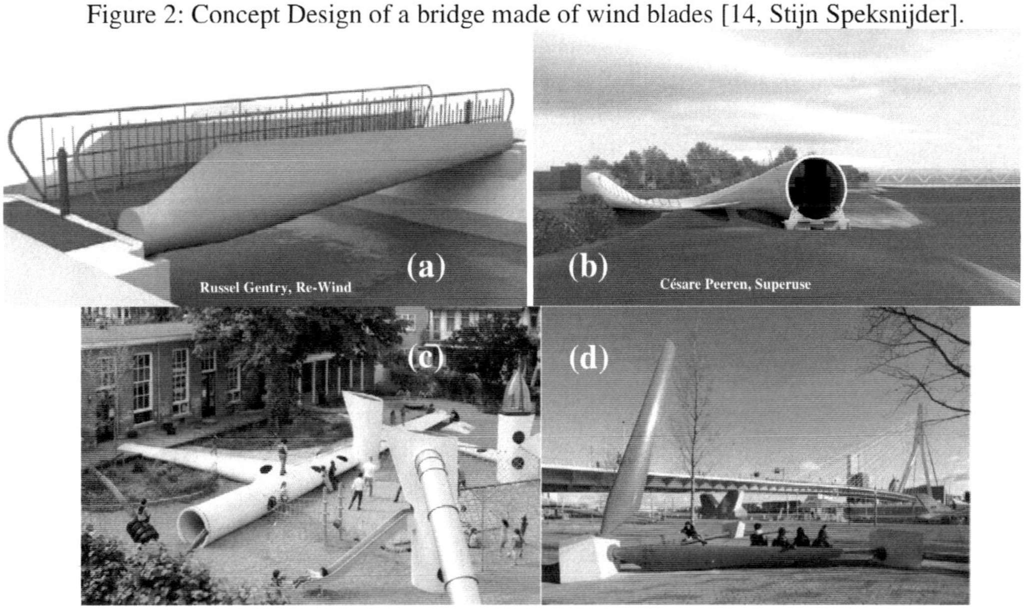

Figure 3: (a) Concept bridge – *Re-Wind project*; (b) Blade bridge in Aalborg – *Wild project*; (c-d) Playground and benches in the Netherlands (Superuse Studios, Denis Guzzo)

Some companies have already realized the potential of "mining" their raw material in the mountain of decommissioned blades to come. Innovative design solution is the key as geometry and properties are already assigned. *Superuse Studios,* a Netherlands based company has developed some interesting designs and construction of structures made of re-used wind turbine blades, see figure 3c-d. These constructions are the only projects within re-using wind turbine blade in new application that have been built. Other projects are mostly at the conceptual design phase [18].

In Sweden, the need to develop knowledge and increase the expertise on recycling processes for FRP composites to reduce the future environmental impacts has also been identified [19]. At the present, the biggest problem for the recycling of FRP is the economic factor, therefore more basic research must be given priority in order to enable the development of new technical solutions that increase resource efficiency and thereby the new players' interests.

In parallel, we observe a rapid growing interest in using FRP for new infrastructure applications, mainly due to a large number of bridge structures which are in need for immediate and costly maintenance work [20]. In Sweden today, about 2500 bridges are in need for major deck rehabilitation or replacement [20]. These figures underline the fact that the low maintenance and sustainability aspects must be central for future bridges.

Several projects investigating FRP bridges and recycling of FRP material have been conducted by the authors during the past years. The recently concluded research project FALCON [13] investigated Future Lightweight Composite Infrastructure in Sweden and created the condition to build Sweden first FRP bridge in Malmö in June 2019. In the pilot project RECYTAL [21], investigations on the potential of recycling / reusing defective and EoL GFRP composite insulators as reinforcement in cement and new composite products were carried out. It was demonstrated in the project that FRP elements can be ground down and used as additive materials both in new composite products and in concrete. However, the economic potential was identified somewhere else, through a short investigation on the re-use of these composite insulators as load-bearing components in the construction industry where several tons of EoL FRP materials can be consumed in one application. Four partners of the FALCON project are today working on the RECINA project [22] based on the result obtained from RECYTAL. The focus is on reusing the decommissioned composite insulators to design a composite pedestrian bridge deck.

Reuse means to either extend the products lifetime, or to reuse it in a new product by taking advantage of its original design purpose. The most promising solution for decommissioned wind turbine blades today is refurbishing [23]. This solution is also consistent with the European waste hierarchy where reuse is the second best alternative. Indeed, the blades show good mechanical properties after 18 years in service according to Sayer et al. [24]. The polish-based company Anmet also performed a full-scale test of decommissioned blades and the results showed that the blades still had a lot of capacity left [25].

The intrinsic mechanical properties and value of decommissioned blade is still high, while handling of waste is expensive for the manufacturers and producers. This clearly creates an incentive to find a new function for the blades in a secondary market. Wind turbine blades are designed to withstand shear and bending stresses along their entire length, a property that is useful in pedestrian bridges. In this study, the focus will be on the development of preliminary conceptual design for pedestrian bridges.

3. Conceptual designs of pedestrian bridges

3.1. Typical wind turbine blade material and structure, and the Sandia blade model

Wind turbine blades are mostly made of glass fibre reinforced polymers (GFRP), but newer blades contain even stiffer carbon fibres that allows the design of longer blades. High strength and stiffness, lightweight and durability are features that differentiate polymer composites from other materials. High costs are usually associated to the use of carbon fibre, but glass fibres are available at very competitive cost. The matrix used is a thermoset of different types (epoxy, polyester, vinylester) depending on the blade manufacturer. The full list of components contained in the blade is almost never disclosed, which complicate the EoL processing. In the case of re-using the blade for structural applications such as pedestrian bridge, one of the challenges will be to determine the quality of the blades and its mechanical properties, without knowing the exact composition.

For most in-service blades today, the construction is made of a shell structure and a load bearing structure. The shell structure is often made of sandwich GFRP, with core materials at different position made of balsa wood or foam material such as PVC. The load bearing stiffeners, that are positioned along the whole length of the blade, give the blade its stiffness and strength. They can be of three different types, mainly single shear web, double shear web or box beam. Adhesive bonding is usually used to attach the shell structure to the stiffeners [14].

The initial idea in the current study was to work with Vestas-V90 which is the most common type of wind turbine in Sweden. However, due to the insufficient information about this blade, it was decided to work with the open-source blade model from Sandia National Laboratories where geometry, material data and FE models are available for research studies. The blade model used in the current study is the SNL100-00 from [26]. This blade has three shear webs and integrated spar caps. To get a comparable size to the Vestas-V90, the model of the SNL100-00 was scaled down to half. Estimations were made on the material thickness, and a parametric study was performed to obtain a range of results (deflection and frequency).

3.2. Brainstorming, ideation and evaluation of potential concepts

To identify as many potential design concepts as possible, a brainstorming with architects was conducted. The pros and cons for each concept were analyzed and the most promising concepts were thereafter investigated more thoroughly. Generally, all the concepts had a 20 m span and a 4 m wide deck as it is the recommended width for a shared pedestrian and bicycle paths.

Table 1. Pros and cons of generated concepts (Figure 4)

Concept 1: Use four blades that carry the load in the edgewise direction. Beams will be placed transversely between the blades to make the blades work as a unit.

Pros: The vertical positioning of the blades provides the largest available bending stiffness. Also, more blades are re-used in this design. This is desirable as finding a purpose for the decommissioned blades is the origin of the problem.	**Cons:** The construction height. The full chord width of the blades needs to be considered, which is approximately four meters. Even if the smallest part of the blades is used, the chord length would still be approximately two meters. The conclusion drawn from this is that the blades should be positioned in the flatwise direction instead to reduce the construction height. The connection between the blades is also anticipated to be complex and time consuming to manufacture.

Concept 2: Use two blades placed in the flatwise direction as beams - bridge deck on top of the beams.

Pros: The main advantage with this concept is a simple construction with a relatively low height. The blades are also positioned in a way that highly utilize their design intention. Furthermore, there is no connection between the blades which simplifies the construction.	**Cons:** As the blades have a varying cross section and are positioned in the opposite direction of each other, the cross section of the bridge will be asymmetric. The blade will deflect more on one side of the bridge deck and torsional forces in the bridge deck are introduced. Furthermore, due to the uneven shape of the blades the connection to the bridge deck becomes even more complicated. This concept also needs uneven level of supports.

Concept 3: A suspension bridge created out of four blades. The blades are leaning towards each other connected at the top to transfer shear forces between the blades and making them work together. In order to transfer the moment forces to the abutments, the blades have a fixed connection at their root.

Pros: Large and spectacular construction, with an innovative design. This construction allows for longer span lengths as the blades can be used as cantilevers as they were designed for.	**Cons:** Probably the most complex of our concepts. Crucial details need to be addressed: The connections between the blades as well as the connections between cables and bridge deck are difficult and expensive to produce, maintain and inspect. In addition, large bending moments needs to be handled in the abutments.

Concept 4: An attempt to create a slender structure. Similar to concept two, although this connects the same section out of four blades to create a more constant and lower cross section.

Pros: This concept has the lowest construction height out of all the concepts. Since the blades are joined together at mid-span, transportation of the blades to the site will also be simplified due shorter components.	**Cons:** Clamped foundations, which are used, are a more complex and expensive solution compared to having a simply supported bridge. The clamped foundation will be on both sides of the bridge which is not optimal from an economical point of view.

Concept 5: The two blades are attached to each other to create one single beam. By removing a part of the shell, the shear webs can be connected directly to each other. Mechanical connectors or adhesive bonding solution could be investigated. The bridge is intended to be simply supported.

Pros: The two blades are working together. This leads to a simpler design for the bridge deck.	**Cons:** Complex joining of the blades, regardless of the method chosen (mechanical or adhesive joint). Adhesive requires planar surfaces, and bolts requires labor inside the blade where space is limited.

Concept 6: Two blades span the whole length. The blades are positioned the wide parts on one side and the narrow on the other. The thick end of the blade is clamped to the abutment while the other end has a roller support.

Pros: This concept is clamped on one side which simplifies the construction. The bending moment curve follows the cross section of the blades.	**Cons:** Clamped foundation is an expensive and complicated solution, with complicated connection to the blade. The two blades are working separately.

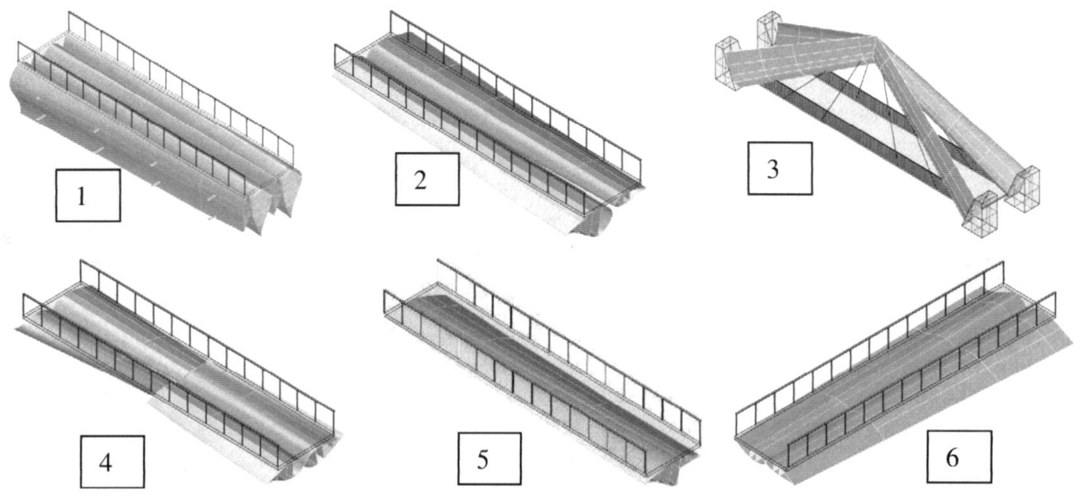

Figure 4: Different bridge concepts resulting from the brainstorming exercise.

When choosing the concept for further studies, several aspects needed to be considered. A life cycle perspective was therefore applied where factors as initial cost, transportation, production method, usage and maintenance were roughly evaluated for each concept.

Concept one was eliminated in an early stage due to the unrealistic construction height. Concept three, the suspension bridge, was also eliminated early due to its complex details. These details would probably have led to an advanced production process as well as high construction and maintenance costs. The remaining concepts, two and four, were then further evaluated and improved to find the most promising alternative. This process resulted in concepts five and six. The connection of the two blades in concepts four and five was analysed as too complicated to construct and thus these concepts were eliminated. Concept six was realized to be the concept with the least complicated details as well as the simplest structural system. This is economically favourable as the production and maintenance costs will be kept low. The detailed design of concept six was therefore conducted.

4. Concept design detailing and preliminary analysis

As the final concept was chosen, all the details needed to be further analysed and designed. This involved how and where the bridge would be assembled, how it would be affected by the transportation and how to handle the boundary conditions as well as all the connections.

During this process, it was realised that the initial clamped foundation was probably not necessary as it is more expensive and complicated than the simply supported solution. Therefore, the bridge was set to be simply supported to create a material efficient and economically attractive solution.

Apart from the blades which have been the focus in the concept design, the bridge deck, railing and connectors have also been investigated:

4.1. Bridge deck

The selection of a bridge deck depends on factors such as structural requirement, weight, initial cost, maintenance requirement and EoL management. Most of the time, the cost factor prevails, especially the initial cost. A pultruded FRP deck with box section was chosen due its strength and lightweight properties, its low maintenance cost and easy prefabricated production process as it simplifies the connections between the deck and the blades. Furthermore, a closed section reduces the wear caused by the surrounding environment and reduces the need for maintenance. On top of the bridge deck an acrylate coating will be used to achieve a suitable walking and biking surface.

Figure 5: Design details for bridge concept with best potential

4.2. Railing
The railing will be made of FRP. This is mainly to keep the self-weight low and the material in the bridge concept consistent. The railing is mounted on top of the bridge deck with a bolted connection.

4.3. Connectors
Wind turbine blades have a very irregular shape and thus a material to fill the void between the flat bridge deck and the curved blades is required (e.g. polyurethane support components). This can be achieved by casting on-site, 3D-scanning the blades and produce the part off-site or cutting a solid block to fit. A bolted connection can efficiently be used to connect the deck to the blade.

4.4. Preliminary numerical analysis
A preliminary numerical analysis was conducted on the blade. Both static and frequency analysis were carried out. The material data from [26] were used.

The analysis for the serviceability limit state showed that the deformation was well below the requirement whereas the fundamental frequency was lower than 5 Hz (4.1 Hz) so further studies on dynamic behaviour need to be carried out. Further analysis on the acceleration needs to be done. The thicknesses are in relation to the full Sandia blade model SNL100-00.

5. Conclusion and Future work
This paper reports the preliminary results of an investigation on the conceptual design study of a pedestrian bridge made of decommissioned wind turbine blades.

Based on the results obtained and knowledge gained, building bridges out of decommissioned wind turbine blades is considered as a promising solution. However, detailed calculations, and further investigations on the remaining capacity of the blades have to be conducted in order to validate the concept. There is especially a need to investigate closer the dynamic loading as the preliminary fundamental frequency was lower than 5Hz.

Even though reusing the blades in bridges is an efficient way of extending the lifetime of wind turbine blades, the problematic of waste management is only postponed during the lifespan of the new bridge. Developing a cost-efficient method to recycle the material in parallel is therefore still needed.

Acknowledgments
The authors would like to thank Vinnova (ReComp – Circular streams from fiberglass composites, 2018-04132) for financial support.

References

[1] Naqvi S R, Mysore Prabhakara H, Bramer E A, Dierkesa W, Akkerman R and Brem G 2018 A critical review on recycling of end-of-life carbon fibre/glass fibre reinforced composites waste using pyrolysis towards a circular economy *Resour. Conserv. Recycl.* **136**, 118–29

[2] Wind Europe – Discussing paper on managing composite blade waste, 2017

[3] European Commission 2015 Report from the Commission to the European Parliament and the Council on the implementation of regulation (EC) no 1013/2006 on shipments of waste, 2015.

[4] Halliwell S 2006 End of life options for composite waste – Recycle, Reuse or Dispose

[5] EuCIA position paper 52816 2011 Glass fibre reinforced thermosets: recyclable and compliant with the EU legislation

[6] WindEurope 2019 -https://windeurope.org/newsroom/press-releases/new-joint-project-between-wind-and-chemical-industry-to-advance-wind-turbine-recycling/

[7] Beauson J, Bech J I and Brøndsted P 2014. Composite recycling: Characterizing end of life wind turbine blade material. *Conf. Proc. ICCM 19*

[8] Ierides M, Reiland J and Dierckx A 2019 Wind turbine blade circularity: Technologies and practices around the value chain *Wind. Conf. Exhib. Work. Bl. Dispos.* 1-38

[9] WindEurope 2020 Accelerating Wind Turbine Blade Circularity

[10] Martin C 2020 Wind Turbine Blades Can't Be Recycled, So They're Piling Up in Landfills, Blomberg green – Right picture: Brendan LaChance, Oil City

[11] Oliveux G, Dandy L O and Leeke G A 2015 Current status of recycling of fibre reinforced polymers: Review of technologies, reuse and resulting properties *Prog. Mater. Sci.* **72** 61-99

[12] Mara V, Haghani R and Harryson P 2014 Bridge decks of fibre reinforced polymer (FRP): A sustainable solution *Constr. Build. Mater.* **50** 190-99

[13] FALCON (2015-2018) FALCON: Future Lightweight Composite Infrastructure

[14] Speksnijder S 2018 Reuse of wind turbine blades in a slow traffic bridge, master thesis, Delft university of technology, faculty of industrial design engineering

[15] Bank L, Arias F, Yazdanbakhsh A, Gentry R, Al-Haddad T, Chen J F and Morrow R 2018 Concepts for Reusing Composite Materials from Decommissioned Wind Turbine Blades in Affordable Housing *Recycling.* **3** 3-14

[16] Suhail R, Chen J F, Gentry R, Taristro-Hart B, Xue Y and Bank L 2019 Analysis and Design of a Pedestrian Bridge with Decommissioned FRP Windblades and Concrete FRPRCS14

[17] Wild project 2019 www.tv2nord.dk/aalborg/vildt-projekt-skrottede-mollevinger-bliver-til-ny-bro-i-lindholm

[18] Wikado och Rewind Willemsplein, SuperUse studio, https://superuse-studios.com/en/

[19] COST345, Procedures Required for Assessing Highway Structures. Working Group 1, Report on the current stock of highway structures in European countries, the cost of their replacement and the annual costs of maintaining, repairing and renewing them, 2004

[20] PANTURA project 2011-2013 Flexible Processes and Improved Technologies for Urban Infrastructure Construction Sites. https://www.chalmers.se/en/projects/Pages/Pantura.aspx

[21] RECYTAL project 2017 RISE, Sweden - Application of recycled/recovered glass fibre-reinforced composite insulators and discarded glass fibres in construction.

[22] RECINA project 2020 RISE, Sweden - Reuse of composite parts for infrastructure applications

[23] Beauson J and Brøndsted P 2016 Wind Turbine Blades: An End of Life Perspective.Citation, 421–432. https://doi.org/10.1007/978-3-319-39095-6{_}23

[24] Sayer F, Bürkner F, Blunk M, van Wingerde A M and Busmann H-G 2009 Influence of Loads and Environmental Conditions on Material Properties over the Service Life of Rotor Blades

[25] Anmet and Rzeszów University of Technology 2020 https://www.thefirstnews.com/article/the-green-team-turning-disused-wind-turbines-into-stylish-street-and-garden-furniture-10543

[26] Griffith T and Ashwill T 2011 SANDIA REPORT The Sandia 100-meter Allglass Baseline Wind Turbine Blade: SNL100-00 Sandia National Laboratories. Albuquerque, New Mexico, US

IOP Conf. Series: Materials Science and Engineering **942** (2020) 012016 doi:10.1088/1757-899X/942/1/012016

Composite wind turbine blade recycling - value creation through Industry 4.0 to enable circularity in repurposing of composites

Raphael Geiger[1,*], Yasser Hannan[1], William Travia[2], Roberto Naboni[1] and Christian Schlette[1]
University of Southern Denmark SDU, Campusvej 55, 5000 Odense, Denmark[1]
École Centrale de Lyon, 36 Avenue Guy de Collongue, 69134 Écully, France[2]
E-mail: rge@iti.sdu.dk

Abstract. Composite waste, especially from wind turbine blades will increase foreseeably. The motivation of this paper is to develop new solutions to repurpose the global material stream coming from wind turbine blades at their end of use. Its suggested to use the composite structure for repurposing, differing from the available recycling methods that aim for a separation of the composite into fibre and matrix fractions. One key factor towards circularity in composite materials is converting significant amounts of composite material into multiple meaningful applications and products. Industry 4.0 can offer a systemic solution utilising robots and digital tools, such as a digital twin of a composite wind turbine blade. The digital handling of relevant material information can offer a new affordable material source to multiple industries, such as building industry or manufacturing businesses. Based on a digital platform the composite material can be made available for architects, designers and developers to repurpose composites into multiple new products.

1. Introduction

Wind turbines consist of various parts, like tower, nacelle, generator, blades etc. In the recycling of wind turbines, blades have been identified to be the most challenging part to recycle, due to their mixed material composition and the nature of thermoset reinforced composites [1]. Over the last two decades the installed wind turbine power grew substantially therefore the amounts of blades reaching end of use are foreseeably growing. Turbine blades have an expected lifetime of about 20-25 years, in case of repowering the wind turbine with new blades this lifetime can be shortened significantly. By 2025 the annual composite waste from wind is estimated to be 66,000 tons [2].

Currently there are various scenarios and technologies dealing with turbine blades after end of use. Landfill is one scenario often used while there are various composite recycling technologies at various technological readiness levels [TRL] available [3]. The available recycling technologies aim either to separate the composite material in a fibre and matrix fraction or recover the materials energy value within thermal recycling processes. With perspective on the waste treatment hierarchy the available wind turbine blade recycling solutions are addressing the two lowest options, recovery and disposal. None of the recycling technologies mentioned in [2] are aiming towards repurpose or recycling of the blade. Yet the composite materials used in blades show excellent material properties and therefore would be ideal for repurpose scenarios.

Some showcases and projects make use of the structural qualities of recycled composite materials in a repurposing or a recycling scenario, for example repurposing the wind turbine blade into a bridge [4] a playground [5] or recycling into outdoor furniture [6]. These projects create value by utilising the structural qualities of reused composite material and integrate the fixed shape of the blade into the repurposed product. Single repurposing projects show the possibilities to climb up the waste treatment hierarchy, yet it might be difficult to match the global amounts of recycled composite materials.

This publication is proposing a scenario for Industry 4.0 to improve composite recycling on a global scale by utilising the composites material properties and fixed shape. A digital twin of the wind turbine blade is gathering relevant data including shape, material quality and mechanical properties. This data can be used to predict the blade lifetime and for maintenance and service. At the end of use, the digital twin data is the basis for material qualification and certification needed for repurpose solutions. Designers and architects use the geometrical data of the blade for their repurpose designs. Within architectural design, digital manufacturing tools allow, using the given shape of the blade and its digital conversion, to design it into new buildings and shapes. Once digitally designed robots disassemble the blade accordingly and reassemble it into its second life. This systemic approach can be used for building industry and others and is allowing for unlimited sustainable solutions in repurposing composites towards a more circular use.

2. Recycling today

Wind turbines consist of various parts, like tower, nacelle, generator, blades etc. In the recycling of wind turbines, blades have been identified to be the most challenging part to recycle, due to their mixed material composition and the nature of thermoset reinforced composites [1]. Over the last two decades the installed wind turbine power grew substantially therefore the number of blades reaching end of use are foreseeably growing. Turbine blades have an expected lifetime of about 20-25 years [7]. In case of repowering the wind turbine with new blades this lifetime can be shortened significantly. By 2025 the annual composite waste from wind is estimated to be 66,000 tons [2]. European Composites Industry Association, European Chemical Industry council and Wind Europe point out in their report "we are committed to promoting a circular economy which reduces environmental impacts throughout product lifecycles." About the current turbine blade recycling situation, the report is stating "while various technologies exist that can be used to recycle blades, these solutions are yet to be widely available and cost-competitive" [8] so, the current recycling solutions are not yet fully applicable for global use.

2.1. Composites in wind turbine blades

Wind turbine blades are considered a composite structure, consisting of various materials with different properties. Although material compositions vary between blade types and blade manufacturers [4]. Typical materials in wind turbine blades are shown in Figure 2:

- reinforced fibres: glass, carbon, aramid or basalt;
- a polymer matrix: thermosets such as epoxies, polyesters, vinyl esters, polyurethane, or thermoplastics;
- a sandwich core: balsa wood or foams such as polyvinyl PVC, PET;
- coatings: PE, PUR;
- metals: copper wiring, steel bolts.

The combination of fibers and polymers, also known as glass fiber reinforced polymer (GFRP) composites, represents the majority of the blade's material composition (60–70% reinforcing

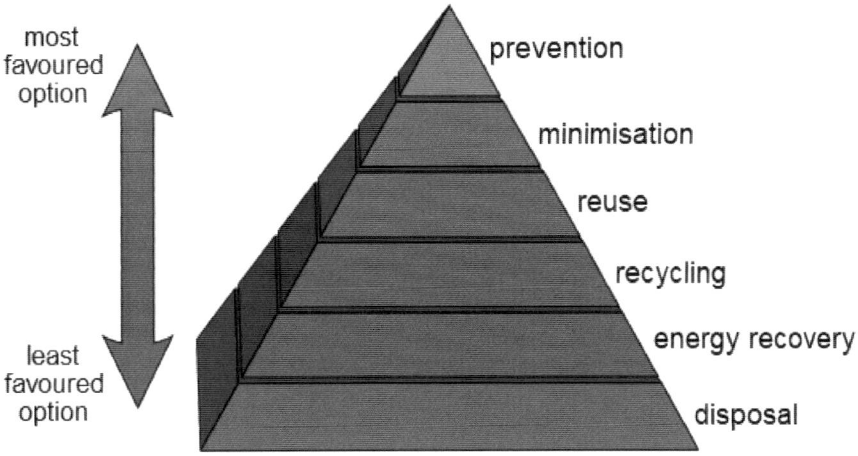

Figure 1. Waste treatment hierarchy from [9] licensed under CC BY-SA 3.0

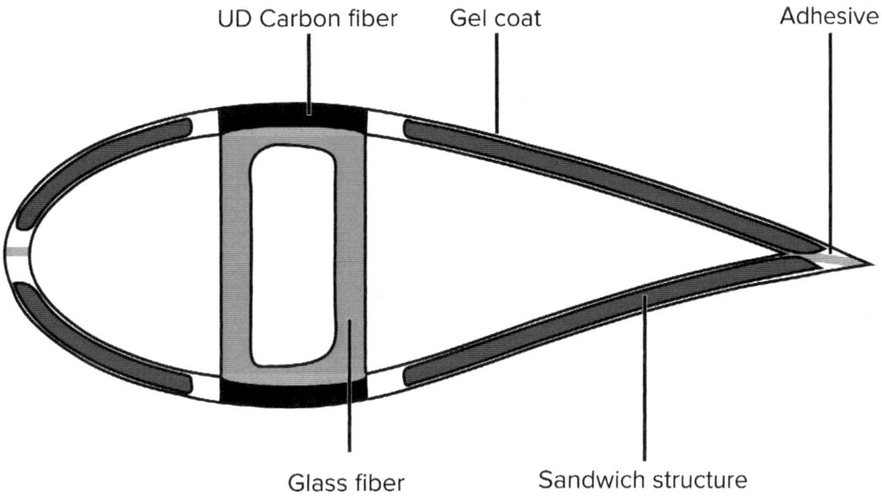

Figure 2. Generic composition of a wind turbine blade

fibers and 30–40% resin by weight) [4]. The shape of the wind blade from the root to the tip transitions rapidly from round to an airfoil shape. These airfoil shapes vary along the length of the blade. A blade can be divided into 4 main parts: shell, caps, shear webs, and root. Shells are designed to transfer lift, created by the shells, to the spar caps, then spar caps transfer the loading to the root. Shear webs keep the caps away from each other, allowing the blade to behave as a beam and retain its global stiffness. This structure is shown decomposed on Figure 3.

The most widely used technology to produce the wind blades is prepreg and resin infusion technology. Prepreg allows the industrial impregnation of fibers, and then forming the impregnated fibers to complex shapes. Currently, vacuum assisted resin transfer molding (VARTM) is the most common manufacturing method for manufacturing of wind turbine rotor blades. 3D woven composites represent a promising alternative to producing fiber reinforced laminates, and this technology allows producing spar caps with higher stiffness and lower weight

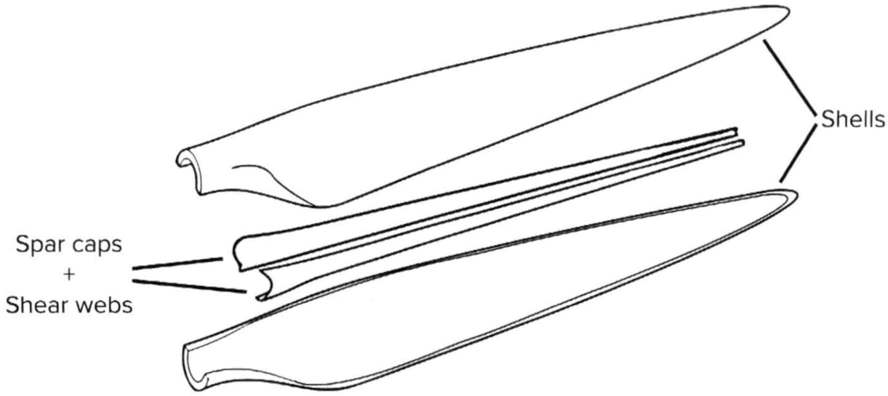

Figure 3. Wind blade structure

[11].

Using automatization technology in the manufacturing process, allow increasing the complexity of the geometry of the blade, enabling to exploit the internal aerodynamic properties, and significantly increase energy [12]. The generic wind blade manufacturing includes 6 steps:

(i) Preparation of the mould;
(ii) Build-up of the dry-layers;
(iii) Resin infusion;
(iv) Adding the webs;
(v) Joining the shells and curing;
(vi) Demoulding, trimming and polishing.

2.2. Available recycling technologies

The main routes for handling end-of-life composite materials are landfill, incineration or recycling to which recycling has a number of approaches: mechanical, thermal and chemical recycling [4]. Currently, mechanical, thermal and chemical approaches have been used to recycle composites. The composite recycling is still an immature area. This means the developing techniques mentioned above still need to be optimized to produce higher quality recyclates and improve resource efficiency [13]. These technologies have different properties, advantages, disadvantages, and technology readiness level displayed in [3].

2.3. Limitations in state-of-the-art methods

Despite the number of recycling technologies presented so far, they all have major drawbacks and limitations, thus making them close to irrelevant on industrial scales. Energy recovery from the incineration of wind turbine blades is made difficult by their high glass fiber content which prevents an effective burning of the parts [14]. Once incinerated, the remaining ashes may be landfilled, when it must be avoided at all costs being the worst end of life option. Mechanically grinding the blades and reusing the shredded parts as fillers for other products is an interesting option. However, all the companies that had implemented industrial scale grinding have ceased operating [14]. The wind industry is pushing the use of the shredded material as filler for cement [8] but it lacks industrial partners to complete the process. It must also be noted that the use of the composite waste for this matter makes no use of its structural properties and therefore significantly lowers the value of the material [14]. A Danish company, Miljøskærm [15], processes shredded GFRP into thermal and acoustic isolation panels. It is however not operating on an industrial scale yet [14].

Thermal recycling processes, such as pyrolysis or fluidized bed, allow only for the recovery of the fiber material, and it comes at the cost of significantly diminished mechanical properties [14]. The Danish company ReFiber made use of pyrolysis to salvage the glass fibers but was terminated in 2007 [1]. In the same way, chemical recycling processes, despite the recovery of both fiber and matrix components with an acceptable quality, are not operated at industrial scale. This is because the processed materials are more expensive than in their virgin condition. Therefore, the state-of-the-art methods for recycling wind turbine blades are either responsible for low value applications of an originally high value material or are more costly than buying virgin material for the producers, while providing limited performances. Hence, solid waste disposal is less costly for wind turbine blades than recycling [16]. In the Netherlands for example, there are no recycling options under the government-fixed threshold of 200 EUR/t, meaning that landfilling is still accepted for wind turbine blades when composite waste landfilling is banned [8].

2.4. Reuse of structural composite parts - best practice

The currently available recycling technologies for wind turbine blades, introduced in the previous sections, have their limitations as they are all within the three least favored levels of the waste treatment hierarchy (Figure 1). They offer only low value applications for the recycled composite parts. It is of a greater interest, on all the plans, to make use of the structural nature of composite wind turbine blades, and therefore achieve higher value recycling. This is characterized by methods belonging to the 2nd and 3rd most favorable levels of the waste treatment hierarchy (Figure 1), Reuse and Repurpose respectively.

2.4.1. Reuse methods Reuse recycling scenarios consist in the refurbishment of decommissioned wind turbine blades which are, mostly, still usable after their service life of around 20 years [14]. It is by far the most desirable recycling option, since it extends the blade's life beyond its intended service life while retaining its original purpose and making full use of its design. However, it must not be forgotten that this solution depends on the blade's condition and that, at some point, it will not be possible to reuse a blade anymore. Then, repurpose methods should be envisaged.

2.4.2. Repurpose methods If it is not possible to reuse the blades, they can be repurposed. Several applications can be found where it is made use of the structural nature of the composite wind turbine blades. The architecture office Superuse Studios has conducted several projects of repurposing blades for urban furniture through the Blade Made initiative [18]. They made use of decommissioned blades to design a playground, bus shelters, public seating and a signpost for a recycling centre. This extensive use of the structural nature of the blades goes to show that they can be repurposed and at the same time replace virgin raw materials that would be used otherwise to build these structures. A similar use of blade parts for urban furniture was made in Aalborg, Denmark, as a bike shed [19].

There have been several other design ideas involving decommissioned blades for civil engineering applications that benefit from the structural properties of the blades. First, affordable housing for the coastal regions of the Yucatan province in Mexico was designed (Figure 4), using blade material for the roof, door and window frames and elevated foundations [17]. Another use envisaged is that of poles in power transmission lines [20]. Finally, blades inspired some bridge designs. Stijn Speksnijder designed a slow traffic Bridge of Blades as his master thesis at the Delft University of Technology [21]. Similarly, a pedestrian bridge has been designed, using blades as girders [22]. Superuse Studios also designed a bridge using two blades for the city of Aalborg, Denmark, and is waiting for its installation [18].

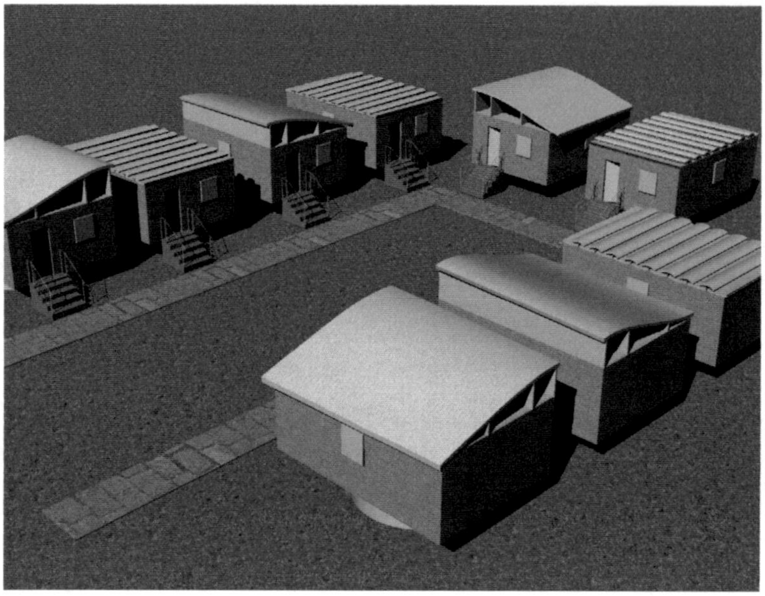

Figure 4. Conceptual housing community from [17] licensed under CC BY 4.0

2.5. Motivation for new approaches

In the perception of the authors none of the currently available recycling technologies are yet able to handle the global composite material waste stream. Cost competitiveness and availability are two of the reasons for this [2]. We further estimate scalability and value of the recycled products from composites as significant factors. Looking at handling large amounts of material within the available composite recycling technologies, this is mostly achieved by a "uniformation" of the composite material. One example is cutting and shredding a composite blade into small parts in order to run pyrolysis efficiently. This uniformation comes with a significant loss of mechanical material properties.

On the other hand, there are several projects repurposing the shape and the mechanical properties of the wind turbine blade to create bridges, playgrounds or furniture. The challenge hereby is scalability, in order to repurpose 14.000 blades predicted for Europe in 2023 [23]. The motivation for this paper is to develop a recycling method utilizing the outstanding mechanical properties of composite materials and being scalable at the same time. The method should be able to create value by utilizing the growing global composite material waste stream and turn it into a circular material economy. The mindset of this principle sees composites as a source for a high-quality material. The idea is applying Industry 4.0 technology to bridge the gap between material specific repurposing and scalability.

3. Industry 4.0

For manufacturers in high wage countries, a major strategy to keep production costs competitive without outsourcing production is to increase the level of automation, i.e. by introducing robots as highly reconfigurable manipulators for product assembly. Such reconfigurable and agile robotic solutions are of particular interest in high mix/ low volume production scenarios, where manufacturers offer many product variants between which production is often switched [24]. The resulting complexity of controlling agile robotic solutions to be fit for manipulating high amounts of product variants is clearly demanding. The paradigm of Industry 4.0 aims to resolve such automation complexities by means of digitalization and the implementation of model-, simulation- and data-based planning and control methods [25].

The essence of the paradigm of Industry 4.0 is to complement the physical scenario with digital representations of the relevant assets (i.e. product and production system) and then to solve planning and control in the digital realm first, before executing the solution in the physical setup. Such an approach, based on the use of Cyber-Physical Systems (CPS) has opened fundamental opportunities in architectural robotics, by shortening the distance between the phases of design and making, and ultimately, has provided a fundamental platform for customization and bespoke manufacturing.

3.1. Digital Twin

Digital Twins were first mentioned in the context of Product Lifecycle Management [26], and later became an integral concept of Industry 4.0 [27]. In 2018, Gartner deemed Digital Twins as one of the five major technological trends of the next decade [28]. However, the term lacks a unique definition, and its interpretations range from "digital interface of physical assets" to "simulation of physical assets" [27]. Based on [29] and [30], we understand Digital Twins to be comprehensive virtual replica of physical robotic equipment regarding the exposed functionality (e.g. to command robot motions identical to the physical robot) as well as the relevant parameters and behaviors (e.g. to set joint velocities or to stop at collisions). Such replica allow for applying various 3D methods of simulation, planning and visualization as well as connecting to the physical robots and sensors for accessing their data from actual operations and controlling them based on the (simulated or real) data processed in the Digital Twin [29].

3.2. Industry 4.0 to enable circularity in repurposing of composites

Next to the obvious challenges due to extraordinary dimensions, production of large structures such as wind turbines and blades can be categorized as "high mix/low volume" production as they tend to be complex few-of-a-kind or even one-of-a-kind designs that are produced and operated in highly dynamic environments, particularly if it comes to offshore installations. Accordingly, also the recycling of wind turbines and blades is complex and can highly benefit from the application of Digital Twins to analyze and then control their recycling processes:

- Wind turbine blades like other large structures (e.g. ships and buildings) consist of an extensive variety of components vastly differing in dimensions, materials and production processes to make them part of the overall product.

- Wind turbine blades may require recycling operations in highly dynamic, often unique environments such as offshore windparks, where operations need to be carefully coordinated in time and space.

- Due to the complexity of the product and the production environment, automation of recycling of wind turbine blades requires flexible, reconfigurable, "smart" production systems based on digitalization and robotization.

Industry 4.0 is combining the Digital Twin approach with the physical repurposing environment, such as decommissioning robots, transport and automation solutions. Figure 5 is outlining a systematic approach of recycling composite wind turbine blades with Industry 4.0. The green arrows, "Reuse", "Refurbish", etc., denote the escalating options of treating the blades in subprocesses, often with various alternatives (red boxes). If reusing and refurbishing is not possible any more, the composite materials of the blades can be cut, the cuts can be shredded, the shreds can be milled and/or treated chemically (e.g. by means of solvolysis). For each of

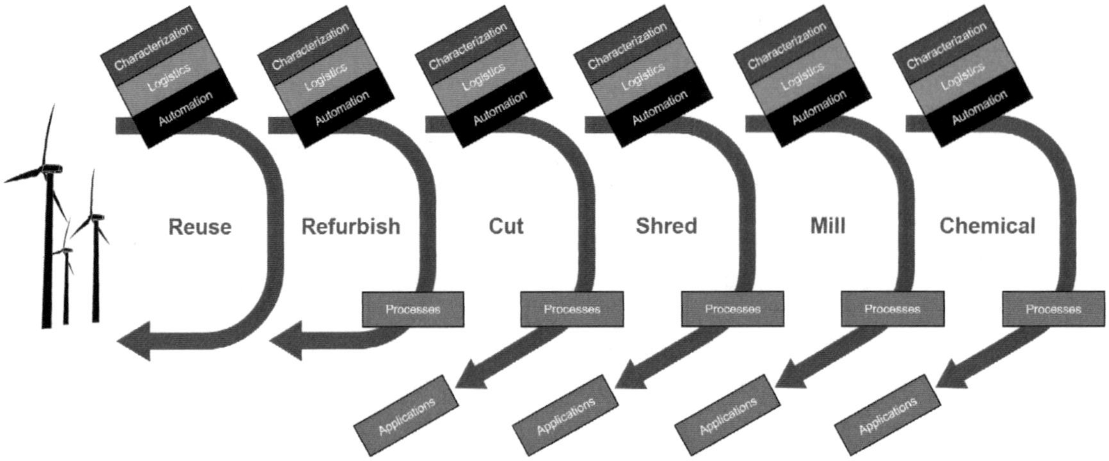

Figure 5. Systematic view on technology options in composite wind turbine blade recycling scenarios.

the subprocesses and alternatives the incoming and outgoing materials need to be characterized and the the logistic alas well as the automation options of each of the subprocesses need to be identified, evaluated and selected (green, yellow and black boxes). The resulting material outputs of cutting, shredding, milling and chemical treatment can be used in various recycling applications (grey boxes).

3.3. Value creation in recycling of composite wind turbine blades

The previous chapters where introducing into the general possibilities of Industry 4.0 applied for large scale composite structure recycling. In order to create value for society and businesses it is benefitable to get up the waste treatment hierarchy shown in Figure 1. A 5-step schematic is outlining the structure of such a recycling method.

(i) Evaluation: geometrical and structural;
(ii) Digitalization: creation of a digital twin;
(iii) Computational design: architectural structures, product design;
(iv) Robotic composite blade disassembly;
(v) Transformation and reassembly: repurposed composite product.

An evaluation of the repurposing wind turbine blade will deliver data on geometry and surface quality, wear and tear as well as structural material parameters. The data can be gathered by scanning, non-destructive testing and production data from the manufacturer. The result of the evaluation is a classification of the repurposing blade. Based on this classification a later repurposing scenario can be determined.

These digital data will help mirroring each specific blade digitally into a digital twin. The digital twin allows to additional data sets as for example economical parameters, logistic data and relevant material certification. A digital twin can already be created while the blade is still operational.

Computational design can utilize the digital twin with the shape constraint composite structures to redesign and repurpose the blade into new products and architecture. At this point the original design will get digitally transformed into its second live usage. This process is possible while the composite blade is still operational.

Based on data from computational design, robots will disassemble the blade in the specific

IOP Conf. Series: Materials Science and Engineering **942** (2020) 012016 doi:10.1088/1757-899X/942/1/012016

segments needed for its circular use. Disassembly could be executed by robotic sawing and milling into pieces of the right size and specific material properties. For logistic reasons disassembly could take place at the wind turbine site.

Once disassembled the different pieces can be easier transported and delivered to its destination. Reassembly into the repurposed product or architectural structure could be executed by robots or manually.

4. Computational Design Approach for Circularity

The recent proliferation of digital tools has offered to the designers the possibility of approaching design with new perspectives. In particular, the use of computation has fundamentally shifted the essence of the design process, by the introduction of a data-driven approach. Through the use of computational means, the designer can easily evaluate a large number of design options in relation to various datasets, he can assess the performance features of a certain design [31] and establish strategies for optimization and differentiation. Moreover, access to computational design has allowed designers to customize workflows, create their own tools and engage with novel processes of construction through the use of digital fabrication tools and robotics. In the context of this work, computational design offers interesting perspectives in the informed manipulation of complex geometry, material and mechanical data, integration of design/simulation into a unique process based on data-rich digital models.

4.1. A computational design process for blade repurposing

One fundamental aspect of a digital twin in this work is providing high-resolution geometrical and material data into the digital space. Data from digital photogrammetry and LiDAR scanning can be nowadays easily converted into geometrical three-dimensional models. High-precision scanning techniques can provide data-intensive geometry, such as point clouds or large mesh, which can be conveniently translated NURBS description.

This fundamental step enables a form of abstraction of the blade geometry, which can be subsequently processed through its fundamental mathematical description. Available tools can be used to inspect existing models and retrieve underlying geometrical information, such as curvature principals and Gaussian curvature of the different areas of a blade's shell. Once a specific design application is proposed with the re-purposing of shell elements starting from existing blades, it is possible to establish a design optimization where an existing geometry can be discretized into parts that match a target design within predefined tolerances.

A second approach can proceed with an initial analysis of the curvature and the extension of the various parts of a blade and suggest useful cutting patterns to obtain composite elements falling into classes of a gaussian curvature. In this case, the various parts can be used as a mechanically reliable substrate for further and more general construction purposes. This can be utilized as a lightweight structural core for certain architectural applications, relying on complementary techniques and materials to further complete the construction.

4.2. From Computational Design to Robotic Fabrication

In the repurposing of existing large-scale elements, the possibility of an integration platform from survey to design to fabrication is fundamental to reduce inconsistencies. Once cutting patterns over a blade shell element are defined, the physical machining of such elements requires the utilization of a large-scale facility that can operate with multi-axial cutting tools. Recent developments of large-scale robotic setups, where industrial robots are integrated with building-scale gantry systems, are offering solutions in this sense. Directly from design information, robots equipped for intensive CNC milling can be operated to cut effectively the shell elements into elements of different sizes and then sorted and stored automatically. Possibly, the same infrastructure can be used for survey and inspection, so that enough precision is ensured in the geometrical analysis. Computer vision can be also integrated for an evaluation of potential material inconsistencies, tracing of existing decay, analysis of potential cracking patterns. This data can be then confronted with mechanical analysis of specimens that are physically extracted on the blade's external body.

5. Conclusion

This paper has described the existing composite recycling technologies and the limitations of the different methods. It has been found that composite recycling methods described in [2] were working based on a principle of "unification" of the recycled composite material. This unification process is coming with a significant loss of mechanical material properties of the composite structure. On the other hand, unification is a principle to deal with the multi material composition coming from composite wind turbine blades. Further unification is offering a possibility of scaling, therefore the handling of larger amounts of composite is possible by scaling up the processes.

In contrast to unification of the recycled composite materials there has been different projects using the mechanical material properties in order to repurpose the composite wind turbine blade into different structures, such as bridges, kids' playgrounds and others. These projects are however singular individual projects with a specific one-time outcome. Therefor a scalability of the solutions seems difficult to realize. These kind of projects may not be the optimal solution to address the global composite waste stream.

This paper is developing a composite recycling scenario that allows using the composite material properties and scalability at the same time by using industry 4.0 and digital fabrication. Industry 4.0 and a digital twin of the composite wind turbine blade is mirroring the fixed shape and the specific material properties of each single wind turbine blade. The specific data of the digital twin can be designed into endless designs for repurposed composite products. Digital manufacturing allowing for the specific decommissioning of each single blade and rebuilding into the repurposed design that has been digitally designed.

The developed Industry 4.0 based repurposing scenario is of theoretical nature and indicates significant benefits in improving composite recycling. It is suggested to investigate the specific value creation in composite recycling. It is further suggested to create a proof of concept in order to verify the theory developed in this paper.

References

[1] PD Andersen, A Bonou, J Beauson and P Brøndsted. 2014. Recycling of wind turbines. *DTU International Energy Report 2014*: 92-7.

[2] ETIP Wind. 2019. How wind is going circular - blade recycling. EUROPEAN TECHNOLOGY & INNOVATION PLATFORM ON WIND ENERGY.

[3] ETIP Wind. 2019. ETIP Wind Roadmap. [Online]. Available from: https://etipwind.eu/publications/. [Accessed 12 05 2020].

[4] JP Jensen, K Skelton. 2018. Wind turbine blade recycling: Experiences, challenges and possibilities in a circular economy. *Renewable and Sustainable Energy Reviews*;**97**: 165-76.

[5] IP Medici. 2011. Human Health Risk Assessment of "Wikado" playground.

[6] E Allameh and M Heidari. 2020. Sustainable Street Furniture. *Periodica Polytechnica Architecture.* Available from: https://doi.org/10.3311/.

[7] J Chen, J Wang, A Ni. 2019. Recycling and reuse of composite materials for wind turbine blades: An overview. *Journal of Reinforced Plastics and Composites*;**38**(12): 567-77.

[8] cefic, EuCiA, Wind EUROPE. 2020. Accelerating Wind Turbine Blade Circularity.

[9] https://en.wikipedia.org/wiki/User:Drstuey, "www.wikipedia.org," [Online]. Available from: https://en.wikipedia.org/wiki/Waste_hierarchy#/media/File:Waste_hierarchy.svg. [Accessed 20 06 2020].

[10] KIRTxThomsen, Bladena. 2019. *WIND TURBINE BLADES Handbook.* 2019 ed. Available from: https://www.kirt-thomsen.com/s/cortir_handbook_2019.pdf [Accessed 18 06 2020]

[11] L Mishnaevsky, K Branner, HN Petersen, J Beauson, M McGugan, B F Sørensen. 2017. Materials for Wind Turbine Blades: An Overview. *Materials*;**2017**(10).

[12] JA Sainz. 2015. New wind turbine manufacturing techniques. *MESIC 2015.*
[13] Y Liu, M Farnsworth, A Tiwari. 2017. A review of optimisation techniques used in the composite recycling area: State-of-the-art and steps towards a research agenda. *Journal of Cleaner Production*;**140**.
[14] J Beauson, P Brøndsted. 2016. Wind Turbine Blades: An End of Life Perspective. *MARE-WINT. New Materials and Reliability in Offshore Wind Turbine Technology*: 421-32.
[15] MILJØSKÆRM. *Recycling of Fiberglass/Composite Materials.* Available from: http://miljoskarm.dk/english/. [Accessed 20 06 2020].
[16] B Fitchett, K Ladwig. 2020. Wind Turbine Blade Recycling - Preliminary Assessment. *ELECTRIC POWER RESEARCH INSTITUTE.*
[17] LC Bank, FR Arias, A Yazdanbakhsh, TR Gentry, T Al-Haddad and J-FCaR Morrow. 2018. Concepts for Reusing Composite Materials from Decommissioned Wind Turbine Blades in Affordable Housing. *Recycling*;**3**(1).
[18] Superuse Studios. 2014. *Blade Made.* Available from: https://issuu.com/2012architecten/docs/blademade. [Accessed 20 6 2020].
[19] H Eilers. 11 02 2020. Vindmøllevinge får nyt liv på Aalborg Havn. *Energy Supply.* Available from: https://www.energy-supply.dk/article/view/699757/. [Accessed 20 06 2020]. *
[20] A Alshannaq, LB David Scott, M Bermek, R Gentry. 2019. Structural Re-Use of De-Commissioned Wind Turbine Blades in Civil Engineering Applications. *34th ASC Technical Conference.*
[21] S Speksnijder. Bridge of blades. Available from: http://www.stijnspeksnijder.com/gallery/bridge-of-blades/. [Accessed 18 06 2020].
[22] R Suhail, J-F Chen, R Gentry, B Taristro-Hart, Y Xue, LC Bank. 2019. Analysis and Design of a Pedestrian Bridge with Decommissioned FRP Windblades and Concrete. *FRPRCS14.*
[23] Wind Europe. 2019. Wind Energy in Europe: Outlook to 2023.
[24] C Brecher, D Özdemir. 2017. *Integrative Production Technology : Theory and Applications.* Aachen: Springer International Publishing.
[25] CJ Bartodziej. 2017. *The concept Industry 4.0.* Wiesbaden: Springer Gabler.
[26] MW Grieves. 2015. Product lifecycle management: the new paradigm for enterprises. *Int J Prod Dev*;**2**(1/2).
[27] German Federal Ministry of Education and Research. The background to Plattform Industrie 4.0. Available from: https://www.plattform-i40.de/PI40/Navigation/EN/ThePlatform/Background/background.html. [Accessed 18 05 2020]
[28] K Panetta. 15 08 2017. Top Trends in the Gartner Hype Cycle for Emerging Technologies, 2017. *Gartner.* Available from: https://www.gartner.com/smarterwithgartner/top-trends-in-the-gartner-hype-cycle-for-emerging-technologies-2017/. [Accessed 19 04 2020].
[29] C Schlette. 2016. Database technologies for long-term/large-scale mental models in robotics. *Robotics in the 21st century: Challenges and Promises.*
[30] M Schluse, M Priggemeyer, L Atorf, J Rossmann. 2018. Experimentable Digital Twins—Streamlining Simulation-Based Systems Engineering for Industry 4.0. *IEEE Transactions on Industrial Informatics*;**14**(4): 1722-31.
[31] T Peters, B Peters. 2018. *Computing the Environment: Digital Design Tools for Simulation and Visualisation of Sustainable Architecture.* Wiley.

Influence of the glass non-crimp fabric intrinsic undulation on the stiffness of the composite ply: A micromechanical approach

A Antoniou[1,3], L P Mikkelsen[2], S Goutianos[2], O Bagemiel[1], I Gebauer[1], R Flindt[1] and F Sayer[1]

[1]Fraunhofer IWES, Am Seedeich 45, 27572 Bremerhaven, Germany
[2]DTU Wind Energy, Frederiksborgvej 399, 4000 Roskilde, Denmark
[3]E-mail: alexandros.antoniou@iwes.fraunhofer.de

Abstract. Load carrying components of modern wind turbine blades are manufactured from composites, consisting of non-crimp fabrics infused with polymer resins. The effective stiffness of the resulting laminate is a combination of the properties of its building blocks i.e. fibers, and matrix as well as from the fabric texture imperfections e.g. fiber undulations. Moreover, ply inherent boundary conditions, e.g. the restriction of the Poisson deformation of the matrix imposed from the adjacent fibers, are determining the in-situ orthotropic performance. Towards modelling the in-plane stiffness of a unidirectional (UD) infused non-crimp fabric, a two-step modular procedure is proposed, accounting for the aforementioned parameters, based only on experimental data and analytical formulations. Initially, a micromechanical model is predicting the stiffness of the ideal UD ply i.e. disregarding fiber undulations. Subsequently, a plate model is generated based on the classical lamination theory, approximating the UD laminate as a multiaxial configuration of ideal UD sub-plies. Each sub-ply thickness and orientation is based on the fiber angle density distribution of dry fabrics and cured laminates. These are derived experimentally with an integrated optical camera system and Computer Tomography scans respectively. The theoretical laminate stiffness is correlating very well with standard and thick UD laminate quasi-static tests.

1. Introduction

Wind turbine blades are subjected in high cycle fatigue loads over their operational life span [1]. Especially in the edge-wise direction these loads are driven from the blade mass [2]. Therefore, the weight optimization is of substantial importance. One of the parameters that are determining the total mass and its distribution along the structure is the specific stiffness of the employed materials, particularly of the UD plies which are the main building blocks for load bearing sub-components. The non-crimp fabrics that are employed in the manufacturing of wind turbine rotor blades are consisted of single fibres, bund into rovings and stitched for handling stability reasons to the so called backing plies. The fibers oriented along the blade length-axis are majorly responsible for its stiffness properties, although off-axis fibers contribute eventually when multiaxial fabrics are employed e.g. biaxial or triaxial. In the case of the UD fabric, due to the stitching of the rovings in the fabric configuration, the backing ply and the laminate stacking sequence, fiber angle often deviates from the absolute 0° roll direction which in turns is resulting in stiffness deviation from the ideal ply.

The in-plane stiffness properties are thoroughly studied in the international literature and several models are proposed particularly for the unidirectional ply, ranging from the rule of mixtures [3] up to micro-mechanical models considering different levels of the composite details [4-6]. Chamis and Sendeckyj [7] are summarizing the predicting theories for the thermoelastic properties of UD materials in a review paper. They are concluding that due to the fabrication process and the statistic variability of the materials, the mechanical properties can be most reliably described with semi-empirical equations. Schürmann [8] proposes a semi-empirical model, considering the matrix Poisson ratio constraint in between the fibers, assuming a serial sequence of them. The model is corrected with a fitting on known test results. Krimmer [9] is stating that the fitting processes can induce errors that potentially could affect the theoretical outcome. Although he proposes similar physical assumptions to the aforementioned model, he alleviates though all fitting parameters by considering the roving geometry configuration in the UD ply, particularly for square or hexagonal pack schemes. This results in an improved prediction of the matrix dominated properties in comparison to all state of the art models. However, in order to incorporate the effect of the in-situ fiber undulations he introduces a knock-down factor for the stiffness of the single-fiber. Similar knock-down factor approaches are adopted in other analytical works [10-11].

In the present research it is attempted to predict the in-plane mechanical properties of the UD E-glass/epoxy laminate with an analytical methodology, implementing for the first time, to the author's knowledge, a modelling procedure with no fitting or knock-down input parameters. The model is considering the matrix in-situ stiffening due to the Poisson deformation restriction, resulting from the adjacent fibers and the fiber in-plane angle distribution of the non-crimp fabric. Therefore, a modular procedure is unfolded in two succeeding steps based on well-established formulations [9] and the classical lamination theory [12]. First, the stiffness of the ideal UD ply i.e. with no fiber misalignment to the laminate 0°-axis, is estimated with a micromechanical model [9]. For this step, the input parameters are the fiber and matrix stiffness's as well as the corresponding fiber volume fraction. The implementation of the intrinsic undulation of the non-crimp fabric is performed in a subsequent step. Therefore, the UD laminate is split into a multiaxial configuration, composed of ideal UD sub-plies that are stacked in variable orientations. The corresponding sub-ply thickness and angles are based on experimental data derived from the fibre in-plane angle distribution about the 0° laminate-axis. The angle deviations of dry UD fabrics and infused laminates, 4- and a 22-layer configurations, are recorded and analysed both with an integrated optical camera system and with micro Computer Tomography (microCT) scans. The optical camera is delivering 2-Dimensional (2D) the in-plane fiber orientation distribution while the microCT 3-Dimensional (3D). In order to enhance the comparison of the fiber-angle probability density derived from the two systems, the microCT results are transformed also in a 2D plane pattern. The theoretical effective mechanical properties of the UD plate model are validated against standard coupon (DIN EN ISO 527-5 [13]) and thick laminate quasi static tests with very good agreement.

The examined fabric consists of a 1134 gr/m^2 glass fibers in roll 0° direction, 36 gr/m^2 backing glass fibers and 12 gr/m^2 stitches (PES). The employed E-glass fibers are the SE2020 from 3B-fibreglass [14] and the infusion epoxy-resin is the RIMH035c/037 from Hexion GmbH [15]. All laminates were infused at 35 °C, cured at 40 °C for 10 hours and post-cured at 80 °C for 7 hours.

2. Methods for prediction of the composite stiffness of the ideal UD ply

A micromechanical approach is implemented for the prediction of the stiffness of the ideal E-glass/epoxy UD laminate manufactured with non-crimp fabrics and vacuum assisted resin infusion molding. For that ideal ply there is no fiber angle deviation from the 0° roll-axis. The model considers the ply in-situ fiber volume fraction, the backing fibers and the stitching material [9].

In the ply level, the orthotropic stiffness parameters and the corr0esponding indexes in the used mathematical formulations are described in Figure 1.

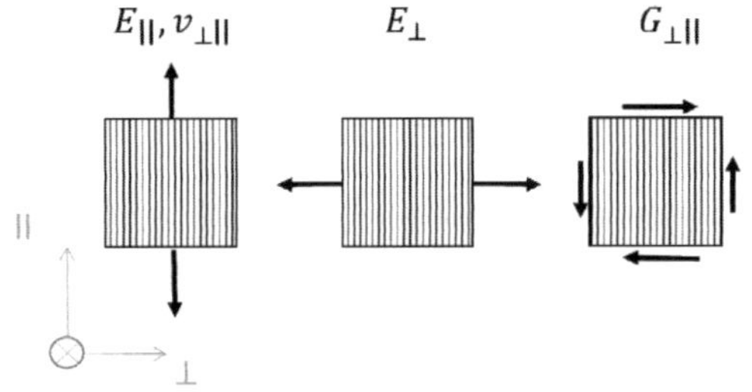

Figure 1. Ply level orthotropic stiffness

2.1. Analytical micromechanical model

In the current research, a micromechanical model is employed, as introduced by Krimmer [9] , towards predicting the in-plane stiffness properties of the ideal UD, i.e. for the ply without fiber undulations. The laminate stiffness along the fibers, $E_{||}$, is estimated based on the rule of mixtures (RoM), by implementing the ply basic building block properties, the stiffness of the matrix E_M and the fibres $E_{F||}$, see eq. (1),

$$E_{||} = (1 - FVF)E_M + FVFE_{F||} \tag{1}$$

where FVF stands for the laminate fiber volume fraction. Diverging from the classical formulation of the RoM, the matrix stiffness E_M, is substituted in eq. (1) from its in-situ property parallel to the fibers, $E'_{M||}$, see eq. (2). The latest is derived when considering the matrix Poisson ratio restriction due to the adjacent fibers, as described in [9],

$$E'_{M||} = \frac{E_{\perp h}(1-v'_{M||})+E_M v'_{M||}}{\frac{E_{\perp h}}{E_M}(1-v'_{M||})+v'_{M||}+2v'^2_{M||}\left(1-\frac{E_{\perp h}}{E_M}\right)} \tag{2}$$

where $v'_{M||}$, eq. (3) stands for the in-situ Poisson ratio of the matrix when pulling parallel to the fibers.

$$v'_{M||} = \left(1 - \frac{v_{F\perp||}E_M}{v_M E_{F\perp}}\right) v_M \tag{3}$$

The model input parameters such as the matrix stiffness E_M and the Poisson ratio v_M are obtained through standard tensile tests [16]. $E_{\perp h}$ stands for the composite stiffness in the direction transverse to the fibers. It is derived analytically in [9] as a function of the rovings pattern in the laminate. The assumption for the fiber-bundle packing geometry in the infused laminate configuration has a direct impact on the in-situ performance of the matrix dominated properties. Therefore, a hexagonal pack scheme is considered and thus the corresponding UD composite ply properties $E_{\perp h}$, $G_{\perp||h}$, (stiffness transverse to the fibers and in-plane shear stiffness), are described in eq. (4) and (5).

$$E_{\perp h} = \frac{2E_M}{\sqrt{3}}\left[\frac{\sqrt{3}}{2} - \sqrt{2\sqrt{3}\frac{FVF}{\pi}} - \frac{\pi}{2\left(1-\frac{E_M}{E_{F\perp}}\right)} + \frac{\frac{\pi}{2} + \arctan\left(\frac{\sqrt{2\sqrt{3}\frac{FVF}{\pi}}\left(1-\frac{E_M}{E_{F\perp}}\right)}{\sqrt{1-2\sqrt{3}\frac{FVF}{\pi}\left(1-\frac{E_M}{E_{F\perp}}\right)^2}}\right)}{\left(1-\frac{E_M}{E_{F\perp}}\right)\sqrt{1-2\sqrt{3}\frac{FVF}{\pi}\left(1-\frac{E_M}{E_{F\perp}}\right)^2}}\right] \tag{4}$$

$$G_{\perp||h} = \frac{2G_M}{\sqrt{3}} \left[\frac{\sqrt{3}}{2} - \sqrt{2\sqrt{3}\frac{FVF}{\pi}} - \frac{\pi}{2\left(1-\frac{G_M}{G_{F\perp||}}\right)} + \frac{\frac{\pi}{2} + arctan\left(\frac{\sqrt{2\sqrt{3}\frac{FVF}{\pi}}\left(1-\frac{G_M}{G_{F\perp||}}\right)}{\sqrt{1-2\sqrt{3}\frac{FVF}{\pi}\left(1-\frac{G_M}{G_{F\perp||}}\right)^2}}\right)}{\left(1-\frac{G_M}{G_{F\perp||}}\right)\sqrt{1-2\sqrt{3}\frac{FVF}{\pi}\left(1-\frac{G_M}{G_{F\perp||}}\right)^2}} \right] \tag{5}$$

Matrix and fibers are assumed to be isotropic materials and therefore their shear stiffness is calculated accordingly, see eq. (6) and (7).

$$G_M = \frac{E_M}{2(1+v_M)} \tag{6}$$

$$G_{F\perp||} = \frac{E_F}{2(1+v_F)} \tag{7}$$

The in-situ properties of the matrix $E'_{M\perp}$, $v'_{M\perp}$ i.e. the stiffness transverse to the fibers and Poisson ratio of the matrix when pulling transverse to the fibers are derived under the same assumptions as described above and are listed for completeness purposes in Appendix A, see eq. (A.1) and (A.3).

2.2. Elastic properties of the multilayer laminate

Based on the fiber angle distribution measurements of the dry fabrics and the cured laminates, the UD configurations can be approximated as the summation of UD sub-plies with a range of fiber orientation. The measurement analysis is presented below in section 4. Due to the symmetric fiber distribution about laminate 0° direction, the multiaxial laminate is considered to be balanced and symmetric. Therefore, any stiffness coupling terms are eliminated through the employed classical lamination plate theory analysis. The axial stiffness matrix $[A_{ij}]_\varphi$ is calculated for each ply with the respective fiber-angle orientation (φ). The summation of the corresponding matrix terms weighted with the probability density of the angle distribution is resulting to the laminate axial stiffness matrix terms, eq. (8).

$$A_{ij}^{Lam.} = \sum_{k=\varphi_{min}}^{\varphi_{max}} A_{ij}^k \cdot PD_k \tag{8}$$

Given the axial stiffness matrix of the laminate, the effective elastic properties are derived as described in [12]. For completeness, the implemented formulas are listed below, see eq. (9-12).

$$E_x = \frac{A_{xx}A_{yy} - A_{xy}^2}{hA_{yy}} \tag{9}$$

$$E_y = \frac{A_{xx}A_{yy} - A_{xy}^2}{hA_{xx}} \tag{10}$$

IOP Conf. Series: Materials Science and Engineering **942** (2020) 012017 doi:10.1088/1757-899X/942/1/012017

$$v_{xy} = \frac{A_{xy}}{A_{yy}} \tag{11}$$

$$G_{xy} = \frac{A_{ss}}{h} \tag{12}$$

3. Input parameters for the modelling procedure

The required fiber input material properties are taken from the literature while the matrix mechanical properties are experimentally obtained in the context of this exercise. For the determination of the fibre angle distribution the materials,. dry fabrics and cured laminates are subjected to optical and microCT scans.

3.1. Constituent materials mechanical properties

The glass-fiber stiffness [14] and Poisson ratio ($\boldsymbol{v_{F\perp||}}$) [9] are taken from the international literature and are listed in Table 1. The glass material is expected to perform isotropic and therefore the stiffness is considered to be equal along and transverse to the fibers, $\boldsymbol{E_{F||}} = \mathbf{E_{F\perp}}$. All properties are assumed to be the same both for the UD and the backing fibers.

Table 1. Fiber mechanical properties

	Young's mod. [GPa]	Poisson [-]
Average Value :	81.0	0.22
Standard deviation:	-	-

Standard DIN EN ISO 527-2 [16] quasi-static tests are conducted for the derivation of the Young's modulus and Poisson ratio of the cured matrix. The tests are performed in a 25kN calibrated universal machine of class 0.5, both for load cell and displacement, according to international standards [17], [18]. The specimens are equipped with extra soft strain gauges in order to avoid local stiffness effects and thus influencing the recorded data. The measured mean values and the respective standard deviations are listed in **Table** 2. The corresponding glass transition and potential (T_g, Tg_∞) temperatures [19] are included for completeness. Detailed mechanical properties of the test specimens are listed in Appendix B.

Table 2. Matrix mechanical properties

	Young's mod. [GPa]	Poisson [-]	T_g [°C]	$T_{g\infty}$ [°C]
Average Value :	3.155	0.366	77.56	83.9
Standard deviation:	0.05	0.02	0.85	1.20

3.2. Fibre distribution of dry non-crimp fabrics and cured laminates

The fiber angle deviation (φ) is mapped for dry UD fabrics and UD infused laminates after curing. Typical fabric in-plane non-conformities i.e. undulations are shown in Figure 2. The measurements are conducted in 4- and 22-layer stacking sequences.

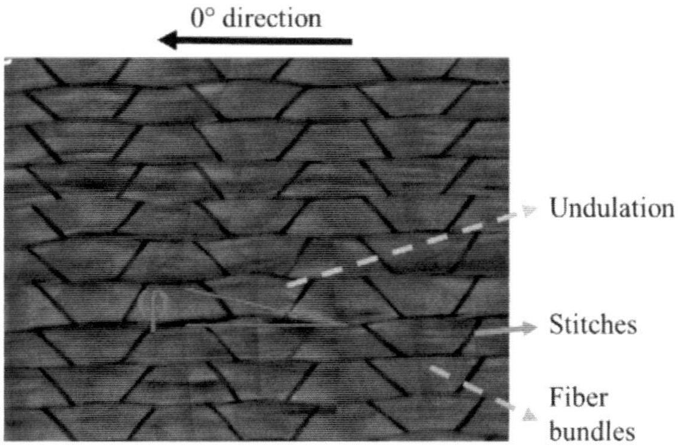

Figure 2. Typical in-plane undulation (φ) of the dry UD 1182 gr/m^2 non-crimp fabric.

For the quantification of the fibre angle distribution of the dry non-crimp fabrics, a camera (HP-C-V3D Apodius Vision System) integrated system is implemented, as shown in Figure 3. It consists of a mobile arm (Hexagon Absolute ARM 8335), a camera head which is installed at the end of the arm and a software (Apodius Explorer 3D) for digital analysis of the acquired photos. Before the infusion and during the preparation of UD stacking sequences, with 4- and 22- layers, each of them is scanned from the top view with an optical camera. Therefore, a 10x10cm measurement field is selected, located in the middle area of each ply. It is assumed that the defined area is representative for the fabric. A red tape line is attached on the table-form as a reference coordinate axis (x), assisting on the alignment of the dry UD fabric fibres and at the same time for the positioning of the Apodius System, see Figure 4. In order to assure the same position of the scanning head for all photos, the supporting arm is fixed on a station next to the infusion table. Because of the symmetric lay-up and the laying process, half of the plies are shot from the UD-fibre side and half from the backing-ply side.

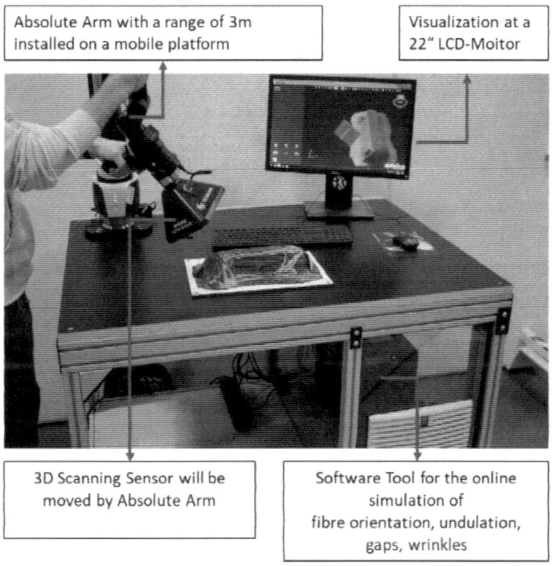

Figure 3. Apodius vision System 3D

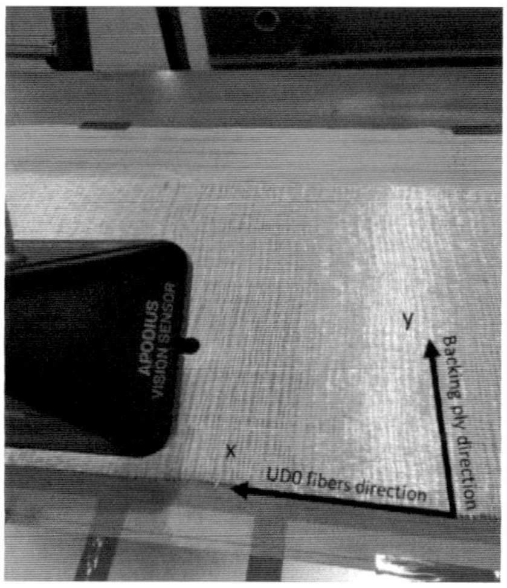

Figure 4. Coordinate system for the scans with the optical camera

The cured laminate is scanned using a Zeiss Xradia 520 microCT scanner with a 2000x2000 detector. The 4-layer and 22-layer laminate is scanned by means of a 20 mm and 13 mm FoV resulting in a voxel size on 9.8 microns and 6.4 microns, respectively. Both scans are based on a 8s exposer time using 6801 projections resulting in 20 hours scan time. The resulting 16 GB reconstructed scanning file was analysed via a structure tensor method as described in [20]. The analysis step took around 1/2 hour for each scan.

The distribution of the fiber angle in the dry plies is averaged for the 4-layer and the 22-layer laminates, respectively. A bell-shaped fitting of the angle distribution probability $\boldsymbol{f(x)}$, see eq. (5), is performed on the experimental data.

$$f(x) = ae^{-\frac{(x-b)^2}{2c}} \tag{13}$$

where x is standing for the corresponding fiber angle and a, b and c bell-shape fitting parameters. All of them are listed in **Table** 3.

Table 3. Fitting parameters of the Bell-shaped angle distributions

	optical camera		
	a	b	c
4-layer	5.3	0.7	8
22-layer	5.4	-0.25	7.8
	microCT		
4-layer	13	0.1	3.1
22-layer	5.3	-0.17	1.8

The experimental in-plane angle distribution for the 4-layer and the 22-layer laminates before and after infusion with the corresponding bell-shaped probability densities are shown in Figure 5 and Figure 6.

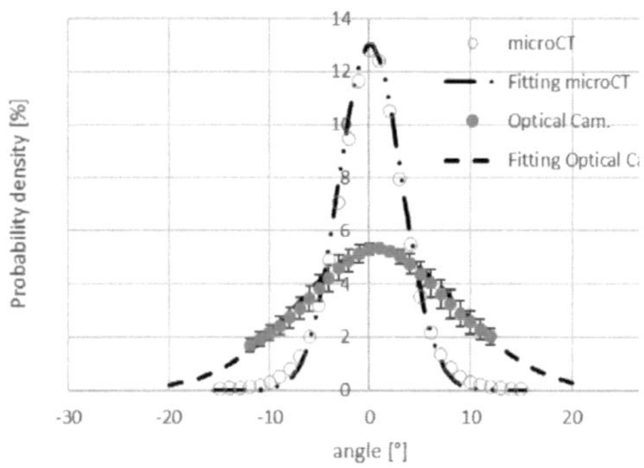

Figure 5. Bell-shaped angle distribution of the 4-layer laminate before and after infusion

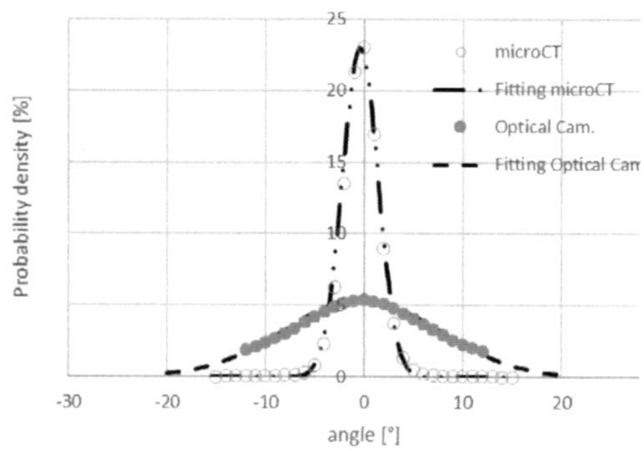

Figure 6. Bell-shaped angle distribution of the 22-layer laminate before and after infusion

3.3. Quasi-static tests for the derivation of UD composite mechanical properties

For the characterization of the principal Young's Modulus $\boldsymbol{E_1}$ and $\boldsymbol{E_2}$, the Poisson ratio $\boldsymbol{v_{12}}$ [13] and the shear stiffness $\boldsymbol{G_{12}}$ [21] of the UD ply, three sets of specimens are tested with 4-layer symmetric layup. The average values and the corresponding standard deviations of the tested coupons are listed in Table 4. For completeness purposes the fiber volume fraction (FVF) and the glass transition temperatures (T_g) [19] are summarized too. The FVF values implemented in the model are derived according to [9] taking under consideration the stitches, the fiber sizing and the backing ply volumes. In brackets are listed the FVF's derived according to the ISO standard [22]. For the derivation of the shear modulus, the shear strain range between 500-2500 µm/m is selected. This deviation from the range

1000-5000 μm/m proposed in the DIN EN ISO 14129:1998-02 standard [21] is chosen in order to account only for the linear deformation in the modulus calculation and eliminate non-linear effects which are observed over the 2500 μm/m limit.

Table 4. UD ply mechanical properties

	E_1[GPa]	E_2[GPa]	v_{12} [-]	G_{12} [GPa]	FVF [%]	T_g [°C]	$T_{g\infty}$ [°C]
Average:	42.35	12.61	0.301	4.59	54* (51.85)	82.25	90.24
St. dev.:	1.19	0.41	0.03	0.150	- (0.29)	1,70	1,41

* for the E_2 estimation, the theoretical FVF is 55%

3.4. *Large scale coupon quasi-static test*

Three large scale specimens are tested under tensile loading till rupture. To record the laminate deformation, four strain gauges are applied on each. Three linear with a measuring grid-length of 10mm are applied on one side, in the load direction at the middle of the coupon length i.e. at L=750 mm. They are located in 25 mm, 50 mm and 75 mm of the coupon width. The fourth is a cross-pattern strain gauge with a 0°- 90° orientation and 8mm measuring grid length applied on the other side of the specimen also at L=750 mm and centred at W=50 mm. The specimens are clamped in a 2,5 MN universal servo hydraulic coupon machine with 224 bar clamping pressure. The testing machine is calibrated with a load cell and displacement class of 0.5 according to [17], [18]. The coupon vertical angle to the machine load axis is measured with a calibrated digital angle recording device, with alignment tolerance of 0.1° (89.9° to 90.1°). The test is displacement controlled with a cross-head speed of 1mm/min. The specimen geometry and the experimental setup are illustrated in **Figure** 7 and **Figure** 8 respectively.

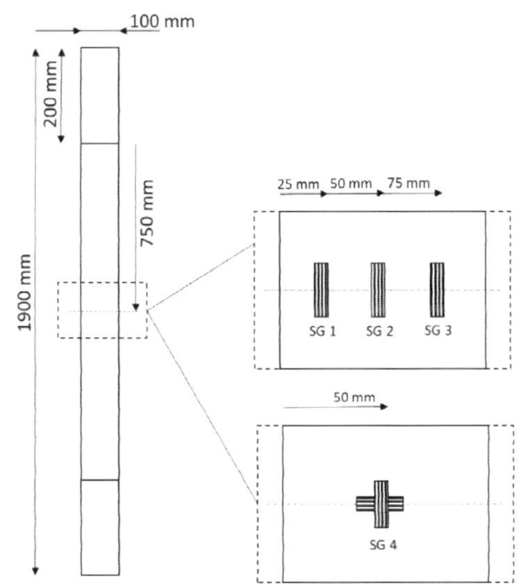

Figure 7. Large scale coupon geometry

Figure 8. Large scale coupon test set-up.

The average Young's Modulus, the Poisson ratio, the glass transition and the potential glass transition temperatures are listed in Table 5.

Table 5. UD thick laminate mechanical properties

	E_1 [GPa]	v_{12} [-]	FVF [%]	Tg [°C]	Tg∞ [°C]
Average:	46.09	0.291	58	77.87	89.27
St. dev.:	1.35	0.006	-	3.00	1,50

4. Analysis results

4.1. Stiffness of the ideal UD ply

The predictions of analytical micromechanical model for the stiffness of the ideal UD ply (without undulations) by implementing the mean values and the standard deviation of the matrix properties are listed in Table 6.

Table 6. Theoretical mechanical properties of the ideal UD ply

	E_1 [GPa]	E_2 [GPa]	v_{12} [-]	G_{12}[GPa]
Prediction:	43.48	12.01	0.26	3.87
Standard Dev.	0.11	0.50	0.01	0.14

4.2. Stiffness prediction of the UD laminate when incorporating the in-plane fiber bundle undulations

The proposed model predictions of the effective mechanical properties of the 4- and the 22-layer UD laminates are summarized below. Therefore, the laminate theory, see section 2.2, the stiffness of the ideal UD ply as presented in the previous section and the fiber-angle distribution as presented in section 3.2 both in the dry plies and the cured specimens, are combined. In Table 7 are listed the model predictions after employing the angle distribution measured on the dry fabrics with the optical camera system. In Table 8 are listed the model predictions after employing the angle distribution recorded with microCT on the cured specimens. Theoretical results are compared to the corresponding experimental.

Table 7. Effective laminate stiffness and difference from experim. (dry fabric angles)

	E_1 [GPa]		E_2 [GPa]		v_{12} [-]		G_{12} [GPa]	
4-layer	42.19	-	11.72	-	0.303	-	4.37	-
22-layer	-	47.03	-	13.44	-	0.301	-	5.11
Stand. Dev.	0.12	0.13	0.49	0.57	0.005	0.005	0.14	0.17
Differ. from exp.	[%]		[%]		[%]		[%]	
4-layer	-0.39	-	-7.05	-	0.70	-	-4.68	-
22-layer	-	2.35	-	-	-	3.52	-	-

Table 8. Effective laminate stiffness and difference from experim. (cured laminate angles)

	E_1 [GPa]		E_2 [GPa]		v_{12} [-]		G_{12} [GPa]	
4-layer	43.15	-	11.77	-	0.272	-	4.00	-
22-layer	-	46.31	-	12.92	-	0.261	-	4.34
Stand. Dev.	0.11	0.11	0.49	0.55	0.007	0.007	0.14	0.17
Differ. from exp.	[%]		[%]		[%]		[%]	
4-layer	1.89	-	-6.63	-	-9.58	-	-12.83	-
22-layer	-	0.49	-	-	-	-10.33	-	-

5. Discussion

To the authors knowledge this is the first attempt to benchmark the optical camera system and the angle density distribution with a microCT technique and vice versa. The comparison is highlighting a difference in the fiber angle probability between the dry fabrics and the cured laminate. This might be originating from the scale measurement level that the two techniques are zooming, assuming that in a macroscopic level higher degrees of angle deviations might be measurable. However, other effects, as fiber alignment after the vacuum application or even straightening due to the infusion and curing process cannot be excluded. Further microCT measurements of the dry fabric are planned in order to investigate the aforementioned issue.

In both cases the proposed model could predict the laminate in-plane stiffness along the fibers direction with very good precision. Nevertheless, the comparison to the matrix dominated properties is moderate. The model variant implementing the cured laminate angles distribution is deviating more especially for the thick laminate. This could be in line with the remark made above about the capacity of the technique to capture higher angle degrees.

When employing the higher end properties of the matrix material (standard deviation), the model predictions are improved approaching the test values in some cases with less than 4% difference for all mechanical properties. It has to be stated though the curing degree of the implemented matrix properties that were derived through static tests, is not totally the same with the composite test specimens, especially with the thick laminates. A curing kinetics model and a correlation model of stiffness with the curing degree could enhance future comparisons.

6. Conclusions

A theoretical model for the prediction of the in-situ stiffness of non-crimp laminates is successfully validated against quasi-static tests of different geometry scales and fiber volume fractions. The model is based exclusively in analytical formulations and well established mechanics without any knock-down or fitting parameters. As input parameters are required only the basic building block mechanical properties and the intrinsic fiber-angle probability distribution. Two fiber-angle distribution techniques are compared to each other, resulting in relative differences to each other. Further investigations have to be performed e.g. microCT scans on dry fabrics in order to conclude on the root causes.

Given these, the effective mechanical properties of any laminate with various fiber volume fraction could be predicted. Both sets of angle distributions results in very good along the laminate 0° direction i.e. less than 2.5%, however the model shows a higher deviation from the experimental results in the matrix dominated properties. These have to be investigated further.

Acknowledgments

Fraunhofer IWES acknowledge the support provided by the German Federal Ministry for Economic Affairs and Energy (BMWi) grant 0324335A and DTU Wind Energy acknowledge the support from by the Danish Energy Agency through the EUDP grant 64018-0068. Both grants is within the ReliaBlade project: Improving Blade Reliability through Application of Digital Twins over Entire Life Cycle.

References

[1] IEC, IEC 61400-1 - Wind Turbines Part 1: Design Requirements for Wind Turbines - Ed. 3, International Electrotechnical Commission, Geneva, Switzerland, 2010

[2] Krimmer A, Leifheit R, and Bardenhagen 2016 A Assessment of quasi-static and fatigue performance of uni-directionally fibre reinforced polymers on the basis of matrix effort, *6th EASN International Conference on Innovation in European Aeronautics Research, Porto, Portugal*

[3] Halpin J C and Kardos J L 1976 The Halpin Tsai equations. A review, *Polym. Eng. Sci.* **16** 5 344-352

[4] Ekvall J C 1961 Elastic Properties of Orthotropic Monofilament Laminates *ASME Aviation Conference* Los Angeles California

[5] Greszczuk L B 1966 Theoretical and Experimental Studies on Properties and Behavior of Filamentary Composites *SPI 21st Conference Chicago Illinois*

[6] Shaffer B W 1964 Stress-Strain Relations of Reinforced Plastics Parallel and Normal to the Internal Filaments *J. AIAA* **2** 348-352

[7] Chamis C C and Sendeckyj G P 1968 Critique on Theories Predicting Thermoelastic Properties of Fibrous Composites *J. Composite Materials* 2 Nr. 3

[8] Schürmann H Konstruieren mit Faser-Kunststoff-Verbunden Berlin, Heidelberg, New York : Springer-Verlag, 2005. – ISBN 3–540–40283–7

[9] Krimmer A 2014 *Mikromechanische Modellierung von Fasergelege-Kunststoff-Verbunden auf Basis von Normprüfungen unter Berücksichtigung der in-situ-Eigenschaften der Matrix Ph.D. thesis* ed Krimmer A (TU Berlin). URL http://opus4.kobv.de/opus4-tuberlin/frontdoor/index/index/docId/4440 chapter 3 pp 43-53

[10] Edgren F Asp LE 2005 Approximate analytical constitutive model for non-crimp fabric composites *Composites: Part A* **36** 173–181

[11] Marklund E Varna J Asp L E 2011 Stiffness and strength modelling of non-crimp fabric composites 52nd *AIAA/ASME/ASCE/AHS/ASC Structures, Structural Dynamics and Materials Conference Denver Colorado*

[12] Reddy J N 2004 *Mechanics of laminated composite plates and shells: theory and analysis*, (CRC press LLC)

[13] DIN EN ISO 527-5:2010-01 Plastics - Determination of tensile properties - Part 5: Test conditions for unidirectional fibre-reinforced plastic composites

[14] 3B-Fibreglass sprl, Technical Data Sheet for SE 2020 Direct Rooving, https://www.3b-fibreglass.com/sites/default/files/products-data-sheets/TDS-SE-2020-DR-for-Epoxy-Resins-2015-sans-trame-LR.pdf, on-line status 19.06.2020

[15] Hexion Specialty Chemicals B.V., Technical Data Sheet for Epikote Resin MGS RIMR 035c/RIMH 037, http://www.hexion.com/en-US/product/epikote-resin-mgs-rimr-035c-and-epikure-curing-agent-mgs-rimr036-0366-037-038-038f, on-line status 19.06.2020

[16] DIN EN ISO, 527-2:2012-06, Plastics - Determination of tensile properties - Part 2: Test conditions for moulding and extrusion plastics

[17] DIN EN ISO 7500-1, Metallic materials - Verification of static uniaxial testing machines - Part 1: Tension/compression testing machines - Verification and calibration of the force-measuring system
[18] DIN EN ISO 9513 Metallic materials - Calibration of extensometer systems used in uniaxial testing
[19] ISO 11357-2:2020(E) - Plastics - Differential scanning calorimetry (DSC) - Part 2: Determination of glass transition temperature and glass transition step height, International Organization for Standardization.
[20] Jeppessen N, Dahl V A, Christensen A N, Dahl A B, Mikkelsen L P Characterisation of the fiber orientations in non-crimp glass fiber reinforced composites using structure tensor, *41st RISOE International Symposium on Materials Science*, 7-10 September 2020
[21] DIN EN ISO 14129:1998-02 Fibre reinforced plastic composites - Determination of the in-plane shear stress/shear strain response, including the in-plane shear modulus and strength, by the ±45 tension test method
[22] DIN EN ISO 1172 : 1998-12 Textile-glass-reinforced plastics —Prepregs, moulding compounds and laminates —Determination of the textile-glass and mineral-filler content —Calcination methods

Appendix A

The analytical formulation of the in-situ matrix stiffness transverse to the UD fibers is described through eq. (A.1). For the formula derivation, a hexagonal-pack fiber geometry is assumed superimposed with the restrain of the matrix Poisson deformation due to the adjacent fibers.

$$E'_{M\perp} = \frac{\tilde{E}_{F||} A \cdot FVF + E_{\perp}(1 - FVF)}{\frac{\tilde{E}_{F||}}{E_M}\left(\frac{E_{\perp}}{E_M}(1 - 3v'^2_{M\perp} - 2v'^3_{M\perp}) + 2(3v'^2_{M\perp} + v'^3_{M\perp})\right) FVF + A(1 - FVF)} \tag{A.1}$$

$$A = \frac{E_{\perp}}{E_M}(1 - v'^2_{M\perp}) + v'^2_{M\perp} \tag{A.2}$$

Implementing the same assumptions as for the matrix transverse stiffness, the in-situ Poisson ratio of the matrix when pulling transverse to the fibers is described through eq. (A.3)

$$v'_{M\perp} = \left(1 - \frac{A \cdot FVF + E_{\perp}(1 - FVF)}{\frac{\tilde{E}_{F||}}{E_M}\left(\frac{E_{\perp}}{E_M}(1 - 3v'^2_{M\perp} - 2v'^3_{M\perp}) + 2(3v'^2_{M\perp} + v'^3_{M\perp})\right) FVF + A(1 - FVF)}\right) \tag{A.3}$$

Appendix B

Table 9. Matrix DIN EN ISO 527-2 test results

Specimen ID	Young's mod.	Poisson's ratio
[-]	[GPa]	[-]
114774_019_09_007	3.13	0.36
114774_019_09_008	3.09	0.35
114774_019_09_009	3.19	0.37
114774_019_09_011	3.21	0.39
114774_019_09_012	3.15	0.37

Table 10. UD ply stiffness test results

Specimen ID [-]	Young's mod. E_1 [GPa]	Poisson's ratio v_{12} [-]	Specimen ID [-]	Young's mod. E_2 [GPa]	Specimen ID [-]	Shear mod. G_{12} [GPa]
*_006_01_001	43.43	0.29	*_007_02_001	12.24	**_010_41_001	4.30
*_006_01_002	44.49	0.32	*_007_02_002	12.38	**_010_41_002	4.62
*_006_01_003	42.03	0.35	*_007_02_003	12.32	**_020_03_001	4.70
*_006_01_004	41.03	0.30	*_007_02_004	12.63	**_020_03_002	4.56
*_006_01_005	42.12	0.30	*_007_02_005	13.36	**_020_03_003	4.64
*_006_01_006	41.49	-	*_007_02_006	12.71	**_020_03_004	4.69
*_006_01_007	41.89	0.25	-	-	-	-

* 114774, **111848

IOP Conf. Series: Materials Science and Engineering **942** (2020) 012018 doi:10.1088/1757-899X/942/1/012018

Exploration of irreversible residual stresses in a carbon/epoxy composite

Johnny Jakobsen* and Fahimeh Shakibapour

Aalborg University, Fibigerstraede 16, 9220 Aalborg, Denmark

*joj@mp.aau.dk

Abstract. Residual stresses in composite material may rise from several sources; difference in thermal straining between fibre and matrix, chemical shrinkage of the matrix material. The presented paper explores an irreversible Eigen-strain in the material which is shown to change and be removed by a thermal treatment above the Tg of the matrix material. This has been observed though TMA measurements of a carbon/epoxy unidirectional sample. The local Eigen-strain is translated into stresses using a visco-elastic constitutive relation. Further exploration indicates that cooling rate influence the generation of residual stresses in the material.

1. Introduction

Residual stresses in composite materials have been investigated by many researchers throughout the past decades and their effect on the mechanical performance. Some of the consequences of residual stresses may be geometrical distortions and altered fatigue endurance.

The build-up of these stresses are rather well understood and may rise due to thermal and chemical straining in the material which often take place during the manufacturing process of the material. This has been investigated by several authors. Hahn [1], Bogetti et al [2] and White et al [3], [4] were among the first to formulate the ground laying theory of residual stresses buildup by the curing process.

Later authors like Baran et al. [5] applied the theory to study residual/cure stresses in pultruded rods, which may be used e.g. in wind turbine blade applications.

The complexity in modelling residual stress build–up increases due to the directional dependence from the fiber [6]. Some material properties show to be highly temperature-depended. In most cases, properties in the fiber direction show less dependence compare to the transverse directions and may often be approximated to be temperature-independed. However, the properties transverse to the fibers are dominated the matrix properties which typically are highly temperature-depended with transition at the glass transitions temperature (Tg). Examples hereof is the modulus, coefficient of thermal expansion, etc.

The thermal expansion in the transverse direction of a fully cured carbon fiber/epoxy composite can be measured using a Thermo Mechanical Analyzer (TMA) and during the first heating material relaxation is observed slightly before the Tg of the epoxy material (see Figure 1). Moreover, it is noticed that during cooling the material contracts in a bi-linear manner with transition at Tg. Further heating and cooling of the sample follows the same bi-linear behavior. This further shows that the sample ends up a slightly contracted shape.

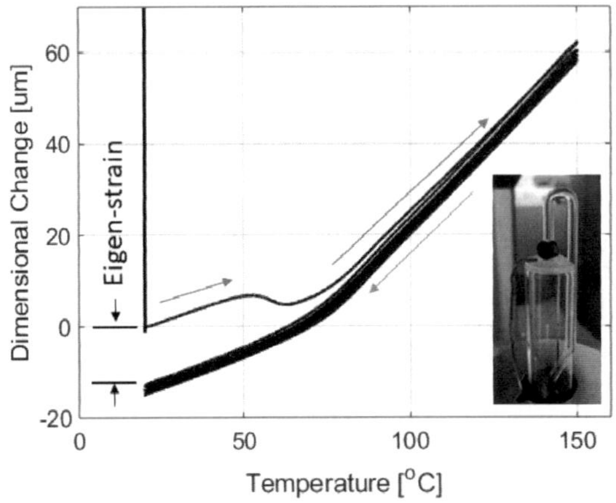

Figure 1. TMA measurement of a small cylindrical carbon/epoxy composite. Measurement conducted transverse to the fibre direction.

This permanent contraction (or eigen-strain) has also been observed by other researcher in the past [7] and [8] and has been explained by some type of material relaxation possibly originating from residual stresses in the material. However, further details on this relaxation seems to be missing to support the argumentation that this is true.

In this paper, the authors attempt to investigate this permanent contraction behavior by applying a visco-elastic model, where the thermal dependence on several properties are taken into account. The model takes a starting point, where the material has reached the state of fully cured and to be cooled rapidly to room temperature. After a holding period at room temperature the material is heated to examine if this sudden contraction can be observed in the model.

In case the model results can confirm that during cooling stresses are build up in the material and later relaxed upon a second heating, it may be a contribution to the understanding the residual stresses in composite material and their ability to relax at elevated temperatures and further support other researchers explanation of this phenomenon.

2. Material

The material used in this study is a carbon fiber epoxy composite, which has been cured at 100C and then rapidly cooled to room temperature. Properties of the material used in the later modelling is as stated in Table 1.

Table 1. Material data used in the later modelling

Model parameter	Value,	Comment
T_g	80 [C]	Tg of the epoxy matrix
T_i	100 [C]	Stress free temperature
v_{23}	0.35 [-]	Poissons ratio
CTE_{subTg}	$100\cdot10^{-6}$ [1/C]	Coefficient of thermal expansion below Tg
CTE_{supTg}	$200\cdot10^{-6}$ [1/C]	Coefficient of thermal expansion above Tg

The visco-elastic behavior of the material is measured by a series of DMA tests and shown in the next section.

2.1. Generation of a master curve

IOP Conf. Series: Materials Science and Engineering **942** (2020) 012018 doi:10.1088/1757-899X/942/1/012018

The stiffness behavior is both time and temperature depended as argued earlier [9]. Therefore, a number of stress relaxation measurements are performed at a number of isothermal temperatures to create a "master curve" for the modulus (see Figure 2).

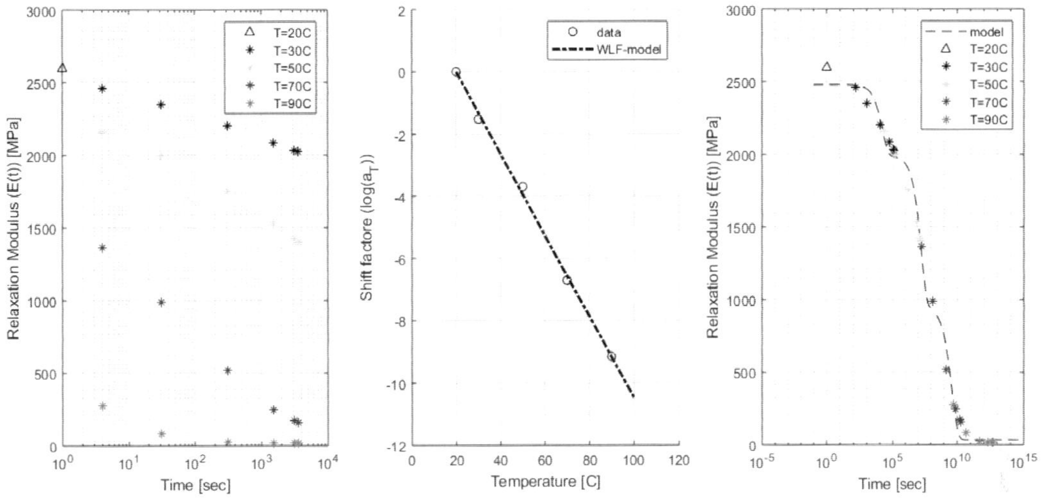

a b c

Figure 2. Stress relaxation data measured at several temperatures are shifted horizontal to form a master curve for the relaxation modulus. a) stress relaxation data, b) applied shift factors and c) generated master curve for T_{ref}=20C.

A Prony-series is then fitted to the master curve data from which the visco-elastic behaviour of the material described in equation (1) [10].

$$E(t) = E_0 + \sum_{i=1}^{3} E_i e^{\left(-\frac{\tau}{t_{Ri}}\right)} \qquad \tau = \frac{t}{a_T} \tag{1}$$

The fitted values obtained are stated in Table 2 and the model is plotted together with data in right side graph of Figure 2.

Table 2. Model parameters for the master curve.

Model parameter	Value, (MPa)	Model parameter	Value, (s)
E_0	35.0	t_{R1}	$1.579\ 10^4$
E_1	490.9	t_{R2}	$1.908\ 10^7$
E_2	1039	t_{R3}	$4.200\ 10^9$
E_3	914.2		

The time-temperature shift (a_T) is modelled using a WLF type equation (2) and similar fitted to data (see Table 3) [10].

$$log(a_T) = \frac{-C_1(T - T_{ref})}{C_2 + T - T_{ref}} \tag{2}$$

Table 3. Model parameters for the time-temperature shift model

Model parameter	Value,	Comment
T_{ref}	20 C	Reference temperature for the master curve
C_1	1002	
C_2	7582 C	

The WLF equation is known to be valid for polymers above its Tg [10] and an Arhenius equation is normally used to express the shift function below Tg. According shift data presented in the middle graph of Figure *2* the WLF equations seems to match the data reasonably well for the entire temperature range. However, the obtained values for the coefficients C1 and C2 are far from their typical values when T_{ref} is set equal to T_g. Therefore the use of equation-2 should merely be considered as good fitting functions for the data presented.

3. Model

A thermo-mechanical model is made in the multi-physics finite element packages Comsol®. The model starts by being in thermal equilibrium at 100C and then cooled rapidly at the outer edge to mimic the experimental cooling process of the sample. This rapid cooling is simulated to occur over a period of 10s and then hold at 20C for a period of 900s. Lastly, the sample edge is heated slowly back to 100C. Symmetry condition are utilized at x=0 and y=0, which means that displacement and heat fluxes normal to these edges are zero. The material includes the visco-elastic behavior as described earlier and temperature depended coefficient of thermal expansion. The model is meshed with second order triangular elements (see Figure 3). The imposed boundary temperature is seen in the graph of Figure 3.

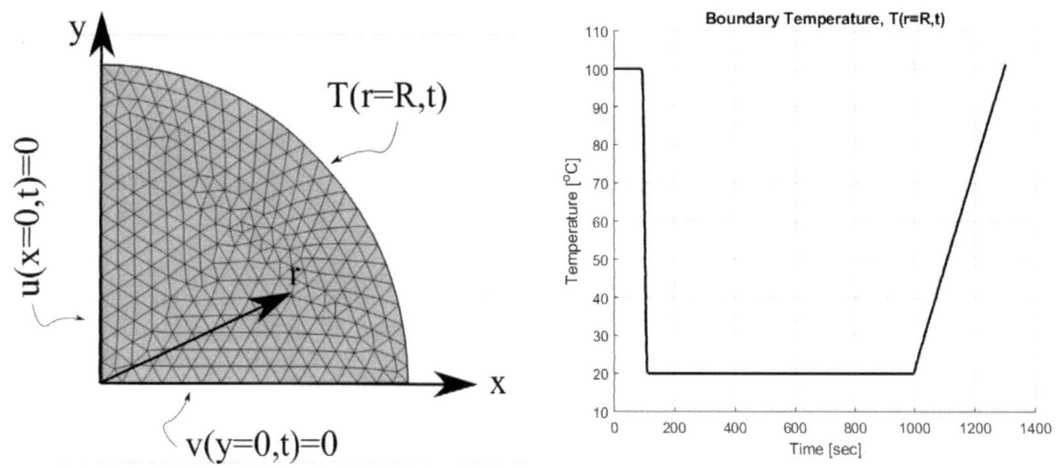

Figure 3. The cylindrical composite sample is modelled as a 2D quarter circle where symmetries are utilized and plane-stress is assumed. The temperature boundary condition applied is shown on the right side graph.

The model has a radius of 2.5mm (R=2.5mm) and a thickness of 2mm. Moreover, plane stress is assumed. Lastly the model presented here is a homogenized representation of the material.

4. Results

The initial rapid cooling of the model leads to rather large temperature differences between outer boundary and center in the magnitude of 50C (see top graph of Figure 4). This temperature gradient results in to thermal stresses in the material, where the outer boundary comes in tension and the center is in compression. As the stresses are relaxed and thermal equilibrium is obtained, the stress state is to some extend reversed such that the boundary comes in a state of cylindrical compression and the center in radial tension (see bottom graph of Figure 4). It is however worth noticing that residual stresses remain in the sample after completing the cooling process and thermal equilibrium is obtained.

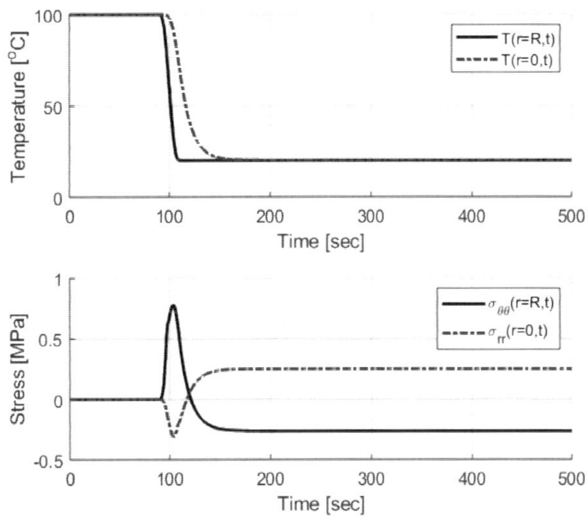

Figure 4. Top graph) the temperature response at the model boundary and in the center. Bottom graph) the stress response at the boundary and in the center. Notice that residual stress remains in the model after the rapid cooling.

It is then speculated if these residual stresses can cause this sudden contraction in the material which are argued earlier and reported by other researchers. It should however be mentioned that the residual stress buildup is a consequence of the cooling rate and the visco-elastic material behavior.

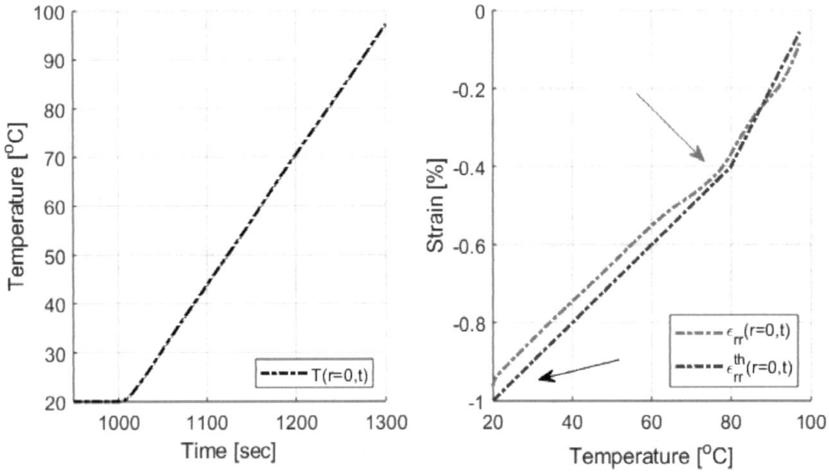

Figure 5. Left graph) subsequent "slow" heating of the model. Right graph) elastic and thermal strain in the center. Notice that stress relaxation occurs when the temperature approach Tg and further that an eigen strain is present in the model at 20C.

Heating the model at a slower rate to above its Tg should then allow the induces stresses to relax. This is also observed to some extend in the resluts presented at right side graph in Figure 5, where both the elastic and thermal strain are presented. The thermal strain displays a bi-linear behaviour which is explained by the difference in coefficient of thermal expansion below and above Tg. Moreover, it is seen that some stress relaxation takes place slightly before reaching Tg and are displayed through a small contraction.

The eigen-strain obtained by the model at 20C is evaluated to 0.05% on the other hand the eigen-strain obtained from the TMA measurement is (~0.30%). Therefore it is still questionably if the sudden contraction observed is purely due to relaxation of residual stresses or if other effects add to pnenomenon.

5. Conclusion

The sudden contraction of a fully cured carbon/epoxy composite cylinder when heated to near Tg is analyzed using a thermo-mechanical model to replicate the experimental measurement. It was found that rapid cooling and visco-elastic relaxation lead to residual stress in the material at room temperature. Upon a slow reheating of the model, residual stress relaxation does to some extend explain the contraction near Tg but the magnitude of contraction was much less than observed from the TMA measurement.

At this stage, it is not clear if the disagreement between model and experimental data is due to lack in precision in the material data (i.e. Prony-series) or if some other phenomenon not accounted for may contribute to this sudden contraction under investigation.

References

[1] Hahn H T, (1976), "Residual Stresses in Polymer Matrix Composite Laminates", Journal of Composite Materials, vol. 10, pp. 266-278.

[2] Bogetti T A and Gillespie J W, (1992), "Process-Induced Stress and Deformation in Thick-Section Thermoset Composite Laminates", Journal of Composite Materials, vol. 26, pp. 626-660.

[3] White S R and Hahn H T, (1992), "Process Modeling of Composite Materials: Residual Stress Development during Cure. Part I. Model Formulation", Journal of Composite Materials, vol. 26, pp. 2402-2422.

[4] White S R and Hahn H T, (1992), "Process Modeling of Composite Materials: Residual Stress Development during Cure. Part II. Experimental Validation", Journal of Composite Materials, vol. 26, pp. 2423-2453.

[5] Baran I and Tutum C C and Nielsen M W and Hattel J H, (2013), "Process induced residual stresses and distortions in pultrusion", Composites Part B: Engineering, vol. 51, pp. 148-161.

[6] Thomsen J S and Pyrz R, (1999), "Creep of carbon/polypropylene model composites—a Raman spectroscopic investigation", Composites science and technology, vol. 59, p. 1375–1385.

[7] Reid R G and Paskaramoorthy R, (2009), "A novel method to measure residual stresses in unidirectional {GFRP}", Composite Structures, vol. 88, pp. 388-393.

[8] Jensen M and Jakobsen J, (2016), "Effect of cure cycle on enthalpy relaxation and post shrinkage in neat epoxy and epoxy composites", Journal of Non-Crystalline Solids, vol. 452, p. 109–113.

[9] Kim Y K and White S R, (1996), "Stress relaxation behavior of 3501-6 epoxy resin during cure", Polymer Engineering & Science, vol. 36, p. 2852–2862.

[10] Brinson H F and Brinson L C, (2015), Polymer Engineering Science and Viscoelasticity - An Introduction (2.ed)", Springer, ISBN: 978-1-4899-7484-6

Micromechanical and spectroscopic characterisation of the curing performance of epoxy resins in the microbond test

David Bryce[1, 2] James Thomason[1] and Liu Yang[1]

[1]Department of Mechanical and Aerospace Engineering, University of Strathclyde, Glasgow, G1 1XJ, UK
[2]Author to whom any correspondance should be addressed

david.bryce.100@strath.ac.uk

Abstract. In the present work, the microscale curing performance of two epoxy resin systems was investigated. It was shown that the degree to which these resins formed cured axisymmetric droplets suitable for microbond testing was influenced by the inclusion of a room-temperature pre-cure stage. Immediate curing at elevated temperature of these systems resulted in the formation of under-cured, soft droplets that deformed under loading. The introduction of a protracted room temperature standing time increased the apparent interfacial shear strength substantially. Fourier-transform infrared spectra generated by a novel technique allowed for the glass transition temperature of microdroplet samples to be determined and detected deviations from stoichiometry. Good correlation was shown between the two methods, in that an increase in IFSS was commensurate with spectra indicating droplets were closer to the ideal stoichiometric ratio, and had both higher degrees of cure and glass transition temperatures.

1. Introduction

The fibre-matrix interface region is critical to the mechanical performance of a composite, thus optimisation of the stress transfer capability of this region is extremely important. Accordingly, it is essential to be able to characterise the interface and associated level of adhesion through a robust methodology in order to fully understand the composite performance. While the use of the microbond test developed by Miller *et al.* [1] to characterise the adhesion between various fibre and polymer systems has been widely reported in the literature, only a very small number of authors have commented on the difference in the behaviour of matrix materials at the microscale and the implications this holds for the applicability of the test. If the microbond test is to be presented as an efficient and robust methodology for effectively characterising the fibre/matrix interface it is essential for this phenomenon to be understood in greater detail.

Analysis of the relevant literature indicates that, while the issue of microdroplet curing is recognised by a relatively small number of authors, there is little general agreement on what the cause behind this observed loss in performance is, with studies in fact providing conclusions entirely contrary to one another. Investigations by Haaksma *et al.* [2], Rao *et al.* [3,4], and Ozzello *et al.* [5] examined an epoxy/ metaphenylene diamine (m-PDA) system and attributed the loss in microdroplet glass transition temperature (Tg) to a stoichiometric imbalance due to evaporation of curing agent. Laurikainen [6,7] and Dirand [8] similarly agree that diffusion of curing agent was the cause of the problem in vinylester systems (due to the large surface-to-volume ratio) though the former states that this could not have been the case for the same effect observed in their epoxy system due to insignificant vapour pressures. Instead it was proposed that a phase separation occurred during sample preparation, or possible interaction with atmospheric moisture. Pyrolysis gas chromatography analysis conducted on under-cured microbond samples showed huge emissions in the thermal desorption step compared to fully cured resin samples, a result indicative of the degree of cure in the microdroplet being significantly lower than that of a bulk matrix sample subjected to the same curing schedule. Charlier *et al.* [9] report that processing and curing conditions had a significant effect on the apparent interfacial shear strength (IFSS) in an acrylic system. Without the addition of paraffin wax, radical polymerisation was not possible at elevated temperatures due to the rapid evaporation of methyl methacrylate (MMA) out of the droplets. Zinck *et al.*[10] refute that evaporation was the cause of the issue in amine-cured epoxy samples and proposed instead that imine formation reducing available amine functions available to react with the oxirane ring was to blame. For an anhydride-cured epoxy they suggested fibre surface moisture or hydrolysis of the anhydride during the pre-cure as the root cause; and, that surface oxidation may also be a contributing factor.

In a glass fibre/cyanate ester system, Bartolomeo *et al.* [11] reported good correlation between the degree of conversion (α) of bulk resin and microdroplet samples after 30 minutes ($\alpha_{resin} = 0.4$, $\alpha_{microdroplet} = 0.35$) and 1 hour ($\alpha_{resin} = 0.52$, $\alpha_{microdroplet} = 0.51$) of isothermal curing at 180°C. At low degrees of conversion measured IFSS remained relatively constant (in the region of 8 MPa) before increasing significantly once the gel point of the system ($\alpha_{gel\ point} = 0.6$) had been reached. Curing at higher temperature produced no significant effect on the adhesion, leading to the assertion that IFSS was dependent only on the degree of monomer conversion. Furthermore, as the values of IFSS at low conversion were the same regardless of fibre used, it was proposed that the IFSS is entirely matrix dependent at low degrees of conversion, while at high degrees of conversion both and fibre and matrix interactions become apparent. Failure mode was dependent on microbond sample processing and resultant degree of monomer conversion present in the sample. Low conversions resulted in a fragile fracture of the matrix while in high-conversion specimens exhibited the expected linear interfacial behaviour and showed "stick-slip" phenomenon characteristic of two-dimensional interfacial behaviour.

Montgomery [12] has reported that both the Tg and degree of cure increased with cure temperature. Furthermore, the measured IFSS increased linearly in relation to Tg. Samples cured at room temperature had a degree of cure of 79% and 91-99% for samples that underwent a post-curing procedure. The author indicates that the change in measured IFSS values was likely a result of changes in residual stress. Peters and Springer [13] conclude that while the curing temperature and degree of cure affected the mechanical properties of resin systems, an increase in the degree of cure did not necessarily correspond to an increase in mechanical properties, and may in fact decrease as degree of cure increases. The tenacity of the fibre/resin bond strength, however, was affected by curing temperature and the maximum interfacial strength occurred at a curing cycle different to that resulting in the maximum bulk matrix strength.

A number of authors have reported on the effect of different curing schedules on matrix microstructure. Patel *et al.* [14] demonstrated that while the inclusion of a standing time (and gelation) had little effect on the decomposition temperature, static flexural tests, or storage modulus, a profound effect was noted in the composite fatigue performance. Mak *et al.* [15] showed that in order for the microdroplets to share a similar microstructure to the bulk sample, a much lower curing temperature must be used while conversely, Charlier *et al.* [16] propose that a higher temperature was needed to fully cure both acrylic and epoxy droplets compared to schedule used for macroscale composite parts. They propose that as a result of the change in scale, phenomena such as auto-acceleration/Trommsdorff's effect do not occur as it would in a bulk sample of several grams and accordingly, does not reach comparable local temperatures resulting in a lessened extent of reaction.

Methods employed in order to overcome this issue are also discussed. Both Rao *et al.* [3] and Ozzello *et al.* [5] saw great improvement by introducing a pre-cure/standing time, allowing the comonomers to react at low temperature and, in theory, preventing vaporisation of curing agent. Zinck *et al.* [10] reported contradictory results in that no change was seen after a 48-hour pre-cure in an amine-cured system and in fact, that it had a hugely detrimental effect on anhydride-cured samples. Rao *et al.* [3] saw a measure of improvement, though still not ideal, by curing the samples in an amine-rich environment though Biro *et al.* [17] employed a similar procedure and saw no discernible difference between the samples cured in the presence of curing agent vapours. Laurikainen [7] reports that curing under a styrene-rich environment alleviated curing problems in vinylester droplets. Zinck *et al.* [10] proposed that amine-cured systems should be cured under inert atmosphere (and at 120°C or below) to prevent loss of Tg from imine formation though Biro *et al.* reported no difference between samples cured in air and nitrogen using an epoxy and an aromatic amine. Laurikainen [6,7] was able to produce cured epoxy microdroplets by allowing the resin viscosity to increase before application, amounting to much the same process as a protracted room temperature pre-cure. Phase separation was thought to be prevented due to the existence of covalent bonds between the resin and amine oligomers.

The above, and the present work, casts reasonable doubt on the abundance of literature published on the utilisation of the microbond test in works where no apparent problem was detected or discussed. Though the testing method has been employed in conjunction with a number of different fibre and matrix materials, whether these works make successful use of the microbond must be carefully considered. This study was undertaken in order to show that, for a number of amine-cured epoxy systems, microdroplet samples exhibit properties that are different from those of the bulk material, and to characterise the effect of these differences as relates to the use of the microbond test. Unanswered questions appear to remain in regards to how much cure shrinkage is effectively locked in place prior to the gel point. It is important that a greater understanding of the relationship between gel point, Tg, residual stress and interfacial shear strength in the microbond test be developed.

2. Experimental

2.1. Materials

Two multiple-component commercial resin systems designed for use in wind turbine blades were selected for investigation. The first was Epotec YD-535 LV cured with Epotec TH7257 hardener at a stoichiometric ratio of 100: 35 pbw. Olin Airstone™ 780E was cured with Olin Airstone™ 785H hardener at a stoichiometric ratio of 100: 31 pbw. In order to minimise the complexity of the interface, experiments were conducted using bare (water-sized) E-glass fibres taken from larger rovings supplied by Owens Corning. For Fourier-transform infrared spectroscopy (FTIR) measurements, the glass fibres were replaced by annealed AISI 302 steel wire of diameter 50 μm purchased from Goodfellow to improve signal clarity.

2.2. Methodology

The basic procedure for the microbond test involves a single filament being pulled from a small droplet of cured matrix while measuring the force required to detach the filament. The procedure for preparation and testing of microbond samples and the development of the specific apparatus has been described in detail in previous literature [18].

Differential scanning calorimetry was performed to characterise the degree of cure according to ASTM Standard E2160-04 *Standard Test Method for Heat of Reaction of Thermally Reactive Materials by Differential Scanning Calorimetry* using a TA Instruments Q20 DSC. Scans from -50 to 300°C were made at a rate of 10°C following a number of room temperature standing times in order to characterise the degree of conversion during the room-temperature pre-cure. Tg of off-stoichiometric compositions were also determined by DSC.

FTIR analysis was performed using a 4100 ExoScan FTIR fitted with a spherical diamond attenuated total reflectance (ATR) interface, adjustable probe and benchtop docking station. Glass fibres were substituted for lengths of 50 μm diameter steel wire in order to create larger droplets of consistent dimensions and to exclude the influence of chemical interactions with fibre sizing. The use of steel wire allowed for accurate application of a single droplet per sample by preventing a capillary effect creating multiple droplets from the initial application as is observed with deposition onto glass or carbon fibres. Signal strength in collecting spectra is improved both by the use of a larger droplet and a reflectance effect of the steel wire, allowing for conventional FTIR apparatus to be used in the analysis of microscale samples. Microbond tests ensured that curing issues were prevalent across both glass fibre ($L_e \approx 120$ μm) and steel wire ($L_e \approx 380$ μm) samples to verify that size effects alone were not responsible for the discrepancies in curing behaviour. Microbond samples were prepared at the stoichiometric ratio, applied to the steel wire, and then either cured (65° 3.5 hr + 75° 7 hr) immediately, or with a room temperature pre-cure ranging between 0.5 and 24 hours followed by post-curing at 65 and 75°C. Samples with altered stoichiometric ratio were created by mixing resin with reduced amounts of curing agent in 10% decrements before syringing a small bead (≈ 10-20 mg) on release film and curing according to the same schedule. Individual droplets were then placed on the ATR interface of the apparatus and the probe fully lowered to ensure maximum contact with the sample. Results were plotted as a normalised average of 30 independent samples. Degree of cure/monomer conversion (α) was quantified using the reduction of the area of the oxirane/epoxy group at 915 cm^{-1} against the invariant peak at 1507 cm^{-1} (C=C stretching) as an internal standard [19].

The experimental configuration is illustrated in Figure 1 and shows a typical optical micrograph of a steel wire-mounted epoxy droplet at 100x magnification.

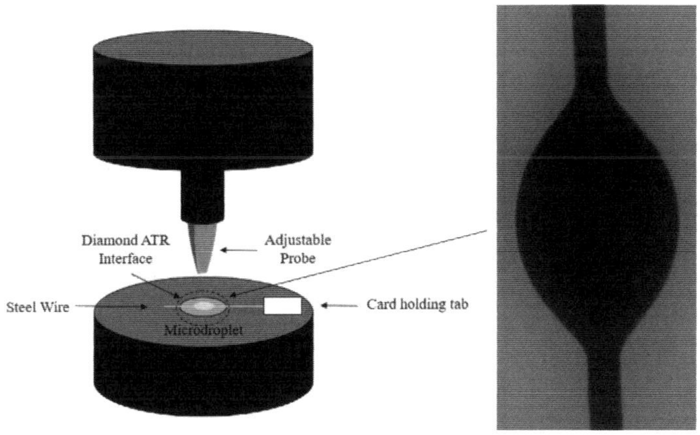

Figure 1: Schematic view of microbond ATR FTIR experimental set-up.

3. Results and Discussion

As illustrated in Figure 2, microbond testing results showed that the room-temperature pre-cure had a profound effect on the apparent IFSS of both the Epotec and Olin resins. It was not possible to accurately measure the IFSS of samples cured immediately due to the plastic deformation of the samples. It was found that a 3-hour room-temperature standing time was sufficient to ensure good curing and interfacial adhesion, and that further standing times of up to 48 hours showed no significant improvement. A standing time of only 1 hour was insufficient to ensure good curing and such samples showed only some tackiness while still deforming under load. Plotted alongside the interfacial adhesion data is the degree of cure to which samples had cured prior to undergoing elevated temperature curing as measured by DSC. Samples did not reach particularly high degrees of cure prior to heating.

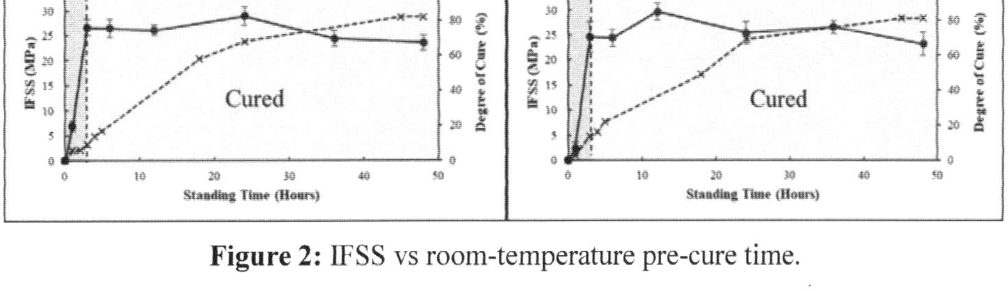

Figure 2: IFSS vs room-temperature pre-cure time.

IOP Conf. Series: Materials Science and Engineering **942** (2020) 012019 doi:10.1088/1757-899X/942/1/012019

The deformation of microbond samples under loading and the increases in IFSS after allowing a pre-cure can be plausibly attributed to whether the samples were allowed to gelate prior to exposure to elevated temperature. The observed deformation of the 0-hour and 1-hour samples would further indicate that the Tg of these samples was sub-ambient. The FTIR microbond method was developed in order to directly measure the Tg of microdroplet samples and to detect deviations from stoichiometry, if any, not present in bulk matrix samples. Previous studies have characterised the properties of microdroplet samples using FTIR microscopy [2], a novel TMA technique using similarly sized droplet clusters [3], modelling thin film thickness as an equivalent to embedded length [10], and pyrolysis gas chromatography [6]. Conventional thermal analysis using DSC was initially applied to microdroplet samples but it was found that sample size was insufficient to generate a signal from which Tg measurements could be made. Using clusters of similarly sized droplets in a similar manner to [3] indicated that, while signal strength was improved, Tg appeared to increase with the number of droplets. In any case, the Tg values measured were about 50°C lower than that of the bulk sample. Furthermore, the FTIR method allowed for the use of samples directly comparable to those used in the microbond test, as thin films did not consider sample geometry or surface-to-volume ratio.

In order to generate a range of spectra for off-stoichiometric composites, samples with curing agent reductions in 10% decrements were assessed. DSC was utilised in order to determine the relationship between amount of curing agent and Tg. Spectra of microdroplets cured after varying lengths of pre-cure were then collected, degree of conversion calculated, and collated with the data below to determine the droplet Tg and how much curing agent had been lost. Table 1 details the DSC Tg and FTIR data for both resin systems. A DSC stoichiometry study was similarly linked to TMA droplet Tg measurements in [3] though Tg was equated between the two methods rather than degree of cure.

Table 1: DSC and FTIR stoichiometry study

	DSC Tg (°C)		**FTIR α**	
Curing Agent Reduction	**Epotec**	**Olin**	**Epotec**	**Olin**
0%	92.3	93.5	0.93	0.89
10%	79.5	75.8	0.90	0.82
20%	60.5	53.2	0.87	0.77
30%	47.7	44.1	0.80	0.75
40%	23.5	21.2	0.72	0.77
50%	16.8	11.1	0.69	0.71
60%	-2.7	-0.8	0.55	0.66
70%	-14.8	-16.5	0.36	0.4
80%	-19.6	-22.0	0.30	0.31

Figure 3 shows the FTIR spectra for the stoichiometry study of the Epotec resin system. Spectra of the Olin resin system samples follow an identical trend. The C-O deformation of the oxirane ring, also known as the epoxy group, is located at 915 cm^{-1} and the C-H stretch of the ring is found at 3050 cm^{-1}. The broad band observed at around 3500 cm^{-1} is assigned to the O-H stretching of hydroxyl group. Bands corresponding to ether linkage are found between 1000-1100 cm^{-1}. Development of the hydroxyl band is a qualitative indicator of extent of reaction, though their use in quantitative analysis is limited as a result of the shape and overlapping bands.

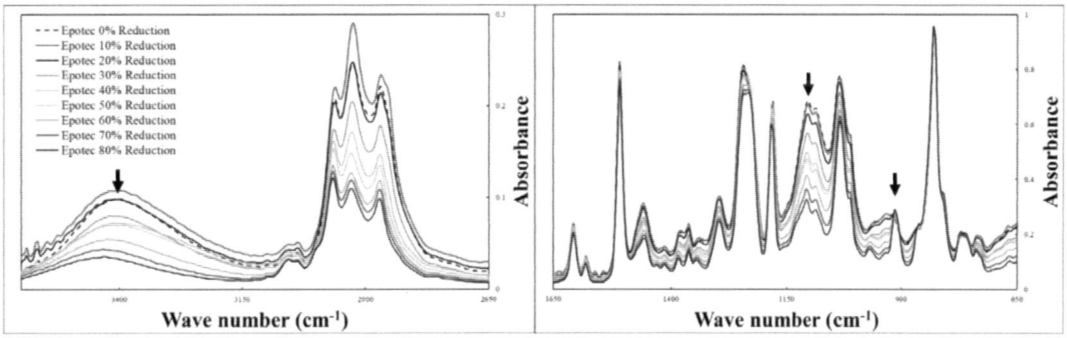

Figure 3: FTIR spectra of stoichiometry study

Changes in the concentration of species related to the curing process in the mIR range is primarily concerned with the broad band between 3000-3400, the sharp peak at 1100 cm^{-1} and the peak at 915 cm^{-1}. As the epoxy curing reaction progresses the oxirane ring opens and the band reduces, then disappears [20]. The reduction and disappearance in the bands located at 3056 and 915 cm^{-1} is an indication of the opening of the oxirane/epoxy ring. Due to the larger peak, and relative separation from neighboring peaks, the reduction in the band at 915 cm^{-1} is most commonly used for quantitative analysis. Furthermore, direct analysis of amine content is generally only possible using nIR methods.

The most apparent changes in microbond FTIR spectra can be observed in the comparison of 0-hour and 24-hour pre-cured samples as is shown in Figure 4. The micromechanical data presented previously would indicate that in the former case, these samples have IFSS values that were too small for accurate measurement and deformed plastically in the former case, and in the latter showed good interfacial adhesion and had an IFSS value in the range of 25-30 MPa.

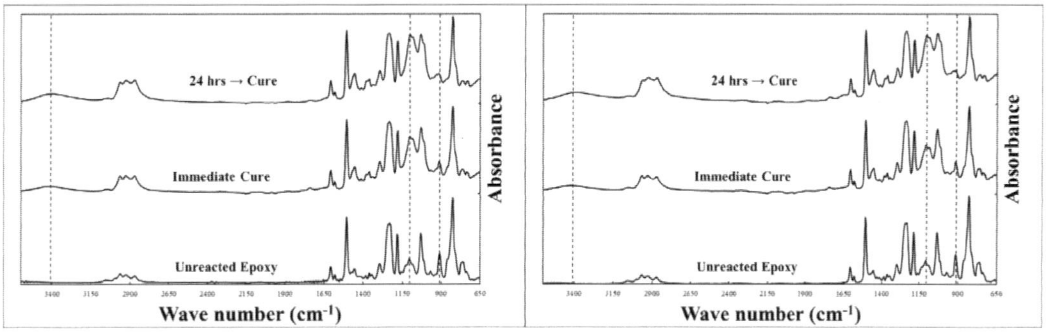

Figure 4: FTIR microbond spectra for Epotec (left) and Olin (right) resins

Samples cured immediately clearly show unreacted epoxy groups, reduced hydroxyl group accumulation and a weaker etherification peak, commensurate with spectra indicating an off-stoichiometric formulation. It is possible that curing agent evaporation following exposure to immediate high-temperature curing is the cause of the issue. Quantitative analysis of the epoxy peak indicated that microbond samples cured immediately had degrees of cure in the region of 0.55, loss of up to 60% of the curing agent, and Tg below zero. Conversely, samples that were allowed to pre-cure at room temperature and reach the gel point prior to heating at elevated temperatures produced spectra much closer to a stoichiometric mixture, high degrees of cure, and Tg of 80°C or greater, comparable to that of bulk matrix samples. The values for the amount of curing agent lost are somewhat higher than those reported by Rao and Drzal [3], though this is possibly due to the use of different curing agents between the studies. The full results of the pre-cure standing time investigation are detailed in Table 2.

Table 2: Summary of FTIR, DSC, and micromechanical data

	α		≈ Tg(°C)		IFSS (MPa)	
Pre-Cure Time (Hrs)	**Epotec**	**Olin**	**Epotec**	**Olin**	**Epotec**	**Olin**
0	0.55	0.55	-2.7	-1	0	0
0.5	0.71	0.69	23	11	-	-
1	0.68	0.62	16	4	6.9	2.33
2	0.87	0.9	60	75	-	-
3	0.87	0.81	60	46	26.6	24.6
6	0.89	0.84	73	49	26.5	24.4
24	0.91	0.93	80	92	28.9	25.5

4. Conclusions

Micromechanical interface testing using the microbond test has shown that in two amine-cured epoxy systems, the apparent IFSS was dependent on the inclusion of a room-temperature pre-cure prior to exposure to elevated curing temperature. A stand time of 3 hours was sufficient to ensure well-cured droplets and increasing the pre-cure to upwards of 48 hours had little significant effect on the IFSS. The exclusion of a pre-cure resulted in the formation of low-IFSS droplets with sub-ambient Tg.

The development and use of a novel FTIR technique appeared to show good correlation with the micromechanical testing results. Spectra generated from individual microbond samples was consistent with a non-stoichiometric epoxy-amine network, a sub-ambient Tg and significant loss of curing agent. Samples that were allowed to stand before curing showed a linear increase in degree of cure and Tg, two parameters which directly contribute to the value of IFSS. A plateau in the IFSS data after 3 hours of pre-cure was consistent with the FTIR measurements in the Epotec system, though this behaviour was slightly less pronounced in the Olin system. High degrees of conversion corresponded to increased Tg and IFSS.

The results presented can be interpreted as possible further evidence for the evaporation of curing agent out of microbond samples in these two systems. Exposure to elevated temperature curing prior to allowing microbond sample gelation resulted in low-IFSS samples that deformed plastically. It is possible that evaporation of the curing agent out of the samples during this early stage accounts for the loss in material properties. It is also possible that alterations to the curing schedule affect the contributions of residual stress present to varying degrees. What does appear to be clear, is that such effects take place early in the curing reaction.

Interfacial testing methods are often employed to measure the influence of factors such as fibre surface treatment, application and screening of sizings, and degradation thereof. Changes in IFSS as the result of such alterations are, however, typically significantly smaller than the influence of the cure state of the matrix microdroplet, thus the route taken in creating microbond samples, and the potential effect of discrepancies in microscale curing must be carefully considered. We have ourselves observed that the curing issues are not prevalent in triethylenetetramine-cured epoxy samples [21] and further research involves extending the study to a range of amine chemistries, curing atmosphere, and oversaturation of curing agent.

References

[1] Miller B, Muri P, and Rebenfeld L 1987 A Microbond Method for Determination of the Shear Strength of a Fiber/ Resin Interface *Compos. Sci. Technol.* vol **28** pp 17–32

[2] Haaksma R A and Cehelnik M J 1989 A Critical Evaluation of the Use of the Microbond Method for Determination of Composite Interfacial Properties *MRS Proc.*

[3] Rao V, Herrera-Franco P, Ozzello A D, and Drzal L T 1991 A Direct Comparison of the Fragmentation Test and the Microbond Pull-out Test for Determining the Interfacial Shear Strength *J. Adhesion* vol **34** pp 65–77

[4] Rao V and Drzal L T 1991 Loss of Curing Agent During Thin Film (Droplet) Curing of Thermoset Material *J. Adhesion* vol **22** pp 245–49

[5] Ozzello A D, Grummon D S, Drzal L T, Kalantar J, Loh I-H, and Moody R A 1989 Interfacial Shear Strength of Ion Beam Modified UHMW-PE Fibers in Epoxy Matrix Composites *Materials Resarch Society Symposium Proc.* vol **153** pp 217–22

[6] Laurikainen P *et al.* 2018 High throughput mechanical micro-scale characterization of composites and the utilization of the results in finite element analysis *Proc. 18th European Conf. on Composite Materials (ECCM 18)* pp 24–8

[7] Laurikainen P 2017 Characterization of the Aging of Glass-Fibre Reinforced Polymers *Master of Scence Thesis* Tampere University of Technology

[8] Dirand X, Hilaire B, Soulier J P, and Nardin M 1996 Interfacial shear strength in glass-fiber/vinylester-resin composites *Compos. Sci. Technol.* vol **56** pp 533–39

[9] Charlier Q, Lortie F, Gerard P, and Gerard J F 2015 Interfacial adhesion between glass fibers and acrylic-based matrices as studied by micromechanical testing *Proc. 20th Int. Conf. Composite Materials (ICCM 20)* pp 19–24

[10] Zinck P, Wagner H D, Salmon L, and Gerard J F 2001 Are microcomposites realistic models of the fibre/matrix interface? II. Physico-chemical approach *Polymer* vol **42** pp 6641–50

[11] Bartolomeo P, Chailan J F, and Vernet J L 1999 Micromechanical and Microscopic Effects of Preprocessing on Interfacial Shear Stress of Glass Fiber-Cynate Ester Resins Composite *Proc. 12th Int. Conf. Composite Materials (ICCM 12)*

[12] Montgomery CB 2018 Multiscale Characterization of Carbon Fiber-Reinforced Epoxy Composites *PhD Thes*is University of Illinois

[13] Peters PWM and Springer GS 1987 Effects of Cure and Sizing on Fiber-Matrix Bond Strength *J. Compos. Mater.* vol **21** pp 157–71

[14] Patel A, Kravchenko O, and Manas-Zloczower I 2018 Effect of curing rate on the microstructure and macroscopic properties of epoxy fiberglass composites *Polymers* vol **10** p 125

[15] Mak M, Lowe A, Jar B, Stachurski Z, and Fitzgerald J 1998 Improvements to the microdroplet single fibre composite test *J.Mater. Sci. Letters* vol **17** pp 645–47

[16] Charlier Q, Lortie F, and Gérard J F 2020 Interfacial adhesion in glass-fiber thermoplastic composites processed from acrylic reactive systems, a multi-scale experimental analysis *Int. J. Adhes. Adhes.* vol **98**

[17] Biro D A, Pleizier G, and Deslandes Y 1993 Application of the Microbond Technique IV: Improved Fiber-Matrix Adhesion by RF Plasma Treatment of Organic Fibers *J. Appl. Polym. Sci.* vol **47** pp 883–94

[18] Yang L and Thomason J L 2012 Development and application of micromechanical techniques for characterising interfacial shear strength in fibre-thermoplastic composites *Polym. Test.* vol **31** pp 895–903

[19] Karayannidou E G, Achilias D S, and Sideridou I D 2006 Cure kinetics of epoxy-amine resins used in the restoration of works of art from glass or ceramic *Eur. Polym. J.* vol **42** 3311–23

[20] Cañavate J, Colom X, Pagès P, and Carrasco F 2000 Study of the curing process of an epoxy resin by FTIR spectroscopy *Polym. Plast. Technol. Eng.* vol **39** pp 937–43

[21] Minty R F, Yang L, and Thomason J L 2018 The influence of hardener-to-epoxy ratio on the interfacial strength in glass fibre reinforced epoxy composites *Compos. Part A* vol **112** pp 64–70

Modelling of chemical shrinkage evolution with curing degree of a filled epoxy adhesive

T Holst[1,2,3], F Sayer[1] and A Antoniou[1]

[1] Fraunhofer Institute for Wind Energy Systems, Am Seedeich 45, 27572 Bremerhaven, Germany
[2] Institute for Wind Energy Systems, Leibniz University Hanover, Appelstraße 9A, 30167 Hanover, Germany

[3] Corresponding author`s e-mail: tobias.holst@iwes.fraunhofer.de

Abstract. An epoxy adhesive system, commonly used in wind turbine blade manufacturing, is experimentally investigated with respect to its chemical shrinkage behavior to show the shrinkage evolution within rotor blade production related process conditions. Therefore, a new test configuration is set up to record the resulting chemical shrinkage of the adhesive at cross-linking temperatures ranging from 20°C to 90°C. The respective conversion evolution is simulated under several temperature conditions through a curing kinetics model. This is generated with state of the art formulations, implementing test results from a differential scanning calorimeter (DSC). Moreover, the transformation of the curing degree to the corresponding glass transition temperature (T_g) is performed through an experimentally based Di-Benedetto curve-fitting. By the combination of curing kinetics and chemical shrinkage test data, a model for the temperature and conversion dependant chemical shrinkage evolution during cure is developed. This is applied to a realistic curing cycle for adhesive materials in rotor blade production.

1. Introduction

To meet today's requirements in the wind energy market, wind turbine rotor blades must embody improved mechanical performance and reliability [1, 2]. Chemical shrinkage effects during curing of bond lines, are one of the main drivers for induced residual stress states after manufacturing, which can influence the mechanical performance of the adhesive by promoting crack initiations [3, 4]. It is therefore important to understand the process related evolution of chemical shrinkage. Material specific shrinkage values are often not known, since the characterization is complicated [3, 5–7].

The above-mentioned chemical shrinkage effect can be explained with decreasing mean gaps within the macromolecule structure in progression of cross-linking [5]. Therefore the adhesive shrinkage behavior is highly related to the chemical structure and additives [3, 5, 6]. For the characterization, isothermal measurement conditions are essential to decouple chemical from thermal shrinkage effects [7]. Previous research showed that the measurement technique according to the principle of Archimedes' can deliver proper test results with accurate resolution [5, 8–10].

A widely common assumption, found in literature, is a linear connection between degree of cure and chemical shrinkage values [3, 5, 8, 11]. In addition, some authors assume a nonlinear shrinkage evolution, which is influenced by gelation and viftrification [9, 12]. Therefore the knowledge of the curing kinetic behavior is of importance to display the relation towards conversion.

Curing kinetics are describing the curing progress respectively to ambient curing conditions. Cure dependent characteristics like reaction heat can be implemented into model environments [13, 14]. To model the curing behavior of epoxy resins with amine hardeners, a common, state of the art approach, found in literature, is introduced by Kamal-Sourour [15]. The named model is an extension of a simple autocatalytic curing correlation, combined with a simple nth-order-reaction [16, 17]. Acceleration effects of the reaction rate due to intermediate reaction products are characteristic for autocatalytic reaction types. More complex curing kinetic models are available. With additional mathematical functions, the diffusion controlled curing process, induced by vitrification effects, can be taken into account [17–19]. Often these models require the conversion dependent development of the glass transition temperature (T_g) [19]. This relationship can be expressed by the Di-Benedetto-equation and is common in literature [19–22].

This scientific work contributes to the characterization of new material-specific properties during cure. For the selected adhesive system the curing kinetic behavior is determined and a model of the chemical shrinkage evolution is generated with new test methods. The shrinkage model is applied with regard to application related curing conditions.

2. Experimental setup

For analyzing the chemical shrinkage, a two-step experimental characterization method was chosen. At first, the curing kinetic behavior of the adhesive system was examined, using DSC. Subsequently, the time-dependent chemical shrinkage was measured in a temperature range from 20 °C to 90 °C and then compared with the time-dependent conversion data.

2.1. Material

Epikote MGS BPR 135 G3 resin in combination with *Epikure MGS BPH 137G* curing agent from Hexion was used in the following research [23]. The selected material is an industry standard epoxy adhesive system for rotor blade production.

2.2. Determination of experimental data for curing kinetic modelling

A *Netzsch DSC 204 F1* was used for the curing characterization procedure. Four different temperature stages (60 °C, 80 °C, 100 °C and 120 °C) were chosen to track the curing behavior under isothermal conditions. Three specimens were tested at each temperature level. The lower limit of the temperature range is defined by the sensitivity of the heat flow sensor, whereas the upper limit is set by exothermal effects of the specimen itself.

Before starting the measurement, uncured adhesive specimen, embedded in aluminum pans, were put in the preheated measuring chamber to achieve immediately the desired temperature level, followed directly by a start of the measurement. It is assumed that during the very short heat-up-phase (below 10 s), the specimen does not undergo any curing. During the described isothermal stages the time dependent heat flow $h_{iso}(t)$ of the curing reaction was monitored until the reaction came to a standstill at t_{end}. Whereas $H_{iso}(t)$ is the integral of the heat flow curve from the start of the reaction at t_0 to the present time t.

After the isothermal stage, a temperature-modulated dynamic ramp from 0 °C to 250 °C was applied on the specimens to track the heat flow during post curing to full cure, H_{res}. The modulated ramp function was used to generate a smooth base line level and isolate effects due to glass transition. A heating ramp of 3 K/min, an amplitude of 0.5 K and a period of 60 s was set for the DSC-measurement.

The development of the degree of cure $p(t)$ was then calculated by correlating the time-depending amount of heat energy during isothermal cure $H_{iso}(t)$ to the total amount of heat release H_{tot},

$$p(t) = \frac{H_{iso}(t)}{H_{tot}} = \frac{\int_0^t h_{iso}(t)dt}{\int_0^{t_{end}} h_{iso}(t)dt + H_{res}} \tag{1}$$

IOP Conf. Series: Materials Science and Engineering **942** (2020) 012020 doi:10.1088/1757-899X/942/1/012020

The Di-Benedetto relation, describing the conversion dependent glass transition temperature, was derived by determining the glass transition temperatures of uncured, partially cured and fully cured samples. The midpoint method according to DIN EN ISO 11357-2 with 20 K/min heating ramp was used for the analysis.

2.3. Chemical shrinkage test rig

The design of the test rig for measuring the chemical shrinkage effect is based on the Archimedean Principle. The used shrinkage test rig consists of a tempered oil bath, which contains the measuring chamber, filled with silicone oil, see Figure 1. To avoid any chemical reactions with the adhesive specimen, the silicone oil has inert chemical properties. The specimen holder, especially developed for pasty adhesive materials, is hanged in the silicon oil below a precision scale. For temperature documentation, each oil bath owns a temperature sensor.

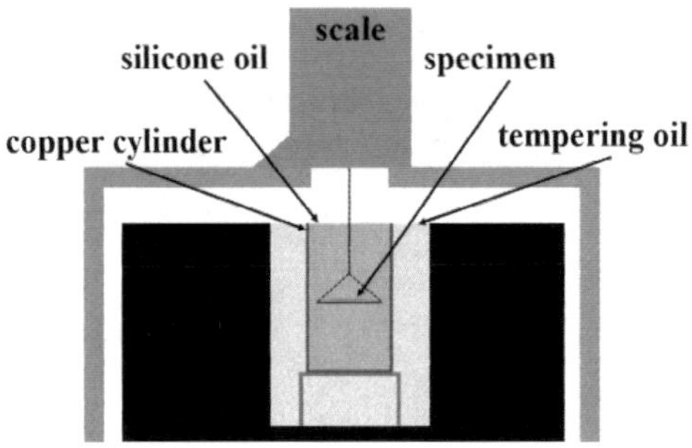

Figure 1. Schematic construction of chemical shrinkage test rig.

Knowing the weight of the specimen in air m_{air} and the density of the silicone oil $\rho_{oil}(T)$ at certain temperatures T, the volume of the specimen $V(t)$ in theory, can be determined any time t by its weight in oil $m_{oil}(t)$, based on the Principle of Archimedes'.

$$V(t) = \frac{m_{air} - m_{oil}(t)}{\rho_{oil}(T)} \tag{2}$$

In the measurement procedure the weight value of the scale measurement $m_S(t)$ in oil consists of the sample and the sample holder. To isolate the effect of the sample holder, its weight in oil $m_{Hoil}(T)$ must be known at respective temperatures T. The volume of the specimen can then be calculated.

$$V(t) = \frac{m_{air} - (m_S(t) - m_{Hoil}(T))}{\rho_{oil}(T)} \tag{3}$$

By comparing the change of the specimen volume $\Delta V(t)$ during cure to the starting volume $V(t_0)$, the chemical shrinkage value S can be calculated.

$$S(t) = \frac{\Delta V(t)}{V(t_0)} = \frac{V(t_0) - V(t)}{V(t_0)} \tag{4}$$

Three specimen were tested each at eight curing temperature levels in a range from 20 °C to 90 °C. For monitoring the specimen temperature, in order to detect unwanted exothermal effects, an additional measurement was made for each temperature with identical specimen geometries without tracking the scale data. Otherwise, the measurement would be highly affected by the temperature sensor due to the high sensitivity of the measuring configuration.

3. Model results and discussion

At first, the modelling results of the curing kinetics are presented. These are then combined with the chemical shrinkage data to show and model the relation with conversion.

3.1. Curing kinetics

The modelling of the curing kinetics was performed by a parameter fit of the model data to the experimental data by using a least square method in a dp/dt-p-diagram. The mean p-values of the three performed measurements at each temperature have been used as the data basis for the fit, see Figure 2. At first a Kamal-Sourour model approach was used, Equation (5). The temperature dependent reaction rate constants k_1 and k_2 can be calculated with the Arrhenius correlation, Equation (7).

The parameter fit leads to a reaction rate constant k_1 (A_1=2075841 1/s and E_{a1}=555214 J/mol) which tends towards zero at variable curing temperatures *T*. Therefore the model approach can be simplified to an autocatalytic model with only one remaining reaction rate constant k_2, Equation (6). The updated model parameters are shown below in Table 1. Since the model environment is implemented in further calculations, a simpler model approach is preferred to reduce the overall calculation effort.

$$\frac{dp}{dt} = (k_1 + k_2 p^m)(1-p)^n \tag{5}$$

$$\frac{dp}{dt} = k_2 p^m (1-p)^n \tag{6}$$

$$k_{1,2} = A_{1,2} * e^{\frac{-Ea_{1,2}}{RT}} \tag{7}$$

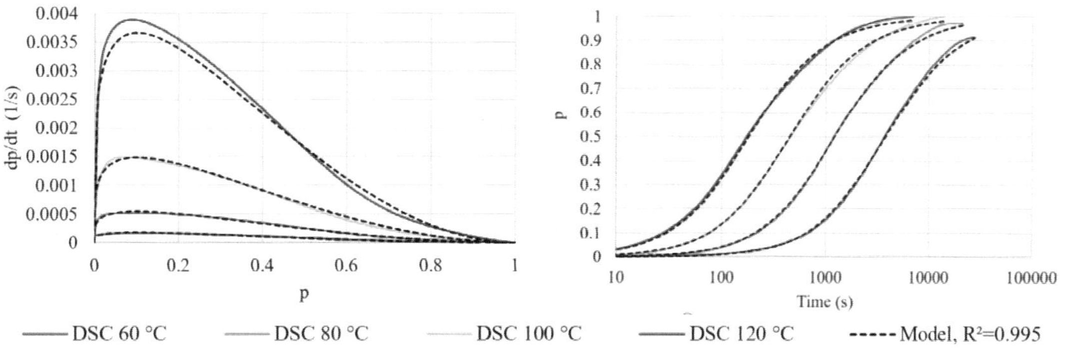

Figure 2. Least square fit of autocatalytic curing model parameters, mean values of measured data.

Table 1. Determined parameters for autocatalytic curing model.

A_2 (1/s)	E_{a2} (J/mol)	*m*	*n*
153965	55055	0.22	1.95

The development of the glass transition temperature is described with the Di-Benedetto-relation, Equation (8). The parameters T_{g0} and $T_{g\infty}$ were determined with glass transition temperature measurements of uncured and fully cured DSC-samples (DIN EN ISO 11357-2 midpoint method, 20 K/min). The Di-Benedetto-parameter λ can be calculated by using a least square data fit as seen in Figure 3. Determined parameters for the Di-Benedetto-equation are listed in Table 2.

$$\frac{T_g - T_{g0}}{T_{g\infty} - T_{g0}} = \frac{\lambda p}{1-(1-\lambda)p} \tag{8}$$

IOP Conf. Series: Materials Science and Engineering **942** (2020) 012020 doi:10.1088/1757-899X/942/1/012020

Table 2. Determined parameters for Di-Benedetto relation.

T_{g0} (°C)	$T_{g\infty}$ (°C)	λ
-40.9	90.9	0.377

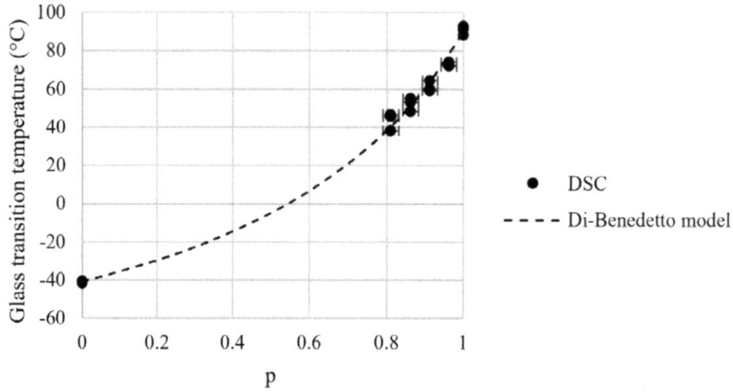

Figure 3. Least square fit of Di-Benedetto model data.
*Note: some error bars not visible, as deviations are very small

By looking at Equation (6) the mathematical forecast for the development of the conversion is reaching full cure after a certain amount of time at arbitrary temperatures. Under real conditions at lower curing temperatures, the curing reaction comes to a standstill because of vitrification effects. Taking this into account an additional simple condition can be implemented into the curing model environment. This condition includes a stop criterion when the rising glass transition temperature of the curing system is more than a certain temperature range T_S above the current curing temperature, Equation (9).

$$\frac{dp}{dt} = \begin{cases} k_2 p^m (1-p)^n \\ 0 \end{cases} \quad for \quad \begin{matrix} T_g(p) - T \leq T_S \\ T_g(p) - T > T_S \end{matrix} \tag{9}$$

To find a suitable value for T_S, adhesive specimens, equipped with temperature sensors, were cured in an oven under variable curing cycles reaching from 40 °C to 100 °C, followed by analyzing the glass transition temperatures $T_g\,(DSC)$ afterwards with DSC-measurements (DIN EN ISO 11357-2 midpoint method, 20 K/min). According to the total square deviation, T_S=30 °C yields the highest fit between model prediction and measured data, Table 3.

Table 3. Adaptation of stop criterion for curing kinetics model.

			Model forecast			
Hold time (*h*)	Oven temperature (°C)	T_g (*DSC*) (°C)	*Ts=20 °C* (°C)	*Ts=25 °C* (°C)	*Ts=30 °C* (°C)	*Ts=35 °C* (°C)
48	40	70.4	60.1	65.1	69.9	74.8
24	50	73.5	70.0	74.3	74.8	75.2
12	60	74.2	73.9	74.5	75.0	75.4
6	70	77.4	73.9	74.5	75.0	75.4
3	80	77.9	72.9	73.5	73.5	74.6
1.5	90	77.7	73.0	73.6	73.6	74.7
0.75	100	76.3	72.4	72.6	72.6	72.7
		Total square deviation (°C²):	**192.1**	**86.2**	**57.3**	**61.3**

The model environment of the curing kinetics gives a good agreement with the measurement data by using a modified autocatalytic model with an implemented stop criterion, which simplifies the effect of diffusion controlled reaction progress. More complex curing models, containing diffusion controlled elements, have also been analyzed and deliver similar results.

3.2. Chemical shrinkage

The following diagram shows the evaluated mean shrinkage values at variable temperature levels plotted against conversion, Figure 4. Visible is a linear connection between shrinkage and degree of cure. Several studies in literature, like Holst [7] observed the same behavior. Minor variations from the linear relation can be explained with uncertainties of the curing kinetics model due to slightly exothermal behavior and vitrification effects at high degrees of cure.

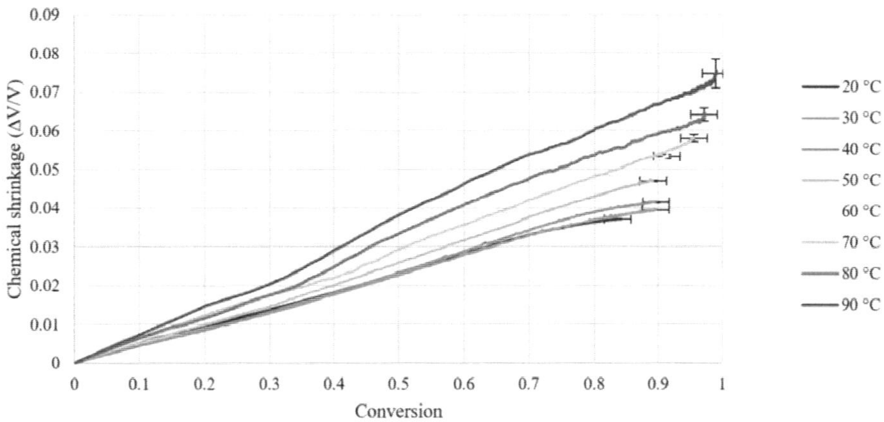

Figure 4. Mean values of observed chemical shrinkage with conversion and temperature. *Note: some error bars not visible, as deviations are very small

Furthermore, the gradients are dependent on the curing temperature. Therefore, at similar degrees of cure, a range of different chemical shrinkage values can be observed. An increasing trend is recognizable with higher curing temperatures. Blumenstock [10] obtained similar results. Here this effect is explained with different amounts of free volumes which are being frozen within the curing process at different temperature levels. This can be described closer with detailed information of the occurring thermal shrinkage effects during cure. Since these details are not known, a simple model approach for describing the temperature dependent chemical shrinkage of the adhesive was chosen.

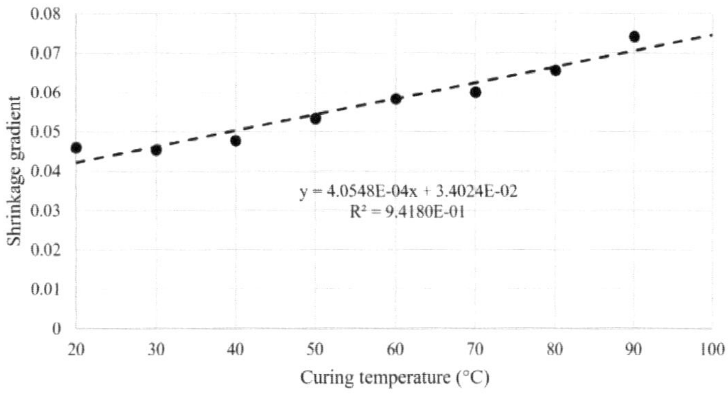

Figure 5. Model approach for chemical shrinkage gradients with temperature dependency.

By plotting the respective straight line gradients of the shrinkage measurements with their corresponding curing temperatures, a linear relationship is visible, Figure 5.

Table 4. Determined parameters for shrinkage gradient modelling.

a (1/°C)	*b*
$4.0548*10^{-4}$	$3.4024*10^{-2}$

The parameters of the linear regression *a* and *b*, Table 4, can be used to derive an equation for predicting the chemical shrinkage development *S* with conversion *p* and ambient curing temperature θ in °C.

$$S = p(a\theta + b) \tag{10}$$

4. Conclusions

The developed models for curing and chemical shrinkage forecast can be used for an application towards a chosen representative curing cycle for the adhesive material in rotor blade production, see Figure 6.

It can be seen that the displayed curing cycle initially remains at a low temperature for the application of the adhesive at 40 °C. After the subsequent joining process, the temperature is increased to accelerate the curing of the adhesive material. At higher temperatures, the reaction rate of the exothermic curing reaction has the effect of a temperature overshoot. In this phase most of the curing progress and also most of the chemical shrinkage take place. After this, a temperature level of 80 °C is maintained constant to achieve the desired degree of cure. At the end of the curing cycle it can be seen that a glass transition temperature of approximately 80 °C has been reached and a chemical volume shrinkage of 7% has taken place in the adhesive material.

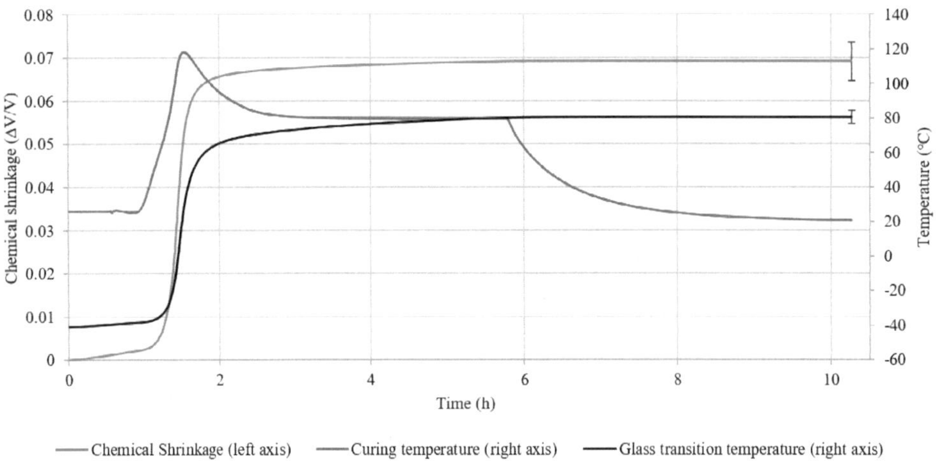

Figure 6. Application of chemical shrinkage- and curing model to a curing cycle.

These illustrations of the mechanisms during the curing process are one of the bases for understanding the development of residual stresses in bond lines during rotor blade manufacturing. In future research works, material specific values for the related thermal shrinkage behavior are of great importance. Besides the knowledge of specific values for further model development towards residual stress prediction, the thermal shrinkage data can also be used to formulate an alternative model approach for describing the temperature dependent chemical shrinkage according to the frozen volume theory. This will contribute to further validate the presented work.

Acknowledgments
We acknowledge the support provided by the German Federal Ministry for Economic Affairs and Energy (BMWi) within the Smart Blades 2.0 project (0324032B/C).

References

[1] Toft H S and Sørensen J D 2011 Reliability-based design of wind turbine blades. *Structural Safety* **33** 333–42.
[2] Hayman B, Wedel-Heinen J, and Brøndsted P 2008 Materials challenges in present and future wind energy. *MRS Bull.* **33** 343–53.
[3] Schümann J-P 2018. *Zur zeit-, temperatur- und umsatzabhängigen Entwicklung polymerphysikalischer Vernetzungsschwindung aminisch vernetzender Epoxide* Dissertation. Technical University of Clausthal, Clausthal.
[4] Jørgensen J B, Sørensen B F, and Kildegaard C 2019 The effect of residual stresses on the formation of transverse cracks in adhesive joints for wind turbine blades. *International Journal of Solids and Structures* **163** 139–56.
[5] Holst M 2001. *Reaktionsschwindung von Epoxidharz-Systemen* Dissertation. Technical University of Darmstadt, Darmstadt.
[6] Haider M, Hubert P, and Lessard L 2007 Cure shrinkage characterization and modeling of a polyester resin containing low profile additives. *Composites Part A: Applied Science and Manufacturing* **38** 994–1009.
[7] Holst M, Schänzlin K, Wenzel M, Xu J, Lellinger D, and Alig I 2005 Time-resolved method for the measurement of volume changes during polymerization. *J. Polym. Sci. B Polym. Phys.* **43** 2314–25.
[8] Khoun L and Hubert P 2010 Cure shrinkage characterization of an epoxy resin system by two in situ measurement methods. *Polym Compos* **31** 1603–10.
[9] Li C, Potter K, Wisnom M R, and Stringer G 2004 In-situ measurement of chemical shrinkage of MY750 epoxy resin by a novel gravimetric method. *Composites Science and Technology* **64** 55–64.
[10] Blumenstock T 2003. *Analyse der Eigenspannungen während der Aushärtung von Epoxidharzmassen* Dissertation. University of Stuttgart, Stuttgart.
[11] Billotte C, Bernard F M, and Ruiz E 2013 Chemical shrinkage and thermomechanical characterization of an epoxy resin during cure by a novel in situ measurement method. *European Polymer Journal* **49** 3548–60.
[12] Wenzel M 2005. *Spannungsbildung und Relaxationsverhalten bei der Aushärtung von Epoxidharzen* Dissertation. Technical University of Darmstadt, Darmstadt.
[13] Hardis R, Jessop J LP, Peters F E, and Kessler M R 2013 Cure kinetics characterization and monitoring of an epoxy resin using DSC, Raman spectroscopy, and DEA. *Composites Part A: Applied Science and Manufacturing* **49** 100–8.
[14] Deutsches Institut für Normung e.V. 2014 DIN EN ISO 11357-5 Plastics - Differential scanning calorimetry (DSC) - Part 5: Determination of characteristic reaction-curve temperatures and times, enthalpy of reaction and degree of conversion.
[15] Sourour S and Kamal M R 1976 Differential scanning calorimetry of epoxy cure: isothermal cure kinetics. *Thermochimica Acta* **14** 41–59.
[16] Yousefi A, Lafleur P G, and Gauvin R 1997 Kinetic studies of thermoset cure reactions: A review. *Polym. Compos.* **18** 157–68.
[17] Karkanas P I, Partridge I K, and Attwood D 1996 Modelling the cure of a commercial epoxy resin for applications in resin transfer moulding. *Polym. Int.* **41** 183–91.
[18] Cole K C, Hechler J-J, and Noel D 1991 A new approach to modeling the cure kinetics of epoxy/amine thermosetting resins. *Macromolecules* **24** 3098–110.
[19] Javdanitehran M, Berg D C, Duemichen E, and Ziegmann G 2016 An iterative approach for isothermal curing kinetics modelling of an epoxy resin system. *Thermochimica Acta* **623** 72–9.

[20] DiBenedetto A T 1987 Prediction of the glass transition temperature of polymers: A model based on the principle of corresponding states. *J. Polym. Sci. B Polym. Phys.* **25** 1949–69.

[21] Yu H, Mhaisalkar S G, Wong E H, and Khoo G Y 2006 Time–temperature transformation (TTT) cure diagram of a fast cure non-conductive adhesive. *Thin Solid Films* **504** 331–5.

[22] O'Brien D J and White S R 2003 Cure kinetics, gelation, and glass transition of a bisphenol F epoxide. *Polym. Eng. Sci.* **43** 863–74.

[23] Hexion. *Technical data sheet. EP-Klebeharz Epikote BPR 135G + Epikure BPH 134G-137GF.* Retrieved May 19, 2020 from https://www.hexion.com/en-US/product/--archive--epikote-resin-mgs-bpr-135g-series-and-epikure-curing-agent-mgs-bph-134g-137gf.

IOP Conf. Series: Materials Science and Engineering **942** (2020) 012021 doi:10.1088/1757-899X/942/1/012021

A basic orthotropic viscoelastic model for composite and wood materials considering available experimental data and time-dependent Poisson's ratios

R Bengtsson[1], R Afshar and E K Gamstedt

Department of Materials Science and Engineering, Division of Applied Mechanics , Uppsala University, Lägerhyddsv. 1, SE-75121 Uppsala, Sweden

E-mail: [1]rhodel.bengtsson@angstrom.uu.se

Abstract. Long-term deformation in creep is of significant engineering importance. For anisotropic materials, such as wood, composites and reinforced concrete, creep testing in several axial directions including shear is necessary to obtain a creep model which is able to predict deformation in the basic orthotropic case. Such a full set of experimental data is generally not available, and simplifying assumptions are typically made to conceive a useful 3D model. These assumptions should preferably be made based on the material behaviour and sound engineering arguments. This problem appears to be addressed in many different ways and sometimes the assumptions are not well justified. In the present study, we examine 3D creep of wood and composite materials. Particular emphasis is made on explaining the choices made in developing the model, considering practicality, incomplete material data and the specific behaviour of wood and composites. An orthotropic linear viscoelastic model is implemented as a material model in a commercial FE software. The constitutive equations are derived in the 1D case using a hereditary approach, then later generalized to the 3D formulation. Guidelines are shown how to implement it into the FE software to predict creep of components and structures. Although the model itself is conventional, the effect of considering time-dependent Poisson's ratios is investigated here, as well an optimization approach when inserting inevitably asymmetric experimental creep data into the model. As far as the authors know, creep of wooden materials have not been defined using this approach before. The model of interest is calibrated against experimental data. Examples using experimental results from solid wood data and a unidirectional fiber composite are demonstrated. The results show that the model is able to capture the orthotropic behaviour adequately. Orthotropy requires symmetry of the creep compliance matrix, which typically is not the case experimentally. It is shown that in rendering the matrix symmetric, one needs to decide which direction is more important. It is also shown that the frequently employed assumption of constant Poisson's ratios should be made with caution.

1. Introduction

If a material is subjected to a persistent load, a gradually increasing deformation of the material is introduced, e.g. a wooden shelf loaded by books may increase its deflection as years goes by. This phenomenon refers to a time-dependent material behaviour also called as creep behaviour. If the time-dependent strains are linearly related to the stresses, it is commonly known as linear viscoelasticity. Over the past years the amount research into the viscoelastic behaviour of composites and timber structures has been significant, given the long-term use of these materials in load carrying structures. A review of previous models is first presented, before showing the advantages of the present approach.

In Refs. [1] and [2] a coincident element method for orthotropic composites is proposed. It utilises two coincident shell elements to model the orthotropic viscoelastic behaviour of a composite laminate. The first element exhibits isotropic and viscoelastic behaviour, while the second element exhibits orthotropic and elastic behaviour. The elements are superimposed in such a way that they are coincident, i.e., the two shell elements share the same nodes and hence deform together. The procedure was used to model a number of published examples of viscoelasticity of laminated composites. It is shown that the coincident element method is a relatively simple and useful tool for modelling orthotropic and viscoelastic response of laminated composites by using a finite element package that only supports isotropic viscoelastic material models. The drawback was that the number of equations to be solved essentially doubled since there are two shell elements describing the same domain. This means, that the elements of each material group will experience the individual stress state under the same displacement field. The resulting stress matrix field in the model is a sum of the stress matrix fields for each group of elements. The model was also restricted to shell problems. Thus the proposed technique is well-suited for the problems of anisotropic viscoelasticity of thin and thick homogeneous and multilayered shells including nonlinear contact problems.

Papers [3], [4] and [5] used the differential approach by combining dashpots and springs in their viscoelastic models for timber structures. These models took into account different phenomena common in wood such as effects of moisture content and mechanosorption. A challenge when using these models is often to find the parameters for the model. In [3], a full 3D constitutive law is proposed for orthotropic nonlinear viscoelastic behaviour. The nonlinearities are taken into account using a generalized Maxwell model with two nonlinear elements. The viscosities in these two Maxwell elements depend on stress level. That parameter is introduced using a danger factor based on the Tsaï-Wu criterion and the calculation of an "elastic multiplier". Viscosity of one of those Maxwell elements also depends on moisture changes. In Ref. [4], in contrary to Ref. [3], a generalized Kelvin model with nonlinear elements was used. In the Kelvin elements, a mechanosorptive dashpot is used for the irrecoverable mechanosorption. The above creep mechanisms are added to the elastic strain and to the hygroexpansion strain. The whole model includes that the mechanical quantities depend on the moisture content. The temperature is considered to be constant during the studied cases.

Papers [6], [7] and [8] used the hereditary approach to exhibit thermoviscoelastic behaviour of composite material to estimate the residual stresses during cool-down. The analysis has been shown to be useful in prediction of residual stresses during production of composite materials and also other thermal environments. It is only needed to characterise the thermo-viscoelastic behaviour of the polymer part of the composite material. The properties of the orthotropic material are calculated in the Laplace domain and used as the input to the subroutine provided to the material model in ANSYS.

3D effects in viscoelasticity are not always strictly accounted for in these models. For instance, none of the above-mentioned models specifically investigated the effect of time-dependent Poisson's ratios on creep behaviour. Papers such as [7] and [8], assume constant Poisson's ratios for their models, although that is not often experimentally the case [9]. Another effect that is largely unaddressed is that experiments typically show asymmetric relaxation and compliance matrices [10][11]. In this paper, the hereditary approach is used to represent the viscoelastic behaviour of orthotropic materials. An approach will been shown on how to take into account the actual time-dependent Poisson's ratios as well as showing how to deal with asymmetric behaviour in a simple manner.

2. Theoretical background

Starting from the fundamentals, the simplified model is derived and explained in this section. As opposed to linear elastic materials, the response of a viscoelastic material depends on the loading history. Performing a relaxation test is a common experimental method to characterize the viscoelastic properties of a material. Characterization may be done by a uniaxial tensile test with a constant strain ε_0. The stress $\sigma(t)$ is then measured as a function of time.

$$\sigma(t) = \int_{0^-}^{t} E(t-\tau)\frac{d\varepsilon}{d\tau}d\tau. \tag{1}$$

Deriving the above equation is called the hereditary approach to linear viscoelasticity, which allows greater freedom when constructing models than the differential approach that relies on the concepts of certain combinations of springs and dashpots. The essential issue in the hereditary approach is that of superposition. Equation (1) can be generalized to 3D by exchanging the the 1D relaxation modulus to a relaxation stiffness matrix using Voigt Notation. Starting from an unloaded and completely relaxed state at $t = 0^-$, the total stress $\boldsymbol{\sigma}$ at time t can be expressed using the convolution integral,

$$\sigma_i(t) = C_{ij}(t)\varepsilon_j(0) + \int_0^t C_{ij}(t-s)\dot{\varepsilon_j}(s)ds. \tag{2}$$

where σ_i and ε_i are stress and strains in index form using the Voigt notation, and C_{ij} are the components of the time-dependent (relaxation) stiffness matrix. Furthermore, C_{ij} is assumed to be described by a Prony series such as

$$C_{ij}(t) = C_{ij}^{\infty} + \sum_{k=1}^{M} C_{ij}^{(k)} e^{\frac{-t}{\tau_{ij}^{(k)}}}, \tag{3}$$

where $C_{ij}^{\infty} = C_{ij}(t \to \infty)$, τ_{ij} and C_{ij}^{k} are material parameters to be determined, and M is the number of Maxwell elements used that is represented by the Prony series.

The stress at an arbitrary time t can be obtained by inserting (3) in (2), which in return can be rearranged such as

$$\sigma_i(t) = C_{ij}^{\infty}\varepsilon_j(t) + \sum_{k=1}^{M} \sigma_{i,viscous}^{(k)}(t) \tag{4}$$

where,

$$\sigma_{i,viscous}^{(k)} = C_{ij}^{(k)} \left(e^{\frac{-t}{\tau_{ij}^{(k)}}} \varepsilon_j(0) + \int_0^t e^{\frac{-t-s}{\tau_{ij}^{(k)}}} \dot{\varepsilon}_j(s)ds \right) \tag{5}$$

2.1. Finite element implementation

To implement a material model in a commercial finite element software, it is often the case to calculate the stress based on a strain increment and on the information of the previous time steps. That is why the authors have used the relaxation form to derive the viscoelastic constitutive equations. For small time steps Δt, the strain rate $\dot{\varepsilon}_j$ may be considered constant, and thus eq. (4) transforms into

$$\sigma_{i,viscous}^{(k)} = e^{\frac{-\Delta t}{\tau_{ij}^{(k)}}} \sigma_{i,viscous}(t-\Delta t) + \frac{1 - e^{\frac{-\Delta t}{\tau_{ij}^{(k)}}}}{\Delta t/\tau_{ij}^{(k)}} C_{ij}^{(k)} \Delta\varepsilon_j \tag{6}$$

From eq. (6), it is now possible to obtain the viscous stresses from each time step Δt. It is noted that the viscous stresses at time t is equal to a fraction of the viscous stress from the previous time step plus a stress increment due to the strain increment $\Delta\varepsilon_j$.

Finally, the total stress is the sum of the elastic and viscous contributions,

$$\begin{aligned}\sigma_i(t) = C_{ij}^{\infty}\varepsilon_j(t) + &\sum_{k=1}^{M} e^{\frac{-\Delta t}{\tau_{ij}^{(k)}}}\sigma_{i,viscous}(t-\Delta t)\\ &+\sum_{k=1}^{M}\frac{1-e^{\frac{-\Delta t}{\tau_{ij}^{(k)}}}}{\Delta t/\tau_{ij}^{(k)}}C_{ij}^{(k)}\Delta\varepsilon_j\end{aligned} \tag{7}$$

In addition to the stress update, it is necessary to provide the consistent tangent stiffness, also known as the material Jacobian, to the FEM software. The consistent tangent stiffness is necessary to derive the constitutive component of the internal virtual work, where the latter is discretized to formulate the FE equations which is later solved via Newton-Raphson iterations to find the corresponding displacements that yield equilibrium. Deriving eq. (7) w.r.t. strain ε_j, the Jacobian is obtained,

$$C_{ij}^{tan} = \frac{\delta\sigma_i(t)}{\delta\varepsilon_j(t)} = C_{ij}^{\infty} + \sum_{k=1}^{M}\frac{1-e^{\frac{-\Delta t}{\tau_{ij}^{(k)}}}}{\Delta t/\tau_{ij}^{(k)}}C_{ij}^{(k)} \tag{8}$$

From eq. (3), it should be noted that this approach will have 12 + 24M parameters that need to be defined in order to fully describe the viscoelasticity of a general 3D orthotropic material. Choosing an adequate number of parameters is always a balance. Adding more and more empirical parameters does not add any significant predictive precision given the generally large scatter in measured data. Too few parameters will not capture the main traits of the time-dependent load-deformation behaviour.

2.2. Determination of the model parameters

Given that creep experiments in all direction have been performed, one may easily determine the parameters of the model. The stiffness matrix $\mathbf{C}$ is the inverse of the compliance matrix $\mathbf{D}$ such as

$$\mathbf{C}^{-1} = \mathbf{D} \tag{9}$$

whereas, assuming orthotropic symmetry,

$$\mathbf{D} = \begin{bmatrix} \frac{1}{E_1} & \frac{-\nu_{12}}{E_2} & \frac{-\nu_{13}}{E_3} & 0 & 0 & 0\\ \frac{-\nu_{21}}{E_1} & \frac{1}{E_2} & \frac{-\nu_{23}}{E_3} & 0 & 0 & 0\\ \frac{-\nu_{31}}{E_3} & \frac{-\nu_{23}}{E_3} & \frac{1}{E_3} & 0 & 0 & 0\\ 0 & 0 & 0 & \frac{1}{G_{12}} & 0 & 0\\ 0 & 0 & 0 & 0 & \frac{1}{G_{13}} & 0\\ 0 & 0 & 0 & 0 & 0 & \frac{1}{G_{23}} \end{bmatrix} \tag{10}$$

It should be emphasized that each element of the compliance matrix, as well as its components, are time dependent. The time dependent Poisson's ratios

$$\nu_{ij} = -\frac{\varepsilon_j(t)}{\varepsilon_i(t)} \tag{11}$$

are determined by the strain in the lateral direction of the creep test, ε_j, and the strain in the loading direction of the same creep test, ε_i. The time dependent Young's moduli

$$E_i(t) = \frac{\sigma_i}{\varepsilon_i(t)} \tag{12}$$

are determined by the constant creep stress σ_i and the aforementioned ε_i. Determining G_{ij} is analogous to eq. (12).

It important to keep $\mathbf{D}(t)$ as a symmetric matrix in order to invert it. Thus one should optimize the off-diagonal elements depending on which of the lateral strains that are more important to describe.

Consider two uniaxial tensile creep tests of where constant stresses, σ_1 and σ_2, are applied in each test respectively. The constant stress σ_1 is applied in direction 1, thus resulting into a creep strain $\varepsilon_1^1(t)$ in the loading direction, and creep strain $\varepsilon_2^1(t)$ in direction 2 due to the Poisson effect. Similarly, the constant stress σ_2 is applied in direction 2, and thus resulting into a creep strain $\varepsilon_2^2(t)$ in the loading direction, and creep strain $\varepsilon_1^2(t)$ in direction 1 due to the Poisson effect. Using eq. (11), $\nu_{12}(t)$ and $\nu_{21}(t)$ and using eq. (12), $E_1(t)$ and $E_2(t)$ can be calculated. Knowing the Poisson's ratios as well as the Young's moduli, will allow us to calculate the off-diagonal elements d_{21} and d_{12} in accordance to eq. (10). It is usually expected that d_{12} and d_{21} are not equal to each other, and thus an optimization has to be made to make the compliance matrix $\mathbf{D}$ symmetric. E.g., one can define a new diagonal element d_{12}^{sym} such as

$$d_{12}^{sym} = xd_{12} + (1-x)d_{21}, \tag{13}$$

where x is decided by minimizing the absolute errors,

$$\epsilon_1 = \| \varepsilon_2^1 - d_{12}^{sym}(x)\sigma_1 \|, \quad \epsilon_2 = \| \varepsilon_1^2 - d_{12}^{sym}(x)\sigma_2 \| \tag{14}$$

In applications, it might be necessary to use some form of weighted absolute errors in cases where there is a specific normal stress component that is dominant throughout the structure. Finally, when $\mathbf{D}(t)$ is fully defined, the elements of the stiffness matrix $\mathbf{C}(t)$ may be fitted by a Prony series described in eq. (3) with an appropriate curve-fitting scheme.

3. Examples of implementation

In this section, the proposed model is implemented in the ANSYS subroutine USERMAT and is used to describe the viscoelasticity of common orthotropic materials. The model parameters are determined by calibrating against experimental data from various experimental studies. Two examples of solid wood and one of a unidirectional fiber composite will be demonstrated.

A 1 [mm] x 1 [mm] x 1 [mm] cube is modelled in Ansys Mechanical APDL. The cube itself is meshed by a single trilinear hexahedral element, denoted as a SOLID185 element in ANSYS Mechanical APDL. To capture the full 3D effects of a load, symmetry boundary conditions are applied onto the cube expressed as different constraints on the eight nodes.

3.1. Fiber reinforced thermoplastic material

Endo et al [11] performed an approximately 3.5 weeks long tensile creep test of a fiber reinforced thermoplastic at a temperature of 23.5 C. Here, two loading directions were performed, parallel and perpendicular relative to the fiber directions. A tensile stress of 12 MPa was applied in both cases. The resulting creep curves are shown in for each loading direction (Fig. 1a and Fig. 1b), showing the creep strains parallel (L) and perpendicular (T) to the fiber direction.

3.2. European beech wood

Ozyhar et al. [10] performed a tensile creep test of European beech wood. In addition to loading in the longitudinal direction, the same test was performed in the tangential and radial directions.The creep tests were carried out in a constant climate test room under controlled environmental conditions, (20C, 65% RH). The resulting creep curves are shown for each loading direction, showing the creep strains in three perpendicular material axes L, T and R in Fig. 2.

It is readily possible to take into account either fully time dependent or constant (time invariant) Poisson's ratios in the model. Additionally, one can choose which specific Poisson's ratios are to be constant or time dependent. This selection can be done by identifying which Poisson's ratio changes the least in comparison to others. In the experiment of Ozyhar et al., it was identified that ν_{RL}, ν_{TL} and ν_{RT} did not change as fast relative to the remaining three Poisson's ratios, perhaps due to the stiffening radial rays present in the hardwood. Thus, the assumption of having these specific ratios to be time-independent was implemented into the model as well.

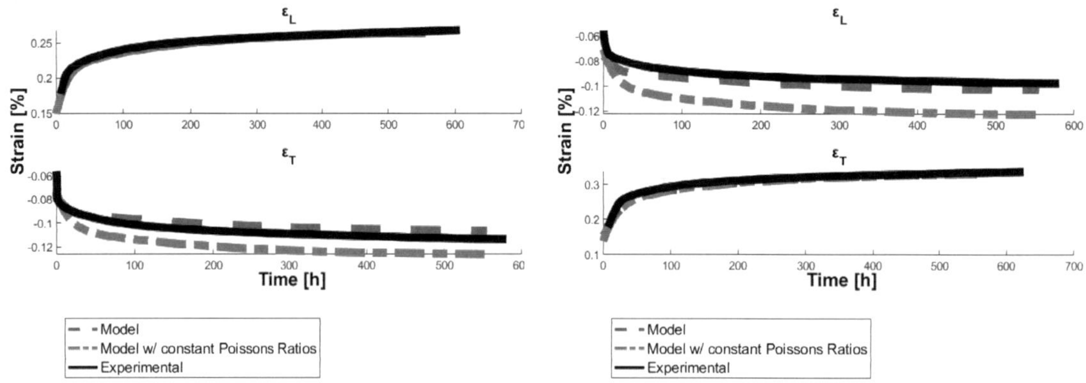

(a) Creep plots from tensile load in L (Endo et al) (b) Creep plots from tensile load in T (Endo et al)

Figure 1: Creep loading according to Endo et al. For (a), the specimen was subjected to a given creep stress of 12 MPa in the longitudinal direction. For (b), The specimen was subjected to a given creep stress of 12 MPa in the perpendicular to the fiber. The resulting creep strains are shown in L and T (perpendicular to the fiber) directions.

4. Discussion

It should be noted that a single 1x1x1 [mm] cube was modelled in FE-software to simulate creep. At this scale, the cellular wood microstructure affects the mechanical behaviour, and the material cannot be regarded as a homogeneous continuum. There is also a risk when using the fitted creep model to simulate larger and complex structures, as the size-effects are not taken into account. For example the polar orthotropy of wood is not considered here, where Cartesian orthotropic behaviour is assumed. The quantification of the error due to size effects is of interest in future studies.

From Fig. 1 and Fig. 2, there is a distinct difference between the assumption of constant and time-dependent Poisson's ratios . However, the trends are different in the two types of material. In the fiber reinforced thermoplastic material, the assumed constant Poisson's ratios resulted into larger transverse creep under longitudinal loading direction, see Fig. 1 (a). In contrast, the solid wood example showed smaller transverse creep instead, as can be seen in Fig.

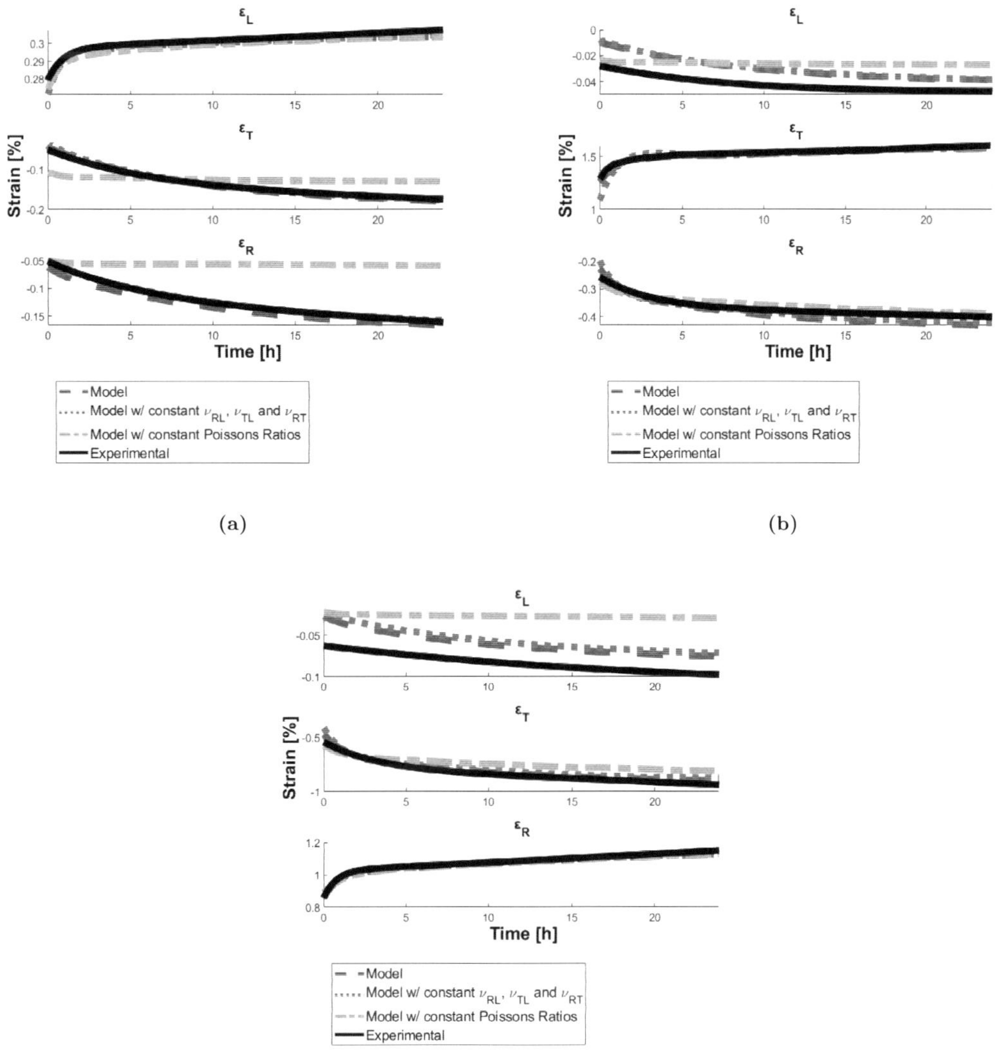

Figure 2: Creep data set from Ozyhar et al. For (a), the specimen was subjected to a given creep stress of 26.2 MPa in the longitudinal direction. For (b), the specimen was subjected with a given creep stress of 5.7 MPa in the tangential direction. For (c), the specimen was subjected to a given creep stress of 11.9 MPa in the radial direction. The resulting creep strains are shown in L, T and R direction for all cases.

2 (a). This difference is of interest and may need some further investigation in the future from a micromechanics viewpoint. Nevertheless, it is clear that applying time-dependent Poisson's ratios to the model gave similar results compared to experimental data. It was also discovered using the solid wood example, that applying constant Poisson's ratio of where the longitudinal direction is perpendicular to the loading direction, results to similar results to the assumption of time-dependent ratios. This is simply due to the fact that the Poisson's ratio is in this case relatively constant comparison to the remaining ones. Thus, it is up to the user of the model to decide which ratios should be taken as constant by comparing the Poisson's effect in the experimental data. The invariance of certain Poisson's ratios should be attributed by

the microstructure of the composite or wood material, and could possibly be addressed by micromechanics.

For the sake of convenience, we used only two exponential terms for each component in the C matrix. However, the model allows to have different number of terms for C_{ij}, thus it is of future interest to study the the optimal combination. Optimal not only in fitting experimental data, but also out of ease of use and how many parameters are motivated to include given the experimental scatter. At this point, the authors have shown the capabilities of the relatively straightforward hereditary approach on the viscoelastic model of orthotropic materials, as well as how to manage asymmetric experimental data.

5. Conclusions

The following conclusions and recommendations can be made:

- It can be recommended to use two exponential terms (m=2) for each component in both transversely isotropic and orthotropic materials, judged by the available experimental data.
- Experimental creep data do generally not comply with a symmetric creep compliance matrix, as required for orthotropic materials. A scheme to render the compliance matrix symmetric is presented in this work.
- If necessary, one may use constant Poisson's ratio if it has been shown that the Poisson's ratios are time-independent in the experimental testing. This facilitates the use of the model.

References

[1] L Nallainathan, X L Liu, W K Chiu, and R Jones (2004) Modelling creep behaviour of orthotropic composites by the coincident element method. *Comp Struct*, **66** 409-13.

[2] V G Martynenko (2017) An original technique for modeling of anisotropic viscoelasticity of orthotropic materials in finite element codes applied to the mechanics of plates and shells. *Mech Mech Eng*, **21** 389–413

[3] E Vidal-Sallé and P Chassagne (2007) Constitutive equations for orthotropic nonlinear viscoelastic behaviour using a generalized Maxwell model: Application to wood material. *Mech of Time Dependent Mat*, **11** 127–42

[4] S Fortino, F Mirianon and T Toratti (2019) A 3D moisture-stress FEM analysis for time dependent problems in timber structures. *Mech Time Dependent Mat*, **13** 333–56

[5] S Hûc and S Svensson (2018) Influence of grain direction on the time-dependent behaviour of wood analyzed by a 3D rheological model. A mathematical consideration. *Holzforschung* **72** 889-97

[6] M Abouhamzeh, J Sinke, K M B Jansen and R Benedictus (2015) A new procedure for thermo-viscoelastic modelling of composites with general orthotropy and geometry. *Comp Struct* **133** 871–77

[7] M A Zocher AND S E Groves (1997) A three-dimensional finite element formulation for thermoviscoelastic orthotropic media. *Int J Num Methods Eng*, **40**, 2267-88

[8] S R White and Y K Kim (1998) Process-induced residual stress analysis of AS41350 1-6 composite material. *Mech Comp Mat Struct*

[9] S Pandini and A Pegoretti (2008) Time, temperature, and strain effects on viscoelastic Poisson's ratio of epoxy resins. *Poly Eng Sci*

[10] T Ozyhar, S Hering, and P Niemz (2013) Viscoelastic characterization of wood: Time dependence of the orthotropic compliance in tension and compression. *J Rheo* **57** 699-718

[11] V T Endo and J C de Carvalho Pereira (2017) Linear orthotropic viscoelasticity model for fiber reinforced thermoplastic material based on Prony series. *Mech Time Dependent Mat* **21** 199–221

[12] ANSYS Mechanical *ANSYS reference guide.*

IOP Conf. Series: Materials Science and Engineering **942** (2020) 012022 doi:10.1088/1757-899X/942/1/012022

Multiaxial stress and strain analysis on laminated plates under different flexural loading rates

S Horta Muñoz* and M C Serna Moreno

Universidad de Castilla La-Mancha, Escuela de Ingeniería Industrial y Aeroespacial de Toledo, Departamento de Mecánica Aplicada e Ingeniería de Proyectos, Instituto de Investigación Aplicada a la Industria Aeronáutica, Real Fábrica de Armas, 45004, Toledo, Spain

*E-mail: sergio.horta@uclm.es

Abstract. The occurrence of events associated with Low-Velocity Impacts (LVI) is analysed by drop-weight impact tests on carbon fibre reinforced polymer laminates, comparing the response in these tests with Quasi-Static Indentation (QSI). A laminate sequence of interest, the angle-ply $[\pm45]_{4S}$, is tested in both testing procedures with a simply supported configuration on a circular fixture. Both methodologies show similarities in the critical points for change of behaviour in terms of energy, while the strain rate effect is verified to influence the higher apparent stiffness of the laminate submitted to the LVI test compared with the QSI test. The effect of multiaxiality in loads existing in these tests is addressed in the quasi-static experiment by applying Digital Image Correlation (DIC), comparing the results with numerical simulations using the Finite Element Method (FEM). The orthotropy of the material generates strain and displacement fields which allow understanding the dissimilar response of the laminate in the different radial directions, also giving explanation to the initiation and evolution of different damage mechanisms.

1. Introduction

The widespread application of fibre polymer composites in highly demanding structural applications, such as aerospace or wind energy sectors, is related to their outstanding specific strength and stiffness. Moreover, the mechanical response of some laminates, for instance pseudo-ductile effects in angle-ply sequences, has been proven to enhance large strain response while increasing energy absorption and damage accumulation under quasi-static loading [1,2]. Nevertheless, a major drawback of the applications of these materials is their limited tolerance to out-of-plane loadings, while impact damages are even more concerning. Low energy strikes may produce internal matrix cracking and delamination, difficult to detect but threatening the structural integrity of the component.

Likewise, the experimental characterization of this type of damage produced by impact on a laboratory scale raises certain difficulties due to the dynamic and short-timed nature, with larger dispersion of results during testing campaigns. Therefore, several authors [3–10] have pursued to establish a relationship between the mechanisms of damage and failure under Low Velocity Impact (LVI) with tests applying a lower deformation rate, i.e. Quasi-Static Indentation (QSI). These works contribute conclusions in certain contradictory points, without reaching a clear consensus of the possible extrapolation of the quasi-static tests. The advantage of lower rate tests resides in greater simplicity in their performance and analysis, compared with the dynamic phenomenology observed under impact loads. Some authors highlighted certain similarities and differences not only in the failure mechanisms and their visualization, but also in parameters recorded in the test such as the force or energy involved in the stages of damage initiation and evolution.

This study proposes is different than the work already existing in bibliography due to the detailed analysis of the multiaxial stress/strain state in laminates with angle-ply orientations. The orthotropy of the material will be studied in depth from the point of view of the radial strain measured through the application of Digital Image Correlation (DIC), combined with numerical simulations using the Finite Element Method (FEM). In addition, the similarities between the tests at different loading rates will be detailed, highlighting the search for a correlation between the critical points of the behaviour of these laminates, in terms of force-displacement and applied energy. In this way, it will be possible to estimate the initiation and evolution of the main failure mechanisms that appear similarly in QSI and LVI with certain independence of the strain rate, until reaching the final rupture of the laminate.

2. Experimental and numerical procedure

The carbon fibre reinforced polymer (CFRP) laminates tested under QSI and LVI methodologies are manufactured using a commercial prepreg ply by Hexcel, with the specific designation of M21E/34%/UD268/IMA-12K. This lamina is composed by a M21E toughened epoxy matrix at 34 wt.% and reinforced with a IMA-12K intermediate modulus carbon fibre at a 268 g/m^2 density [11]. Additionally, the average cured ply thickness is equal to 0.25 mm. The manufacturing process consists of hand lay-up of the plies with the proposed stacking sequence, followed by the curing cycle recommended by the manufacturer, which is performed on a LabEcon 300 hot platen press. The fabricated laminates are cut with a circular saw, producing specimens in the form of square plates with 80 mm nominal side dimensions.

The elastic properties of the lamina in material principal directions, characterized in previous studies [1,12–15], is summarized in Table 1, where elastic moduli (E), shear moduli (G) and Poisson's ratios (ν) are included.

Table 1. Elastic properties of the CFRP lamina in principal directions [1,12–15].

E_1 (GPa)	E_2 (GPa)	E_3 (GPa)	G_{12} (GPa)	G_{13} (GPa)	G_{23} (GPa)	ν_{12}	ν_{13}	ν_{23}
177.56	11.84	11.84	5.42	5.42	3.10	0.39	0.39	0.36

For both testing procedures, i.e. QSI and LVI, the same specimen dimensions and boundary conditions are applied (see Figures 1 and 2). A non-standardized procedure is developed, in order to perform a test in which a squared plate is simply supported in a 74 mm diameter span fixture, while a semi-spherical indenter with 10 mm radius applies the load at the centre of the specimen. In the case of the QSI testing, displacement is applied at a fixed rate of 0.8 mm/s by means of a MICROTEST MAEFH electromechanical machine (Figure 1a). During this test, strain values are acquired at different gauge zones, that is, the displacement fields at the upper compressive face are recorded by means of a 3D DIC system (namely LaVision StrainMaster). Simultaneously, strains at the centre of the bottom surface are acquired by means of a foil strain gauge rosette. Specifically, KYOWA KFGS triaxial rosette with 0°, 45° and 90° stacked gauges are utilised, with 5x1.4 mm gauge length and width, respectively.

(a) (b)

Figure 1. Experimental set-up for QSI testing: (a) Testing facility showing DIC installation. (b) Detail of loading application over simply supported specimen.

Regarding the impact testing, an instrumentalized drop-weight tester INSTRON CEAST 9340 is utilised (Figure 2a). The LVI tests are performed with an impact velocity of 2.0 m/s to avoid total penetration of the specimen and the kinetic energy applied was 89.9 J. The utilised indenter is the same than in QSI testing (Figure 1b), fitted to an impact hammer with the mass adapted to deliver the required impact velocity and energy. Regarding the supporting fixture (Figure 2b), this is specifically designed to be fitted in both the electromechanical and impact testing machines.

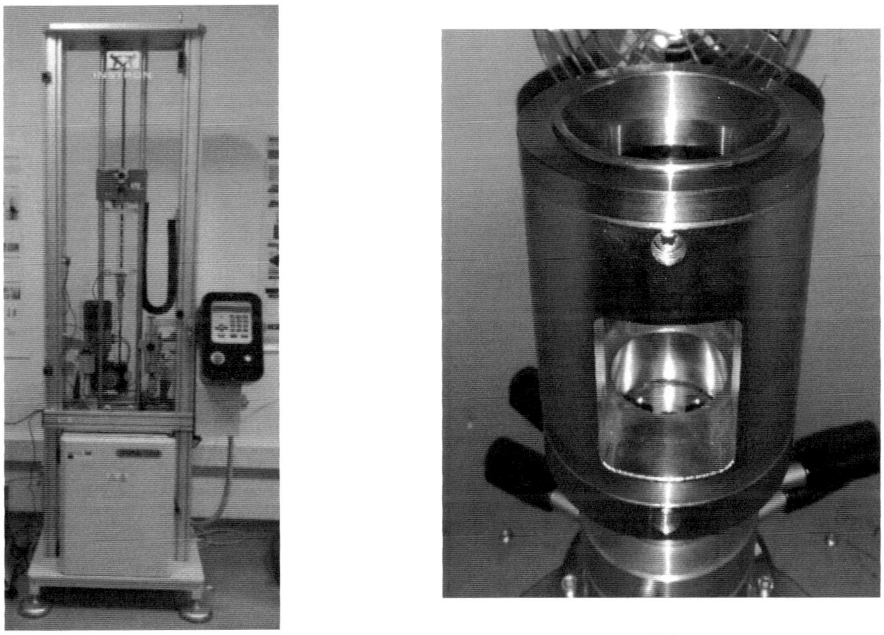

(a) (b)

Figure 2. Experimental set-up for LVI testing: (a) Drop-weight tower. (b) Novel fixture with 74 mm diameter circular window.

3. Results and discussion

This section describes the experimental results obtained in the QSI and LVI tests, analysing the similarities in the recorded data and detailing the failure modes observed in the fractographies for the tested specimens. Furthermore, the analysis of the multidirectional strain state is completed with the comparison of the normal strain contours obtained by DIC and FEM.

3.1. Experimental results

The results obtained for two groups of $[\pm45]_{4S}$ specimens respectively tested under QSI and LVI are detailed below. The force-displacement and energy-displacement curves obtained for two representative tests are shown in Figure 3, although note that this study collects the numerical data from a campaign of repetitive tests. Figure 3a depicts a significant difference in the apparent stiffness of the laminate, leading to an average rigidity that is 36.3% higher in the case of LVI. This change can be explained by the strain rate effects observable in polymer matrix composites, as is extensively reported in literature [16,17]. Due to this, similarities between both testing methods are not found in terms of force or displacement.

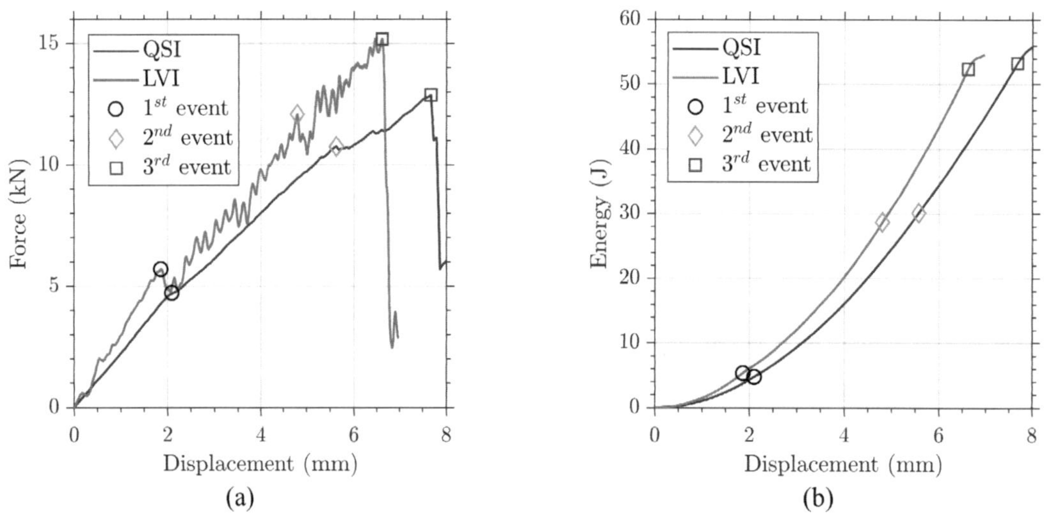

Figure 3. Representative curves for the $[\pm45]_{4S}$ under QSI and LVI, featuring the most notable events: (a) Force-deflection and (b) Energy-deflection

However, neglecting the dynamic component of the LVI test, both curves have a significant qualitative similarity. It is verified how there are similar points of change in behaviour in both evolutions, so it is possible to foresee the appearance of similar damage mechanisms. If the events highlighted in Figure 3a are compared in terms of instantaneous internal energy, it is possible to find a relevant correlation between both tests at different velocities. The internal energy values (Figure 3b), calculated as the integral of the force-displacement curve, differs only 6.5%, 3.0% and 1.8% between both tests (in average values), for the three marked events, respectively. These correlated values of energy allow us to indicate a relationship between the appearance and evolution of damage phenomena, mainly macrocracks, independently of the applied load rate.

This idea is supported by the optical observations of the tested specimens, which helps to estimate the evolution of the typically expected failure mechanisms in CFRP laminates. Figure 4 depicts the photographs of both sides, upper and lower faces, of a specimen subjected to LVI, and the inspection of a plate subjected to QSI is very similar when observed under similar internal energy levels. Taking into account that the plate is subjected to bending in all radial directions, it is expected that, for instance, the bottom-side will develop tensile stresses and strains in all radial directions, these being the maximum through-thickness values. The absolute maxima are located in the mid-span, where bending moment is also the highest. Out-of-plane shear stresses and strains are also expected, which are maximum in the neutral fibre of the specimen, i.e. the midplane in case of similar tensile and compressive behaviour of the laminate. In addition, stress/strain concentrations appear in the vicinity of the indenter, so the first failures at the upper layer are observed in this region. In Figure 4b it is possible to observe the indentation produced in the central area of the plate.

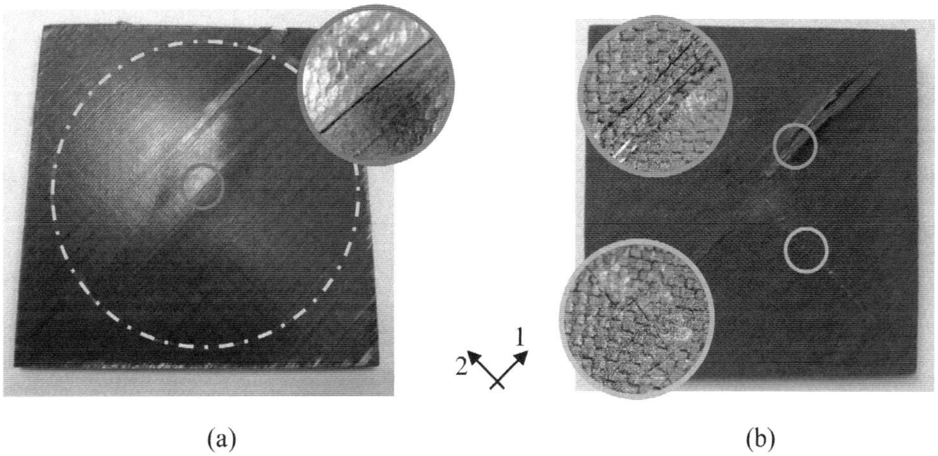

Figure 4. Failure modes of the angle-ply laminate under LVI. (a) Bottom/tensile face view. (b) Upper/compressed face. The fibre orientation at the outer ply is symbolised with the material principal directions.

If the most-tensioned surface is observed in detail (Figure 4a), the previous explanation indicates that the central zone is the region of maximum normal tensile stress. The highest local elastic modulus (ply-level) is found on the fibre direction, i.e. at 45° in Figure 4, and fibre-matrix debonding is expected to be the initial failure mode, as this weak interface is subjected to higher stresses. This microcracks propagate originating a larger matrix crack that advances through the diagonal, but with less intensity than in the central zone (see detail in Figure 4a). The apparition of large matrix cracking is reflected in Figure 3 with the first highlighted event. This failure mode extends also to internal plies, but alternating the orientation with the fibre direction, and decreasing in length due to the lower bending moment when moving towards the neutral fibre. This damage mechanism is combined with the delamination of the plies, producing larger surfaces which are reflected in the loss of rigidity, associated to an internal energy liberation.

When observing the compressed surface (Figure 4b), the failure modes are not only matrix dominated, but also certain fibre kinking could be observed (green magnification in Figure 4b). Nevertheless, the initial failure mechanism continues to be the matrix-dominated, due to the lower strength of the matrix, giving rise to large cracks that propagate again on the diagonal at 45°, associated with the fibre-matrix debonding described above. The larger cracks propagate up to reaching the proximities of the boundary conditions, which has been related to the second event in Figure 3 [10]. Finally, out-of-plane shear stresses are relevant in the propagation of the final failure (third event in Figure 3), producing the greatest drop in stiffness of the plate. This results in an abrupt opening near to the laminate mid-surface.

Finally, post-mortem photographs for the QSI test are shown in Figure 5, in a similar manner to Figure 4. Comparable failure modes are observed, with a significant difference in the severity of the matrix splitting (Figure 5a) and the depth of the indentation mark (Figure 5b), due to the higher deflection reached in the quasi-static test.

IOP Conf. Series: Materials Science and Engineering **942** (2020) 012022 doi:10.1088/1757-899X/942/1/012022

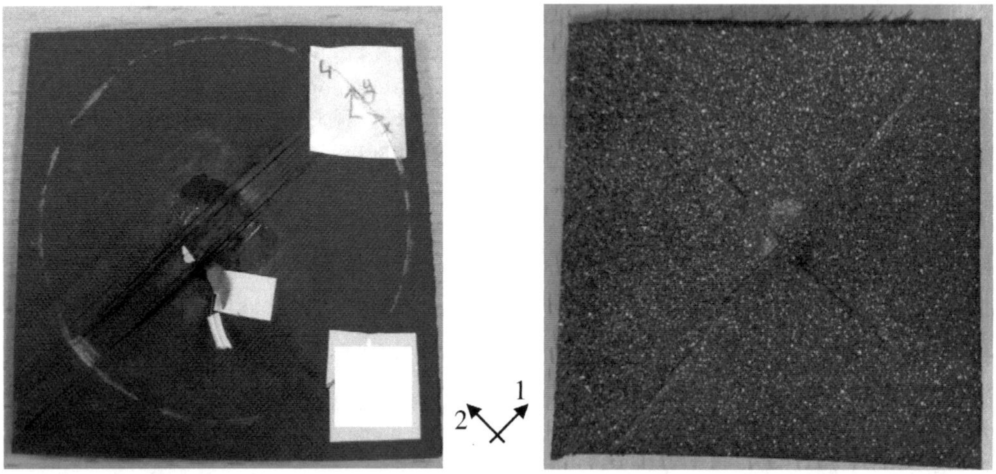

(a) (b)

Figure 5. Failure modes of the angle-ply laminate under QSI. (a) Bottom/tensile face view. (b) Upper/compressed face. The fibre orientation at the outer ply is symbolised with the material principal directions.

3.2. Comparison of numerical and experimental strain fields

Numerical simulations performed using Abaqus/Standard FEM software lead to a better understanding of the results from the quasi-static bending test. The model carried out is based on a mesh consisting of solid linear elements with 8 nodes and reduced integration (S8R), in which the symmetry at 45° is used to generate only half of the actual geometry (Figure 6a). A static non-linear analysis is carried out, applying an orthotropic material model, and defining the elastic properties of the ply and its orientations.

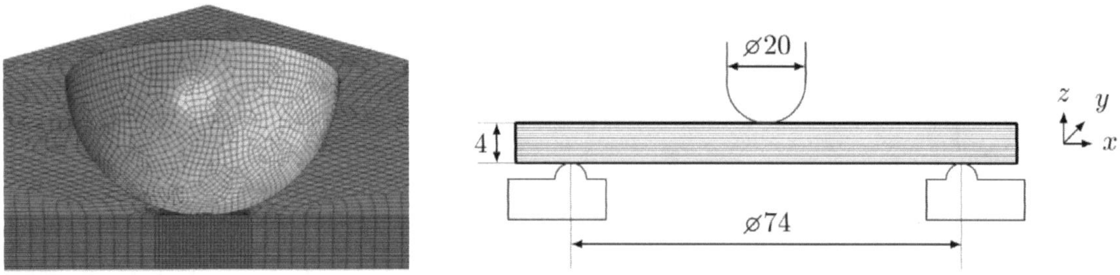

(a) (b)

Figure 6. (a) Detail of the mesh applied in numerical simulations of QSI testing. (b) Schematic drawing of the fixture, laminate and indenter (dimensions in mm, not to scale)

The testing facility is represented in Figure 6b. The indenter is modelled as a rigid solid, on which a 1 mm displacement is applied perpendicular to the laminated plate. The indenter-laminate interaction is defined as frictionless surface-to-surface contact [10]. The boundary conditions imposed by the support tool are reproduced preventing perpendicular displacements in a circumference of diameter 74 mm. The results of this analysis are compared (Figure 7) with the experimental strain contours obtained by DIC at the same level of applied displacement, before damage initiates and the response changes. Due to limitations of the image acquisition with the DIC cameras, i.e. range of view constrained by indenter, supports and load application system, only half of the top surface can be measured.

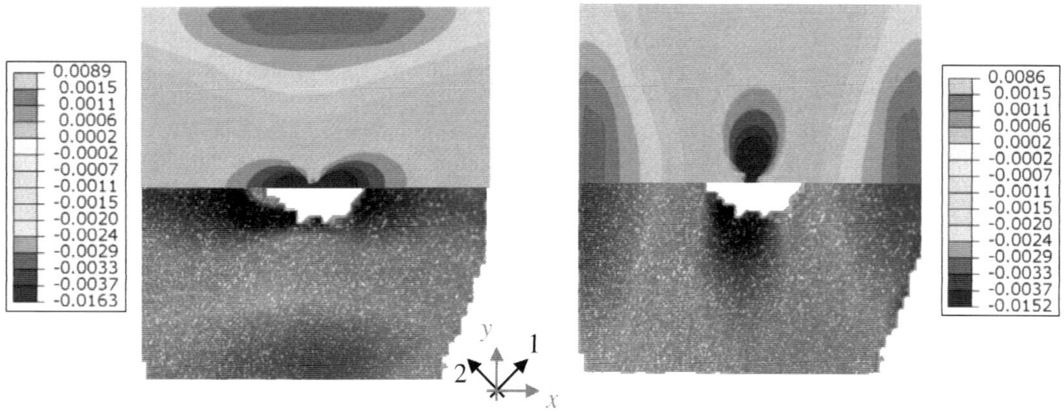

(a) (b)

Figure 7. Comparison of experimental (top) and numerical (bottom) strain fields at uppermost compressed ply: (a) ε_{xx} and (b) ε_{yy}. Local coordinate system for the observed ply is included.

The similarities between both approaches are noticeable, in terms of numerical values and shape of the strain fields. It can be noted the lack of symmetry regarding the x and y directions, due to the flexural orthotropy of the laminate. In other words, the $[\pm45]_{4S}$ laminate experiments a non-negligible orthotropy in the flexural stiffness when observed at the material principal directions, developing the maximum apparent flexural stiffness at the direction aligned with the fibre of the outermost ply. Therefore, the strain gradients, with reference to the centre of the plate, moves forward with an angle close to the 0° and 90°, but slightly deviated to the maximum stiffness direction (corresponding to the *1* principal direction in Figure 7). Additionally, the effect of the circular boundary conditions is reflected on the maximum tensile strains appearing at the edges of the plate.

4. Conclusions

The thorough analysis of the QSI and LVI tests instrumented for the acquisition of the strain states, applied forces and accumulated strain energy in each stage of the test has allowed us to find similar situations under both methodologies. That is, the events where the behaviour changes are most notable occur at similar levels of internal energy regardless of the strain rate applied, despite the notable apparent stiffness difference under different velocities. Damage initiation and evolution stages are correlated, and the more progressive quasi-static methodology helps to describe the failure mechanisms.

5. Acknowledgements

This work was financially supported by the Ministry of Economy and Competitiveness of Spain and the European Regional Development Fund under the grant DPI2016-77715-R. In addition, the Government of Castilla-La Mancha and the European Social Fund have funded S. Horta Muñoz through the predoctoral grant SBPLY/16/180501/000263.

References

[1] Serna Moreno M C, Horta Muñoz S, Romero Gutiérrez A, Rappold C, Martínez Vicente J L, Morales-Rodríguez P A and López Cela J J 2018 Pseudo-ductility in flexural testing of symmetric ±45° angle-ply CFRP laminates *Compos. Sci. Technol.* **156** 8–18

[2] Horta Muñoz S and Serna Moreno M C 2020 Numerical modelling of the pseudo-ductility

effect in ±45° angle-ply laminates under biaxial loading ECCM 2018 - 18th European Conference on Composite Materials (Applied Mechanics Laboratory)

[3] Lee S M and Zahuta P 1991 Instrumented Impact and Static Indentation of Composites *J. Compos. Mater.* **25** 204–22

[4] Kwon Y S and Sankar B V. 1993 Indentation-flexure and low-velocity impact damage in graphite epoxy laminates *J. Compos. Technol. Res.* **15** 101–11

[5] Spronk S W F, Kersemans M, De Baerdemaeker J C A, Gilabert F A, Sevenois R D B, Garoz D, Kassapoglou C and Van Paepegem W 2018 Comparing damage from low-velocity impact and quasi-static indentation in automotive carbon/epoxy and glass/polyamide-6 laminates *Polym. Test.* **65** 231–41

[6] Nunes S G, Reichwald L G G, Júnior W F A, Manes A and Amico S C 2018 Quasi-Static Indentation and Low-Velocity im pact response of aramid/epoxy *Composites* **323**–9

[7] Wagih A, Maimí P, Blanco N and Costa J 2016 A quasi-static indentation test to elucidate the sequence of damage events in low velocity impacts on composite laminates *Compos. Part A Appl. Sci. Manuf.* **82** 180–9

[8] Sun X C, Wisnom M R and Hallett S R 2016 Interaction of inter- and intralaminar damage in scaled quasi-static indentation tests: Part 2 - Numerical simulation *Compos. Struct.* **136** 727–42

[9] García-Rodríguez S M, Costa J, Bardera A, Singery V and Trias D 2018 A 3D tomographic investigation to elucidate the low-velocity impact resistance, tolerance and damage sequence of thin non-crimp fabric laminates: effect of ply-thickness *Compos. Part A Appl. Sci. Manuf.* **113** 53–65

[10] Serna Moreno M C and Horta Muñoz S 2020 Mechanical response of ±45° angle-ply CFRP plates under low-velocity impact and quasi-static indentation: Influence of the multidirectional strain state *Compos. Sci. Technol.* **194** 108145

[11] Hexcel 2015 HexPly® M21 - Product Data Sheet - EU Version Hexcel 1–6

[12] Serna Moreno M C, Romero Gutiérrez A and Martínez Vicente J L 2016 Different response under tension and compression of unidirectional carbon fibre laminates in a three-point bending test *Compos. Struct.* **136** 706–11

[13] Serna Moreno M C, Romero Gutiérrez A and Martínez Vicente J L 2016 First flexural and interlaminar shear failure in symmetric cross-ply carbon-fibre laminates with different response under tension and compression *Compos. Struct.* **146** 62–8

[14] Serna Moreno M C, Romero Gutiérrez A and Martínez Vicente J L 2016 Flexural testing on carbon fibre laminates taking into account their different behaviour under tension and compression *IOP Conf. Ser. Mater. Sci. Eng.* **139** 012047

[15] Serna Moreno M C and Horta Muñoz S 2020 Elastic stability in biaxial testing with cruciform specimens subjected to compressive loading *Compos. Struct.* **234** 17–8

[16] Cui H, Thomson D, Pellegrino A, Wiegand J and Petrinic N 2016 Effect of strain rate and fibre rotation on the in-plane shear response of ±45° laminates in tension and compression tests *Compos. Sci. Technol.* **135** 106–15

[17] Fotouhi M, Fuller J D, Longana M, Jalalvand M and Wisnom M R 2019 The high strain rate tension behaviour of pseudo-ductile high performance thin ply composites *Compos. Struct.* **215** 365–76

IOP Conf. Series: Materials Science and Engineering **942** (2020) 012023 doi:10.1088/1757-899X/942/1/012023

Rain droplet erosion behavior of a thermoplastic based leading edge protection system for wind turbine blades

T.H. Hoksbergen*, I. Baran, R. Akkerman

University of Twente, Engineering Technology, Drienerlolaan 5, 7522NB, Enschede, NL

E-mail: t.h.hoksbergen@utwente.nl

Abstract. This paper discusses the rain droplet erosion mechanisms of an acrylonitrile butadiene styrene (ABS). Rain droplet impingement was modeled based on a coupled smoothed particle hydrodynamics and finite element method (SPH/FEM). Using linear elastic material parameters at low strain rates, the dynamic stress behavior was studied and the location of damage initiation was predicted. Experiments using a pulsating jet erosion tester were performed and the resulting erosion behavior was analyzed using confocal microscopy. The damage was expected to initiate at the surface and remain superficial during propagation. It was shown that a pitting behavior occurred at the surface after the first few impacts. This pitting continued until 100.000 impacts. After this, the pits connected through a cracking mechanism and finally, at 300.000 impacts, cratering was observed which led to the onset of material loss. The depth of these craters was observed to be approximately 80μm, which was relatively low as compared to the material thickness of 4mm, indicating superficial damage. The resulting volume loss curve showed an initial period where no volume loss occurred, called the incubation period, followed by a linear relation between the volume loss and the number of impacts. This behavior agreed well with behavior found for other materials in literature. The surface roughness parameters were determined for each amount of impacts and the mean roughness value corresponded well to the volume loss behavior. Earlier stages of damage could be detected by analyzing the skewness value.

1. Introduction

Sustainable energy production is an important topic concerning society, economics and the environment. Wind energy is of vital importance to the energy transition and large investments in wind energy are made globally. Currently, a cumulative capacity of 205GW wind energy is installed in Europe, of which 22GW is offshore wind energy [1]. This offshore wind energy market is growing fast with an installed capacity of 3.6GW in 2019 only. Onshore turbines had an average power rating of 3.1MW while offshore turbines had an average power rating of 7.8MW with the largest being the 12MW GE Haliade-X prototype [2] which has a rotor diameter of 220m. Not only the size of the turbines increases, but also the tip speeds of the blades, currently exceeding 100ms^{-1}. These high tip speeds lead to interactions of the blade material with rain droplets causing high dynamic stresses in the blades, especially at the leading edge of the blade. These stresses cause surface fatigue damage, ultimately leading to a material loss [3]. The eroded surface of the leading edge has a negative effect on aerodynamics and might lead to structural failure of the blade, which is mainly made of glass fiber reinforced thermosetting polymer composites. To prevent a possible failure of the leading edge of the blade, regular, expensive maintenance has to occur.

Currently, technologies to limit leading edge erosion consist of protective polyurethane coatings, tapes or shells as well as thermosetting gelcoats [4]. These are added on top of the blade structure, influencing the blade aerodynamics and reducing the erosion damage development. The research field on leading edge protection (LEP) systems is very active and currently involved with the influence of interphases between the LEP and the composite blade structure [5], the use of thermoplastic materials [6] and the optimization of (multilayer) microstructures of materials [7]. The recent research project InLEP (Integrated Leading Edge Protection for Offshore Wind Turbine Blades at High Speed) carried out in the Netherlands, focuses on an integrated leading edge protection concept based on thermoplastic materials to reduce the erosion mechanisms further without compromising aerodynamic efficiency of the blade. In order to design this InLEP concept, a fundamental understanding of the erosion and bonding mechanisms for different types of thermoplastics is required.

When looking at the erosion damage development, the intensity of the damage is often represented by the cumulative mass-loss and the time by either the cumulative number of impacts or exposure time. This results in a mass-loss curve such as in Figure 1. It can be seen that up to a certain number of impacts there exists no mass loss which is defined as the incubation period. Afterwards, a linear relation between mass-loss and exposure time is observed which is followed by a random behavior corresponding to a catastrophic failure of the blade. This relation is found to be similar for most LEP materials although the underlying damage mechanisms may differ. The end of incubation, which is the transition from the incubation period to the linear mass loss progression, is usually defined as the critical point for assessing the rain erosion performance of the LEP.

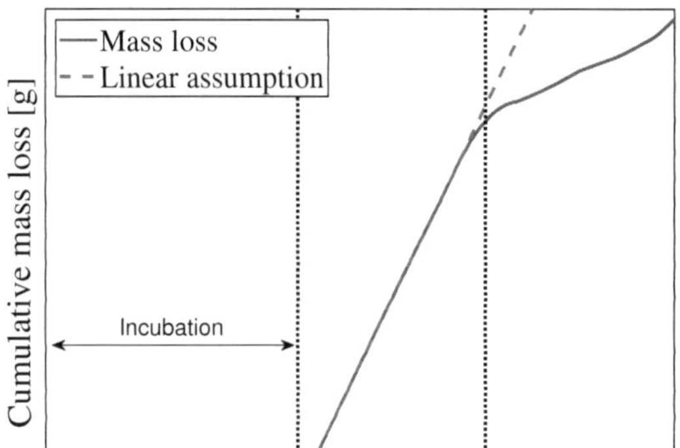

Figure 1. Typical mass-loss curve for leading edge erosion.

Rain droplet impacts cause elastic waves in the material. These highly dynamic waves interact with interfaces, defects and each other leading to complex stress situations. Understanding how these waves propagate as a function of not only material parameters but also geometric and impact parameters is therefore of vital importance when designing LEP systems. Initially, analytic models using the (modified) waterhammer pressure were proposed [8]. These models are still widely used and adapted to more advanced materials. Recent developments have been made in numerical modeling to obtain a more accurate description of the stresses. Often, a smoothed particle hydrodynamics (SPH) approach for the fluid is used in combination with a finite element method (FEM) approach for the solid [3, 9]. Some work has used a combined Eulerian/Lagrangian mesh for the fluid structure interaction and the results corresponded well with SPH/FEM approach [10]. Simulation of damage is very challenging due to the dynamic waves and microstructures of the materials. Combining SPH/FEM simulations with experimental observations can be useful to understand damage initiation and development [11]. More difficulties lie in predicting the long term performance based on these stress wave fields. So far, fitting of experimental results to a fatigue model is still required [12, 13].

Understanding the rain erosion behavior and validating the models requires methods to do accelerated testing of material coupons. Generally, two different types of experiments exist, being rotating arm and jet based erosion tests [14]. In rotating arm rigs, a material coupon or small leading edge section is rotated through an artificial rain field at high velocity [15]. During the test, the mass loss can be evaluated by drying and weighing the specimen at set intervals and the end of incubation and breakthrough points are identified. With this data, the Wöhler curve can be made and the Springer model can be fitted [8]. These tests provide a fast method to characterize different materials for long term erosion behavior, however the insight gained in the stress and damage development under rain droplet impact is limited. To overcome this, jet based systems are generally used [16]. In these systems, a stationary sample is impacted by a high velocity jet or jet segments. Because the sample is stationary, it is easier to perform measurements on the sample and the number of impacts is controlled in a better way. The material damage due to rain droplet impact however is very local in jet based setups, therefore a validation of these tests based on erosion behavior observed in the field is relatively difficult. Some efforts have been made to correlate both test methods [17], including surface topology parameters after damage [18]. A third test method using elastomeric balls as projectiles is also under development [19] and shows high potential for accelerated testing of coating systems. The visco-elastic effects in the balls do however differ from the compressive effects in the water droplets.

In this paper, a pulsating jet erosion test (PJET) is performed on an acrylonitrile butadiene styrene (ABS) material. The rain droplet erosion damage initiation and propagation are studied and compared with modeling results. A novel method to determine the volume loss of material after the PJET is discussed and the results are compared with the surface roughness parameters.

2. Model

A coupled SPH/FEM method was used to model the dynamic fluid structure interaction between the water droplet and the ABS LEP system. The model was interpreted in ABAQUS and solved for a linear elastic material. The results gave insight in stress locations and magnitudes as a function of time. This was used to predict damage initiation locations. The model was solved using the parameters in Table 1.

Table 1. SPH/FEM model input parameters.

Property	Density [kg/m^3]	E [GPa]	ν [−]	C_p [m/s]	μ [Pas]	V [m/s]	Diameter [mm]
Droplet	997	-	-	1481	8.9e-4	160	2
LEP	1050	2.45	0.408	-	-	-	-

3. Experiments

The experimental setup shown in Figure 2 was based on a pulsating jet erosion test and developed in house at the University of Twente, The Netherlands. A high pressure pump ejected water from a reservoir through a nozzle. The resulting continuous jet was sliced using a slotted rotating disk resulting in jet segments that impacted a stationary sample. The water was captured and filtered and flowed back into the reservoir. The setup could reach pressures of 180bar and flow rates of 19l/min. This allowed for jet velocities of up to 190ms^{-1} at nozzle diameters of up to 1.4mm. At larger nozzle diameters, the maximum velocity decreased, as expected. For the tests described in this report, a jet velocity of 160ms^{-1} was used with a nozzle diameter of 1mm. The jet length was determined by the width of the slots in the rotating disk and the angular velocity. The latter was set to a constant value of 600rpm and the used slot width theoretically resulted in a jet length of 47mm. The used disk had 20 slots and therefore the impact frequency was 200Hz. For ABS, this frequency did not result in heating of the specimen and the resulting impact waves of two consecutive impacts did not interact with each other as proofed by piezoelectric sensors.

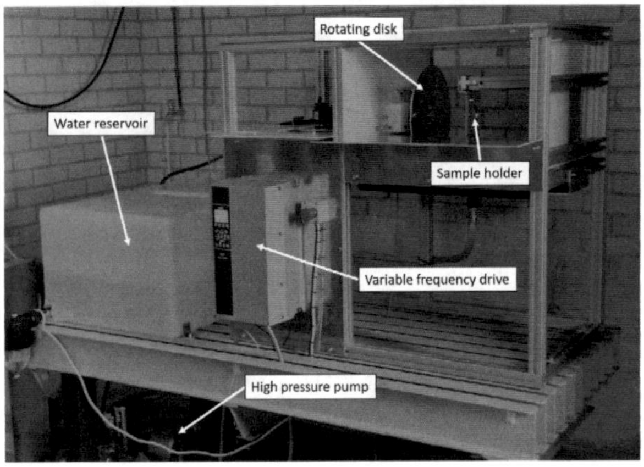

Parameter	Value
Nozzle diameter	0.6 – 1.6mm
Jet velocity	0 – 190m/s
Impact frequency	$\leq$ 200Hz
Impact angle	90 – 15deg

Figure 2. Overview of the pulsating jet erosion tester at the University of Twente.

The material used was a 4mm thick ABS plate which was clamped between two aluminum plates with a round hole of 55mm diameter. The impacts occurred in the center of this plate. The spread of the droplet impact locations was found to be smaller than the droplet diameter confirmed by the damage pattern and location sensing using piezo electric sensors. Several tests were performed with increasing numbers of impacts. Confocal microscopy was performed after testing to visualize the damaged area and determine the volume loss and roughness parameters.

A Keyence VK-9700 laser scanning confocal microscope with a 10× lens was used to obtain the 3D height data of the surface. A single image was used to determine the roughness parameters. A stitched image was made to obtain an overview of the damaged area. From this overview image, it was possible to determine the volume loss. To gain more detailed information of the damage mechanisms, lenses with higher zooms (50×, 100× and 150×) could be used. These lenses however did not give an image with a representative surface and could therefore not be used for roughness studies. The measured data was stored as a height profile and post-processed by using MATLAB.

To determine the volume loss, the height profile was extracted from the microscope as shown in Figure 3 (a). A reference surface was fitted with respect to the undamaged material at the sides of the image (red regions). The total deviation of the height profile with respect to this reference surface was taken as the volume loss in the distance to mean method. This method was improved by reducing the noise using the standard deviation (STD) of the surface roughness. A second plane was defined as the reference plane lowered by the standard deviation. Only the points beneath this plane were considered for the volume loss with respect to the initial reference plane. This improved method considered only the dark regions in Figure 3(a) and reduced the influence of noisy measurements. It could also be seen that the considered region for the volume loss corresponded well with the microscopic image in Figure 3 (b). It was therefore expected that the STD corrected method resulted in a higher accuracy compared with the distance to mean method.

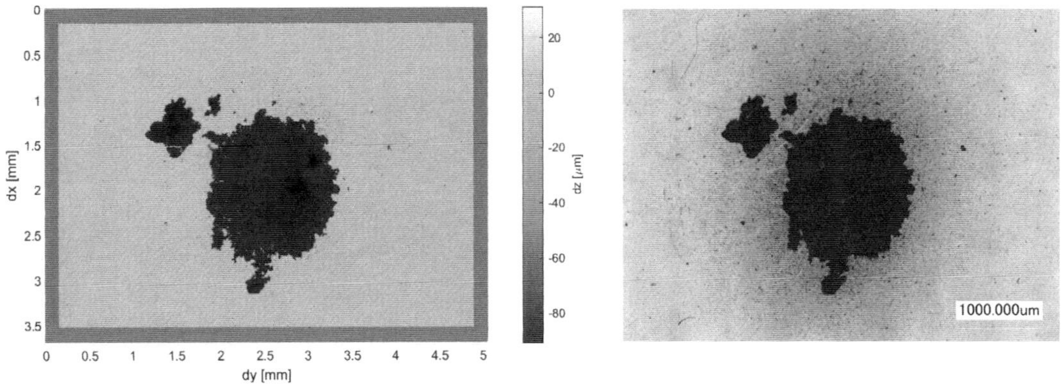

a - Volume loss algorithm b - Confocal microscope image

Figure 3. The volume loss algorithm (a) and an overview image of the damage after 500.000 impacts (b).

4. Results

The results from the coupled SPH/FEM model are shown in Figure 4 for two different time instances. It can be seen that the regions with the highest stresses were very small and remained superficial. These locations were expected to also be the locations where damage initiated during testing. The high stresses were observed only during the initial stages of the rain droplet impact, which emphasized the importance of the compressive effects in the fluid. After 0.9μs lateral jetting occurred and the compressive pressure in the droplet was released which led to reduced contact pressures.

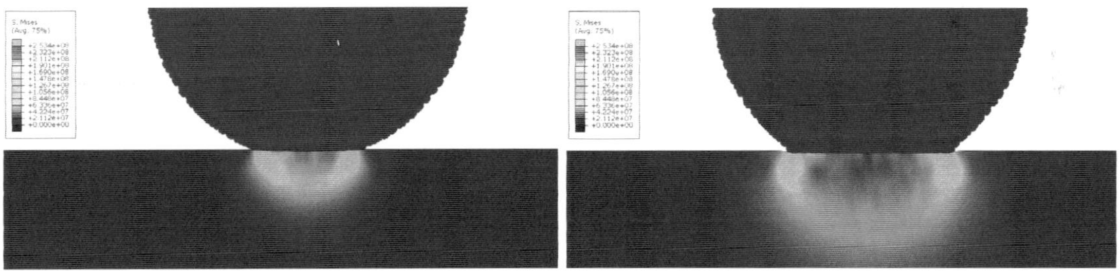

a - 0.5μs b - 0.9μs (onset of lateral jetting)

Figure 4. SPH/FEM results for two different time steps.

The surfaces of the eroded ABS specimens for different amounts of impacts are given in Figure 5. It can be seen that the damage intensity increased with a higher number of impacts. It started with the creation of small pits at the surface that were connected by cracks at 200.000 impacts. Material loss occurred in the form of cratering after 300.000 impacts. According to the analyzed surface roughness height, the depth of these craters was less than 0.1mm, which was found to be low as compared to the material thickness of 4mm.

Figure 6 shows a more detailed view of the pitting, cracking and cratering mechanisms. It can be seen that with increasing pitting densities, the pits started to connect by the cracking mechanism. This appeared between 150.000 and 200.000 impacts. Later, these cracks connected and this resulted in craters which led to the material loss. As seen in Figure 6(c), the cracks

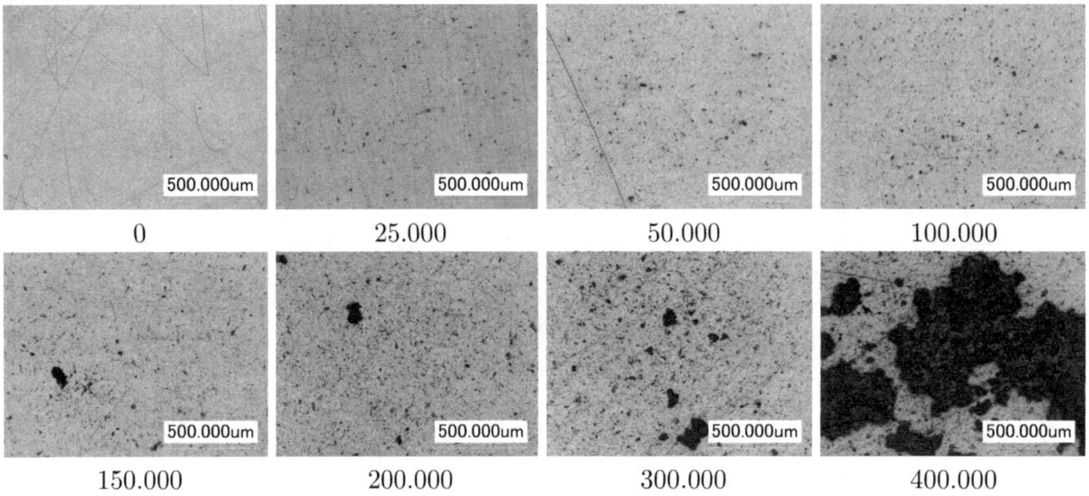

Figure 5. Confocal microscopy (10× lens) images of eroded ABS specimens with increasing numbers of impacts.

extended from the craters and more material was released which expanded the crater size or connected multiple craters together.

Figure 6. Detailed confocal microscopy images of the damage (50× lens).

Figure 7 shows the roughness values and the volume loss for increasing numbers of impacts for the ABS specimen. It can be seen from both the roughness and the volume loss curves that a transition region was found to be present between 300.000 and 400.000 impacts. Other observations were the increase of the skewness value in the early stages and the small differences between the volume loss calculated by using the distance to the mean and the standard deviation corrected methods.

5. Discussion

As seen from the model results, the highest stresses were very local and remained superficial. When the damage patterns were observed, it was found that damage initiated locally and at the surface. This indicated a good correlation between the model and the experiments. The depth of the craters was around 80μm, indicating that damage did not penetrate far into the material and a thin layer could be studied for LEP applications. It should however be noted that in this case, interfaces between the LEP and the blade structure start to play an important role. The

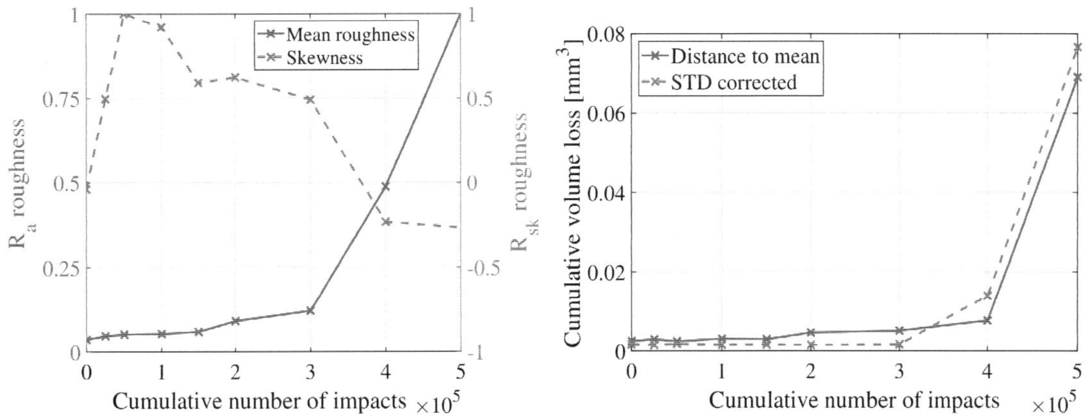

a - Normalized roughness values b - Volume loss

Figure 7. Roughness values and volume loss of eroded ABS specimens for increasing numbers of impacts.

combined method of modeling and experiments could be used to study multiple materials and predict their erosion behavior as a function of not only material parameters, but also impact and geometric parameters. Incorporation of damage in the model is very challenging due to the high strain rates for which material parameters are generally not known. To solve this, it is recommended to test promising materials at high strain rates or to fit a damage model with rain erosion experiments (eg. elasto-viscoplasticity).

As seen from the volume loss graph, a higher number of impacts resulted in more volume loss. For lower numbers of impacts, surface pitting and cracking was observed indicating that damage was accumulated during the incubation stage but no volume loss was measured. Traditionally, the end of incubation point (start of volume loss) is considered when lifetime predictions are made, however damage might occur earlier as seen from these results. In order to develop new materials for LEP systems, it is important to understand what happens during the incubation stage. In order to quantify this, the surface roughness parameters were analyzed.

It was seen that the skewness showed a high dependency on the pitting intensity. A positive skewness value indicated that a small amount of pits was present, a negative skewness indicated that a lot of pits or craters were present and a zero skewness indicated that there was as much high as low points or the surface was undamaged. Because of these characteristics, the skewness could be used to describe the surface during incubation. After incubation, the mean roughness seemed to increase significantly following the same trend as the volume loss which corresponded to literature for other materials [20]. This showed that using the skewness and arithmetic mean roughness values, the damage progression during incubation as well as the volume loss after incubation could be described.

The two methods that were used to determine the volume loss showed a minor deviation, this has to do with a small amount of noise on the measurements. By looking at the surface roughness development and the initiation of craters, it could be seen that it was likely that volume loss occurred after 300.000 impacts. This indicated that the STD corrected method was slightly more accurate than the distance to mean method.

6. Conclusions

ABS specimens were eroded by consecutive water droplet impacts using a pulsating jet erosion tester. The resulting samples were analyzed using a Keyence VK9700 laser confocal microscope.

A novel method for the determination of volume loss using the height profile was developed and successfully used to identify the end of incubation point. This method can be used to compare jet based erosion tests to whirling arm based erosion tests in terms of mass loss and volume loss.

It was seen that initial damage for this particular material occurred during incubation in the form of pitting at the surface. It was shown that this could be analyzed using the skewness parameter. The damage initiation at the surface was also in agreement with the model, that showed high local superficial stresses. The pitting density increased with prevailing impacts to the point at which they were connected by cracks. After this, loss of material occurred in the form of cratering. The observed material loss could be measured by using the arithmetic mean surface roughness parameter as well as the developed volume loss algorithms. It is of vital importance to understand the combination of these damage mechanisms and how they develop in order to study rain erosion resistance of materials and develop sophisticated leading edge protection systems.

Future work includes testing of ABS at lower velocities to verify the damage mechanisms and establish the Wöhler curve for comparison to other LEP materials as well as development of a sophisticated damage model using high strain rate material parameters to predict long term erosion performance.

Acknowledgments
This project is financed by TKI-Wind op Zee Topsector Energy subsidy from the Ministry of Economic Affairs of the Netherlands with the reference number TEWZ118008. The authors would like to thank Udaya Aditya Ghantasala for his support during the measurements and the microscopic analyses of the presented results.

References
[1] Komusanac I, Brindley G and Fraile D 2020 *Wind Europe*
[2] Ramírez L, Fraile D and Brindley G 2020 *Wind Europe*
[3] Keegan M H, Nash D H and Stack M M 2013 *Journal of Physics D: Applied Physics* **46** 383001
[4] Herring R, Dyer K, Martin F and Ward C 2019 *Renewable and Sustainable Energy Reviews* **115** 109382
[5] Cortés E, Sánchez F, O'Carroll A, Madramany B, Hardiman M and Young T M 2017 *Materials* **10** 1146
[6] Slot H M, IJzerman R M, le Feber M, Nord-Varhaug K and van der Heide E 2018 *Wear* **414-415** 234–242
[7] Mishnaevsky L 2019 *Wind Energy* **22** 1636–1653
[8] Springer G S, Yang C I and Larsen P S 1974 *Journal of Composite Materials* **8** 229–252
[9] Verma A S, Castro S G P, Jiang Z and Teuwen J J E 2020 *Composite Structures* **241** 112096
[10] Doagou-Rad S and Mishnaevsky L 2019 *Meccanica* **55** 725–743
[11] Mishnaevsky L, Fæster S, Mikkelsen L P, Kusano Y and Bech J I 2020 *Wind Energy* **23** 547–562
[12] Slot H M, Gelinck E R, Rentrop C and van der Heide E 2015 *Renewable Energy* **80** 837–848
[13] Eisenberg D, Laustsen S and Stege J 2018 *Wind Energy* **21** 942–951
[14] Bartolomé L and Teuwen J 2019 *Wind Energy* **22** 140–151
[15] Ibrahim M E and Medraj M 2020 *Materials* **13** 157
[16] Zhang S, Dam-Johansen K, Nørkjær S, Bernad P L and Kiil S 2015 *Progress in Organic Coatings* **78** 103–115
[17] Tobin E F, Young T M, Raps D and Rohr O 2011 *Wear* **271** 2625–2631
[18] Tobin E F, Rohr O, Raps D, Willemse W, Norman P and Young T M 2015 *Wear* **328-329** 318–328
[19] Fraisse A, Bech J I, Borum K K, Fedorov V, Frost-Jensen Johansen N, McGugan M, Mishnaevsky L and Kusano Y 2018 *Renewable Energy* **126** 1102–1112
[20] Tobin E F and Young T M 2017 *Materials Performance and Characterization* **6** MPC20160085

IOP Conf. Series: Materials Science and Engineering **942** (2020) 012024 doi:10.1088/1757-899X/942/1/012024

Combatting rain erosion of offshore wind turbine blades by co-bonded thermoplastic-thermoset hybrid composites

Jamal Seyyed Monfared Zanjani*, Ismet Baran and Remko Akkerman

Faculty of Engineering Technology, University of Twente, 7500AE Enschede, The Netherlands

*j.seyyedmonfaredzanjani@utwente.nl

Abstract. An integrated thermoplastic-thermoset hybrid leading edge protection system is developed based on the co-bonding process. Co-bonding is a joining method in which a prefabricated part joints with a thermoset composite during the curing process. In such a multi-material hybrid design, the reliability of the bonding between the prefabricated protection layer and the main body of the blade is of crucial importance to prevent any delamination failures. Nevertheless, the adhesion of prefabricated thermoplastics to the thermoset remains a challenge as the interphase between two dissimilar materials is prone to form defects and irregularities. Such interface defects may lead to early failure and reduced structural integrity of the components. Therefore, the focus of this study is on achieving a strong, and reliable bonding between the prefabricated thermoplastic leading edge protection system and thermoset main body of the blade. In this study, the effect of processing temperature on the interphase quality and thickness during the co-bonding process is investigated. Next, mechanical characteristics and microstructure of the interphases are examined by Vickers' microhardness tests. The effect of processing condition on the fracture toughness of structure is examined by climbing drum peel tests (CDP). Finally, fractography investigations are used to provide an understanding of failure mechanisms and its correlations with interphase morphology and microstructure.

1. Introduction

The latest manufactured wind turbine blades have reached a length of over a hundred meters where the tip speed can be higher than 100 m/s. The high tip speed of blades and extremely harsh environment of offshore sites present a new challenge in terms of leading-edge erosion by the impact of objects such as raindrops. Rain erosion damages at the leading-edge of wind blades significantly reduce the aerodynamic efficiency and consequently jeopardize the overall integrity of the blades and also decrease the energy production efficiency. The reduction in energy production imperils the price competitiveness of the wind-energy to non-sustainable energy sources. Therefore, a more reliable and durable protection layer that overcomes the rain erosion in wind blades is urgently needed to increase the uptime and productivity of off-shore wind farms and eliminate the costly blade repairs.

Engineering thermoplastics have shown to be more damage tolerant than their thermoset counterparts in many applications for the harsh environment [1]. Therefore, the main objective of this study is to integrate a thermoplastic protection layer as a durable leading edge protection system in the leading edge of the wind turbine blades as schematically presented in Figure 1. Nevertheless, the joining of the thermoplastics leading edge protection to the thermoset of the main body of the blade is a challenging task due to their chemical, physical and mechanical dissimilarities. In such a multi-material hybrid design, the reliability of the bonding between the protection layer and the main body of the blade is of crucial importance to prevent any delamination failures as shown in Figure 1. Such defected interphase may lead to early failure and reduced structural integrity of the components. Therefore, the focus of this study is on achieving a strong, defect-free, and reliable joining between thermoset and thermoplastic materials.

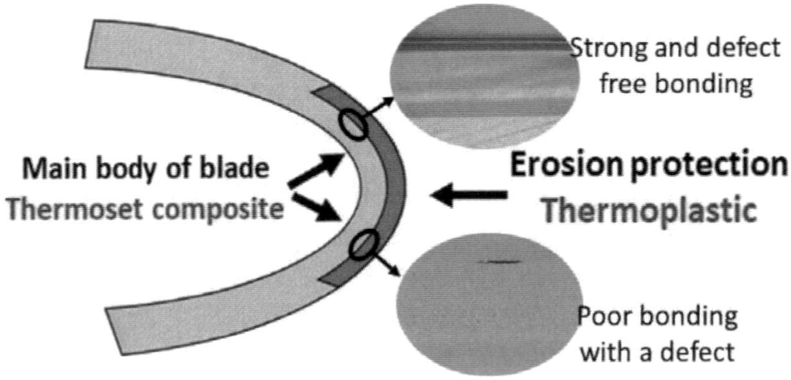

Figure 1. Integrated leading-edge protection by thermoplastic with defected and defect-free bonding.

Various methods have been developed for the joining of polymeric composites including mechanical fastening, welding, adhesive bonding, co-curing, and hybrid joints. Nevertheless, all these joining methods have disadvantages in terms of being unreliable, labor-intensive, time-consuming and, expensive [2]. An alternative and effective joining method for composites is co-bonding where an uncured part joins with one or more cured parts through a single step process [3, 4]. The co-bonding method can be applied for jointing both similar or dissimilar materials including thermoplastics to thermosets. Co-bonding eliminates the needs for mechanical fasteners, and adhesives resulting in a more effective joining for many applications.

It is known that the interface between two parts in the co-bonding process is controlled by their thermodynamic affinity and physical interactions in between and also the curing process of the reactive resin [5]. Nonetheless, the complete understanding of the interdiffusion of a reactive resin into the thermoplastic structure during the co-bonding process remains largely uncovered due to the several challenges and complexities concerning the changes in the properties of the thermoset resin, and unclear diffusion mechanisms.

In this study, the effect of processing temperature on the interphase quality formed between acrylonitrile butadiene styrene (ABS) as the thermoplastic leading edge protection system and unsaturated polyester resin (UPR) as the thermoset during the co-bonding process is investigated. Initially, the effect of three different manufacturing temperatures of 25 °C, 35 °C and 50°C on the interphase thickness are studied. Interphase formation mechanisms are identified and correlated to interphase thickness through the curing kinetics, and rheological changes of the resin. Next, mechanical characteristics and microstructure of the interphases are examined by Vickers' microhardness tests. The effect of processing condition on the bonding quality and the fracture toughness of structure is examined by climbing drum peel tests (CDP). Finally, the fracture surface analyses are performed and insight on various crack propagation mechanisms contributing to bonding quality are recognized.

2. Experimental and Methods

2.1. Materials

An industrial medium reactive orthophthalic unsaturated polyester resin (UPR) system designed for resin transfer molding and vacuum assisted resin transfer molding processes was used as the thermoset resin. A room temperature activated liquid peroxide system specialized for the unsaturated polyester with long working time (gel time) was used as an initiator. A glass fabric with a 750 g/m^2 unidirectional roving reinforced (660 g/m^2) with random filament and stitching (90 g/m^2) on the back was used as the reinforcement. Acrylonitrile butadiene styrene (ABS) in the sheet form was used as thermoplastic material in this study with the grade name of VIKUREEN ABS PLAAT GLANS WIT 0291 known to have a strong resistance to corrosive chemicals and/or physical impacts.

2.2. Composite manufacturing and sample preparation

The effect of processing temperature on the interphase thickness was studied by vertically placing ABS specimens with dimensions 15 mm× 15 mm× 3 mm in the middle of a cylindrical embedding mold with a diameter of 25 mm by using a couple of metallic clamping rings. Next, degassed UPR resin mixture was poured onto the ABS specimens until the resin completely covered the specimen. Prepared specimens and molds were immediately placed in a preheated oven fixed at the set temperatures of 25°C, 35 °C, and 50 °C for 24 hrs followed by a post-curing at 60 °C for 24 hrs.

Next to the infusion of the pure UPR onto the ABS, three different hybrid sandwich composites were manufactured to analyze the effect of process temperature on the fracture toughness by employing a Climbing Drum Peel (CDP) test. Figure 2 represents the stacking sequence of the composites where initially a layer of UD glass fabric was placed on the mold surface followed by placing an ABS sheet with a thickness of 3 mm and another fabric layer on top. Then the stack was impregnated with degassed UPR through vacuum-assisted resin transfer molding. The curing was performed at three different temperatures of 25°C, 35 °C, and 50 °C for 24 hrs followed by a post-curing at 60 °C for 24 hrs.

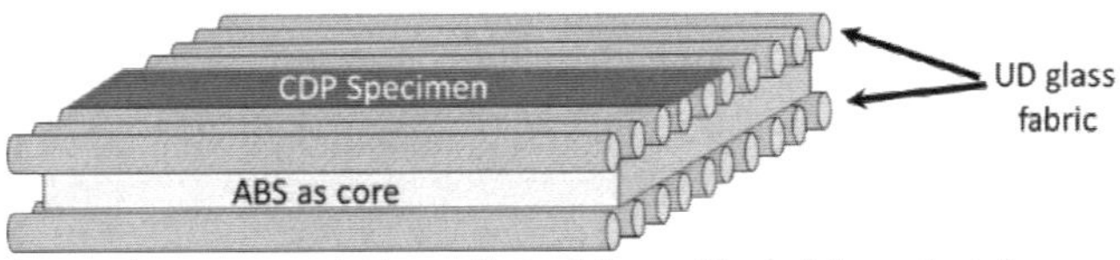

Figure 2. Schematic representation of stacking of hybrid sandwich composite plates.

2.3. Characterizations

The polished cross-section of interphases formed between the co-bonded ABS and UPR systems at various processing conditions were analyzed by using a Keyence VHX-5000 digital microscope equipped with a VH-100UR lens. The interphase thicknesses were measured for each temperature at mid-section of interface far from the clamps and obtained from at least three specimens.

The viscosity and gel time of the UPR were measured using the Anton Paar-Physica MCR 501 rheometer in "plate–plate" mode. All the measurements were performed by using circular aluminum plates of 25 mm diameter in oscillatory mode at a 0.5% strain and a 1 Hz with a plate-plate spacing of 0.3 mm. The viscosity of resin at different temperatures were measured for UPR resin without mixing with the initiator to eliminate the cure effect on viscosity. The temperature-dependent gel time of the resin, as an indication of the cure kinetics of UPR resin, was estimated from the crossing point of storage modulus (G') and loss modulus (G") of UPR mixture including the liquid peroxide [6].

The Vickers microhardness test was used to probe the mechanical response at the UPR-ABS interphase. Vickers microhardness mapping was conducted using the LECO LM100AT micro-hardness tester by applying a load of 10 gr and an indent spacing of 100 μm. The Vickers diamond pyramid hardness number, HV, was defined as the ratio of the applied load, P, to the pyramidal contact area of the indentation, $H_v=\alpha\ P/d^2$,where d was the diagonal length of the resultant impression, and α was defined as 1.8544 for Vickers indenter [7].

Climbing Drum Peel (CDP) test based on ASTM Standard D1781was employed to evaluate the quality of bonded joints between the UPR skin (laminate) and ABS core in the sandwich hybrid composites. The test was performed using a 1 kN load cell at a displacement rate of 25 mm/min. The specimens were cut as 300 mm in length, 10 mm in width as shown in Figure 2. In this test, the critical strain energy release rates were calculated by using Eq 1 as:

$$G_c = \frac{F_d - F_w}{w}\frac{r_2 - r_1}{r_1} \tag{1}$$

where F_w was the force required to overcome the drum weight and to wind the peel arm around the drum, F_d was the force contribution related to the propagation of debonding, w was the width of the specimen and r_1 and r_2 were the radii of the drum and the flanges as 100 mm and 130 mm, respectively. Further details of CDP can be found in [8]. Afterward, fracture analyses of the fracture surfaces after the CDP test were performed using a Keyence VHX-5000 digital microscope.

3. Results and discussion

3.1. Effect of processing temperature on the interphase

Figure 3a reveals the morphology of ABS-UPR interphases fabricated at 25 °C and Figure 3b exhibits the average interphase thickness formed at different temperatures of 25 °C, 35 °C and 50 °C. It is seen that thickness of ABS-UPR interphase prepared at 25 °C was about 710±20 μm while an increase in temperature to 35 °C decreased the interphase thickness to 635±10 μm. This reduction in the interphase thickness by an increase in the temperature from 25 °C to 35 °C was correlated with the acceleration in the cure kinetics of the resin and consequently limited time available for diffusion and interphase formation before solidification of the thermoset resin. To elaborate it is seen in Table 1 that by an increase in temperature from 25 °C to 35 °C, the gel time of the UPR considerably decreased from 12600 S to 3960 S associated with 8640 S shorter gel time. It is worth noting that gel time is an indication of cure kinetics of the resin and it defines the time in which resin switches from a viscous liquid to a semi-solid gel. Herein, gel time was assumed as the time where the diffusion process stopped. Therefore, such a reduction in gel time restricted the diffusion time and resulted in a shorter interphase region. On the other hand, the increase in temperature let to a drop in viscosity of the resin from 1.51 Pa.S at 25 °C to 0.78 Pa.S at 35 °C which favors the increase in the interphase thickness. However, the observed decrease in the interphase thickness showed that the reduction of gel time is more effective than the reduction of resin viscosity on interphase thickness by increasing the temperature from 25 °C to 35 °C. Nonetheless, a further increase in temperature form 35 °C to 50 °C resulted in an increase of interphase thickness to 665±10 μm. This increase in the interphase is associated with the significant reduction of resin viscosity at elevated temperature which reached to 0.35 Pa.S which enhanced the diffusion process and increased the interphase thickness while gelation time only decreased by 3120 S. Thus, the extent of changes in both viscosity and also gelation time are determinant factors of the interphase thickness.

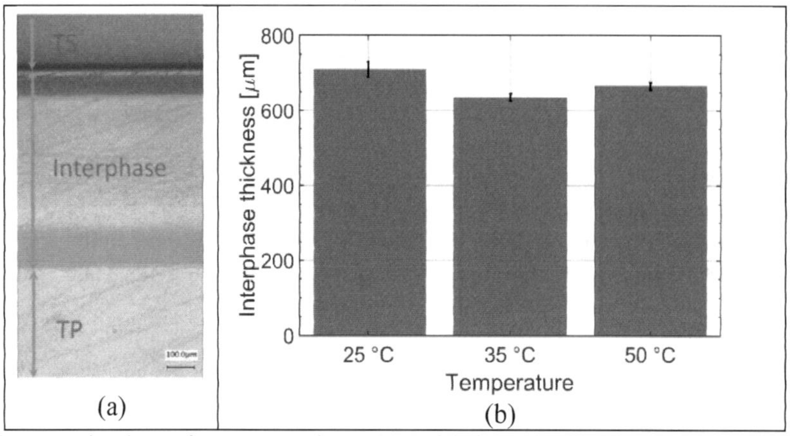

Figure 3. Microscopic view of ABS-UPR interphase fabricated at 25 °C and thickness variation by temperature.

Table 1. Gel time and viscosity of UPR at different temperature.

	25 °C	35 °C	50 °C
Gel time (S)	12600	3960	840
Viscosity (Pa.S)	1.51	0.78	0.35

3.2. Interphases mechanical response by microhardness

Figure 4a shows the variations of Vickers microhardness in the vicinity of the interphase prepared at 25 °C. It was observed that the hardness at the interphase was lower than both cured UPR and ABS. The softer interphase uncovered the plasticizer effect of unreacted UPR diffused into ABS. It also revealed that a significant amount of diffused resin into ABS remained unreacted due to the limited access to reaction components at the interphase area. Figure 4b discloses the average Vickers hardness value at interphase region for specimens fabricated at different temperatures. It was noticed that the average Vickers hardness value at interphase was correlated with the processes temperature and an increasing temperature resulted in an increase in interphase hardness. To elaborate, the average Vickers hardness at the interphase for specimens fabricated at 25 °C measured to be 6.39±0.90 while the increase of processing temperature to 35 °C and 50 °C resulted in hardness at the interphase of 7.28±0.82 and 7.93±0.94, respectively. This increasing trend in hardness value at interphase for specimens by increasing the manufacturing temperatures is associated with the concentration of resin at the interphase area which was decreased by an increase in temperature due to the acceleration in resin cure kinetics and diffusion kinetics.

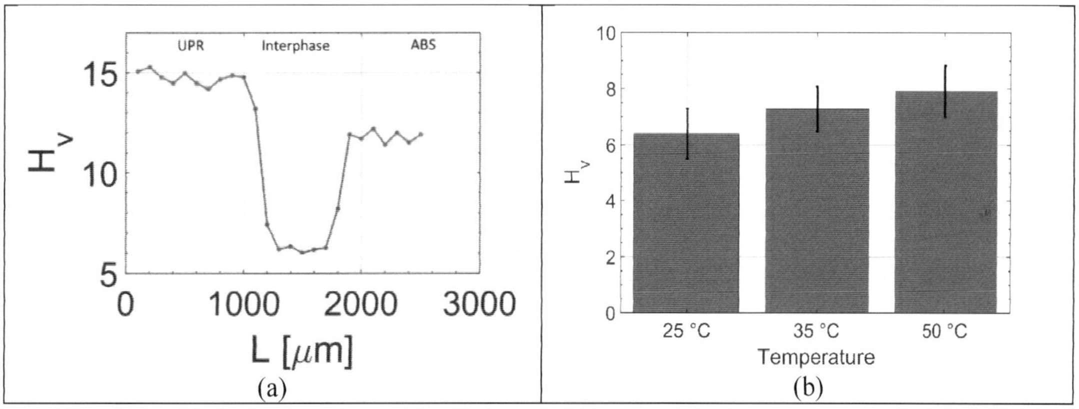

Figure 4. (a) Vickers microhardness for UPR-ABS prepare at 25 °C, and (b) variation of interphase hardness by manufacturing temperature.

3.3. Fracture toughness by climbing drum test

Figure 5a displays the typical CDP test curve obtained for a specimen prepared at 25 °C. The F_w is measured to be 160 N and the average force for crack propagation was measured in the crack propagation area as shown in Figure 5a. Figure 5b exhibits the fracture toughness measured for specimens fabricated at three different temperatures of 25 °C, 35 °C and 50 °C. It was observed that the highest fracture toughness was measured for specimens manufactured at 25 °C as 3.45±0.3 N/mm. However, an increase in manufacturing temperature to 35 °C reduced the fracture toughness to 1.44 ±0.2 N/mm. A further increase in processing temperature to 50 °C slightly increased the fracture toughness to 1.8 ±0.2 N/mm.

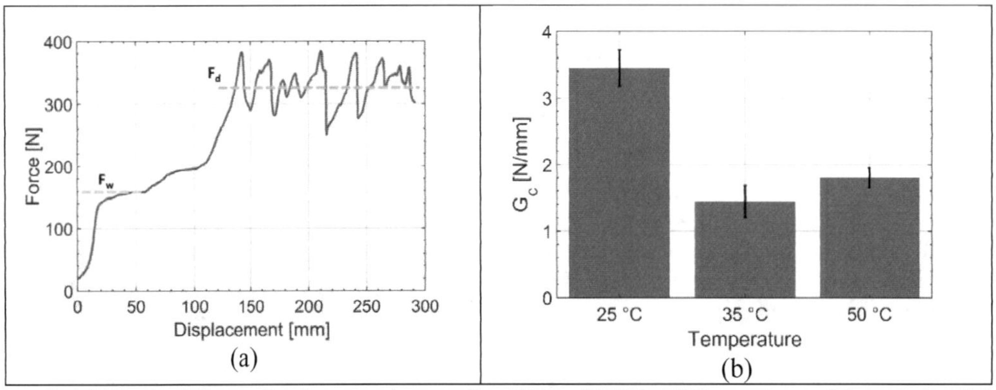

Figure 5. (a) Typical force-displacement curve for CDP of UPR-ABS specimen fabricated at 25 °C, and (b) fracture toughness for specimens prepared at different temperatures.

3.4. *Fractography analysis*

Figure 6. depicts the fracture surface of glass fabric reinforced thermoset skin after CDP tests. It is seen that two different mechanisms were contributing to crack propagations namely interface failure and cohesive failure. In the interface failure, skin and core were debonded at their interface without any sign of their attachment. On the other hand, in the cohesive failure area residues of ABS were remained on the skin surface. Since the cohesive strength of ABS is much higher than the UPR-ABS interfacial strength, cohesive failure enhanced the fracture toughness measured by CDP tests. Figure 6 reveals that the fracture surface of the specimen manufactured at 25 °C contained a high level of cohesive failure which was complied with its comparative higher fracture toughness as observed in Figure 5b compare to other specimens. Furthermore, it was observed in Figure 3b that the specimen manufactured at 25 °C formed the thickest interphase and therefore higher level of interdiffusions which reflected in higher cohesive failure and consequently higher fracture toughness for these specimens. On the other hand, the fracture surface of specimen manufactured at 35 °C, showed a very low amount of cohesive failure which is also coordinated with its lowest fracture toughness and the thinnest interphase thickness compared to other specimens. Alternatively, specimen prepared at 50 °C, showed a moderate level of cohesive failure compared to other specimens in which the number of cohesive failures was alike to specimen fabricated at 25 °C while their size was much smaller which was in correlation to its intermediate interphase thickness and fracture toughness values.

Figure 6. Fracture surface on thermoset skins after CDP test.

4. Conclusions

An integrated leading edge protection system based on co-bonding of an engineering thermoplastics to the thermoset main body of blades is proposed for rain erosion inhibition. In the proposed design, the effect of processing conditions on the bonding quality of ABS as the thermoplastic protection to polyester composite as the main body of the blade were studied during the co-bonding process. The bonding quality was correlated with the cure kinetics and interphase thickness of the resin. It was observed that the interphase size changed with the processing temperature and it was linked to the thermal changes in the viscosity and cure kinetics of UPR. Microhardness tests showed that the hardness of interphase was lower than both ABS and UPR regions due to the plasticization effect of UPR molecules at the interphase. An increasing trend in hardness value at interphase by increasing the manufacturing temperatures were observed. CDP tests were performed and maximum fracture toughness was obtained for specimens fabricated at 25 °C while specimens manufactured at 35 °C present the minimum fracture toughness. Fracture surface analyses revealed that two different mechanisms were contributing to crack propagations namely interface failure and cohesive failure. Consequently, it was concluded that fracture toughness of specimens was closely related to the level of cohesive failure and specimen with thicker interphase resulted in higher cohesive failure and fracture toughness whereas thinner interphase led to lower cohesive failure and consequently lower fracture toughness.

Acknowledgment

Authors gratefully acknowledge the research program Talent Scheme (Veni) with project number 15897, which is (partly) financed by the Netherlands Organization for Scientific Research (NWO) and TKI-Wind op Zee Topsector Energy subsidy from the Ministry of Economic Affairs, the Netherlands with the reference number TEWZ118008. Authors are also thankful to LM Wind Power for consultations during the project and supply polyester resin.

References

[1] Seyyed Monfared Zanjani J, Saner Okan B and Menceloglu Y 2016 Manufacturing of multilayer graphene oxide/poly(ethylene terephthalate) nanocomposites with tunable crystallinity, chain orientations and thermal transitions, *Mater. Chem. Phys.* **176**, 58-67

[2] Karakaya N, Papila M and Özkoç G 2020 Overmolded hybrid composites of polyamide-6 on continuous carbon and glass fiber/epoxy composites: 'An assessment of the interface' *Compos. Part A Appl. Sci. Manuf.* **131**, 105771

[3] Baran I 2017 Warpage prediction in over-infusion process of glass/polyester composite laminates. *21st International Conference on Composite Materials,* 2017,

[4] Leone C and Genna S 2018 Effects of surface laser treatment on direct co-bonding strength of CFRP laminates, *Compos. Struct.* **194**, 240-51

[5] Stavrov D and Bersee H E N 2005 Resistance welding of thermoplastic composites-an overview *Compos. Part A Appl. Sci. Manuf.* **36**, 39-54

[6] Hsu C and Lee L J 1993 Free-radical crosslinking copolymerization of styrene/unsaturated polyester resins: 3. Kinetics-gelation mechanism *Polymer,* **34,** 4516-23

[7] Gong J, Wu J and Guan Z 1999 Examination of the indentation size effect in low-load vickers hardness testing of ceramics. *Eur. Ceram.* **19**, 2625-31

[8] Daghia F and Cluzel C 2015 The Climbing Drum Peel Test: An alternative to the Double Cantilever Beam for the determination of fracture toughness of monolithic laminates *Compos. Part A Appl. Sci. Manuf.* **78**, 70-83

IOP Conf. Series: Materials Science and Engineering **942** (2020) 012025 doi:10.1088/1757-899X/942/1/012025

Mechanisms and computational analysis of leading edge erosion of wind turbine blades

Leon Mishnaevsky Jr.[1], **Søren Fæster**[1] **and Saeed Doagou Rad**[1]

[1]Department of Wind Energy, Technical University of Denmark, Roskilde, Denmark

lemi@dtu.dk

Abstract. Microstructural characterization and computational simulations are combined to study the micromechanisms of leading edge erosion of two different coating systems. Scanning electron microscopy (SEM) and X-ray tomography investigations were performed and micromechanical models of the two coatings were developed which take their micro- and nanoscale structures into account. The computational unit cell models are compared to the microscopy studies and both show that the heterogeneities, particles and voids in the protective coatings have critical effect on the crack initiation in the coatings under multiple liquid impact.

1. Introduction

Leading edge erosion of wind turbine blades is one of the most common reasons of reduction of power generation by wind turbines [1]. Leading edge erosion is responsible for more than 5% reduction of annual energy production for a utility-scale wind turbines [1], [2]. The erosion of wind turbines leads also to an increase in drag coefficient by 80-200% [3], and decrease in lift coefficient for higher angles of attack [4]. In this paper, the damage mechanisms of leading edge erosion, methods of computational modelling of the erosion and possibilities of development of new protection coatings are discussed.

2. Damage mechanisms

In order to develop the blade protection against erosion, understanding the microscale erosion mechanisms is necessary. Two different industrially used leading edge coating systems were investigated by electron microscopy and X-ray tomography after a rain erosion test performed on an R&D rain erosion tester [5]. Two typical coatings were investigated, provided by leading companies in this area, one with relatively thin (coating A, thickness of top layer ~100 μm, filler layer 240 μm) and another two times thicker (coating B, thickness of top layer ~200 μm, filler layer 400 μm). The tested specimens were 3-mm thick, flat and U-shaped, consisting of a coating on a glass fibre reinforced polymer laminate manufactured by vacuum infusion. The two coating systems were tested in an R&D rain erosion tester with a drop size of 3.5mm and a max speed of 168.9 m/s until erosion was observed. Figures 1 and 2 are scanning electron microscopy (SEM) images that show typical structures of coating system A and B respectively. The figures show the heterogeneity of the coating and putty layers. Voids, heterogeneities, particles can be seen on images. Large round Si particles and large elongated particles of Al, Si and K were observed. It can be seen that the manufactures have used completely different coating systems, which also lead to completely different damage mechanisms when subjected to a rain erosion test.

Coating system A contain many elongated particles which mainly is oriented parallel to the surface. One can see the cracks along these elongated particles close to the surface, see Figure 3, supposedly due to a difference in stiffness between the elongated particle and the matrix. This type of cracks are not observed in coating system B. Coating system B contain a lot of bubbles both in the top coat and in the putty, see Figure 2 and Figure 3. X-ray tomography was performed on coating system B after testing using a Zeiss Xradia 520 Versa. The scanning was carried out with a polychromatic X-ray beam with energies up to 40 kV originating from a tungsten target. The sample was rotated ±180° while 1601 projections were acquired. Reconstruction was performed using a Feldkamp cone beam reconstruction algorithm. The voxel size in the reconstruction is 2.0 μm which is sufficient to resolve cracks within the material. Figure 3 show two different virtual slices through the reconstruction. Cracks can be observed between the surface and the bubbles in the top coat and the cracks extend from these bubbles further into the putty.

The electron microscopy investigations and X-ray tomography showed that the leading edge erosion is controlled to a large degree by heterogeneities of the coating systems and manufacturing defects. Voids and weak interfaces heterogeneities trigger therefore damage mechanisms (cracks from voids, debonding particles, debonding between layers etc.). So, one of the keys to the improvement of anti-erosion coating lies in improving the coating manufacturing technology and better quality control.

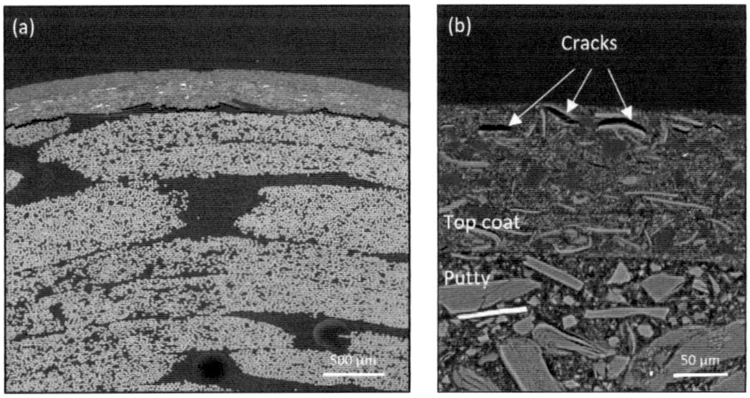

Figure 1. *SEM images of a cross sections through the coating systems A. (a) Overview image and (b) close up of the top coating and the putty after testing showing dark cracks close to elongated white filler particles.*

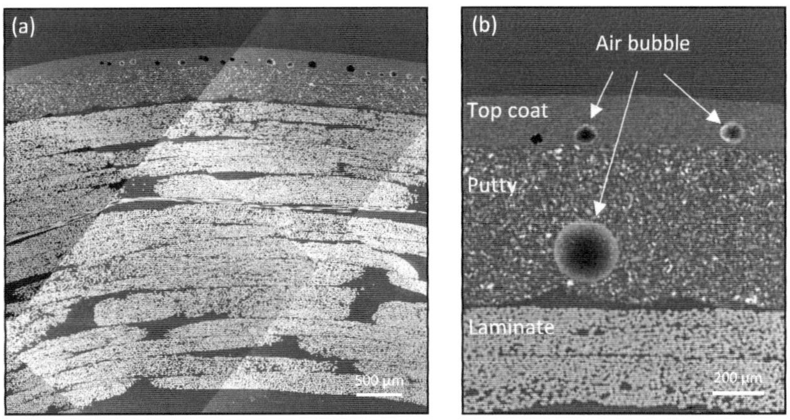

Figure 2. *SEM images of a cross sections through the coating systems B. (a) Overview made as a collage of SEM images (b) Close up of the top coating and the putty showing air bubbles before testing.*

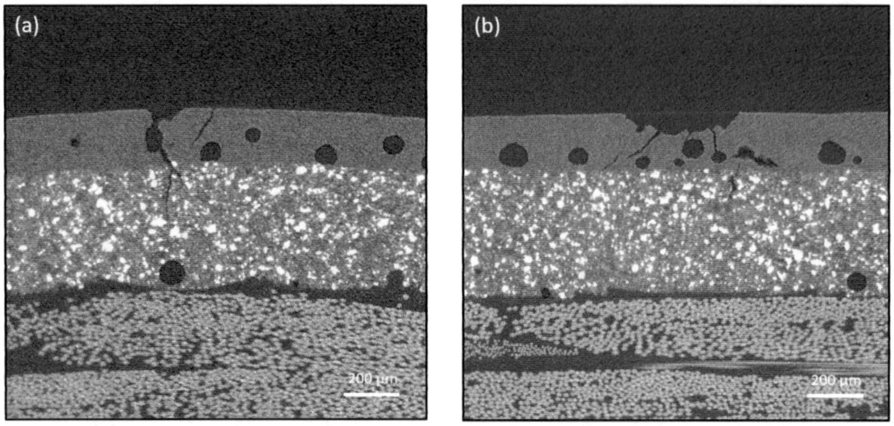

Figure 3. *Virtual slice through a tomographic reconstruction of coating system B after testing showing cracks between the surface and bubbles within the top coat.*

3. Computational modelling: Drop/Surface interaction

In order to develop the required protective coatings against rain erosion, predictive computational models of the material degradation under erosion are required. Main steps of leading edge erosion modelling include the evaluation of loading (rain density, droplet size distribution, dust, flow velocity), modelling of impact contact between droplet/particle and surface (pressure on the surface, time, deformation and damage initiation; wave distribution, Rayleigh waves), modelling of materials degradation over time (coating cracking, debonding, cracks in composite fatigue, material loss, roughening of surface, estimation of lifetime). The schema of the rain erosion modelling is presented on Figure 4.

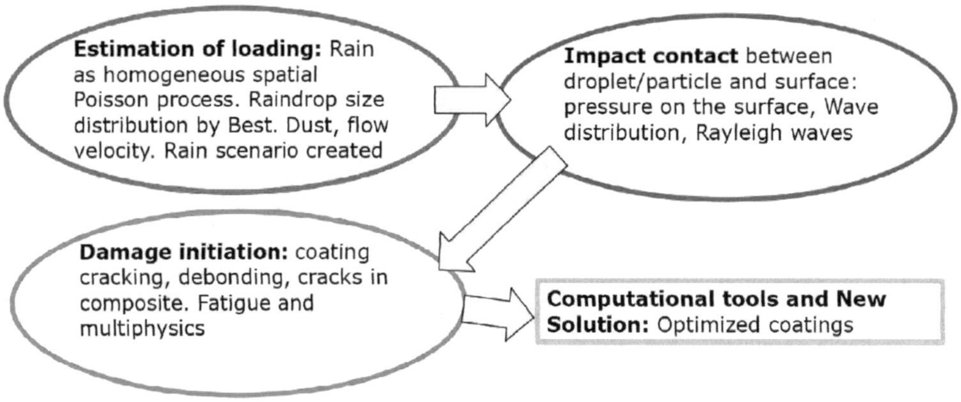

Figure 4. *Schema of the rain erosion modelling.*

Numerical models of surface erosion of wind turbine blades are expected to provide both predictions of performance, lifetime and maintenance requirements of repaired structures, and serve as a tool for the optimization of the protection and repair technologies. A number of quite complex multiscale and multi-physical models have been developed, from good phenomenological formulas (water hammer, damage threshold velocity, see [9]) to complex multistep models [6], including the full model chain from turbulent flow, real rain statistics via the liquid impact [6], up to the fatigue damage studies [7].

A new research project "DURALEDGE - Durable leading edges for high tip speed wind turbine blades" has started at the Technical University of Denmark, to optimize the structure of protective coatings of wind turbine blades on the basis of computational models [10]. Figures 5 -7 show some computational models and results of the numerical modelling of rain droplet impact, obtained in the framework of Duraledge project.

A parametric 3D model of raindrop hitting the coated laminate has been developed in [11]. For the simulation of rain drop fluid behavior, both Smoothed Particle Hydrodynamics (SPH) and Coupled Eulerian Lagrangian (CEL) were used and compared. It was observed that both methods tend to predict similar trends of stress evolution during the full period of contact. The coupled Eulerian-Lagrangian method is more accurate in capturing the details, but more computationally expensive.

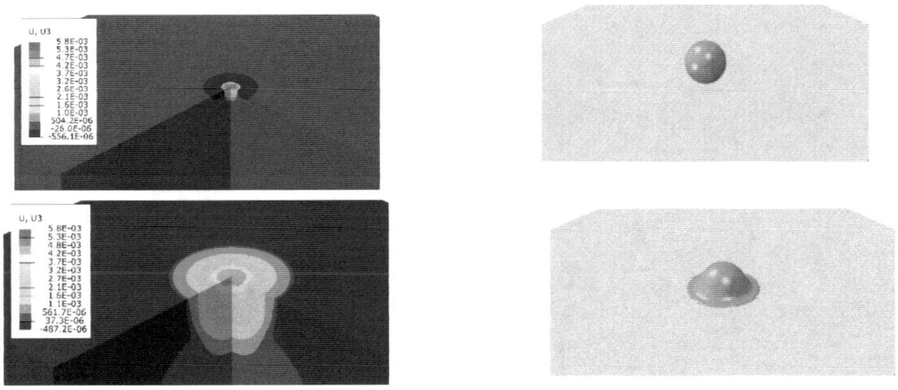

Figure 5. *Simulated stress field under impact and droplet simulated with SPH. (Reprinted from [11] with kind permission from Springer)*

A crucial element in erosion behavior of the coatings protecting wind turbine blades is the surface characteristics of the applied coatings, in particular, roughness. Figure 6 (b) shows the model to investigate the influence of surface roughness on the stress fields within the coating. The surface is dented randomly with half-spheres with the various radii of 5-20 μm. The results showed that the high stress concentration sites around the surface irregularities, that could lead to crack initiation.

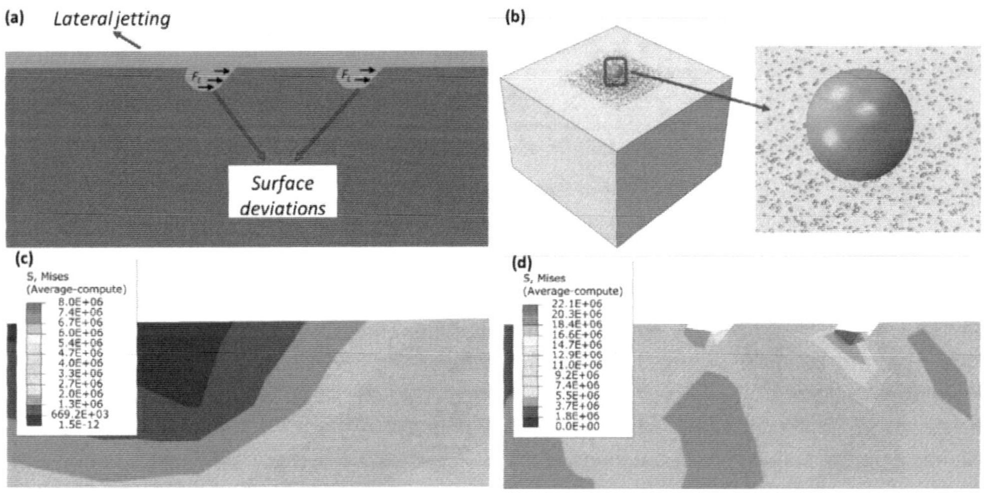

Figure 6. *Finite element rough surface model. Stress fields at (c) smooth versus (d) rough surface. Reprinted from [11] with kind permission from*

Manufacturing defects have also strong influence on the erosion. One of the common manufacturing defects are bubble [5], [12]. In order to investigate the sensitivity of the erosion behavior to the presence of the bubbles, bubbles with different sizes were included within the coatings [11]. The simulation results on these defected coatings showed that even minuscule bubbles could aggravate the coating erosion resistance. While presence of 40 μm bubbles within the coating intensified the initial peak stress, larger sizes of bubbles increased the initial stress and heightened the stress levels afterwards. Even presence of 100 μm bubbles could lead to at least 10 % shortening of the lifetime of the coatings. For the analysis of the fatigue degradation of the coatings, software, based on the critical plane models (I.e,, assuming that fatigue damage accumulates on a specific plane in the material, denoted the "critical plane"), Miner-Palmer fatigue rule and the rainflow counting was developed [13]. The critical plane model allows simulation of multiaxial fatigue. Figure 7 shows the output of this software code [13] in which the load cycles are presented and rainflow histogramm is shown.

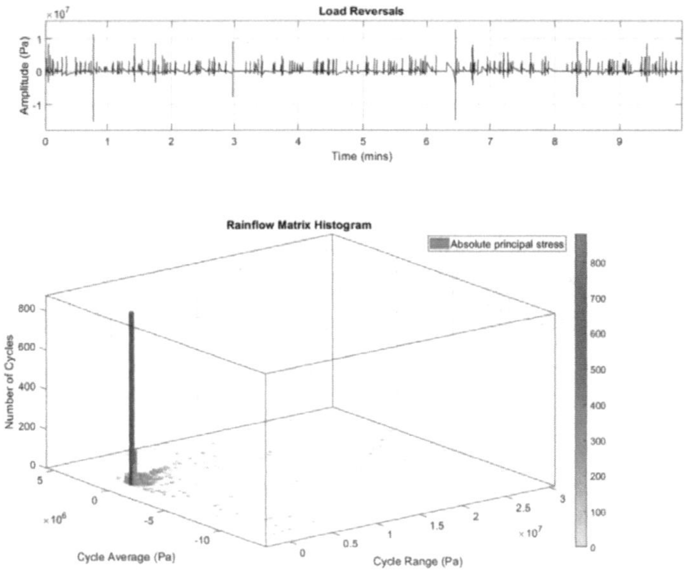

Figure 7. (a) Principal stress cycles, and (b) rainflow histogram.

4. Directions of optimization of coatings

As a first evaluation of the materials potential for erosion protection, various predictors for the erosion resistance of coatings were proposed, an easily tested value. Among them, resistance against abrasive wear: [14], viscoelastic moduli at given frequency [14], fracture energy in tensile tests [15], rebound resilience [16], reduction of storage modulus of coating [17], recovery of material after impact [17], [18] are considered.

Multilayer coatings is a promising option to enhance the protection of laminates, since they allow combining advantages of various materials in different layers, and exploiting interfaces for wave reflection. The multi-layered coatings with alternating stiff and ductile layers found applications in many protective systems. Engel et al [19] demonstrated that multilayer polyurethane/PUR coatings ensure at least better, and more stable erosion protection than two layer PUR coatings, which in turn ensure at least better and more stable protection than single coatings. Cortes and colleagues [20], [21] studied polycarbonatediol-polyurethane based coating (hybrid polyurethane-urea technology) for erosion protection of wind turbines, and reported good performance of these coatings. In-mould coatings were painted or sprayed onto the mould tool. Mohagheghian and colleagues [22] considered the polymer–metal laminates (with thin polyethylene coating). Also, in British project Leading Edge for Turbines (LEFT), the integration of electroformed metallic leading-edge protection solutions (nickel cobalt alloys) into the erosion protection systems is investigated. Herring, Dyer et al. [23] proposed electroformed nickel shields, located in higher speed regions of the wind blades and thermoplastic coatings (which are lighter and cheaper) located at the intermediate, closer to roots areas. Another mechanisms of wave dissipation is the scattering on inclusions. An interesting type of materials is the materials with interpenetrating polymer networks /IPN. For instance, damping properties of polyurethane can be improved by introducing epoxy to form polyurethane/epoxy interpenetrating polymer network [27]. The advantages of IPN as compared with pure polyurethane include better heat resistance, higher strength, and extend the temperature range with sufficiently high damping peaks [28], [29]. In a number of works, particle (microscale and nanoscale) reinforcements in the polymer coatings have been studied. Armada et al. [31] developed epoxy coatings with surface treatedceramic nanoparticles, prepared by sol-gel method. They observed that the erosion resistance clearly increases with increasing the nanoparticles content (both for FunzioNano and SiC). Gou et al. [32] developed carbon nanofiber paper based coatings, with embedded nickel nanostrands. As with other applications of nanoengineering, the positive (or negative) output of nanomodification strongly depends on specific technology: manufacturing and distribution of nanoparticles, functionalization, surface modification, etc. [33], [34]-[37]. For instance, Gohardani et al. [38] explored the effect of CNT (carbon nanotube) reinforcement in resins on the erosion under multiple liquid impact and did not observe sufficient improvement of CNT reinforced resin as compared with pure resin.

5. Conclusions

Micromechanisms of leading edge erosion of coated wind turbine blades are studied with the use of scanning electron microscopy and X-ray tomography and computational micromechanics simulations. Heterogeneities of the coating systems and manufacturing defects such as air bubbles are found to initiate leading edge erosion. In the improvement process of anti-erosion coatings it is therefore vital to improve the coating manufacturing technology and to ensure a better quality control

Acknowledgement: The authors gratefully acknowledge the financial support of the Innovation Foundation of Denmark in the framework of the Grand Solutions project DURALEDGE, Durable leading edges for high tip speed wind turbine blades, File nr.: 8055-00012A.

References

[1] Mishnaevsky Jr. L., 2019 *Renewable Energy*, **140**, 828-839
[2] Mishnaevsky Jr L., Branner K., Petersen H.N., Beauson J., McGugan M. and Sørensen B., 2017 *Materials*, **10**, 1285
[3] Sareen A. et al, 2012, *Wind Engineering*, **36**, 525-534
[4] K. Chamberlain, 2017, *New Energy Updates*
[5] Mishnaevsky Jr. L., Fæster S., Mikkelsen L. P., Kusano Y.,Bech J.I., 2019, *Wind Energy*, 1–16
[6] Amirzadeh B., Louhghalam A., Raessi M., Tootkaboni M., J. 2017, *Wind Engineering and Industrial Aerodynamics*, **163**, 33–43
[7] Slot H.M., Gelinck E.R.M., Rentrop C., van d. Heide E., 2015 *Renewable Energy*, **80**, 837-848
[8] Cortés E.et al, 2017, *AIP Conference Proceedings* 1896, 030023
[9] Eisenberg D., Laustsen S., Stege S., 2018,*Wind Energy*, **21**, 942-951
[10] IFD project DURALEDGE, www.duraledge.dk
[11] Doagou-Rad S., Mishnaevsky Jr. L. *Meccanica*, doi:10.1007/s11012-019-01089-x
[12] Mishnaevsky Jr. L., Sütterlin J., 2019, *Polymer Degradation and Stability* **166**, 283-289
[13] Doagou-Rad S., Mishnaevsky Jr. L., Bech J., 2020, *Wind Energy*, https://doi.org/10.1002/we.2515
[14] Arena G.et al, 2015 eXPRESS *Polymer Letters* **9**:166-176
[15] Rutherford K.L., Trezon R.I., Ramamurthy A.C., Hutchings I.M., 1997,*Wear*, **203–204**, 325-334
[16] Hutchings, I. M.; Deuchar, D. W. T.; Muhr, A. H. 1987, *J Materials Science*, 22, **11**, 4071-4076
[17] O'Carroll A. , Hardiman M., Tobin E.F., Young T.M. , 2018,*Wear*, **412–413**, 38-48

[18] Busch H. et al, 1966. Philos. Trans. R. Soc. Lond. Ser. A, *Math. Phys. Sci.*, **260** (1110) 168-181

[19] Engel O.C.et al, 1974, Investigation of composite coating systems for rain erosion protection, *Naval Air Systems Commands*, Florida Atlantic University

[20] Project "AEI-010500-2015-373 Design, validation and manufacture of a prototype equipment to test and check a new polycarbonatediol-polyurethane based coating to protect the leading edge of wind turbine blades", http://www.aerox.es/projects/

[21] Cortés E. et al, *AIP Conference Proceedings* **1896**, 030023 (2017); https://doi.org/10.1063/1.5008010

[22] Mohagheghian I., McShane G.J., Stronge W.J., 2016, *Int J Solids Structures*, **88–89**, 337-353

[23] R.Herring, K.Dyer, P.McKeever, F.Martin, Integration of thermoplastic/metallic erosion shields a into wind turbine blades to combat leading edge erosion, Topic: Offshore wind turbines and components

[24] Beltzer A.I., Bert C.W., Striz A. G., 1983, *Int J Solids and Structures* **19**, 785-791

[25] Kinra V.K., Petraitis M.S., Datta S.K., 1980, *Int. J. Solids Structures*, **16**, 301-312

[26] Kinra V.K., Anand, A. 1982. *Int. J. Solids Structures*, **18,** 367-380

[27] Kausar, A. 2017. Polyurethane/Epoxy Interpenetrating Polymer Network, Aspects of Polyurethanes, Chapter: 1 Publisher: InTech-Open Access, Eds. F. Yilmaz http://dx.doi.org/10.5772/67678

[28] Culin J., 2016, Polimery, 3 (157—228) dx.doi.org/10.14314/polimery. 2016.159

[29] Chern Y. C., et al, 1999, *J Applied Polymer Science*, **74**, 328 –335

[30] Sophiea D. et al, 1994, *Advances in Chemistry*, 239

[31] Armada S., Bjørgum A., Knudsen O.Ø., Simon C., Pilz M., NOWITECH, https://www.sintef.no/projectweb/nowitech/, 2010

[32] Gou J. et al. 2010, *Composites B* **41**(2):192-198

[33] Dashtkar A. et al, Rain erosion resistant coatings for wind turbine blades – A Review, *Polymers and Polymer Composites* (in print)

[34] Dai, G.M. Mishnaevsky Jr. L., 2014, *Computational Materials Science*, **95**, 684–692

[35] Mishnaevsky Jr. L., 2015, *Composites B*, **68**, 75–84

[36] Zhou H.W., et al,, 2016, *Composites B*, **88**, 201-211

[37] Dai G.M., Mishnaevsky Jr L., 2015 *Composites B*, **78**, 349-360

[38] Gohardani, O; Williamson, D. M.; Hammond, D. W. 2012 *Wear*, 294

IOP Conf. Series: Materials Science and Engineering **942** (2020) 012026 doi:10.1088/1757-899X/942/1/012026

A practical approach for the peel stress prediction in the trailing-edge adhesive joint of wind turbine blades

M Rosemeier[1], T Gebauer[2] and A Antoniou[1]
[1]Department of Rotor Blades, Fraunhofer IWES, Fraunhofer Institute for Wind Energy Systems, Am Seedeich 45, 27572 Bremerhaven, Germany
[2]P. E. Concepts GmbH, Wiener Straße 5, 28359 Bremen, Germany

E-mail: malo.rosemeier@iwes.fraunhofer.de

Abstract. Wind turbine blades consist of thin-walled cylindric and airfoil-shaped structures, which are prone to "breathing" or "pumping" when subjected to cyclic loading. The "pumping" induces a peel stress in the adhesive layer of the trailing-edge bond line. To take account of this peel stress in the design phase, adequate models are required. State-of-the-art blade finite element (FE) models are usually implemented using shell elements. The trailing-edge joint is often represented by solid elements that are connected with the shell elements. The peel stress peak of interest occurs at the edge of the adhesive joint, which is, subject to a singularity, however. This study proposes a practical approach to estimate the peel stress peak in the adhesive joint with the help of the analytical solution which approximates and extrapolates the FE results. Moreover, different modeling techniques are benchmarked in respect of the peel stress prediction.

1. Introduction

The trailing-edge joint in wind turbine blades is subjected to a fatigue loading which results into a multiaxial stress state in the adhesive layer. The magnitude of each stress component depends on the position along the blade length. Cyclic peel stress amplitudes occur especially in the inboard blade region around maximum chord length. Peel stresses are caused by the "pumping" effect (Fig. 1*a*). [1]

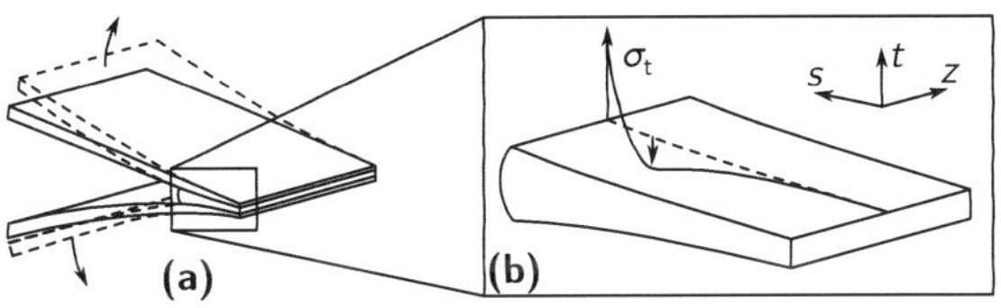

Figure 1. "Pumping" deformation of the blade's trailing-edge panels (*a*) and peel stress σ_t along the width of the adhesive layer (*b*).

Although the presence of peel stresses can provoke several failure modes of the joint [2], e.g., an adhesive failure or a fiber-tear failure, the present research investigates the cohesive failure in the adhesive layer, called a tunneling crack [1, 3].

The stress exposure, as described by Puck [4], in combination with stress-life curves, e.g., according to Basquin [5], can be used to estimate the load cycles in respect of tunneling crack initiation as a consequence of multiaxial fatigue loading, as shown by Krimmer et al. [6]. To this end, all stress components are calculated by means of finite element (FE) simulations to obtain the equivalent stress according to [7], for example.

Rosemeier et al. [1] used a state-of-the-art FE implementation of the adhesive layer with the aid of solid elements which share nodes with the blade shell elements (Fig. 4*d* and Fig. 2*a*). Haselbach [8] compared other modeling techniques of the trailing-edge bond line, e.g., a multi-point constraint (MPC) connecting the shell and the solid element (Fig. 4*e*).

The peel stress peak occurs at the edge of the adhesive represented by solid elements. This edge is subject to a singularity that typically occurs at a geometric or a stiffness discontinuity. That means stress values tend toward infinity when the mesh density is increased. To overcome this singularity, the stress value can be extracted at a defined distance from the edge, c.f. [9]. This method, however, may underestimate the actual stress peak value.

Spies [10] was the first to present a theoretical analysis of the mechanics of elastic peeling. He considered a substrate strip, which is modeled as an elastic beam, embedded into an elastic adhesive layer, and loaded by a peel force which acts perpendicular to the adhesive layer plane, i.e., at an angle of 90°. A similar model was presented by Bikerman [11]. Where Spies' and Bikerman's models assume small deflection angles, which is valid for relatively stiff substrate strips, other models take into account large deflections of the strip, viscoelasticity of the adhesive, and peeling angles greater or less than 90°, c.f. [12–14].

Hinopoulus et al. [15] were able to capture the typical peel stress distribution in a relatively detailed solid FE model of a T-peel joint.

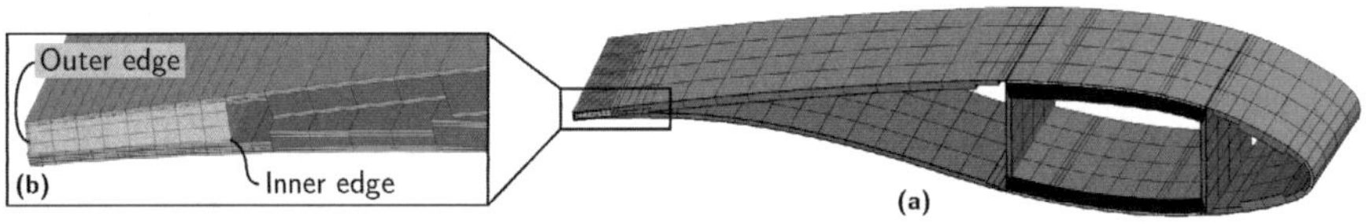

Figure 2. Finite element model showing extruded shell elements of the trailing-edge cell of the blade (*a*), and zoom into the trailing-edge bond line modeled with solid elements (*b*).

In this work, a practical approach is proposed to estimate the peel stress peak in the adhesive layer of an FE model with the help of the analytical solution. Moreover, different FE modeling techniques are investigated on a plate level to provide a thorough representation of the trailing-edge joint subjected to "pumping". The analytical solution by Spies [10] serves as reference. We focus on suitable models that can be implemented into a state-of-the-art 3D FE blade model at a reasonable computational cost.

2. Methods

2.1. Analytical model

The suction side (SS) and the pressure side (PS) of the trailing-edge joint (Fig. 3*a*) are modeled using Spies' [10] assumptions to calculate the peel stress within an adhesive layer that joints a substrate strip with a rigid wall, this peel stress being a consequence of a moment M and a shear force F acting at one strip edge (Fig. 3*b*):

(i) The strip is assumed to be straight.

(ii) The strip extends to infinity in the positive x-direction.

(iii) Normal stresses in the x-direction of the strip are neglected.

(iv) The adhesive is homogeneous and obeys Hooke's law.

(v) The influence of the strip length b in the z-direction is neglected.

(vi) The adhesive layer is assumed to be fixed at the lower edge to an absolutely rigid wall.

(vii) Deformations in the xt-plane resulting from normal and shear loads in the strip are neglected.

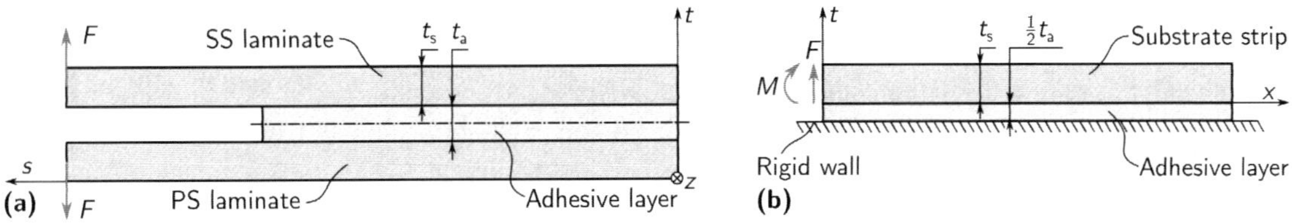

Figure 3. Cross-section view of the trailing-edge joint subjected to a peel force (a); diagrammatic representation of Spies' model (b), based on [10].

The peel stress distribution along the x-coordinate at the PS edge or SS edge of the adhesive layer is calculated as

$$\sigma_{\mathrm{t}} = \sigma_{\mathrm{t}}^{0}\, e^{-\frac{x}{c}} \left(\cos\frac{x}{c} - \frac{M}{M+Fc} \sin\frac{x}{c} \right), \tag{1}$$

where the peak stress at the inner edge, i.e., $x = 0$, is calculated as

$$\sigma_{\mathrm{t}}^{0} = \frac{M+Fc}{b\,c^{2}}, \tag{2}$$

where

$$c = \sqrt[4]{\frac{t_{\mathrm{a}} B}{2E_{\mathrm{a}}}}. \tag{3}$$

B denotes the bending stiffness of the substrate in the transverse s-direction about the z-axis, which is equivalent to the matrix element D_{22} of the bending stiffness matrix, as explained in [16]. E denotes Young's modulus and t the thickness, and index s stands for substrate strip and index a for adhesive. In the case of isotropic substrates, (3) simplifies to:

$$c = t_{\mathrm{s}} \sqrt[4]{\frac{1}{6} \frac{\frac{E_{\mathrm{s}}}{E_{\mathrm{a}}}}{\frac{t_{\mathrm{s}}}{t_{\mathrm{a}}}}}. \tag{4}$$

The positions of the extrema are calculated by setting the first derivative of (1) equal to zero and solving for x:

$$x_n = 2c \left(n\,\pi + \arctan \frac{1 - \frac{M}{M+Fc} + \sqrt{2\left(1 + \frac{M}{M+Fc}^{2}\right)}}{1 + \frac{M}{M+Fc}} \right), \tag{5}$$

where $n \in \mathbb{Z}$. x_0 defines the distance from the edge to the first extreme value, i.e., a minimum or maximum point, and x_1 defines the beginning of the flat plateau (Fig. 5a).

2.2. Finite element plate models

Four two-dimensional sandwich-like plate models were implemented in ANSYS APDL (ANalysis SYStem Parametric Design Language) [17] to generically represent the trailing-edge joint:

- A high fidelity "n-sliced solid" model wherein the nth layer of the bond line laminate is represented by a homogeneous SOLID186 element (Fig. 4*b*).
- A "3-sliced layered solid" model wherein the adhesive layer is represented by a homogeneous SOLID186 element covered on both the PS and the SS with a layered SOLID186 element, which is attributed with the properties of the respective bordering laminae via the SECDATA command (Fig. 4*c*).
- A "solid/shell share" model wherein a homogeneous SOLID186 element represents an adhesive layer which is as thick as the whole laminate, and two SHELL281 elements represent the bordering laminae (Fig. 4*d*). The shell elements share their nodes with the solid. The shell nodes are offset with respect to the outer surface in such a way that the laminae overlap the solid.
- A "solid/shell MPC" model wherein a homogeneous SOLID186 element represents the adhesive layer, and two SHELL281 elements represent the bordering laminae (Fig. 4*e*). The shell nodes are offset with respect to the outer surface in such a way that the laminae protrude into the gap between the solid and shell. Multi-point constraints (MPC) bond the surfaces of the solids and shell with each other.

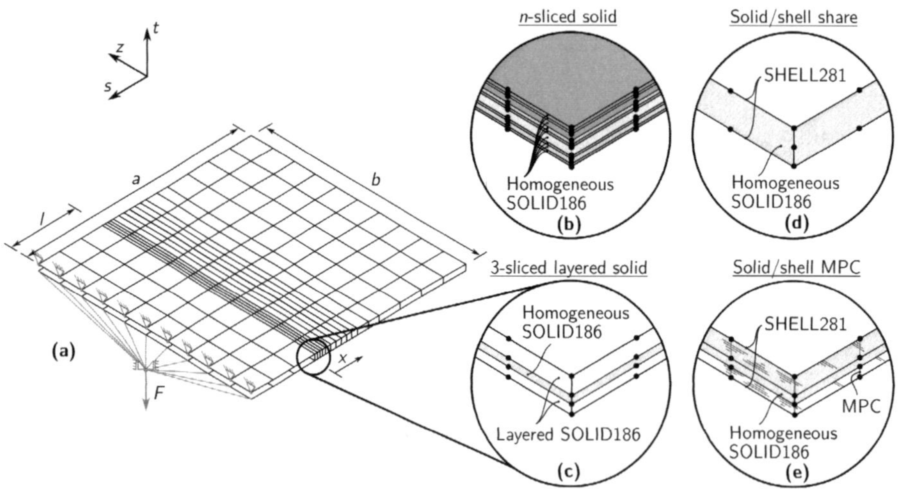

Figure 4. Double cantilevered finite element plate model (a); zooms into different modeling techniques, i.e., n-sliced homogeneous solid (b), 3-sliced layered solid (c), homogeneous solid which overlaps the shell while sharing nodes (d), and homogeneous solid and shell connected with multi-point constraints (MPC) (e).

The SS edge along the z-direction is constrained in all translational degrees of freedom. A rigid body element of type CERIG connects the PS edge with a master node, which can translate in the t-direction. A unit load F is applied to the master node.

The bordering laminae on the SS and PS sides overlap the adhesive layer over the length $l = 1/3a = 0.33\,\text{m}$. The adhesive layer is meshed between the two sides over a width of $2/3a = 0.67\,\text{m}$, a length of $b = 1.0\,\text{m}$, and a mesh size from 16 to 64 elements along the width. The element size in the longitudinal z-direction is 100 mm.

2.3. Mesh density and peak stress calculation

The mesh density along the width of the adhesive layer increases toward the inner edge where the peel stress peak of interest is expected. The required distance x_0 from the edge to reproduce the characteristic peel stress distribution is estimated with the help of the analytical solution (5).

The peel stress distribution is derived from nodal FE stresses along the SS and PS edge along the width (x-direction) of the adhesive layer. By varying M, F, and b, the analytical equation (1) approximates the peel stress distribution of the FE model using the least-squares method [18]. The first node at the inner edge is neglected since it is subject to a singularity: The higher the mesh density, the more the first node tends toward infinity. Finally, the peel stress peak of the approximated solution obtained with (2) is used as the result.

In particular, for the approximation of the "solid/shell share" FE results, the adhesive thickness t_s which is input for c according to (4) is corrected in order to take into account the decrease in stiffness of the elastic embedding, which results from the overlap of the adhesive and the substrate:

$$t_\mathrm{a}^\mathrm{cor} = \frac{1}{2} t_\mathrm{a} + t_\mathrm{s} \tag{6}$$

After approximating the "solid/shell share" FE results with the corrected thickness according to (6), the peel stress peak σ_t^0 obtained with (2) is again corrected in order to obtain comparable results to those of the analytical solution. This correction is derived by relating the stress peak $\sigma_\mathrm{t}^{0,\mathrm{cor}}$ according to (2), which is obtained by taking the corrected thickness $t_\mathrm{a}^\mathrm{cor}$ according to (6) into account, to the stress peak σ_t^0 obtained by taking the actual thickness t_a into account:

$$\sigma_\mathrm{t}^{0,\mathrm{cor}} = \sigma_\mathrm{t}^0 \sqrt{\frac{t_\mathrm{a} + 2t_\mathrm{s}}{t_\mathrm{a}}}. \tag{7}$$

3. Results

3.1. Sensitivity to input parameters

A simple reference steel-adhesive joint is subjected to a sensitivity study using the analytical model. To this end, the input parameters of the reference (Fig. 5*a*) are varied. Their impact on the peel stress peak and the edge-distance to the extreme value x_0 is investigated. The decrease in stiffness of the substrate E_s toward typical values of uni-directional and bi-directional glass fiber-reinforced polymers (GFRPs) increases the peel stress peak and shifts x_0 closer to the edge (Fig. 5*b*). The same effect is observed when either the substrate thickness t_s or the adhesive thickness t_a is decreased (Fig. 5*c* and Fig. 5*d*). The opposite effect is observed when the thickness increases. An increased peel moment M/b increases only the stress peak but has no effect on x_0 (Fig. 5*e* and Fig. 5*f*). The peel force F/b increases by a factor of ten, slightly increases the peak, and slightly shifts x_0 away from the edge.

3.2. Effect of modeling technique

Three FE plate modeling techniques are investigated in respect of the prediction of the peel stress peak and are compared to the analytical model (Fig. 6*a*). The reference steel-adhesive joint is implemented with a mesh density of 64 elements along the adhesive layer width.

The approximation of the "3-sliced layered solid" model (indicated as "fit" in Fig. 6a) is closest to the analytical peak stress (−10.4%), followed by the "solid/shell MPC" model (−11.2%). The fit of the "solid/shell share" model exhibits the lowest peak (−22.3%). When the peak stress of the approximated stress distribution is corrected according to (7), the "solid/shell share" model reveals the peak closest to the analytical solution, which is overestimated by 4.2%.

The first node in all FE models overestimates the analytical solution. Discontinuities are observed in the nodal FE values of the "solid/shell MPC" model at $x \approx 2.5\,\mathrm{mm}$ and $x \approx 5.5\,\mathrm{mm}$.

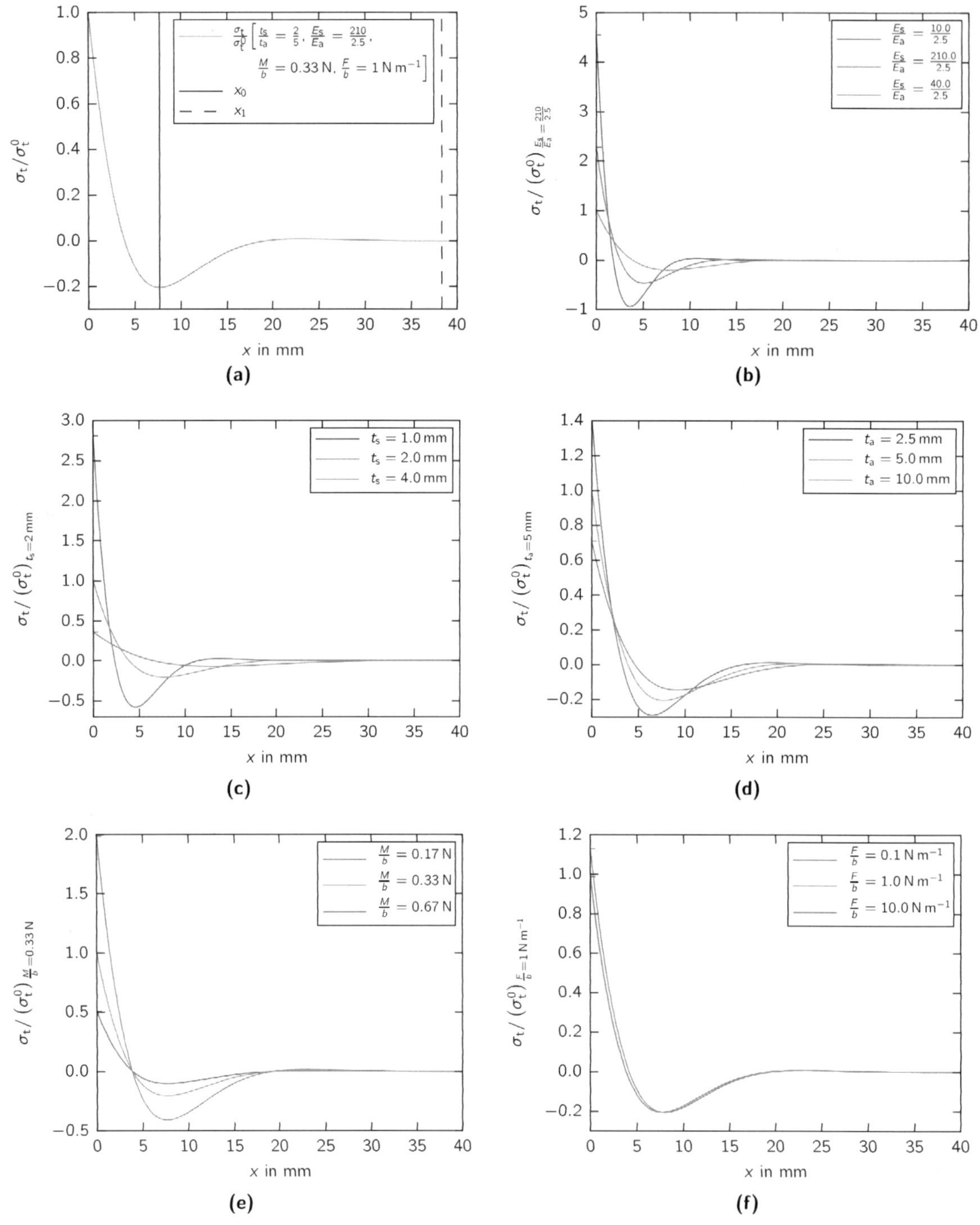

Figure 5. Peel stress distribution along the adhesive layer (a); sensitivity to stiffness fraction E_s/E_a (b), substrate thickness t_s (c), adhesive thickness t_a (d), peel moment M/b (e), and peel force F/b (f).

3.3. *Effect of mesh density*

The element discretization of the "solid/shell share" model is varied between $N_e = 16$ and $N_e = 64$ elements along the adhesive layer width, i.e., in the x-direction (Fig. 6*b*). An element spacing ratio of $2N_e/1$ between the element width at the edge ($x = 0$ mm) and the center ($x = 333$ mm) is set up such that the mesh density increases toward the edge.

At least three nodal results within the distance $2x_0$ from the edge, including the first node, are required to capture a stress peak that is close to the analytical solution using the method of approximation and extrapolation.

The higher the mesh density, the better the match between the analytical peel stress distribution and the modeled one. This trend is confirmed when the peak stress is corrected. The peak stress obtained from the first node, however, increases with the mesh density toward infinity. This is attributed to a singularity.

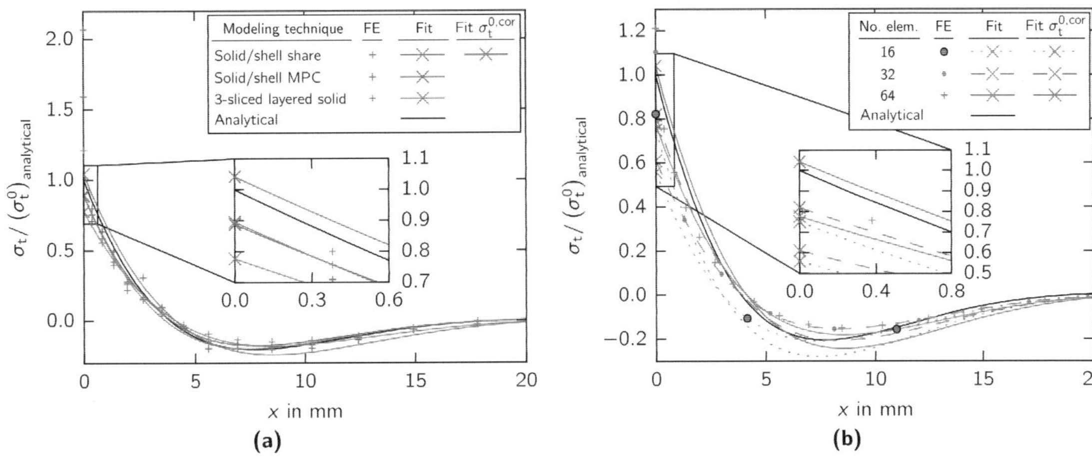

Figure 6. Peel stress distribution along the adhesive layer width; comparison of modeling techniques (*a*); mesh density study for "solid/shell share" model (*b*).

4. Discussion

The approach proposed in this paper circumvents the dilemma that the position of the peel stress peak value of interest coincides with a singularity in an FE model: An analytical model approximates the nodal stress results and extrapolates the stress peak of interest. Since the force and moment acting at the substrate edge cannot be extracted directly from a full 3D blade model with the modeling techniques presented in this work, the approach varies M/b to fit the stress peak and the amplitude of the valley (Fig. 5*e*), and F/b to fit the distance of the valley from the edge x_0 (Fig. 5*f*).

When the peel stress peak in the "solid/shell share" model-fit is corrected, this peak is closest to the analytical solution when compared to the other models. Moreover, this modeling technique is implemented with the least effort in a state-of-the-art full 3D FE shell model. Therefore, this model seems suitable for the design analysis of a trailing-edge joint of a wind turbine blade.

A sufficient number of elements need to be placed between the edge and the materially and geometrically dependent parameter x_0, which is estimated using the analytical solution (5).

The proposed approach represents the more conservative option to predict the stress peak in comparison to the method of extracting the stress value at a defined distance from the edge, as used in [9]. Moreover, this approach is computationally more efficient since the mesh density can be determined with the analytical model by calculating the distance of the extrema from the edge using (5).

The assumption made in [1] that the peel stress distribution and peak may be affected by the connection between the shell and solid elements, i.e., that the rotational degrees of freedom of the shell elements are not adequately transferred to the solid elements, is disproved in this work. All modeling techniques benchmarked in this work are able to capture the stress response of the analytical solution with the proposed approach.

5. Conclusions

In this work, we propose and verify a practical approach to estimate the peel stress peak at the inner edge of the trailing-edge joint with the help of an analytical model which approximates and extrapolates nodal peel stress results. The approach circumvents the dilemma that the position of the peel stress peak value of interest coincides with a singularity in an FE model. The computational efficiency of this approach can be optimized by calculating the distance of the extrema analytically, and thus the region required to have a higher mesh density. Moreover, it is suggested that the "solid/shell share" modeling technique can be implemented with the least effort in a state-of-the-art full 3D FE shell model when compared to the other techniques presented.

In a future work, the analytical model can be extended to consider asymmetric adhesive joints. Furthermore, this approach can be applied to other loading situations of the adhesive joint such as normal or shear stressing.

Supplemental material

The data used for the figures are available at `https://doi.org/10.5281/zenodo.3937258` [19].

Acknowledgments

We acknowledge the support provided by the German Federal Ministry for Economic Affairs and Energy (BMWi) within the ReliaBlade project (0324335A).

References

[1] Rosemeier M, Krimmer A, Bardenhagen A and Antoniou A 2019 *AIAA J.* **57** 5462–5474
[2] ASTM International 1999 *ASTM D5573-99: Standard practice for classifying failure modes in fiber-reinforced plastic (FRP) joints*
[3] Jørgensen J B, Sørensen B F and Kildegaard C 2018 *Eng. Fract. Mech.* **189** 361–376
[4] Puck A 1996 *Festigkeitsanalyse von Faser-Matrix-Laminaten: Modelle für die Praxis* (Hanser)
[5] Basquin O H 1910 *ASTM Proc.* vol 10 p 625
[6] Krimmer A, Leifheit R and Bardenhagen A 2016 URL `https://doi.org/10.5281/zenodo.3878156`
[7] Beltrami E 1885 *Il Nuovo Cimento* **18** 145–155
[8] Haselbach P 2017 *Compos. Struct.* **180** 521–530 ISSN 0263-8223
[9] Rosemeier M, Krimmer A and Antoniou A 2020 *J. Phys.: Conf. Series* **1452** 012060
[10] Spies G J 1953 *Aircr. Eng. Aerosp. Technol.* **25** 64–70 ISSN 0002-2667
[11] Bikerman J 1957 *J. Appl. Phys.* **28** 1484–1485
[12] Inoue Y and Kobatake Y 1959 *Appl. Sci. Res., Sec. A* **8** 321–338
[13] Chang F S 1960 *J. Rheol.* **4** 75–89
[14] Heise O 1963 *Die Ermittlung der spezifischen Anrißkraft beim Winkelschälversuch* (Dt. Forschungsanstalt f. Luft-u. Raumfahrt eV (DFL))
[15] Hinopoulos G and Broughton W 1999 *NPL Report CMMT (A)* **207**
[16] Reddy J N 2004 *Mechanics of laminated composite plates and shells: theory and analysis* (CRC Press)
[17] Swanson J A 2014 *ANSYS Mechanical APDL, version 15.0*
[18] Legendre A M 1805 *Nouvelles méthodes pour la détermination des orbites des comètes* (F. Didot)
[19] Rosemeier M, Gebauer T and Antoniou A 2020 Supplemental Material to Article "A practical approach for the peel stress prediction in the trailing-edge adhesive joint of wind turbine blades"

On the potential of particle engineered anti-erosion coatings for leading edge protection of wind turbine blades: Computational studies

K M Jespersen[1*], G Monastyreckis[2] and L Mishnaevsky Jr[1]

[1]Department of Wind Energy, Technical University of Denmark, Frederiksborgvej 399, 4000 Roskilde, Denmark

[2]Department of Mechanical Engineering, Kaunas University of Technology, Studentu str. 56, 51424 Kaunas, Lithuania

E-mail: kmun@dtu.dk*

Abstract. The potential of particle and fiber reinforced anti-erosion coatings for the protection of wind turbine blades is explored through computational modelling. A hypothesis that stiff disc-shaped particle or fiber reinforcements embedded in viscoelastic coatings ensure better erosion protection is validated numerically, and mechanisms of this effect are analyzed. A computational unit cell model of coatings with embedded fibers (fiber pulp) or disc particles subject to rain droplet impact is developed, and series of computational experiments is carried out. The distribution and scattering of stress waves from the rain droplet impact and damping properties are analyzed for homogeneous viscoelastic polyurethane coatings, coatings with disc-shaped particles, and fiber pulp. It is shown that the stress waves are increasingly scattered, and the damping is increased with higher volume percentage of the fibers. The mechanism of such increased energy dissipation is found to be related to the high local viscoelastic deformation in the regions between closely located fibers and the higher stiffness of the unit cell. The current work demonstrates the high potential of fiber engineered coatings for the improvement of anti-erosion protection of wind turbine blades.

1. Introduction

Leading edge erosion of wind turbine blades is one of the most common mechanisms of the blade degradation. The repair of eroded wind turbine blades, especially off-shore, is a quite expensive and labour consuming process [1]. With the blades becoming longer to increase the power output, the tip speeds increase and as a consequence, rain droplets and particles hit the blades at higher speeds. Significant efforts were put into developing new leading edge protection technologies to reduce the need for repair. Ideas have ranged from applying metal shields [2] and protective tapes [3] on the blade to developing advanced multi-layer coatings with particles [4]. Protective coatings can be applied in-mold and have practically no negative impact on the aerodynamic properties of the blade [2]. However, coatings which can last the entire 20-30 years of life of wind turbine blades have yet to be developed, and pushing the blade length to new limits will require even stronger erosion-resistant coatings in the future. Therefore, it is of great importance to continuously improve the coating solutions for wind turbine blades.

In [4], different effects and mechanisms to enhance the erosion resistance of blade coatings are reviewed, among them, the damping effect caused by the viscoelasticity of coatings, attenuation, dispersion and scattering of stress waves by inhomogeneities, mode conversion at interfaces, redirection, and conversion of deformation energy to heat by molecular relaxation processes. Dissipation of stress waves in composite materials can be caused by the viscoelastic nature of the resin and by scattering on inclusions. At high frequencies, the wavelength is not very large compared to the reinforcements size, and thus, the wave propagation can be influenced by particles. Multiple scattering of waves on particles lead to frequency-dependent velocity and attenuation of coherent waves. Several approaches and estimations were developed to model the wave scattering on the particles: scattered field expressions and quasi-crystalline approximation, models based on the causal differentiation method and Kramers-Kronig expressions between phase velocity and attenuation [5], the generalized self-consistent model by Christensen and Lo [6], and others [7, 8, 5, 9].

An interesting type of coating materials is the materials with Interpenetrating Polymer Networks (IPN). For instance, the damping properties of polyurethane can be improved by introducing epoxy to form a polyurethane/epoxy interpenetrating polymer network. The advantages of IPN as compared with pure polyurethane include better heat resistance, higher strength, and extention of the temperature range with sufficiently high damping peaks [10, 11]. Bicontinuous materials with high specific area and interlocked topology can make the material system defect-insensitive and enhance energy absorption of the materials. In several works, the optimization of IPN structures has been carried out to achieve the best damping properties [10, 11, 12]. Following [4], the assumption was formulated that the stiff distributed particles embedded in the coatings can enhance the erosion protection by scattering the impact wave from the rain droplet. Further, it is assumed that the embedded interconnected fibers (pulp) can also enhance the erosion resistance by creating an additional wave propagation mechanism (along the interfaces), additional damping mechanisms (pulp deformation) and also wave scattering. These assumptions are explored in this work using computational simulations.

2. Computational models

In order to explore the potential of structural modifications of anti-erosion coating materials, a series of computational studies have been carried out. Figure 1 shows examples of three considered damping mechanisms: (a) viscoelasticity, (b) disc-shaped particle reinforcement, (c) pulp (multiple elastic fibers) induced damping of the engineered coatings. Small unit cells models similar to that presented in Figure 2(b) were established. This section describes the materials and particle structures used for the modelling, along with the details of the modelling approaches.

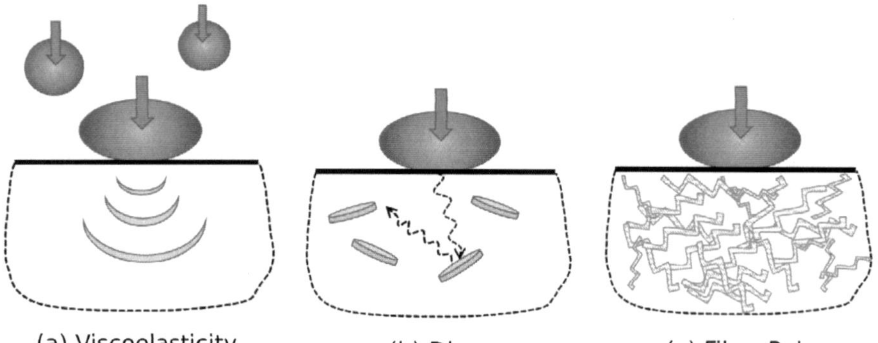

Figure 1: Considered damping mechanisms: (a) viscoelastic damping, (b) disc particle reinforcement induced damping, (c) fiber pulp enhanced damping.

2.1. Simplification of 3D model droplet impact

In the previous work by Doagou-Rad et al. [13, 14] the rain droplet impact on a polyurethane (PU) coated surface was simulated, however without taking the particles distributed in three dimensions (3D) into account. Figure 2(a) shows a droplet impact model similar to that presented in [13]. The material below the water droplet is subjected to stress waves travelling away from the impact location, and therefore the loading of a given material point will be different depending on location and time. To study the scattering of stresses, energy dissipation, and similar mechanisms that can improve the erosion resistance of coatings for different particles in detail while still keeping the problem relatively simple, small volumes (unit cells) were considered. Figure 2(b) shows one of the unit cells studied in this paper with 8 volume percentage (vol%) fibers. It is assumed that the considered volume is sufficiently small so that the impact load would be nearly homogeneously distributed on the surface.

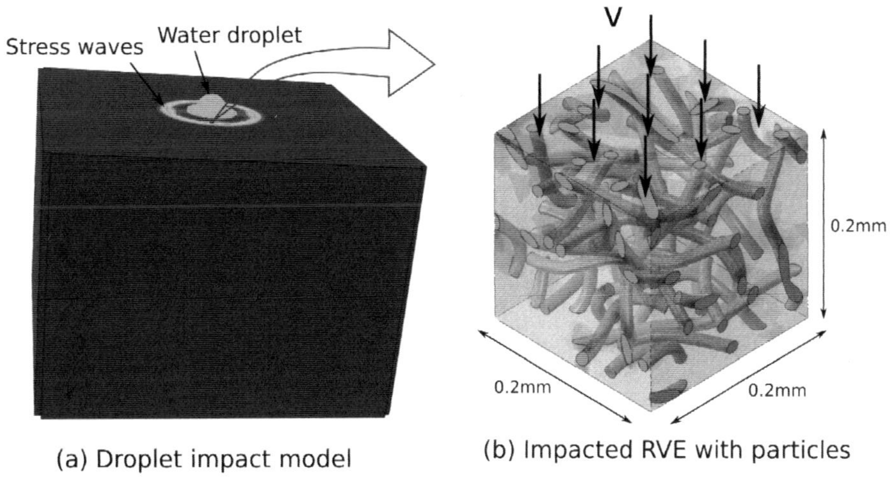

Figure 2: Principle of modelling simplification where a small material point of the full droplet model (a) is considered as a separate unit cell (b) with applied homogenized impact load.

2.2. Materials

Two different types of particles; discs with varying thickness and curved fibers (pulp) are modelled in this paper. Nepheline Syenite particles and Kevlar pulp were used as an example in this study. However, as several other types of particles with similar properties could be used as well, the names "discs" and "fiber pulp" will be used in the rest of the paper. The fiber pulp is modelled with a length of 0.5 mm and a constant diameter of $d = 12\mu m$. The discs were modelled with a diameter of 12 μm and varying thickness of 1-6 μm, giving 2-12 aspect ratio values, respectively. PU was used as the matrix in the models and depending on the analysis was modelled as either linear elastic or hyper- and viscoelastic behaviour. The elastic material properties used for the particles and matrix are summarized in Table 1. The hyper- and viscoelastic behavior of the PU matrix used in this work was the same as that presented by Doagou-Rad et al. [13].

2.3. Explicit finite element impact models

Two types of models (A & B) utilizing different meshing approaches and workflow were used in the present work as listed below.

Table 1: Elastic material properties and density

	Young's modulus [MPa]	Poisson's ratio [-]	Density [g/mm^3]
Polyurethane	300	0.475	0.00118
Kevlar	70500	0.36	0.00144
Nepheline Syenite	150000	0.2	0.0026

(A) Model with high time-resolution of initial impact (0-0.001ms)

(B) Model of over-time behaviour including damping from viscoelasticity (0-0.025ms)

For both model A and B, the model geometry was generated using the software Digimat. Figure 3 shows examples of the unit cells considered in the current study. The finite element modelling was carried out using Abaqus Explicit with an initial mass scaling factor of 2. All models were done with X, Y and Z-symmetry conditions applied to the unit cell sides and using 4-node tetrahedral elements (C3D4). Doagou-Rad et al. [13] reported that the von Mises stresses reached a maximum at approximately 0.001ms after droplet impact, and this was applied as the impact time for all the unit cell models. It was chosen to use a maximum deformation of 0.005 mm, giving a maximum strain of $\varepsilon = 0.025$ for all the modelled unit cells.

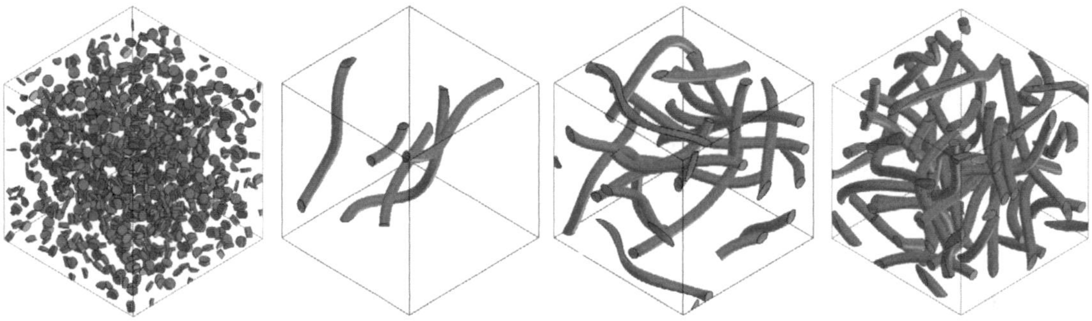

(a) 4 vol% discs (b) 1 vol% fibers (c) 4 vol% fibers (d) 8 vol% fibers

Figure 3: Examples of unit cells generated in Digimat used in the current study.

For model A, the model geometry generated in Digimat was imported as a step file into Abaqus where linear elastic material properties, boundary conditions and mesh were applied. The geometry was meshed in Abaqus which gave an overall number of elements ranging between 1.5-2.5 million, with a size of around 3 μm.

For model B, the unit cells were both generated and meshed in Digimat to be able to study a range of different fiber structures and volume percentages. After generating and meshing the unit cell in Digimat, the input file for Abaqus was exported. The input file was then changed into an explicit impact model by modifying the material properties, step, boundary conditions, history and field output directly in the file. This approach gives less freedom when meshing than using Abaqus as for model A, however it requires less manual work, making it suitable for parametric studies with focus on the overall behaviour.

Model B was carried out using visco- and hyperelastic material behavior for the PU introducing damping into the system. The fibers were modelled as linear elastic. The unit cells were impacted (pressed) and then released using a separate load step in Abaqus. The average displacement of the impacted surface of the unit cell was measured to be able to compare the damping of the different unit cells. The number of elements was up to around 2,500,000 and down to 230,000 depending on the internal structure of the unit cell. For model B, all the analyses were carried out on a computer cluster, where a single model took up to 12 hours using 200 cores.

3. Results and discussion

3.1. Particle/impact wave interaction

In order to understand the stress distribution inside PU coatings reinforced by fiber pulp and discs, a high-resolution impact model (model A) was performed with a step time of 0.001 ns. Unit cell models of pure PU, 4 vol% discs, along with 4 and 8 vol% fibers were compared at the time of 0.2 ns, where the first stress wave was observed. Figure 4(a) shows the pure PU model with an average stress of 1.1 MPa, while the model with 4 vol% discs (b) shows much higher average stress values of 3.8 MPa and maximum stress peak of 142.3 MPa. The model with 4 vol% fibers (c) shows a similar average stress of 4.1 MPa, but the maximum stress peak reaches up to 667.5 MPa. The model with 8 vol% fibers (d) has stresses distributed throughout the whole unit cell model, an average stress of 6.7 MPa, and a stress peak of 693.7 MPa.

To further discuss the micromechanical stress-transfer mechanism, one can notice from Figure 4 that for both discs and fibers the stress is transferred from particle to particle, with higher stresses at shorter neighbor particle distances. The stresses build up inside the matrix in the range of 20-30 μm around the fibers (pulp diameter - 12 μm), while the stress around the discs

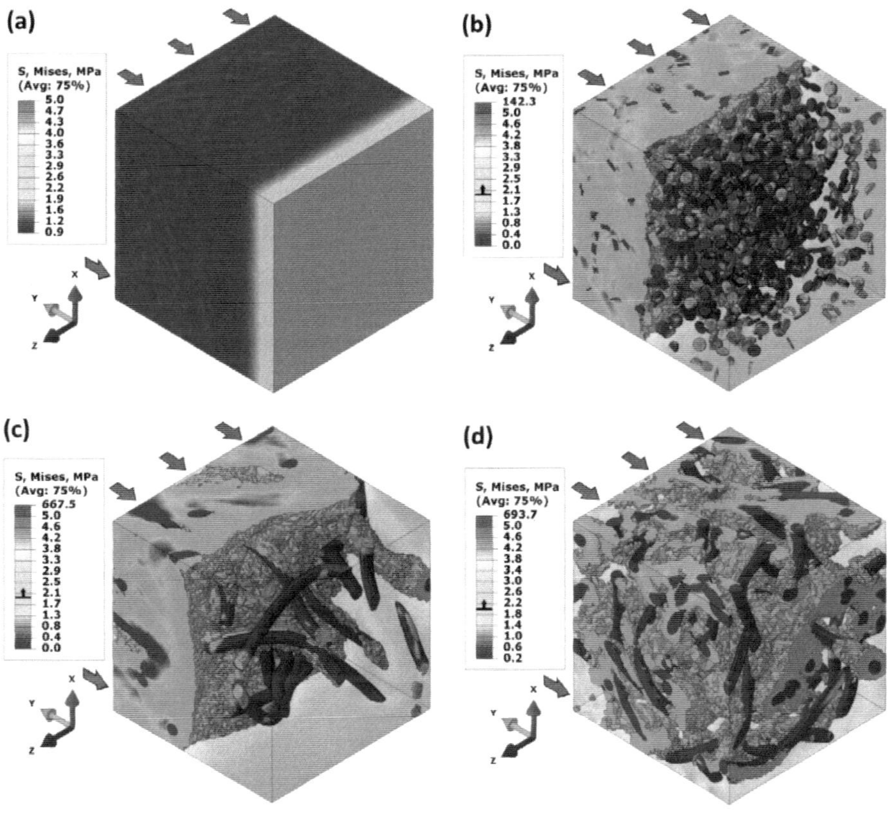

Figure 4: Initial impact models at the same time step (t=0.2 ns). (a) Pure PU, (b) 4 vol% discs, (c) 4 vol% fiber pulp, (d) 8 vol% fiber pulp. Stresses below 2 MPa have been hidden for clarity in (b)-(d).

penetrates only up to 5-10 μm. Additionally, as a single particle of the fiber pulp penetrate far into the depth of the unit cell, they transfer the stresses into the matrix further down at a shorter time than for the discs. In other words, although the overall stress wave is similar, the stresses are locally transferred into the matrix faster for the fibers than for the discs. As stresses are building up in a larger region for the fiber pulp than for discs, one can also expect an increased energy dissipation due to local viscous effects.

3.2. Effect of fiber pulp volume percentage on energy dissipation
To study the damping properties of the fiber pulp in more details, the movement (bouncing) of each unit cell after impact was modelled considering a longer time frame (0-0.025ms) including viscoelastic material properties (model B). Figure 5(a) shows the average displacement of the impacted surface as a function of time for different volume percentages of fibers compared to pure PU. Considering the average displacement, an increasing volume percentage seems to increase the damping and frequency of the unit cell. Increasing the volume percentage of stiff fibers also increases the overall stiffness of the unit cell, which according to vibrations theory will increase the damping and oscillation frequency. Additional local viscous effects are discussed in the following.

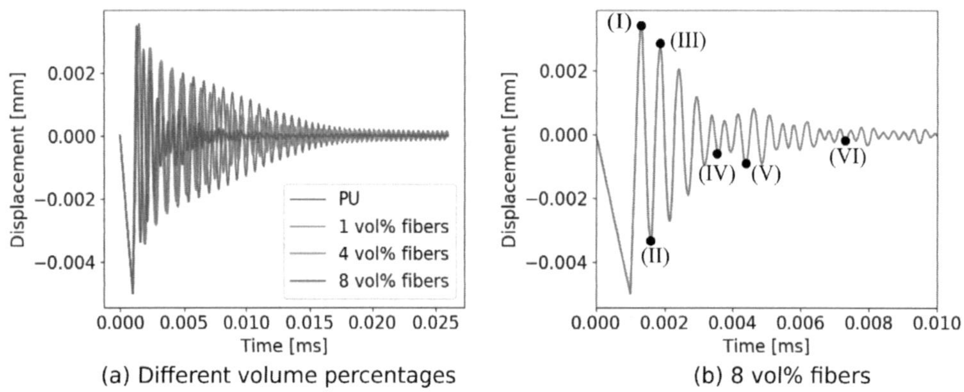

Figure 5: Average displacement of the impacted surface over time for (a) different volume percentage and (b) 8 vol%.

Figure 5(b) shows the time-displacement response for 8 vol% fibers with six marked points in time (I, II, III, IV, V, VI) and Figure 6 shows the highest matrix stresses in the unit cell for each of these time points. In Figure 6, all regions with stresses below 35MPa have been hidden, thus showing only the most stressed regions. Note that compared to Figure 4 which also included the fiber stresses, Figure 6 includes only the matrix stresses. It is seen that at the early stage ((a) and (b)) high stresses are present mainly between fibers located close to one another and along some specific fibers. The regions showing high stresses are relatively similar for when the unit cell is in tension (a) and compression (b). Shortly after impact, the unit cell starts to bounce diagonally as can be seen from the slightly increased stresses in the corners of the unit cell already in (c) and later more significantly in (d) and (e). After some time, the high-stress regions decreases significantly as shown in (f) indicating that most energy has been dissipated. Compared to pure PU where the stresses are evenly distributed, the presence of the fibers lead to regions with more localized stresses where additional energy is dissipated due to the viscoelastic properties of the PU matrix. This is also supported by Figure 7 which shows the energy dissipated by viscous effects for the whole model over time extracted from Abaqus for 8 vol% fibers and pure PU. The fiber case is seen to faster dissipate energy than pure PU.

IOP Conf. Series: Materials Science and Engineering **942** (2020) 012027 doi:10.1088/1757-899X/942/1/012027

In other words, the improved damping properties seem to be both a result of the increased unit cell stiffness and due to local viscous energy dissipation from the fiber structure.

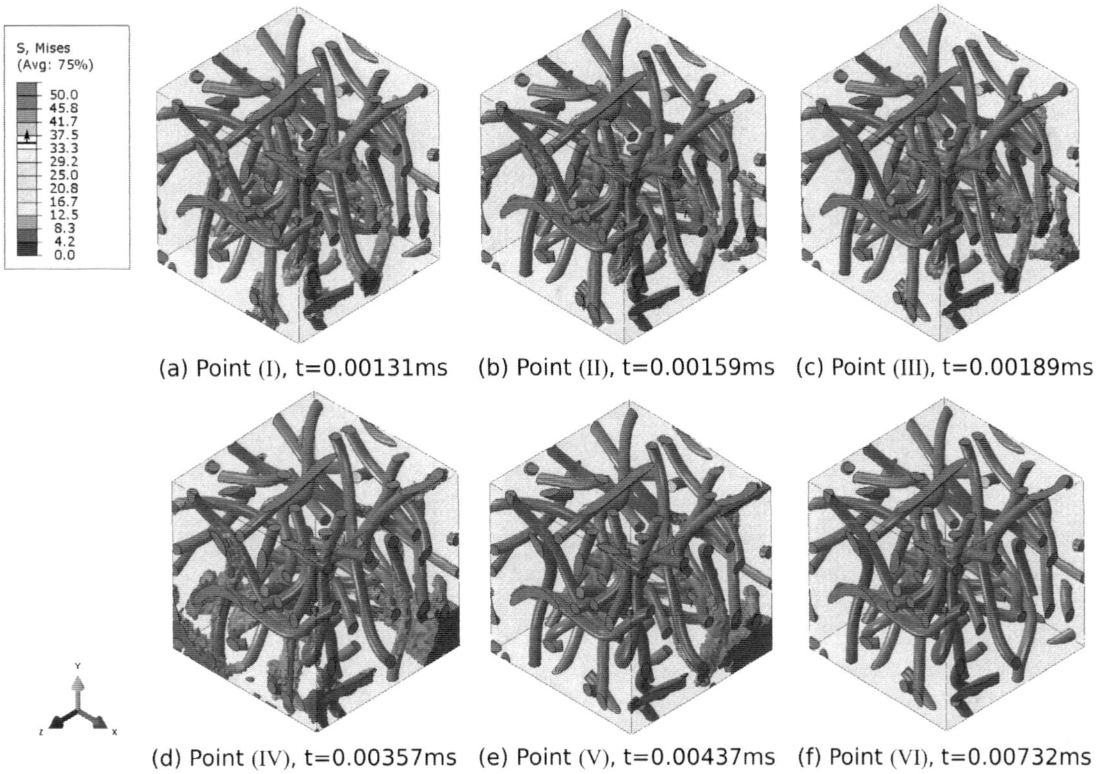

(a) Point (I), t=0.00131ms (b) Point (II), t=0.00159ms (c) Point (III), t=0.00189ms

(d) Point (IV), t=0.00357ms (e) Point (V), t=0.00437ms (f) Point (VI), t=0.00732ms

Figure 6: Von Mises stresses in the matrix above 35 MPa for different time points referring to Figure 5(b) after impact on the top surface in the negative Y-direction.

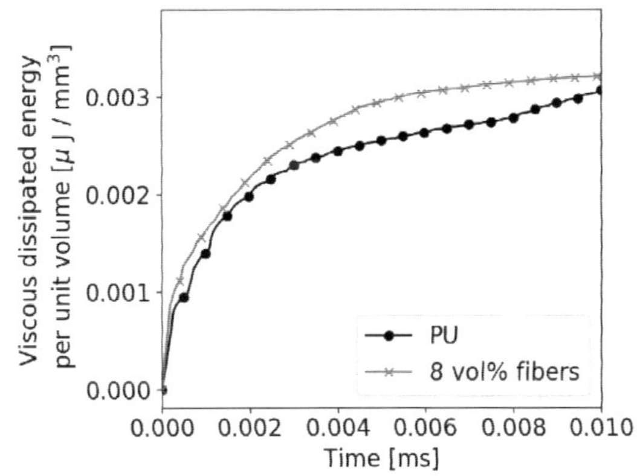

Figure 7: Energy dissipated by viscous effects per element volume in Abaqus models for 8 vol% fibers and pure PU.

3.3. Potential of fiber engineered coatings

As discussed in the introduction, the erosion resistance of blade coatings can be improved by exploiting the viscoelastic damping effect and scattering of stress waves by inhomogeneities. The current work has shown that the presence of fiber pulp (elastic curved fibers) in a viscoelastic resin leads to enhanced scattering of the stress waves, which in turn increases with higher volume percentages (Figure 4(d)). Furthermore, studying the time-displacement response of the unit cells after impact revealed locally higher energy dissipation in regions where stresses build up, particularly near fibers and in regions where the fibers lie close to one another. In addition, even further energy dissipation effects are expected when considering larger network of clustered fibers and the effect of interface behavior between the fiber and matrix. Therefore, the results show good potential of fiber engineered coatings as playing an important role in new future erosion-resistant coating systems for wind turbine blades.

4. Conclusion

In this study, the potential of particle engineered reinforcement of anti-erosion polyurethane coatings for better protection of wind turbine blades against rain erosion was investigated. In a series of computational studies, the effect of discs and fiber pulp reinforcements on the stress wave propagation, wave scattering, and damping in the coating under rain droplet impact was studied. It was demonstrated in the numerical experiments that the coating with an embedded network of stiff elastic curved fibers show enhanced damping and energy dissipation, and therefore better erosion protection can be expected. The embedded particles cause additional wave scattering mechanisms, and viscoelastic energy dissipation was observed in highly stressed regions between closely located fibers. Therefore, particularly for higher volume percentage of fibers, the impact energy of a rain droplet is expected to be dissipated faster for the particle reinforced coatings than for pure polyurethane coatings. Scattering of stresses and increased damping properties serve to improve the erosion resistance of coatings and therefore it was concluded that fibers or similar particles have great potential as a part of new improved leading edge erosion solutions.

Acknowledgments

The authors gratefully acknowledge the financial support of the Innovation Foundation of Denmark in the framework of the Grand Solutions project DURALEDGE, Durable leading edges for high tip speed wind turbine blades, File nr.: 8055-00012A.

References

[1] Mishnaevsky Jr L 2019 *Renew. Energy* **140** 828–839 ISSN 18790682
[2] Herring R, Dyer K, Martin F and Ward C 2019 *Renew. Sustain. Energy Rev.* **115** ISSN 18790690
[3] Sareen A, Sapre C and Selig M 2012 *Wind Eng.* **36** 525–534 ISSN 0309524X
[4] Mishnaevsky Jr L 2019 *Wind Energy* **22** 1636–1653 ISSN 10991824
[5] Beltzer A and Brauner N 1987 *Mech. Mater.* **6** 337–345
[6] Christensen R M and Lo K N 1979 *J. Mech. Phys. Solids* **27** 315–330
[7] Sabina F J and Willis J R 1988 *Wave Motion* **10** 127–142 ISSN 01652125
[8] Varadan V K, Ma Y and Varadan V V 1989 *Pure Appl. Geophys.* **131** 577–603 ISSN 00334553
[9] Yang R B and Mal A K 1994 *J. Mech. Phys. Solids* **42** 1945–1968 ISSN 00225096
[10] Čulin J 2016 *Polimery/Polymers* **61** 159–165 ISSN 00322725
[11] Chern Y C, Tseng S M and Hsieh K H 1999 *J. Appl. Polym. Sci.* **74** 328–335 ISSN 00218995
[12] Sophiea D, Klempner D, Sendijarevic V, Suthar B and Frisch K C 1994 *Interpenetrating Polymer Networks as Energy-Absorbing Materials* (ACS Publications)
[13] Doagou-Rad S and Mishnaevsky L 2020 *Meccanica* **55** 725–743 ISSN 15729648
[14] Doagou-Rad S, Mishnaevsky L and Bech J I 2020 *Wind Energy* 1–15 ISSN 10991824

IOP Conf. Series: Materials Science and Engineering **942** (2020) 012028 doi:10.1088/1757-899X/942/1/012028

Micro-mechanical investigation of glass fibre/resin interface failure in mode I and mode II

Peter G Jenkins*, David Bryce, Georgios Xypolias and James L Thomason

Department of Mechanical & Aerospace Engineering, University of Strathclyde, James Weir building, 75 Montrose Street, Glasgow, G1 1XJ, United Kingdom

*peter.jenkins@strath.ac.uk

Abstract. Fibre reinforced composites are experiencing a period of consistent and significant growth which is predicted to continue, due to the sustained global increase in demand across a number of industrial sectors. One of the critical elements in the success of such composites in both the short and longer term is the interface/interphase between reinforcement fibres and matrix. Poor adhesion will generally lead to composites with insufficient mechanical performance. Degradation of the interface/interphase region during service life will lead to a decrease in performance, eventually to a point where the part can no longer fulfil its intended function. Micro-mechanical test methods can be an advantageous approach to analysing the fibre-matrix interface as they reduce material consumption and simplify the analysis compared with composites. In this paper some of the disadvantages of the commonly used microbond test are discussed, particularly difficulties related to testing of vinyl esters. A new micro-mechanical interface test is introduced which induces failure at the interface in mode I as opposed to the shear failure that occurs with microbond and several other contemporary interface tests.

1. Introduction

The global use of fibre reinforced composite materials continues to grow year on year, across numerous industries. This is driven by their universally attractive properties such as low density, high strength and stiffness and resistance to corrosion, as well as a wide applicability due to the range of reinforcement fibres and matrix materials on the market. Fibre reinforced polymers (FRPs) form a large percentage of the composites market and are relatively mature from a technological viewpoint. A huge array of matrix materials, both thermosets and thermoplastics, are available depending on the requirements of the application. The most widely used reinforcement fibre is glass fibre (GF) due to its high specific properties and low cost. Other important fibres include carbon, basalt, aramid and some polymer fibres. In addition to these generally uniform manufactured fibres there is also a growing market for non-uniform plant-derived natural fibres.

The interface between matrix and reinforcement fibre is one of the most important factors in the success of any composite and significant research effort has been made to understand its role [1–4]. The vast majority of fibres require surface modification to improve compatibility with matrix materials and to produce a strong interface. The precise nature of the interface, or interphase region more generally, is challenging to define and so its discussion is often simplified to one of adhesion. Over the past several decades, a number of micro-mechanical tests have been created and subsequently improved to measure adhesion. In polymer composites research the microbond test has been applied relatively widely, as have the single fibre fragmentation, pull out and push out tests. Advantages of such tests include provision of direct data about interfacial adhesion and relatively small material consumption compared with composite macro-tests.

All of the interface tests listed above share some important features; for example, test specimens are considered symmetric around the longitudinal fibre axis and the failure induced at the interface is assumed to be of a mode II shear type. The most common data output from these tests is a value of apparent interfacial shear strength (IFSS); although more complex versions have been formulated to obtain parameters such as interfacial fracture energy and shear sliding stress [5]. In the literature, however, some critiques have been levelled at all of these micro-mechanical tests [6, 7]. These include that the centro-symmetric form of failure is unrealistic and that results may be significantly influenced by factors such as fibre surface roughness and cure shrinkage during sample manufacture.

Another challenging aspect of these tests can be the effect of the relatively small scale on sample preparation. The microbond test, for example, has been applied relatively widely to different matrix materials [3, 8, 9] but it is not unusual to encounter effects that must be mitigated against to obtain reliable test data. For example, the apparent adhesion varies due to degradation of thermoplastics [10], with relatively small changes in testing temperature [11, 12] or due to thermoset mixing ratios away from stoichiometry [13, 14].

In this paper, specific challenges related to the micro-mechanical adhesion testing of a vinyl ester (VE) resins system are presented and discussed. The procedure developed to successfully perform adhesion testing between VEs and reinforcement glass fibre is described. Following this, a new micro-scale interface test is introduced which addresses some of the criticism that has been made of the most commonly used micro-scale tests for measurement of adhesion. Some early results from this new single fibre peel test are presented.

2. Experimental

2.1. Materials

Two types of glass fibre were utilised in the work described in the paper. A commercially available Advantex E-glass fibre with 17 μm nominal diameter was provided as a single end continuous roving. The fibre was sized at point of manufacture by a contact roller then dried for 24h as a 20kg package. The size applied was for general compatibility with thermoset resins (including vinyl esters) and constituted approximately 0.6 wt% of the total product weight. It is designated GF_{com} hereafter. A smaller 10 kg roving was supplied by Owens-Corning; it was cooled by water spray following fibre drawing but had no surface size applied. The fibres are, therefore, referred to as being bare or unsized, GF_{us}.

A sample of lab-scale manufactured glass fibre was provided by RWTH Aachen Institute of Textile Technology, produced using a small experimental bushing. Fibres were of E-glass formulation; they were not coated with any size but were cooled by water spray before being wound onto cardboard tubes. 203 individual filaments were provided with continuous length of approximately 50 m. The fibre thickness was about 35 μm; these unsized fibres are referred to as GF_{ITA}.

A commercially available vinyl ester resin (VE_{vac}) designed for vacuum infusion with approximate styrene content of 50 wt% was supplied. Curing was achieved using the peroxide based catalyst PBC-21 from United Initiators GmbH by mixing at a ratio of 97.5 : 2.5 by weight respectively. The cure and post-cure procedures and their effect on results form part of the research presented in this paper; the manufacturer recommends a cure cycle of 24 h at room temperature followed by 24 h at 60 °C.

Epoxy resin DER 332 and triethylenetetramine were sourced from Sigma Aldrich and mixed at the stoichiometric ratio. The resin was mixed and degassed under vacuum for approximately 10 mins before application. The cure cycle utilised was heating to 60 °C for one hour followed by a further two hour isothermal at 120 °C, with intermediate heating rates of 2 °C/min.

2.2. *The microbond test*

The adhesion between reinforcement fibres and VE/epoxy resin was measured using an in-house designed microbond rig. The experimental setup is based on the original conception of Miller et al. [15] and has been described in detail in the literature [16]. A schematic of the system used is shown in Figure 1.

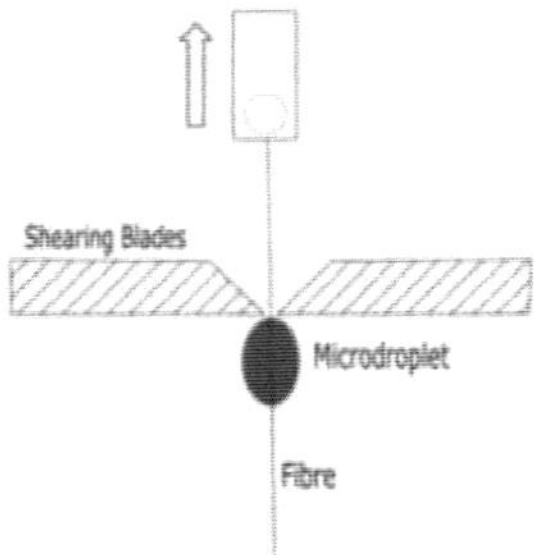

Figure 1: Image and diagram showing parts of the microbond rig used

Tests were carried out at a constant extension rate of 0.1 mm/min and only droplets within a diameter range of approximately 70-120 μm were selected. The apparent interfacial adhesion of each sample was calculated using (1)

$$\tau = \frac{F_{max}}{\pi d_f l_e} \qquad (1)$$

In this equation F_{max} is the peak of the load deformation plot during debonding, d_f the fibre diameter and l_e the embedded length of the fibre within the droplet. At least 20 individual tests were carried out for each condition to obtain an average apparent interfacial shear stress (IFSS).

The sample preparation process comprises several steps which are of particular importance in this work due to the influence they can have on the value of IFSS that is obtained. First, individual fibres are isolated and affixed to card frames with an adhesive so that resin can later be applied in a controlled manner. The standard adhesive used for this purpose is a cyanoacrylate gel superglue which is given 24 hours to fully dry. Next, a VE or epoxy resin was mixed following the relevant directions and applied to fibres using a thin piece of steel wire to produce droplets in the preferred size range. The desired cure cycle was then initiated immediately.

3. Microbond testing of vinyl ester resin

A significant volume of data regarding the microbond testing of various fibres with epoxy resins and with many different thermoplastic polymers can be found in the literature. Research on polyesters and vinyl esters, in particular, are less prevalent although some can be found [9, 17].

The following sub-sections discuss results acquired using the fibre GF_{com} and resin VE_{vac}. Several factors were found to exert a significant influence on the apparent IFSS values obtained, all of which must be controlled in order that meaningful results are generated, therefore allowing comparisons to be made between fibre sizes and resin formulations, for example.

3.1. *Proper cure of samples: atmospheric control and post-cure*

Vinyl ester resins are capable of achieving similar mechanical performance as some epoxies, but often with more straightforward processing requirements. The expected apparent IFSS for glass fibres with many epoxy resins is in the range 40-50 MPa [18] when proper cure is achieved. Values obtained for commercially available GF and vacuum infusion VE resin are given in Figure 2.

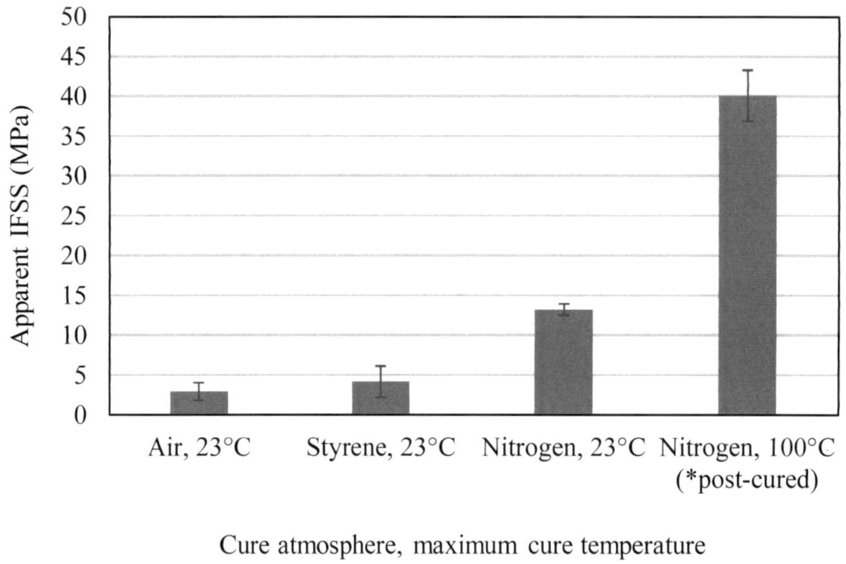

Figure 2: Apparent IFSS between GF_{com} and VE_{vac} dependent on cure atmosphere and maximum post-cure temperature

Very low values of apparent IFSS were obtained when initial curing for 24 h was carried out under air or a styrene-rich atmosphere. In reality droplets do not cure and solidify sufficiently to bear load, so there is no measurable interface by this method. Performing the initial 24 h cure under inert nitrogen allowed sufficient curing of droplets such that they could bear load during the microbond test, yielding a value of 13.2 MPa. Addition of a post-cure procedure at elevated temperature significantly increased the IFSS to around 40 MPa. The post-cure utilised was adapted from the manufacturers recommendation for vacuum infused panels: following 24 h at room temperature (under nitrogen) samples were post-cured in a convection oven at 60 °C for 24 h, 80 °C for 3 h and finally 100 °C for 1 hour.

The most critical step in the preparation of VE test specimens for microbond is initial curing under an inert atmosphere such as nitrogen. In the data presented in Figure 2 this was achieved using a vacuum oven to replace the air with dry nitrogen from a cylinder. Doing so significantly reduces the oxygen inhibition effect [19] which acts against the network forming polymerisation reaction.

The post-cure at significantly elevated temperature is required in order to increase the degree of cure of droplets. A standard DSC temperature ramp program was used to investigate the excess enthalpy of samples of VE resin which experienced different post-cure temperatures. Examples of these results are shown in Figure 3. The samples used were not taken directly from microbond specimens; they were prepared alongside them from the same batch of resin and distributed on release film to produce droplets with mass of 10-15 mg.

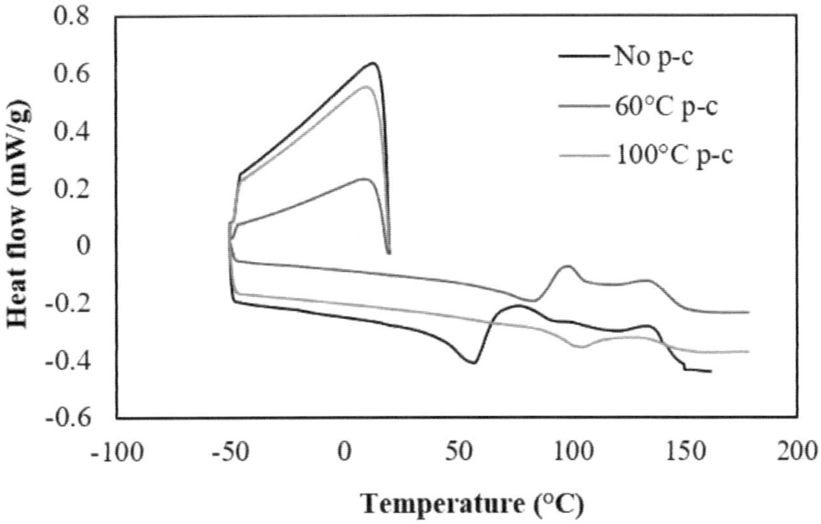

Figure 3: DSC curves (first temperature ramp) of VEvac samples cured at room temperature for 24 h under nitrogen, followed by (a) no post-cure (b) 60 °C for 24 h (c) 60 °C for 24 h, 80 °C for 3 h, 100 °C for 1 h

Elevated temperature following the initial room temperature cure induces further cross-linking within the VE network and this exothermic reaction can be observed in the data in Figure 3. Application of post-curing increases the initiation temperature for the exotherm and reduces the enthalpy of the reaction as expected. Without post-cure the exotherm initiated at 58 °C and excess enthalpy was 70.2 J/g. When a maximum post cure temperature of 60 °C or 100 °C was applied this shifted to enthalpy 31.2 J/g at 85.6 °C and 7.87 J/g at 105.6 °C respectively.

Testing of interfacial adhesion between glass, or other reinforcement, fibres and vinyl ester resins is possible using the microbond test. However, care must be taken to ensure that specimens are properly cured and failure to do so may lead to misleading results that underestimate the level of adhesion. In addition to initial curing under an inert atmosphere, it is necessary to post-cure samples to maximise degree of cure. The conditions required to achieve sufficient cure can be found from straightforward thermal analysis of the resin. The scale at which this analysis is carried out should, ideally, be as close to that of the specimens themselves; in the case of microbond this is very challenging as droplets have mass of the order of micro-grams.

3.2. Effect of fibre adhesive

The role of the fibre adhesive in microbond testing is only to restrain the individual filaments on a frame (usually made of card) so that droplets of the resin under study can be applied in a controlled manner. It is not considered that it should have any influence on the test, because it is applied and allowed to fully dry before the resin is introduced with which adhesion is to be measured. The fibre adhesive is always used in sufficient volume that the fibre will not slip during testing; the embedded length of fibre within this adhesive is many times larger than the embedded length within the resin being tested. Results shown in Figure 4, however, somewhat contradict this prevailing view.

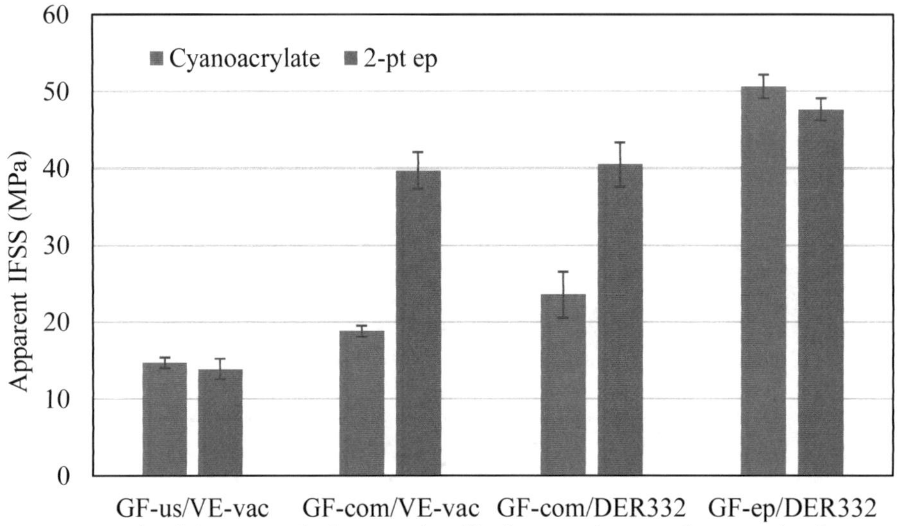

Figure 4: Apparent interfacial shear strength of various glass fibre/resin combinations from microbond testing. Fibre 'GF-ep' is a commercially sized product specifically designed for compatibility with epoxy resins

When using unsized GFs the adhesion obtained is very low between 12-14 MPa. Relatively low adhesion is expected as there is no silane or size present to enhance chemical bonding between the fibre and matrix; instead it is likely a combination of forces from cure shrinkage of the resin and friction. No difference in apparent adhesion was found by changing the fibre fixing adhesive. Similarly, microbond data for an epoxy-compatible GF (GF_{ep}) with a simple epoxy resin were unaffected by the fibre adhesive. The value of approximately 50 MPa is typical for DER332 epoxy resin on fully sized glass fibres.

A very clear effect, however, was found when microbond testing of the multi-resin compatible fibre GF_{com} was carried out. The two central pairs of columns in Figure 4 show that low values of apparent IFSS were obtained when fibres were fixed in place with cyanoacrylate 'superglue' adhesive; which has been the standard selection for microbond testing of resins and thermoplastics. By changing only the fibre adhesive, to a 2-part room temperature cure epoxy, significantly higher values of IFSS were measured. The data strongly suggest that this effect is due to some interaction with the fibre size of GF_{com}: it was not observed for different sizing formulation on GF_{ep} or for unsized fibre GF_{us}.

A satisfactory explanation for this effect has yet to be found; in many ways the result is a curiosity as the use of cyanoacrylate based superglues does not appear to present similar problems when mircobond testing with other reinforcement fibres. Currently, it is believed to be a peculiarity of the fibre GF_{com}; therefore, it may be the result of a reaction with an element or elements within this glass fibre size formulation. It is known that application of cyanoacrylate adhesive produces some vapours which can preferentially deposit [20] and initial experiments suggest that this is being deposited on the surface of the GFs. It is not yet understood, however, why this only effects the particular size on GF_{com} fibres.

4. Single fibre peel test (SFPT) – a mode I interface test

4.1. *Introduction*

As outlined in the introduction, some criticisms have been made of how representative the microbond, and other mode II micro-mechanical interface, tests are in relation to the mechanisms of failure within real polymer composites. A new interface test could, therefore, be of significant benefit to the study of adhesion in composites. In fact, such a method was proposed and demonstrated in a limited capacity within the literature over 20 years ago. A peel test based on a single fibre was described and results presented using an epoxy resin and several fibre types including glass, carbon and aramid [21, 22].

Peel testing is commonplace for the study of numerous materials at the macro-scale, such as adhesive joints and pressure-sensitive tapes, and testing is governed by numerous standards. Common methods include 90°, 180°, T-peel and climbing drum [23–26]. It has also found some utility in the study of internal interfaces within natural fibres [27]. At a basic level peel testing is simple: an average debonding force and peel length (normal to the direction of peel) must be measured, and from this the energy of fracture may be calculated.

4.2. Test specimens

In previous work, small capsule of resin were allowed to gel for several hours; then fibres were 'floated' on top to ensure they did not sink too deeply in the resin and become embedded by more than 50 % of their circumference [22]. To reduce sample preparation time, an alternative method has been developed. Approximately 1 ml of mixed resin was placed at the centre of a glass or PMMA microscope slide and immediately spun using a Laurell WS-650-23 Spin Coater. Spinning at 9000 rpm for 30 s produced a thin and consistent layer of VE or epoxy region in the central portion of the slide. A single glass fibre was then manually placed on to the resin, parallel to the slide width.

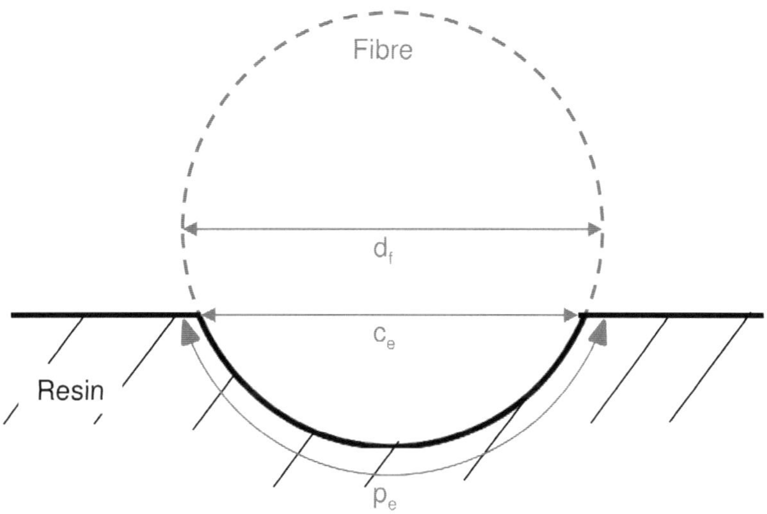

Figure 5: Diagram showing key geometry of a part embedded fibre

A model sample is shown in Figure 5 with the key dimensions that are required in order to calculate the embedded perimeter p_e of the fibre: d_f the fibre diameter and c_e the embedded chord of the fibre. By spin coating a thin, controlled layer of resin it is possible to ensure that (depending on diameters of fibres selected) fibres will be embedded by less than 50 % of their circumference.

4.3. Theoretical background

The energy of fracture for peel test specimens is calculated from the average force of debonding divided by the sample width over which it acts. For large and regular shaped materials this dimension may be well defined and easily measured, but for the single fibre peel test it is more challenging. The embedded perimeter, which is an arc of a circle, must either be measured directly or calculated from other dimensions. This was achieved by obtaining d_f and c_e (Figure 5): fibre diameter was obtained using optical microscopy and c_e by measuring the profile of the depression left behind after a fibre was peeled. This was achieved using a stylus profilometer and importing the profile into Origin to measure. A typical profile is shown in Figure 6.

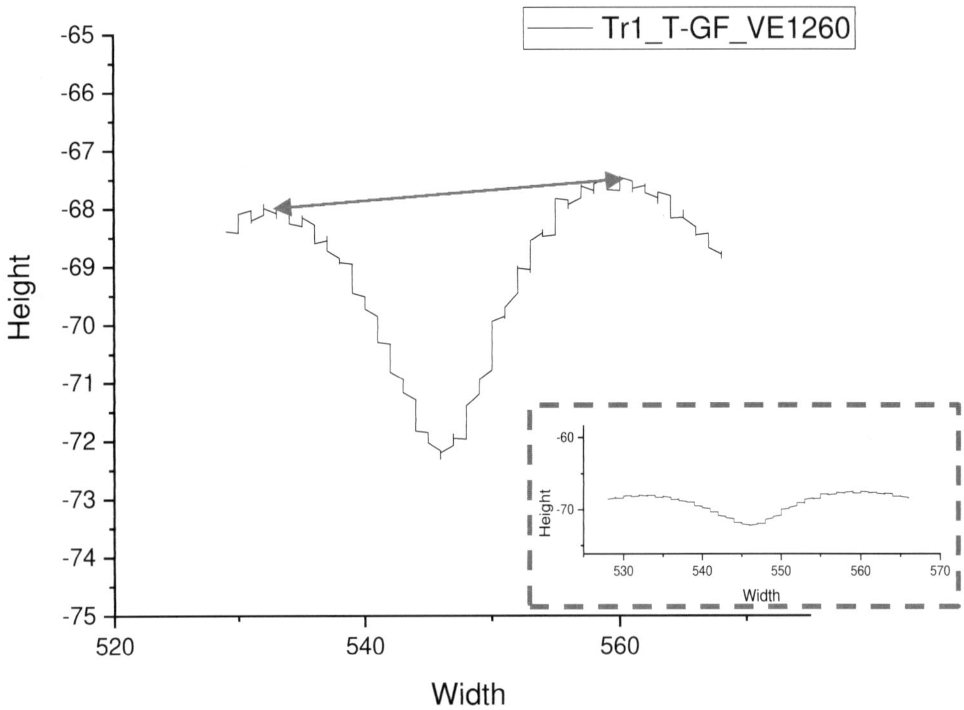

Figure 6: Typical profilometry data of GF_{ITA} (diameter approx. 34 μm) after fibre debonding. Inset (dotted line) shows 'true' proportions of the profile on equally scaled x and y axes (dimensions μm)

With these dimensions the embedded perimeter of the fibre specimen can be calculated at a number of points along the length for which average peel force ($\bar{F}$) is known and stable, according to (2)

$$l_e = \frac{\pi d_f}{180} sin^{-1}\left(\frac{0.5c_e}{r}\right) \quad (2)$$

The fracture energy G_I^c for peeling at 90° is found from (3) or if an oblique peel angle φ° is used then according to (4).

$$G_I^c = \frac{\bar{F}}{l_e} \quad (3)$$

$$G_I^c = \frac{\bar{F}(1 - cos\varphi)}{l_e} \quad (4)$$

4.4. Testing procedure

Fibre peel tests were performed at extension of 1 mm/min and an oblique angle of 30°, in order to maximise the detectable debond force. This force increases with decreasing angle but is unreliable once φ < 30°[22]. Testing was carried out using a single column Testometric M250-2.5CT tensile machine with a 5 N load cell. An aluminium base plate with dovetail was manufactured: this allowed different test fixtures with a range of angles to be easily substituted. Test fixtures were made from PLA using a 3D printer.

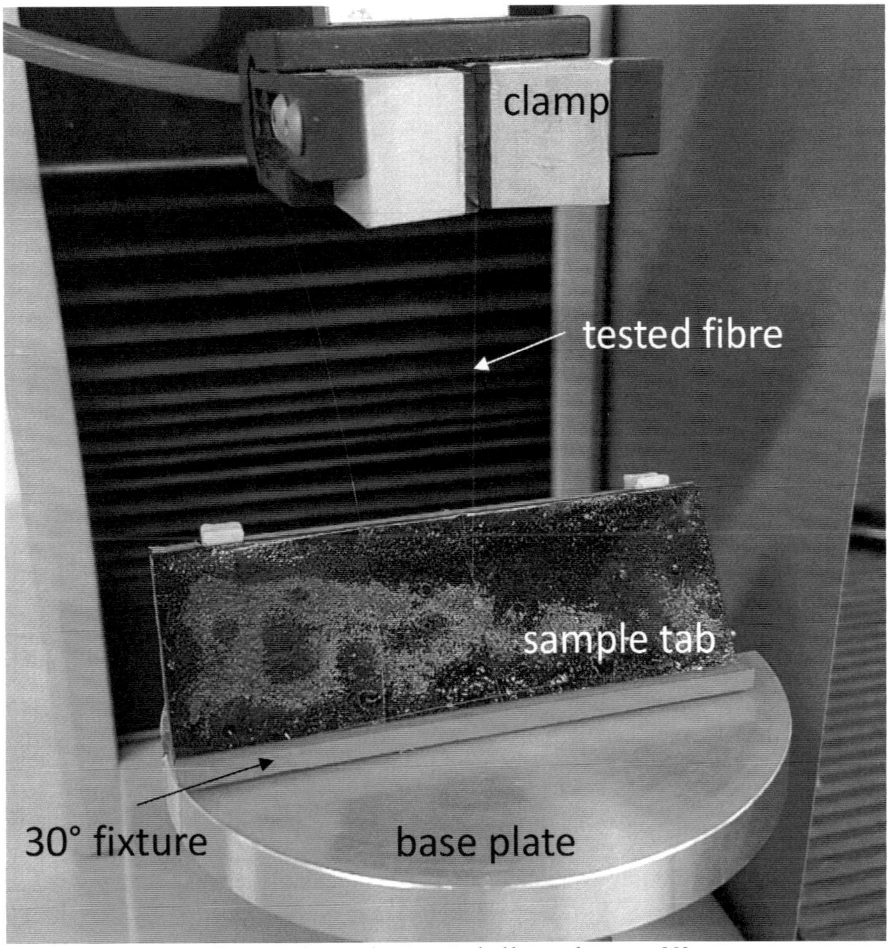

Figure 7: Image showing single fibre peel setup at 30°

The image in Figure 7 shows the 30° fixture with a sample mounted; in this case DER 332 was spun on to a black PMMA slide. In the image the tested fibre can be observed; it is clamped directly in the rubberised pneumatic grips of the instrument. Due to the very small debonding forces involved it does not appear that fibre slippage during testing is a problem. This is a significant advantages as it allows direct fibre gripping rather than the use of a slip foil and adhesive to be applied to each test sample before loading. A diagram showing the progression of the test is shown in Figure 8.

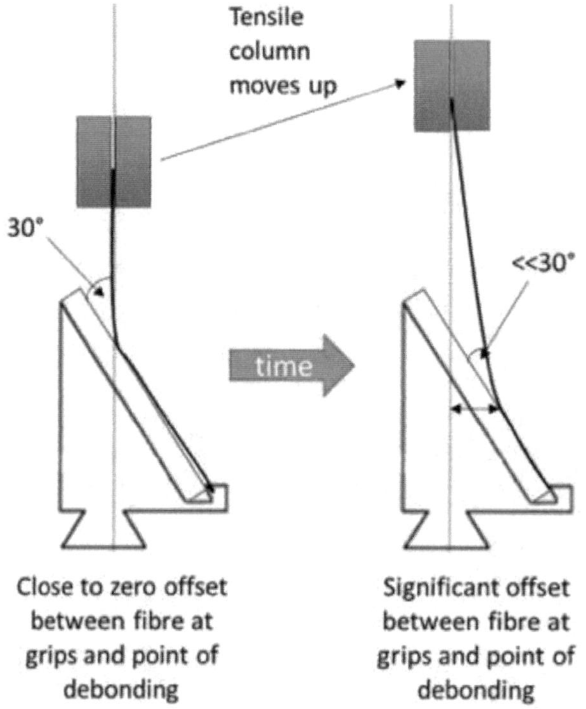

Figure 8: Progression of SFPT experiment showing change in angle between fibre and resin surface over time

The current test setup does not allow for compensatory movement of the sample in sync with the vertical displacement of the fibre/tensile column. Therefore, the debond angle changes gradually throughout a test. In the 30° case, for example, fibre slack that is required during mounting must first be displaced in the early moments of the test. During this time the angle exceeds 30°. The specimen is observed and, when the fibre is parallel with the tensile column, the absolute (machine) displacement is recorded. Average force data from ± 0.5 mm of this point only are considered. The right hand part of Figure 8 shows that, as the test progresses further, the angle between fibre and matrix decreases below that defined by the test fixture.

4.5. Initial results

The load extension plot for an unsized glass fibre (GF_{ITA}) is shown in Figure 9. This plot shows some typical features that can occur during single fibre peel testing. There is an initial period in which slack in the fibre is taken up before it bears any load. The 'middle' portion of the test is of most interest as the useable data are found here. In this case, the fibre was observed to be parallel with the column of the tensile machine at extension of approximately 3 mm. The force data around this point, therefore, were used for analysis of the debond energy.

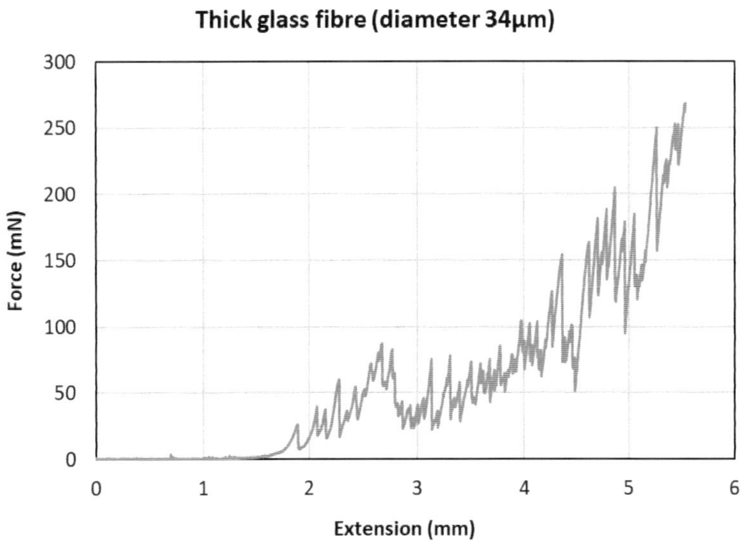

Figure 9: Load-extension plot of GF_{ITA} in DER 332 peeled at 30°

The average force from 3-3.5 mm was 44.2 mN and the fibre was estimated to be 50% embedded from images obtained by optical microscopy. From (3) the fracture energy was estimated as 27.7 J/m^2. Images of the fibre taken post-test, however, suggest that there is likely significant error in this value. Residual matrix was observed along its length; it is possible, therefore, that both debonding at the fibre-matrix interface and matrix cracking contributed to the force detected.

Future investigations using this test method will focus on approaches to minimise this effect. The resin distribution method has been optimised to ensure a layer is created which is thinner than the fibre radius. Testing at additional angles of 45, 60 and 90° will also be carried out.

5. Conclusions

Micro-mechanical testing of the interface between fibres and polymer matrices remains a useful approach in spite of the known (and at times unanticipated) challenges. It was shown that application of the microbond test to investigation of glass fibre/vinyl ester specimens presents several issues which must be overcome in order to achieve accurate and representative values of apparent interfacial shear strength. Initial cure of samples must be performed under an inert atmosphere and, thereafter, post-curing at elevated temperature is also necessary. An unexpected interaction was discovered between the size on a resin-compatible glass fibre and the cyanoacrylate based adhesive commonly used to restrain fibres for specimen preparation. Deposition of the adhesive from its vapour state on to the fibre appears to be the cause; however, the mechanism by which this significantly reduced the adhesion between fibre and resin is not yet understood.

A new micro-mechanical test method for characterisation of interfacial adhesion in mode I, the single fibre peel test, was introduced. It is intended to address some criticisms and limitations of the pre-existing micro-mechanical interface tests which induce mode II failure.

Acknowledgment

This work was supported by the DACOMAT project from the European Union's Horizon 2020 research and innovation programme 5 under GA No. 761072.

References

[1] Drzal LT, Madhukar M (1993) Fibre-matrix adhesion and its relationship to composite mechanical properties. *J Mater Sci* **28**:569–610. doi: 10.1007/BF01151234
[2] Jones FR (2010) A review of interphase formation and design in fibre-reinforced composites. *J Adhes Sci Technol* **24**:171–202. doi: 10.1163/016942409X12579497420609
[3] Dey M, Deitzel JM, Gillespie JW, Schweiger S (2014) Influence of sizing formulations on glass/epoxy interphase properties. *Compos Part A Appl Sci Manuf* **63**:59–67. doi: 10.1016/j.compositesa.2014.04.006
[4] Zhandarov S, Mäder E (2014) An alternative method of determining the local interfacial shear strength from force-displacement curves in the pull-out and microbond tests. *Int J Adhes Adhes* **55**:37–42. doi: 10.1016/j.ijadhadh.2014.07.006
[5] Sørensen BF (2017) Micromechanical model of the single fiber fragmentation test. *Mech Mater* **104**:38–48. doi: 10.1016/j.mechmat.2016.10.002
[6] Piggott MR (1997) Why the fibre/polymer interface can appear to be stronger than the polymer matrix. *Compos Sci Technol* **57**:853–857. doi: 10.1016/S0266-3538(96)00151-0
[7] Piggott MR (1997) Why interface testing by single-fibre methods can be misleading. *Compos Sci Technol* **51**:965–974.
[8] Graupner N, Rößler J, Ziegmann G, Müssig J (2014) Fibre/matrix adhesion of cellulose fibres in PLA, PP and MAPP: A critical review of pull-out test, microbond test and single fibre fragmentation test results. *Compos Part A Appl Sci Manuf* **63**:133–148. doi: 10.1016/j.compositesa.2014.04.011
[9] Kim IC, Yoon TH (2000) Enhanced interfacial adhesion of carbon fibers to vinyl ester resin using poly(arylene ether phosphine oxide) coatings as adhesion promoters. *J Adhes Sci Technol* **14**:545–559. doi: 10.1163/156856100742735
[10] Yang L, Thomason JL, Zhu W (2011) The influence of thermo-oxidative degradation on the measured interface strength of glass fibre-polypropylene. *Compos Part A Appl Sci Manuf* **42**:1293–1300. doi: 10.1016/j.compositesa.2011.05.011
[11] Thomason JL, Yang L (2011) Temperature dependence of the interfacial shear strength in glass–fibre polypropylene composites. *Compos Sci Technol* **71**:1600–1605. doi: 10.1016/j.compscitech.2011.07.006
[12] Thomason JL, Yang L (2014) Temperature dependence of the interfacial shear strength in glass-fibre epoxy composites. *Compos Sci Technol* **96**:7–12. doi: 10.1016/j.compscitech.2014.03.009
[13] Minty R, Yang L, Thomason JL (2018) The influence of hardener-to-epoxy ratio on the interfacial strength in glass fibre reinforced epoxy composites. *Compos Part A Appl Sci Manuf* **112**:64–70. doi: 10.1016/j.compositesa.2018.05.033
[14] Petersen H, Minty R, Thomason JL, et al. (2018) The amine:epoxide ratio at the interface of a glass fibre/epoxy matrix system and its influence on the interfacial shear strength. *Compos Interfaces* 1–13. doi: 10.1080/09276440.2018.1511107
[15] Miller B, Muri P, Rebenfeld L (1987) A Microbond Method for Determination of the Shear Strength of a Fiber/Resin Interface. *Compos Sci Technol* **28**:17–32. doi: 10.1016/0266 3538(87)90059-5
[16] Yang L, Thomason JL (2012) Development and application of micromechanical techniques for characterising interfacial shear strength in fibre-thermoplastic composites. *Polym Test* **31**:895–903. doi: 10.1016/j.polymertesting.2012.07.001
[17 Bénéthuilière T, Duchet-Rumeau J, Gérard JF, et al. (2015) Physico-chemistry of vinylester/glass fiber interfaces used in SMC composites. I*CCM Int. Conf. Compos. Mater.* 2015-July:
[18] Bryce D, Yang L, Thomason JL (2019) Microscale curing performance of epoxy resin in the microbond test [presentation]. *5th Int. Conf. Mech. Compos.*, Lisbon, Portugal
[19] Scott TF, Cook WD, Forsythe JS (2002) Kinetics and network structure of thermally cured

vinyl ester resins. *Eur Polym J* **38**:705–716. doi: 10.1016/S0014-3057(01)00244-0
[20] Wargacki SP, Lewis LA, Dadmun MD (2007) Understanding the chemistry of the development of latent fingerprints by superglue fuming. *J Forensic Sci* **52**:1057–1062. doi: 10.1111/j.1556-4029.2007.00527.x
[21] Alimuddin MA (1999) Peel Test for the Evaluation of Environmental Effects on Fiber-Polymer Interface. University of Toronto
[22] Alimuddin MA, Piggott MR (1999) Fracture Toughness of Fiber-Polymer Interfaces Estimated From Single Fiber Peel Tests. *Polym Compos* **20**:655–663.
[23] D6862-11 (2010) Standard Test Method for 90 Degree Peel Resistance of Adhesives. Annu B ASTM Stand. doi: 10.1520/D3167-10R17.1
[24] D903–98 (2003) Standard Test Method for Peel or Stripping Strength of Adhesive Bonds. Annu B ASTM Stand 1–3. doi: 10.1520/D0950-03R11.responsibility
[25] D1876-08 (2015) Standard Test Method for Peel Resistance of Adhesives (T-Peel Test). Annu B ASTM Stand. doi: 10.1520/D1876-08R15E01.Copyright
[26] D1781–98 (2012) Standard Test Method for Climbing Drum Peel for Adhesives. Annu B ASTM Stand 1–4. doi: 10.1520/D1781-98R12.2
[27] Réquilé S, Le Duigou A, Bourmaud A, Baley C (2018) Peeling experiments for hemp retting characterization targeting biocomposites. *Ind Crops Prod* **123**:573–580. doi: 10.1016/j.indcrop.2018.07.012

Characterisation of fibre/matrix interfacial fracture energy using the single fibre peel experiment

Daan J Hottentot Cederløf and Bent F Sørensen

DTU Wind Energy, Technical University of Denmark, Risø Campus, 4000 Roskilde, Denmark

E-mail: dace@dtu.dk

Abstract. An experimental method for determining the interfacial fracture energy of a single fibre undergoing peeling is presented. Peeling of a partially embedded single fibre is observed under scanning electron microscopy. The fracture energy of the fibre/matrix interface is determined by analysis of the measured curvature of the fibre near the crack tip. This study serves as a demonstration of concept for the characterization of fibre/matrix interfaces through the single fibre peel experiment. A glass fibre/vinylester interface is used as an initial test case, from which obtained interfacial fracture energies was found to be in the range from $2\,\mathrm{Jm^{-2}}$ to $14\,\mathrm{Jm^{-2}}$.

1. Introduction

The utilization of fibre reinforced polymer (FRP) composites is widespread in industries such as aerospace, automotive and wind energy. FRP composites are made up of long aligned fibres which provide high strength and stiffness, whereas the polymer (referred to as matrix) serves to protect and support the fibres. The ability to tailor component properties by controlling the fibre orientation, combined with high specific stiffness and strength in the fibre direction, means that lightweight structures may be realised. However, the out-of-plane and transverse properties are inferior in comparison with the properties in the fibre direction.

Delamination is a typical out-of-plane damage mechanism observed in composite structures, where cracks propagate in-between layers of fibres. Propagation of delamination cracks can cause loss of structural stiffness and lead to structural failure [1, 2]. Delamination growth however may be slowed, or even arrested, by fibre bridging; a phenomenon where single fibres, or ligaments of fibres, cross over between the two delaminating surfaces [3]. As the bridging fibres strain they transfer tractions between the crack faces and act as an energy absorbing (toughening) mechanism, reducing the delamination crack growth rate [4, 5]. However, this toughening mechanism is not fully understood or controlled at the micro-mechanical level, meaning it cannot be reliably incorporated during design of critical composite structures. In the case of fibre bridging under delamination, the peeling fracture energy of the fibre/matrix interface is expected to be a factor governing whether a bridging fibre breaks, or stays intact and acts as an energy absorbing mechanism [3]. It is thus of great importance to be able to characterize the mechanical properties of fibre/matrix interfaces.

Numerical modelling of fibre bridging in wood was demonstrated by Mirzaei et al. [6]. Bridging elements (single fibres and ligaments of fibres) were modelled as cross linking beams. Mirzaei et al. mentioned briefly that if the fibre was only slightly embedded that a peeling action would be a source of energy dissipation and thus a peeling fracture energy is needed. A reference is made to the work by Kinloch et al. ([7]), with no further explanation for how such a fibre/matrix peeling fracture energy value is to be obtained. Kinloch et al. [7] studied the adhesive fracture energy of peeling flexible laminates. By accounting for internal strain energy, energy dissipated by bending and tensile deformation, they determined the geometry independent fracture energy of the interface between the peeling arm and substrate. This was concluded to be a material specific property due to the insensitivity to geometry, however its scope is limited to laminae interfaces.

1.1. Background and Motivation

The interface between the fibre and matrix is one of the most critical aspects of composite materials. Interfacial properties are controlled by the *size* (or *sizing*); a surface coating applied to the fibre during production [8]. Sizing is applied to glass fibres to facilitate bonding the fibre with a specific resin type [8]. Given the fact that size formulae are closely guarded industry secrets and that no test standards exist, the fibre matrix interface remains one of the least understood aspects of composite materials [8, 9].

Existing micro-mechanical experiments, related to the fibre/matrix interface, focus on the shearing between fibre and matrix, in some form of pull-out or push-out of the fibre ([10, 11]. However, stress concentrations, full circumferential fibre/matrix contact and fibre/matrix friction may give erroneous results [12]. At the very least they present significant challenges when pursuing a pure interfacial fracture energy because two processes are occurring at the interface simultaneously; debonding (fracture) and frictional sliding. In the debonding process, chemical bonds are broken at the fibre/matrix interface (fracture energy measured in Jm^{-2}) whereas in the sliding process shear stress (measured in Nm^{-2}) is introduced due to friction [10]. As evidence of the difficulty of isolating these two parameters: three different studies (using either analytical or numerical models) of the same test data (single fibre fragmentation test of glass/epoxy [13]) resulted in three different interfacial fracture energies: $120\,\mathrm{Jm}^{-2}$ [13], $12\,\mathrm{Jm}^{-2}$ [14] and $27\,\mathrm{Jm}^{-2}$ [10]. Furthermore, these shear based experiments are not applicable in the case of fibre bridging where fibres are exposed to both peeling and pulling.

A single fibre peel test was proposed by Alimuddin and Piggott [12] to determine the fracture energy of a fibre/matrix interface. Fibres (glass, carbon and aramid) were placed in resin that had gelled so that they sank to approximately half their diameter. After curing the single fibres were peeled out of the resin whilst placed on a microbalance that allowed both force and displacement to be measured. These measurements were used to calculate the fracture energy for the fibre/matrix interface. The fracture energy was determined to be $140 \pm 50\,\mathrm{Jm}^{-2}$ for the glass/epoxy interface of a $22\,\mu\mathrm{m}$ thick fibre. The testing of carbon/epoxy and aramid/epoxy interfaces resulted in fracture energies of $60 \pm 20\,\mathrm{Jm}^{-2}$ and $250 \pm 40\,\mathrm{Jm}^{-2}$, respectively. It was noted that the aramid fibre showed fibrilation, i.e. the fibre itself experienced splitting, which would contribute to the higher apparent fracture energy of $250 \pm 40\,\mathrm{Jm}^{-2}$. Furthermore, fibres of all three types may have been overembedded in the epoxy resin, which could increase the apparent fracture energy due to plastic deformation in the resin covering the fibre sides.

Liechti and Chai [15] performed mixed mode fracture mechanics experiments on pure glass/epoxy interfaces at different phase angles. The phase angle, ψ, is related to the complex stress intensity factor used in linear elastic fracture mechanics of interfaces, see Rice [16] and Hutchinson and Suo [17]. In the case of interface cracks, the crack tip stress field does not separate into pure normal (Mode I) and tangential (Mode II) fields. In simplified terms, a phase angle of around 0 degrees indicates a predominantly normal opening whereas a phase angle of 90 degrees indicates a predominantly tangential opening at the crack tip. Liechti and Chai found interfacial fracture energies ranging from 4 to $36\,\mathrm{Jm}^{-2}$ for $-50 < \psi < 88$ deg, respectively. They noted that for $0 < \psi < 45$ deg the fracture energy is relatively independent of mode mixity.

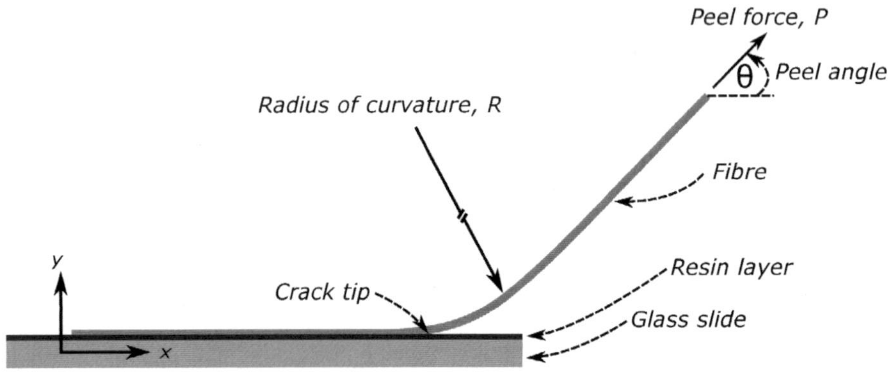

Figure 1: Illustration of test specimen.

An interesting finding by Kinloch et al. [7] observed that the mode-mixity near the crack front of thin films under peel is similarly insensitive to the applied peeling angle. This is further supported by Thouless and Jensen [18] who found that the phase angle for a thin films, with high stiffness, undergoing delamination (peel) is relatively insensitive to the peel angle. The phase angle reported by Thouless and Jensen remained constant at -37.9 deg for all peel angles, which lies in the range of phase angles described by Liechti and Chai [15] where fracture energy is insensitive to mode mixity.

McDaniel et al. [19] utilized a single fibre peel experiment to study the splitting of single ultra high molecular weight polyethylene (UHMWPE) fibres i.e. the fibre itself is split in two. Their sample preparation included partially embedding single fibres onto glass slide spin-coated with epoxy. An energy balance method, inspired by the work of Kinloch et al. [7] was used to determine the fracture energy of a splitting fibre under different fracture modes. The fibre matrix interface was however not explored.

Kawashita et al. [20] developed a method that uses image analysis to derive the radius of curvature and the root rotation angle of a laminate undergoing peel. Laminate arms with thickness in the order of 1 mm were investigated using high quality digital photography. This is several orders of magnitude greater than a single glass fibre with a diameter in the order of 17 µm. Resulting fracture energy values (obtained by global energy analysis similar to Kinloch et al. [7]) were in good agreement with their analytical results.

The present paper presents a novel, scaled down adaptation of the experimental method shown by Kawashita et al. [20] by analyzing the peeling of a single fibre inside a scanning electron microscope (SEM). The general idea of the single fibre peel experiment is shown in fig. 1. A single fibre is peeled with peel force, P, at an angle, θ. The fibre/matrix interfacial fracture energy is determined by measuring the curvature of the fibre near the crack tip. This experiment may be used as a means of validating fibre bridging models or to serve as a screening tool for fibre/matrix adhesion.

2. Method

2.1. Materials

Samples were prepared by spin-coating a glass microscope slide (25x75 mm) with approximately 1 ml commercial vinyl ester resin (VE-1260, Polynt Composites UK Ltd., Stallingborough, UK). The sample preparation was performed at the University of Strathclyde [21]. The spin coater (WS-650-23 Spin Coater, Laurell Technologies Corp., city of purchase, USA) was operated at 9000 RPM for a duration of 30−50 s to obtain a resin layer thickness (H_r) of 8 µm, approximately equal to the fibre radius, r, (see fig. 2). The device was accelerated to operating rotational speed

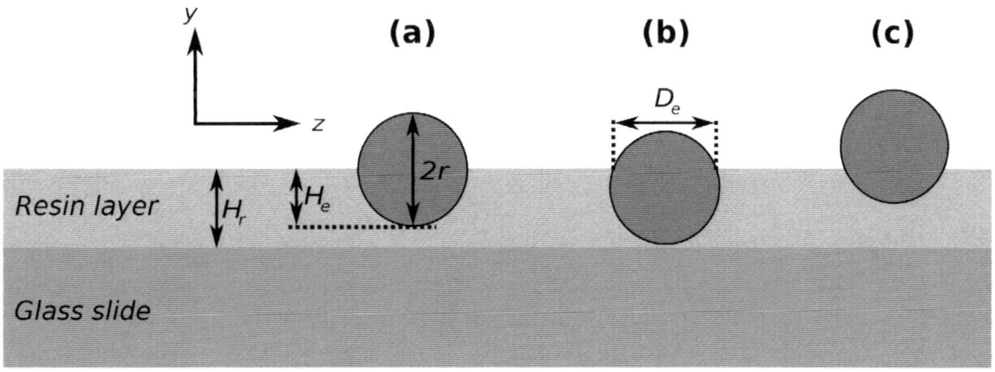

Figure 2: Illustration of test specimen cross section showing three levels of 'embeddedness': (a) fibre embedded to a height (H_e) of exactly its radius; (b) an over-embedded fibre; (c) an under-embedded fibre. The terms H_r and H_e refer to the resin height and the embedded height/depth of the fibre, respectively. The term D_e is the visible diameter when looking onto the fibre from a top view perspective (i.e. in the $x-z$ plane).

in 10 s and decelerated within 2−3 s. Commercial sized glass fibres (SE3030, 3B Fibreglass sprl., Battice, Belgium) with diameter 16−18 µm were hand placed onto the spin-coated microscope slide and cured according to supplier specifications. Final specimens were obtained by cutting the microscope slide to fit in the SEM fixture and sputter coating with gold (ca. 7 nm layer thickness).

2.2. *Experimental setup*

Specimens were tested and observed inside a scanning electron microscope (SEM) (EVO 60, Carl Zeiss A.G., Oberkochen, Germany) whilst mounted to a custom made fixture with a movable stage, as shown in fig. 3. The movable stage is actuated by a fine thread shaft which in turn is driven by a stepper motor. This results in a precise and accurate displacement control of the movable stage with roughly 100000 steps per mm of stage displacement. In order to observe the crack tip, the fixed and movable stages were tilted at an angle of $\varphi = 10°$ with respect to the electron beam originating from the LaB_6 filament (SEM gun), see fig. 3. This ensures that the edge of the microscope slide and any unevenness in the vinyl ester resin layer do not hide the embedded section of the fibre.

The SEM was operated at an accelerating voltage of 15 kV and a working distance of 10 mm. In order to capture a frame as close to the occurrence of fracture as possible every scan of the SEM was saved in a video recording. By employing a high scan speed with minimal noise reduction a frame rate of 2-2.5 frames per second was achieved.

2.3. *Video analysis*

A Matlab [22] script was prepared to extract and analyze still frames from the SEM video. After filtering and binarization, the top edge of the deformed fibre is extracted as a series of (x,y) coordinates. The deflection of the fibre was fitted in three parts: a straight line (the embedded part), a curved line and a straight line away from the detachment point. In fig. 4 these are labelled: linear fit embedded, non-linear fit and linear fit detached, respectively. The lower and upper bound of the non-linear fit are determined by the residuals of the two linear fits. The bounds are defined as the point where the residual (the distance between the fitted curve and the fibre edge) of the linear fit exceeds 1 µm. Note that the lower bound is not the same as the crack tip location. The x co-ordinate of the crack tip location is determined by visual inspection

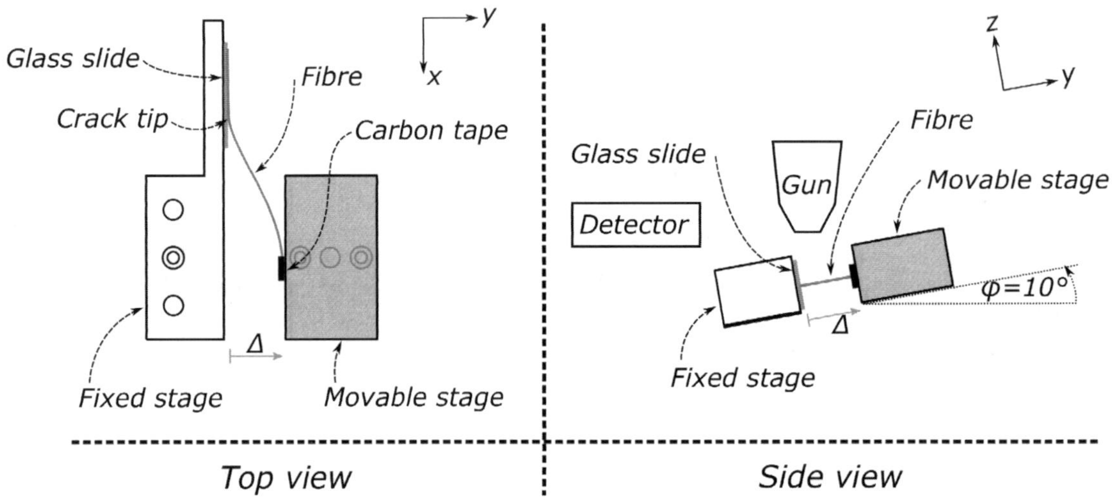

Figure 3: Illustration of test setup in the SEM. The top view is the view from the SEM gun. The movable stage translates in the y-direction a displacement Δ to peel the fibre from the glass slide mounted on the fixed stage.

of SEM images. Kawashita et al. [20] proposed an alternative method with a piecewise linear fit, this however requires extensive smoothing of the obtained fibre edge coordinates. The radius of curvature in the non-linear segment is obtained from differential geometry (eq. (1)) [23]. A detailed description of the image analysis method is detailed in the appendix.

$$\frac{1}{R(x)} = \frac{|y''(x))|}{[1 + (y'(x))^2)]^{\frac{3}{2}}} \tag{1}$$

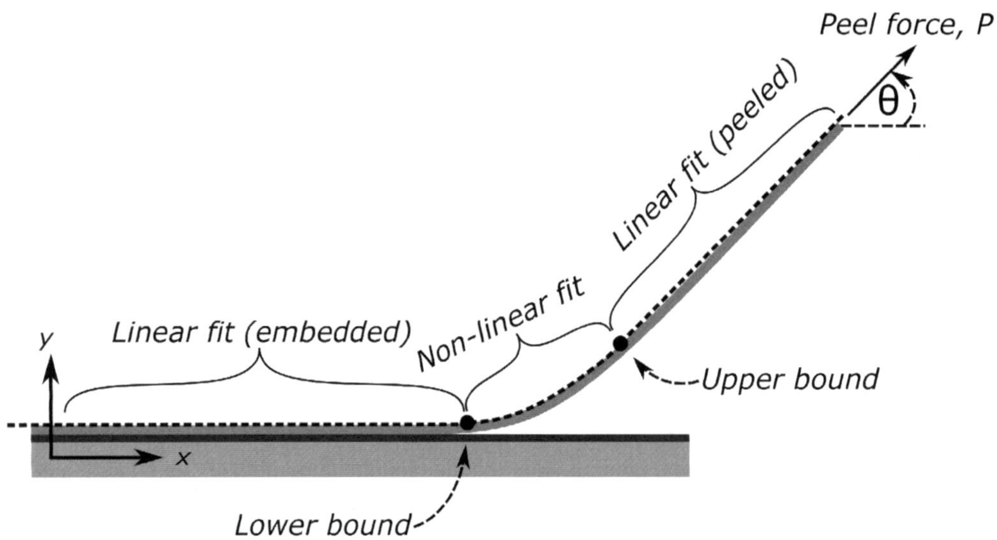

Figure 4: Illustration of the fitting procedure of the top edge of the fibre.

2.4. Fracture energy analysis

The fracture energy is determined by the curvature of the single fibre near the crack tip, the fibre modulus, E, and the fibre radius, r.

We start with the energy release of a beam under pure moments determined by the J-integral as given by [24]

$$J = \frac{M^2}{2BEI} \tag{2}$$

where M and B are the moment at the crack tip and beam width, respectively, E and I are the Young's modulus and second moment of area. In the case of the a single fibre the beam width is equal to two times the fibre radius, r (eq. (3)).

$$B = 2r \tag{3}$$

The second moment of area for a circular cross section is given by

$$I = \frac{1}{4}\pi r^4 \tag{4}$$

From classical beam theory [25] we obtain the relation between moment, M, and beam radius of curvature, R:

$$M = \frac{EI}{R} \tag{5}$$

Finally, we assume that the increase in fracture area to be the product of half the fibre circumference and the incremental crack growth, da, as shown in eq. (6). This implies that the embedded depth (H_e in fig. 2) is equal to the fibre radius. The validity of this assumption may be checked by microscopy of the substrate after complete peeling off of the fibre or by viewing the fibre from a top view ($x - z$ plane) prior to peeling to check how deep the fibre is embedded. The measurement of 'embeddedness' prior to peeling is demonstrated in fig. 5.

$$dA = \pi r da \tag{6}$$

We may then set the energy release to be equal to the energy consumed over a small crack growth, resulting in the relationship given by eq. (7), where G_c is the fracture energy.

$$JBda = \pi G_c r da \tag{7}$$

Isolating G_c in eq. (7) and inserting eqs. (2) and (3) gives eq. (8):

$$G_c = \frac{M^2}{2\pi r EI} \tag{8}$$

Further insertion of eqs. (4) and (5) into eq. (8) results in a solution for G_c purely in terms of fibre modulus, fibre radius and the radius of curvature, as given by eq. (9):

$$G_c = \frac{Er^3}{8R^2} \tag{9}$$

Table 1: Curvature and fracture energy results from single fibre peel analysis. An elastic modulus of 70 GPa was used to calculate the fracture energy (see eq. (9)).

Sample	Meas. No.	Radius of curvature R [m]	Fibre radius r [µm]	Fracture energy G_c [Jm^{-2}]
S3-02	1	7E-04	8.1	9
S3-02	2	6E-04	8.1	14
S3-02	3	8E-04	8.1	8
S3-03	4	12E-04	8.1	3
S3-03	5	14E-04	8.1	2

3. Results and observations

Two samples of SE3030/VE1260 were successfully tested in the SEM. The measured radii of curvature and calculated interfacial fracture energy are presented in table 1. It was possible to obtain multiple measurements from the same specimen. The data points in table 1 were taken from the frame before fracture occurred.

In fig. 5 a top view is presented of a single embedded fibre (specimen S7-03) and the same fibre un-embedded ca. 10 mm away. Measurement of the visible fibre diameter (16.7 µm) and measurement of the same fibre un-embedded 17.4 µm are indicated. The visible fibre diameter of the embedded portion of the fibre is 0.7 µm smaller showing that the fibre in this section was overembedded ($H_e > r$ in fig. 2) by 4% of the fibre diameter.

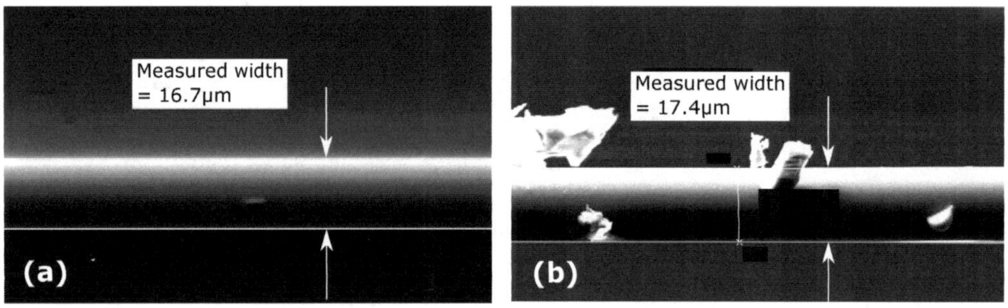

Figure 5: Top view ($x - z$ plane) width measurement of sample S7-03 (not tested in peel). **(a)** Measurement of the un-embedded fibre shows a fibre diameter, $2r$, of 17.4 µm. **(b)** Measurement of the same fibre embedded in resin shows a visible diameter (D_e) of 16.7 µm indicating that the fibre here is over-embedded (i.e. $H_e > r$ in fig. 2).

4. Discussion

Preliminary results show the feasibility of sample preparation and in-situ SEM observation of a single fibre undergoing peeling. However, a large scatter in the limited number of results are observed; between the two samples and between the measurements of the same sample. A lower radius of curvature (i.e. higher interfacial fracture energy) was systematically observed for sample S3-02 than for S3-03.

It is possible that the fibre 'embeddedness' is varying causing a measurable difference in fracture energy. If the fibre is over-embedded, the failure mode changes from a pure peeling action to also include elastic (and possibly plastic) shearing of the overlaying matrix, increasing the energy required to peel out the fibre. Sample S7-03 was checked for embeddedness, it was found that 96% of the fibre was visible. In comparison with the 'embeddedness' of up to 40%

reported by Alimuddin and Piggott [12], this is deemed acceptable. Future peel experiments must be initiated with an analysis of embeddedness prior to mechanical testing to ensure that the assumption of a perfectly half-embedded fibre holds.

The fracture energies obtained (between 2-14 $\mathrm{Jm^{-2}}$) are comparable to the mode I fracture energy values measured by Liechti and Chai (ca. 4 $\mathrm{Jm^{-2}}$) [15] for glass/epoxy interfaces but much lower than the fracture energy obtained by Alimuddin and Piggott for a glass/epoxy system (160 $\mathrm{Jm^{-2}}$) [12]. Two factors may account for that large discrepancy. Firstly, Alimuddin and Piggott reported that their fibres were over-embedded, with down to just 60% of the fibre diameter visible. This means that as the fibre is peeled out, the matrix that sits around the top of the fibre must be plastically deformed, thereby increasing the apparent toughness. Secondly, they used epoxy resin systems which are generally a tougher matrix system than vinyl ester [26, 27].

Several challenges present themselves when working with in-situ testing inside a scanning electron microscope. A single high resolution image typically takes in the order of 10 or more seconds to capture. This does not allow for dynamic events such as the peeling action to be captured. Resolution and sharpness were traded in favour of framerate to obtain a maximum of 2.5 frames per second. The images obtained were sufficient to determine the radius of curvature, however it did not allow for the crack front to be accurately located. This was further influenced by the charging of the fresh fracture surface, since it was not coated with a conductive material, resulting in white spots near the crack tip.

The limited working space within the scanning electron microscope is a primary constraint for this experiment. However, possible validation data for the obtained interfacial fracture energy results could be obtained through measurement of the tensile force in the fibre, as demonstrated by Alimuddin and Piggott [12]. Furthermore, a robust method for determining the crack front must be established, with one possibility being the method described by Kawashita et al. [20] using the location of minimum radius of curvature i.e. minimum of eq. (1).

As stated in the introduction, the present paper serves as a demonstration of the concept and test setup. A much larger experimental regime is required, with different material systems, for any conclusions to be drawn on the interfacial fracture energies obtained from this test.

5. Summary

An experimental method to characterize the fracture energy of fibre/matrix interfaces is presented. This includes manufacturing of specimens, peel testing under in-situ SEM observation followed by image analysis to determine the radius of curvature of peeling fibres. The radius of curvature at the crack tip of the fibre undergoing peel is used as input in a fracture mechanics based approach to derive the interfacial fracture energy of the fibre/matrix interface. It is shown that the interfacial fracture energy may be derived using this method. However, the robustness of this technique must be explored by further experimental testing.

The experimental method presented may be utilized either as a screening method for fibre/matrix compatibility or as input for fibre bridging models which include discrete fibres undergoing bridging and peel-off. It may supplement existing micro-mechanical experiments aimed at determining the fracture mechanical properties of the fibre/matrix interface.

Acknowledgements

This research is conducted as part of the DACOMAT project. The DACOMAT project has received funding from the European Union's Horizon 2020 research and innovation programme 5 under GA No. 761072. Thanks to the research technicians: Gitte Christiansen and Jonas Heinige for preparation of the microscopy specimens. Special thanks to Erik Vogeley for patiently assisting in setup preparation, microscopy and providing sanity checks. Finally, the collaborative work and sample preparation by DACOMAT partners at the University of Strathclyde is gratefully acknowledged.

References

[1] Overgaard L, Lund E and Thomsen O 2010 *Compos. Part A Appl. Sci. Manuf.* **41** 257–270
[2] Lee H G, Kang M G and Park J 2015 *Compos. Struct.* **133** 878–885
[3] Sørensen B F, Gamstedt E K, Østergaard R C and Goutianos S 2008 *Mech. Mater.* **40** 220–234
[4] Suo Z, Bao G and Fan B 1992 *J. Mech. Phys. Solids* **40** 1–16
[5] Yao L, Alderliesten R and Benedictus R 2016 *Compos. Struct.* **140** 125–135
[6] Mirzaei B, Sinha A and Nairn J 2016 *Compos. Sci. Technol.* **128** 65–74
[7] Kinloch A J, Lau C C and Williams J G 1994 *Int. J. Fract.* **66** 45–70
[8] Thomason J 2019 *Compos. Part A Appl. Sci. Manuf.* **127** 105619
[9] Thomason J 2020 *Polym. Test.* **85** 106421
[10] Sørensen B F 2017 *Mech. Mater.* **104** 38–48
[11] Yang L and Thomason J L 2010 *Compos. Part A Appl. Sci. Manuf.* **41** 1077–1083
[12] Alimuddin M and Piggott M 1999 *Polym. Compos.* **20** 655–663
[13] Kim B W and Nairn J A 2002 *J. Mater. Sci.* **37** 3965–3972 ISSN 00222461
[14] Graciani E, Varna J, Mantič V, Blázquez A and París F 2011 *Procedia Eng.* **10** 2478–2483 ISSN 18777058
[15] Liechti K M and Chai Y S 1992 *J. Appl. Mech. Trans. ASME* **59** 295–304 ISSN 15289036
[16] Rice J R 1988 *Am. Soc. Mech. Eng.* **55** 98–103 ISSN 04021215
[17] Hutchinson J W and Suo Z 1992 *Adv. Appl. Mech.* **29** 63–191 ISSN 0065-2156
[18] Thouless M and Jensen H 1992 *J. Adhes.* **38** 185–197
[19] McDaniel P B, Deitzel J M, Gregory D, Polakovic T and Gillespie J W 2018 *J. Appl. Polym. Sci.* **135** 1–11
[20] Kawashita L F, Moore D R and Williams J G 2005 *J. Mater. Sci.* **40** 4541–4548
[21] Jenkins P G, Bryce D, Xypolias G and Thomason J L 2020 *Proceedings of the 41st Risø international symposium on materials science*
[22] The Mathworks Inc 2018 *MATLAB (R2018b)* (Natick, Massachusetts)
[23] Stewart J 2008 *Calculus, Early Transcendentals* 6th ed (20 Davis Drive, Belmont, CA, USA: Brooks/Cole) ISBN 9780871702807
[24] Rice J R 1968 *Journal of Applied Mechanics* **35** 379
[25] Timoshenko S P and Goodier J N 1951 *Theory of Elasticity* (New York, USA: McGraw-Hill) ISBN 9780871702807
[26] Orozco R 1999 *Effects of Toughened Matrix Resins on Composite Materials for Wind Turbine Blades* Msc thesis Montana State University
[27] ASM International 1988 *Engineered Materials Handbook Vol 2: Engineering Plastics* (Materials Park, OH, USA: ASM International) ISBN 9780871702807

Appendix A. Image analysis method

In fig. A1 a flowchart is presented, giving an overview of the various steps taken in the script. Each step is detailed in the bullet list below.

- *Length conversion:* The fibre diameter is measured in pixels and converted to µm.
- *Image prep:* A frame is selected manually (one frame before debond crack propagation) and extracted from the video input. After cropping and rotating it is converted to a binary image using the ForegroundPolarity option (threshold of ca. 0.6) in the Matlab function: imbinarize.
- *Features:* The top edge of the fibre is extracted as a series of (x, y) coordinates (using the coordinate system shown in figs. 1 and 4).
- *Linear fit:* Two linear fits are made on the straight segments of the fibre as shown in fig. 4 using the matlab toolbox: cftool. Next, the upper and lower bounds for the non-linear fit (see fig. 4) are determined to be where the error of the linear fits exceeds a threshold of 1 µm.
- *Non-linear fit:* The curved section of the fibre is fitted with a quadratic function, $y(x)$, using the matlab toolbox cftool.

- *R and G_c:* From the non-linear fit parameters the radius of curvature, R, is calculated as a function of x using eq. (1), where y' and y'' are the second and first derivatives of the non-linear fitting function y.

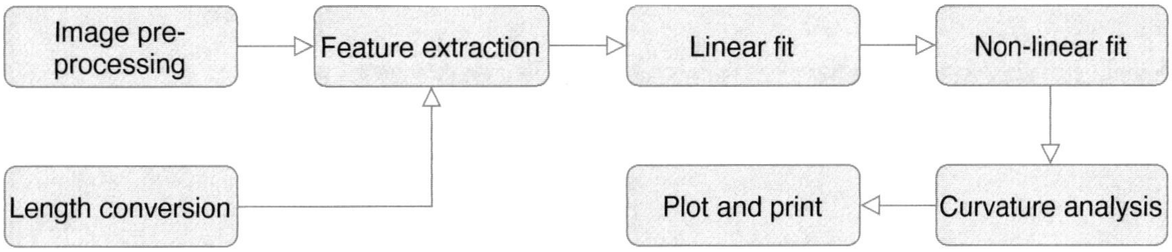

Figure A1: Flowchart of Matlab program used to perform video/image analysis.

The full Matlab code is available upon request.

IOP Conf. Series: Materials Science and Engineering **942** (2020) 012030 doi:10.1088/1757-899X/942/1/012030

Air-to-air atmospheric pressure plasma treatment – perspective for composite manufacturing

Cheng Fang[1,2], Daan Jonas Hottentot Cederløf[1], Alexander Bardenshtein[3] and Yukihiro Kusano[1,3*]

[1] Department of Wind Energy, Technical University of Denmark, Denmark
[2] Harbin Institute of Technology, People's Republic of China
[3] Danish Technological Institute, Denmark

*: corresponding author
e-mail: yuk@teknologisk.dk

Abstract. Fibre-reinforced polymer composites are gaining increasing attention in various applications for constructing mechanical structures such as wind turbine blades. The interface between fibres and a polymer matrix should be optimally designed to promote the mechanical performance of the composites. Plasma treatment shows obvious advantages over conventional approaches, since it has the characteristic of environmental friendliness, low-cost, and easy operation. A plasma can be favourably generated at atmospheric pressure. One of the most commonly used atmospheric pressure plasmas is a dielectric barrier discharge (DBD). In the present work, an air-to-air DBD is introduced. The DBD was generated in a gas mixture of helium and fluorocarbon between a rod-shaped water-cooled powered electrode covered with alumina and a one-dimensionally movable ground aluminium plate. Polyethylene terephthalate films were used as model specimens, and attached on the aluminium plate for the surface modification. The results indicate that specimen surfaces can be oxidized or fluorinated, depending on the conditions, and that the gap between the electrodes and gas flowrates significantly affect the treatment effect.

1. Introduction

Glass fibre reinforced polymer (GFRP) composites are widely used due to high strength-to-weight ratios, mechanical and corrosion resistance properties [1]. Examples of applications include sporting equipment, vehicles, architectures, and wind turbine blades.

It is indicated experimentally [2] and theoretically [3] that fracture toughness of GFRPs can be improved not by strong interactions at the interfaces between fibres and polymer matrices, but by optimally designed interfaces. More specifically, creation of multiple cracks in GFRPs due to locally distributed domains with low adhesion can improve overall fracture toughness of GFRPs. Therefore, at the initial stage of polymer composite manufacturing, surface treatment of fibres is important for controlling the interfacial properties of GFRPs. Motivated by this idea, fluorination of sized glass fibres is studied aiming at reducing initial interaction of glass fibre surfaces with polymer matrices [4,5]. Specifically, it is expected that fluorination of sized glass fibre surfaces can pronounce de-wetting of uncured matrix polymers, delay interaction between the sizing and the matrix, and that subsequent interfacial interaction can be lowered. Among the reported surface treatment techniques, atmospheric pressure plasma treatment is attractive due to its high treatment efficiency, environmental friendliness and easy operation with simple setups [6,7]. Commonly used atmospheric pressure plasmas include corona discharge, dielectric barrier discharge (DBD) [8], cold plasma torch [9,10], and gliding arc [11-17]. DBD is generated by applying an alternating-current (AC) voltage between electrodes separated by one or more dielectrics. Simple and robust configuration of the DBD allows for design flexibility including air-to-air type continuous plasma treatment systems [18,19].

Adhesion improvement of material surfaces using plasma treatment [6,7] is commonly demonstrated by oxidation for acid-based interactions [20] and/or surface roughening for mechanical interlocking effects for the purpose of improving interactions at interfaces. By contrast, recent publications [4,5] propose designed lower interactions between glass fibre fabrics with a polymer matrix by creating a polytetrafluoroethylene (PTFE) like surface by substitution reactions between hydrogen and fluorine using a DBD in a helium/tetrafluoromethane (He/CF_4) gas mixture. However, it is reported that use of fluorocarbons such as hexafluoropropylene (HFP. C_3F_6) and octafluorocyclobutane (OFB. c-C_4F_8) in a DBD exhibits better hydrophobic effects by plasma polymerization to synthesize a PTFE like coating than use of CF_4 [21]. It is noted here that if plasma polymerization on a material surface is appropriately carried out, the surface property of the coating is generally independent of the material to be coated.

In the present work, an air-to-air DBD is introduced for investigating hydrophobic plasma surface modification in a He/OFB gas mixture. Polyethylene terephthalate (PET) films were used as model specimens to study the modification effects, since PET is easily available, and its properties are well-known. Contact angle measurements were carried out using deionized water and glycerol as test liquids for studying wetting characteristics. X-ray photoelectron spectroscopy (XPS) was used to characterize the elemental analysis of the modified PET film surfaces.

2. Experimental methods

A photo image and a diagram of the DBD setup are shown in Fig. 1. The powered electrode is water-cooled cylindrical metallic tube covered with an alumina tube. Inner and outer diameters of the alumina tube are 12 and 16 mm, respectively. The lower ground electrode is an aluminium plate (280 mm x 400 mm). The gap between the alumina tube and the aluminium plate was adjusted to 0.6 mm or 2.0 mm. He and OFB were used as a dilute gas and a reactive gas, respectively. A flowrate of He was adjusted between 14.5 and 29 standard litre per minute (SLM). OFB flowrate was adjusted between 0.066 and 0.262 SLM. A gas mixing ratio (MR) is defined in equation (1).

$$MR = F_{OFB} / (F_{He} + F_{OFB}) \times 100 \text{ (vol. \%)}, \quad (1)$$

where F_{OFB} and F_{He} are the flowrates of OFB and He, respectively.

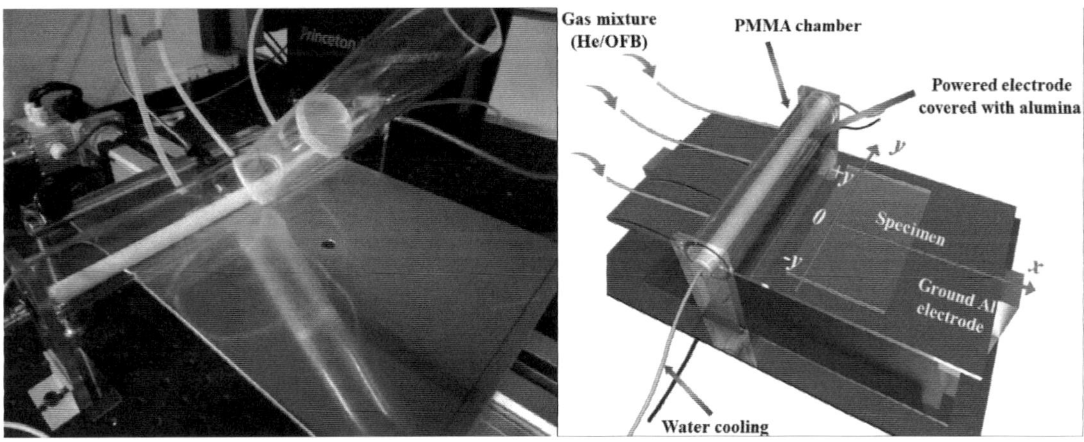

Fig. 1 A photo image and a diagram of the air-to-air atmospheric pressure DBD setup.

IOP Conf. Series: Materials Science and Engineering **942** (2020) 012030 doi:10.1088/1757-899X/942/1/012030

He gas was pre-mixed with OFB, and then introduced into a poly-methyl-methacrylate (PMMA) chamber, separating the gas mixture from the surrounding atmosphere. The bottom part of the PMMA chamber facing the aluminium plate is open to air so that the DBD was generated between the alumina tube and the aluminium plate. The DBD was driven by an AC power supply (Generator 6030, SOFTAL Electronic GmbH, Germany) at a frequency of approximately 40 kHz. Untreated general-purpose PET films with thickness of 65 μm were used as specimens. They were attached on the aluminium plate at positions in a y-axis direction of "-y" (approximately 5 cm from the edge of the aluminium plate), "0" (centre) and "+y" (approximately 5 cm from the edge of the aluminium plate). The aluminium plate was moved forward and backward in an x-axis direction at a speed ranging from 1 to 100 mm/s. The specimens were exposed to the DBD once or four times. The average power input was obtained by measuring voltage and current with a high-voltage probe and a 10 Ω current viewing resistor, respectively. The power of each treatment was adjusted to 100 W.

Static contact angles of deionized water and glycerol on the PET films before and after the treatment were measured in air at room temperature using a contact angle measurement system (CAM100; Crelab Instruments AB, Sweden). Here, deionized water is characterized by its significantly polar nature, and is the most commonly used test liquid for the contact angle measurement. Glycerol is also often used as a test liquid, exhibiting both polar and non-polar nature with high viscosity. Glycerol is considered to be an appropriate test liquid for screening adhesion properties of polymer surfaces due to similarities of physical properties to general adhesives and uncured resins [22].

XPS data were collected using a monochromatic Al K_{α} X-ray source with a lateral resolution of 30 mm (K-alpha; ThermoFischer Scientific, UK) to study the changes of the elemental composition at the PET film surfaces. Atomic concentrations of each element were calculated by determining the relevant integral peak intensities subtracting a linear background.

3. Results

A photo image of the DBD in the He/OFB gas mixture is exemplified in Fig. 2. The discharge looks a mixture of filamentary and glow-like discharges. It was confined between the electrodes, and spread along the powered electrode in the y-axis direction. The photoemission on the right side in Fig. 2, corresponding to +y direction, was more intense than at the centre and at the left-hand side. The difference of the photoemission is related to the uniformity of the plasma and treatment, which will be discussed below.

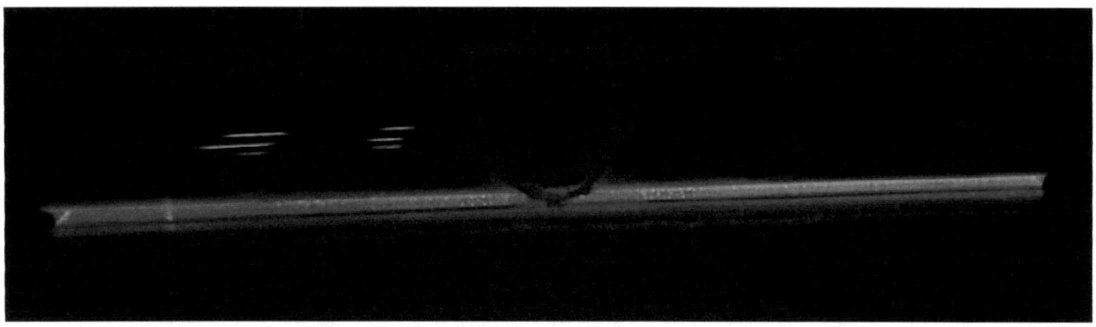

Fig. 2. A photo of the DBD in He/OFB gas mixture (gap: 2.0 mm, MR value: 0.45 %). The purple colour indicates photo-emission from the DBD between the cylindrical powered electrode and the ground electrode.

The voltage and current waveforms of the DBD in the He/OFB gas mixture are shown in Fig. 3. The voltage waveform was sinusoidal. The phase shift was approximately π/4, indicating that the DBD is primarily a capacitively coupled plasma. Generation of the DBD in each half period deformed the current waveform from the sinusoidal feature.

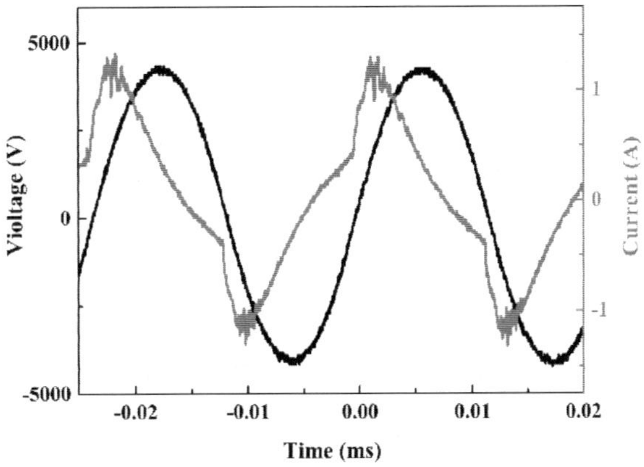

Fig. 3. An example of voltage (black) and current (red) waveforms of the DBD in He/OFB gas mixture (gap: 2.0 mm, MR value: 0.45 %).

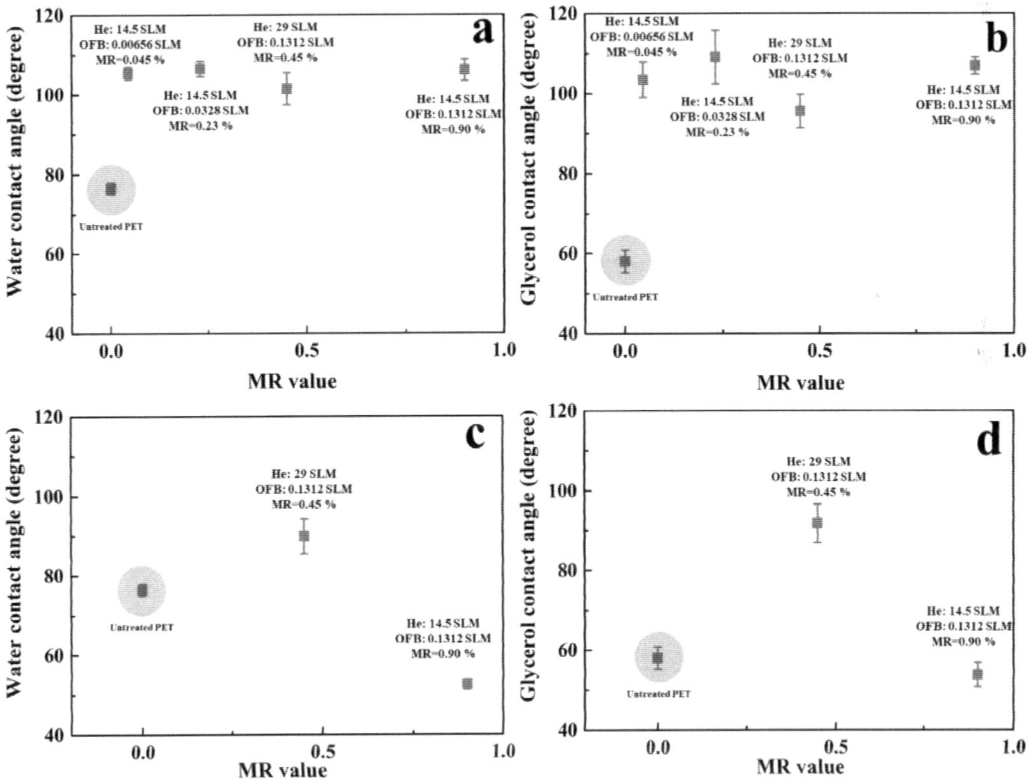

Fig. 4. Contact angles of deionized water and glycerol with the gaps of 0.6 mm (a, b) and 2.0 mm (c,d). (exposure: 4 times, MR value: 0.045 %, 0.23 %, 0.45 %, and 0.90 %). The contact angles for the untreated PET are also plotted in each figure.

In order to investigate the influence of different plasma process parameters on the wettability of the PET surfaces, static contact angles of deionized water and glycerol were measured after treatment. PET films without treatment were also tested for comparison and the values for deionized water and glycerol were 76.4 ± 1.5 and 58.0 ± 2.8 degrees, respectively. The gas content in a plasma is an important parameter for atmospheric pressure plasma processing. It depends on flowrates of the gases, the MR value, introduction of air from the surrounding, and consumption of the reactive gas by polymerization and/or decomposition in the plasma. The gap between the two electrodes, which affects the introduction of atmospheric air into the DBD, was first set to the minimum value of 0.6 mm. The MR value was set to 0.045 %, 0.23 %, 0.45 %, or 0.90 %. PET films were placed at the "0" position on the aluminium plate (see Fig. 1) and exposed to the DBD four times. Results of the contact angle measurement of the PET films treated under different conditions are summarized in the Fig. 4 (a,b). The contact angle increased greatly after each treatment.

Next, the gap was set to 2.0 mm and similar experiments were carried out. Fig. 4 (c,d) illustrates the results of the contact angle measurement, showing a different trend from the smaller gap in Fig. 4 (a,b). When the MR value was set from 0.045 to 0.45 %, the contact angles of deionized water and glycerol increased, but were lower than the values achieved at the gap of 0.6 mm. When the MR value was 0.9 % by decreasing the He flowrate from 29 SLM to 14.5 SLM, the contact angles were even lower than those of the original PET films.

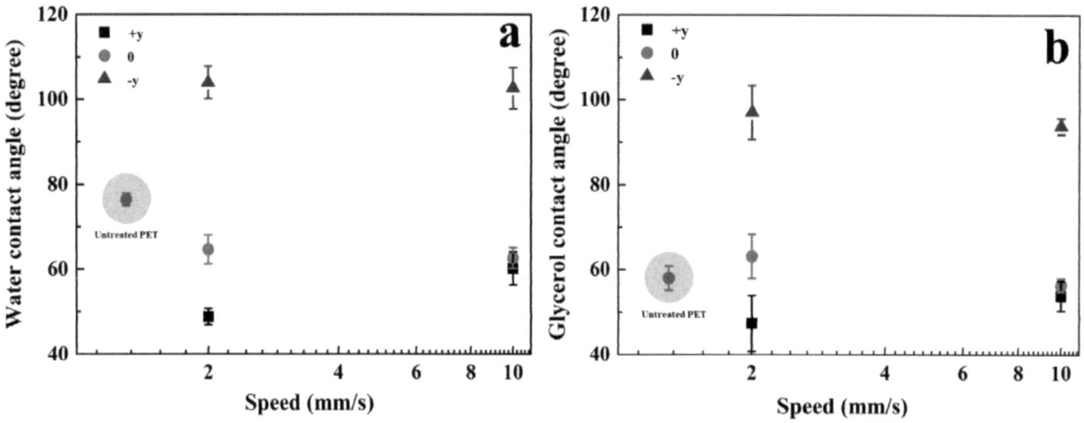

Fig. 5. Contact angles at position of –y, 0, and +y. (a) deionized water, (b) glycerol (exposure: once, gap: 2.0 mm, MR value: 0.45 %, He flowrate: 29 SLM). The contact angles for the untreated PET are also plotted in each figure.

Since the DBD can treat an up to A4 size specimen (210 mm x 300 mm), it is worth investigating the treatment uniformity. In addition, time for treatment is an important parameter for practical applicability of the processing. It can be changed by the number of the exposures and the moving speed of the aluminium plate. In the following experiments, the number of exposures is fixed to one time. In addition, the PET films were fixed at the positions of "-y", "0" and "+y", the moving speed was varied between 1 and 100 mm/s, and the gap and the He flowrate were fixed at 2.0 mm and 29 SLM, respectively.

The MR value was fixed at 0.45 %, and contact angles of deionized water and glycerol were measured after the treatments as shown in Fig. 5. The result indicates that the treatment was significantly uneven. The "-y" position became highly hydrophobic. On the other hand, at the "0" and "+y" positions, the water contact angle was slightly lowered from the original value of 76.4 degrees, and the glycerol contact angle did not change significantly from the original value of 58.0 degrees.

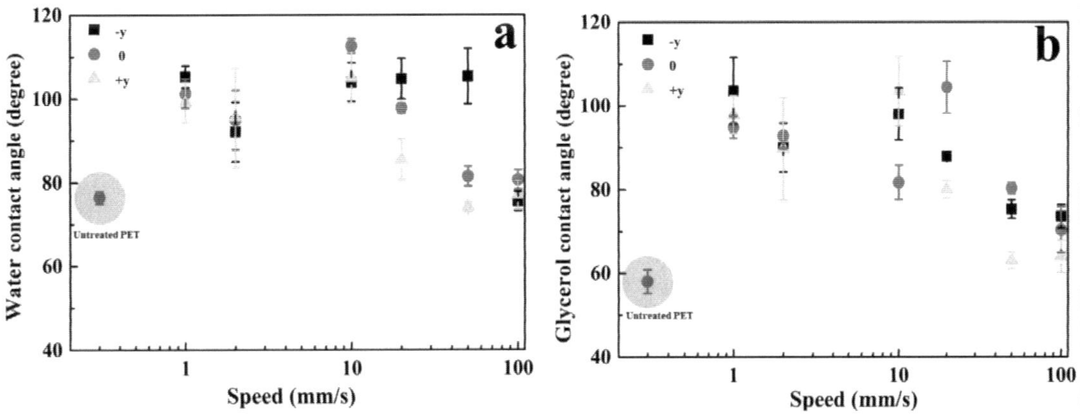

Fig. 6. Distributions of contact angles at positions –y, 0, and +y. (a) deionized water, (b) glycerol (gap: 2.0 mm, exposure: once, MR value: 0.90 %, He flowrate: 29 SLM). The contact angles for the untreated PET are also plotted in each figure.

The flowrate of OFB was increased from 0.131 SLM to 0.262 SLM with a fixed He flowrate of 29 SLM, and the similar experiment was carried out (MR value: 0.90 %). The results of the measured contact angles are summarized in Fig. 6. The uniformity of the treatment was generally improved, compared with the results in Fig. 5. At higher speeds, the contact angle approaches the values of the untreated PET films.

Table 1. XPS elemental analysis of the PET films (gap: 2.0 mm, exposure: once, MR value: 0.90 %).

Speed (mm/s)	*Position*	*Atomic content (at.%)*			*Contact angle (°)*	
		C	F	O	Water	Glycerol
-	-	75.8	0.0	24.2	76.4 ±1.5	58.0± 2.8
1	-y	45.5	52.4	2.1	105.3 ±2.6	103.6± 8.0
	0	48.5	25.6	26.0	101.3 ±3.5	94.9± 2.7
	+y	36.6	36.6	26.8	99.1 ±4.8	97.8± 4.5
10	-y	45.5	39.9	14.6	104.8 ±4.6	98.0± 6.2
	0	55.2	13.2	31.6	112.6 ±1.8	81.7± 4.1
	+y	52.9	17.5	29.6	105.1 ±6.1	103.5± 8.3

XPS measurements were carried out to analyze the elemental composition of the PET film surfaces before and after the plasma treatments. Table 1 summarizes the results XPS together with contact angles. The plasma conditions were the same as those in Fig. 6, and the moving speed was 1 or 10 mm/s. The untreated PET film surface was dominated by carbon (C) and oxygen (O) atoms without fluorine (F) atoms. After each treatment, F atoms bonded with C were introduced at the surfaces. The F content on the film surfaces tended to be higher at the slower speed. Compared with the "0" and "+y" positions, the F content at the "-y" position was significantly higher at each speed. Specifically, the F content of the film was 52 at.% when the speed was 1 mm/s. At the "0" and "+y" positions, the O content was higher than the untreated one. In other words, both fluorination and oxidation simultaneously occurred at these positions.

4. Discussion

The discharge mode of the DBD is discussed by comparing the visual observation of the plasma in Fig. 2 with the voltage current waveforms in Fig. 3. The overall deformation in the current waveform from the sinusoidal feature suggests an increase in the electrical impedance by the generation of a bulk plasma, while the complex spiky current waveform indicates formation of filamentary micro-discharges, which agrees with the observation of the DBD in Fig. 2.

Measurement of wetting characteristics gives direct indication of surface treatment effects. Fig. 4 (a,b) indicates that the DBD treatment could effectively increase the hydrophobicity of the PET surfaces when the gap was 0.6 mm. On the other hand, when the gap was 2.0 mm, the surfaces did not show hydrophobicity when the He flowrate was lowered as shown in Fig. 4 (c,d). The results demonstrate that the electrode gap has a great impact on the treatment effect. Furthermore, it is suggested that oxidation and/or etching on the PET film surfaces would take place with the gap of 2.0 mm. When the gap becomes larger, the gas flowrates should be increased to account for extra leakage.

The measured result in the unevenness of the wettability shown in Fig. 5 can be related to the uniformity of the gas mixture in the DBD. The photo in Fig. 2 was captured in the same condition. The intense photoemission at "+y" position indicates the lower OFB content in that region. It can be due to insufficient gas mixing process of He and OFB, or uneven local consumption of OFB. It is therefore worth increasing the flowrate of OFB or the MR value as demonstrated in Fig. 6. The result confirms that the treatment can be more uniform by the higher MR value at the high He flowrate.

It is interesting to compare the XPS result with the result of the contact angle measurement in Table 1. It seems that the measured contact angles would be rather insensitive against the significant difference of the F and O contents at different positions. One possible explanation of the discrepancy is that each position would be sufficiently fluorinated to be hydrophobic, and that the differences of the F and O contents may not necessarily affect the contact angles. However, additional investigation will be necessary for further improving the uniformity of the plasma treatment.

5. Conclusion

An air-to-air DBD plasma in a He/OFB gas mixture can introduce fluorine and promote hydrophobicity of the PET films. The flowrates of the gases and the gap of the electrode played important roles for the treatment effects, attributed to the gas content in the plasma. The measured wetting characteristics of hydrophobicity was rather insensitive to the difference in elemental composition of the PET surfaces. The technique presented can be used for continuous air-to-air surface modification of sheet-like specimens including glass fibre fabrics, indicating industrial feasibility of the technology for fibre composite manufacturing.

Acknowledgements

The DACOMAT project has received funding from the European Union's Horizon 2020 research and innovation programme 5 under GA No. 761072. Cheng Fang's work was supported by the Harbin Institute of Technology Scholarship Fund. Research technician Jonas Kreutzfeldt Heininge is acknowledged for design and construction of the air-to-air DBD.

References

[1] Information on https://www.sintef.no/projectweb/dacomat/ "DACOMAT - Damage Controlled Composite Materials". The European Union's Horizon 2020 research and innovation programme under GA No. 761072. Coordinator: SINTEF, Norway. The website inspected on 24th August 2020.

[2] Rask M, Sørensen B F 2012 Eng. Fract. Mech. **96** 37–48

[3] Goutianos S, Sørensen B F 2016 Eng. Fract. Mech. **151** 92–108

[4] Cederløf D J H, Fæster S, Kusano Y 2020 J. Adhesion **96** 2-12

[5] Kusano Y, Cederløf D J H, Fæster S 2020 Key Eng. Mater. **843** 159-164

[6] Kogoma M, Kusano M, Kusano Y 2011 Nova Science Publishers

[7] Kusano Y 2014 J. Adhesion **90** 755-777

[8] Kogelschatz K U 2003 Plasm. Chem. Plasm. Proc. **23** 1-46

[9] Mortensen H, Kusano Y, Leipold F, Rozlosnik N, Kingshott P, Sørensen B F, Stenum B, H Bindslev 2006 J. Appl. Phys. **45** 8506-8511

[10] Lee B J, Kusano Y, Kato N, Horiuchi T, Koinuma H 1997 Jpn. J. Appl. Phys. Pt **36** 2888-2891

[11] Fridman A, Nester S, Kennedy L A, Saveliev A, Mutaf-Yardimci O 1999 Prog. Energy Combustion Sci **25** 211–231

[12] Zhu J J, Ehn A, Gao J L, Kong C D, Aldén M, Larsson A, Salewski M, Leipold F, Kusano Y, Li Z S 2017 Optics Express **25** 20243-20257

[13] Kusano Y, Berglund L, Aitomäki Y, Oksman K, Madsen B 2016 Mater. Sci. Eng. **139** 012027

[14] Kusano Y, Salewski M, Leipold F, Zhu J J, Ehn A, Li Z S, Aldén M 2014 Eur. Phys. J. D **68** 319

[15] Kusano Y 2009 Surf. Eng. **25(6)** 415-416

[16] Zhu J J, Gao J L, Ehn A, Aldén M, Li Z S, Larsson A, Kusano Y 2017 Phys. Plasmas **24** 013515

[17] Kusano Y, Zhu J J, Ehn A, Li Z S, Aldén M, Salewski M, Leipold F, Bardenshtein A, Krebs N 2015 Surf. Eng. **31(4)** 282-288

[18] Teodoru S, Kusano Y, Rozlosnik N, Michelsen P K 2009 Plasm. Proc. Polym. **6** S375-S381

[19] Kusano Y, Andersen T L, Michelsen P K 2008 J. Phys. Conf. Ser. **100** 012002

[20] Mittal K L 1977 Polym. Eng. Sci. **17** 467-473

[21] Kusano Y, Yoshikawa M, Naito K, Okazaki S, Kogoma M 1994 SPSM **7** 77-81

[22] Kusano Y, Madsen B, Berglund L, Oksman K 2019 Celulose 26 7185-7194

The effect of environmental ageing in the interphase of a glass fibre/vinyl ester composite designed for wind turbine applications

G Xypolias[1], J Thomason and L Yang

Mechanical & Aerospace Engineering, University of Strathclyde, James Weir building, 75 Montrose Street, Glasgow, United Kingdom

georgios.xypolias@strath.ac.uk[1]

Abstract. The use of advanced fibre reinforced polymer (FRP) composites in the renewable energy sector has significantly increased over the last decade. However, the durability and damage tolerance of such materials when exposed to environmental conditions still remains under investigation. This creates some uncertainty for their introduction and development in the sector. The scope of the present work is the assessment of the durability of composite laminates and structures through providing a fundamental understanding of the micromechanical performance of the glass fibre/vinyl ester fibre-matrix interphase as a function of the environmental history. An accelerated ageing study has been conducted in order to create a correlation between the moisture sorption at the interphase and its effect in the micromechanical performance of the composite system. The composite interphase performance has been assessed by measuring the Interfacial Shear Strength (IFSS) using the microbond test. An estimation of the moisture sorption of each microbond sample has been obtained by the gravimetric analysis of thin film neat resins specimens of similar thickness to the microbond sample diameter. These results have also been coupled to results obtained by Fourier-transform Infrared Spectroscopy (FTIR). Additionally, moisture induced degradation has been characterised by recording the change in the Tg of the vinyl ester resin as a function of environmental ageing by a number of thermal analysis techniques.

1. Introduction

The popularity of FRPs in wind turbine applications has significantly increased over the last decade. During their service life their contact with environmental degradation agents, such as moisture and thermal cycling is inevitable. Regardless of the close attention to details and the employment of optimum design methods, the long-term exposure of FRPs to the use-environment will likely shorten the durability of the materials and eventually trigger irreversible degradation mechanisms which will affect their properties. Moisture uptake is known to cause a series of degradation phenomena in thermoset polymers and FRPs, such as swelling and plasticization of the matrix, hydrolysis and loss of adhesion at the fibre/matrix interphase [1]. A method of accurately determining the effect of the in-service environmental conditions to the exposed material is yet to be established. This creates a level of uncertainty about the durability of composites for infrastructure applications, where clients and stakeholders need to "know" how the material will perform in over timescales of many decades.

A real-time assessment of in-service composite structures is a time-consuming process involving high costs. These parameters can be lowered by employing in-lab accelerated ageing conditions. The basic concept of accelerated ageing is to increase the severity of the degradation conditions, such as temperature or humidity, to a higher level than that achieved in real-life operating environments [1].

It has been suggested in previous studies that the fibre/matrix interphase is a crucial region of FRPs [2, 3]. In fact, the condition of the interphase can define the reliability and therefore the durability of composite materials or structures during their service life. The hydrothermal conditioning of the interphase directly can significantly accelerate ageing times. Even though the transferability of the properties of the fibre/matrix interphase to the bulk laminate is complex, it has been established that a strong interphase can potentially contribute to an FRP with high mechanical and chemical properties, and an interphase failure is likely to compromise the performance of the bulk material and even lead to its failure [1, 2, 3]. Studies have also shown that whenever there is any improvement in the macroscopic properties of the composite, there was a corresponding improvement in the micromechanical properties of the composite as well [3, 4]. The integrity and properties of the interphase are particularly sensitive to hygrothermal ageing. In fact, moisture exhibits the ability to force itself between the fibre and the matrix and therefore the interphase is a potential passageway for the permeation of moisture into polymer composites. Moisture ingression is known to cause hydrolysis of the bonds at the composite interphase, resulting in reduced load transfer capability between the matrix and the fibre [1, 3].

The objective of this study was to create a correlation between the moisture uptake of a glass fibre/vinyl ester interphase and its translation into the micromechanical properties of the system. This work is an initial approach in predicting the hydrothermal response of the GFRP intended for wind turbine blade applications and is a part of a greater study aiming to develop more durable composites for use in large infrastructures. Thin film specimens have been used as model for the moisture uptake in the interphase, while bulk resin specimens have been conditioned hydrothermally too to allow a comparison between the micro and macro scales. A great part of this work is focused on the change of neat resin properties, which in turn can have a great effect in the glass fibre vinyl ester interphase.

2. Experimental

2.1. Materials

The experiments were conducted using a low viscosity bisphenol epoxy based vinyl ester resin (DION® VE-1260), with a styrene content of 48 – 52%. The resin was supplied by Polynt Composites Ltd. The resin was mixed with a UN 3103 Organic peroxide type C initiator (1.95%) and was initially cured under a nitrogen environment for 24 h at room temperature and post-cured at 60 °C for 24 hours, 80 °C for 3 hours and 100 °C for 1 hour. The nitrogen environment was chosen due to the susceptibility of microdroplets (high surface area to volume ratio) to oxygen, which can act as a copolymer and result in incomplete matrix cure. The glass fibre used was a Boron-free Advantex® ECR glass fibre a commercial glass vinyl ester compatible sizing (se3030). The fibre was provided by 3B Fibreglass in the form of a direct roving.

2.2. Ageing Parameters

The ageing conditions were selected as a starting/reference point in order to produce initial results and examine the durability of a glass fibre/vinyl ester interphase in a wet environment. A water temperature within the range of room temperature (23 °C) was selected, as well as a temperature of 50 °C as a means of accelerating the ageing process. Specimens were aged by full immersion in deionized water using two different water baths.

2.3. Sample Preparation

2.3.1. Microbond Sample Preparation. A high grade stainless steel rig was designed and manufactured in order to accommodate microbond specimens and allow their exposure in humid environments. The single fibre specimens were adhered on a stainless steel washer by the means of araldite glue. The microbond sample preparation procedure has been previously described by Yang and Thomason [5].

IOP Conf. Series: Materials Science and Engineering **942** (2020) 012031 doi:10.1088/1757-899X/942/1/012031

Acceptable droplet dimensions were considered: 90-180 μm in embedded length and 40-80 μm in diameter.

2.3.2. Neat resin sample preparation. Thin films were used as a model of prediction of moisture uptake in microbond droplets. Film thickness was kept as close as possible to the embedded length of the microbond droplets. The fabrication of films was carried out with the means of a seam roller by applying several resin layers on a silicone rubber substrate of 170 x 100 mm and a thickness of 5 mm. The resin was left at ambient laboratory conditions for 30 minutes to become more viscous and facilitate rolling before use. All cured film sheets were removed from the substrate were cut into 25 x 25 mm specimens, which were dried under vacuum for 48 h before use.

Generally, the manufacturing of thin neat resin film was a challenging task, due to the nature and the physical properties of the thermoset (i.e. low viscosity), as well as the difficulty to produce specimens of such thickness. Several techniques and various substrates were used until the optimum result was achieved. The films used for the moisture uptake measurements were 150±30 μm and 220±30 μm for the 50 °C and 23 °C hydrothermal exposure, respectively. The bulk neat resin specimens were manufactured by using open silicone moulds providing specimen dimensions of 80x80 mm and a thickness of around 3.9 mm.

A semi-sphere droplet-like specimen with a flat bottom was also manufactured for thermal analysis by piping resin (mixed with initiator) onto a non-porous release film. The thickness of the droplet was around 2 mm. Cured semi-sphere specimens are illustrated in Figure 1. The purpose of these specimens was to allow accurate Tg measurements (DSC) according to ASTM 1356-08.

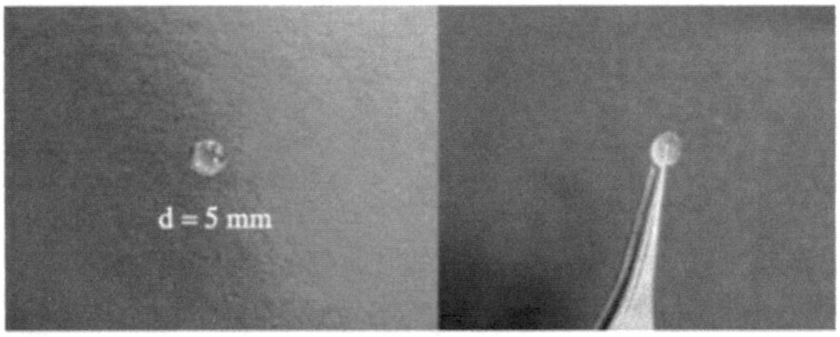

Figure 1. Neat resin semi-sphere droplet-like specimens used for thermal analysis.

2.4. Microbond Test

The microbond test was used to assess the change of IFSS of a single fibre glass fibre/vinyl ester system as a function of short-term, full immersion exposure in deionized water at 23 °C and 50 °C. The microbond test used was built in house and is described in detail by Yang and Thomason [5]. Upon ageing specimens remained in a desiccator of 10% RH at room temperature for 0.5 h prior to testing.

2.5. Gravimetric Analysis and Fourier-transform Infrared Spectroscopy (FTIR)

The gravimetric analysis on neat resin specimens was carried out by removing the specimens from the exposure medium, drying them with paper and weighing them at the nearest analytical balance. Assuming Fickian diffusion, diffusion coefficients were determined for the bulk resin specimens. According to those, equilibrium times were calculated for the film specimens and were compared to experimental data obtained. All equations used were taken from ASTM D5229/D5229M - 20.

An FTIR analysis was carried out for the coupling of the gravimetric measurements. Preliminary results have been included only for the film specimens exposed at 23 °C. A total of 9 specimens were analysed. All measurements were taken using an Agilent Technologies 4100 Exo Scan FTIR. A 45° specular reflectance interface was used for the collection of the spectra. This is a non-destructive interface and allowed multiple measurements of the same specimens at different ageing intervals. A thin and light-weight stainless steel plate was placed on top of the thin film specimen producing a transflectance effect. This was required due to the partial transparency of the specimen, which can cause the light exerted the interface to dissipate. Spectra were run with background scans between each run at 64 scans and 8 cm^{-1} resolution resulting in a spectral acquisition time of under 30 seconds. All measurements were taken from the centre point of the films and all spectra were averaged and normalized used Origin software per batch. Spectragyrph software was used in order to calculate the difference in areas between signals and therefore moisture gain. The integration with baseline tool was used.

2.6. Thermal Analysis

A DSC analysis was carried out to provide insight on the Tg of the matrix at different thicknesses. All results included in this paper are representative of the dry state of the matrix material prior to ageing. A Q20 differential scanning calorimeter with a cooling unit was used for all DSC tests. The instrument was manufactured by TA instruments. The following heating programme was used for each specimen:

- Equilibrate at -10 °C and held isothermally for 10 min (cycle 1)
- Temperature ramp at 10 °C /min up to 180 °C and held isothermally for 10 min (cycle 1)
- Temperature ramp at 20 °C /min to -10 °C and held isothermally for 10 min (cycle 2)
- Temperature ramp at 10 °C /min up to 180 °C (cycle 2)

A cool air environment supplied at 50 ml/min was used for all tests. The glass transition temperature could then be assigned from a clear step change visible on the resultant thermal curve of heat flow plotted against temperature.

3. Results

3.1. Gravimetric Analysis and FTIR

The gravimetric measurements for the bulk resin specimens exposed at 23°C and 50°C in deionized water are shown in Figure 2.

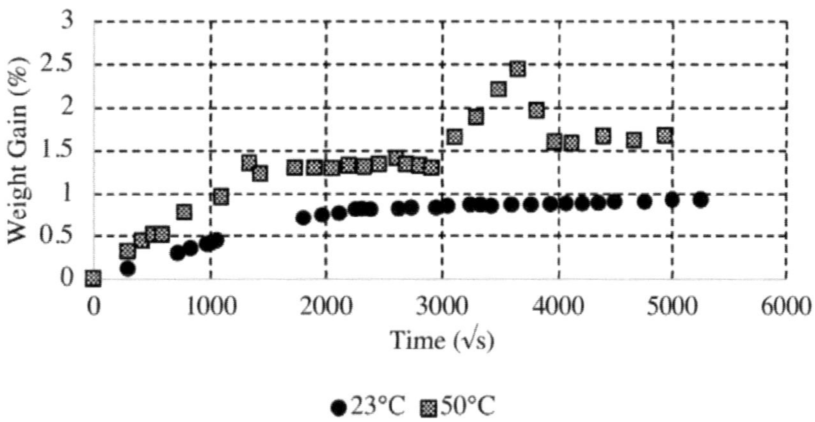

Figure 2. Gravimetric measurements of bulk resin specimens at 23°C and 50°C.

The specimen exposed at 23°C followed a Fickian-like behaviour and reached equilibrium after approximately 2 months (≈2300 √s) absorbing approximately 0.87% of water. The specimen exposed at 50°C exhibited a different response to hydrothermal ageing. Even though a stabilisation in the weight gain of the matrix was observed after 1 month (≈1800 √s) of exposure at around 1.3% gain, indicating equilibrium, an abrupt increase in weight was observed after around 3.3 months (3000 √s). This was attributed to hydrolytic degradation mechanisms starting to take place within the polymer network of the specimens. Both specimens were re-dried after exposure. It was observed that the specimen exposed at 50°C experienced a weight loss of 0.6% from its original weight, while the specimen exposed at 23°C experienced no weight change. Assuming Fickian diffusion for both specimens (until initial equilibrium for the specimen at 50°C), the diffusion coefficients were determined and are shown in Table 1. Using the diffusion coefficients of the bulk specimens, a prediction of time to equilibrium for the thin film specimens was carried out and all resulting values are contained in Table 1.

Table 1. Diffusivity values of bulk resin specimens and estimation of equilibrium times for thin films

	Diffusivity (mm^2/s)	Time to equilibrium (√s)	Time to equilibrium (days)	Thickness (mm)
23°C	4.9E-07	275	0.9	0.2
		344	1.4	0.25
50°C	1.37E-06	123	0.2	0.15
		165	0.3	0.2

The gravimetric measurements for the film specimens aged at 23°C and 50°C are shown in Figure 3. Confidence limits of 95% have been included for all gravimetric measurements. The moisture gain measured by the FTIR analysis for the films exposed at 23°C has also been plotted in Figure 3. As shown, the equilibrium time obtained through gravimetric measurements for the film exposed at 23°C was between 2 and 4 days (≈416 - 588 √s) experiencing a moisture gain of 1.28%. Most values measured by FTIR are within the error bars of the gravimetric data, except for the values recorded the first 24 h (≈293 √s) of exposure. The films treated at 50°C reached equilibrium almost immediately after they were exposed in the aqueous medium. Strong fluctuations in the moisture gain of the films were observed throughout their exposure. These changes could be attributed to a combination of moisture induced degradation phenomena, such as plasticization, hydrolysis and leaching.

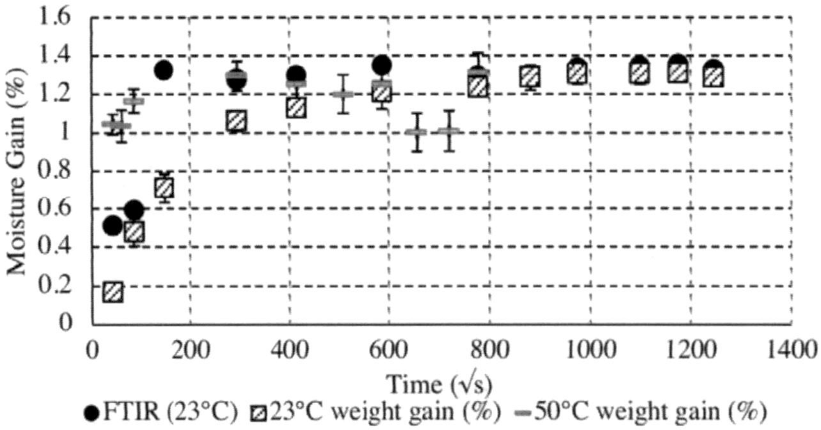

Figure 3. Gravimetric measurements of aged films and FTIR quantitative measurements for film at 23°C.

Hydrolytic degradation and leaching were confirmed by an apparent weight loss of re-dried film specimens. More specifically, a 0.8% reduction from the weight of the virgin specimen was observed after 72 and 96 h of hydrothermal ageing at 50°C and 1.1% after 168 h. This is a clear indication of chemical changes in the polymer structure.

3.2. *Thermal Analysis*

The Tg values obtained for the bulk resin and films of different thicknesses are contained in Table 2. Both DSC cycles employed are shown in the Table. It was evident that the Tg of the thin films shown from the 1st DSC cycle was within the 59 – 69°C range, depending on the thickness of the film. However, re-heating the film (after heated isothermally at 180°C) resulted to a step change in temperature and therefore a possible Tg at around 107°C This value was found to be close to the Tg value of the bulk resin specimen.

Table 2. Tg change vs thickness of dry matrix. Tg 1 represents the step change in temperature in cycle 1 and Tg (2) in cycle 2

Thickncss (mm)	Tg (1) (°C)	Tg (2) (°C)
0.15 ± 0.2	59	109
0.2 ±0 .2	69	108
0.25 ± 0.2	64	108
2 ± 0.2[a]	100	107
3.9 ± 0.2	107	110

[a]semi-sphere shaped droplet - flat bottom

3.3. *Microbond Test*

The IFSS of the glass fibre/vinyl ester microcomposites are presented graphically in Figure 4. Confidence limits of 95% are included for each experimental value shown. It was evident that a reduction in the IFSS of the system (≈25%) was not present until 72 h of ageing. All the IFSS values recorded prior to 72 h are the same within the range of the 95% confidence limits. On the other hand, an immediate decrease of approximately 28% was observed in the IFSS of the microcomposites conditioned at 50°C. A further drop of 37 - 40% from the dry value started taking place after 48 h. A comparison between a dry and aged (24 h at 50°C) microbond droplet is shown in Figure 5 and is indicative of the moisture induced degradation at 50°C. "Flaked-off" resin components are apparent in the surface of the aged droplet, which may be indicative of leaching, which in turn can contribute to reduced adhesion.

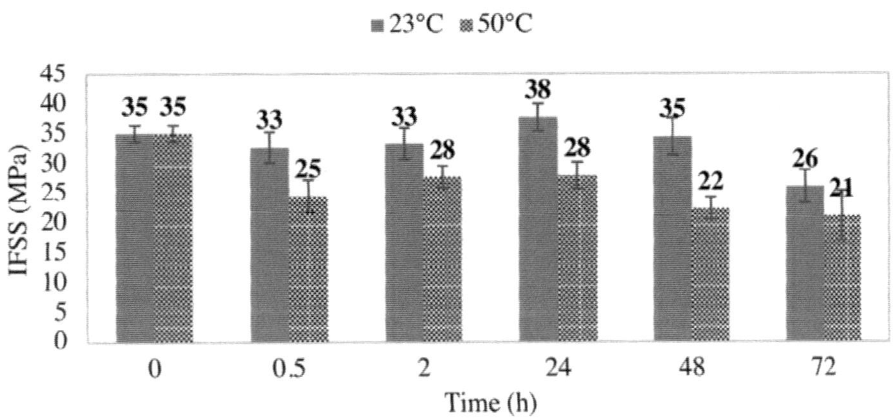

Figure 4: Interfacial Shear Strength as a function of ageing.

Figure 5. Morphology comparison of dry (LHS) and aged microbond droplet (RHS) on glass fibre at 50°C in deionized water. Images obtained by the means of an optical microscope at 200x magnification.

4. Discussion

Moisture induced degradation was evident from the presented results for both specimens immersed at deionized water of 23°C and 50°C. An unexpected increase in the moisture gain of the bulk specimen aged at 50°C was observed after initial equilibrium. The effect was attributed to probable swelling and plasticization, which are moisture induced degradation phenomena commonly present in thermoset resins [1]. Leaching, which is usually induced by hydrolysis was also confirmed, since the specimen experienced a weight loss of 0.6% from its dry state. A difficulty in extracting all moisture from the specimen was present. A decaying weight loss was observed while the specimen was dried under vacuum for around 2 weeks and after it was firstly induced to a temperature of 50°C for 72 hours, until the weight was stabilised. On the other hand, the specimen at 23°C was fully dried within 72 h under vacuum and no further weight changes took place. The difficulty of drying the specimen was an indication of water molecules trapped within microcavities and capillaries, found in the polymer network of the thermoset making it susceptible to hydrolysis. This effect has been previously studied by S.O. Han [6] and is an indication of chemical ageing. A Fickian-like behaviour was observed for the bulk resin specimen conditioned at 23°C. Physical ageing (reversible degradation) was observed since the specimen was dried out to its original weight after 72 h under vacuum.

Even though the estimation of equilibrium times of the film specimens by using the bulk specimen diffusion coefficients was close to the experimental values recorded, the weight gain behaviour of the films was found to vary from that of the bulk specimens. More specifically, the total averaged weight gain of the film specimens exposed at 23°C was found to be higher than that of the bulk specimen. Nevertheless, the film exhibited a Fickian-like behaviour, similar to the one of the bulk specimen. This was attributed to potential incomplete cure and a variation in the properties of the material, which was justified by the thermal analysis. It was apparent that the Tg of the resin varied significantly with a change in the thickness of the material – the Tg of the films exposed at 23°C was notably lower than that of bulk resin. A similar effect has been presented by P. Zinck and J. Gérard on polyepoxide films [7].

The 50°C aqueous environment was found to be aggressive for the vinyl ester film even after short-term exposure. Saturation was reached by the polymer almost immediately after being exposed to water and strong fluctuations in moisture gain were observed. Similar to the bulk specimen, irreversible ageing took place with the polymer losing approximately 0.8% of its weight after 72 and 96 h of ageing and around 1.1% after 168 h of ageing. Due to the continuous fluctuation in weights and the large error bars it was difficult for an average, maximum weight gain to be identified. However, a relatively stable weight gain plateau was observed between 24 h (≈293 $\sqrt{s}$) and 96 h (≈587 $\sqrt{s}$) of ageing and a weight gain of around 1.2%, which is close to the 1.3% weight gain value exhibited by the bulk resin specimen at its initial equilibrium. The chemical attack induced by the elevated temperature of the aqueous medium can be justified considering that the film specimens were actually exposed to a temperature close to their Tg (60°C).

Incomplete polymerization is often apparent in thermosets such as vinyl ester resins cured in low or moderate temperature regimes. Incomplete polymerization may potentially increase the susceptibility of the polymer to hydrolysis, plasticization and further deterioration, such as leaching when exposed to hydrothermal environments [8–10]. Thermal analysis carried out on vinyl ester films (≤250 um) was an indicator of incomplete cure of the resin. The Tg obtained in the 1st heating cycle (DSC) ranged from 59 to 69°C depending on the film thickness. However, re-heating the film to a value above its initial Tg increased the Tg of the resin to the same magnitude as the Tg of the bulk specimens.

The link between the moisture level in the composite interphase and its effect on the micromechanical performance of the microcomposite was established through the microbond test. A significant reduction in the IFSS of the specimens aged at 23°C was present after 72 h of ageing, which is the stage when moisture gain equilibrium was achieved in the interphase according to the thin film model. However, no significant changes were identified during the initial stages of moisture uptake. A probable indication is that water is penetrated through the sample thicknes and not the interphase. The sharp weight increase of the film treated at 50°C was translated into an immediate compromise in the micromechanical performance of the system. In addition, degradation increased with time and a further decrease of around 40% was apparent after 48 and 72 h. The leaching effect confirmed by the weight loss of the film model and the changes in the microbond sample morphology are indicators that chemical changes are induced in the fibre/matrix interphase. However, microscopy images of a higher resolution as well as a screening of re-dried IFSS values would contribute to the understanding of the degradation effects in more depth.

5. Conclusion

The effect of hydrothermal ageing was found to be severe for the glass fibre/vinyl ester interphase. The increase in temperature not only resulted in accelerated moisture gain but also led to degradation effects for the system. Chemical changes in the polymer were induced regardless of its scale and a reduction in the adhesion between the fibre and matrix was evident after short-term exposure. A reduction in the IFSS at 23°C was observed upon moisture equilibrium. The ageing environmental of 50°C was found to be aggressive for the microbond specimens and a sharp moisture gain of the model films correlated with a sharp decrease in the IFSS. There is significant doubt about the transferability of material properties from the microscale to macroscale when it comes to mechanical performance. Moreover, the chemical structure and the moisture uptake kinetics of vinyl ester films were found to be significantly different from those of the bulk matrix. FTIR was found to be a useful tool for quantifying moisture gain in films – nevertheless – a more thorough study to fully quantify moisture uptake in the material would be beneficial. The recording of the films Tg change as a function of hydrothermal ageing will also be useful to provide more insight into the moisture induced degradation in vinyl esters.

Acknowledgements
This work has received funding from the European Union's Horizon 2020 Research and Innovation Programme under Grant Agreement No 761072 (DACOMAT).

References
[1] Maxwell A S, Broughton WR 2017 Survey of long-term durability testing of composites adhesives and polymers (UK: NPL Laboratory) 1–8.
[2] Thomason J L 1995 *Composites* **26** 477-85.
[3] Sethi S, Ray B C 2015 *Adv. Colloid Interface Sci.* **217** 43–49.
[4] Višnjić D, Lalić H, Dembitz V, et al 2014 *Period. Biol.* **116** 37–43.
[5] Yang L, Thomason J L 2012 *Polym. Test.* **31** 895–903.
[6] Han S O, Drzal L T 2003 *Eur. Polym. J. J* **39** 1791–99.
[7] Zinck P, Gérard J F 2008 *Compos. Sci. Technol.* **68** 2028–33.
[8] Laurikainen P 2017 *Characterization of the ageing of glass fibre-reinforced polymers* 3-5.
[9] Chu W, Karbhari V M 2005 *J. Mater. in Civ. Eng.* **17** 63–71.
[10] Nakashima K, Shirakawa T 2003 *AIP Conf. Proceedings* **680** 517–21.

IOP Conf. Series: Materials Science and Engineering **942** (2020) 012032 doi:10.1088/1757-899X/942/1/012032

Design of lamina orientation for biaxially loaded off-axis tunnelling cracks

Ashish K Bangaru, Lars P Mikkelsen and Bent F Sørensen

Department of Wind Energy, Technical University of Denmark, Risø Campus, DK-4000 Roskilde, Denmark

E-mail: akba@dtu.dk

Abstract. The primary objective of this work is to demonstrate how the condition for biaxially loaded off-axis tunnelling crack growth can be reproduced in a modified laminate under uniaxial loading. More specifically, laminates with the stacking sequences $[0/\theta/0/-\theta]_s$ and $[0/\alpha/0/-\alpha]_s$ have been considered for the biaxial and uniaxial loadings, respectively. A steady-state crack growth condition is analysed for an isolated off-axis tunnelling crack. The crack tip stress state is expressed in terms of a through thickness averaged mode mixity and energy release rate. The presented framework uses an energy accounting method with the crack opening displacements extracted from a finite element analysis. Results shows that the tunnelling crack tip stress state of the modified laminate under uniaxial loading matches with the biaxially loaded laminate. With a modified off-axis lamina orientation and the additional load amplitude from the biaxially loaded laminate, a simplified uniaxial testing laminate can be designed that yields the same biaxial tunnelling crack tip conditions.

1. Introduction

Over the past few decades, fibre reinforced polymer composites (FRP) have been an excellent choice of material in several industry sectors like aerospace, automotive, marine and wind energy. This is primarily due to their high strength and stiffness properties, combined with the low density. Long aligned FRP composites are laminated materials in which the individual plies are tailored to orient in one single or multiple directions depending on the direction of loads. Laminates consisting of only 0° oriented plies are called unidirectional laminates, otherwise, when stacked up with different ply angles like $\pm45^\circ$, 90° etc. they are called multi-directional laminates. However, these off-axis plies which are inclined at an angle to the primary load direction are always prone to the very first mode of damage during the service [1].

A typical wind turbine blade has load carrying composite laminates in spar caps [2]. Often, these laminated composites are made from non-crimp fabrics which has secondary oriented fibre bundles with respect to the main load carrying fibres, alongside few off-axis plies in between the quasi-unidirectional plies [3]. Cracks in off-axis plies or bundles conventionally called tunnelling cracks, propagate along the direction of off-axis fibres. The multiple tunnelling cracks not only cause the initial stiffness loss but also leads to other failure modes like fibre failures at the crack tips [4] and delaminations [5] at the interface of the neighbouring load carrying unidirectional laminates. Thus, it is necessary to understand and control the formation of the first mode of damage in order to prolong the fatigue life of the composite laminate.

1.1. Background
Generally, the onset of a tunnelling crack occurs either from a flaw like void within the ply or from free edges of the test specimen. Tunnelling cracks with an off-axis angle 90°, called transverse cracks, have been first studied in the late 1970s [6, 7]. From the experimental investigations, it was found that thickness of the transverse ply largely influences the crack spacing [6, 7]. In later years, the effect of multiple cracks and the crack spacing in transverse ply has been studied by finite element models with respect to the changes in the laminate stiffness [8, 9]. In one of the study by Nakamura and Kamath [10], it was shown that the energy release rate of a thin film crack bonded to a rigid substrate and oriented perpendicular to the loading direction reaches a steady state value (remains constant) when the crack length is approximately twice the ply thickness.

For tunnelling cracks propagating in steady state condition, Beuth [11] and Ho and Suo [12] presented an energy accounting method to calculate the energy release rate. Moreover, the cracking condition in transverse plies is pure Mode I [12]. In such conditions, the local energy release along the crack front remains constant and is equal to Mode I fracture energy, $G = G_{Ic}$. Further, the steady-state energy release rate found to be proportional to square of the applied stress and thickness of the lamina [12].

In case of off-axis angles other than 90°, the cracking occurs under mixed mode condition i.e., in both Mode I and Mode II. Besides, the fracture energy is found to depend on the mode mixity [14], defined as the phase angle of the stress intensity factors.

An interesting observation by Nakamura and Kamath [10] was that the far field stress state from the crack tip is found to be independent of the actual crack front shape. Therefore, the energy accounting method [12] which uses the stress state far ahead of the crack tip and crack opening displacements far behind the crack tip will therefore not depend on the actual crack front shape. This approach for calculating energy release rate was further extended to off-axis plies by Quaresimin et al. [15]. Following the studies of Quaresimin et al. [15] and Mikkelsen et al. [16], the present study uses this approach to calculate the average energy release rate.

Existing biaxial test methods are classified based on the loading system [17]. Briefly, these methods consists of either applying in-plane loads along the two perpendicular arms of a cruciform specimen or tension-torsion loads to tubular specimens. However, a good testing greatly relies on a good design of biaxial specimen. Additionally, to overcome the limitations associated with the biaxial testing [21], an elegant approach is demonstrated in terms of laminate design such that the same crack tip stress state is mimicked under uniaxial loading.

1.2. Motivation
The stress state in many composite structures is multiaxial. Therefore, a first step moving from a uniaxial fatigue testing is to do a fatigue testing under biaxial loading and there have been a number of testing strategies ([18, 19, 20]). Nevertheless, the fatigue damage initiation for many fibre composites observed to be governed by the evolution of the tunnelling cracks. The tunnelling crack growth condition during a biaxial fatigue testing can be reproduced using a uniaxial fatigue testing on a laminate with a modified layup. Thereby, a complicated biaxial fatigue testing aiming for the characterization of the tunnelling cracks damage can be replaced by a much simpler uniaxial fatigue testing series.

The present study illustrates a novel and scaled down approach of analysing the biaxially loaded off-axis tunnelling cracks under uniaxial loading. For a non-interacting crack under steady-state propagation, the specific mode mixity of biaxially loaded crack tip can be obtained by changing the orientation of the off-axis lamina under uniaxial loading. Furthermore, the uniaxial load applied on the new laminate can be specified to acquire the same energy release at the crack front. A specific biaxial loading scenario is considered on the laminate layup $[0/\theta/0/-\theta]_s$ to determine the crack tip stress state. The aim is to get same crack tip conditions in the modified laminate $[0/\alpha/0/-\alpha]_s$ subjected to uniaxial loading.

2. Theory of tunnelling cracks

2.1. Cracks in orthotropic materials

Considering the linear elasticity in orthotropic materials, the relationship between Mode I and Mode II energy release rates and the corresponding stress intensity factors is given by

$$\begin{Bmatrix} G_I \\ G_{II} \end{Bmatrix} = \left(\frac{1+\rho}{2E_{11}E_{22}}\right)^{1/2} \begin{Bmatrix} \lambda^{-1/4}K_I^2 \\ \lambda^{1/4}K_{II}^2 \end{Bmatrix} \tag{1}$$

where G_I and G_{II} are the Mode I and Mode II energy release rates, K_I and K_{II} are the Mode I and Mode II stress intensity factors. The two non-dimensional parameters, λ and ρ, are related to the engineering elastic constants ([22, 23]) (see Table 1) as

$$\lambda = \frac{E_{22}}{E_{11}}, \quad \rho = \frac{\sqrt{E_{11}E_{22}}}{2G_{12}} - \sqrt{\nu_{12}\nu_{21}} \tag{2}$$

Using the stress intensity factors at the crack tip for a given loading condition, the linear elastic fracture mechanics mode mixity is defined as ([13])

$$\psi = \arctan\left(\frac{K_{II}}{K_I}\right) \tag{3}$$

In general for off-axis tunnelling cracks, the mode mixity at the crack front varies along the position of the crack front. Furthermore, the fracture energy which is dependent on the mode mixity, $G_c = G_c(\psi)$, also varies along the crack front. However, in the event of steady-state crack propagation, the crack front will realign its shape such that for all points across the front, the local energy release rate is identical to the fracture energy of the associated local mode mixity. Moreover, the energy dissipation at the crack front that of the steady-state crack remains unchanged, even for an unknown crack front shape. As the crack front shape is not known, averaged properties (averaged over lamina thickness) are used to quantify the stress state of the crack front, defined similarly to eqs. (1) and (4). The average mode mixity then becomes

$$\bar{\psi} = \arctan\left(\frac{\bar{K}_{II}}{\bar{K}_I}\right) = \arctan\left(\lambda^{-1/4}\sqrt{\frac{\bar{G}_{II}}{\bar{G}_I}}\right) \tag{4}$$

here $\bar{G}_I$ and $\bar{G}_{II}$ represent average Mode I and Mode II energy release rates respectively, calculated by the formulae presented in the next section.

2.2. Calculation of the energy release rate

To calculate the steady-state energy release rate of an off-axis tunnelling crack, we use the energy accounting method ([15, 16]) as shown in Fig. 1. For an isolated crack in the thick lamina of height $2h$, the average steady-state energy release rate is given by the expression

$$\bar{G}_{ss} = \frac{1}{2h}\int_0^h \sigma_{22}(x_3)\,\delta_n(x_3)\,dz + \frac{1}{2h}\int_0^h \sigma_{12}(x_3)\,\delta_t(x_3)\,dz \tag{5}$$

here x_3 is through-the-thickness coordinate and varies from 0 to the ply thickness h. The stress components σ_{22} and σ_{12} are the normal and shear stresses in the cracked lamina far ahead of the crack tip, whereas δ_n and δ_t are the normal and tangential crack opening displacements far behind the crack tip. In eq. (6) the first integral expression is assumed as Mode I energy release

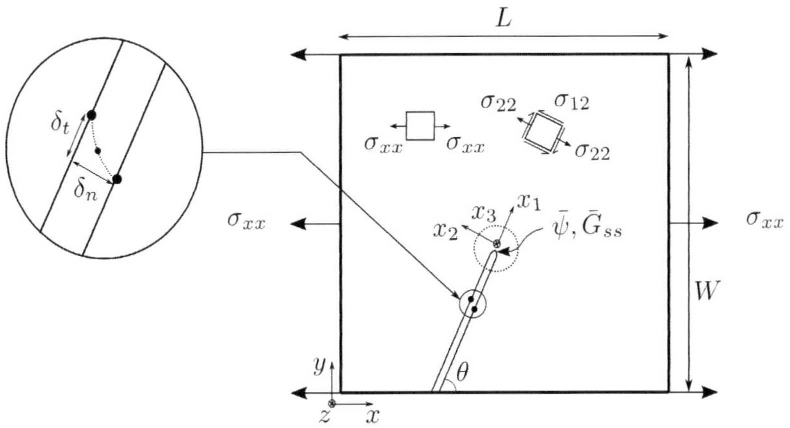

Figure 1: Illustration of the energy accounting method for an off-axis tunnelling crack from the crack normal opening and tangential opening displacements.

rate while the second expression is assumed to be the Mode II energy release rate. Additionally, because of the constant (far field) stress components σ_{22} and σ_{12} along the thickness direction of the off-axis lamina, the eq. (6) can be rewritten as

$$\bar{G}_I = \frac{1}{2}\sigma_{22}\,\bar{\delta}_n, \quad \bar{G}_{II} = \frac{1}{2}\sigma_{12}\,\bar{\delta}_t \tag{6}$$

where $\bar{\delta}_n$ and $\bar{\delta}_t$ represents average normal and tangential crack opening displacements respectively,

$$\bar{\delta}_n = \frac{1}{h}\int_0^h \delta_n(x_3)\,dz, \quad \bar{\delta}_t = \frac{1}{h}\int_0^h \delta_t(x_3)\,dz \tag{7}$$

therefore, the average total energy release rate is now given by

$$\bar{G}_{ss} = \bar{G}_I + \bar{G}_{II} \tag{8}$$

3. Problem definition and modelling

3.1. Geometry and boundary conditions

Tunnelling crack growth in the present study is analysed in the context of linear elastic fracture mechanics. Symmetric and balanced laminates with the layup a) $[0/\theta/0/-\theta]_s$ for biaxial loading and b) $[0/\alpha/0/-\alpha]_s$ for uniaxial loading has been considered for the steady state crack growth analysis. In case of biaxial loading, the laminate is a square plate with a side length, L_s, while in uniaxial loading the laminate is a rectangular plate with a length, L_r, and a width, W_r, as shown in Fig. 2. Both the plates are assumed to have identical total thickness, $2H$, and individual ply thickness, h. Here, the tunnelling crack is assumed to propagate along the fibre orientation of the thick central lamina (thickness $2h$) in the respective laminates. We operate on two Cartesian coordinate systems, a global coordinate system represented by the axes (x, y, z) along the two edges and thickness of the laminate respectively and a local coordinate system on ply level denoted by the axes (x_1, x_2, x_3), representing fibre direction in the plane, normal to the fibre direction and the out-of-plane thickness direction.

Due to symmetry, only lower half of the laminates (H) are modelled. Boundary conditions include the symmetric conditions applied on the mid-plane of the laminates along the z-direction i.e., all the nodes positioned at $z = H$ are constrained in the z-direction, $u_3^s = u_3^r = 0$. For biaxial

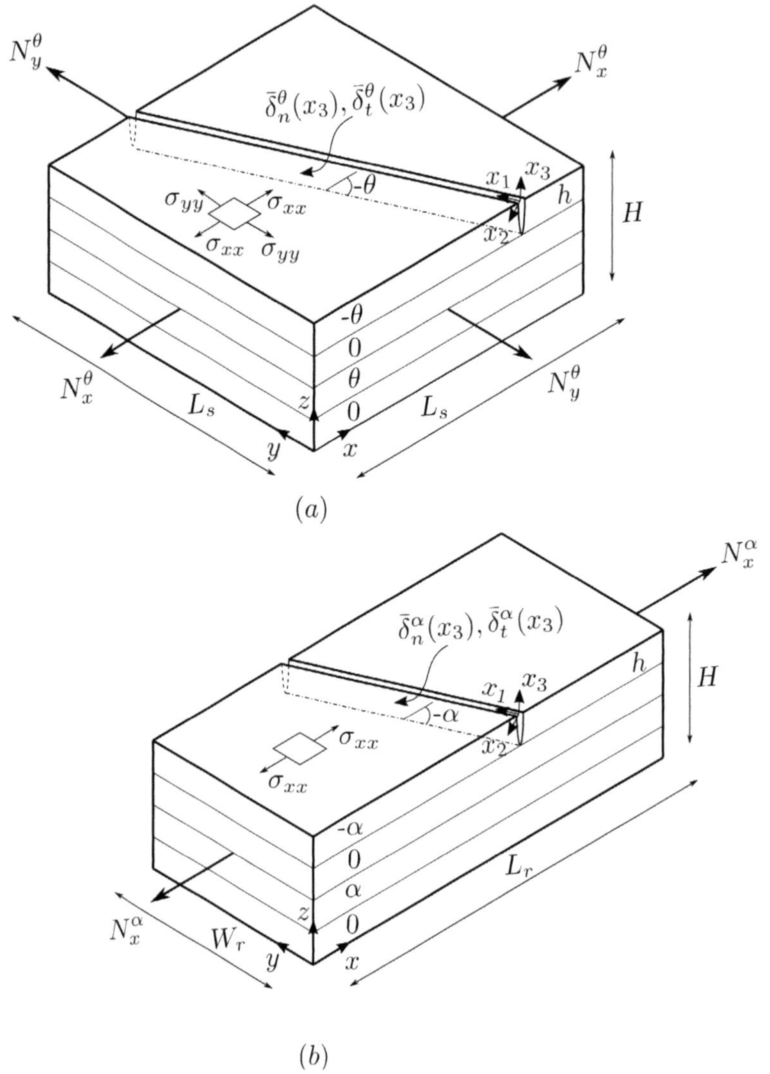

Figure 2: Geometries and boundary conditions of the symmetric laminate models along with the full-width cracks shown for a) biaxial loading of $[0/\theta/0/-\theta]_s$ and b) uniaxial loading of $[0/\alpha/0/-\alpha]_s$. The average crack opening displacements are extracted at the centre of the full-width crack for the respective model.

loading, the displacements were applied to the edges of the plate, along x and y-directions. This means, the nodes located at $x = 0$ and $x = L_s$ are subjected to prescribed displacements, $u_1 = \mp u_x^b$ with u_2 unconstrained, likewise, the nodes at $y = 0$ and $y = L_s$ are subjected to displacements in the $y-$direction, $u_2 = \mp u_y^b$ with u_1 left free. While for uniaxial loading, the displacements are applied to the nodes located $x = 0$ and $x = L_r$ as $u_1 = \mp u_x^u$ with u_2 left free.

3.2. Finite element modelling

Following Mikkelsen et al. [16], a non-crack tip model denoted as a full-width crack model is used in the present study as shown in Fig. 2. Each of the laminate model has only one full-width crack to analyse isolated and non-interacting crack growth condition.

For the two loading scenarios, the average normal and tangential crack opening displacements are extracted at the center of the respective laminate model and thereby the average mode mixities and the average energy release rates are calculated. The dimensions of the finite element model used for biaxial loading are $L_s \times L_s = 180h \times 180h$, whereas for uniaxial loading, $L_r \times W_r = 140h \times 40h$. Both the models use 8 noded brick elements (C3D8R) with a refined mesh size of $0.04h \times 0.04h \times 0.01h$ focussed at the center of the full-width crack, similar to what is used in Mikkelsen et al. [16]. The biaxial model consists of 6.5 million elements while the uniaxial model has 3.5 million elements. Since the relative difference between the stress, σ_{xx} (σ_{22} and σ_{12}), far away from both the edges and the crack of the finite element and the stress calculated from classical laminate theory is less than 1%, stresses from the latter are used in the energy accounting eq. (7). Linear analysis on the finite element models were performed using a commercial finite element code (Abaqus standard, 2017).

4. Results

4.1. Material properties

Glass fibre composite made up of unidirectional E-glass and epoxy resin is analysed with the elastic properties given in Table 1. The laminate is modelled as a homogeneous orthotropic material with each ply possessing the given elastic properties.

Table 1: Elastic constants of unidirectional E-glass/epoxy composite

Property	Value
Longitudinal modulus, E_{11}	40.8 (GPa)
Transverse modulus, E_{22}	12.25 (GPa)
In-plane Poisson's ratio, ν_{12}	0.27 (-)
In-plane shear modulus, G_{12}	4.62 (GPa)
Out-of-plane Poisson's ratio, ν_{23}	0.40 (-)

[a] Assumption of transverse isotropy: $E_{22} = E_{33}$, $\nu_{12} = \nu_{13}$, $G_{12} = G_{13}$ and $G_{23} = E_{22}/(2\,(1+\nu_{23}))$.
[b] $\nu_{21} = (E_{22}/E_{11})\,\nu_{12}$.
[c] $\lambda = 0.30,\ \rho = 2.25$.

Following Mikkelsen et al. [16], the finite element solutions of the biaxial and uniaxial loadings are presented in non-dimensional form. The results include the crack opening displacements, δ_n, δ_n, the local stress components σ_{22}, σ_{12}, and the energy release rates, G_I and G_{II}. Parameters used for normalization are cracked ply thickness, $2h$, global stress, σ_{xx}, and global strain, ε_{xx}, in the $x-$ direction for each loading case. defined as follows:

$$\bar{\delta}_n^* = \frac{\bar{\delta}_n}{\varepsilon_{xx}\,2h}; \quad \bar{\delta}_t^* = \frac{\bar{\delta}_t}{\varepsilon_{xx}\,2h}; \quad \sigma_{22}^* = \frac{\sigma_{22}}{\sigma_{xx}}; \quad \sigma_{12}^* = \frac{\sigma_{12}}{\sigma_{xx}}; \quad \bar{G}^* = \frac{\bar{G}}{\sigma_{xx}\,\varepsilon_{xx}\,2h} \tag{9}$$

4.2. Biaxial loading of $[0/\theta/0/-\theta]_s$

With the aim to analyse the effect of biaxial loads on the steady-state crack propagation, the average mode mixity and the average energy release rate was calculated in $-\theta$ layer of the layup, $[0/\theta/0/-\theta]_s$, under investigation. For this, we define load biaxiality ratio as transverse load to horizontal load, N_y^θ/N_x^θ, acting on the laminate. Additionally, positive (tensile) loads are investigated with the condition: $N_y^\theta < N_x^\theta$. In the present study, an example of biaxial loading case with specific N_y^θ/N_x^θ ratio is simulated on the laminate as:

Biaxially loaded condition: $\theta = 30°$ in $[0/\theta/0/\theta]_s$ and $N_y^{30°}/N_x^{30°} = 0.12$

The normalized average crack opening displacements and the stress components are found to be

$$\begin{aligned} \delta_n^* &= 0.4980, \quad \delta_t^* = 0.6985 \\ \sigma_{22}^* &= 0.1522, \quad \sigma_{12}^* = 0.1673 \end{aligned} \tag{10}$$

now substituting above values in eq. (7) gives an estimate for the average Mode I and Mode II energy release rates as

$$\bar{G}_I^* = \frac{1}{2}\sigma_{22}^*\bar{\delta}_n^* = \frac{1}{2}\cdot 0.1522 \cdot 0.4980 = 0.0379 \tag{11}$$

$$\bar{G}_{II}^* = \frac{1}{2}\sigma_{12}^*\bar{\delta}_t^* = \frac{1}{2}\cdot 0.1673 \cdot 0.6985 = 0.0584 \tag{12}$$

when substituted the energy values into eqs. (5) and (9) gives the average mode mixity and the average total energy release rate for as

$$\bar{\psi}^\theta = 59.2^\circ \text{ and } \bar{G}_{ss}^\theta = 0.21 \text{ N/mm (absolute)} \tag{13}$$

4.3. Uniaxial loading of $[0/\alpha/0/-\alpha]_s$

To match the average mode mixity of approx. 59° at the crack front of an off-axis tunnelling crack under uniaxial loading, lamina orientation α in $[0/\alpha/0/-\alpha]_s$ was altered [24]. Further, the uniaxial load N_x^θ applied to the laminate is varied in order to attain the same average energy release rate at the steady-state crack front [24].

Uniaxially loaded condition: $\alpha = 50^\circ$ in $[0/\alpha/0/-\alpha]_s$ and $N_x^{50^\circ}/N_x^{30^\circ} = 0.57$

For this case, the average crack opening displacements and the stress components and in normalized form are

$$\begin{aligned} \delta_n^* &= 0.6471, \quad \delta_t^* = 0.9530 \\ \sigma_{22}^* &= 0.4425, \quad \sigma_{12}^* = 0.4706 \end{aligned} \tag{14}$$

using eq. (7) again, the Mode I and Mode II energy release rates are calculated as

$$\bar{G}_I^* = \frac{1}{2}\sigma_{22}^*\bar{\delta}_n^* = \frac{1}{2}\cdot 0.4425 \cdot 0.6471 = 0.1431 \tag{15}$$

$$\bar{G}_{II}^* = \frac{1}{2}\sigma_{12}^*\bar{\delta}_t^* = \frac{1}{2}\cdot 0.4706 \cdot 0.9530 = 0.2242 \tag{16}$$

and thus from eqs. (5) and (9), the average mode mixity and the average energy release rate are estimated as

$$\bar{\psi}^\alpha = 59.4^\circ \text{ and } \bar{G}_{ss}^\alpha = 0.21 \text{ N/mm (absolute)} \tag{17}$$

Thus, it can be seen that the average mode mixity and the average energy release rate of a steady-state tunnelling from the biaxial loading case can be recreated in the uniaxial loading, $\bar{\psi}^\theta = \bar{\psi}^\alpha$ (with less than 1% discrepancy) and $\bar{G}_{ss}^\theta = \bar{G}_{ss}^\alpha$.

5. Summary

Using three dimensional finite element analyses, steady-state crack growth condition of an off-axis tunnelling crack is analysed. The energy accounting method is applied to the isolated cracks in two different laminate layups, $[0/\theta/0/-\theta]_s$ and $[0/\alpha/0/-\alpha]_s$, subjected to the biaxial and uniaxial loading respectively. It is shown that the average mode mixity and the average energy release rate of the biaxially loaded tunnelling crack tip can be attained on the modified laminate in the uniaxial loading. By altering the orientation angle of the off-axis lamina and the uniaxial load amplitude, mode mixities and the energy release rates at the crack tip can be varied accordingly.

The approach of performing uniaxial testing to understand tunnelling cracks under biaxial loads may be seen as an alternative to the complicated biaxial testing.

Acknowledgements
This research is carried out as a part of the CASMaT (Center for Advanced Structural and Material Testing) Initiation Project.

References

[1] Quaresimin M, Carraro P A and Maragoni L 2016 Early stage damage in off-axis plies under fatigue loading *Compos. Sci. Technol.* **128** 147-54
[2] Mishnaevsky L, Branner K, Petersen H N, Beauson J, McGugan M and Sørensen B F 2017 Materials for wind turbine blades an overview *Materials* **10** 1285
[3] Vallons K, Adolphs G, Lucas P, Lomov S V and Verpoest I 2013 Quasi-UD glass fibre NCF composites for wind energy applications: a review of requirements and existing fatigue data for blade materials *Mech. Ind.* **14** 175-89
[4] Jamison R D, Schulte K, Reifsnider K L, Stinchcomb W 1984 Characterization and analysis of damage mechanisms in tension-tension fatigue of graphite/epoxy laminates *J. ASTM Int.* 21
[5] Talreja R 1981 Fatigue of composite materials: damage mechanisms and fatigue-life diagrams *P. Roy. Soc. A-Math Phy.* **378** 461-75
[6] Garrett K W and Bailey J E 1977 Multiple transverse fracture in 90 cross-ply laminates of a glass fibre-reinforced polyester *J. Mater. Sci.* **12** 157-68
[7] Parvizi A, Garrett K W and Bailey J E 1978 Constrained cracking in glass fibre-reinforced epoxy cross-ply laminates *J. Mater. Sci.* **13** 195-201
[8] Hashin Z 1985 Analysis of cracked laminates: a variational approach *Mech. Mater.* **4** 121-36
[9] Varna J and Berglund L 1991 Multiple transverse cracking and stiffness reduction in cross-ply laminates *Compos. Sci. Technol.* **13** 97-106
[10] Nakamura T and Kamath S M 1992 Three-dimensional effects in thin film fracture mechanics *Mech. Mater.* **13** 67-77
[11] Beuth Jr J L 1992 Cracking of thin bonded films in residual tension *Int. J. Solids Struct.* **29** 1657-75
[12] Ho S and Suo Z 1993 Tunneling cracks in constrained layers *J. Appl. Mech* **60** 890-94
[13] Hutchinson J W, Suo Z 1991 Mixed mode cracking in layered materials *Adv. Appl. Mech.* **29** 63-191
[14] Sørensen B F and Jacobsen T K 2009 Characterizing delamination of fibre composites by mixed mode cohesive laws *Compos. Sci. Technol.* **69** 445-56
[15] Quaresimin M, Carraro P A, Mikkelsen L P, Lucato N, Vivian L, Brøndsted P, Sørensen B F, Varna J and Talreja R 2014 Damage evolution under cyclic multiaxial stress state: A comparative analysis between glass/epoxy laminates and tubes *Compos. B. Eng.* **61** 282-90
[16] Mikkelsen L P, Klitgaard S J, Niordson C F and Sørensen B F 2020 Tunneling cracks in arbitrary oriented off-axis lamina *Int. J. Fract.* Accepted
[17] Smits A, Ramault C, Makris A, Hemelrijck V D, Clarke A, Williamson C, Gower M, Shaw R, Mera R, Lamkanfi E and Paepegem V W 2007 A review of biaxial test methods for composites *In Experimental Analysis of Nano and Engineering Materials and Structures* 933-34
[18] Radon J C, Wachnicki C R 1985 Biaxial fatigue of glass fiber reinforced polyester resin *Multiaxial Fatigue ASTM Int.*
[19] Smits A, Hemelrijck V D, Philippidis T P and Cardon A 2006 Design of a cruciform specimen for biaxial testing of fibre reinforced composite laminates *Compos. Sci. Technol.* **66** 964-75
[20] Quaresimin M and Carraro P A 2013 On the investigation of the biaxial fatigue behaviour of unidirectional composites *Compos. B. Eng.* **54** 200-8
[21] Makinde A, Thibodeau L, Neale K W 1992 Development of an apparatus for biaxial testing using cruciform specimens *Exp. Mech.* **32** 138-44
[22] Suo Z 1990 Delamination specimens for orthotropic materials *J. Appl. Mech* **53** 627-34
[23] Suo Z, Bao G, Fan B and Wang T C 1991 Orthotropy rescaling and implications for fracture in composites *Int. J. Solids Struct.* **28** 235-48
[24] Bangaru A K, Sørensen B F, Mikkelsen L P and Legarth B N. Approach for analysing off-axis tunnelling cracks in biaxially loaded laminates. In preparation.

IOP Conf. Series: Materials Science and Engineering **942** (2020) 012033 doi:10.1088/1757-899X/942/1/012033

Understanding the mechanical response of glass and carbon fibres: stress-strain analysis and modulus determination

Rajnish Kumar*, Lars P Mikkelsen, Hans Lilholt and Bo Madsen

Division of Wind Energy Materials and Components, Department of Wind Energy, Technical University of Denmark, 4000 Roskilde, Denmark

*E-mail: rakum@dtu.dk

Abstract. Accurate characterization of fibres is crucial for the understanding the properties and behaviour of fibre-reinforced composite materials. Fibre properties are key parameters for composite design, modelling and analysis. In this study, characterization of mechanical properties of glass and carbon fibres has been performed using a semi-automated single-fibre testing machine. Based on a sample set of 150 glass and carbon fibers fibres, engineering and true stress-strain curves are analyzed. Different modulus determination methods are discussed based on true stress-strain and tangent modulus-strain relationships. For glass fibres, the true stress-strain based tangent modulus is found to be independent of applied strain, whereas for carbon fibres, a tendency of tangent modulus to increase with applied strain is observed. The modulus of glass fibres is found to be independent of fibre diameter, whereas carbon fibres with smaller diameter show higher modulus compared with carbon fibres with larger diameters.

1. Introduction

Fibre-reinforced composites are seeing a rapid growth in their development and application in wind energy, aerospace and automotive industries. It is due to their excellent mechanical properties combined with low density. The use of fibres as reinforcement with high levels of specific modulus and specific strength imparts desirable mechanical performance to composites. Reliable characterization of the reinforcing fibres is critically important for reliable predictions of the mechanical behaviour of composites, as they are the principal load-bearing members.

Different methods are used to characterize the mechanical properties of fibres like single-fibre tensile test and fibre bundle test. The single-fibre tensile test is the most commonly used method [1]. Individual fibres are tested one by one by subjecting them to an increasing tensile load until failure. The traditional card frame method is a labour intensive and time-consuming process. Due to the small diameter of fibres, measurement of their cross-sectional area becomes challenging. In addition, a careful processing and handling technique is required, as this might induce new flaws and/or change the existing ones on the fibres. Most studies on fibre characterization have been limited to small number of samples due to very challenging and time-consuming testing method [2]. A number of studies have highlighted that for a reliable analysis, a large number of single-fibre data is needed [3, 4, 5]. Swolfs et al. reported that several hundred of fibres should be tested to obtain less than 10 % of variation in modelling predictions of composite materials [6].

Glass and carbon fibres are the two most industrially important reinforcement fibres for composites [7]. Both of them are brittle in nature and exhibits scatter in strength. Due to their high modulus to weight ratio, they attract wide applications where minimum structural weight is required, e.g. wind turbine blades.

A literature survey on single glass and carbon fibre testing reveals that the analysis of single-fibre tensile test results are mostly restricted to strength distribution using different probabilistic models, where the Weibull model is by far the most widely used [8, 9]. There is a lack of studies analysing the basic mechanical response of single fibres.

In this study, we characterised the mechanical response of glass and carbon fibres using a semi-automated single-fibre testing machine. This equipment enabled us to test large number of fibres with minimized manual handling. The tensile test results were analysed using both engineering and true stress-strain curves, to present a complete picture. In addition, we investigated different modulus determination methods. All the data is available as a Zenodo repository [19].

2. Materials and methods

2.1. Materials

Boron-free R-glass fibres (HiPer-tex™ W 3030, 3B Fiberglass Norway AS) supplied by Owens Corning, Belgium were investigated in this study. The roving had a filament count of 4,000 and a nominal single fibre diameter of 17 μm.

PAN-based carbon fibres (PYROFIL™ TRW40 50L, Mitsubishi Chemical Corporation, Japan) were investigated in this study. The roving had a filament count of 50,000 and a nominal single fibre diameter of 7 μm.

2.2. Single fibre tensile testing

Single-fibre tensile properties of glass and carbon fibres were determined using a semi-automated single-fibre testing machine (TexTechno, Favimat+, Germany). The fibre cross-sectional area (A, m^2) was measured by the built-in vibroscope system. Initially, after a fibre is gripped between the two clamps in the machine, a small tension force is applied to the fibre and it is excited by acoustic waves to measure its resonance frequency. Based on the measured resonance frequency, the fibre linear density (T, kg/m) is determined by using a recently developed iterative analysis method [10]. The method corrects for the fact that the fibre is not perfectly flexible (i.e. the factor E/L^2F is not neglected, where E is fibre modulus, L is fibre length and F is fibre pre-tension), and this leads to values of linear density that are slightly larger (1-2 %) than the values determined by assuming a simply supported boundary at the clamps. Using the fibre density (ρ, kg/m^3), the fibre cross-sectional area can be calculated ($A = T/\rho$).

The density of glass and carbon fibres was measured based on gas pycnometry method. A gas pycnometer (Ultrapyc 1200E; Quantachrome Instruments, USA) was used to measure the volume of fibres. The density of glass and carbon fibres were measured to be 2.597 ± 0.013 g/cm^3 and 1.774 ± 0.002 g/cm^3, respectively. The measured density values are very close to the manufacturer data sheet values of 2.6 g/cm^3 and 1.80 g/cm^3 for glass and carbon fibres, respectively. The density values based on the manufacturer data sheet were used to determine the fibre cross-sectional area in the vibroscope method described above.

The tensile tests were carried out with a crosshead speed of 1 mm/min. The approximate tensile test-time for each fibre was 1-2 minutes. In order to determine the machine and gripping compliance, glass and carbon fibres were tested at different gauge lengths (initial fibre length) in the range of 40-80 mm, and the data was analysed by a standard method for determination of compliance [11]. In this study, the analysis of the mechanical response of glass and carbon fibres was performed based on 150 fibres tested at a gauge length of 60 mm.

2.3. Stress-strain curve analysis

The mechanical response of the glass and carbon fibres were analysed using engineering and true stress-strain curves. The *engineering* stress-strain measures of nominal stress (σ_n) and engineering strain (ε_e) are determined from the measured load (P), initial cross-sectional area (A_0), initial length (L_0) and current length (L) as follows:

$$\sigma_n = \frac{P}{A_0} \tag{1}$$

$$\varepsilon_e = \frac{L - L_0}{L_0} \tag{2}$$

The *true* stress-strain measures of true stress (σ_t) and logarithmic strain (ε_{ln}) are determined from the measured load (P) and current cross-sectional area (A) as follows:

(Assuming constant volume dV=0, $A_0/A = L/L_0$)

$$\sigma_t = \frac{P}{A} = \frac{P}{A_0} \cdot \frac{L}{L_0} \tag{3}$$

$$\sigma_t = \sigma_e(1 + \varepsilon_e) \tag{4}$$

The increment of strain is the incremental increase in displacement (dL) divided by the current length (L):

$$d\varepsilon_{ln} = \frac{dL}{L} \tag{5}$$

Integrating both sides

$$\int_0^{\varepsilon_{ln}} d\varepsilon_{ln} = \varepsilon_{ln} \tag{6}$$

$$\int_{L_0}^{L} \frac{dL}{L} = \ln\frac{L}{L_0} \tag{7}$$

Using equation (2)

$$\ln\frac{L}{L_0} = \ln(1 + \varepsilon_e) \tag{8}$$

$$\varepsilon_{ln} = \ln(1 + \varepsilon_e) \tag{9}$$

The tangent modulus as a function of strain was determined from the stress-strain data for each tested fibre using the following method:

1) Range of strain was selected. Starting point of the range was the first point of measured strain (%) + 0.1 %. Final point of the range was the failure strain (%) - 0.1 %. This range of strain was divided into 1000 strain values.
2) Linear fitting was performed to calculate the corresponding tangent modulus value at each strain values with a window size of 0.20 % i.e. 0.10 % on each side of the strain value.

2.4. Observation of fibre cross-sectional shape

Samples with a rectangular area of 25×25 mm^2 were cut from unidirectional glass and carbon composites manufactured in another study [12] with the same type of glass and carbon fibres used in this study. The samples were cast into cylindrical blocks of epoxy resin. Edges of the blocks were ground and polished. The polished surface of the blocks were sputtered with carbon particles with thickness ≈10 nanometres.

The cross-sectional shape of glass and carbon fibres was examined using scanning electron microscope (VEGA3 TESCAN, Czech Republic) at 1.38 kx magnification.

3. Results

3.1. Mechanical response of glass fibres

Single-fibre tensile test results of glass fibres have been analysed using engineering and true stress-strain curves as shown in Figure 1. The stress-strain curve for brittle materials like glass fibres is generally believed to be linear over the full range of strain, without showing appreciable plastic deformation. However, as shown in Figure 1a, the engineering stress-strain curves begin to deviate from linearity with increasing strain. In true stress-strain curves, as shown in Figure 1b, the glass fibres can be observed to behave linearly over the full range of strain.

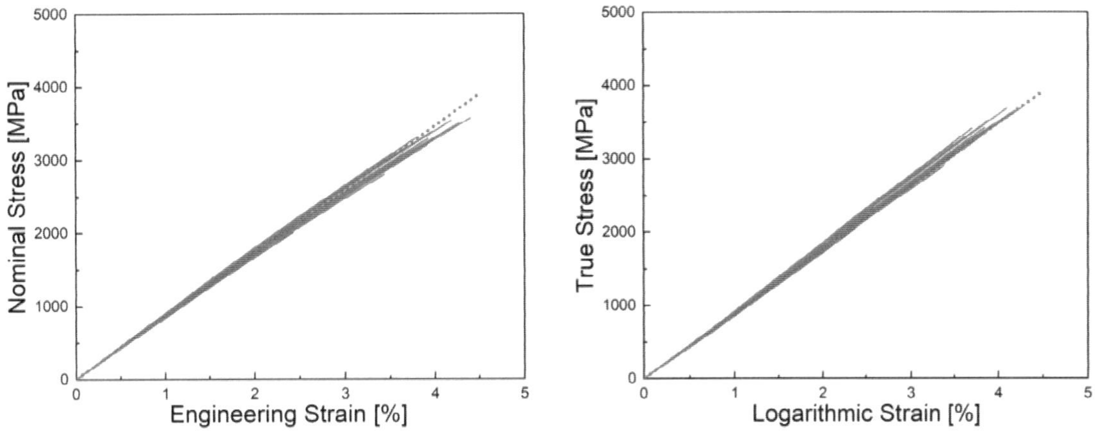

(a) (b)

Figure 1. Stress-strain curves of single glass fibres: (a) engineering σ_n-ε_e, (b) true σ_t-ε_{ln}. Dotted line represents a constant modulus line calculated between strain 0.05 % and 0.25 %.

This can also be seen in the determined tangent modulus-strain curves, again comparing engineering and true stress-strain measures, as shown in Figure 2. In Figure 2a, the slope of the linear fit line was calculated to be -3.0 GPa/%. This trend demonstrates that the modulus is decreasing with increasing strain. In Figure 2b, modulus tends to remain almost constant with increasing strain, with the slope of the linear fit line being -0.6 GPa/%.

The modulus of glass fibres has been determined using the following five methods, based on the *engineering* stress-strain curves, the *true* stress-strain curves and the *true* tangent modulus-strain curves. The results are summarized in Table 1.

1) Modulus based on using linear regression of the engineering stress-strain curve between strain 0.05 % and 0.25 %. This is the conventional method for determination of materials modulus according to the ISO 527 standard [13].
2) Modulus based on using linear regression of the true stress-strain curve between strain 0.05 % and 0.25 %.
3) Modulus based on the midpoint value of the linear fit line for the engineering tangent modulus-strain curve between strain 0.05 % and 0.25 %.
4) Modulus based on the midpoint value of the linear fit line for the true tangent modulus-strain curve between strain 0.05 % and 0.25 %.

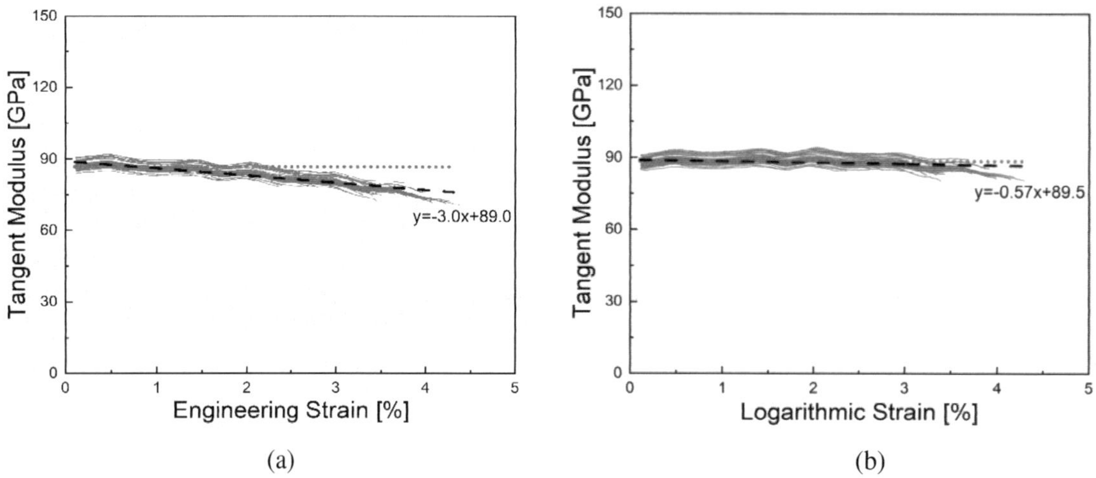

Figure 2. Tangent modulus-strain curves for single glass fibres: (a) engineering ε_e, (b) logarithmic ε_{ln}. Dotted line (grey) represents a constant modulus line (based on stress-strain curves) calculated between strain 0.05 % and 0.25 %, and dashed line (black) represents a linear fit for the tangent modulus-strain curves.

5) Modulus based on the linear fit line for the true tangent modulus-strain curve at 1.5 % strain.
6) Modulus at 0 % strain determined as the intersection with the y-axis of the linear fit line for the tangent modulus-strain curve.

Table 1. Summary of determined modulus values of glass fibres.

Modulus determination methods	Modulus (GPa)
Linear regression of the engineering stress-strain curve (0.05 % - 0.25 %)	86.6 ± 1.5
Linear regression of the true stress-strain curve (0.05 % - 0.25 %)	87.0 ± 1.6
Linear fit of the engineering tangent modulus-strain curve (0.05 % - 0.25 %)	88.5 ± 1.3
Linear fit of the true tangent modulus-strain curve (0.05 % - 0.25 %)	89.4 ± 1.6
Linear fit of the true tangent modulus-strain curve (at 1.5 %)	88.7 ± 1.6
Linear fit of the true tangent modulus-strain curve (at 0 %)	89.5 ± 1.7

In Table 1, it can be seen that there, as expected, is a very little difference between the young modulus found performing a linear regression on the stress-strain curve of the engineering or true stress-strain measure. This is because, the strains in the 0.05% to 0.25% strain range is so small that the two stress-strain measure approximate each other. On the other hand, testing thin fibers, it can be a challenge to get a precise measure of the load and strain in this range. During the initial loadings, the influence of an initial alignment of the fibers may affect this precision. Nevertheless, observing a linear relation between the strain and tangent modulus measure, it can be seen that this linearity can be used to make a much more robust stiffness measurement. Making a linear fit of the tangent modulus versus strain in the full tested strain range, a good agreement between the linear fit and the tangent modulus variation is found, see Figure 2a-b. Based on the linear fit of the tangent modulus versus strain curve, modulus values of 88.5 ± 1.3 GPa for the engineering stress-strain case and 89.4 ± 1.6 GPa for the true stress-strain case was determined. Both values significant larger than the 86.6 ± 1.5 GPa and 87.0 ± 1.6 GPa found using a linear regression of the stress-strain curve in the 0.05-0.25% range. As it is not expected that the glass-fiber material should behave significant softer for smaller strains, those smaller stiffness values is judge to be influenced by this small measurement precision at the lower strain and load values. Therefore, a modulus of 88.5 ± 1.3 GPa in the engineering stress-strain domain and 89.4 ± 1.6 GPa in the true stress strain domain is found to be more trustworthy.

In addition, a rather constant slope of the stress-strain curve is found using the true stress-strain measure, see Figure 1b and 2b. The modulus values determined at 0 % and 1.5 % are 89.5 ± 1.7 GPa and 88.7 ± 1.6 GPa, respectively. Comparing the average values, modulus of glass fibre decreases by 0.9 % going from 0 % to 1.5 % strain. Considering the standard deviation values, it can be said that the measured modulus value of glass fibre is independent of the applied strain. In the commercial finite element code Abaqus, a linear elastic material will be linear in the true stress-strain measure when including geometrical non-linear effect [14]. It is therefore interesting to see that the glass-fibers, as a good approximation be considered as a linear elastic material using this true stress versus logarithmic strain measure.

3.2. Mechanical response of carbon fibres

Single-fibre tensile test results of carbon fibres have been analysed using engineering and true stress-strain curves as shown in Figure 3. Non-linear behaviour is observed for both engineering and true stress-strain curves. There is an appreciable increase in modulus with increasing strain, resulting in the curves to bend upwards. Curtis et al. [15] first observed this behaviour.

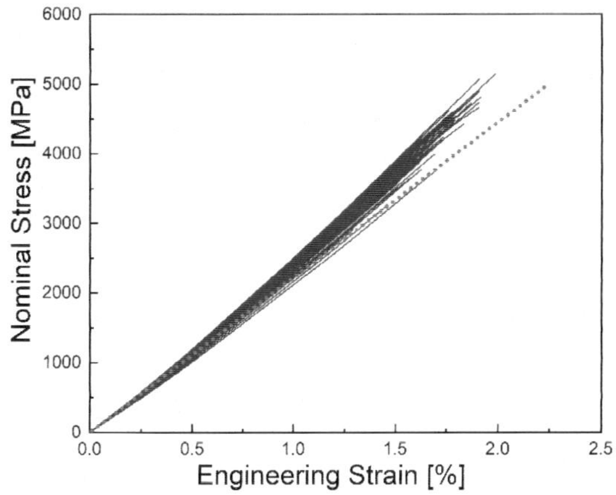

(a)

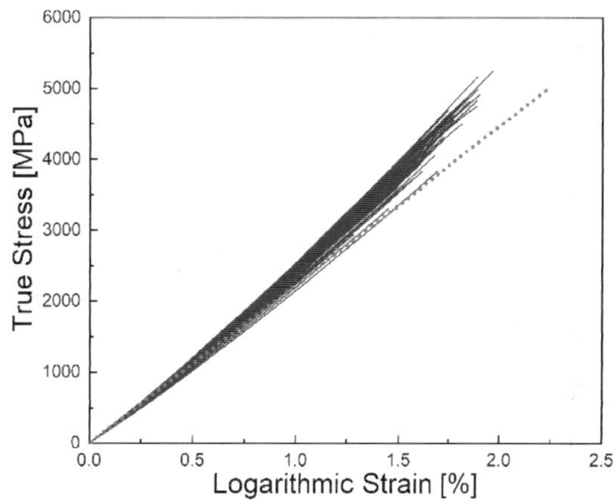

(b)

Figure 3. Stress-strain curves of single carbon fibres (a) engineering σ_n-ε_e, (b) true σ_t-ε_{ln}. Dotted line represents a constant modulus line calculated between strain 0.05 % and 0.25 %.

Engineering and true tangent modulus-strain curves are shown in Figure 4. The dashed line represents the linear fit for the tangent modulus-strain curves of all the fibres. In Figure 4a and 4b, the increasing tendency of modulus with increasing strain can be observed, with the slope of the linear fit line being 36.5 GPa/% and 44.3 GPa/% for engineering and true curves, respectively.

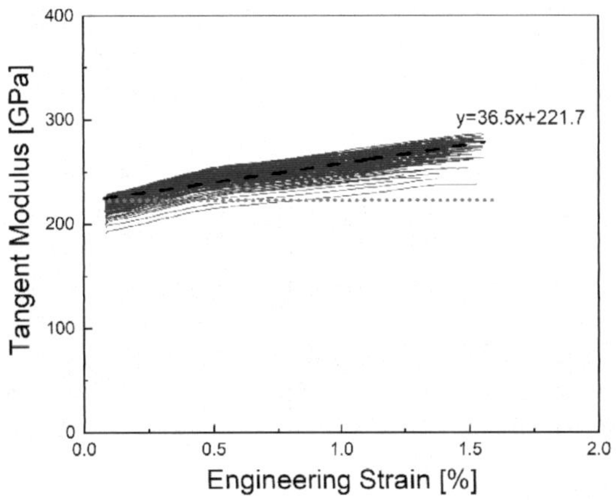

(a)

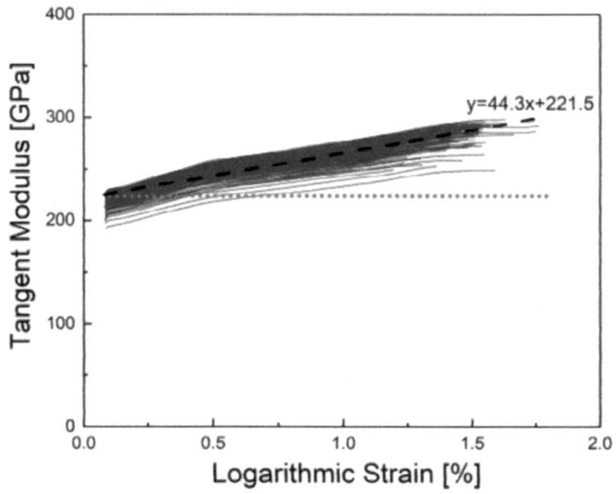

(b)

Figure 4. Tangent modulus-strain curves for single carbon fibres: (a) engineering ε_e, (b) logarithmic ε_{ln}. Dotted line (grey) represents a constant modulus line (based on stress-strain curves) calculated between strain 0.05 % and 0.25 %, and dashed line (black) represents the linear fit for the tangent modulus-strain curves.

As can be seen in Figure 4, the increase in modulus of carbon fibres with applied strain appears to follow a two-step increment, and performing a single linear fitting of both these regions will give an incorrect estimate for modulus. As shown in Figure 5, the green dotted line represents the linear regression line from 0 % to 0.4 % strain, having a slope of 69.5 GPa/%. Similarly, the red dotted line represents the linear regression from 0.6 % strain until failure, having a slope of 38.1 GPa/%. The intersection of these two linear fit lines gives a transition point at 0.47 % strain.

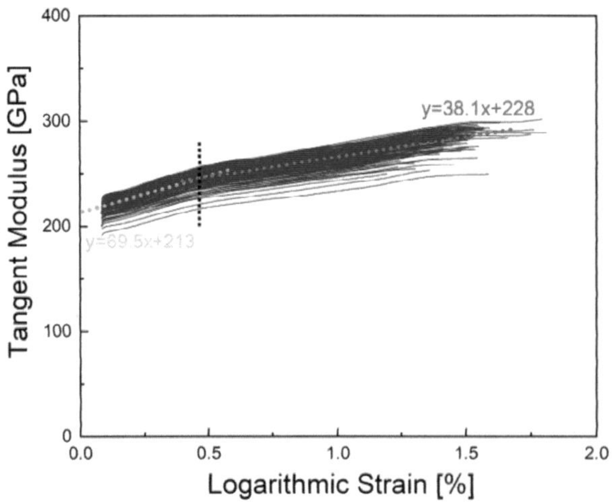

Figure 5. True tangent modulus-strain curves for single carbon fibres (same as in Figure 4b). The two linear fit lines "green" and "red" represents the two-step tangent modulus curves.

The modulus of carbon fibres has been determined by the same five methods used for the glass fibres, based on the *engineering* stress-strain curves, the *true* stress-strain curves and the *true* two-step tangent modulus-strain curves. The results are summarized in Table 2.

Table 2. Summary of determined modulus values of carbon fibres.

Modulus determination methods	Modulus (GPa)
Linear regression of the engineering stress-strain curve (0.05 % - 0.25 %)	223 ± 6
Linear regression of the true stress-strain curve (0.05 % - 0.25 %)	224 ± 6
Linear fit of the engineering tangent modulus-strain curve (0.05 % - 0.25 %) *	223 ± 7
Linear fit of the true tangent modulus-strain curve (0.05 % - 0.25 %) *	224 ± 7
Linear fit of the true tangent modulus-strain curve (at 1.5 %) **	284 ± 9
Linear fit of the true tangent modulus-strain curve (at 0 %) *	213 ± 6

* Analysis based on first linear fit line "green". ** Analysis based on second linear fit line "red"

The modulus values determined at 0 % and 1.5 % are 213 ± 6 GPa and 284 ± 9 GPa, respectively. Comparing the average values, modulus of carbon fibre increases by 33 % going from 0 % to 1.5 % strain. This two-step tangent modulus-strain relation is based upon initial observations. Further in-depth analysis of this trend will be done to verify its existence.

3.3. Relationship between fibre diameter and modulus

In Figure 6, microscopic images of the cross-section of glass and carbon fibres are presented. It can be observed that glass fibres have an almost circular cross-section, while carbon fibres have clearly non-circular kidney-shaped cross-section. The average cross-sectional areas of the fibres were determined with the vibroscopy method using the measured linear density and density of fibres. The average cross-sectional area for glass and carbon fibres are 249 ± 43 μm^2 and 42 ± 3 μm^2, respectively. A representative diameter assuming the fibres to be circular in cross-section was calculated, and is only used here as a representative value for the cross-section area. The average diameters of glass and carbon were determined to be 18 ± 1 μm and 7 ± 0.3 μm, respectively.

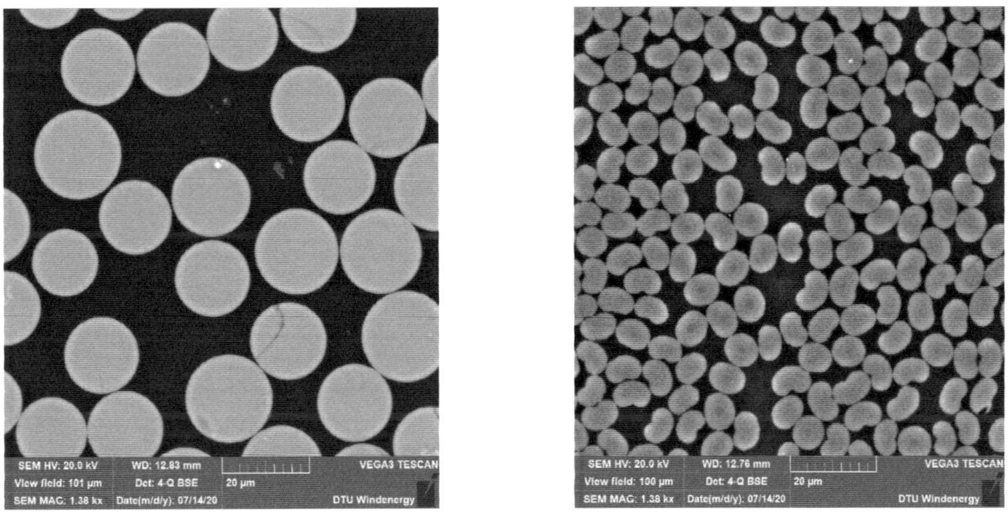

(a) (b)

Figure 6. SEM images showing the cross-section shape for (a) glass fibres and (b) carbon fibres.

Figure 7a shows the relation between modulus (linear regression of the true stress-strain curve (0.05 % - 0.25 %)) and diameter for glass fibres. The measured modulus appears to be independent of the fibre diameter. In Figure 7a, the dashed line represents the linear fit for modulus vs fibre diameter. The slope of the linear fit line was calculated to be 0.12 GPa/μm. It implies that a similar modulus was measured irrespective of the difference in the fibre diameter for glass fibres.

Figure 7b shows the relation between modulus and diameter for carbon fibres. The slope of the linear fit line for modulus vs fibre diameter relation was calculated to be -15.2 GPa/μm. It shows a tendency that the fibres with smaller diameters have higher modulus. During the manufacturing of PAN-based carbon fibres, there is a processing step called fibre spinning. The precursor fibres are drawn in a series of stages, in order to increase polymer chain alignment in the fibre axial direction and results in diameter reduction. In general, the precursor fibre tensile modulus will continue to increase as the total draw ratio increases [16]. Then, the conversion of the precursor fibre into carbonized structure requires two heat treatment processes, namely stabilization and carbonization. The fibres are again subjected to an applied force, which is an important processing parameter to influence mechanical properties. The process parameters are optimized to minimize fibre-to-fibre diameter variation and mechanical property variance. Due to the highly anisotropic nature of the graphite elements, the apparent modulus increases as the structural units reorients [17]. Carbon fibres with smaller diameters might have increased alignment of the crystallites, thus showing a higher modulus, as observed in Figure 7b.

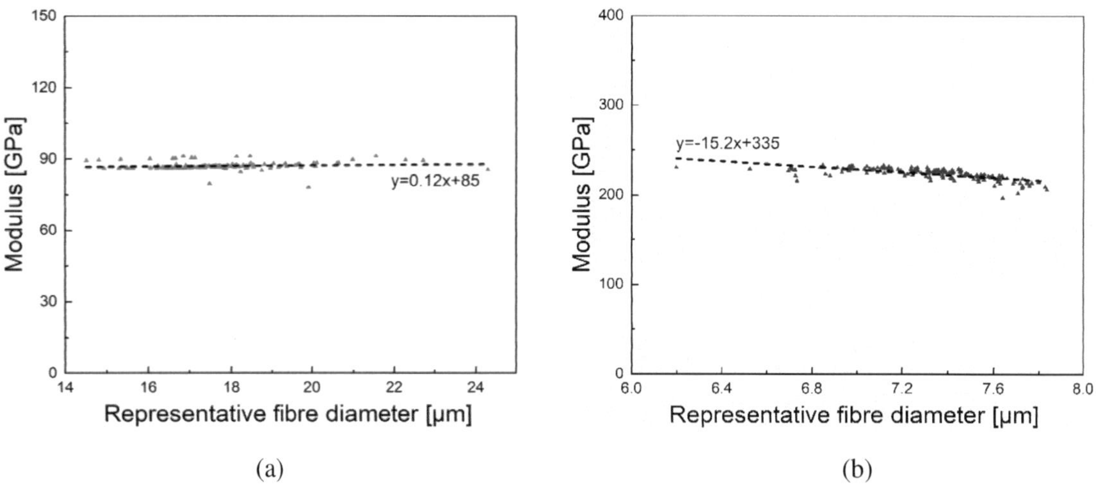

Figure 7. Relationship between modulus and fibre diameter for (a) glass fibres and (b) carbon fibres.

4. Discussion

By implying a second-order polynomial relation between stress (σ) and strain (ε), a linear relation between tangent modulus and strain can be established as [18]:

$$E_t = \frac{d\sigma}{d\varepsilon} = \alpha \cdot \varepsilon + E_0 \tag{10}$$

where E_t (= $d\sigma/d\varepsilon$) is the tangent modulus at a given value of ε, E_0 is the initial stiffness ($\varepsilon = 0$), and α is a so-called stress-strain curvature coefficient, respectively.

Integrating the equation (10) with respect to ε (for intercept at origin) the second-order polynomial relation between stress (σ) and strain (ε) can be established:

$$\sigma = \frac{1}{2} \cdot \alpha \cdot \varepsilon^2 + E_0 \cdot \varepsilon \tag{11}$$

Equation (11) can be directly used for glass fibres. Whereas for carbon fibres, based on the two-step modulus curves, Equation (11) can be written as:

$$\sigma = \begin{cases} \frac{1}{2} \cdot \alpha_{(1)} \cdot \varepsilon^2 + E_{0(1)} \cdot \varepsilon & \text{for } \varepsilon < \varepsilon_{tr.} = 0.47\ \% \\ \frac{1}{2} \cdot \alpha_{(2)} \cdot \varepsilon^2 + E_{0(2)} \cdot \varepsilon & \text{for } \varepsilon > \varepsilon_{tr.} = 0.47\ \% \end{cases} \quad (12)$$

Based on the true tangent modulus-strain curves for glass and carbon fibres in the present study, values for initial modulus E_0 at 0 % strain and α (= slope) can be found from the linear fitting lines, and they are summarised in Table 3.

Table 3. Summary of initial modulus and curvature coefficient of glass and carbon fibres.

	$E_{0\,(1)}$ (GPa)	**$\alpha_{(1)}$ (GPa)**	**$E_{0(2)}$ (GPa)**	**$\alpha_{(2)}$ (GPa)**
Glass	89.5 ± 1.7	-57 ± 30	-	-
Carbon	213 ± 6	+6950 ± 360	228 ± 7	+3810 ± 340

Based on the initial modulus E_0 and curvature coefficient α, stress can be calculated at any particular value of strain. Using these values, the whole stress-strain curve can be modelled.

5. Conclusions

1) The semi-automated single-fibre testing machine (TexTechno, Favimat+) made it possible to test large sample size of glass and carbon fibres, with reduced manual handling and inferences, thus enhancing the accuracy and reliability of the obtained fibre properties.
2) The true stress-strain curve of glass fibres exhibits linear behaviour, with constant modulus over the range of applied strain. The modulus values determined at 0 % and 1.5 % were 89.5 ± 1.7 GPa and 88.7 ± 1.6 GPa, respectively. This linear behaviour makes it possible to determine a reliable fibre modulus based on a larger strain range.
3) The measured modulus of the glass fibres in the 0.05 %-0.25 % strain range was found to be significantly lower compared with the modulus based on a larger strain range. The measurement based on 0.05 %-0.25 % strain range may include some larger measurements error due to smaller displacement and force values and an influence of the initial aligning of the fiber during the initial loading and initial deformation of the rubber grips.
4) The stress-strain curve of carbon fibres exhibits non-linear behaviour, with increasing modulus with applied strain. The modulus values determined at 0 % and 1.5 % are 213 ± 6 GPa and 284 ± 9 GPa, respectively
5) The study of relationship between fibre diameter and modulus revealed a similar modulus for glass fibres irrespective of the difference in the diameter. For carbon fibres, the fibres with smaller diameters showed a higher modulus.
6) The second-order polynomial relation between stress (σ) and strain (ε) based on the initial modulus E_0 and curvature coefficient α is useful tool to calculated stress at any particular value of strain and to model the whole stress-strain curve.

Acknowledgements

The research leading to these results has been performed within the framework of the HyFiSyn project and has received funding from the European Union's Horizon 2020 research and innovation programme under the Marie Skłodowska-Curie grant agreement No 765881. The research was conducted using single-fibre testing machine (TexTechno, Favimat+) at DTU Center for Advanced Structural and

Material Testing (CASMaT). The authors would like to thank Fiberline A/S, Denmark for suppling the carbon fibres used in this study.

References

[1] Swolfs Y, Verpoest I and Gorbatikh L 2016 A review of input data and modelling assumptions in longitudinal strength models for unidirectional fibre-reinforced composites *Compos. Struct.* **150** 153

[2] Islam F, Joannès S, Bucknell, S, Leray Y, Bunsell A and Laiarinandrasana L 2020 Investigation of tensile strength and dimensional variation of T700 carbon fibres using an improved experimental setup *J. Reinf. Plast. Compos.* **39** 144

[3] Mesquita F, Swolfs Y, Bucknell S, Leray Y, Lomov S V and Gorbatikh 2019 Tensile properties of single carbon fibres tested with automated equipment ICCM 22 (Melbourne, Australia) 1336-1342 pp.

[4] Berger M H and Jeulin D 2003 Statistical analysis of the failure stresses of ceramic fibres: dependence of the Weibull parameters on the gauge length, diameter variation and fluctuation of defect density *J. Mater. Sci.* **38** 2913

[5] Bunsell A R, Joannès S and Marcellan 2018 *A Handbook of properties of textile and technical fibres* (Cambridge UK: Woodhead Publishing)

[6] Swolfs Y, Verpoest I and Gorbatikh L 2015 Issues in strength models for unidirectional fibre reinforced composites related to Weibull distributions, fibre packings and boundary effects *Compos. Sci. Technol.* **114** 42

[7] Thomason J L 2013 On the application of Weibull analysis to experimentally determined single fibre strength distributions *Compos. Sci. Technol.* **77** 74

[8] Weibull W A 1939 *Statistical Theory of the Strength of Materials* (Royal Swedish Academy of Engineering Sciences)

[9] Weibull W A 1951 A statistical distribution function of wide applicability *J. Appl. Mech.* **18** 293

[10] Madsen B, Høgsberg J B and Lilholt H. Determination of fibre cross-sectional area by the vibroscope method. In preparation.

[11] Yang L and Thomason J L 2013 Effect of Silane coupling agent on mechanical performance of glass fibre *J. Mater. Sci.* **48** 1947

[12] Kumar R, Mikkelsen L P, Lilholt H and Madsen B. Accurate experimental failure strain determination of unidirectional carbon composite. In preparation.

[13] ISO 527-5:2009 Plastics – Determination of tensile properties – Part 5: Test conditions for unidirectional fibre-reinforced plastic composites (International Organization for Standardization, Switzerland)

[14] Abaqus 2019 *Theory manual online documentaion help* (Dassault Systèmes, france)

[15] Curtis G J, Milne J M and Reynolds W N 1968 Non-hookean behaviour of strong carbon fibres *Nature* **220** 1024

[16] Newcomb B A 2016 Processing, structure, and properties of carbon fibers *Compos .Part A Appl .Sci. Manuf.* **91** 262

[17] Kant M and Penumadu D 2014 Dynamic mechanical characterization for nonlinear behavior of single carbon fibers *Compos .Part A Appl .Sci. Manuf.* **66** 201

[18] Markussen C M, Madsen B, Lilholt H and Løgstrup Andersen T 2015 Proc. of the 20th International Conference on Composite Materials ICCM20 (Copenhagen, Denmark)

[19] Kumar R, Mikkelsen L P, Lilholt H and Madsen B, Single fiber tensile test - Glass and Carbon fibers [Data set], Zenodo, http://doi.org/10.5281/zenodo.4033353

IOP Conf. Series: Materials Science and Engineering **942** (2020) 012034 doi:10.1088/1757-899X/942/1/012034

Investigations of the carbon fibre cross-sectional areas and their non-circularities by means of laser diffraction

Jonas J. Huether[1] **and Wilfried V. Liebig**[1]

[1]Karlsruhe Institute of Technology, Institute for Applied Materials IAM-WK, Kaiserstrasse 12, 76131 Germany

E-mail: `Jonas.huether@kit.edu`

Abstract. Laser diffraction is a commonly used tool to measure the fibre diameter of carbon fibres prior to mechanical testing. However, non-circularities of carbon fibres need to be considered in order to minimise measuring errors. As the work at hand demonstrates, using a single measurement of the fibre diameter may cause deviations as high as 30% from a computationally determined value. It appears that the error can be minimised by acquiring a data set of several apparent diameters as a function of the angle around the fibre axis. Based on this data, the cross-sectional area can be calculated as a circle with an averaged diameter or as an ellipse by applying an elliptical fitting procedure.

1. Introduction

Carbon fibres offer outstanding mechanical properties in relation to their low density. Thus, demand and markets for carbon fibre reinforced polymers (CFRP) are still growing and CFRP are used in various lightweight structures, for instance in the automotive, aviation and aerospace industry. However, when it comes to carbon fibre testing, the focus often lies upon composite properties rather than on single fibre testing. Nonetheless, single fibre testing is the method of choice when a better understanding of the fibre properties and how they are affected, for instance by recycling, ageing and production processes themselves, is sought.

2. State of research

Single fibre testing has its roots in the cotton industry [1–3] and several methods and measurement protocols have been transferred to technical high-modulus fibres without assuring their scientific accuracy. For instance, strains are commonly not measured on fibre scale and fibre diameters are assumed to be circular. The latter cause of error is investigated in the work at hand. Considering non-circular carbon fibre shapes is expected to allow for enhanced single fibre testing.

2.1. Carbon fibre shapes

Assuming that all types of carbon fibres are shaped perfectly circular oversimplyfies the reality. Instead, various fibre shapes co-exist in dependence of the manufacturer, the processing temperatures and other process parameters. Reviewing publications concerned with carbon fibres, four main categories of fibre shapes emerge. The list presented here serves as a general classification and makes no claim to completeness.

a) **circular**: cross-sections of these fibres resemble a perfect circle. Exclusively for fibres of this category it is legitimate to calculate the cross-sectional area based directly on the fibre diameter [4–14].
b) **oval** and **elliptical**: fibres within this category are convex but deviate from circular shapes. Their apparent diameter varies as a function of the viewing angle [4, 14–19].
c) **kidney-shaped**: kidney-shaped fibres are, mathematically speaking, not convex. The are oval-like but possess a distinct indentation along the fibre axis [4, 6–9, 11, 14, 20–22].
d) **complex** and/or **hollow**: This category contains the largest spectrum of different fibre shapes. Published representations of fibres include trilobal, cross-like, rectangular-like and hollow (circular) fibre shapes [13, 23–31].

The four shapes are schematically presented in Figure 1 a) to Figure 1 d). A direct determination of the cross-sectional area is difficult to accomplish in tensile tests. More convenient is the measurement of an apparent fibre diameter which is then commonly used to calculate the cross-sectional area of a circle, neglecting the variety of different fibre shapes. It is evident that such a simplification is deficient for all non-circular fibres, particularly for hollow and kidney-shaped fibres as presented in Figures 1 c) and Figure 1 d).

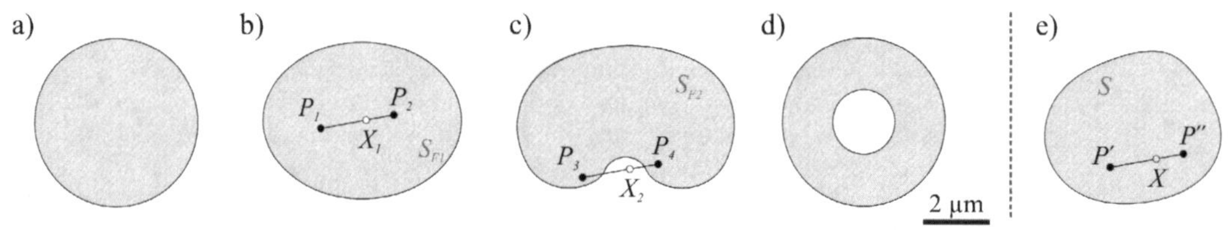

Figure 1: Categorisation of carbon fibre shapes: a) to d) are simplified shapes based on a literature review and are declared as a) circular, b) oval, c) kidney-shaped and d) complex and/or hollow. e) is a tracing of a real fibre. Shapes a), b) and e) fulfil the definition of convex sets, while c) and d) are non-convex.

The fibre shapes shown in Figure 1 a) and Figure 1 b) fulfil the definition of being convex sets, while those in Figure 1 c) and Figure 1 d) do not. With reference to Figure 1e), a set S is defined as being convex if and only if for all arbitrary points P' and P'' in S and all points X on the line segment connecting P' and P'', the points X also belong to the set S [32, 33]. If any point X does not belong to the set, the set is non-convex. Figure 1 gives an illustration with X_1 being included in the set S_{F1}, while X_2 is excluded from set S_{F2}. To apply laser diffraction techniques correctly, convexity of the fibre shape is a mandatory boundary condition.

2.2. Measurement of cross-sectional areas of carbon fibres

To obtain the cross-sectional area of carbon fibres, a range of measurement techniques exist and most of the procedures are classified in standard BS ISO-11567 [34]. In the standard it is also emphasised that the term diameter may refer to a true diameter in case of a perfectly circular fibre and to an apparent diameter when observing solely the side view of a non-circular fibre.

Laser diffraction is one technique proposed in standard BS ISO-11567 and it was first used for determining the diameters and eccentricity of metal wires [35–37] but is nowadays also applied for investigating carbon fibres. Details of the working principle of laser diffraction were published by Fedorov, Wang and Valdivia-Hernandez and Koedam [38–40], but a general explanation shall be given in accordance with Figure 2b): Here, parallel beams of light from a laser hit the carbon fibre (i). Given the fineness of the fibre, it can be described as a single slit (ii), which is diffracting the incoming light. The resulting diffraction pattern is mathematically transformed to an apparent fibre diameter d' as a function of the angular rotation (β). If only one measurement is taken, this reading is commonly considered to calculate a circular cross-section. If a series (> 2) of individual apparent diameters is measured in different angles, a circular cross-section can be computed based on an average apparent diameter. It is advisable to rotate the system in angular increments β_i, starting at $0°$ and completing at $(180° - \beta_i)$, to take irregular fibre shapes into consideration without giving more weight to a certain direction [11, 34, 41]. If a series of diameter readings is recorded, additional evaluation principles emerge. They are expected to be superior to the simplified circular assumption as non-circularities are considered: for instance, a procedure to approximate the carbon fibre shape by assuming an elliptical instead of a circular shape was presented by the authors in [15, 42] and is also applied in the work at hand.

Particularly, the closed measuring system LDS0200 by *CERSA-MCI* which is also incorporated in the LEX820/LDS0200 tensile tester by *Dia-Stron* is commonly used for determining diameters and/or cross-sectional areas of carbon fibres. Published examples are studies on carbon fibre tensile strength [15, 41], on the comparison between virgin and recycled carbon fibres [15, 43, 44] and on the manufacturing of carbon fibres based on cellulose/lignin precursors [45–47]. In the LEX820/LDS0200 apparatus the ends of the fibre are embedded in resin and the fibre diameter is measured once while a pretension is applied. A rocking movement of $\pm$ 8° is executed to receive an optimised laser signal. Yet, the fibre rotation and thus the rotational angle between fibre and laser source are arbitrary. The factory-set calibration of the LDS0200 is based on stainless steel wires with an uncertainty of 0.01 µm, according to the manufacturer. However, several research groups expressed doubts concerning the utilisation and calibration of laser diffraction sensors to measure carbon fibres. Particularly, the laser diffraction signal can be affected by different surfaces of the fibre [42, 48–50]. Details of standardised calibration procedures of laser diffraction sensors as well as other testing methods discussed in standard BS ISO-11567 [34] are beyond the focus of the present paper and shall neither be discussed nor used in the work at hand.

3. Experiments

All experiments were run on *Panex 35* carbon fibres manufactured by *Zoltek Toray Group*. The experiments are based on principle of laser diffraction and a computational procedure to mimic laser diffraction measurements. Both were applied to determine apparent diameters and to calculate the cross-sectional area either by assuming elliptical [15] or circular fibre shapes [34].

Figure 2a) illustrates the geometrical approach based on fibre micro sections. Two parallel tangents, $\overline{PQ}$ and $\overline{RS}$, rotated by an angle β from the horizontal axis are positioned in a manner that they coincide with the outermost curvature of the fibre section. This procedure provides an apparent diameter d as a function of the angular rotation. It mimics an experiment based on laser diffraction as shown in Figure 2b), explained in Section 2.2.

In the following, the experimental procedure based on laser diffraction and an approach based on computational geometric shapes are described. In both investigations, apparent diameters were obtained in angular steps of ($\beta_i = 10°$), from 0° to 140° in the real experiment, and from 0° to 170° in the computational approach. An angle of 140° was the maximum coverage feasible with the experimental setup described in [15] and used here.

Based on the series of individual measurements of d' as a function of the rotational angle β', a non-linear fit of least squares is run to provide the semi-major axis a and the semi-minor axis b of an approximated ellipse [15]:

$$d' = \sqrt{(2\,a\cos\beta')^2 + (2\,b\sin\beta')^2} \tag{1}$$

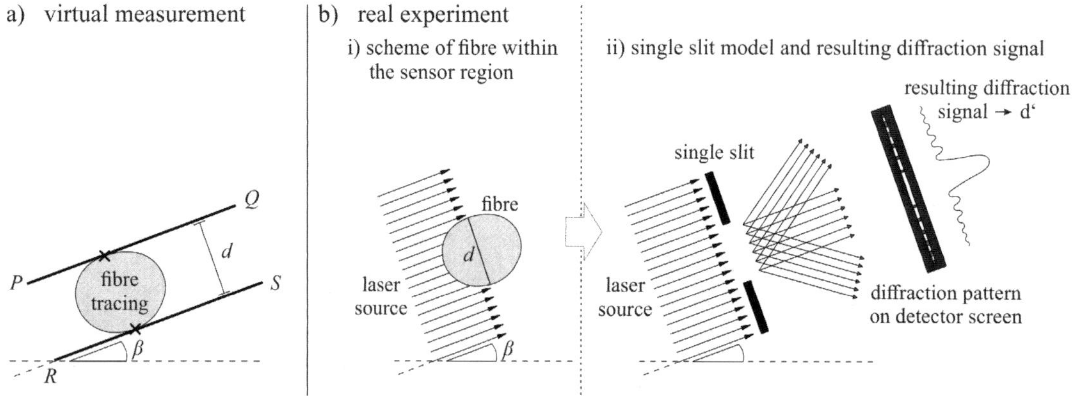

Figure 2: Laser diffraction principle: a) a set of the two parallel tangents $\overline{PQ}$ and $\overline{RS}$ imitates the measurement based on laser diffraction when the tangents touch the outermost part of a fibre tracing. b) in laser diffraction, a fibre can be seen as single slit and the occurring diffraction pattern is used to derive apparent diameters.

The geometric derivation and mathematical relationships are shown in Figure 3. With both axes, the area of the elliptical cross-section (A_{fibre}) can be calculated by:

$$A_{fibre} = \pi \, a \, b \tag{2}$$

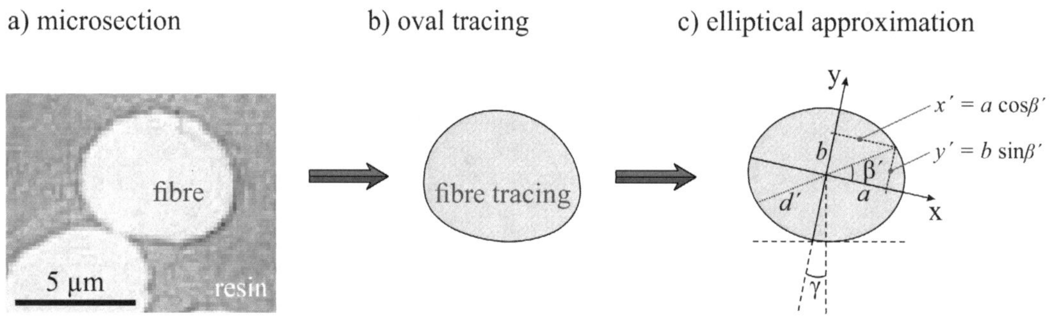

Figure 3: Elliptical approximation of an oval carbon fibre. Geometric derivation and mathematical relationships to approximate the cross-sectional area of a carbon fibre.

3.1. Laser diffraction measurements on carbon fibres

For the experimental investigations of carbon fibres, the fibres were strained and aligned with a parallel gripping system. Laser diffraction measurements were executed with the closed measuring system LDS0200. The laser diffraction sensor worked with the aforementioned factory-set calibration with steel wires. To adapt the sensor to carbon fibres, an empirically determined correction function was applied [42]: all measured apparent diameters d' were transformed to corrected diameters d'_c with Equation (3):

$$d'_c \;=\; 1.03774 + 0.96497 \cdot d' \tag{3}$$

All diameters used for the evaluations within the work at hand are corrected diameters d'_c. The laser diffraction sensor could be rotated in angular steps of 10° in respect to the fibre axis. In the investigation of real carbon fibres, no alternative procedures to verify the results could be executed, as it is impossible to evaluate the very same position along the fibre with another procedure, for instance metallographical investigations.

3.2. Computational geometric approach

To gain a database for the computational approach, fibre tracings were generated on base of microsections of ten arbitrarily chosen carbon fibres, shown in ascending cross-sectional area as geometric shape 1 to 10 in Figure 4. For these shapes, the 0°-direction is arbitrarily chosen and exemplary shown for fibre 1.

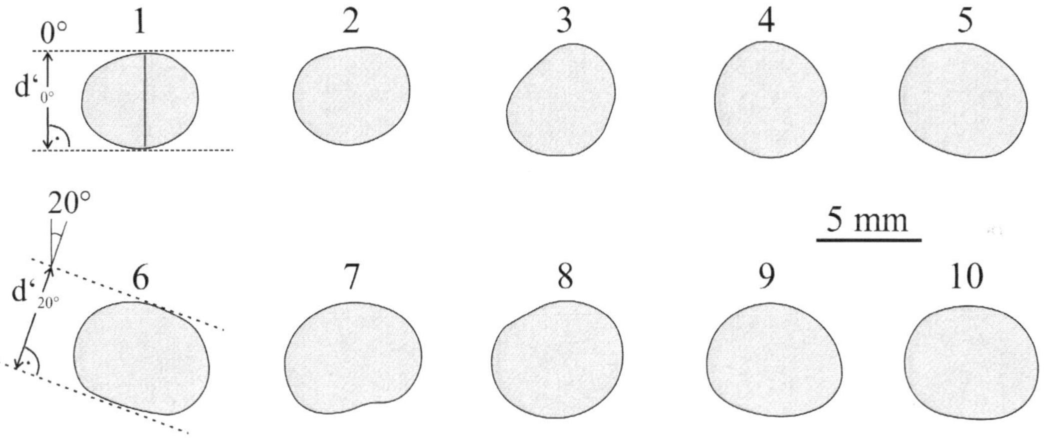

Figure 4: Ten fibre tracings sorted by ascending cross-sectional area.

Apparent diameters were determined as follows: the *true* cross-sectional area was determined computationally by image analysis to serve as comparison. A set of two parallel tangents was positioned in a way that they coincided with the outermost curves of the shape, starting at an initial angle of 0°. The tangents were then rotated in steps of 10°, providing a series of apparent diameters as a function of the angle. For each shape, 18 measurements were taken from $d'_{0°}$ to $d'_{170°}$ to fully cover the geometric shape. 18 measurements were the database for different procedures to calculate the cross-sectional area of each fibre tracing.

3.3. Nomenclature of different calculation procedures

The following calculations were used:

- **ellipse 140** was calculated by an elliptical fit according to [15]. Apparent diameters were taken in steps of 10° from 0° to 140°. This corresponds to the constraints caused by the experimental setup used in the work at hand for tensile tests on fibres.
- **ellipse 170** is analogous to procedure *ellipse 140* but angles up to 170° were considered. This corresponds to a half circle, which is sufficient to cover the complete shape by the fitting procedure. This calculation was solely feasible for the computational approach.
- **circle random**, **circle random 0** and **circle random 40** connote that one apparent diameter for each fibre tracing was randomly picked to calculate its circular cross-sectional area. Fortuitousness was ensured by the fact that the fibres were not specifically angled before mechanical testing or metallographic investigation.

- **circle max** and **circle min** signify that the maximum or the minimum apparent diameter of each fibre tracing were used to calculate a circular cross-sectional area.
- **circle 0/60/120** connotes that three apparent diameters were averaged to calculate a circular cross-sectional area, with the 0° -direction being the starting point and then rotating the set of tangents by 60°.
- **circle 0-140** was calculated by averaging the same apparent diameters as for *ellipse 140* but instead of using the described elliptical fit, 15 apparent diameters were used to calculate a circular area.

4. Results

In the following, results of the experimental and computational approach are presented.

4.1. Laser diffraction

200 carbon fibres were measured with the laser diffraction sensor according to the procedure detailed in Section 3.1. Based on 15 apparent diameters per fibre, cross-sectional areas were calculated with either the elliptical approach or based on circular assumptions. Results are plotted in Figure 5: Figure 5 a) displays the complete data set, Figure 5 b) highlights the interval from fibre number 85 to 115. The dashed lines are the envelope curve obtained by calculating the circular area based on the maximum or minimum apparent diameter, respectively.

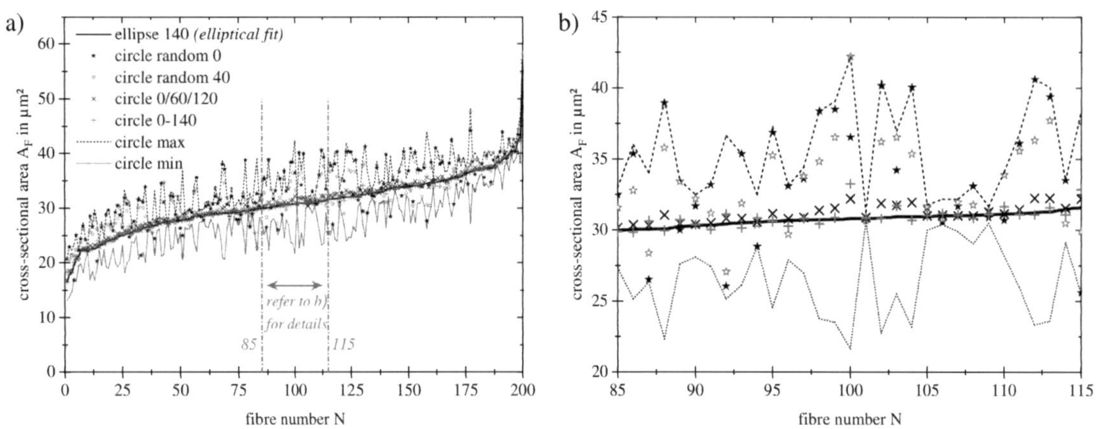

Figure 5: a) Cross-sectional areas of 200 fibres were determined based on different calculations. The dataset is sorted by ascending cross-sectional area of the elliptical fitting procedure (ellipse 140). For each fibre number N, the differently calculated results are plotted. b) For better clarity, the segment between fibre number 85 and 115 is shown enlarged.

4.2. Computational approach

Ten fibre tracings were measured according to the procedure presented in Section 3.2. Based on 18 individual apparent diameters per fibre tracing, the cross-sectional areas were calculated in different ways. Results are plotted versus the computational (*true*) value in Figure 6 a). In addition, Figure 6 b) displays the calculations for each fibre tracing, analogously to Figure 5 b). As Figure 6 a) highlights, all procedures depicted there tend to overestimate the cross-sectional area.

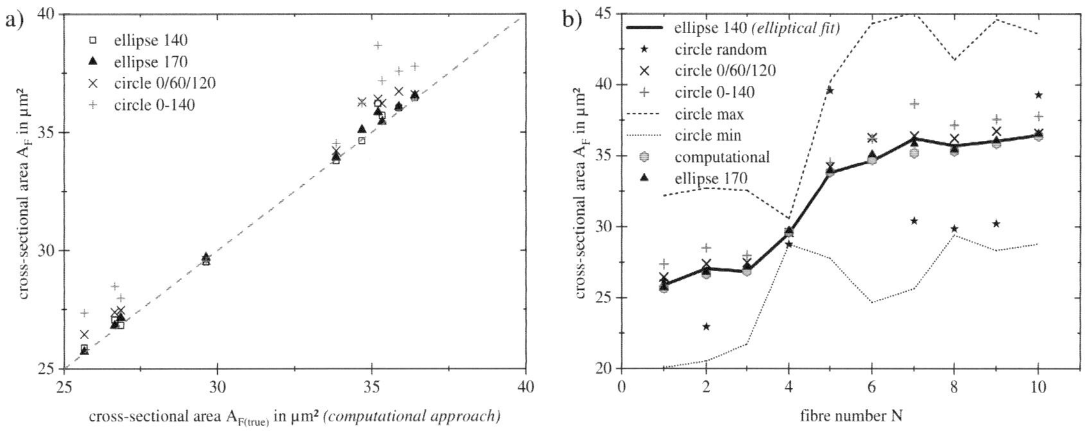

Figure 6: a) Cross-sectional areas calculated based on different procedures plotted versus the cross-sectional area determined computational (*true* value). b) Different calculation approaches for each fibre tracing; illustrated analogous to Figure 5 b).

5. Discussion

The work at hand has its focus on evaluating to what extend laser diffraction sensors provide reliable measurements of the cross-sectional area of carbon fibres. Laser diffraction is a widely used principle to determine the cross-sectional areas of carbon fibres. However, only few publications considered the non-circularity of fibre shapes, by either recording and averaging a series of individual apparent diameters in various rotational angles to increase the statistical certainty [11, 41] or by hypothesizing elliptical shapes [15, 42].

In Section 4 it was shown that calculations of the cross-sectional area of irregularly shaped carbon fibres vary significantly depending on whether circular or oval shapes are assumed, for both, the experimental measurements on carbon fibres and the computational approach on fibre sections.

This effect shall be further described in accordance with Figure 7. The value equivalence line represents the values obtained by the elliptical fitting procedure based on 15 apparent diameters (*ellipse 140*) for both, the experimental and the computational approach. Procedure *ellipse 140* was chosen for comparative purposes, as its values are available in both, the experimental and the computational approach. Here, cross-sectional areas calculated based on the circular assumptions are plotted as small symbols for the experimental approach (200 fibres) and as large symbols for the computational approach (10 fibre tracings, shown in Figure 4). Again it becomes obvious that a cross-sectional area based on one (randomly picked) apparent diameter causes large deviations from the elliptical fitting procedure. As shown in Figure 6 b), values obtained by the elliptical fitting procedures differ only slightly from the *true* values determined computationally. Moreover, data in Figure 7 indicates that circular calculations based on laser diffraction measurements tend to overestimate the cross-sectional area: linear fits of the experimental data are shifted towards larger values compared to the value equivalence line. This effect is most prominent for the randomly picked apparent diameters. Listed in Table 1 are the slope, the intersection with the axis of ordinates and the Pearson correlation coefficient. The correlation coefficient and the slope for the linear fits of *circle 0/60/120* and *circle 0-140* are close to unity, and the deviation from the origin is low (0.5 and 0.1, respectively). When picking a single random value (*circle random*), the correlation is rather low (0.811) and slope and origin deviate significantly from unity. In general, the overestimation caused by laser diffraction measurements can be explained by the fact that crenellations and indentations along the fibre

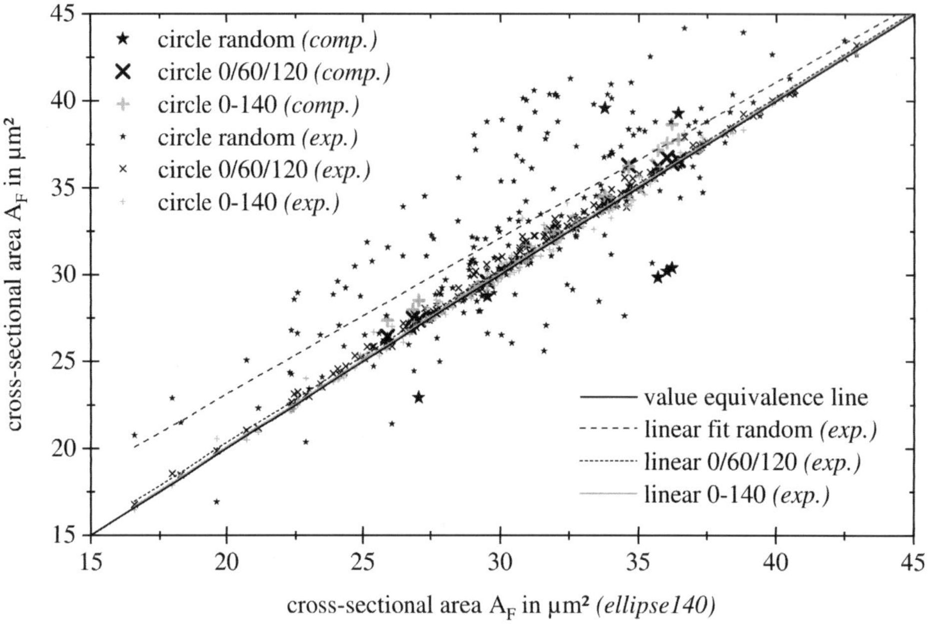

Figure 7: Calculations of circular cross-sectional areas of carbon fibres and their tracing compared to the elliptical fitting procedure. Computational investigations of fibre tracings are labelled *(comp.)*, experimental measurements on carbon fibres *(exp.)*. Procedure *ellipse 140* serves as comparison as it was available for both, the experimental and the computational approach.

cannot be detected by the laser. This effect is expected for both, the circular and the elliptical assumption. However, it tends to be more prominent when assuming circular shapes, a trend depicted in Figure 6 b) where all computational values (obtained by image analysis) are below the circular assumptions *circle 0/60/120* and *circle 0-140* and the elliptical procedures *ellipse 140* and *ellipse 170*.

Table 1: Evaluation of the linear fits of experimental data in Figure 7.

	circle random	circle 0/60/120	circle 0-140
slope	0.898	0.994	1.001
intersection with axis of ordinates in µm^2	5.2	0.5	0.1
Pearson correlation coefficient	0.811	0.998	0.996

The deviation from the elliptical fit shall be further discussed for the experimental results on 200 carbon fibres: relative deviations between elliptical and circular assumptions were as high as 38 %. Both, *circle 0-140* and *circle 0/60/120* show mostly deviations below 4 %, with the former showing some deviations higher than 4 % compared to the values obtained by the elliptical fit. Opposed to that, picking a random or the maximum apparent diameter to calculate a circular cross-sectional area yields a wide range of deviations up to more than 30 %. For the randomly picked apparent diameters, half of the data set (100 fibres) deviates 7 % or more in cross-sectional area, for the maximum apparent diameters 100 fibres deviate 10 % or more. As elucidated, laser diffraction is a common tool to investigate the cross-sectional area of carbon fibres [11, 15, 41–47]. The data provided in the work at hand clearly indicates that when applying laser diffraction, picking a single apparent diameter might cause significant measuring errors. Instead, it is recommended to use a series of individual measurements in equidistant angular steps and an averaged diameter to calculate the cross-sectional area. Considering non-circular shapes by applying an elliptical fitting procedure proved to be a purposeful approach, as this procedure is more suitable when the laser diffraction sensor cannot be rotated completely around the fibre axis [15, 42]. Without detailed quantification, some additional aspects that might decrease the quality and reliability of determining the cross-sectional area by means of laser diffractometry shall be listed, partly in accordance with Figure 8. These effects can occur separately or collectively. The likelihood and severity of erroneous readings is expected to rise from a) to e). If the fibres exhibit perfect circular shapes (a), the limiting factor is the accuracy of the laser diffraction sensor. As shown in the work at hand, oval shapes (b) demand for recording a series of individual diameter measurements [15, 42]. Kidney-shaped fibres (c: [4, 6–9, 11, 14, 20–22]), which do not fulfil the definition of being convex sets, will affect the measurement as the size of undercuts cannot be detected by laser diffraction. Crenellations along the fibre (d) are an additional deviation from the convex shape and the cross-sectional area will be overrated. Debris like dust (e) may also interfere with the laser beam and deteriorate the measurement. Not shown graphically but nonetheless relevant sources of inaccuracies are listed at what follows: misalignments between the laser beam and the fibre may influence the measurement. Furthermore, laser diffraction sensors average the apparent diameter along a certain increment of the fibre, for instance approximately 2 mm in case of the LDS0200. According to the manufacturer, *Diastron*, this device has a resolution of 0.01 μm which is another limiting factor of the measurement accuracy. The device is calibrated with steel wires, and as explained earlier, the calibration needs to be adjusted to carbon fibres [42].

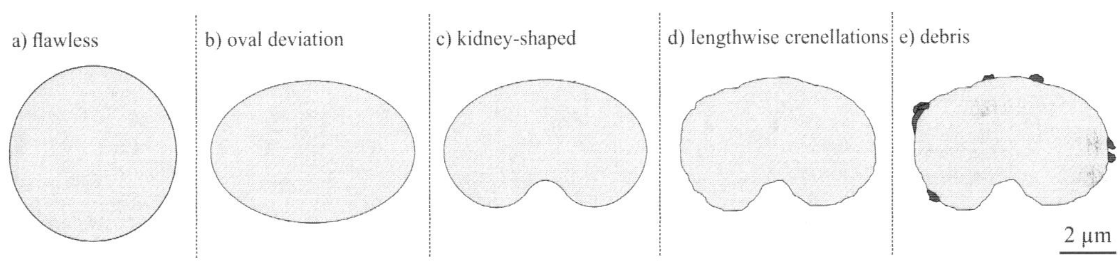

Figure 8: Effects on the quality of laser diffraction measurements: the more irregular the fibre shape and the higher the amount of undercuts and debris, the lower the quality of laser diffraction measurements.

6. Conclusion

It can be concluded that when applying laser diffraction techniques to determine the diameter of carbon fibres, the non-circularity of carbon fibres should be considered. Firstly, the fibre population should be investigated microscopically to detect the predominant fibre shape. Secondly, a series of apparent diameters can be averaged to increase statistical certainty. Here, the measuring sensor has to be rotated at constant angular steps in relation to the fibre, while ensuring that none direction is given more weight to. Thirdly, calculations of cross-sectional areas can be improved by considering oval or elliptical shapes instead of circular fibre shapes.

7. Acknowledgement

The authors would like to thank Michael Seitz for his scientific contribution to this article.

References

[1] 2012 *ASTM standard D1445/D1445M – 12: Standard Test Method for Breaking Strength and Elongation of Cotton Fibers (Flat Bundle Method)* (West Conshohocken, PA)

[2] Osumi G and Kato E 1937 *Journal of the Textile Institute Transactions* **28** T129–T144 ISSN 1944-7027

[3] Pressley E H 1942 *American Society for Testing and Materials Bulletin* **118** 13–17

[4] Naito K, Tanaka Y, Yang J M and Kagawa Y 2008 *Carbon* **46** 189–195 ISSN 00086223

[5] Naito K, Yang J M, Tanaka Y and Kagawa Y 2012 *Journal of Materials Science* **47** 632–642 ISSN 0022-2461

[6] Naito K, Tanaka Y and Yang J M 2017 *Carbon* **118** 168–183 ISSN 00086223

[7] Kumar S, Anderson D P and Crasto A S 1993 *Journal of Materials Science* **28** 423–439 ISSN 0022-2461

[8] Endo M 1988 *Journal of Materials Science* **23** 598–605

[9] JDH Hughes 1986 *Carbon* **24** 551–556 ISSN 00086223

[10] Cho T, Lee Y S, Rao R, Rao A M, Edie D D and Ogale A A 2003 *Carbon* **41** 1419–1424 ISSN 00086223

[11] Tzeng S S and Chang F Y 2001 *Materials Science and Engineering: A* **302** 258–267 ISSN 09215093

[12] Vezie D L and Adams W W 1990 *Journal of Materials Science Letters* **9** 883–887 ISSN 0261-8028

[13] Liu X, Wang R, Wu Z and Liu W 2012 *Materials Letters* **73** 21–23 ISSN 0167577X

[14] Kobets L P and Deev I S 1998 *Composites Science and Technology* **57** 1571–1580 ISSN 02663538

[15] Huether J, Esse D, Rupp P, Seitz M and Weidenmann K 2019 *Measurement Science and Technology* **30** 1–12 ISSN 0957-0233

[16] Saburow O, Huether J, Maertens R, Trauth A, Kechaou Y, Henning F and Weidenmann K A 2017 *Procedia CIRP* **66** 265–270 ISSN 22128271

[17] Li Z, Zhang D, Qin W and Geng D 2017 *Proceedings of the Institution of Mechanical Engineers, Part B: Journal of Engineering Manufacture* **231** 2347–2358 ISSN 0954-4054

[18] Paris O, Loidl D, Müller M, Lichtenegger H and Peterlik H 2001 *Journal of Applied Crystallography* **34** 473–479 ISSN 0021-8898

[19] Fitzer E 1989 *Carbon* **27** 621–645 ISSN 00086223

[20] Daumit G and Ko Y 1986 *High tech: The way into the nineties – Proceedings of the 7th International Conference of the Society for the Advancement of Material and Process Engineering European Chapter* **7** 201–213

[21] Marcuzzo J, Otani C, Polidoro H and Otani S 2013 *Materials Research* **16** 137–144 ISSN 0033-4545

[22] Johnson J W and Thorne D J 1969 *Carbon* **7** 659–661 ISSN 00086223

[23] Curtis P T and Travis S 1999 *Plastics, Rubber and Composites* **28** 201–209 ISSN 1465-8011

[24] Liu Y, Chae H, Choi Y and Kumar S 2015 *Journal of Materials Science* **50** 3614–3621 ISSN 0022-2461

[25] Wang C Y, Li M W, Wu Y L and Guo C T 1998 *Carbon* **36** 1749–1754 ISSN 00086223

[26] Xie W, Cheng H, Chu Z, Chen Z and Long C 2011 *Ceramics International* **37** 1947–1951 ISSN 02728842

[27] Park S J, Seo M K, Shim H B and Rhee K Y 2004 *Materials Science and Engineering: A* **366** 348–355 ISSN 09215093

[28] Gulgunje P, Newcomb B, Gupta K, Chae H, Tsotsis T and Kumar S 2015 *Carbon* **95** 710–714 ISSN 00086223

[29] Hunt M, Saito T, Brown R, Kumbhar A and Naskar A 2012 *Advanced materials (Deerfield Beach, Fla.)* **24** 2386–2389

[30] McMahon P E 1973 Graphite fiber tensile property evaluation *Analysis of the Test Methods for High Modulus Fibers and Composites* ed Whitney J M (West Conshohocken, PA: ASTM International) pp 367–367–23

[31] Eom S Y and Ryu S K 2010 *Korean Journal of Chemical Engineering* **27** 1592–1595 ISSN 0256-1115

[32] Bonnesen T and Fenchel W 1934 *Theorie der konvexen Körper* (Berlin, Heidelberg: Springer) ISBN 9783540062349

[33] Schneider R 1993 *Convex bodies: The Brunn-Minkowski theory* (*Encyclopedia of mathematics and its applications* vol 44) (Cambridge and New York: Cambridge University Press) ISBN 9780521352208

[34] British Standards 2018 BS-ISO 11567:2018: Carbon fibre - Determination of filament diameter and cross-sectional area: Carbon materials

[35] Blohm W, Sikora H and Beining A 1999 Durchmessermessung mit Beugungssäumen sowie elektronische Verschmutzungskorrektur - European Patent Office - EP 0924493 B1: Europäische Patentschrift

[36] Khodier S 2004 *Optics & Laser Technology* **36** 63–67 ISSN 00303992

[37] Troll J and Baker C 1972 US Patent 3659950A – Laser apparatus for detecting the size and form of filamentary material by measuring diffracted light

[38] Fedorov M E 2015 *IOP Conf. Ser.: Mater. Sci. Eng.* **81** 1–10 ISSN 1757-8981

[39] Wang H and Valdivia-Hernandez R 1995 *Measurement Science and Technology* **6** 452–457 ISSN 0957-0233

[40] Koedam M 1966 *Philips technical review* **27** 208–211

[41] Islam F, Joannès S, Bucknell S, Leray Y, Bunsell A and Laiarinandrasana L 2020 *Journal of reinforced plastics and composites* **39** 144–162

[42] Huether J 2020 - in print *Disseration, Schriftenreihe des Instituts für Angewandte Materialien, Band 84*

[43] Pickering S J, Turner T A, Meng F, Morris C N, Heil J P, Wong K H and Melendi-Espina S 2015 Developments in the fluidised bed process for fibre recovery from thermoset composites *2nd Annual Composites and Advanced Materials Expo, CAMX 2015; Dallas Convention CenterDallas; United States* pp 2384–2394

[44] Fischer H and Schmid H 2013 *Kunststoffe* **103** 88–91

[45] Kleinhans H and Salmén L 2016 *Journal of Applied Polymer Science* **133** 119 ISSN 0021-8995

[46] Nowak A, Hagberg J, Leijonmarck S, Schweinebarth H, Baker D, Uhlin A, Tomani P and Lindbergh G 2018 *Holzforschung* **72** 81–90

[47] Bengtsson A, Bengtsson J, Olsson C, Sedin M, Jedvert K, Theliander H and Sjöholm E 2018 *Holzforschung* **72** 1007–1016

[48] Li C T and Tietz J 1990 *Journal of Materials Science* **25** 4694–4698 ISSN 0022-2461

[49] Meretz S, Linke T, Schulz E, Hampe A and Hentschel M 1992 *Journal of Materials Science Letters* **11** 1471–1472 ISSN 0261-8028

[50] Perry A, Ineichen B and Eliasson B 1974 *Journal of Materials Science* **9** 1376–1378

IOP Conf. Series: Materials Science and Engineering **942** (2020) 012035 doi:10.1088/1757-899X/942/1/012035

Evolution of CFRP stress cracks observed by *in-situ* X-ray refractive imaging

A Kupsch[1]*, V Trappe[1], B R Müller[1], and G Bruno[1,2]

[1] BAM Federal Institute for Materials Research and Testing, 12200 Berlin, Germany
[2] University of Potsdam, Institute of Physics and Astronomy, 14476 Potsdam, Germany

* E-mail: andreas.kupsch@bam.de

Abstract. Modern air-liners and wind turbine rotor blades are made up primarily of fiber reinforced plastics. Failure of these materials heavily impairs the serviceability and the operational safety. Consequently, knowledge of the failure behavior under static and cyclic loads is of great interest to estimate the operational strength and to compare the performance of different materials. Ideally, the damage evolution under operational load is determined with *in-situ* non-destructive testing techniques. Here, we report *in-situ* synchrotron X-ray imaging of tensile stress induced cracks in carbon fiber reinforced plastics due to inter-fiber failure. An in-house designed compact tensile testing machine with a load range up to 15 kN was integrated into the beamline. Since conventional radiographs do not reveal sufficient contrast to distinguish cracks due to inter-fiber failure and micro cracking from fiber bundles, the Diffraction Enhanced Imaging (DEI) technique is applied in order to separate primary and scattered (refracted) radiation by means of an analyzer crystal. This technique allows fast measurements over large fields-of-view and is ideal for *in-situ* investigations. Imaging and the tensile test are run at the highest possible frame rate (0.7 s^{-1}) and the lowest possible strain rate ($5.5 \cdot 10^{-4}$ s^{-1}). For 0°/90° non-crimp fabrics, the first inter-fiber cracks occur at 380 MPa (strain 0.8 %). Prior to failure at about 760 MPa (strain 2.0 %), we observe the evolution of nearly equidistant (1 mm distance) cracks running across the entire sample in the fully damaged state.

1. Introduction

Fiber reinforced plastics, FRPs, reinforced with carbon fibers or glass fibers, are polyvalent: they are an established material in aircraft construction, they are widely used in wind turbine rotor blades and increasingly used in automotive engineering. The low weight together with the high strength make this class of materials very attractive for modern lightweight construction applications. In general, FRPs are of increasing interest for structural components in lightweight constructions, whose failure wholly or partially impairs serviceability and operational safety.

Knowledge of the spatial and temporal evolution of cracks is valuable for both the manufacturer and the customer of composite materials in order to estimate the material's lifetime. In the case of transparent glass fiber reinforced plastics (GFRP), the behavior prior to and during fracture has been successfully recorded by common optical systems with mega-pixel resolution [1, 2]. While the occurrence of the first cracks in GFRPs can be inspected with visible light, it is not suited to characterize the evolution of micro cracks at high load levels or due to cyclic loading. Since the (integral) intensity of transmitted light is not proportional to the internal surface per volume caused by micro cracks, the number of cracks has to be evaluated by image analysis [3, 4]. Additionally, the increasing intensity of transmitted light due to increasing crack opening at a constant number of cracks can be mistaken for increased inner surface.

In a comparative study, the inter-fiber failure in GFRP has been investigated by means of X-ray-refraction topography [5, 6]. This technique overcomes the disadvantages of visible light absorption techniques, depending on the type and size of cracks. Separated cracks could be imaged, and the internal (specific) surface was determined by calculating the co-called X-ray refraction value [5]. Even dry textiles or impregnated laminate can be quantitatively analyzed with this technique, for instance for different mean fiber filament diameters even at identical fiber volume content.

2. Experimental

2.1. Samples

The CFRP specimens were made from 2 layers of a 400 g/m² twill style textile and 4 layers of a 200 g/m² NCF, each made with Tenax-E HTA40 E1, 6K yarn (400tex) (figure 1). In both cases the stacking sequence was symmetric and balanced. In the non-crimp fabric, the carbon fiber rovings were fixed by a 34tex glass fiber yarn. The matrix system used was Huntsman Araldite® LY 556 / Aradur® 917 / Accelerator DY 070. Carbon fiber reinforced plastics, CFRP, plates were manufactured in a resin transfer moulding technique, and flat specimens with a length of 140 mm, a width of 10 mm (0°/90°-laminate and -45°/+45°-laminate) and a thickness of 1.1 mm were cut out and tested. Tab reinforcement was applied onto the 0°/90°-specimens for clamping. Due to the high Poisson's ratio of the -45°/+45°-laminate, the fracture occurs in the middle of the specimens even without the tabs. 0°/90°- and -45°/+45°-laminates of each textile reinforcement were investigated. The fiber volume content is 41% in both laminates.

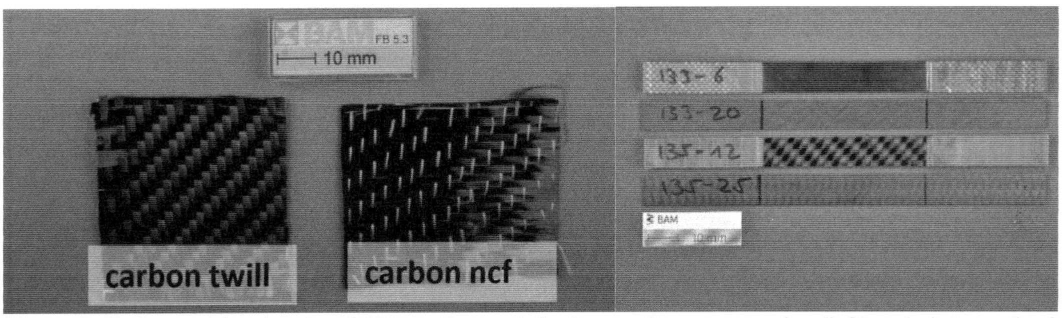

Figure 1. Carbon twill and carbon NCF textiles used for the specimen preparation (left) and photograph of the produced specimens (right): NCF (top two) and twill (bottom two). Specimens with 0°/90° fiber orientation were additionally prepared with tab reinforcements.

Table 1. Compilation of the fabrics investigated. The specimen names are assigned to the type of fabrics and fiber orientation relative to the direction of tensile stress.

	0°/90°	-45°/+45°
non-crimp fabric (NCF)	133-3, 133-11	133-17, 133-19
2 × 2 twill fabric (TF)	135-10, 135-14	135-18

2.2. X-ray refraction

X-ray refraction techniques were introduced a few decades ago [7] and have been successfully used for both material characterization and non-destructive testing [8]. These techniques are used to obtain the amount of the relative internal specific surface (*i.e.*, surface per unit volume, relative to a reference state) of a specimen, and are therefore effective in the investigation of inhomogeneities within fiber reinforced plastics such as cracks, fibers, and fiber debonding.

X-ray refraction occurs whenever X-rays interact with interfaces between materials of different density, as in the case of cracks, pores, and particles in a material. This is analogous to the behavior of visible light in transparent materials, such as lenses or prisms. It should be noted that the X-ray refraction index is smaller than unity. Thus, the optical behavior is inverted [9], so structures such as pores act rather as focusing objects. The difference in the refraction indices between the two interfacing materials, the so-called refraction decrement, determines the refraction angle at the interface (angle between the propagation direction of the primary and the refracted beam). Since the refraction decrement for X-rays is of the order of 10^{-7} to 10^{-5}, X-ray optical effects can only be observed at very small deflection angles ranging from a few seconds to a few minutes of arc. X-ray refraction techniques detect pores and cracks as soon as they exceed a certain size or opening of a few X-ray wavelengths. Since the typical X-ray wavelengths are approximately 0.1 nm, the smallest detectable object size is down to the nanometer range. This is not to be confused with the spatial resolution or the size of the objects that can be *imaged*. The spatial resolution of the technique is limited by the pixel size of the detector system.

It must be emphasized that, because of the inevitable background noise, it is impossible to conclusively detect one single defect. A certain population of objects is necessary to yield an integrated signal above the background noise. Thus, X-ray refraction is primarily used in radiographic mode with thin specimens (platelets) and yields 2.5D information about the microstructure of the sample (*i.e.*, integrated over the whole specimen thickness); the 2D lateral resolution corresponds to the number and size of the detector pixels. This results in the detection and imaging of a population of defects and not in the imaging of single defects. If the lateral size of defects or structures exceeds the spatial resolution of the detector system, they can be localized by the X-ray refraction contrast. This analysis can be conducted on samples of macroscopic size of a few cm, where the field of view of X-ray refraction techniques is much larger than that of high-resolution X-ray computed tomography and of classic scanning electron or even optical microscopy. Furthermore, X-ray refraction techniques are sensitive to defect orientation, thereby allowing different kinds of defects to be identified. The refraction signal of an isotropic inhomogeneity, such as spherical voids, is present at any orientation of the specimen. For cracks or elongated pores, on the other hand, the signal decreases when the defect surface normal is parallel to the scattering vector of the detection system (the bisectrix between the incident and the refracted beam directions). The X-ray refraction signal has been *quantitatively* correlated to microstructural changes, which have been rationalized in micromechanical models [10-17].

2.3. X-ray refraction radiography by Diffraction Enhanced Imaging

Synchrotron X-ray refraction radiography (XRRR) measurements were carried out at the BAM synchrotron laboratory BAM*line* at the Helmholtz-Zentrum Berlin, Germany [18, 19]. A double crystal (Si (111)) monochromator (DCM) was used to extract a highly collimated monochromatic X-ray beam with an energy band width of 0.2 %. The beam energy was set to 25 keV. A flat panel detector in combination with a lens system and a 50 µm thick CWO scintillator screen provided a pixel size of 7.2 × 7.2 µm^2, capturing a field of view of about 14 × 7.2 mm^2. The incident beam was narrowed to the field of view by a slit system in order to avoid detector backlighting [20, 21]. The exposure time for each image was 1 second (plus 0.4 s read out time). In order to achieve a sufficient count rate per pixel we applied a 4 × 4 pixel binning, resulting in an effective pixel size of 28.8 × 28.8 µm^2. The X-ray beam was oriented along the specimen thickness direction (t = 1.1 mm).

In contrast to transmission-based radiographic measurements, a Si (111) analyzer single crystal was placed in the beam path between the specimen and the camera system, shown in figure 2a, to perform refraction radiographs (Diffraction Enhanced Imaging, [22]) in two different modes. In mode 1 the crystal is positioned in the X-ray beam in such a manner that the Bragg condition is fulfilled (Bragg angle θ_B = 4.537° at 25 keV, between beam direction and crystal surface) and the free X-ray beam (without sample) is completely reflected to the detector. In mode 2 the crystal is slightly tilted (by some $10^{-4\circ}$), the Bragg condition is not fulfilled for the primary radiation. Here, we show images obtained in mode 1, exclusively. X-rays, refracted by the sample's internal surface such as cracks violate the Bragg condition. Hence, they are not reflected and create an additional attenuation to the sample's true absorption properties.

We performed complementary experiments with the samples 133-11 and 135-14 in mode 2, which revealed the same material behavior as their sister samples 133-3 and 135-10. In mode 2, only those beam portions are reflected by the analyzer, which fulfill the *actual* Bragg condition. The different contrasts are shown in figures 2b and 3.

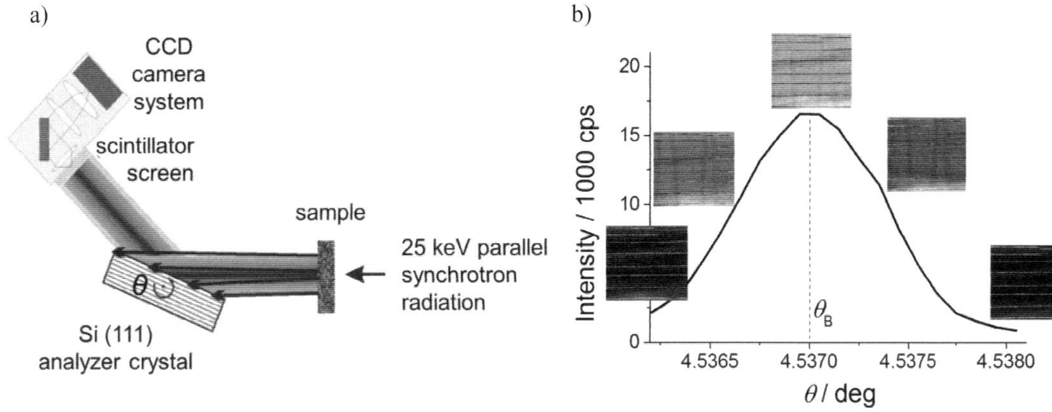

Figure 2. (a) Set-up of synchrotron X-ray refraction radiography. In the sketch, the dark shaded (attenuated) primary beam portions are diffracted by the analyzer crystal. The black arrowed rays (scattered by the sample) do not fulfill the Bragg condition, and thus are not reflected by the crystal (and not detected by the camera system). (b) The crystal's rocking curve provides X-ray images of changing contrast. In the center of the rocking curve (mode 1), all refracting beam portions are discarded and the primary beam is imaged, only. Thus, the cracks appear dark (low intensity). Off the center (mode 2), the primary beam is discarded. Only rays of a predefined refraction angle are accepted; the cracks appear as bright (high intensity) features.

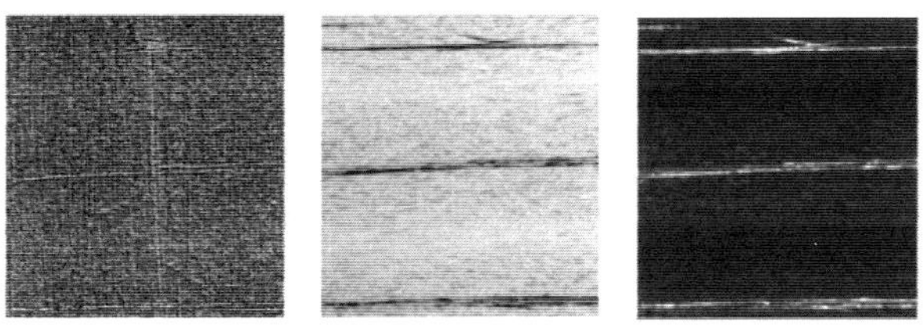

Figure 3. Comparison of absorption (left) and refraction contrast of a pre-loaded NCF CFRP sample ((σ_{max}=530 MPa). The middle image is obtained from the rocking curve center, the right image from an off-center rocking curve position.

2.4. *In-situ tensile tests*

A tensile rig with an electro-mechanical actuator and a hydraulic clamping system was integrated into the DEI set-up at BAM*line* (figure 4a). The compact load rig was designed and manufactured specifically for *in-situ* tensile and cyclic loading in a laboratory X-refraction set-up. The maximum load capacity is +/-15 kN. In the laboratory set-up, the loading of the FRPs must be performed in load steps, since each scan of the specimen takes more than one hour. At BAM*line*, the tensile tests were done with a typical actuator speed of 2 mm/minute, as is common in quasi static tensile tests standards [23]. The typical stress strain curves obtained during the tests are shown in figures 4b and 4c. Table 2 summarizes some mechanical characteristics of the individual samples (obtained from tensile testing).

Imaging and the tensile test were started separately. Roughly 15 seconds after continuous imaging had started, the tensile stress machine was set into displacement-controlled mode. It recorded the displacement (strain) and the applied force as a function of time. With the known time stamp, the single images were assigned to the respective stress/strain by computing residual images (figure 5a), *i.e.*, the difference of subsequent images. The variance of these residual images clearly indicated the start of the tensile test and the rupture (figure 5b). It should be noted that the generated cracks are clearly visible in the residual images, but their local contribution to the variance of the full image is too low to cause significant changes.

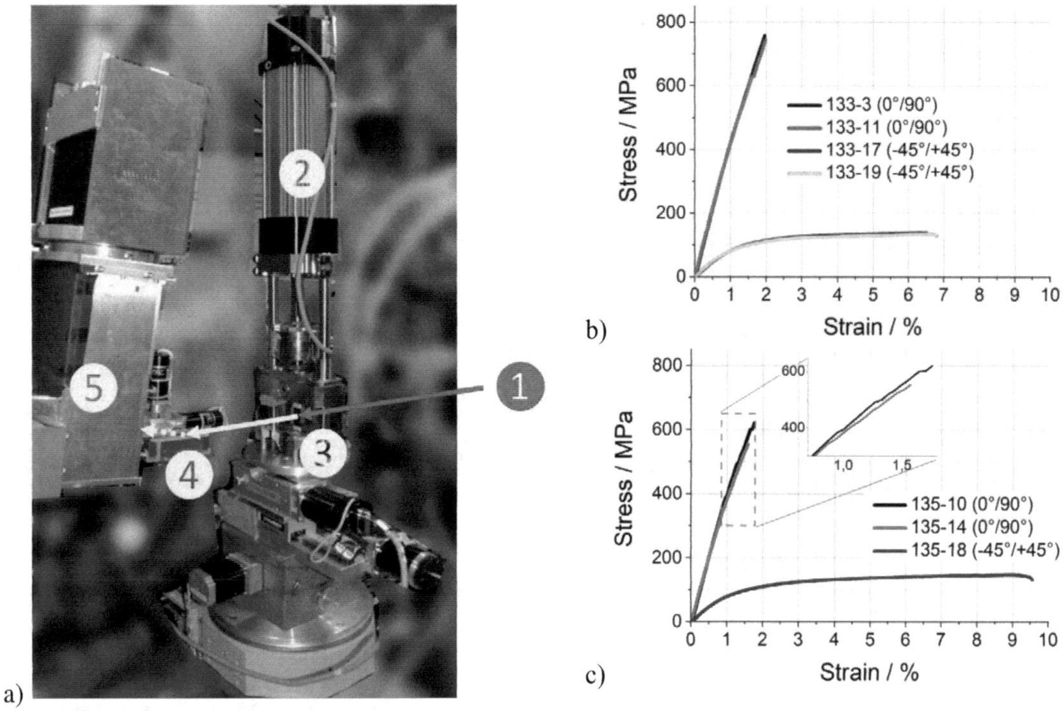

Figure 4. (a) Main components of the experimental set-up - The primary beam (red arrow) is emitted from (1) the synchrotron source and monochromatized by a DCM. It is absorbed and scattered by (3) the (loaded) sample; the tensile rig, including the electro-mechanical actuator, the load cell, and (2) the hydraulic clamping . The transmitted and scattered beam (solid yellow arrow) is diffracted at the analyzer crystal (4). The diffracted beam (dotted yellow arrow) is recorded by (5) the flat panel detector. Stress strain curves (b) of the non-crimp fabrics and (c) the twill fabrics reveal the larger stiffness of the NCF samples. The magnified detail in (c) visualizes the occurrence of transverse cracking and its impact on the slight decrease of the 0°/90° samples' stiffness ("Puck's knee" [24]).

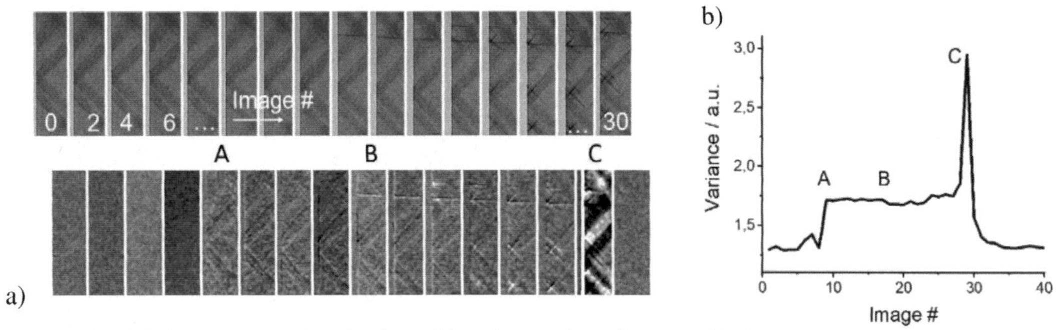

Figure 5. (a) Subsequent synchronization of imaging and tensile test with the example of an image detail of a -45°/+45° non-crimp fabric (sample 133-17). The as-measured radiographs (top row) are used to compute residual radiographs (bottom row). Points A, B, and C indicate the start of the tensile test, the appearance of first cracks, and the sample rupture, respectively. (b) Computing the variance of the residual radiographs reveals clear indication of A and C, while the evidence of B is less pronounced.

Table 2: Compilation of the tensile test results, where *UTS* is the ultimate tensile strength, ε_r the strain at rupture, and *E* is Young's modulus.

sample	type	Fiber orientation	*UTS* / MPa	ε_r / %	*E* / GPa
133-3	NCF	0°/90°	760	2.0	44.5
133-11	NCF	0°/90°	740	2.0	44.4
133-17	NCF	-45°/+45°	138	6.3	9.7
133-19	NCF	-45°/+45°	135	6.6	9.8
135-10	twill	0°/90°	620	1.8	41.2
135-14	twill	0°/90°	550	1.6	41.4
135-18	twill	-45°/+45°	148	8.9	9.6

3. Results

3.1. Non-crimp fabrics

Figure 6 shows selected images of the tensile test on the 0°/90° non-crimp fabric (specimen 133-3). Since the specimens are unnotched and the field of view is about 10% of the specimen's free length, it is a coincidence if the rupture occurs within the image.

The 35tex glass fiber binding yarn is not detectable by naked eye inspection (figure 1) when impregnated with the epoxy matrix. However, the density of the glass (ρ = 2.55 g/cm³) is significantly higher than the carbon fiber epoxy (according to fiber volume content of 41 % - ρ = 1.4 g/cm³). Hence, the glass fiber yarn causes an absorption contrast. Due to the orientation sensitivity of X-ray refraction, the contrast of the horizontal glass fiber yarns is enhanced compared to the perpendicular-oriented ones (absorption and refraction confounded *vs.* pure absorption). The specimens strain becomes visible by the displacement of the binding yarns.

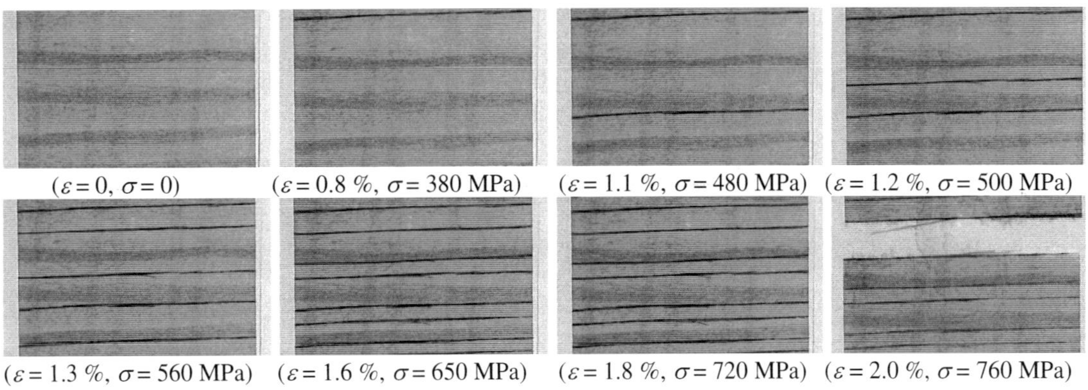

Figure 6. Selected XRRR images of the tensile test of the NCF sample 133-3 covering the start of the test, the subsequent occurrence of 90° inter-fiber cracks until reaching a saturated crack density, followed by rupture. The images correspond to a size of 11 × 7.2 mm². The respective strain (ε) and stress (σ) is indicated below each image.

Before the first cracks appear (σ = 380 MPa), the strain is evidenced from stretching the glass fiber yarns (figure 5a). The sample roughly follows a linear-elastic stress-strain behavior (Young's modulus 44.4 GPa [23]). However, due to the increasing number of transverse cracks, the stiffness of the specimen decreases slightly to a secant modulus of about 38 GPa short before rupture. Only 90° inter-fiber cracks are observed. Once such cracks are generated, they immediately cross the entire width of the NCF sample. (The time resolution does not allow locating the site of initiation.) In an intermediate state, they form an irregular pattern of 1 mm and 2 mm distance. Since subsequent cracks fill the 2 mm gaps symmetrically, before rupture (which is by chance in the field of view) we observe a regular 1 mm distance pattern, which corresponds to the saturated crack density for quasi static loading [2]. The transverse strain is negligible. After the rupture due to 0°-fiber failure, the cracks remain visible with lower contrast, indicating reduced crack opening due the relaxed external load. However as discussed above, a transverse crack opening of a few nm is sufficient to generate the refraction contrast. A second measurement on specimen 133-11 (cf. figure 4b, table 1) confirmed the results obtained for 133-3.

Figure 7 shows selected images of the tensile test on the -45°/+45° non-crimped fabric (sample 133-17). Again, before the first cracks appear (at 123 MPa) the strain is evidenced from the stretching of the glass fiber yarns. Here, however, the translation of the yarn (due to displacement) is accompanied by fiber rotation by about 2° (figure 8a), which is due to the high Poisson's ratio (of about 0.7). The -45°/+45° NCF laminate shows a non-linear viscoelastic-plastic stress-strain behavior [1] (Young's modulus 9.7 GPa [23]) due to the strong influence of the polymer matrix up to the rupture strain (of about 6.3 %). Figure 8b demonstrates that the local remaining plastic deformation in the field of view is rather low.

The initial cracks observed are 90° cracks near the free edges and cross about 10 % of the sample width. Presumably, such cracks are generated in the matrix at the intersection of fiber bundles, where the epoxy matrix agglomerates. At higher strain (about 5 %), the generation of 45° inter-fiber cracks sets in; such cracks stop near the specimen's edges. Until rupture, we observe no cracks crossing the entire width of the sample in the field of view. A second measurement on sample 133-19 (cf. figure 4b, table 1) confirmed the results obtained on 133-17 but it shows that one particular 45° crack crosses the entire width at rupture in the field of view.

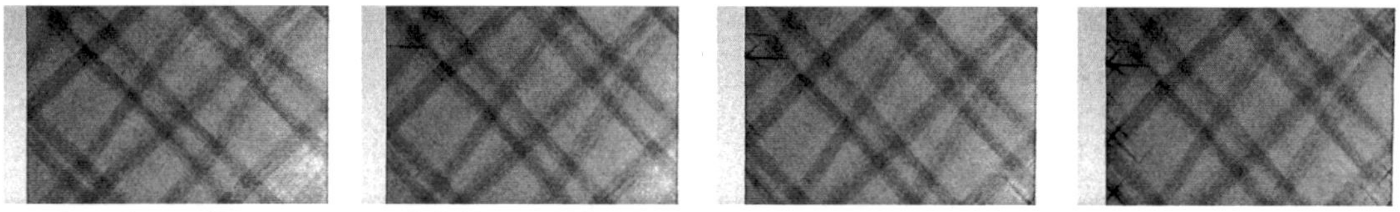

($\varepsilon = 0$, $\sigma = 0$) ($\varepsilon = 2.5$ %, $\sigma = 123$ MPa) ($\varepsilon = 4.9$ %, $\sigma = 135$ MPa) ($\varepsilon = 6.3$ %, $\sigma = 138$ MPa)

Figure 7. Selected XRRR images of the tensile test of the NCF sample 133-17 showing the start of the test, the first 90° crack, the first 45° inter-fiber cracks and the damage before rupture.

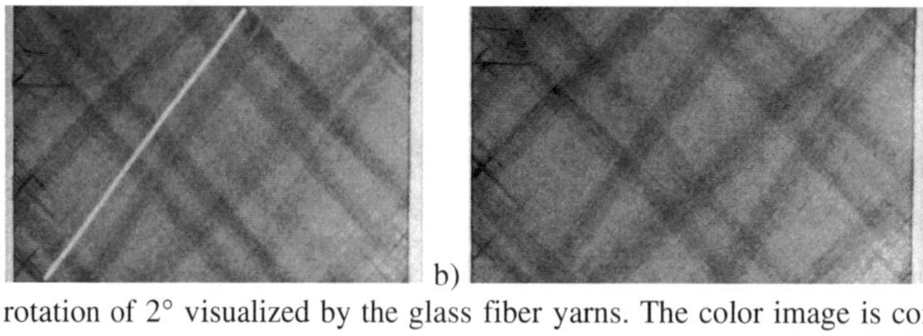

a) b)

Figure 8. (a) Fiber rotation of 2° visualized by the glass fiber yarns. The color image is composed of single XRRR images at the start of the tensile test (green picture) and short before rupture (red picture). The two solid lines indicate the translation and rotation of one and the same yarn in the two states. The light green areas on both sides of the sample indicate the transverse contraction (of about 5 %). (b) The low plastic deformation is evidenced by superimposing the start of the tensile test (green) with the image after rupture (red). The glass fiber yarns rotate back to their initial orientation, and the transverse contraction is nearly reversed (2 % residual contraction) in the field of view.

3.2. Twill fabrics

Figure 9 shows selected images of a 0°/90°-twill specimen (133-10) under tensile load. The slight horizontal whitening comes from the horizontal carbon fiber bundles and is due to the refraction contrast. The perpendicular fibers are not detectable [5, 6]. The first crack occurs at 340 MPa in the center of a fiber bundle. Subsequent cracks are also generated within the fiber bundles, where the local fiber volume is high. In contrast to the NCF-laminate, the cracks do not suddenly cross the total specimen width due to the presence of the perpendicular woven fiber bundles. With increasing load, about half of the cracks grow across the entire sample width. The crack density in the saturated state is lower than in the NCF-

IOP Conf. Series: Materials Science and Engineering **942** (2020) 012035 doi:10.1088/1757-899X/942/1/012035

laminate. However, in the roughly linear-elastic stress-strain behavior, the stiffness slightly decreases with the increasing transverse cracking by about 15 % to 35 GPa (detail in figure 4c), as observed with the NCF 0°/90° laminate. Finally, the twill textile shows a more damage tolerant inter-fiber failure behavior accompanied by a slightly lower total strength and stiffness compared to the NCF-textile. It should be noted that in this particular case, the stress-strain curve shows at least three “Puck knees” [24], *i.e.*, discontinuities, at about 400, 500, and 600 MPa. These kinks are a footprint of the occurrence of transverse cracks crossing the sample as the sample’s compliance increases.

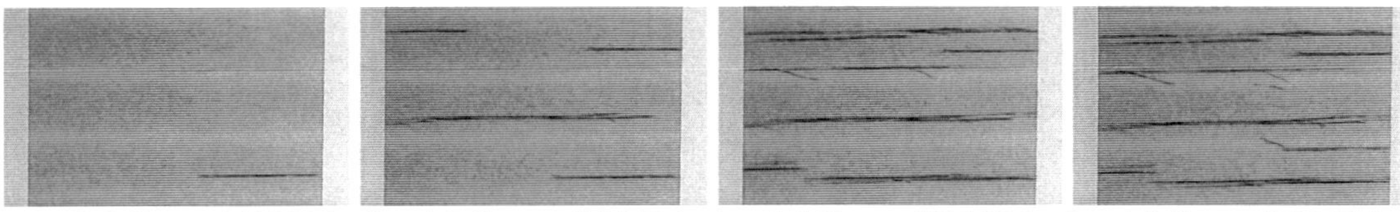

($\varepsilon = 0.8$ %, $\sigma = 340$ MPa) ($\varepsilon = 1.0$ %, $\sigma = 420$ MPa) ($\varepsilon = 1.4$ %, $\sigma = 520$ MPa) ($\varepsilon = 1.8$ %, $\sigma = 620$ MPa)

Figure 9. Selected XRRR images of the tensile test of the twill sample 135-10 showing the first 90° crack, the first crack crossing the sample width, and the crack pattern at 520 MPa, which remains nearly unchanged until rupture at 620 MPa.

Cracking in the -45°/+45° twill specimen (figure 10, sample 135-18) starts at about 5% strain. The first cracks observed are 45° inter-fiber shear cracks of limited length at the free edges. With increasing load, 90°-cracks are generated in matrix-rich areas near the edges, preferably, at the intersections of former 45° cracks. However, they do not cross the entire sample width until rupture. The total crack density is low compared to the 0°/90°-laminates. In the -45°/+45°-twill-laminate, the refraction effect occurs in both fiber orientations (proportional to 1-cos$^2\alpha$, where α is the angle of fibers relative to the scattering direction, here: vertical [5]). Hence, the -45°/+45°-fiber bundles become visible.

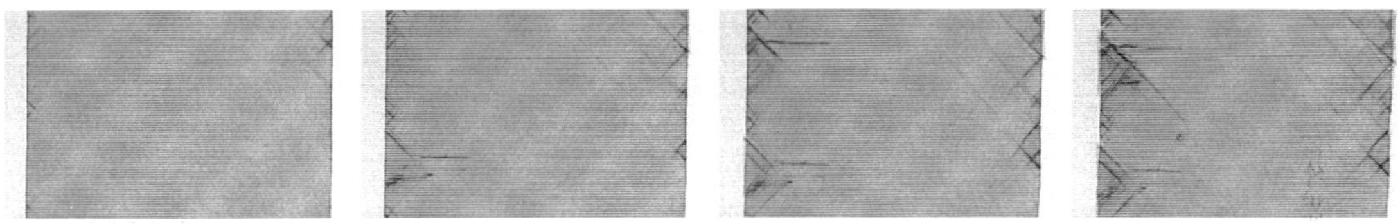

($\varepsilon = 5.0$ %, $\sigma = 137$ MPa) ($\varepsilon = 6.6$ %, $\sigma = 142$ MPa) ($\varepsilon = 8.9$ %, $\sigma = 148$ MPa) ($\varepsilon = 9.5$ %, $\sigma = 131$ MPa)

Figure 10. Selected XRRR images of the tensile test of the twill sample 135-18, showing the first 45° inter-fiber cracks, the first 90° crack, and a grown crack pattern, which remains nearly unchanged until rupture at 131 MPa.

4. Conclusions

We have tracked the crack evolution in differently prepared carbon fiber reinforced plastics (CFRPs) by *in-situ* X-ray refraction imaging without the need to use contrast agents. The different samples showed significantly different mechanical behavior mainly due to fiber orientation: 0°/90° oriented fabrics show a linear elastic behavior and break at about 2 % strain, while -45°/+45° oriented ones show a viscoelastic-plastic behavior and break at considerably higher strain. It could be shown that crack generation is correlated to discontinuities in the stress-strain curves for 0°/90° samples. Crack generation and growth is mainly governed by the type of fabric: in non-crimp fabric (NCF) several cracks propagate across the sample, while twills stop cracks with the presence of perpendicular yarns. Hence, twill fabrics exhibit a lower strength and stiffness but higher damage tolerance. The comparatively low stiffness of the 0°/90° fabrics is due to the low fiber volume content of 41 %. The observed inter-fiber-failure is qualitatively similar to the phenomena observed with glass fiber reinforced plastics, GFRP. The cracks remain visible after the rupture, although at reduced contrast. This demonstrates that XRRR is sensitive even to cracks of very small opening, which are not detectable using X-ray absorption techniques at comparable spatial resolution.

IOP Conf. Series: Materials Science and Engineering **942** (2020) 012035 doi:10.1088/1757-899X/942/1/012035

Acknowledgements
The authors are indebted to T. Wolk, R. Britzke, S. Hickmann, and M. Harwardt (BAM) for assistance with the measurements. V.T. gratefully acknowledges financial support by the Deutsche Forschungsgemeinschaft (DFG) in the Priority Program 1466 "Life∞" under grant numbers TR499/3-1 and TR499/3-2.

References
[1] Trappe V and Harbich K W 2006 *Int. J. Fatique* **28** 1187.
[2] Müller A, Trappe V, Hickmann S and Ortwein H P 2018 Investigation of the infinite life of fibre-reinforced plastics using X-ray refraction topography for the *in-situ*, nondestructive evaluation of micro-structural degradation *Fatigue of Materials at Very High Numbers of Loading Cycles* ed. H J Christ (Wiesbaden: Springer Spektrum) pp. 417-439.
[3] Adam T J and Horst P 2018 Very high cycle fatigue testing and behavior of GFRP cross and angle-ply laminates *Fatigue of Materials at Very High Numbers of Loading Cycles* ed. H J Christ (Wiesbaden: Springer Spektrum) pp. 511-532.
[4] Adden S and Horst P 2006 *Compos. Sci. Technol.* **66** 626.
[5] Trappe V, Hickmann S and Sturm H 2008 *Mater. Test.* **50** 615.
[6] Trappe V, Hentschel M P and Ivers H 2006 *Proc. 16th Europ. Conf. of Fracture (Alexandroupolis)* (New York: Springer) p. 900.
[7] Hentschel, M P, Hosemann R, Lange A, Uther B and Brückner R 1987 *Acta Cryst.* A **43** 506.
[8] Müller B R and Hentschel M P 2013 Micro-diagnostics: X-ray and synchrotron techniques *Handbook of technical diagnostics - Fundamentals and application to structures and systems.* ed. H Czichos (Berlin, Heidelberg: Springer) pp. 287-300.
[9] Evsevleev S, Müller B R, Lange A and Kupsch A 2019 *Nucl. Instrum. Meth.* A **916** 275.
[10] Kupsch A, Lange A, Hentschel M P, Onel Y, Wolk T, Staude A, Ehrig K, Müller B R and Bruno G 2013 *J. Ceram. Sci. Tech.* **4** 169.
[11] Kupsch A, Müller B R, Lange A and Bruno G 2017 *J. Eur. Ceram. Soc.* **37** 1879
[12] Müller B R, Cooper R C, Lange A, Kupsch A, Wheeler M, Hentschel M P, Staude A, Pandey A, Shyam A and Bruno G 2018 *Acta Mater.* **144** 627.
[13] Nellesen J, Laquai R, Müller B R, Kupsch A, Hentschel M P, Anar N B, Soppa E, Tillmann W and Bruno G 2018 *J. Mater. Sci.* **53** 6021.
[14] Laquai R, Gouraud F, Müller B R, Huger M, Chotard T, Antou G and Bruno G 2019 *Materials* **12** 1017.
[15] Mack E, Laquai R, Müller B R, Helle O, Sebold D, Vaßen R and Bruno G 2019 *J. Am. Ceram. Soc.* **102** 6163.
[16] Erdmann M, Kupsch A, Müller B R, Hentschel M P, Niebergall U, Böhning M and Bruno G 2019 *J. Mater. Sci.* **54** 11739.
[17] Schob D, Roszak R, Sagradov I, Sparr H, Ziegenhorn M, Kupsch A, Léonard F, Müller B R and Bruno G 2019 *Arch. Mech.* **71** 507.
[18] Müller B R, Lange A, Harwardt M and Hentschel M P 2009 *Adv. Eng. Mater.* **11** 435.
[19] Görner W, Hentschel M P, Müller B R, Riesemeyer H, Krumrey M, Ulm G, Diete W, Klein U and Frahm R 2001 *Nucl. Instrum. Meth.* A **467-468** 703.
[20] Lange A, Hentschel M P, Kupsch A and Müller B R 2012 *Int. J. Mater. Res.* **103** 174.
[21] Al-Falahat A A M, Kupsch A, Hentschel M P, Lange A, Kardjilov N, Markötter H and Manke I 2019 *Rev. Sci. Instrum.* **90** 125108.
[22] Chapman D, Thomlinson W, Johnston R E, Washburn D, Pisano E, Gmür N, Zhong Z, Menk R, Arfelli F and Sayers D 1997 *Phys. Med. Biol.* **42** 2015.
[23] ISO 527-1:2012-06 2012 Plastics - Determination of tensile properties - Part 1: General principles (Berlin: Beuth).
[24] Puck A and Schürmann H 1998 *Compos. Sci. Technol.* **58** 1045.

IOP Conf. Series: Materials Science and Engineering **942** (2020) 012036 doi:10.1088/1757-899X/942/1/012036

Comparison of the damage evolution in glass fiber-reinforced polyurethane and epoxy in the HCF and VHCF regimes investigated by intermittent in situ X-ray computed tomography

D Hülsbusch[1,*], R Helwing[1], S Mrzljak[1] and F Walther[1]

[1]Department of Materials Test Engineering (WPT), TU Dortmund University, Baroper Str. 303, D-44227 Dortmund, Germany

E-mail: daniel.huelsbusch@tu-dortmund.de

Abstract. The aim of this work is the comparative characterization of the fatigue and damage behaviors of GFR-polyurethane and GFR-epoxy with application-relevant quasi-isotropic layer setup in the high cycle and very-high cycle fatigue regimes. Therefore, a high-frequency test method based on a resonant testing system (1 kHz) has been further developed and assessed with special consideration of self-heating. In intermittent test procedures, the damage state has been explored by in situ X-ray computed tomography analysis after certain numbers of cycles. It was shown that the overall damage state in the VHCF regime is reduced by a factor of three compared to the HCF regime and accompanied by delayed initiation and propagation of delamination. The latter was proven to be the main reason for a decreased inclination of the S/N-curve in the VHCF regime by 50-60%.

1. Introduction and state of the art

Glass fiber-reinforced polymers (GFRP) are increasingly used in components that are subjected to fatigue. In major applications, e.g., wind turbines, the structures have to last more than 1E9 cycles. However, there is still insufficient knowledge about the fatigue and damage behaviors in the very-high cycle fatigue (VHCF) regime, which are essential for providing operational reliability of the structures. Furthermore, recent developments and changes in polymer matrices, e.g., polyurethane-based instead of epoxy-based GFRP used in rotor blades of wind turbines e.g. in China, lead to the need for determination of the polymer influence on the fatigue properties, i.e., damage evolution.

In the last couple of years, a lot of investigations were carried out focusing on the damage evolution in GFRP. In this context, mechanical test methods were combined with advanced techniques like X-ray computed tomography (CT), to obtain information of the internal volume structure. The potential of portraying FRP structures with high-resolution CT and its advantages is shown by various studies. In this instance, two techniques are available: conventional CT and in situ CT. Conventional CT has been used in a variety of methods to determine characteristics of FRP damage evolution. Constant amplitude tests were carried out on GFRP, in which in between test sections [1] and specimen failure [2] CT scans were used to locate proceeding damages. In situ CT is applied in an expanded way to explore the nature of FRP behavior by scanning at different levels of load [3]. Studies on FRP using in situ CT show promising potential, since they lead to significant improvements regarding result reliability by minimizing crack closure effects, which occur because of elastic behavior. In the case of FRP, CT images carried out on an in situ loaded specimen show up to more than 200% increased delamination area and 30 to 100% more cracks than the same specimen scanned via conventional CT [4]. In this way, damage initiation and propagation can be successfully monitored. [5]

Investigations of damage evolution using in situ CT are almost exclusively limited to the low cycle fatigue (LCF) to high cycle fatigue (HCF) regime. This is mainly due to practical and economical reasons, such as the enormous test duration under standard test conditions for investigating the VHCF regime, e.g., more than 110 days for 1E8 cycles at 10 Hz. Previous approaches in the VHCF regime were therefore based on unconventional test methods with increased frequencies. However, many of these test methods are of limited use, as they are accompanied by extensive restrictions, for example with regard to maximum load. In the near past, the ultrasonic test method has been used by several researchers, which was applied on FRP in 3P bending [6] as well as axial [7]. This test method is limited to a frequency of 20 kHz. It is expected that an extensive self-heating takes place, which impairs the comparability with results from the HCF regime determined in accordance with the standard. Therefore it can be concluded that a holistic, suitable experimental method for the investigation of the VHCF regime still needs to be developed.

In principle, it has to be discussed whether a characterization of the fatigue properties in the VHCF regime is necessary at all. According to Talreja [8], FRP exhibit an endurance limit represented by region III in the fatigue-life diagram. In this case, the strain amplitude is too small to initiate matrix cracks and/or to be able to trigger a crack growth rate sufficient for failure. In investigations of Wu et al. [9], Hosoi et al. [10], and Adam and Horst [11] the theory of Talreja is supported. For example, Hosoi et al. did not detect any damage in quasi-isotropic GFRP up to 3E8 cycles at a relative maximum stress of 0.2, which is why a limit value with regard to damage initiation and thus endurance limit was assumed. Adam and Horst investigated the damage evolution in GFRP up to 1.5E8 cycles. At an initial strain of 46-51% of the static strain to inter-fiber breakage, no or marginal delamination occurred. Based on this, the authors assumed a limit value of the initial strain representing an endurance limit.

In recent test series, findings contrary to those of the research groups listed above can be found. In tests on CFRP [12], GFRP [7] and flax fiber-reinforced EP [13], no endurance limit or trend in this respect can be identified. Weibel et al. [12] detected a significant increase in crack density under 3P bending up to 1E9 cycles and failure occurred at approx. 2E9 cycles on additional specimens. Investigations on unidirectional GFRP under axial tensile loading show similar trends [7]. Although the S/N curve shows a flattening from the HCF to the VHCF regime, failure still occurs in tests up to approx. 1.6E9 cycles, accompanied by an initial strain amplitude of only approx. 0.11%. With reference to the theory of Talreja, this proves that even at low strain amplitudes failure of FRP structures can still occur. Although it cannot be ruled out that an endurance limit may be present at even lower strain amplitudes and/or other structural properties, recent research results show the need to investigate the VHCF regime up to at least 2E9 cycles.

This work focusses on the comparative characterization of the fatigue and damage behavior of GFR-polyurethane and GFR-epoxy with application-relevant quasi-isotropic layer setup in the HCF and VHCF regimes. Therefore, an intermittent test procedure has been applied to determine the damage evolution and interaction stepwise after certain numbers of cycles. For the investigations in the VHCF regime, an unconventional high-frequency test method based on a resonant testing system (1 kHz) has been further developed and assessed with special consideration of self-heating.

2. Material

In this study, GFR-polyurethane (GFR-PU) and GFR-epoxy (GFR-EP) are comparatively investigated. Both GFR-PU and GFR-EP are based on an application-relevant quasi-isotropic layer setup of $[45/-45/0/90]_{2S}$ with a fiber volume fraction of 46%. The fiber material is Hex-Force 07781 1270 (continuous E-glass, 9 μm fiber diameter) in 8h satin fabric. The fibers are showing a Silan TF970 sizing, which is suitable for both polymers.

In Table 1 the properties of the neat resins are listed. The EP is the Hex Ply 914 matrix system certified for use in the aviation industry. The GFR-EP was manufactured by FACC (Austria) in an autoclave process at 7 bar and 175 °C for 1 hour. The newly PU thermosetting resin was developed by Rühl (Germany) and is based on a multi-stage modification of existing resin systems suitable for a vacuum RTM manufacturing process. The final configuration used in this work is a combination of the slowly reacting polyol puropreg 185-2L IT with internal release agent and the low-viscosity isocyanate puronate 905. In the production process of GFR-PU at the Institute for Material Technologies and Plastic Processing (Switzerland), both components are mixed in a high pressure device and injected with a rate of 7 g·s^{-1} for 41 s into the high pressure tool, in which a preform is fixed. The curing process takes place at 35 bar and 85 °C for 300 s. Further details on the manufacturing process can be found in [14].

Table 1: Properties of the neat resins of polyurethane and epoxy [14].

	Young's modulus E (GPa)	Tensile strength σ_M (MPa)	Strain to failure A (10^{-2})	Glass trans. temperature Tg (°C)
Polyurethane	3.1	63	9.1	130
Epoxy	3.9	48	1.5	190

3. Experimental methods

For the investigations of the damage evolution, an intermittent test procedure was applied, in which mechanical tests were stopped after certain numbers of cycles to determine the damage state via in situ CT. Two different kinds of mechanical testing systems for the LCF to HCF regime and the VHCF regime were used and will be presented in the following subsections. A special CT specimen geometry (Fig. 1) is necessary due to restrictions of the in situ CT-stage. The dimensions allow a CT-scan of the whole measurement length at a time and the shape of the specimen results in an almost completely homogeneous stress distribution in the measurement length, shown for an elongation of 10 μm in Fig. 1.

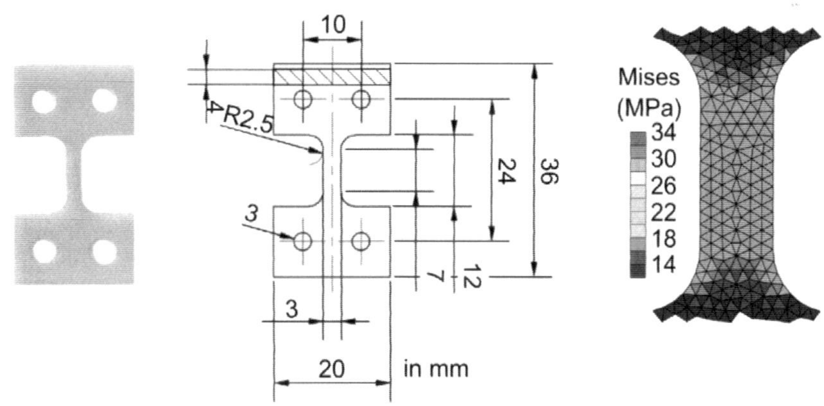

Fig. 1: Specimen geometry and simulation of the stress distribution for an elongation of 10 μm.

3.1. Fatigue tests in the LCF to HCF regime

The investigations in the LCF to HCF regime were carried out stress-controlled with the use of a servo-hydraulic testing system (Shimadzu EHF-UV100, F_{max} = ±100 kN) with a sinusoidal load time function under tension-tension loading (R = 0.1) in a climatic chamber at 23°C. Due to the CT specimen geometry, a special adapter was used (Fig. 2). With the adapter, the specimen is mechanically clamped by combining friction and pinning.

The frequency was varied stress-related between 1.5 Hz (200 MPa) and 30.0 Hz (100 MPa) to realize a constant induced energy rate and thereby a similar self-heating. More information regarding the principle of induced energy rate can be found in [15]. The fatigue tests for investigating the damage evolution were carried out at a relative maximum stress of $\sigma_{max,r} = 0.55$ related to the tensile strength.

Fig. 2: Test setup for fatigue tests in the LCF to HCF regime [5].

3.2. Fatigue tests in the VHCF regime

In view of economical and practical reasons, investigations in the VHCF regime need to be carried out at higher frequencies. Therefore, in this study, a resonant testing system (Rumul Gigaforte, F_{max} = ±25 kN) was used, which works at a frequency of about 1 kHz. The further test parameters and the specimen adapter were similar to the ones applied in the HCF regime.

The use of this testing system leads to an improved comparability between the results in the HCF and VHCF regimes compared to the use of an ultrasonic testing system, which was applied in multiple previous studies, e.g., [7, 12], due to the lower frequency (1 kHz vs. 20 kHz). However, there is still a big difference between the frequencies in the HCF and VHCF regime. That is why special attention was paid to the self-heating. Therefore, specimens made of GFR-EP were provided with thermocouples on the surface as well as embedded in the core. The results have been used to determine a suitable stress range, in which a sufficient comparability of the test results is assumed.

Figure 3a) shows the stress-dependent development of the surface and core temperature at simultaneous air cooling. The surface temperature remains almost constant with increasing maximum stress. The core temperature increases significantly by about 40 K from an initial temperature of 30 °C between 60 and 140 MPa, showing a linear behavior. Thus, the difference between surface and core temperature increases with increasing stress. In order to evaluate the influence of air cooling, additional tests were carried out without air cooling (Fig. 3b). In this case, the core temperature increases exponentially, so that a temperature of well over 100 °C (> 50% Tg) is already reached at σ_{max} = 90 MPa. In contrast to the tests with air cooling, the surface temperature shows a qualitatively comparable course with only a slight increase in the temperature difference from the core temperature.

From the investigations it can be concluded that the concept of evaluating the specimen temperature on the basis of surface temperature measurements with simultaneous air cooling must be considered improper. On the basis of the above-mentioned findings, the low self-heating detected by Flore et al. [7], among others, is to be questioned with regard to their validity. It can be assumed that the core temperature is significantly higher. Lower fatigue strengths compared to investigations with conventional test methods, such as those determined by Weibel et al. [12], can therefore probably be justified by the temperature increase in the core.

For the resonant testing system, the temperature development in the core confirms that this test method can only be used validly on GFRP in a low stress range (VHCF). In this range, the temperature development in the core can be limited to, e.g., approx. 20 K for σ_{max} = 90 MPa by using air cooling.

IOP Conf. Series: Materials Science and Engineering **942** (2020) 012036 doi:10.1088/1757-899X/942/1/012036

The testing system was therefore used exclusively in a stress range of $\sigma_{max,r} = 0.25$ to 0.35, taking into account the temperature development and following VHCF-specific threshold values of Hosoi et al. [16] and Jeannin et al. [13]. For the investigations of the damage evolution in the VHCF regime, a relative maximum stress of $\sigma_{max,r} = 0.275$ was applied, resulting in a calculated temperature increase in the core of approx. 23 K, which is considered permissible for FRP in Vieille and Taleb [17]. Furthermore, the maximum temperature stays below 35% Tg for GFR-PU and 25% Tg for GFR-EP, respectively, and is thereby significantly below the threshold value of 50% Tg defined by Backe et al. [6].

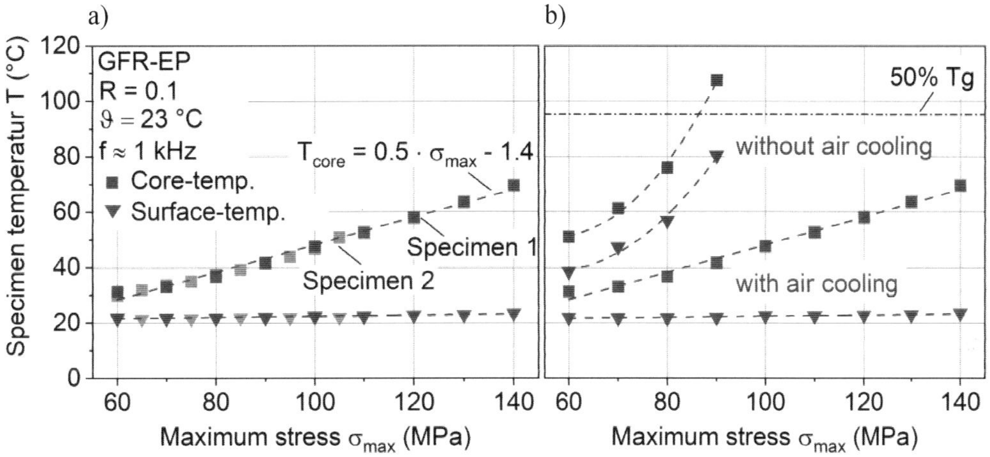

Fig. 3: Stress-dependent self-heating of GFR-EP specimens at a frequency of 1 kHz. a) Comparison of surface and core temperatures and b) influence of air cooling.

3.3. *In situ CT investigations of the damage evolution*

CT investigations were carried out with Nikon XT H 160. The system contains an X-ray tube with a maximum voltage of 160 kV and a maximum power of 60 W and allows, with the 1008^2-pixel detector, voxel sizes down to 3 µm. In situ load application was performed using an in situ CT stage (Deben Uk CT5000, $F_{max} = \pm 5$ kN). A glass-like carbon tube with a wall thickness of 3 mm transmits the force between the lower and upper clamping device. X-rays can pass through this homogeneously constructed tube under a small, uniform damping. [5]

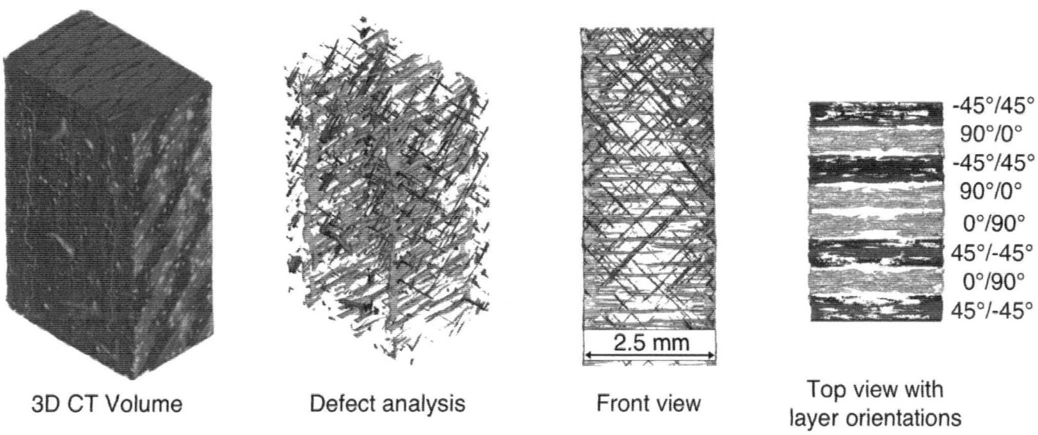

Fig. 4: Procedure of in situ CT defect analysis [5].

The defect analysis is done using in situ acquired 3D CT volumes containing the damage states of the specimens. For an illustration of the 3D damage volume, the specimen material is hidden, and only the contrast agent is visible (Fig. 4). Besides, the contrast agent residues on the surfaces of the front and back are hidden to provide an insight into the specimen. The front and top view of the 3D damage volume are used to analyze the damage state and distribution with the usage of single layer defect analysis for all layer orientations, to separate the occurring damages and their locations. [5]

4. Results

4.1. Fatigue behavior

Figure 5 presents the material-specific S/N curves. The S/N curve of GFR-PU (Fig. 5a) shows a similar shape to GFR-EP (Fig. 5b), although the tensile strength is reduced by approx. 15%. In recent publications various approaches for modelling the S/N curve are used. In this study, the S/N curve is separated into three regions, which are representing the (I) LCF, (II) HCF, and (III) VHCF regime. As a demonstration of the insufficient description by simpler approaches, a power function is shown in a grey dotted line across all regions as well as across regions II and III. The entire power function leads to an underestimation of fatigue strength in the LCF regime and an overestimation in the VHCF regime, which is critical with respect to a conservative design. With a power function for the joint description of regions II and III, fatigue strength in the VHCF regime is consistently underestimated, which does not allow an ideal utilization of the material potential. The separation of the S/N curve into three regions, on the other hand, leads to a very good agreement of region-specific power functions with the test results and is therefore recommended for a comprehensive description of the fatigue strength.

$$\sigma_{max} = a \times {N_f}^{n}; n = k^{-1} \tag{1}$$

For GFR-PU, the ratio of the inclination index k (Eq. 1) for regions II and III ($k_{II}/k_{III} \approx 10.5/15.9 \approx 1/1.5$) is almost identical to GFR-EP ($10.9/17.5 \approx 1/1.6$). The inclination index k presents the slope of the S/N curve described by Basquin's law. It can be summarized that for quasi-isotropic GFRP there is a transition region from the HCF to VHCF regime, which is accompanied by a change in the slope of the S/N curve of 50 to 60%. The course of the S/N curve is thus comparable to that of face-centered cubic metals whereby their change in the slope is more pronounced and entails a change in the failure mechanism from primarily surface damage to internal damage. A deviating or delayed development of damage in the VHCF regime was also suspected for FRP as a reason for the changing inclination from region II to III [6, 11] and will be demonstrated in the following subsection. The results do not show any trend regarding an endurance limit, which confirms recent findings in other studies [7, 12].

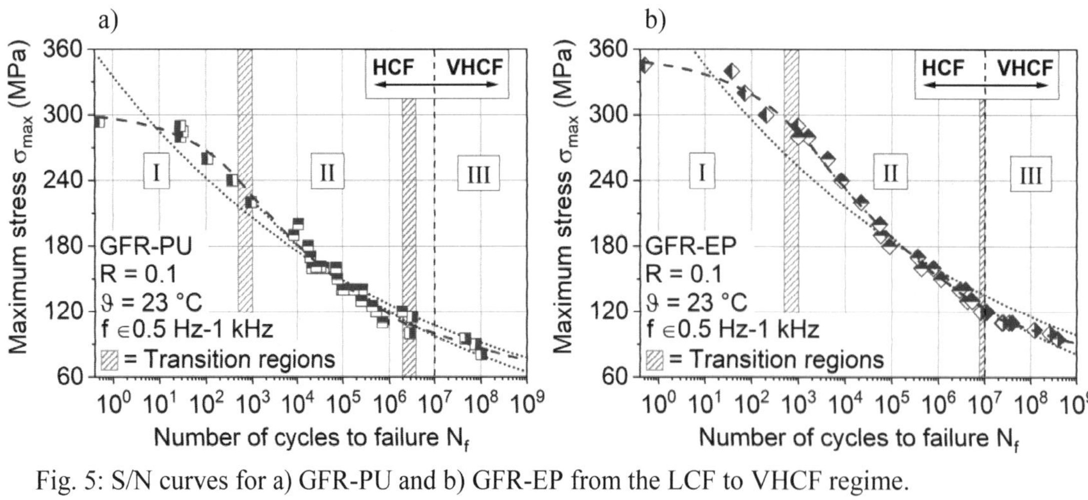

Fig. 5: S/N curves for a) GFR-PU and b) GFR-EP from the LCF to VHCF regime.

4.2. Damage evolution

The damage evolution is exemplarily illustrated in Fig. 6. Generally, the damage evolution is characterized by a gradual development. In the HCF regime, directly after the test start, inter-fiber cracks form in the 90° layers and subsequently in the ±45° layers. In the further course of the test, these damage mechanisms proceed, accompanied by intra-layer delamination in the 0°/90° layer, which originate from meta delamination. In the following, existing cracks and intra-layer delamination grow. Additionally, the amount of meta delamination increases significantly in the 0°/90° layers and first inter-layer delamination occur between the 0°/90° and ±45° layer. The delamination proceed over the entire measuring length and spread towards the center of the specimen, indicating the subsequent failure.

GFR-EP consistently shows a more extensive damage state. In particular, the number of inter-fiber cracks is significantly increased, which can be attributed to the lower ductility of EP compared to PU. However, the fatigue strength is higher than that of GFR-PU. This is due to the fact that the formation of delamination is reduced. GFR-PU has a significantly higher proportion of delamination, which are more critical in terms of fatigue properties than inter-fiber cracks. The reduced fatigue strength can therefore be attributed to a decreased interlayer strength, which results, amongst others, out of a pore content of approx. 1.2 vol.-% (GFR-PU) compared to 0.0 vol.-% (GFR-EP).

In the VHCF regime, the damage mechanisms and the general damage evolution are similar to those in the HCF regime, although a reduced crack density is formed. This results in a reduction of the damage volume by a factor of three compared to the HCF regime. However, the main difference in the damage evolution in the VHCF regime is represented by a delayed initiation and propagation of delamination. In the HCF regime, first delamination occur after approx. 10% of the fatigue life. In the VHCF regime, on the other hand, no delamination can be detected in GFR-PU up to 79% and in GFR-EP, after 89% of the fatigue life first delamination formation is visible. It can therefore be assumed that delamination will probably also be initiated in GFR-PU in the further course, which will initiate the failure. Based on the findings it is reasonable to assume that the delayed initiation of delamination, as a critical damage mechanism, leads to the decrease of inclination of the S/N curve in the VHCF regime by 50-60%.

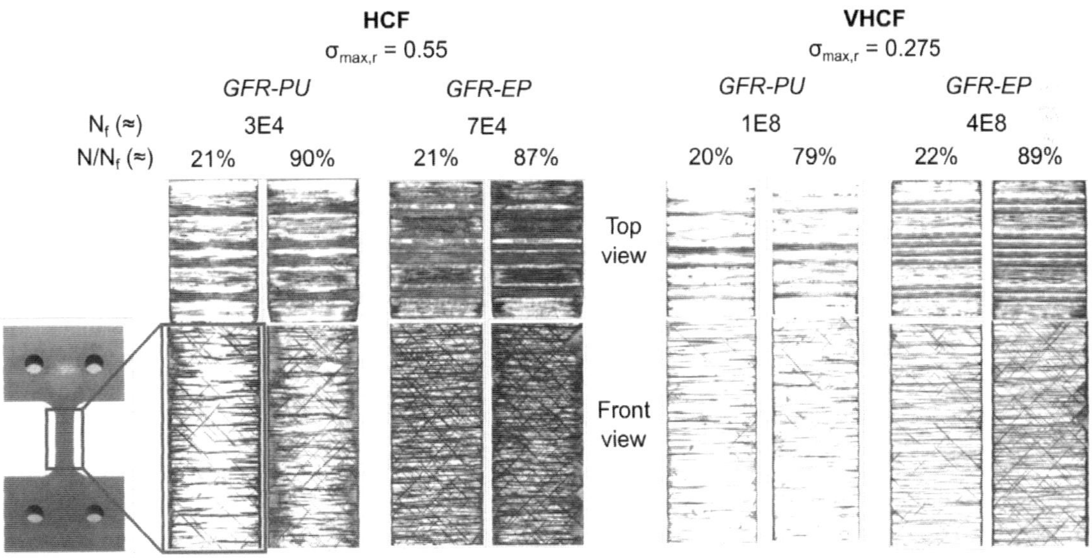

Fig. 6: Comparison of the damage evolution in GFR-PU and GFR-EP in the HCF and VHCF regimes.

5. Conclusions

In this study the necessity for improved alternative test procedures in the VHCF regime was proven. In particular, the influence of increased frequencies has to be assessed by paying special attention to the self-heating. For the 1 kHz resonant testing system, a suitable stress range was identified by measuring the stress-dependent increase in temperature with the use of embedded thermocouples. In intermittent test procedures, the damage state has been determined by in situ 3D X-ray computed tomography. It was shown that GFR-PU exhibits decreased fatigue strength compared to GFR-EP, although the overall damage state of GFR-EP is significantly higher. This contradiction has been explained by the increased proportion of delamination in GFR-PU, which are much more critical than the higher amount of inter-fiber cracks in GFR-EP. In the VHCF regime the general damage evolution is similar to the one in the HCF regime, although much less extensive and delayed. The damage volume is reduced by a factor of three compared to the HCF regime. However, the most important difference was proven to be the delayed initiation and propagation of delamination, which was considered the main reason for the change in the slope of the S/N curve of 50 to 60% from the HCF to VHCF regime. Although the slope of the S/N curve changes, a trend regarding the existence of an endurance limit could not be identified.

References

[1] Yu B, Blanc R, Soutis C and Withers P 2006 Evolution of damage during the fatigue of 3d woven glass-fibre reinforced composites subjected to tension-tension loading observed by time-lapse x-ray tomography *Compos. Part A: Appl. Sci. Manuf.* **82** pp. 279-290

[2] Yu B, Bradley R, Soutis C, Hogg P and Withers P 2015 2d and 3d imaging of fatigue failure mechanisms of 3d woven composites *Compos. Part A: Appl. Sci. Manuf.* **77** pp. 37-49

[3] Hufenbach W, Boehm R, Gude M, Berthel M, Hornig A, Rucevskis S and Andrich M 2012 A test device for damage characterisation of composites based on in situ computed tomography *Compos. Sci. Technol.* **72** pp. 1361-1367

[4] Böhm R, Stiller J, Behnisch T, Zscheyge M, Protz R, Radloff S, Gude M and Hufenbach W 2015 A quantitative comparison of the capabilities of in situ computed tomography and conventional computed tomography for damage analysis of composites *Compos. Sci. Technol.* **110** pp. 62-68

[5] Hülsbusch D, Mrzljak S and Walther F 2019 In situ computed tomography for characterization of the damage propagation in glass fiber-reinforced polyurethane *Mater. Test.* **61** pp. 821-828

[6] Backe D, Balle F and Eifler D 2015 Fatigue testing of CFRP in the very high cycle fatigue (VHCF) regime at ultrasonic frequencies *Compos. Sci. Technol.* **106** pp. 93-99

[7] Flore D, Wegener K, Mayer H and Oetting C 2017 Investigation of the high and very high cycle fatigue behaviour of continuous fibre reinforced plastics by conventional and ultrasonic fatigue testing *Compos. Sci. Technol.* **141** pp. 130-136

[8] Talreja R and Singh CV 2012 *Damage and Failure of Composite Materials* Cambridge University Press (Cambridge, UK)

[9] Wu T, Yao W and Xu C 2018 A VHCF life prediction method based on surface crack density for FRP *Int. J. Fatigue* **114** pp. 51-56

[10] Hosoi A, Sato N, Kusumoto Y, Fujiwara K and Kawada H 2010 High-cycle fatigue characteristics of quasi-isotropic CFRP laminates over 10^8 cycles *Int. J. Fatigue* **32** pp. 29-36

[11] Adam T J and Horst P 2018 Very high cycle fatigue testing and behavior of GFRP cross- and angle-ply laminates *Fatigue of Materials at Very High Numbers of Loading Cycles* Springer Spektrum (Wiesbaden, Germany) pp. 511-532

[12] Weibel D, Beck T and Balle F 2019 Fatigue properties and damage of CF-PPS from high to very high cycles *Proc. 22nd Int. Conf. Compos. Mater.* (Melbourne, Australia) pp. 1-7

[13] Jeannin T, Gabrion X, Ramasso E and Placet V 2019 About the fatigue endurance of unidirectional flax-epoxy composite laminates *Compos. Part B: Eng.* **165** pp. 690-710

[14] Hülsbusch D, Jamrozy M, Frieling G, Müller Y, Barandun GA, Niedermeier M and Walther F 2017 Comparative characterization of quasi-static and cyclic deformation behavior of glass fiber-reinforced polyurethane (GFR-PU) and epoxy (GFR-EP) *Mater. Test.* **59** pp. 109-117

[15] Hülsbusch D, Kohl A, Striemann P, Niedermeier M, Strauch J and Walther F 2020 Development of an energy-based approach for optimized frequency selection for fatigue testing on polymers – Exemplified on polyamide 6 *Polym. Test.* **81** 106260

[16] Hosoi A, Takamura K, Sato N and Kawada H 2011 Quantitative evaluation of fatigue damage growth in CFRP laminates that changes due to applied stress level *Int. J. Fatigue* **33** pp. 781-787

[17] Vieille B and Taleb L 2019 High-temperature fatigue behavior of woven-ply thermoplastic composites *Fatigue Life Prediction of Composites and Composite Structures* Woodhead Publishing (Duxford, UK) pp. 195-238

IOP Conf. Series: Materials Science and Engineering **942** (2020) 012037 doi:10.1088/1757-899X/942/1/012037

Characterization of the fiber orientations in non-crimp glass fiber reinforced composites using structure tensor

N Jeppesen[1,3], V A Dahl[1], A N Christensen[1], A B Dahl[1] and L P Mikkelsen[2]

[1]Department of Applied Mathematics and Computer Science, Technical University of Denmark, Lyngby, Denmark

[2]Department of Wind Energy, Technical University of Denmark, Roskilde, Denmark

E-mail: [3]niejep@dtu.dk

Abstract. The mechanical properties of composite fiber materials are highly dependent on the orientation of the fibers. Micro-CT enables acquisition of high-resolution 3D images, where individual fibers are visible. However, manually extracting orientation information from the samples is impractical due to the size of the 3D images. In this paper, we use a Structure Tensor to extract orientation information from a large 3D image of non-crimp glass fiber fabric. We go through the process of segmenting the image and extracting the orientation distribution step-by-step using structure tensor and show the results of the analysis of the studied non-crimp fabric. The Jupyter notebooks and Python code used for the data-analysis are publicly available, detailing the process and allowing the reader to use the method on their own data. The results show that structure tensor analysis works well for determining fiber orientations, which has many useful applications.

1. Introduction

Fiber reinforced polymer matrix composites are used in many structural applications due to their excellent stiffness, strength, and fatigue properties and relatively low weight. Nevertheless, these properties are very sensitive to the fiber orientations, and therefore characterizing fiber orientations is an important part of quality control. However, the microscopic nature of the fibers makes it difficult to bridge the gap between the fiber/matrix scale on the micro-meter scale to the structural component on the meter scale and thereby examine representative samples, while maintaining adequate resolution. With modern micro-CT scanners, it is possible to capture and stitch together high-resolution data for larger samples, which results in large volumetric data-sets.

To extract orientation information about the fibers, one approach is to track every fiber individually [1]. This can provide detailed information about the fibers, but often requires some user interaction and can be very computationally demanding, making it less attractive for large data-sets. In addition, it is often not the location of the individual fibers but just the fiber orientation in each material point which is of relevance. An approach addressing this is the voxel-based Structure Tensor method [2, 3, 4]. This method is relatively simple to use as it requires only two scalar parameters, σ and ρ. Furthermore, distributing the structure tensor calculations on modern multi-core CPU and GPU systems, allows the method to scale well, even with large volumetric data-sets.

In Sections 2 and 3 we introduce the Structure Tensor method and the data-set, respectively. Section 4 describes the procedure of using structure tensor analysis to extract orientation information from the volumetric data-set. A procedure which is demonstrated in the Jupyter notebook *StructureTensorFiberAnalysisDemo* which can be downloaded from [5]. In Section 5, we demonstrate the procedure on a realistic sized composite specimen, a full cross-section of a tension/tension fatigue test sample. The data and Jupyter notebook for this analysis can also be downloaded from [5].

2. Structure Tensor

In the context of volumetric (3D) image analysis, a structure tensor is a 3-by-3 matrix, which summarizes orientation in a small neighborhood around a point in space. The structure tensor is computed from the gradients of the 3D image. More precisely, if $\nabla V = [V_x \;\; V_y \;\; V_z]^T$ denotes the gradient of the 3D volume V, we can compute the structure tensor as

$$\mathbf{S} = \sum \nabla V \, (\nabla V)^T , \tag{1}$$

where summation runs in a certain neighborhood around the point. Nice properties of the structure tensor in respect to the scale of the analyzed structures are obtained if a Gaussian window is used for integration, and a Gaussian derivative for computation of the gradient [6], leading to the following formulation

$$\mathbf{S} = K_\rho * \left(\nabla V_\sigma \, (\nabla V_\sigma)^T \right) . \tag{2}$$

Here, the parameter σ is the standard deviation of the Gaussian derivative kernel used for computing the gradient ∇V_σ, while ρ is the standard derivation of the Gaussian kernel K_ρ used for integrating by convolution. The parameter σ is called the *noise scale*, and indicates the size of the structures filtered out while computing the gradient. The parameter ρ is called the *integration scale* and reflects the size of the window where the orientation is analyzed. Therefore, ρ should be chosen based on the size of the structures to be analyzed.

Given a unit vector $\mathbf{u}$, the product $\mathbf{u}^T \mathbf{S} \, \mathbf{u}$ gives the squared change in intensity for a small displacement in direction $\mathbf{u}$. Therefore, finding the predominant orientation amounts to minimizing $\mathbf{u}^T \mathbf{S} \, \mathbf{u}$, which is achieved through eigendecomposition of $\mathbf{S}$. Being symmetric and positive semi-definite, $\mathbf{S}$ yields three positive eigenvalues $\lambda_1 \leq \lambda_2 \leq \lambda_3$ and mutually orthogonal eigenvectors $\mathbf{v}_1$, $\mathbf{v}_2$ and $\mathbf{v}_3$. The eigenvector $\mathbf{v}_1$, corresponding to the smallest eigenvalue, is an orientation leading to the smallest variation in intensities, which indicates a predominant orientation in the volume. In this work, we focus on the information obtained from $\mathbf{v}_1$.

3. Data-set

A stitched data-set based on three 3D X-ray CT scans is used in this demonstration of using the structure tensor method for a fiber orientation characterization. The specimen that is scanned is a so-called butterfly-shaped tensile test-sample see e.g. [7] with a cross-section of approximately $15 \times 4\,\text{mm}^2$ and a uniform shaped gauge length of 60 mm. Three 16.5 mm field of view (FoV) binning 1 scans are reconstructed and stitched into a single Transmission X-ray Microscopy (TXM) file of 31 GB covering 32 mm of the sample length. Cropping away the air around the sample results in a 9 GB file, which is saved in the NIfTI format. Both files can be obtained from [5]. The laminate used in the test-sample contains four unidirectional non-crimp fabrics surround two biaxial non-crimp fabrics resulting in the following layup $[\mathrm{b_{biax}/biax, b_{UD}/UD, b_{UD}/UD}]_\mathrm{s}$ where the "b" indicates the location of the backing layer, which is orientated in the transverse

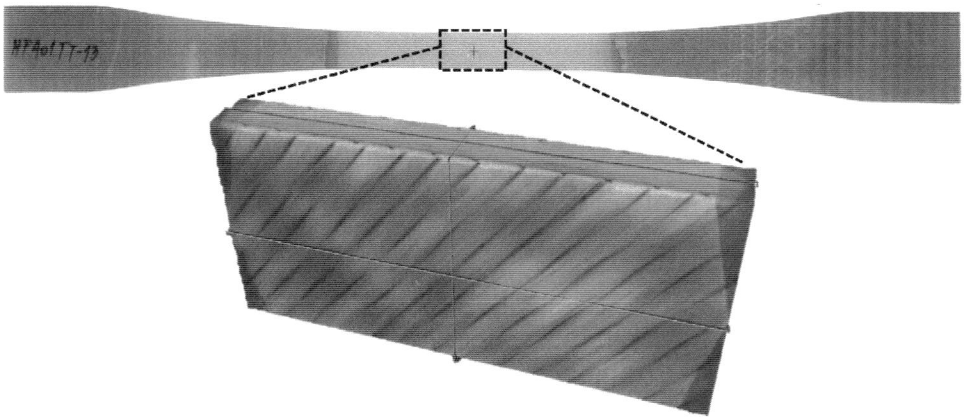

Figure 1. The test-sample and the scanned part of the gauge-section with the dimension $32 \times 15 \times 4.5\,\text{mm}^3$.

direction. The UD and biax layers in the fabric have an area weight of $1134\,\text{g/m}^2$ and $384\,\text{g/m}^2$, respectively, and consist of 17 µm diameter glass fibers. The backing layers, b_{UD} and b_{biax}, have an area weight of $36\,\text{g/m}^2$ and $4\,\text{g/m}^2$, respectively, with a glass fiber diameter of 9 µm. With a sample thickness of 4.4 mm in the scanned area and a glass fiber density on $\rho = 2600\,\text{kg/m}^3$ this corresponds to an average fiber volume fraction in the composite laminate of $V_f = 54\%$. The fraction of fibers in the different directions is calculated later in table 2. The scan settings and resulting voxel size are listed in table 1.

Table 1. X-ray CT scan information (stitched size: 32 mm×16.5 mm×16.5 mm)

Sample ID	Optical mag.	Voltage	Exposer time	Projec-tions	Bins; Stitch	Scan Time	FoV	Voxel Size
HF401TT-13	0.4×	50 keV	24 s	3201	1;3	3×24h	$3 \times 16.5\,\text{mm}^3$	8.08 µm

4. Procedure

We will now go through the steps needed to analyze the data using structure tensor in Python. With this paper, we release three Jupyter notebooks and a Python script file/module with helper functions. The first notebook, *StructureTensorFiberAnalysisDemo*, goes through the structure tensor analysis step by step, while the second, *StructureTensorFiberAnalysisAdvancedDemo*, offers a more advanced approach for speeding up computations on large volumes using multi-CPU or GPU systems. The third, *HF401TT-13_FoV16.5_Stitch*, can be used to recreate the results of the paper. The Python module file, *structure_tensor_workers.py*, contains various helper functions, including functions for parallel structure tensor computations across many CPUs and GPUs. The notebooks and helper functions can be obtained from [5].

Step 0: Pre-process data. Depending on available tools and experience it may be useful to pre-process data. This includes cropping and rotating data, as well as converting it to a suitable format for reading (see next step).

Step 1: Reading data. In our case the original data is stored in the TXM file format. To read this data in Python we can use the `dxchange` [8] package. However, we prefer either RAW format, which can be read directly with NumPy [9] `memmap`, or NIfTI, which can be read using the `nibabel` [10] package, due to the excellent functionality and performance of the `numpy` and `nibabel` packages. Another popular format is TIFF for which `scikit-image` and `tifffile` packages are useful. For large data-sets, using a memory-mapped file to access data is often more practical than reading all data into memory at once.

Step 2: Prepare data. Unless data has already been prepared before reading it, now is the time to crop, rotate, or otherwise transform the data. Depending on the size of the data-set and the amount of memory in our system, you can either keep all data in memory while doing this or use memory-mapped files. It is often convenient to save the prepared data to disk.

Step 2.5: Partition data. For large datasets, calculating the structure tensor for the full data-set at once is often infeasible due to hardware memory constraints. One way around this is to partition the data-set into blocks as shown in figure 2 and calculate the structure tensor, $\mathbf{S}_b$, for each block separately, before merging the results. Here b is the block index. This is not only more memory efficient but also allows for parallel computations. To ensure consistent results when calculating $\mathbf{S}$ in blocks, the blocks must be padded appropriately. The padding consists of actual voxels from the data-set, shown as the area between the inner and outer red box in figure 2(c). The values $\mathbf{S}_b$, which correspond to voxels located inside the padding, will not be correct and have to be discarded afterward. The size of the padding depends on the σ and ρ parameters, as they determine the size of the Gaussian filter kernels used to calculate $\mathbf{S}_b$. Because the kernel is discrete, the kernel radius is rounded to the nearest integer number. In our case the largest kernel radius is $\lfloor 4\rho \rceil = \lfloor 4 \cdot 2.96 \rceil = \lfloor 11.84 \rceil = 12$. Thus, for a value in $\mathbf{S}_b$ to be valid, the 12 neighboring voxels in all six directions must be included in the same block. Values in $\mathbf{S}_b$, where the block does not include the 12 neighboring voxels in each direction will vary depending on the partitioning of the blocks. Since $\mathbf{S}$ should not depend on the partitioning of blocks, these values have to be thrown away. The *structure_tensor_workers.py* file includes functions for creating and merging blocks (referred to as "crops" in the code) with proper padding. In our implementation,

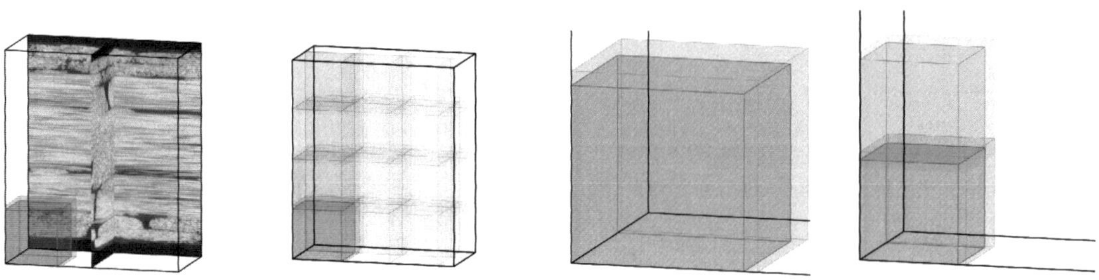

(a) Block 1 (b) All blocks (c) Block 1 (zoomed) (d) Block 1 & 2

Figure 2. We partition the volume (black box) into blocks (colored boxes) as shown in (a) and (b). The outer red box, more clearly visible in (c), represents block 1, which contains the data required to calculate the structure tensor for block 1, $\mathbf{S}_1$. The inner red box represents the voxels for which the values of $\mathbf{S}_1$ are valid. The overlapping of the blocks needed to calculate a valid $\mathbf{S}$ for the complete volume is more clearly visible in (d).

the user specifies the size of the blocks using the `crop_size` parameter. The maximum size of a block is a cube with sides of length `crop_size` plus the kernel radius. Of course, blocks may be smaller to ensure they fit inside the volume.

Step 3: Compute structure tensor and eigendecomposition. Using the `structure-tensor` package, we can calculate **S** for a volume using the `structure_tensor_3d` function. Although **S** is in theory a 3-by-3 matrix, because it is symmetric, we only need to store six values per voxel. As a result, given a volume with the shape (X, Y, Z), the array returned by `structure_tensor_3d` function has the shape (6, X, Y, Z). To get the eigenvalues and vectors, we pass **S** to the `eig_special_3d` function. By default, given an input with shape (6, ...), the values and vectors returned by `eig_special_3d` will have shape (3, ...), where (...) is an arbitrary shape. The values returned are the three eigenvalues in ascending order. For the vectors, only the vector for the smallest eigenvalue is returned by default. If the other two vectors are needed, setting `full=True` will return vectors as a (3, 3, ...) array instead where the first dimension corresponds to the vectors for each of the three eigenvalues.

In our experiments we choose σ and ρ as

$$\sigma = \frac{r_f}{\sqrt{2}} = 0.74 \qquad \rho = 4\sigma = 2.96, \tag{3}$$

where $r_f = 1.05$ is the UD fiber radius in voxels. Decreasing σ much further can lead to numerical instability due to truncation, numerical precision used in the implementation, and the discrete nature of the data. For ρ, we want to include a large enough area to determine structural "direction", while keeping the integration area small enough to avoid to much "bleeding" between structures (fibers/bundles).

For computing **S**, the `structure-tensor` package relies on the SciPy [11] `scipy.ndimage.gaussian_filter` function. To speed up computation using GPUs, the `structure-tensor` package uses the CuPy library [12], which implement GPU targeted versions of large parts of the NumPy and SciPy libraries.

Step 3.5: Saving results. Depending on the purpose of the analysis and workflow, it may be relevant to save the computed results. For instance, if we want to view the eigenvalues or vectors in another application, or if we want to save them for later analysis. Again, saving the data as NIfTI or RAW is simple with `niabel` or `numpy` and the formats are easily loaded in many applications.

Step 4: Orientation metrics. Since we focus on orientations, we will only be using the eigenvectors, $\mathbf{v}_1$. The eigenvalues have interesting use cases as well, but we will not discuss those in this paper.

Our analysis of orientations is based on the four directions, in which fiber bundles in our material are oriented. We represent each of the four directions as a class with a corresponding class unit vector, $\mathbf{c}_o$, where o is the counter-clockwise rotation of the bundle in the xy-plane from the x-axis in degrees. The four class vectors are $\mathbf{c}_0 = [1, 0, 0]$, $\mathbf{c}_{45} = [0.707, 0.707, 0]$, $\mathbf{c}_{-45} = [0.707, -0.707, 0]$ and $\mathbf{c}_{90} = [0, 1, 0]$. The class vectors can and should be specified to match material composition. At least one class vector must be set.

The first class vector is considered the primary vector, which means it is the direction in which the orientations are calculated relative to. The remaining class vectors are used solely to label the voxels, also known as segmentation. In our notebooks we use the `calculate_angles` function from the `structure_tensor_workers` module to get a label (class), stiffness (η_o), angle (θ), rotation in xy-plane and rotation out of the xy-plane for each eigenvector in $\mathbf{v}_1$.

The `calculate_angles` function works by first determining the class of each vector. Since each vector belongs to a specific voxel in the volume this results in a complete segmentation of the volume. The segmentation is done by calculating the angle between the eigenvectors, $\mathbf{v}_1$, and each class vector, $\mathbf{c}_o$. A vector is assigned the class of the class vector with which is it most aligned (has the smallest angle). The angle between $\mathbf{v}_1$ and the primary class vector, in our case $\mathbf{c}_0$, is the value θ, which is between 0 and 90 degrees. Based on θ we calculate the stiffness estimate, η_o, which is between 0 and 1. The last two metrics are the orientations in and out of the xy-plane. Both are between -90 and 90 degrees.

Step 5: Choosing fiber threshold. One issue with the segmentation from step 4 is that it labels all voxels as belonging to one of the fiber classes. However, the volume is not only fiber material but also includes epoxy, which have a lower density than the glass fiber but higher density than air. Segmenting epoxy (or air) into one of the four orientation classes or calculating the orientation of epoxy voxels does not make sense. Therefore, we create a background class and use a simple intensity threshold to separate the foreground (fibers) and background (non-fiber). Using a histogram, it is fairly easy to choose a sensible threshold value. Another option is to use a statistical method, such as Otsu's threshold, which we also demonstrate in our notebooks. When choosing our threshold, we make sure not to include values outside the sample or near the edge in our histogram. This is because we want to separate foreground and background inside the sample and intensities outside the sample may be significantly different.

Step 6: Plotting segmentation and orientations. Using the results of steps 5 and 6, we can plot segmentation and orientations on top of the data to verify the correctness of our calculations. Here, choosing appropriate color maps for orientations is important for meaningful figures. We demonstrate this in both our notebooks.

Step 7: Orientation distributions. After qualitatively assessing our results by plotting, we calculate the class fractions, which tell us how much of the fiber material belongs to each class. This is easily comparable to the material specifications. Afterward, we can use the segmentation labels to create histograms for each of the orientation metrics for each class, as well as combined for all classes. Along with the histograms, we can also calculate variables such as the median, mean, and standard deviation for the distributions. However, it is important to note that the orientation metrics are *not* normally distributed, so we have to be careful with treating them as such. A more appropriate distribution for describing orientation data would probably be the Bingham distribution [13], but that is outside the scope of this paper.

5. Results

The results shown here are also available in the *HF401TT-13_FoV16.5_Stitch* Jupyter notebook and can be reproduced by running the code in the notebook [5].

The composition of the fabric is critical to its mechanical properties. Table 2 show the expected fiber distribution across the four classes, as well as the segmented distribution in the scanned region, estimated using the previously described procedure. The expected fiber distribution is based on the reported fiber distribution for the individual non-crimp fabric used for the composite. The segmented fraction estimates are less than 2% from the expected fractions, indicating both that the material follows the specification closely and that the segmentation is accurate.

To qualitatively validate our results, we can look at the segmentation results for the analyzed region of the sample, shown in figure 3. We see that the segmentation appears accurate, although some voxels near the class interfaces may be mislabeled. This is not surprising, as orientations

Table 2. Volumetric fractions of the fiber orientations

$[\mathrm{b/biax, b/UD, b/UD}]_s$	0°	45°	−45°	90°
Area weight $[g/m^2]$	4×1139	2×384	2×384	4×36+2×4
Calculated fractions	0.729	0.123	0.123	0.024
Segmented fractions	0.746	0.122	0.116	0.016

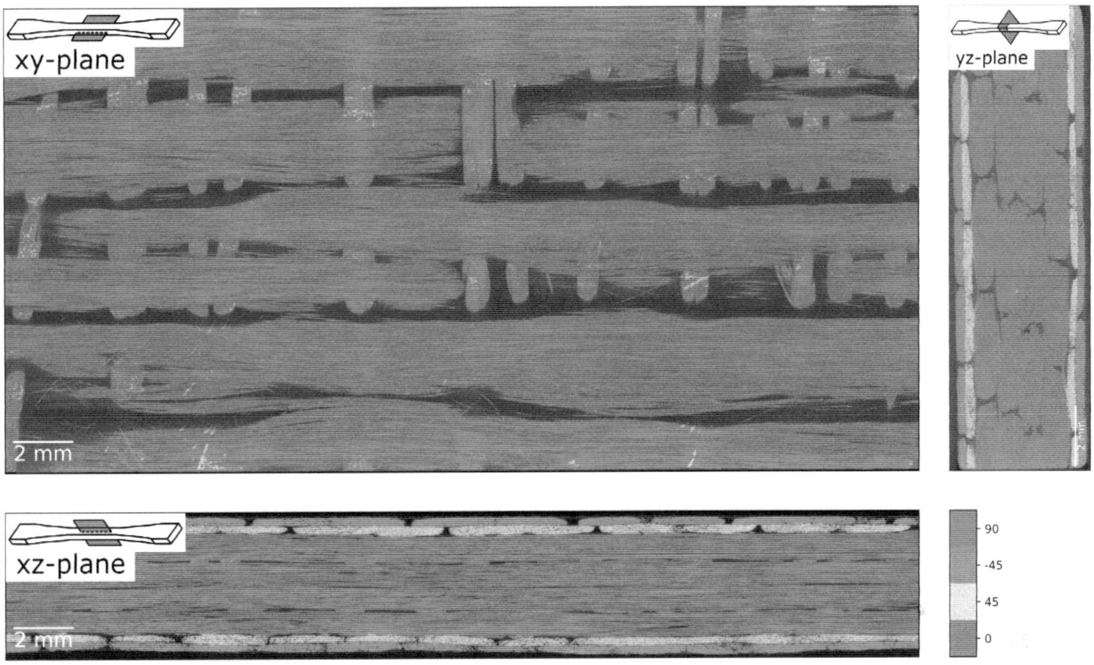

Figure 3. Segmentation based on fiber orientations of the non-crimp fabric shown in the center of the sample in the xz and yz-plane and in the xy-plane in a plane with backing bundles. The four different fiber classes are represented by four different colors, while the background is gray. The segmentation is shown on top of the original grayscale images.

near the interfaces are integrated over data from both sides of the interface. For instance, this could cause a voxel near the 0°/90°-interface to be classified as −45° or 45°. We see this in the figure 3 xy-plane image as blue or green "bleeding" between the red and pink areas. One way to correct for this is to remove/reassign small components, as done in [2]. However, as the number of mislabeled voxels is relatively small, it should not affect the orientation distributions much. One exception could be the least represented class, 90°, where the mislabel voxels near the interfaces could be contributing to the estimated fraction (1.6%) being smaller than the expected fraction (2.4%). In general, we suspect that the bleeding effect, which depends on the integration scale, ρ, could result in slightly underestimated fractions for the less frequent classes, which is exactly what we see. However, we do not know at this point, if this is actually the case here. Either way, the estimated fractions and visual inspection of segmented slices makes us quite confident in the accuracy of our segmentation and estimated orientations.

To distinguish the matrix-rich regions without a material orientation from the fiber-rich

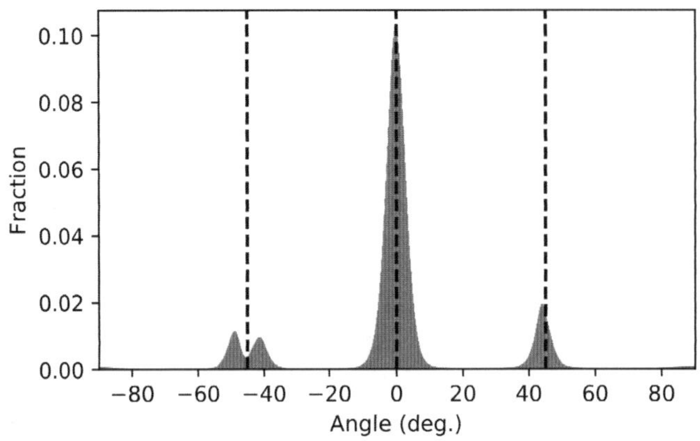

Figure 4. Distribution of the fiber orientations projected onto the xy-plane.

regions with a material orientation, the fiber regions are segmented using a threshold value corresponding to the non-black regions in figure 3. In our experiments we use a histogram to choose a reasonable threshold value. In theory, we can use the threshold to calculate the fiber volume fraction, V_f. However, with a voxel size of $8\,\mu m^3$ and a fiber diameter of $17\,\mu m$ for the dominating (UD) fibers, this is unlikely to give a good estimate of V_f. With this resolution, we cannot represent the small pockets of epoxy between the fibers inside the bundles, which means the majority of the voxels inside the bundles are counted as fiber material. Thus, in our experiment, the threshold value results in $V_f = 75\%$, which is much higher than the specified 54% for the material. We could increase the threshold value to get closer to the expected V_f, however inspecting the intensity distribution and the segmentation qualitatively, this does not seem like the right solution. For instance, a higher threshold value appears to favor the thicker UD fibers, thereby over-representing the 0° class. Although V_f is not accurate, we should still be able to trust the estimated ratio between the different fiber orientation classes, as long as the fiber volume fraction inside the bundles is approximately the same for all the orientation classes.

To quantitatively evaluate the fabric and bundles (fiber rowings) used for building the fabric, we can plot the in-plane fiber orientation distributions. Figure 4 shows the distribution orientations of the fibers projected onto the xy-plane for all fibers in the analyzed sample. In figure 5 the same distribution has been separated for the four classes: 0°, −45°, 45°, and 90°. From figures 5(b) and 5(d) we see that both 0° and 45° fibers appear to align with the expected orientation, with means being off by less than a degree and standard deviations of less than four degrees. However, the −45° class is another story. It appears that the fibers in this class follow two different distributions, shifted around four degrees in opposite directions.

To investigate the −45° class further, we have separated the orientation data for the two −45° surface biax-plies in the sample. Figure 6 clearly shows that the two plies correspond to the two peaks in 5(c). The two plies appear to be rotated about four degrees in opposite directions. It can, therefore, be concluded that there is a mismatch between the expected $+45°/-45°$ biax layup and the actual $+45°/(-45{\pm}4)°$. This difference could either come from the manufacturing of the non-crimp stitched fabric or have been introduced by shearing the fabric during the layup.

Figure 7 shows the out-of-plane fiber orientations for the xy-plane. Here, all classes should have the same 0° orientation, which also appears to be the case, with all means being within one degree of the expected orientation. Again the UD (0°) fibers appear to align almost perfectly, while the biax-ply fibers diverge slightly more. We also note that the −45° and 45° classes

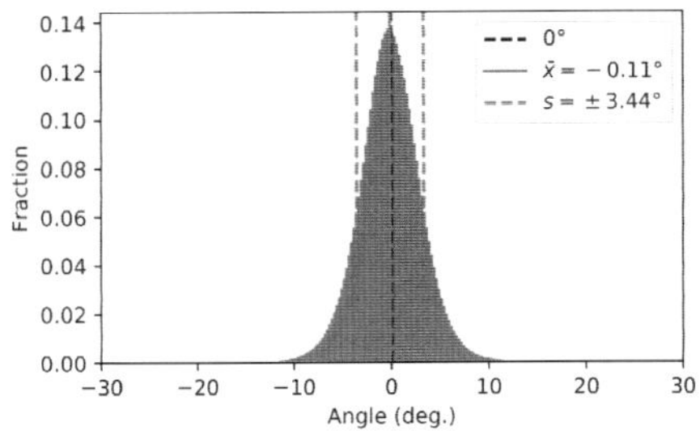

(a) Distribution of 0°

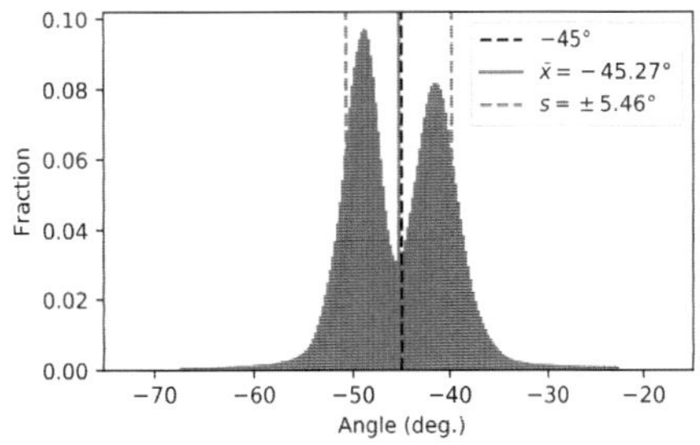

(b) Distribution of −45°

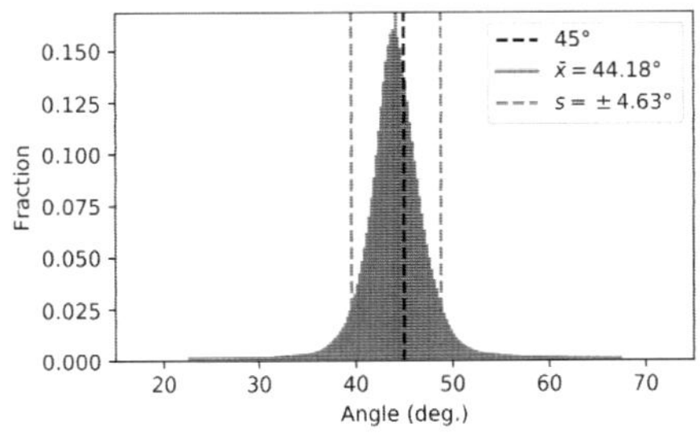

(c) Distribution of +45°

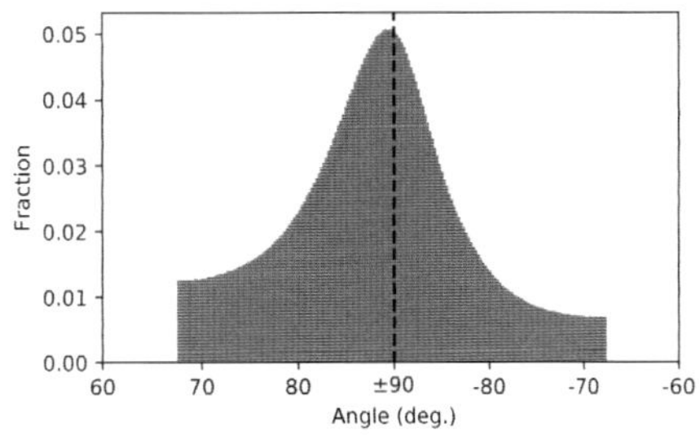

(d) Distribution of 90°

Figure 5. Distribution of the in-plane fiber orientations for the xy-plane. The angle is the orientation of the eigenvectors projected onto the xy-plane. (a), (b), (c) and (d) shows the distribution for the four classes, 0°, −45°, 45°, and 90°, with corresponding class vectors, $\mathbf{c}_0$, $\mathbf{c}_{-45}$, $\mathbf{c}_{45}$ and $\mathbf{c}_{90}$. For the first three distribution the mean, $\bar{x}$, and standard deviation, s, are also included.

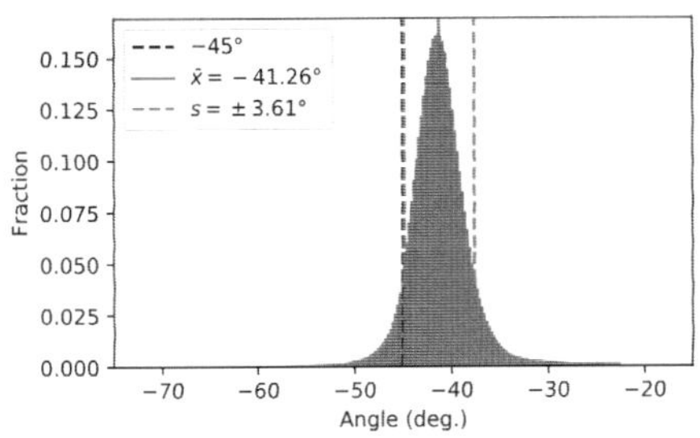

(a) Distribution of the upper ply

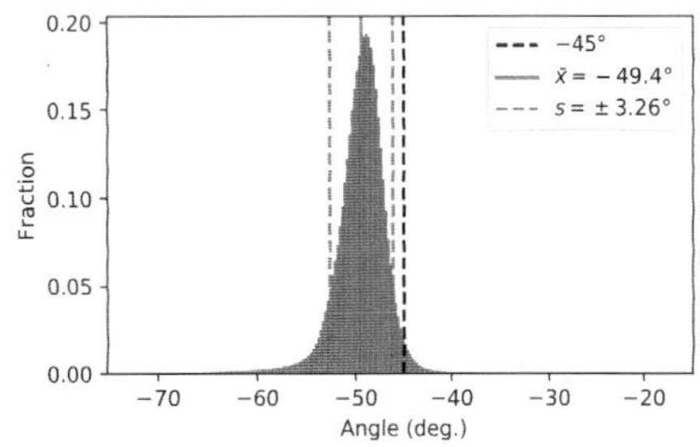

(b) Distribution of the lower ply

Figure 6. Distribution of the in-plane fiber orientations for the xy-plane for the two −45° plies in the sample. Together these should be almost equal to the distribution in 5(c).

IOP Conf. Series: Materials Science and Engineering **942** (2020) 012037 doi:10.1088/1757-899X/942/1/012037

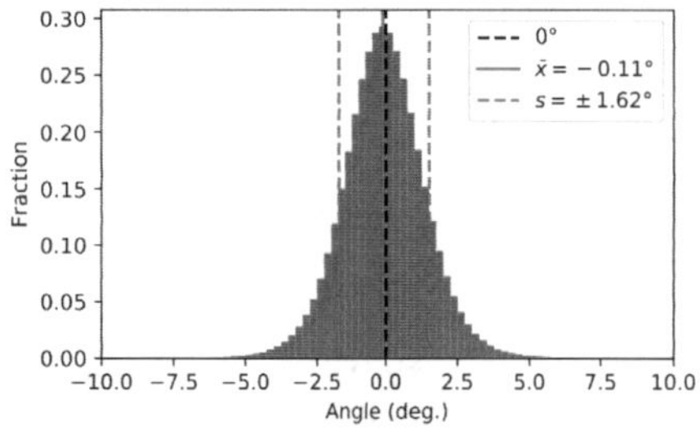

(a) Distribution of 0°

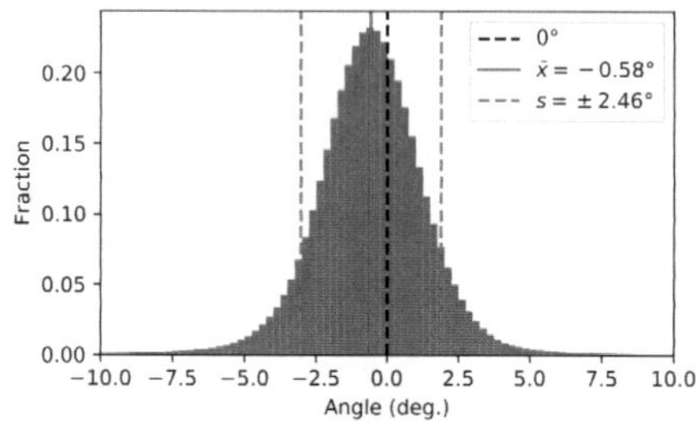

(b) Distribution of −45°

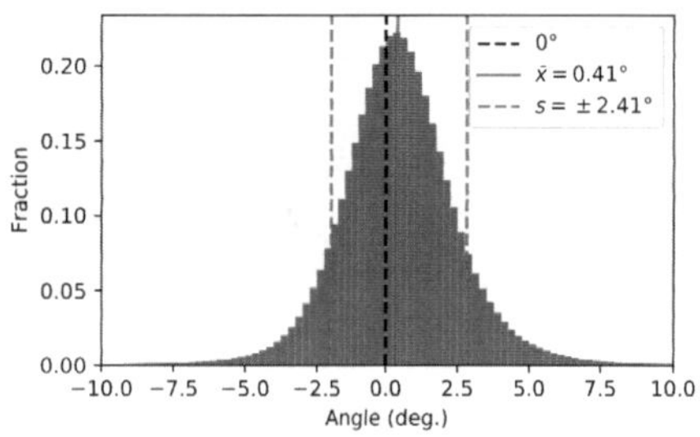

(c) Distribution of 45°

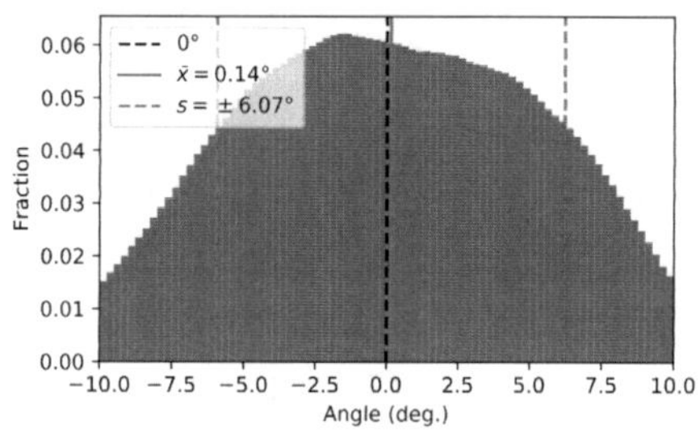

(d) Distribution of 90°

Figure 7. Distribution of the out-of-plane fiber orientations for the xy-plane. This is the angles between the eigenvectors and the xy-plane. (a), (b), (c) and (d) shows the distribution for each of the classes. For each distribution the mean, $\bar{x}$, and standard deviation, s, are also included.

appear to be oriented about half a degree in opposite directions. However, this small deviation could be an artifact from the structure tensor analysis, rather than the fibers actually being misoriented. While the 90° class is also oriented as expected, the orientation distribution for the class has a large variance. The most likely explanation is simply that it is by far the least represented class and thus the one most affected by noise and "bleeding" artifacts. Introducing a fifth class containing fibers, which fit poorly into all the four known classes could remove some of the noise from the distribution. Trying different values for ρ and σ could also help to reduce both noise and bleeding.

For a given load direction, the fiber orientation distribution can be quantified with a so-called fiber orientation efficiency factor [14],

$$\eta_o = \sum_i a_i \cos^4 \theta_i \tag{4}$$

where θ_i is the orientation with respect to the loading direction of the i-th fraction a_i. The loading direction is here given as the x-axis. A value of $\eta_o = 1$ corresponds to an ideal uni-directional composite with all fibers in the loading direction while $\eta_o = 0$ corresponds to a composite with all fibers orthogonal to the loading direction. For our sample, the fiber efficiency factor for fibers belonging to the 0° class is found to be $\eta_o(0°) = 0.991$, while the overall efficiency factor, including all fibers for all four classes, is $\eta_o = 0.802$.

6. Conclusion

We have shown that structure tensor analysis can be used to acquire important information about fiber orientations from micro-CT data of glass fiber fabrics. We have described how to do this step by step and included three Jupyter notebooks and a Python module demonstrating the procedure.

The results show that structure tensor information can be used for volume segmentation, as well as to extract important metrics, such as fiber class fractions and orientations. Despite relativity low resolution with a voxel-size only half the dominating fiber diameter, a good agreement is obtained with the expected volumetric fractions for the four different orientation classes. In addition, the method has shown its value as a non-destructive quality control procedure, in this case, making it possible to identify the $\pm 4^\circ$ misorientation of the -45° class. Furthermore, the method can produce orientation distributions for each class, which amongst other things, can be used to calculate the fiber orientation efficiency factor. Such mechanical properties are valuable for quality control, and the scalability of structure tensor on modern hardware allows acquisition of these properties to be done quickly, even for large samples. In this paper, we have studied a glass fiber-based composite, but the method should work equally well for other fiber materials.

The presented segmentation method is simple and works well for cases, where the fiber classes are known beforehand. That said, there are many ways to improve the accuracy of the orientation distributions and segmentation. Firstly, methods for noise reduction, normalization, and image sharpening could enhance the data quality. This could make both threshold-based segmentation and estimated orientations more accurate. Secondly, using a connected component approach to eliminate isolated falsely classified voxels, as in [2], and/or introducing another class for "noisy" fiber voxels, could improve segmentation accuracy. However, we show that a very simple segmentation method can produce sufficiently accurate results when combined with structure tensor analysis.

Acknowledgments

N Jeppesen's work is supported by FORCE Technology. The material system used and L P Mikkelsen's work is supported by the Danish Energy Agency through the Energy Technology Development and Demonstration Program (EUDP), grant no. 64018-0068. The supported project is RELIABLADE: Improving Blade Reliability through Application of Digital Twins over Entire Life Cycle. This work is partly supported by The Center for Quantification of Imaging Data from MAX IV (QIM) funded by The Capital Region of Denmark.

References

[1] Emerson M J, Jespersen K M, Dahl A B, Conradsen K and Mikkelsen L P 2017 Individual fibre segmentation from 3D X-ray computed tomography to study the misalignment in unidirectional composite materials *Compos. Part A Appl. Sci. Manuf.* **97** 83–92 ISSN 1359835X URL `http://dx.doi.org/10.1016/j.compositesa.2016.12.028`

[2] Straumit I, Lomov S V and Wevers M 2015 Quantification of the internal structure and automatic generation of voxel models of textile composites from X-ray computed tomography data *Compos. Part A Appl. Sci. Manuf.* **69** 150–58 ISSN 1359835X URL `http://dx.doi.org/10.1016/j.compositesa.2014.11.016`

[3] Straumit I, Hahn C, Winterstein E, Plank B, Lomov S V and Wevers M 2016 Computation of permeability of a non-crimp carbon textile reinforcement based on X-ray computed tomography images *Compos. Part A Appl. Sci. Manuf.* **81** 289–95 ISSN 1359835X URL `http://dx.doi.org/10.1016/j.compositesa.2015.11.025`

[4] Nguyen N Q, Mehdikhani M, Straumit I, Gorbatikh L, Lessard L and Lomov S V 2018 Micro-CT measurement of fibre misalignment: Application to carbon/epoxy laminates manufactured in autoclave and by vacuum assisted resin transfer moulding *Compos. Part A Appl. Sci. Manuf.* **104** 14–23 ISSN 1359835X URL `https://doi.org/10.1016/j.compositesa.2017.10.018`

[5] Jeppesen N, Dahl V A, Christensen A N, Dahl A B and Mikkelsen L P 2020 Characterization of the fiber orientations in non-crimp glass fiber reinforced composites using structure tensor [data set] *Zenodo* URL `http://dx.doi.org/10.5281/zenodo.3877522`

[6] Weickert J 1998 *Anisotropic Diffusion in Image Processing* (Teubner Stuttgart) URL `https://www.mia.uni-saarland.de/weickert/Papers/book.pdf`

[7] Jespersen K M and Mikkelsen L P 2017 Three dimensional fatigue damage evolution in non-crimp glass fibre fabric based composites used for wind turbine blades *Compos Sci Technol* **153** 261–72 ISSN 02663538 URL http://linkinghub.elsevier.com/retrieve/pii/S0266353817301811
[8] De Carlo F *et al.* 2014 Scientific data exchange: a schema for HDF5-based storage of raw and analyzed data *J. Synchrotron Radiat.* **21** 1224–30 URL https://doi.org/10.1107/S160057751401604X
[9] van der Walt S, Colbert S C and Varoquaux G 2011 The numpy array: A structure for efficient numerical computation *Comput Sci Eng* **13** 22–30
[10] Brett M *et al.* 2020 nipy/nibabel: 3.1.0 URL https://doi.org/10.5281/zenodo.3757992
[11] Virtanen P *et al.* 2020 SciPy 1.0: Fundamental Algorithms for Scientific Computing in Python *Nat. Methods* **17** 261–72
[12] Okuta R, Unno Y, Nishino D, Hido S and Loomis C 2017 *Proceedings of Workshop on Machine Learning Systems (LearningSys) in The Thirty-first Annual Conference on Neural Information Processing Systems (NIPS)* URL http://learningsys.org/nips17/assets/papers/paper_16.pdf
[13] Bingham C 1974 An antipodally symmetric distribution on the sphere *Ann. Stat.* 1201–25
[14] Krenchel H 1964 *Fibre reinforcement. Theoretical and practical investigations of the elasticity and strength of fibre-reinforced materials* (Akademisk Forlag)

IOP Conf. Series: Materials Science and Engineering **942** (2020) 012038 doi:10.1088/1757-899X/942/1/012038

X-ray tomography based numerical analysis of stress concentrations in non-crimp fabric reinforced composites - assessment of segmentation methods

R M Auenhammer[1,2,*], L P Mikkelsen[2], L E Asp[1] and B J Blinzler[1]

[1]Department of Industrial and Materials Science, Chalmers University of Technology, Hörsalsvägen 7A, SE-41296 Göteborg, Sweden

[2]Department of Wind Energy, Technical University of Denmark, Roskilde, Denmark

E-mail: *robaue@chalmers.se

Abstract. In this study two automated segmentation methodologies of an X-ray computer tomography based numerical analysis are compared. These are then assessed based on their influence on the stress distribution results of finite element models of glass fibre reinforced composites made out of non-crimp fabrics. Non-crimp fabrics reinforced composites are commonly used for wind turbine blades due to their high stiffness to weight ratio for the dominating bending load. Finite element modelling based on X-ray computer tomography allows the reduction of the cost and can accelerate the development process of the key material parameters of wind turbine blades. Recent research progress in the last years has laid the basis for such a procedure. Those processes must be easy applicable, fast and accurate. The main challenge in current methodologies is the segmentation part. The segmentation methods applied for this study have overcome this issue by being automated. This allows for a comparatively fast transfer from X-ray computer tomographic data to finite element results.

1. Introduction

High performance polymer matrix composites are the standard materials used for wind turbine blades due to their durability and the high stiffness to weight ratio. A good fatigue behaviour is especially essential due to the huge amount of load cycles (between 50 and 300 million) [1]. In addition, with increasing blade length, higher tensile stiffness and compression strength are also necessary [2]. The improvement of the composite structure is expensive since it is mostly based on experiments. Finite element modelling based on X-ray computer tomographic data allows to account for real material behaviour

IOP Conf. Series: Materials Science and Engineering **942** (2020) 012038 doi:10.1088/1757-899X/942/1/012038

after manufacturing of the composite structure. Misaligned and displaced fibres, fibres bundles as well as void formations can be implemented in the model [3].

1.1. X-ray computer tomography

Developed originally for the medical diagnostics field, X-ray computer tomography has expanded its application field to non-medical research and industry in the last decades [4]. This technology allows to create images of the inside of the scanned sample while being non-destructive [5]. The resolutions of modern tomographs can visualise single fibres [6]. X-ray computer tomography scans of composites can be used to assess manufacturing processes, tensile failure, compressive failure and kinking, fatigue as well as impact damage [7].

1.2. X-ray computer tomography image based modelling

X-ray computer tomography image based modelling can improve the quality of finite element analysis [8]. Recent improvements in the images acquisition and analysis allow enhanced applications for evaluating material models and structural improvements of fibre reinforced composites [9]. Increased usage of finite element modelling will reduce the development and production cost for industrial applications. Automated segmentation is in particular important since huge amount of data (in the range up to dozens of Gigabytes), that is generated during X-ray computer tomography, needs to be processed to serve as basis for finite element modelling. The volume of data creates the necessity for automated algorithms which allow accurate, repeatable and cost and time efficient processing. An implementation of the fibre orientation is fundamental for modelling of fibre reinforced composites. X-ray computer tomography can additionally supply the fibre orientation as well. Karamov et al. [10] compared two high-fidelity methods, fibre tracking with the commercial software package Avizo™ and ellipsometry, with a structural tensor approach. Depending on the scanning resolution, high accuracy of the calculated fibre orientation was achieved for all methods in their investigation. The study of Jespersen et al. [11] on the other hand did not include fibre tracking. They focused more on the generation of surface geometries of the single fibre bundles with the application of the commercial software packages Avizo™ and Geomagic Wrap®. The interpolation between single 2D cross-sections resulted in a 3D volume. After the mesh creation, the orientations of the lay-up were used to calculate the material parameters with the Classical-Laminate-Theory. Blinzler et al. [12] conducted a similar study. In contrast to Avizo™ and Geomagic Wrap® they used Solidworks™ to create a 3D surface geometry. With the help of a generated centroid spline for the 3D fibre bundles, the waviness could be accounted for, which is advantageous compared to the study of Jespersen et al. [11]. Both studies however, are manual, time intensive, and depend on the experience of the operator.

1.3. Objective

The goal of this investigation is to assess the influence of the automated segmentation procedures for non-crimp fabrics reinforced composites on the fibre bundle scale on the finite element analysis results. Two methods, called *Physical Boundary* and *Thresholding* in the following, differing in the representation of the boundary between fibre bundles will be compared in their ability to model stress concentrations.

2. Approach

Glass fibre reinforced composites are used for the highly automated X-ray computer tomography aided process. The fundamental parts for such a process are image acquisition, segmentation and the finite element modelling.

2.1. Material system

Non-crimp fabric glass fibre reinforced composites were manufactured with the architecture Jespersen et al. described in [13]. Four uni-directional bundle layers are placed on each other and each layer is supported by a layer of backing bundles. The composite was manufactured with the vacuum assisted resin transfer moulding technique. An unsaturated polyester was used as the resin. The composite layup with $[b/0, b/0]_s$ was used where the term "b/0" denotes a uni-directional bundles layer oriented at 0° on top of a layer of backing bundles oriented at ±80° (Figure 1).

For the material properties in the finite element analysis, a linear transverse orthotropic

Figure 1: Rendered reconstructed X-ray computer tomographic data of the specimen with a size of 376 × 936 × 1708 pixels and a voxel size of 10.94 μm displayed in the commercial software package Avizo™ reflecting a physical size of $4.11 \times 10.2 \times 18.7\ mm^3$.

model is chosen. A precis value of the fibre volume fraction is essential for predicting the material parameters. A 2D scanning electron microscopy image of the centre plane of the cross-section was taken of the specimen after it was scanned in the X-ray computer tomograph. The fibre volume fraction was determined based on a image analysis using a threshold operation and taking the ratio of the high intensity glass-fibre pixels to the total amount of pixels in fibre bundle part of the cross-section. The material parameters were calculated based on the fibre volume fraction of 61 % inside the fibre bundles with the

IOP Conf. Series: Materials Science and Engineering **942** (2020) 012038 doi:10.1088/1757-899X/942/1/012038

rule of mixtures for the stiffness along fibre direction and Halpin-Tsai for the transverse and shear properties [14]. In Table 1 the values normalised with the longitudinal stiffness can be seen.

Table 1: Calculated material parameters with the Rule of Mixture and Halpin-Tsai normalised with the value of the longitudinal stiffness E_{11}.

E_{11} [-]	E_{22} [-]	E_{33} [-]	G_{12} [-]	G_{13} [-]	G_{23} [-]	ν_{12} [-]	ν_{13} [-]	ν_{23} [-]
1.00	0.265	0.265	0.0775	0.0775	0.0575	0.258	0.258	0.254

2.2. Image acquisition
The central part of the gauge section of a tensile test specimen was scanned at DTU Wind Energy with a 3D X-ray computer tomograph Zeiss Xradia Versa 520 scanner, that has a 2000 × 2000 pixel detector. A so-called binning of 2 was used. With this setting 2 × 2 pixels are merged to one pixel. This results in a shorter scanning time but also signifies a lower resolution. Three single scans were stitched together to cover a larger field of view. Further, with the exposure time 4 s, the number of projections 3601, the accelerating voltage 50 kV, the optical magnification 0.4×, the voxel size 10.94 μm resulted in a scanning time of approximately 22 h. The voxel size with 10.94 μm is too coarse to resolve single fibres with diameters of 9 and 17 μm. The scanned field of view covers the entire width and thickness of the tensile test specimen. In the main fibre direction, the sample needed to be cropped to reduce the scanning and modelling time. The result is visualised in Figure 1.

2.3. Segmentation method
To investigate the influence of the segmentation method two automated approaches were used. They differ in their complexity and how the boundaries between uni-directional and backing bundles are modelled. Yet, both are comparatively simple considering the goal they need to fulfil. To assure that the segmentation outcomes can be used for accurate finite element modelling, a fibre orientation must be either provided or the assignment possibility must be given. Fundamental to achieving this objective in our approach is a fibre tracking, which was carried out in Avizo™. Based on the correlation of the image data with cylinder templates fibre centre lines are generated [15]. Those fibre centre lines are exported as a comma separated value file and the orientation is later mapped on the finite elements. This material orientation mapping allows for one of the simplest segmentation methods, the so called thresholding. Thresholding assigns pixels/voxels based on their grey-scale values to different groups. In this case fibres and matrix can be separated with thresholding. Combining thresholding and fibre tracking creates a fast, automated and very accurate method serving as basis for finite element modelling. This method is called *Thresholding* in this study (Figure 2a). Applying the method *Thresholding* for finite element modelling will deliver accurate stiffness results

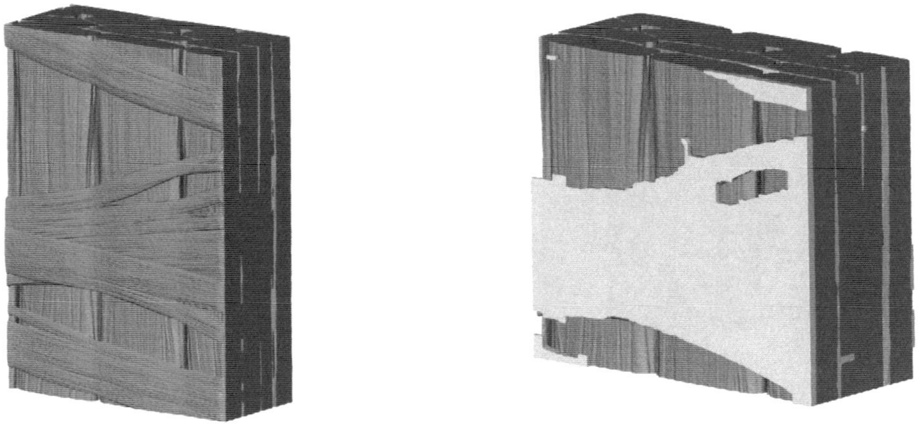

(a) *Thresholding* (b) *Physical Boundary*

Figure 2: Automatically segmented surfaces of the methods *Thresholding* and *Physical Boundary* with quasi uni-directional (blue) and backing bundles (yellow) depicted in the commercial software package Avizo™

but can create the drawback that no physical boundary between quasi uni-directional and backing bundles is generated. The resulting stress state at the interface between quasi uni-directional and backing bundles might not be resolved correctly. In order to overcome this issue, advanced morphological operators are applied on the fibre centre lines. With the Avizo™ operators *Dilation*, *Closing* and *Subtract Image* independent backing bundle surfaces are created. This method is called *Physical Boundary* in this study (Figure 2b). Since the operations are not only based on the image data, they can adulterate the actual volume of the uni-directional and backing bundles. The influences of those segmentation choices will be discussed later on in the results of the finite element analysis.

2.4. Modelling

The surface data of both automatically generated segmentations are imported in ANSA™. In this pre-processor mesh quality issues and intersections are removed and a volume mesh of the bundles is created. The meshing of the *Physical Boundary* method takes longer, since penetration issues arise between uni-directional and backing bundles. This segmentation method separates bundles which are very close together. In order to be able to fit in solid elements for the matrix between those bundles a shrinkage of the backing bundles is required. The solid mesh for the matrix surrounding the bundles is generated with the ANSA™ function "Include solid facets in detection". The obtained fibre orientation from AVIZO™ is mapped element-wise onto the solid elements. This ensures accurate stiffness and stress results since the ratio between longitudinal and transverse stiffness with approximately four (Table 1) generates a significant influences of even small misalignments. Figure 3 shows a comparison of the final models.

IOP Conf. Series: Materials Science and Engineering **942** (2020) 012038 doi:10.1088/1757-899X/942/1/012038

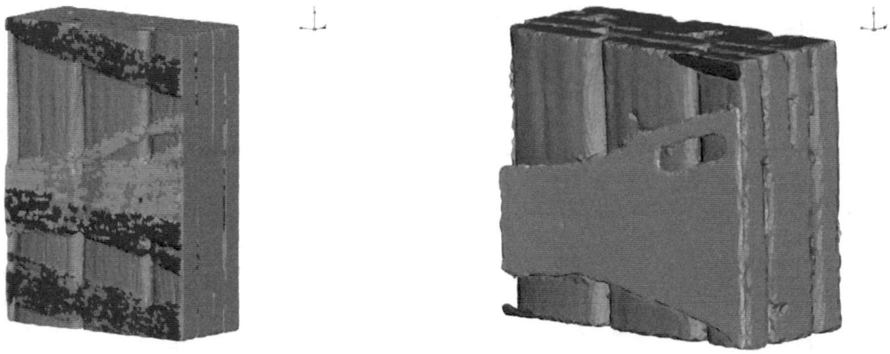

(a) *Thresholding* (b) *Physical Boundary*

Figure 3: Meshed models based on the *Thresholding* and *Physical Boundary* segmentation methods with quasi uni-directional (marine-blue) and backing bundles (yellow and brown) depicted in the commercial software package ANSA™. The matrix mesh is masked.

3. Results

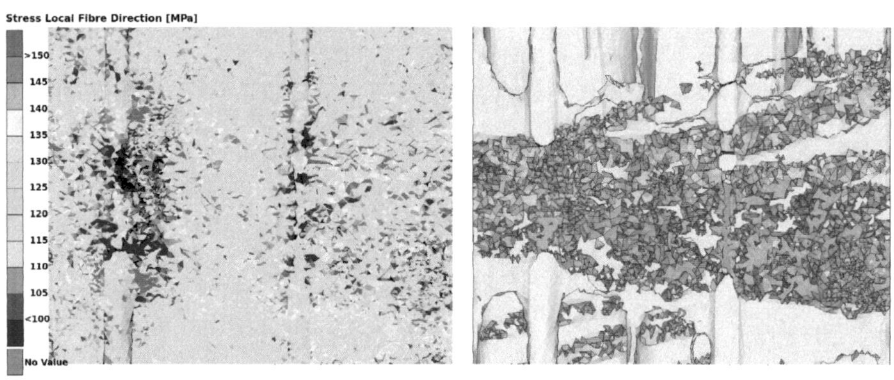

Figure 4: Plane cut of the *Thresholding* model through the volume showing the stresses in fibre direction inside the uni-directional bundles on the left and the position of the overlapping backing bundles on the right.

Both models (*Physical Boundary* and *Thresholding*) were loaded in the main fibre direction with a strain of 0.25 %. Regarding the obtained stress in fibre direction, it becomes clear that there are stress localisations, where backing bundles are crossing uni-directional bundles, depicted in Figure 4. This effect is occurring at all locations where backing bundles and uni-directional bundles are overlaying. Jespersen et al. [16] showed in their study that in this area the fatigue damage process of tension-tension

IOP Conf. Series: Materials Science and Engineering **942** (2020) 012038 doi:10.1088/1757-899X/942/1/012038

loaded specimens with a similar material system is originating. They assumed stress concentrations to be the reason for this initial fibre failure. Their theory can be supported by these simulation results. Similar finite element analysis by Jespersen et al. [11] and Blinzler et al. [12] have also shown stress localisations in the overlapping area.

Both models (*Physical Boundary* and *Thresholding*) are able to predict these stress concentrations. There is not significant difference in the stress distribution visible (Figure 5), though areas with higher stresses are more spread in the *Thresholding* model. The

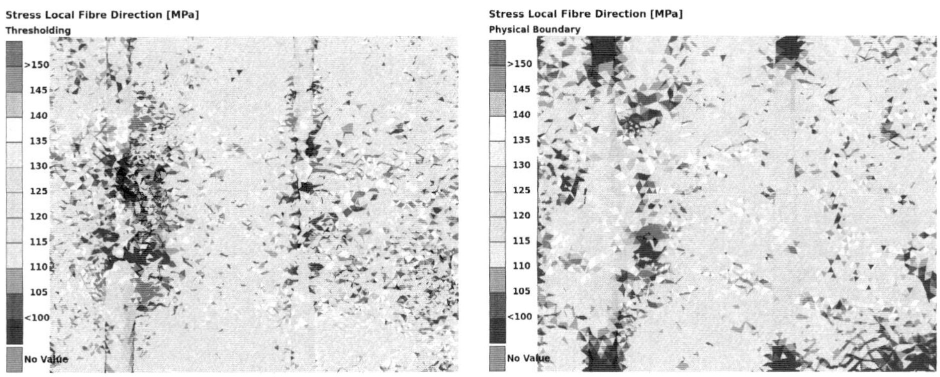

Figure 5: Plane cut through the volume showing the stresses in fibre direction inside the uni-directional bundles. On the left the model based on the segmentation method *Thresholding* is shown, the model on the right bases on the method *Physical Boundary*.

model with the physical boundary representation allows for a smother surface mesh, resulting in a more uniform stress distribution. The model is not able to resolve an effect on the single fibre scale with an average mesh size of 120 μm.

Nevertheless, even on larger scales, stress concentrations are occurring. The stiffness differences between matrix, backing bundles and uni-directional bundles could cause those stress concentrations. By assigning matrix material for the backing bundles in the model based on the segmentation method *Thresholding*, this influence was examined. In other words, the fibres have been virtually removed from the backing bundles. The result is shown in Figure 6. The stress concentrations remain, they are even more severe at this position in the composite. At the other positions of stress concentrations through the volume nearly no differences between the model with backing bundles and the model without backing bundles become visible. By this analysis it becomes evident that the stiffness properties of the backing bundles have no significant influence on the stress localisations inside the uni-directional bundles.

In order to exclude the option, that the observed stress localisations are induced by the mesh, the backing bundles were deleted in the model *Physical Boundary* and the accrued volumes were combined and remeshed as solid matrix elements. The results are depicted in Figure 7. The effect seen in Figure 6 where the concentrated stresses are more spread are visible in this analysis as well. This signifies that, even without the presence of backing bundles in the finite element model, stress concentrations occur.

IOP Conf. Series: Materials Science and Engineering **942** (2020) 012038 doi:10.1088/1757-899X/942/1/012038

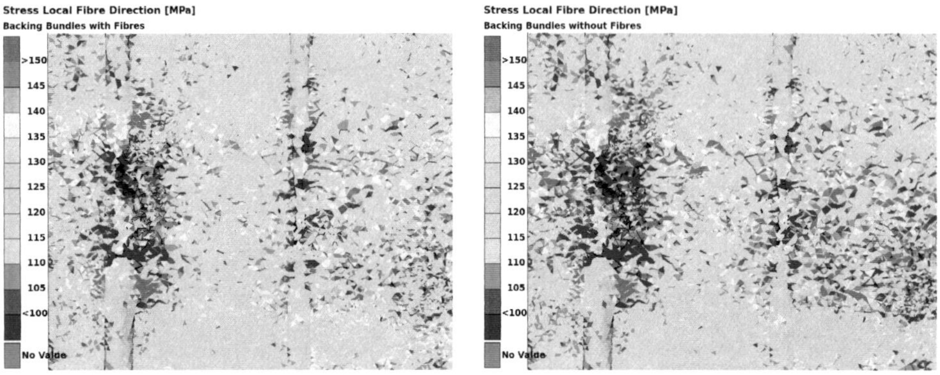

Figure 6: Plane cut through the volume showing the stresses in fibre direction inside the uni-directional bundles. Both models base on the segmentation method *Thresholding*. For the model on the right the material was changed to the matrix material, the model on the left shows the result of the original model with the correct backing bundle stiffness properties (Figure 5 left model).

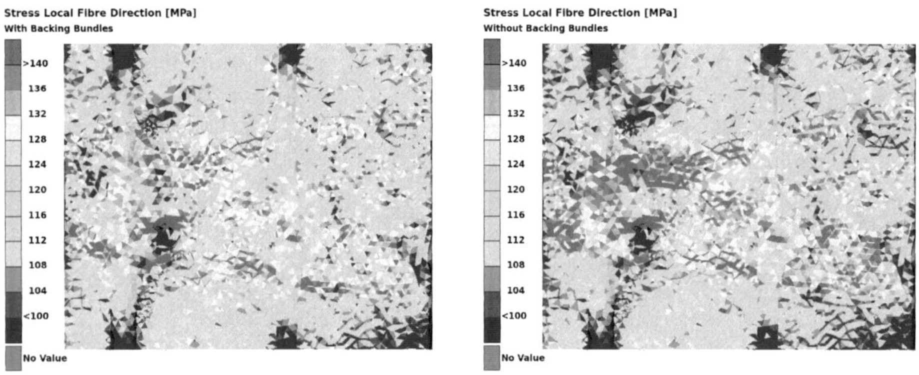

Figure 7: Plane cut through the volume showing the stresses in fibre direction inside the uni-directional bundles. Both models are based on the segmentation method *Physical Boundary*. However, for the model on the right the backing bundles were removed and the accrued volumes were new matrix solid elements. The model on the left shows the result of the original model with the backing bundles included and the correct stiffness properties.

Nevertheless, the waviness of the uni-directional bundles induced due to the backing bundles was not removed. Consequently the waviness is assumed to be the reason for the localised stresses.

It can be concluded that the stress localisations are caused by the backing bundles.

However, not due to different stiffness differences but due to the induced waviness of the uni-directional bundles.

4. Conclusion

In this study the influence of two automated segmentation approaches for X-ray computer tomographic data on the results of a finite element analysis have been compared. As test specimen, a composite made of non-crimp fabrics comprising four uni-directional bundles stacked on top of each other and held together by smaller off-axis backing bundles, was used. The automated segmentation methodologies were developed in Avizo™ and differ in the generation of a boundary between backing and uni-directional bundles.

It has been shown that both models show stress localisations in the overlapping area of backing and uni-directional bundles which has been observed in experimental studies on a similar material system. Based on further investigations presented in this study, it can be concluded that the stress concentrations originate at least partly from the induced waviness due to the backing bundles. The influence of the single fibre scale was not investigated due to the focus on the fibre bundle level in the finite element model.

Since no significant influences of the segmentation method on the finite element analysis has been detected, it can be concluded that the developed automated segmentations are robust. The method *Thresholding* is regarded favourable due to a higher time efficiency and reduced complexity. In future work, the proposed methodology needs to be tested on other composite layups as well. Further research applying the process of transferring X-ray computer tomography images to finite element models is necessary to investigate the origin of the stress concentration on the single fibre scale.

Acknowledgements

This study was funded by EU Horizon 2020 Marie Skłodowska-Curie Actions Innovative Training Network: MUltiscale, Multimodal and Multidimensional imaging for EngineeRING (MUMMERING), Grant Number 765604 and the Erna and Victor Hasselblad foundation grant for female scientists. Financial support from VINNOVA (the Swedish Innovation Agency) via LIGHTer Academy is also gratefully acknowledged. The computations were performed on resources provided by Chalmers Centre for Computational Science and Engineering (C3SE). The X-Ray computer tomography data was acquired using the Zeiss Xradia 520 Versa from the DTU (Technical University of Denmark) Centre for Advanced Structural and Material Testing (CAS-MAT), grant no. VKR023193 from Villum Fonden. At DTU in Roskilde the scanning electron microscopy images were taken. The support of Dr. Eng. Ioannis Nerantzis form BETA CAE Systems SA, Greece and Dr. Jan Giesebrecht from Thermo Fisher Scientific GmbH, Germany is acknowledged.

References

[1] Mikkelsen L P 2020 The fatigue damage evolution in the load-carrying composite laminates of wind turbine blades. *In Fatigue Life Prediction of Composites and Composite Structures* 2nd ed (Elsevier Ltd., Duxford, United Kingdom) 569–603 URL http://dx.doi.org/10.1016/B978-0-08-102575-8.00016-4

[2] Mikkelsen L P, Emerson M J, Jespersen K M, Dahl V A, Conradsen K and Dahl A B 2016 X-ray based micromechanical finite element modeling of composite materials *29th Nordic Seminar on Computational Mechanics, Gothenburg, Sweden*

[3] Sharma R, Mahajan P and Mittal R K 2013 Elastic modulus of 3D carbon/carbon composite using image-based finite element simulations and experiments *Composite Structures* **98** 69–78 URL `https://10.1016/j.compstruct.2012.11.019`

[4] Villarraga-Gómez H, Herazo E L and Smith S T 2019 X-ray computed tomography: from medical imaging to dimensional metrology *Precision Engineering* **60** 544–69 URL `https://doi.org/10.1016/j.precisioneng.2019.06.007`

[5] du Plessis A, le Roux S G and Guelpa A 2016 Comparison of medical and industrial X-ray computed tomography for non-destructive testing *Case Studies in Nondestructive Testing and Evaluation* **6** 17–25 URL `http://dx.doi.org/10.1016/j.csndt.2016.07.001`

[6] Maire E and Withers P J 2014 Quantitative X-ray tomography *International Materials Reviews* **59** 1–43 URL `https://10.1179/1743280413Y.0000000023`

[7] Garcea S, Wang Y and Withers P 2018 X-ray computed tomography of polymer composites *Composites Science and Technology* **156** 305–19 URL `https://10.1016/J.COMPSCITECH.2017.10.023`

[8] Auenhammer R M, Mikkelsen L P, Asp L E and Blinzler B J 2020 Automated x-ray computer tomography segmentation method for finite element analysis of non-crimp fabric reinforced composites *[Submitted]*

[9] Sencu R M, Yang Z, Wang Y C, Withers P J, Rau C, Parson A and Soutis C 2016 Generation of micro-scale finite element models from synchrotron X-ray CT images for multidirectional carbon fibre reinforced composites *Composites Part A: Applied Science and Manufacturing* **91:1** 85–95 URL `https://10.1016/j.compositesa.2016.09.010`

[10] Karamov R, Martulli L M, Kerschbaum M, Sergeichev I, Swolfs Y and Lomov S V 2020 Micro-CT based structure tensor analysis of fibre orientation in random fibre composites versus high-fidelity fibre identification methods *Composite Structures* **235** 111818 URL `https://10.1016/j.compstruct.2019.111818`

[11] Jespersen K M, Asp L E, Hosoi A, Kawada H and Mikkelsen L P 2018 X-ray tomography based finite element modelling of non-crimp fabric based fibre composite *18th European Conference on Composite Materials, Athens, Greece*

[12] Blinzler B J, Wilhelsson D, Asp L E, Jespersen K M and Mikkelsen L P 2018 A Systematic Approach to Transforming Composite 3D Images Into Meso-Scale Computational Models *18th European Conference on Composite Materials, Athens, Greece*

[13] Jespersen K M, Zangenberg J, Lowe T, Withers P J and Mikkelsen L P 2016 Fatigue damage assessment of uni-directional non-crimp fabric reinforced polyester composite using X-ray computed tomography *Composites Science and Technology* **136** 94–103 URL `https://10.1016/j.compscitech.2016.10.006`

[14] Agarwal B, Broutman L and Chandrashekhara K 2018 *Analysis and performance of fiber composites* 4th ed (Wiley, Hoboken/NJ, USA)

[15] User's guide avizo software 2019 Accessed: 2020-02-25 URL `https://assets.thermofisher.com/TFS-Assets/MSD/Product-Guides/users-guide-avizo-software-2019.pdf`

[16] Jespersen K M, Glud J A, Zangenberg J, Hosoi A, Kawada H and Mikkelsen L P 2018 Uncovering the fatigue damage initiation and progression in uni-directional non-crimp fabric reinforced polyester composite *Composites Part A: Applied Science and Manufacturing* **109** 481–97 URL `https://10.1016/j.compositesa.2018.03.002`

IOP Conf. Series: Materials Science and Engineering **942** (2020) 012039 doi:10.1088/1757-899X/942/1/012039

Observing the evolution of fatigue damage and associated strain fields in a correlative, multiscale 3D time-lapse study of quasi-unidirectional glass fibre composites

Anuj Prajapati[1], Ali Chirazi[2], Lars P. Mikkelsen[3], Timothy Burnett[1] and Philip J. Withers[1,*]

[1] Henry Royce Institute, Department of Materials, The University of Manchester, U.K.

[2] Visualization Sciences Group, Thermo Fisher Scientific, Bordeaux, France.

[3] Composite Mechanics and Structures, DTU Wind Energy, Denmark

* p.j.withers@manchester.ac.uk

Abstract. This research is focused on studying the tension-tension fatigue behaviour of a unidirectional (UD) glass-fibre wind turbine composite. The damage features, their progression and the associated strain fields are tracked in a representative volume by employing a novel correlative approach bringing together x-ray computed tomography (XCT) and digital image correlation (DIC). The focus is on studying ex situ the evolution of damage features (fibre breaks and micro cracks) in an interrupted time-lapse manner. The major drops in stiffness are correlated to the number and location of the damage features in the bulk (XCT) and at the surface (DIC). Results from XCT highlight a localized cluster of fibre breaks and matrix cracks near backing bundles along with axial macro-cracks, while DIC shows that the backing bundles cause regions of higher strain. This highlights the relation between the damage features and strain localisation and their effect on the progressive degradation in stiffness during high cycle fatigue (HCF) cycling.

1. Introduction

With the global call for increasing investment in, and exploitation of, renewable resources owing to environmental damage manifest as climate change, wind energy remains a strong candidate for the shift of power generation from fossil fuel-based resources to renewable and more environmentally sustainable ones [1]. Over recent years there has been a strong increase in the installation and production capacity of wind farms worldwide [2]. As the wind power generated by a turbine is proportional to the swept area by the rotor blades, there is a clear drive towards longer, larger and thus heavier blades, increasing the strength and fatigue life requirements.

The spar caps, which are the main load-bearing components in the blades, are made of quasi-unidirectional glass and carbon fibre reinforced composites, which experience a high-cycle fatigue (HCF) loading regime, having a service life as high as 30 years encompassing 10^8-10^9 cycles [3]. These materials are selected because of their optimum balance between cost, high stiffness-weight ratio, fatigue resistance and design flexibility. Furthermore considerable performance improvements have been achieved for such composites over the last decade increasing their range of applications [4].

The blades are subjected to repeat edge-wise and flap-wise oscillating loading from the weight plus the centrifugal rotation and wind dynamics respectively, which essentially engage the load bearing glass fiber composites in a fatigue regime.

As a result of the fatigue damage accumulated the blade gradually loses its working stiffness which in extreme cases can even lead to collision with the supporting tower or even catastrophic failure. Consequently, the focus of this study is to understand how loads representative of such cycles affect the mechanical properties over the service life, observing the effect it has on the microstructure and how that builds up over time to affect the macro mechanical response of these materials. This will help us isolate key factors in the microstructure which are critical to this behavior and help us optimize them to improve the mechanical response, both in terms of extending the safe service life and achieving a higher load bearing capacity.

The fatigue behavior of unidirectional composites has been an active area of research since the 70s, inspired initially by the metal fatigue experience, but then slowly moving to accommodate other important aspects of the microstructure, especially the interfacial phenomena between the fibers and the matrix [5], [6]. 'Fatigue life diagrams' [7] can be used to lay out the progressive and non-progressive governing mechanisms in the fatigue regime, thereby helping to identify which zones of distinct damage features in the diagram are critical to failure. In general, the dominant damage features on the micro scale include, among others, fibre breaks, fibre-matrix debonding, matrix yielding and cracking, and transverse ply cracking. This leads to complex modes of damage progression which are challenging to predict accurately, especially for different layups. Quasi-unidirectional (UD) non-crimp composites offer excellent unidirectional strength and stiffness and are widely used in wind turbine applications. They have the majority of fibres running in the load-bearing axis direction clustered in bundles (tows), while a small number of backing fibres run in off-axis directions. The UD bundles are tied to the backing bundles by means of stitching threads to aid integrity during laying up and manufacturing.

It has been shown [8] that in the early and intermediate stages of fatigue life, the drop in stiffness can be related to the crack densities in the matrix for materials with unidirectional fibre bundles with backing layers at $\pm 45^{\circ}$ and 90°. The final stage near failure involving 0° fibre rupture exhibits a rather stochastic behavior, believed to be due to stress concentrations near the tip of the intra ply matrix cracks.

It has been proposed [9,10] that the damage starts in, and near, the off-axis bundles, especially in regions which are in crossover between the UD and backing bundles, as matrix cracks and fibre breaks, and then progress towards the thickness in a diffuse manner [9,11]. As the progression is complex and essentially stochastic in the microstructure, it is difficult to predict accurately and reliably.

Recently, combined advanced imaging and characterization methods at the microscale have been put forward which are delivering new insights into the classic initiation, propagation and failure by damage evolution [12].These include not only imaging, but other critical complementary pieces of information regarding stress-strain fields, morphology and chemistry, among others. On this front, x-ray computed tomography (XCT) has been proven particularly successful for investigating this microscale behavior since this is the scale of the onset of sensible damage, and this technique can provide 3D information non-destructively [13,14]. It can provide key insights to help resolve the complex morphology and sequencing of the damage by monitoring the damage in UD composites in a time-lapse manner during fatigue [9,10].

Digital image correlation (DIC) has been developed [15] around the 70's in and has been used to investigate deformation in fibre composites as early as the 90's [16]. DIC involves the tracking of a speckle pattern on the sample specimen surface to infer the displacement and associated strain fields.

In this paper these techniques are used in a complementary workflow to the study the fatigue behavior of UD composites across varying length scales.

2. Materials and Experimental Methods

2.1. Materials systems

The materials have been supplied by Saertex GmBH, comprising UD E-glass fibre bundles stitched to backing bundles (90°) and impregnated in epoxy using vacuum-assisted resin transfer molding, with a layup of $[[0/90]/[90/0]_s]$ with the backing bundles adjacent to each other and none in the center sandwiched between 100 micron thick layers of cross woven fabric on either surface.

Butterfly geometry fatigue test specimens of length 284mm were cut out using waterjet with a parallel sided gauge length of 44 mm, as shown in Fig. 1.

The smallest width at the gauge was 10mm wide, with the plate thickness of 3.8 mm. End tabs of length 120mm were stuck on to avoid crushing the ends, tapered over a length of 60mm, to reduce stress concentrations.

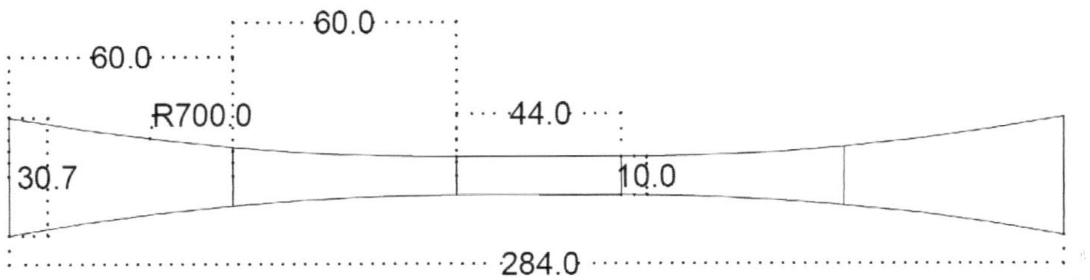

Figure 1. Sample specimen cut out in a butterfly geometry. All the dimensions are in mm.

2.2 Experimental workflow

A workflow involving time-lapse x-ray CT and digital image correlation was used to identify the progression of damaged regions and their associated strain levels through the fatigue test. Fig. 2 shows the characterization workflow aligned with the observed degradation in stiffness degradation. The central focus is to investigate the strain distribution and progression of damage features over the whole gauge as the number of cycles increase. Region I describes the initial drop in stiffness upon fatigue cycling, followed by stable response in region II, followed by a steep decrease in stage III just before failure. The fatigue tests were interrupted for DIC (in-situ) and CT (ex-situ) at 3 intermediate stages in addition to the first stage in the timeline. The region of interest for CT scanning were chosen from the strain hotspots identified from the DIC images. The sample was tested to failure but rather fatigue cycling was stopped when around 10% stiffness was lost.

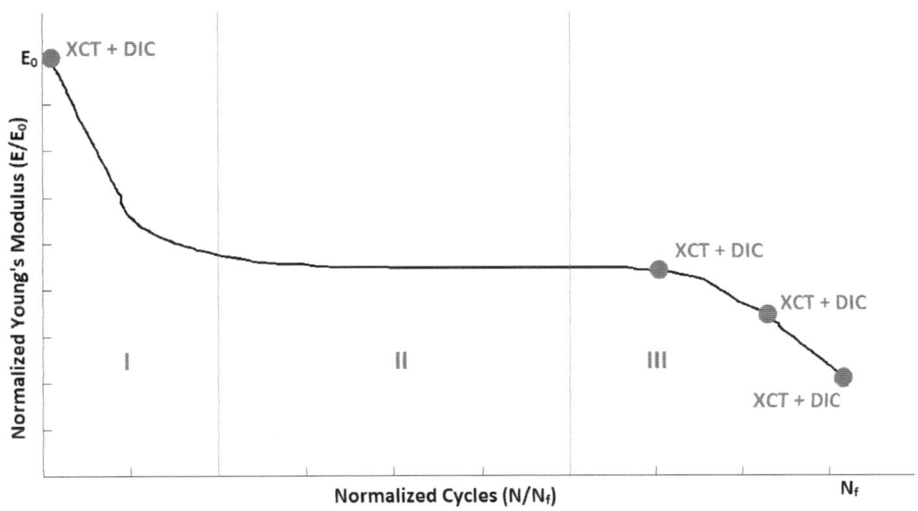

Figure 2. Schematic showing the experimental workflow scheduled around the degradation in stiffness, E, and number of cycles, N. The initial stiffness is E_0 and total number of cycles is N_f.

2.3. Mechanical testing
The fatigue tests were carried on an Instron Servohydraulic 8802 with a 100kN load cell at maximum strain of 1.0% with a stress ratio of R=0.1, in load control with a sinusoidal waveform and 4 Hz to avoid self- heating.

The strain was monitored with both a clip-on extensometer (5mm extension/12.5 mm gauge). The stiffness was continuously tracked while the DIC images were acquired at a cycle frequency of 0.125 Hz to minimize blurring and the imaging frequency of 10 frames per second, over the whole gauge length of 44mm, for a full single cycle at all the 4 stages. The sample gauge was 'speckled' using a black and white spray paint, to obtain sufficient contrast between the black and white points for correlation. LaVision Strainmaster systems were used both in terms of the software and hardware for the DIC acquisition and processing.

2.4. X-ray computed tomography
XCT experiments were carried out on a Zeiss Xradia Versa 520 scanner with the sample mounted on a custom-made holder fixtured to enable it to be mounted as close in the identical location for each scan in the sequence. As has been demonstrated from [9,10], the damage starts to occur near the backing bundles, so a RoI (region of interest) containing a backing bundle was scanned. The projections were not binned. The scan settings are summarized in Table 1.

Table 1. X-ray CT scan conditions used on the Zeiss Versa 520.

Source to sample distance	Detector to sample distance	Optical magnification	Pixel range	Exposure time	Number of projections/ tomogram	Accelerating voltage	Pixel size
42 mm	150 mm	4x	16-bit	20s	3201	70kV	0.7μm

3. Results and discussion
The degradation in stiffness over time is shown in Fig. 3, with the key points marked. The degradation in stiffness can be divided into 3 regions in accord with [9,11,17]. Region I includes the initial stiffness drop, while Region II relates to the period over which the stiffness remains level. Region III leads up to final degradation and the ensuing failure. The curve has been piecewise smoothened with a 4th degree polynomial to filter out the heavy load cell related noise in the stiffness calculations. The co-ordinate system for the sample specimen as: x in width, y in thickness and z in axial length.

As mentioned, the test was interrupted at 3 points (for XCT acquisition) in the loading timeline, at points 1A and 2A and 3A. There is a slight decrease in the modulus after reinstating the test after scans 1 and 2 marked by points 1B and 2B. This behavior has been observed previously [9] and can be attributed to the re-calibration and the sensitivity of the extensometer between the interrupted steps. There is no strong reason to believe that the low level of x-rays exposure degrades the mechanical properties of the composite, in fact x-ray exposure has been found to increase the matrix stiffness elsewhere [18].

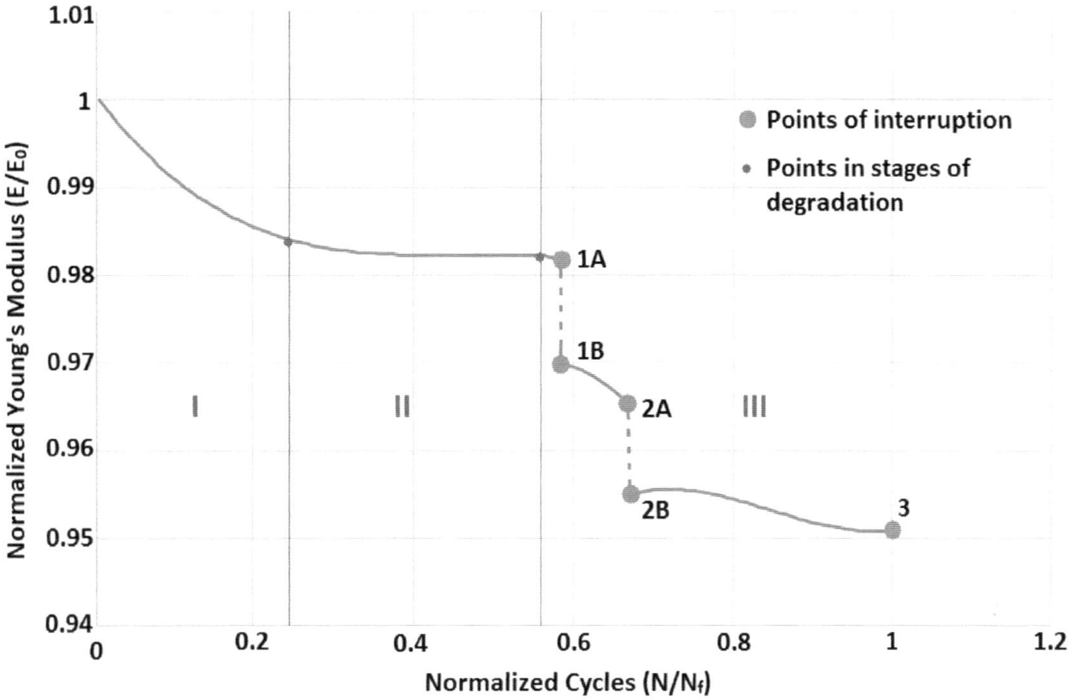

Figure 3. Stiffness degradation curve during fatigue cycling with points demarcating interruption for characterization. Region I and III show steady falls in stiffness, while the stiffness is stable during region II. N_f=120,000 cycles

From the DIC maps in Fig. 4 we can clearly see hotspot regions that experience higher strain than the maximum gross strain of 1% imparted on the gauge. This is attributed to the presence of the 90° backing bundles running across the material, as can be confirmed in the CT volumes. At hotspots, the strain is as high as 2% which is very close to the tensile failure strain of the plate (2.25%), individual fibres (2.7%) and the matrix (4%). The regions at the periphery, have been masked out for DIC analysis because they go in and out of the field of view invalidating DIC analysis. The linear hotspots in strain (marked by rectangles in Fig. 4) correspond to the location of the backing bundles, while the point hotspots (marked by purple arrows in Fig. 4) are regions of crossover where the difference in waviness between UD and matrix is higher, which leads to coupled regions of higher and lower compliance.

Not only do the backing bundles perturb the 'homogeneity' of the microstructure, they also cause the UD bundles to be slightly wavy in the regions of the crossover, which leads to bending and a change in local orientation. This may be why some of the glass fibres fracture locally giving rise to other mixed mode damage modes of debonding, matrix cracking and transverse and longitudinal cracks.

As witnessed previously [9], the damage starts predominantly near the region of crossover between the UD bundles and the backing bundles, although in the previous work it was difficult to resolve any matrix cracks due to reasons of scale and resolution. The mixed mode damage propagation is mostly localized in clusters as seen in Fig. 5. Individual fibre breaks such as that seen in Fig. 6 and the associated debonding changes the local stress state leading to strain concentrations and breaks in neighbouring fibres, as discussed elsewhere [19].

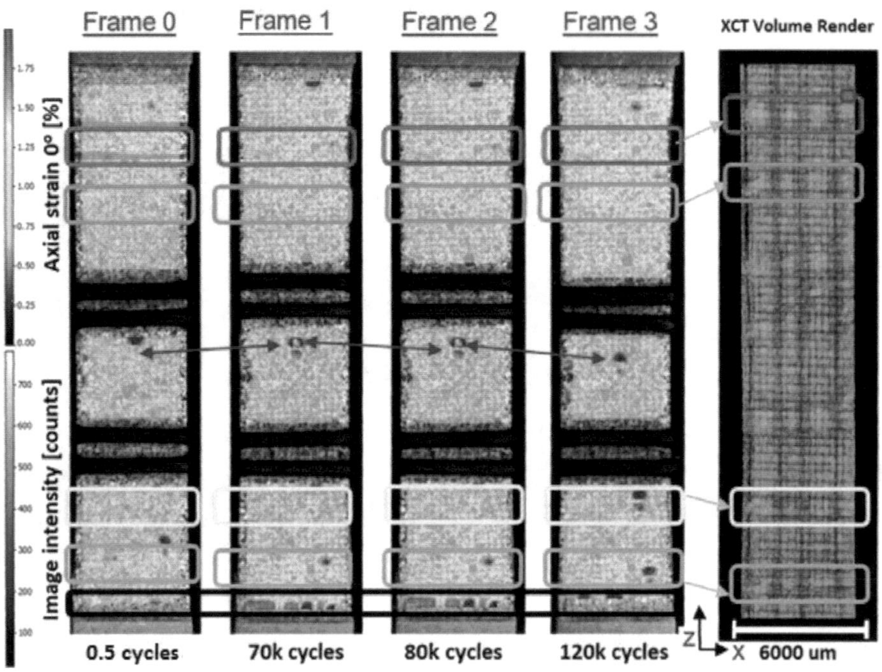

Figure 4. DIC maps (between images at 0% strain and peak strain of 1%) showing hotspots in strain broadly correlated with the location of the backing bundles in the XCT volume render. The blue square near the top of the XCT volume shows the region of interest for corresponding XCT scans. The two pair of black bands on each map are O-rings from the extensometer. It should be noted that the alignment of the 90° bundles (red, blue, yellow and orange) are not precisely correlated with the hotspots, because the backing bundles in the bulk (not shown here) are not at the same z position as the ones on the surface shown here. The black box highlights regions which did not correlate correctly due edge effects.

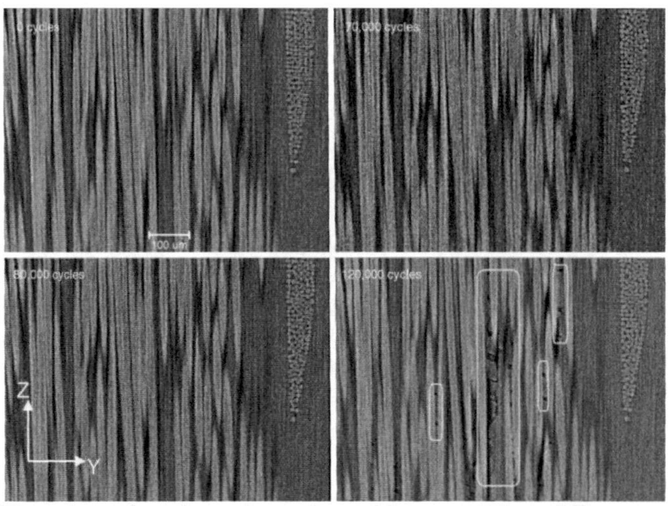

Figure 5. A virtual YZ section taken from the region of interest CT at different stages in the fatigue cycling showing a localized damage cluster (boxed) involving matrix cracking and debonding, located in proximity to the backing bundles.

The x-ray CT scans do not appear to show clear evidence of classical progressive damage; instead, damage clusters originated and progressed only after the third scan. This does not definitely rule out the presence of damage in the microstructure prior to the third scan, as the first cracks in the backing bundles have previously been observed after as little as 480 cycles [9]. Indeed it is highly likely that damage has occurred in other regions of the test-piece which were not CT scanned in accordance with the drop in stiffness, as has been found in a number of studies [3,8,9]. Despite this, lack of evidence for progressive damage accumulation a transverse-longitudinal (YZ) macro-crack (Fig. 7) already present in the apparently pristine sample did undergo progressive damage evolution, more so in terms of opening up the crack, combined with more matrix cracking at higher cycles. The width of the macro-crack appears to increase progressively, although there has been no significant progression observed in the length of the crack, which seems to be arrested from the first XCT scan. Although it is probably due to the machining of the specimens and/or manufacturing and/or the first cycle over which the DIC strain map was collected, it is noteworthy that it is arrested at the backing bundle.

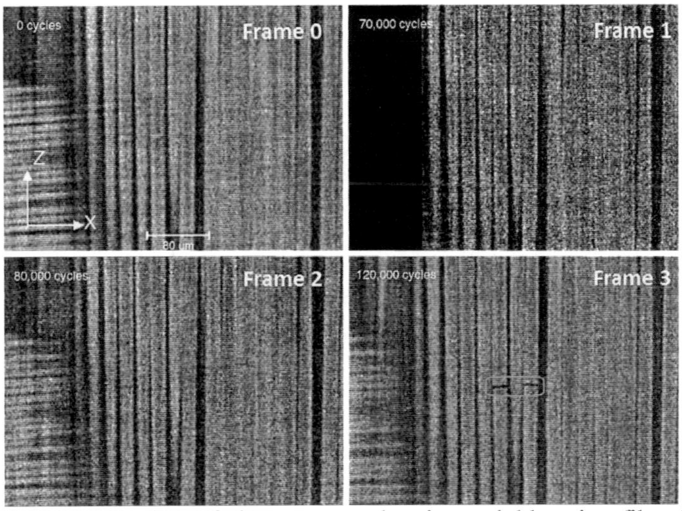

Figure 6. Virtual ZX section for the 4 fatigue stages showing neighbouring fibre breaks (boxed) in regions of crossover between the UD bundle and the 90° backing bundles.

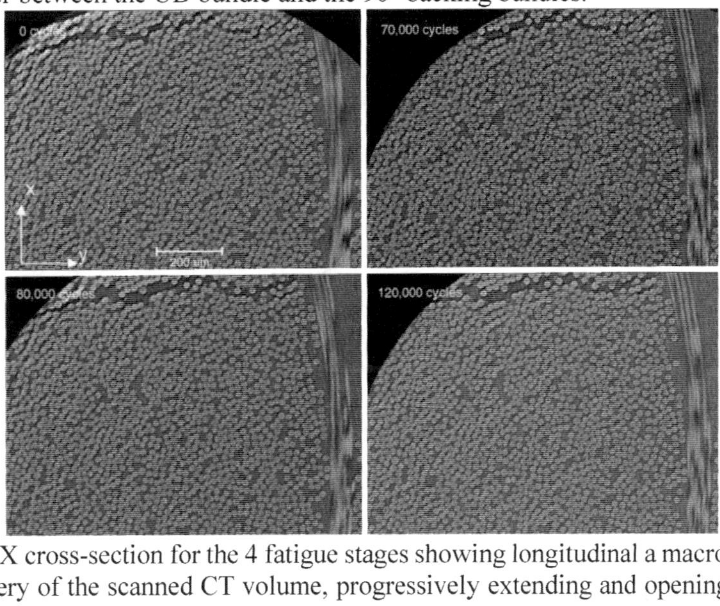

Figure 7. Virtual YX cross-section for the 4 fatigue stages showing longitudinal a macro-cracks running through the periphery of the scanned CT volume, progressively extending and opening throughout the fatigue regime but being arrested at the backing bundle (boxed).

4. Conclusions

This study has shown that the strain field that develops in UD composite during fatigue loading is affected by the presence of the backing bundles. By correlating the strain field maps as a function of the number of fatigue samples with ex situ x-ray CT imaging of the composite it was possible to correlate the resulting stress amplification with localized damage initiation and ultimately to damage propagation and final failure. Advantages of this correlative method include the ability to directly relate hotspots in strain concentration with the underlying microstructure as a function of fatigue life to study the progress of damage accumulation. In this case most of the damage was not observed until just prior to fatigue failure. Next steps of this work will be to focus on designing experimental workflows to follow damage evolution at the microscale and combine it with digital volume correlation (DVC) to study the stress amplification in 3D.

5. Acknowledgements

This study was funded by EU Horizon 2020 MSCA Innovative Training Network 'MUMMERING' Grant Number 765604. We thank Stuart Morse from The University of Manchester for help in fatigue testing. Support from Saertex for material supply is gladly acknowledged. Access to Thermo Fisher Scientific Avizo™ is acknowledged. Beamtime was kindly provided by the Henry Moseley X-ray Imaging Facility (HMXIF), which was established through EPSRC grants EP/F007906/1, EP/ I02249X/1 and EP/F028431/1. HMXIF is a part of the Henry Royce Institute for Advanced Materials, established through EPSRC grants EP/ R00661X/1, EP/P025498/1 and EP/P025021/1. PJW acknowledges a European Research Council Advanced Grant Correl-CT No. 695638.

References

[1] Taylor M 2012 Wind Power Renewable Energy Technologies: Cost Analysis Report *International Renewable Energy Agency*

[2] Ohlenforst K 2018 Annual Wind Energy Report 2018 *Global Wind Energy Council*

[3] Highsmith A L and Reifsnider K L 1982 Stiffness-reduction mechanisms in composite laminates *Damage in composite materials: basic mechanisms, accumulation, tolerance, and characterization* (ASTM International)

[4] Agarwal B D, Broutman L J and Chandrashekhara K 2017 *Analysis and Performance of Fiber Composites.* (John Wiley & Sons, Incorporated)

[5] Talreja R 2008 Damage and fatigue in composites - A personal account *Compos. Sci. Technol.* **68** 2585–91

[6] Hancock JR 1973 Fatigue of Composite Materials, *ASTM* **569**

[7] Talreja R 1981 Fatigue of Composite Materials: Damage Mechanisms and Fatigue-Life Diagrams. *Proc. R. Soc. London, Ser. A Math. Phys. Sci.* **378** 461–75

[8] Gagel A, Lange D and Schulte K 2006 On the relation between crack densities, stiffness degradation, and surface temperature distribution of tensile fatigue loaded glass-fibre non-crimp-fabric reinforced epoxy *Compos. Part A Appl. Sci. Manuf.* **37** 222–8

[9] Jespersen K M 2017 Fatigue Damage Evolution in Fibre Composites for Wind Turbine Blades *DTU Wind Energy PhD*

[10] Wang Y, Mikkelsen L P, Pyka G and Withers P J 2018 Time-lapse helical X-ray computed tomography (CT) study of tensile fatigue damage formation in composites for wind turbine blades *Materials (Basel).* **11** 2340

[11] Zangenberg J, Brøndsted P and Gillespie J W 2014 Fatigue damage propagation in unidirectional glass fibre reinforced composites made of a non-crimp fabric *J. Compos. Mater.* **48** 2711–27

[12] Burnett T L and Withers P J Completing the picture through correlative characterization *Nat. Mater.* ***18*** 1041–49

[13] Yu B, Bradley R S, Soutis C and Withers P J 2016 A comparison of different approaches for imaging cracks in composites by X-ray microtomography *Philos. Trans. R. Soc. A Math.*

Phys. Eng. Sci. **374** 20160037
[14] Garcea S C, Wang Y and Withers P J 2018 X-ray computed tomography of polymer composites *Compos. Sci. Technol.* **156** 305–19
[15] Anuta P E 1970 Spatial Registration of Multispectral and Multitemporal Digital Imagery Using Fast Fourier Transform Techniques *IEEE Trans. Geosci. Electron.* **8** 353–68
[16] Russell S S 1990 Image Correlation Nondestructive Evaluation of Impact Damage in a Glass Fibre Composite *Second Conference on NDE for Aerospace Requirements* (NASA)
[17] Reifsnider K L and Jamison R 1982 Fracture of fatigue-loaded composite laminates *Int. J. Fatigue* **4** 187–97
[18] Startsev O V, Krotov A S and Golub P D 2006 *Effect of climatic and radiation ageing on properties of VPS-7 glass fibre reinforced epoxy composite* **41**, *263-73*
[19] Gamstedt E K 2000 Effects of debonding and fiber strength distribution on fatigue-damage propagation in carbon fiber-reinforced epoxy *J. Appl. Polym. Sci.* **76** 457–74

IOP Conf. Series: Materials Science and Engineering **942** (2020) 012040 doi:10.1088/1757-899X/942/1/012040

Effect of backing fibers on the thermoelastic stress analysis of multi-directional glass/epoxy laminates during fatigue loading

A Quinlan[1] , JM Dulieu-Barton[2] and O Castro[3]

[1]Department of Civil Engineering, Technical University of Denmark, Brovej 118, 2800 Kgs Lyngby, Denmark

[2] Department of Mechanical Engineering, University of Bristol, Queen's Building, University Walk, Bristol BS8 1TR, United Kingdom

[3] Department of Wind Energy, Technical University of Denmark, Frederiksborgvej 399, 4000 Roskilde, Denmark

E-mail: [1]`alequ@byg.dtu.dk` [2]`janice.barton@bristol.ac.uk` [3]`osar@dtu.dk`

Abstract. A preliminary study was conducted regarding Thermoelastic Stress Analysis (TSA) for use in monitoring stiffness degradation of multi-directional GFRP laminated specimens during tension-tension fatigue loading. The specimens were made from a non-crimp fabric consisting of UD fibers stitched to a layer off-axis backing fiber bundles. The thermoelastic responses of the UD surface and backing fiber surface were compared. It was observed that the thermoelastic response differed between two faces of the fabric and was influenced by the presence of backing fibers on the surface.

1. Introduction

Laminated fiber-reinforced polymer composites (FRPs) have become an attractive choice for weight critical engineering applications because their layup can be optimized for the design loads. But this positive attribute also makes FRPs difficult to characterize, especially for fatigue behavior. Fatigue in FRPs [1] occurs in fundamentally different mechanisms than metals, and FRP S-N curves are limited in application to the specific laminate studied. The current FRP design process involves extensive coupon testing to characterize a laminate's fatigue performance before prototype structures are created [2]. The time and cost of this process might be reduced by better understanding the micro-mechanics of FRP damage and fatigue and by performing more tests on FRP components and structures.

1.1. Damage in FRPs

One of the first damage mechanisms to occur in a laminate is matrix cracking [3]. While it is not a catastrophic damage mode, matrix cracking can lead to delamination [4], fiber breakage, and property degradation that all negatively affect a FRP structure's performance. Modeling and predicting the initiation, growth [5], and effects [6, 7] of matrix cracking is an active field of research. Some models [8] use micro-mechanics to predict the degradation of thermo-mechanical properties of a laminate containing matrix cracks. Experimental study is a critical part of developing, calibrating, and verifying these fatigue models. While this typically occurs on material coupons, a multi-scale approach [2,9] is needed to evaluate the applicability of these models on components and sub-structures. Measuring the behavior of sub-components is more challenging than measuring coupon behavior, as the increase in both scale and complexity require additional consideration. In the case of evaluating applied stresses, the typical coupon method of calculating the nominal stress from the load-cell output and specimen cross sectional area cannot be applied, so an alternative method of "measuring" stress is thermoelastic stress analysis.

1.2. Thermoelastic Stress Analysis

Thermoelastic Stress Analysis (TSA) is a full-field non-contact measurement technique that assesses the stress amplitude on a material surface by obtaining a temperature change during cyclic loading using infrared imaging [10, 11]. The thermoelastic effect [12] states that the relationship between stress and temperature for an orthotropic material experiencing cyclic loading under adiabatic conditions is:

$$\Delta T = \frac{T_0}{\rho C_p}(\alpha_1 \Delta\sigma_1 + \alpha_2 \Delta\sigma_2) \tag{1}$$

where ρ is the density, C_p is the specific heat under constant pressure, T_0 is the temporal mean of the temperature signal, α_1 and α_2 are the linear coefficients of thermal expansion, σ_1 and σ_2 are the stresses, and the material's principal directions are notated as subscript 1 and 2. Rather than individually obtaining the exact values ρ, C_p, α_1 and α_2, it is often preferable to combine these into a quantity known as the thermoelastic constant, K, where

$$K_i = \frac{\alpha_i}{\rho C_p} \tag{2}$$

and substitute the K values into equation (1) to yield the following equation:

$$\Delta T = T_0(K_1\sigma_1 + K_2\sigma_2) \tag{3}$$

Procedures for calibrating K are documented in the literature [13, 14]. The calibration is carried out experimentally because the thermal properties of FRP laminates can vary between layups and production batches.

The application of TSA has increased in recent decades as thermal cameras have improved and reduced in cost. Studies have expanded from coupons to composite sub-components and structures [2, 11, 14–18]. To perform TSA, an infrared camera captures a series of image frames at a high rate so that the temperature waveform can be correlated with the load signal and extracted using a sine-fitting or lock-in algorithm. The temperature range and mean are then used in equation (3) to determine the stress values. Temperature is a scalar quantity, therefore it is necessary to *calibrate* ΔT to account for the orthotropic nature of the material by using some additional information from other experimental techniques.

The literature sources present varying interpretations of *which* stresses are analyzed during TSA. Three types of stresses considered are 1) surface resin stresses, 2) surface ply stresses, and 3) substrate ply stresses. Pitaressi et al [19] found that the surface resin rich layer (SRRL) acts as a strain witness and that the thermoelastic response is independent of the surface ply stresses. Bakis and Reifsnider [20] and Sambasivam et al [21] found that laminates with different surface ply orientations loaded to the same global strains do indeed have different thermoelastic responses. Crump and Dulieu-Barton [14] found that the thermoelastic response of a carbon fiber structure is influenced by the substrate plies during low frequency loading. These findings are not contradictory, but illustrate that the methodology for TSA will differ according to the material, surface coating, stress amplitude, and loading frequency.

1.3. Backing fiber bundles in wind turbine blade materials

The material under investigation in this paper is a glass fiber non-crimp fabric. Non-crimp fabrics have been used in the wind energy sector since the 1980's [22]. While some FRP components are made from pre-preg materials cured in an autoclave, composite structures that exceed the size of autoclaves, like wind turbine blades, are made by infusing resin into multi-directional (MD) or uni-directional (UD) fiber fabrics. Backing fibers are often used in UD fabrics to support the UD tows during processing [23] and are usually a small percent of the fiber weight, offering a minor contribution to the fabric's mechanical strength in the UD fiber direction. While useful for manufacturing, the backing fibers have been found as initiation sites for damage [23–27].

The objective of this work is to explore the effect of the backing fibers when using TSA to obtain a stress metric in a multi-directional laminate made with non-crimp fabric and to observe possible changes during fatigue loading. The work described in the paper provides initial steps in developing a TSA methodology for use on components and sub-structures made from a similar material under similar fatigue loadings.

2. Methodology

2.1. Specimens

Two types of specimens were made from E-glass and epoxy using UD non-crimp fabric with backing fibers running at 90°, 45°, and -45°angles relative to the UD tows. The backing fiber bundles cover only one surface, referred to here as the *backing fiber surface* (BFS), which is shown in figure 1.a. The opposite side contains only UD fibers and polyester threading, which is referred to here as the *uni-directional surface* (UDS) and is shown in figure 1.b. While all the fibers in the laminate are E-glass, the backing fibers are of a smaller diameter than the UD fibers. The fiber components of the non-crimp fabric are detailed in Table 1. While the total backing fiber weight is only 10% of the total, figure 1.a shows that about 60% of the fabric surface is covered by backing fiber bundles.

For specimen production, the non-crimp fabric was arranged according to the layups listed in Table 2, where the *b* indicates the location of the backing fibers surface. Vacuum assisted resin transfer molding (VARTM) was used to infuse the fabric with epoxy, which was then cured for 19 hours at 40°C, followed by 5 hours at 80°C. Rectangular specimens measuring 25 x 250 mm were made according to the ASTM guidelines [28]. The 8-ply laminates were 3.8 mm thick and had an approximate fiber volume fraction of 51%. The two specimens are referred to by their outer surfaces: the BFS specimen and the UDS specimen. The specimens were painted matte black to increase the emissivity of the surface.

2.2. Experimental procedure

The tests were performed using a 100 kN hydraulic axial test machine. The specimen temperature was monitored using an infrared camera with the specifications detailed in Table 3. An extensometer was used to measure the strain in the loading direction. Specimens were loaded to 8.25 ± 6.75 kN at a frequency of 5.1 Hz. The tests ran for 5000 cycles with sets of

Table 1. Fibers in Non-Crimp Fabric

Fiber Type	Fiber Diameter	% by Weight	% of Surface Area
UD	17 μm	90.5 %	40 %
90°	9 μm	5.5 %	60 %
± 45°	9 μm	4 %	

Table 2. Laminate layup configurations

Laminate	Layup
BFS	$[_b0°/_b60°/_b0°/_b\text{-}60°]_s$
UDS	$[0°_b/_b60°/_b0°/_b\text{-}60°]_s$

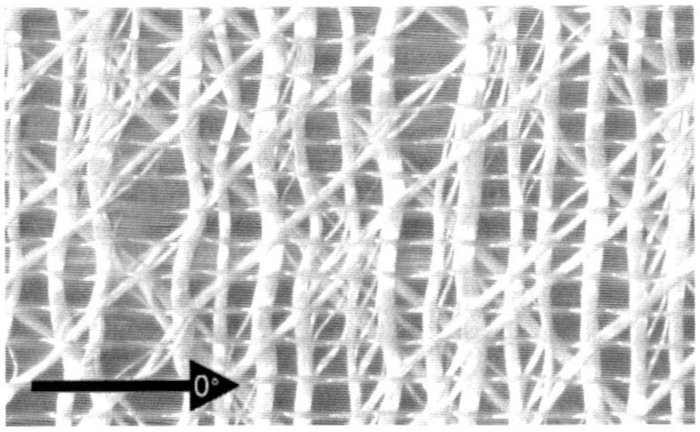

(a)

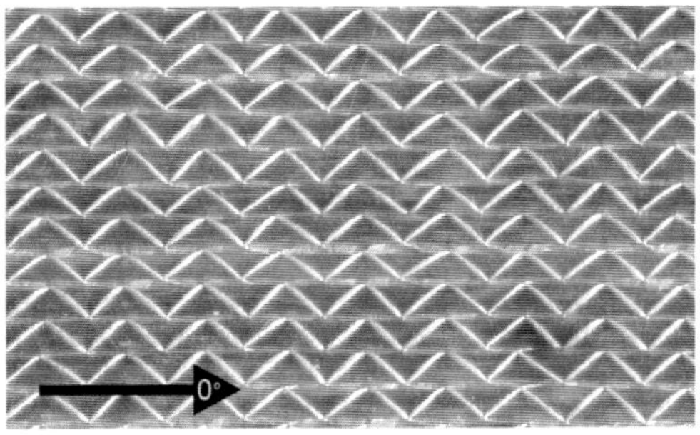

(b)

Figure 1. Two surfaces of the UD non-crimp fabric with backing fibers: (a) BFS, (b) UDS

infrared images captured at intervals throughout the test.

Table 3. Thermal Camera Specifications

Make	Model	Resolution	Sample Rate	Sample Size
FLIR	SC6540	0.172 mm/pix	250 Hz	1500 frames

A motion compensation program was applied to the captured infrared image series. The motion compensated images were analyzed pixel by pixel using a least-squares algorithm developed at the University of Southampton. A sinusoid was fit to the data to compute the temperature's mean, amplitude, and phase as follows:

$$T(x, y, t) = T_0(x, y) + 0.5\Delta T(x, y)\cos(2\pi f_0 t + \phi(x, y)) \quad (4)$$

where x and y are the pixel coordinates, t is time, f_0 is the loading frequency, and ϕ is the phase angle.

3. Results and Discussion

The plots in figure 2 show the thermo-mechanical responses of the specimens as they undergo fatigue loading. Both the BFS specimen and the UDS specimen show similarly shaped stiffness degradation curves in figure 2.a, though the UDS specimen has a slightly higher initial modulus in the longitudinal direction. Consequentially, the UDS specimen experiences a slightly lower longitudinal strain in figure 2.b. The bulk heating of the laminates during the fatigue tests are similar in figure 2.c, with the temperature difference between the specimens remaining below 1°C throughout the test. Large changes in temperature affect the specific heat of GFRP, which is estimated to increase by about 3% over the duration of this test according to the equations provided in [29]. However, the temperature difference between the two specimens at a given cycle count is small enough that the specific heat can be assumed to be the same and the $\frac{\Delta T}{T_0}$ terms can be directly compared.

3.1. Thermoelastic response evolution

Although the nominal stress remains the same throughout the loading, the stress on the 0°plies is expected to increase as the ±60°plies degrade. Using the rough assumption that matrix damage occurs only in the ±60°plies and that no fiber breakage occurs in the 0°laminate, classical laminate theory [30], CLT, suggests that the observed 7% drop in global stiffness would

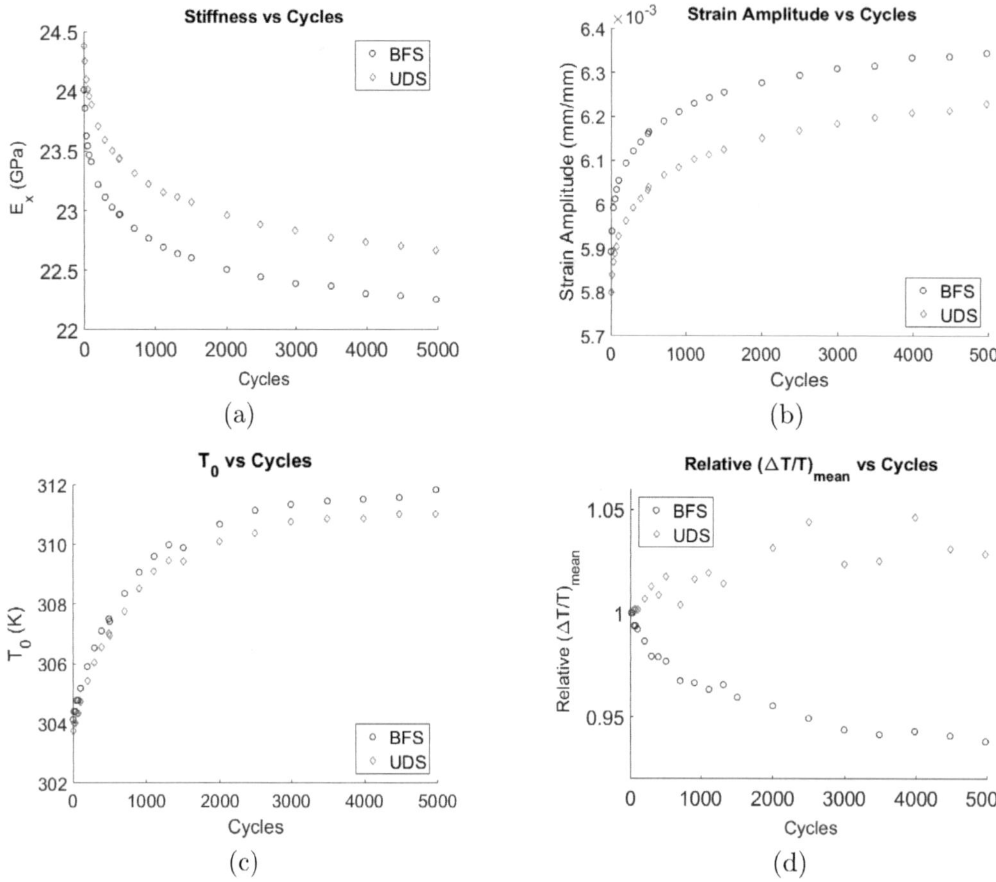

Figure 2. Evolution of the thermo-mechanical responses of the Backing Fiber Surface laminates and UD Surface laminates

correspond with an increase in the UD ply stresses by 7% in the longitudinal direction, while the transverse stresses remain close to 0. According to equation (3), these stress increases would result in a 7% increase in the thermoelastic response of the surface, assuming that the K values are unchanged.

Figure 2.d shows the mean $\frac{\Delta T}{T_0}$ term taken over the central region of the specimen. The average thermoelastic response of the UDS specimen increases by about 2% over the 5000 cycles, but the BFS specimen exhibits a decrease in thermoelastic response. This discrepancy suggests that the CLT approach might not fully account for the thermoelastic response.

Considering the surface resin rich layer (SRRL) approach [19], the SRRL should have a stress that increases proportionally with the laminate strains, assuming that laminate strains are fully transferred into the SRRL. Because the longitudinal strain increased by about 7% over the duration of the test, the thermoelastic response is also expected increase.

3.2. Surface texture thermoelastic response

The thermoelastic responses ($\frac{\Delta T}{T}$) from both specimens are shown in figure 3. As shown in figures 3.b and 3.d, the UDS specimen does not have clearly recognizable regions, but shows variation in thermoelastic response that is suspected to be caused by the surface roughness. In contrast, figures 3.a and 3.c show clear patterns on the BFS specimen that match with the regions of the surface. The most clear features are the horizontal lines corresponding to the

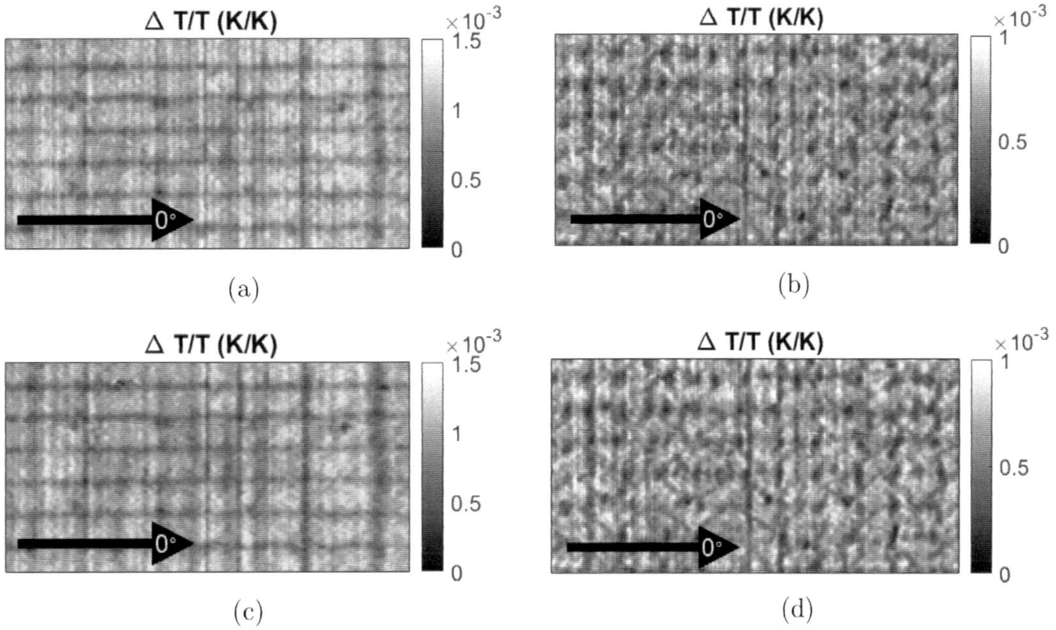

Figure 3. Thermoelastic responses of the specimen surfaces: (a) Test start - BFS, (b) Test start - UDS, (c) Test end - BFS, and (d) Test end - UDS. The horizontal lines are resin rich regions and the vertical lines are backing fibers.

matrix rich regions between the tows. The vertical lines correspond to the 90°backing fibers and become darker as the number of cycles increase. The thermoelastic response field from the BFS specimen shows agreement with Sambasivam [21] in that the texture of the surface ply is detectable through the surface resin layer. This, in addition to the mean thermoelastic response decreasing as the global strain increases, suggests that SRRL cannot account for the entirety of the thermoelastic response in these specimens. The possible lack of conformity to either the SRRL or CLT predictions could be attributed to many factors. First, the material properties of the laminate can change during the test: the specific heat of GFRP could be changing around 3% and the CTE can change during matrix cracking. Also, the surface resin layer may have uneven thickness across the fabric texture. The influence of the sub-surface layers and other non-adiabatic effects could also influence the thermoelastic response.

To further investigate the BFS's mean $\frac{\Delta T}{T}$ evolution in figure 2.d, the response of smaller regions are shown in figure 4. The backing fiber region shows a large decrease in $\frac{\Delta T}{T}$ over the first 500 cycles and then a downward trend afterwards, while the UD region stays relatively flat initially before exhibiting a slight downward trend. As the backing fibers were covering about 60% the surface, it appears that it was their stress changes that were responsible for the decrease in the mean thermoelastic response. Initially, the stiffness in the loading direction of the backing fibers was an order of magnitude lower than the UD tows, so a large stress difference between the features was expected. By the end of the test, cracks were observed in the backing fiber bundles, but not in the UD fiber tows. This may have caused an increase in the stress difference between the backing fibers and UD tows, resulting in a larger contrast in thermoelastic response between the regions.

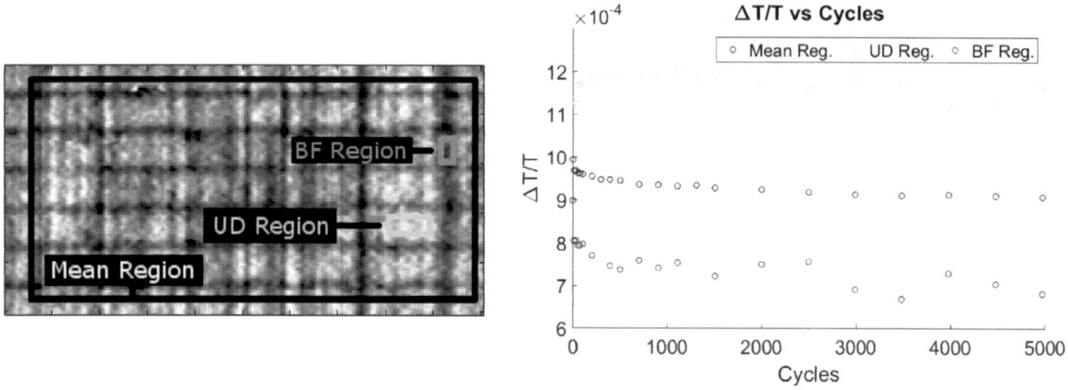

Figure 4. Evolution of thermoelastic response in different regions of the BFS specimen. The black, green, and blue sub-regions on the left correspond with the plot colors on the right.

3.3. Effect of cracks in BFS specimens

After 5000 cycles, many more matrix cracks were observed in the 90°backing than the 0°UD fibers. A transilluminated white light image is presented in figure 5 where the surface cracks are clear at 90°and ±45°, along with the 60°cracks of the sub-surface ply, while there are little to no cracks in the 0°tows on the surface ply. Matrix cracking occurs under transverse [8] and shear loads because of mode I and mode II crack opening [5], so the global longitudinal load acted normal to the 90°backing fiber bundles and facilitated matrix crack growth. In contrast, the longitudinal stress that dominated in the UD tows did not cause crack growth. Any cracking in the UD tows under this longitudinal load would have been caused by the transverse stress developed by the mismatch in Poisson's ratio between the surface ply and the laminate. When matrix cracking occurs, the effective thermoelastic properties decay [8], including the stiffness and the coefficient of thermal expansion α, which are specifically important to TSA. Because the different orientations of fibers had different crack densities, the stresses and thermoelastic response of the BFS specimen surface diverged across the features.

4. Conclusion

Two mechanically similar specimens were tested, with one having it's backing fibers layer exposed and the other having the surface layer flipped so that the UD fibers were exposed. The two specimens exhibited similar stiffness degradation behaviors, but had different thermoelastic responses. The response of neither laminate matched the expectations from the simplified Classical Laminate Theory or Surface Resin Rich Layer methods, which suggests that further work is needed to account for additional effects. The presence of backing fibers on the BFS specimen influenced its thermoelastic response evolution in a non-homogeneous manner as the damage level increased. This was caused by the different levels of stress induced in the UD and backing fibers of different angles. This non-homogeneous nature changed during fatigue loading, likely due to ma-

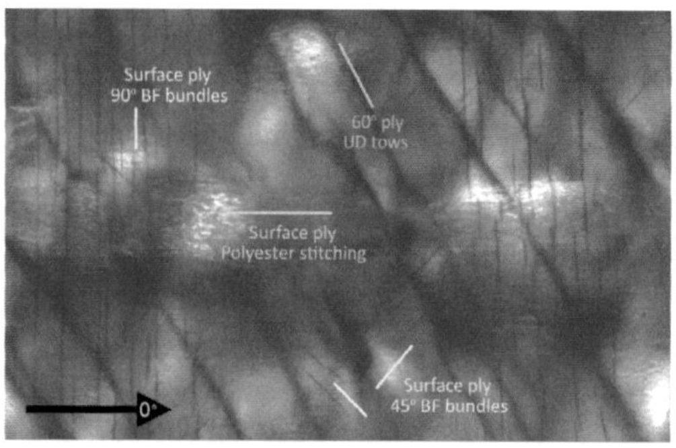

Figure 5. Photograph of a back-lit specimen after 5000 cycles, with surface cracks in 90°and ±45°backing fibers, and sub-surface cracks in ±60°plies.

trix crack development in the backing fiber bundles. Therefore, the UDS configuration appears to be preferable for use in future structural tests, as it avoids the potential influence of backing fibers on the TSA results.

The results presented in this paper encourage future work to both better understand the effects of backing fibers on TSA and how fatigue damage affects the material properties that influence the thermoelastic response. Examples of future work include:

- Testing specimens without paint on the surface so that the matrix crack density evolution can be recorded and compared with the TSA results
- Measuring full-field strains with Digital Image Correlation
- Testing uni-directional specimens with the layups $[_b0/_b0]_s$ and $[0_b/_b0]_s$ so the influence of the $\pm60°$plies will be removed
- Testing specimens at different loading amplitudes and frequencies to observe possible non-adiabatic effects

Acknowledgments

This study was conducted as part of the CASMaT Initiation Project undertaken as part of Villum Center for Advanced Structural and Material Testing (grant no. VKR 023193).

References

[1] Van Paepegem W and Degrieck J 2002 *Int J Fatigue* **24** 747–62
[2] Bull D, Barton J, Thomsen O, Butler R, Rhead A, Fletcher A and Potter K 2017 *ICCM21*
[3] Quaresimin M 2018 *Comprehensive Composite Materials II* ed Beaumont P W and Zweben C H (Oxford: Elsevier) pp 249–74 ISBN 978-0-08-100534-7
[4] Carraro P A, Novello E, Quaresimin M and Zappalorto M 2017 *Compos. Struct.* **176** 420–32
[5] Carraro P, Maragoni L and Quaresimin M 2017 *COMPOS SCI TECHNOL* **145** 24–39
[6] Quaresimin M, Susmel L and Talreja R 2010 *Int J Fatigue* **32** 2–16
[7] Castro O and Branner K 2019 *ICCM22*
[8] Varna J 2016 *Modeling Damage, Fatigue and Failure of Composite Materials* Woodhead Publishing Series in Composites Science and Engineering ed Talreja R and Varna J (Woodhead Publishing) pp 243 – 87 ISBN 978-1-78242-286-0
[9] Talreja R 2006 *J. Mater. Sci.* **41** 6800–12
[10] Dulieu-Barton J M 2012 *Optical Methods for Solid Mechanics: A Full Field Approach* pp 345–66
[11] Dulieu-Barton J and Crump D 2011 *Experimental and Applied Mechanics* vol 6 (Springer) pp 259–66
[12] Stanley P 2008 *Strain* **44** 285–97
[13] Emery T, Dulieu-Barton J, Earl J and Cunningham P 2008 *COMPOS SCI TECHNOL* **68** 743–52
[14] Crump D and Dulieu-Barton J 2012 *Experimental mechanics* **52** 829–42
[15] Fortunato I J, Bull D J, Dulieu-Barton J M, Thomsen O T *et al.* 2019 *ICCM22*
[16] Salerno A and Costa A 2010 *ICEM 2010* vol 6
[17] Paynter R and Dutton A 2003 *Strain* **39** 73–8
[18] Waugh R C, Dulieu-Barton J M and Quinn S 2015 *ICEM 2014*
[19] Pitarresi G, Conti A and Galietti U 2005 *Applied Mechanics and Materials* vol 3 pp 167–72
[20] Bakis C E and Reifsnider K L 1991 *J COMPOS MATER* **25** 809–30
[21] Sambasivam S, Quinn S and Dulieu-Barton J 2009 *ICCM17*
[22] Adolphs G and Skinner C 2011 *Non-Crimp Fabric Composites* (Elsevier) pp 481–95
[23] Zangenberg J, Brøndsted P and Gillespie Jr J W 2014 *J COMPOS MATER* **48** 2711–27
[24] Jespersen K M, Glud J A, Zangenberg J, Hosoi A, Kawada H and Mikkelsen L P 2018 *Compos. Part A Appl. Sci. Manuf.* **109** 481–97
[25] Jespersen K M and Mikkelsen L P 2017 *COMPOS SCI TECHNOL* **153** 261–72
[26] Jespersen K M, Zangenberg J, Lowe T, Withers P J and Mikkelsen L P 2016 *COMPOS SCI TECHNOL* **136** 94–103
[27] Mikkelsen L P 2020 *Fatigue Life Prediction of Composites and Composite Structures* (Elsevier) pp 569–603
[28] ASTM 2017 *ASTM International*
[29] Kalogiannakis G, Van Hemelrijck D and Van Assche G 2004 *J COMPOS MATER* **38** 163–75
[30] Daniel I M, Ishai O, Daniel I M and Daniel I 1994 *Engineering mechanics of composite materials* 2nd ed (Oxford university press New York)

IOP Conf. Series: Materials Science and Engineering **942** (2020) 012041 doi:10.1088/1757-899X/942/1/012041

Effect of maximum load on cyclic crack growth in UD E-glass/vinyl ester composites with constant and alternating stitch density

Rubén I Erives Anchondo* and Stergios Goutianos

Technical University of Denmark, Department of Wind Energy, Risø Campus, Frederiksborgvej 399, DK-4000 Roskilde, Denmark

E-mail: *riea@dtu.dk

Abstract. DCB specimens, manufactured from composites with constant and alternating stitching density, are tested under pure mode I for static and cyclic loading. The tests are carried out for different maximum load values, but keeping the same R-ratio. The crack growth is recorded from pictures at different cycle intervals, and the energy release rate (ERR) is computed using an approach based on the J-Integral. Along with the crack-tip, the length of the fracture process zone (FPZ) is recorded with the number of cycles. A curve relating the crack growth rate (CGR) and the cyclic fracture resistance, ΔJ, which can be directly applied in the design/repair process of composite structures, is derived.

1. Introduction

Composite materials are widely used in many modern engineering lightweight structures. Their high specific mechanical properties (e.g. specific strength, and stiffness) coupled with their good fatigue performance make them preferable to metals for some applications such as aircraft and wind turbine blades. Currently, damage tolerant designs of composite structures are being evaluated. However, such designs can only be implemented if accurate damage propagation prediction tools exist. The separation of initially bonded adjacent layers, which comes from the inherently weak interfaces (out-of-plane strength) of layered materials, is a common damage mechanism known as delamination. Such a process can lead to large stiffness and load-carrying capacity losses, which is why delamination is widely acknowledged as one of the most important failure mechanisms [1]. Most engineering structures are subject to cyclic loading, and thus cyclic crack growth is more relevant.

Accurate predictions of fatigue delamination largely depend on the accuracy of the fracture parameters of the material at hand. Predictions of fatigue delamination in composite structures are commonly based on numerical simulations either using a linear elastic fracture mechanics (LEFM) framework or using the cohesive zone model (CZM) concept originally proposed (independently) by Dugdale [2] and Barenblatt [3]. Cohesive models have become more widely adopted given that they circumvent some of the inherent limitations of LEFM, such as specific mesh requirements at the stationary crack tip, while also providing a framework to account for large scale fracture process zones. Cohesive models require fracture material properties such as critical fracture toughness, critical opening displacements or tractions as input. These fracture properties are commonly derived for each material system typically from experiments like the double cantilever beam (DCB) [4, 5], end-notched flexure (ENF) [6], and the mix-mode bending (MMB) specimens [7], among others. The present work lays out a methodology to obtain the mode I fracture toughness and cyclic crack-growth rate from based on experimental data from DCB specimens. To demonstrate the approach, several tests with different maximum loads (but constant R-ratio) are carried out on laminates manufactured from constant and alternating stitching density. The effects of the maximum load, P_{max} and the stitching density of the composite laminates in the crack growth rate (CGR) and the cyclic fracture resistance ΔJ are discussed.

2. Methods

2.1. Materials and DCB specimens

The laminate used for the DCB specimens is an E-glass/vinyl ester laminate manufactured by vacuum infusion with a curing cycle of 24 hrs at room temperature followed by 24 hrs at 60 °C. The E-glass fabric consisted of nearly unidirectional bundles, and the laminates had a volume fraction of approximately 55%. Two laminates were manufactured using the same fibers, and matrix, but different stitching density. Both configurations use glass yarns as support threads and a tricot stitching pattern, but one has a constant number of stitch ends per inch equal to 6 (6 gg), and the other has a stitching density alternating between 6 gg and 2 gg. The support threads are arranged so that they face each other, leaving alternating interfaces free of threads and thus exclude the effect of support threads in the cyclic crack growth rate. Prior to vacuum infusion, the middle interface (free of support threads) is used for introducing an artificial crack initiation through a release foil of 35 ±15 μm (See Fig. 1).

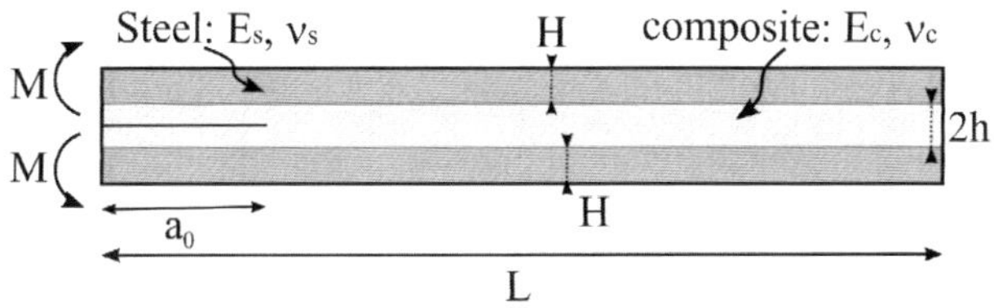

Figure 1. DCB speceimen schematic

The DCB specimen dimensions are; Length, L, equal to 500 mm, width, B, equal to 30 mm, and height, 2h, of approximately 2 mm. The DCB specimen was glued to steel beams on the top and bottom of the beams to add flexural rigidity in order to prevent large rotations. The steel beams have the same length, L, and width, B, but a thickness, H, of 6.67 mm. A schematic of the sandwich structure is shown in Fig. 1

2.2. Test configuration

Both quasi-static and cyclic tests were carried out by applying equal, but opposite constant moments at the top and bottom ends of the DCB specimen. The constant moment is applied through an adjustable lever-arm for both fatigue and static tests as shown in Fig. 2. The load, P, is measured by two load cells and the average value is used to calculate the applied moment, M. The approach of applying constant moment instead of constant load was introduced by Sørensen et al. [8, 9] and it offers the advantage of producing stable crack growth as this no longer depends on the crack length. More importantly, the approach produces more accurate measurements in the case of large scale FPZ typically observed in composite materials.

The static and fatigue test are carried out on the same specimen sequentially, starting with the static test, and then followed by cyclic loading. This allows for more testing per specimen and

IOP Conf. Series: Materials Science and Engineering **942** (2020) 012041 doi:10.1088/1757-899X/942/1/012041

also ensures a sharp crack at the beginning of the cyclic loading. For the static test, the normal opening is measured with an extensometer mounted on the end top and bottom of the DCB specimen (See Fig. 2a). For the cyclic test, instead of the normal opening, the crack growth is measured using digital pictures taken from the same location at different cyclic intervals. The post-processing is carried out manually, which means that it is subject to reading errors and interpretation. The presence of large scale bridging made it so that the definition of the start and end of the FPZ is not trivial. The start of the FPZ, i.e. the crack-tip, is defined as the first observable colour discontinuity, whereas the end of the FPZ is defined as the beginning of the last observable fibre bridge (farthest from crack tip, see Fig. 8). Reading errors are accounted to an extent by carrying multiple measurements and using mean crack length values.

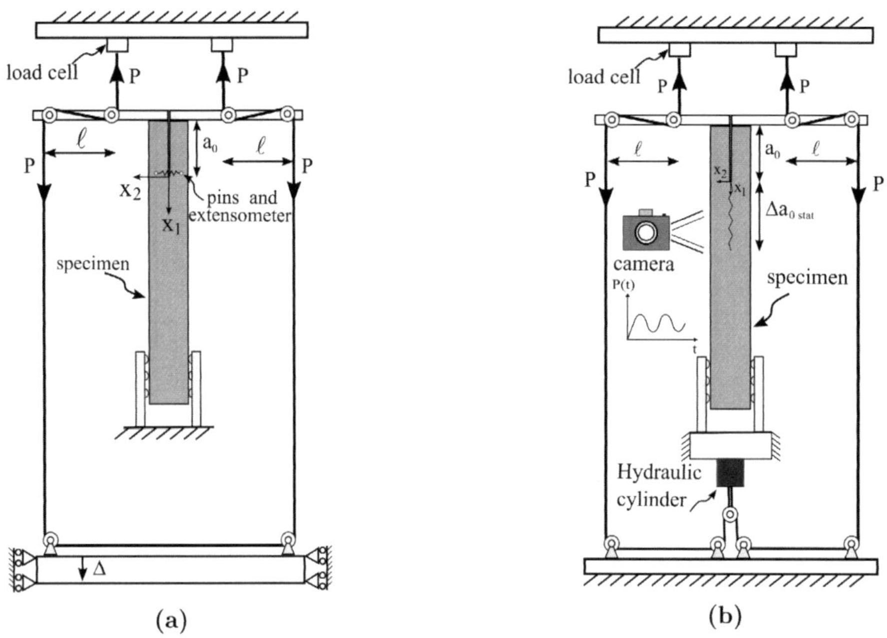

Figure 2. Schematic of test configuration: a) Static test-stand, b) Cyclic test-stand

2.3. Calculation of Fracture Resistance

The strain energy release rate of DCB specimens is commonly calculated using principles of LEFM (see for instance the ASTM D5528-01 [5]), however, LEFM is no longer valid when large scale bridging is present. For non-negligible FPZ cases, the J-integral method [10] is a powerful tool to obtain the energy release rate. Due to the relatively high width to height ratio plain strain condition is assumed. Under such condition, the J-integral for an orthotropic sandwich DCB specimen loaded with constant end moments is [8, 11]:

$$J = \frac{1-\nu_s^2}{E_s}\frac{M^2}{B^2H^3\eta^3 I} \tag{1}$$

with,

$$I = \frac{1}{3}\left\{\frac{1}{\eta^3} + 3\frac{\Delta}{\eta}\left(\Delta - \frac{1}{\eta}\right) + \Sigma\left[1 + 3\left(\Delta - \frac{1}{\eta}\right)^2 - 3\left(\Delta - \frac{1}{\eta}\right)\right]\right\} \tag{2}$$

$$\Delta = \frac{1 + 2\Sigma\eta + \Sigma\eta^2}{2\eta(\Sigma\eta + 1)} \quad , \quad \Sigma = \frac{E_c}{E_s}\left(\frac{1-\nu_s^2}{1-\nu_c^2}\right) \quad , \quad \eta = \frac{h}{H} \tag{3}$$

3. Results and discussion

The static fracture resistance, J_R, as a function of the end normal opening mode, δ_n, calculated using Eq. (1) is shown in Fig. 3. The upper portion of the curve corresponds to the loading part and the bottom to the unloading. The initiation of delamination corresponds to point 1 in Fig. 3, whereas point 2 indicates the steady-state delamination growth. It is observed that the initiation J_R value in the static case (point 1, approximately 200 N/m) is larger than any of the J_{max} values from the cyclic tests shown later in Table 1.

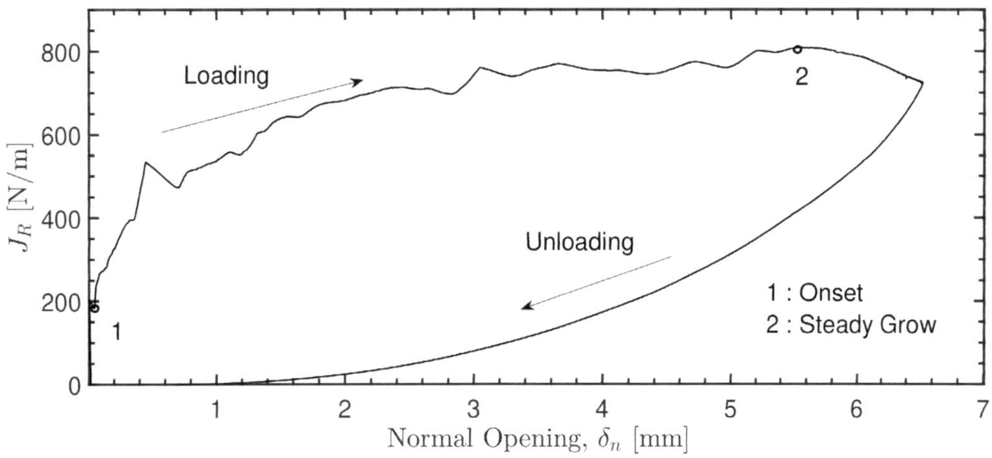

Figure 3. Static fracture resistance of a DCB specimen from alternating stitching density fabric

For the cyclic test, the cyclic fracture resistance ΔJ is calculated instead of J_R, where ΔJ is defined simply as the difference of J_{max} and J_{min}. In Table 1 all of the tested cases are listed with their corresponding maximum load, R-ratio, moment, and ΔJ. It is to be noticed that given the applied moment remains constant through the cyclic loading, the computed ΔJ corresponds to that of a single cycle.

Table 1: List of tested cases and with their maximum load and corresponding ΔJ

Stitching density [gg]*	**P max.** [N]	**P min.** [N]	**R-ratio** [-]	**M max.** [N-m]	**M min.** [N-m]	**J_R max.** [N/m]	**J_R min.** [N/m]	**ΔJ** [N/m]
	500	50	0.1	42.5	4.3	109.2	1.1	108.1
6	400	40	0.1	34.0	3.4	69.9	0.7	69.2
	350	35	0.1	29.8	3.0	53.5	0.5	53.0
	400	40	0.1	34.0	3.4	69.9	0.7	69.2
6-2	350	35	0.1	29.8	3.0	53.5	0.5	53.0
	375	37.5	0.1	31.9	3.2	61.4	0.6	60.8
	385	38.5	0.1	32.7	3.3	64.7	0.6	64.1

* Number of stitches per inch

Readings from the first cycle and close to the last one are shown in Fig. 4. By using the markings in the figure as a reference it can be clearly observed how the crack has propagated from cycle 1 to 22000. As previously mentioned several readings were carried out at each acquired picture to estimate the reading error. For all of the crack-tip measurements, the maximum standard deviation is 1.74 mm and the mean standard deviation is 0.19 mm.

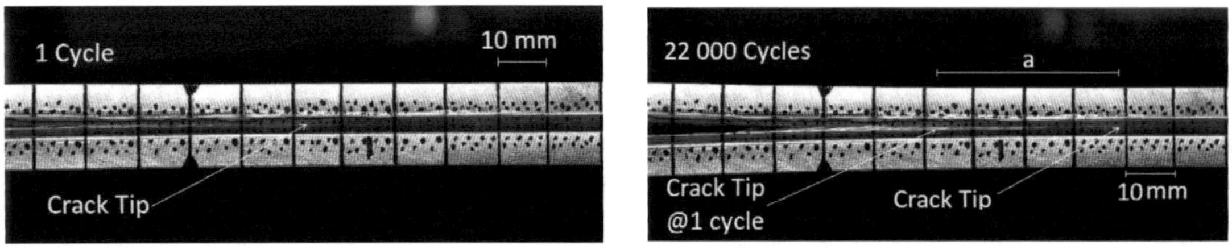

(a) (b)

Figure 4. Crack-tip growth for DCB under $P_{max} = 500N$ and constant 6 gg at, a) 1 cycle, and b) 22000 cycles

The crack growth as a function of the number of cycles is shown in Fig. 5a for different maximum loads of the constant stitching specimens. Note that the crack tip growth is tracked only during the cyclic loading, as such throughout the present paper the term crack growth refers to the growth due to cyclic loading. From Fig. 5a it is observed that as the load increases the crack growth rate increases. As the crack grows it can take up to two orders of magnitude more cycles for the case with a $P_{max} = 350N$ to reach the same crack length as the case with $P_{max} = 500N$. On the right side of Fig. 5, a similar plot shows the comparison of results from the DCBs made with different stitching density. From Fig. 5b, it can be noticed that for the same load (400 N, and 350 N) the constant stitching density specimens are more resilient to fracture than those with alternating stitching density. This observation is somehow expected as the constant 6 gg is more dense than the alternating 2-6 gg. This dependency on stitching density, however, contradicts the observations from Mouritz in [12], where it is concluded that fatigue is not dependent on stitching density.

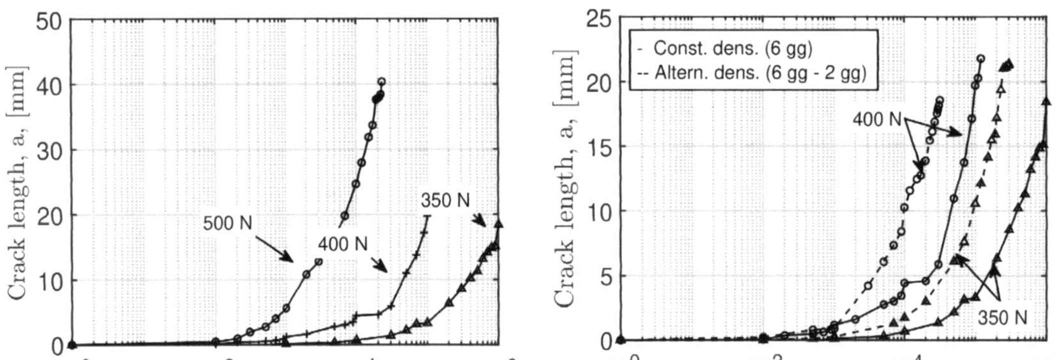

(a) (b)

Figure 5. Crack length vs number of cycles: a) different maximum load (constant stitching density), b) constant vs alternating stitching density

A plot of the Paris region (linear in a log-log plot) is shown in Fig. 6. From visual inspection of the crack growth, it was observed that once the crack reached about 1 mm length its growth behave (close to) linear in a log-log plot. This value was used as a threshold for data to be included in the linear regressions shown in the plot.

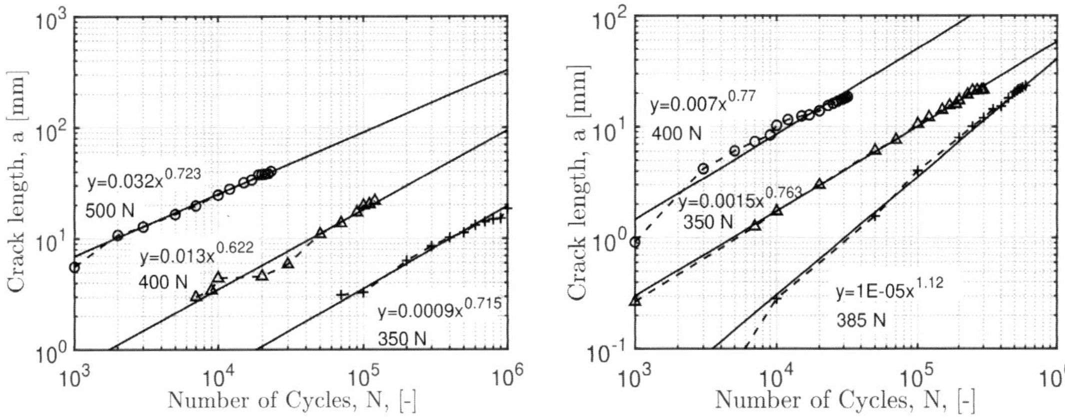

(a) Constant stitching density (b) Alternating stitching density

Figure 6. Linear part of the crack growth

The crack growth rate, da/dN, versus ΔJ is shown in Fig. 7 for each of the studied cases. The crack growth rate is calculated using the 2 and 3-point secant method, and the scatter reduction method proposed by Zheng et al. in [13], however, the 3-point method was selected because it yielded the least amount of scatter. Historically, secant methods have shown to yield large scatter data [13], however, they are easily implemented for any test configuration where the crack growth and the number of cycles are monitored. For the case with the constant stitching density, the three different loads lie almost in the same line, which is expected given that they have the same R-ratio. In the case of the alternating stitching density, one of the data sets seems to deviate. With only three points it is hard to make assertions, however, a likely scenario is that the CGR from the point with $P_{max} = 385N$ is under-estimated as it shows a decrease from the previous point, which does not make sense physically. Linear regression is performed to fit the data in the Paris' relation as shown in figure 7. The values of the obtained Paris' constants are given in Table 2. Both configurations, constant and alternating stitching densities, present similar slopes (constant n), but the intersection constant C is affected (see Table 2).

Table 2: Paris constants values

Stitching density [gg]	**Log(C)** $[\frac{mm}{cycle}/(MPa * \sqrt{m})^n]$	**n** [-]
6	4736.33	6.47
2 - 6	3035.06	6.04

As it is common in DCB tests of composite materials with 0°interfaces, all of the specimens showed large scale fibre bridging (LSB). This made the reading of the FPZ more challenging, which is reflected by the measurements statistics. The maximum standard deviation and mean standard deviation were 4.25 mm and 0.53 mm respectively, for the measurements of the FPZ. Fig. 8 shows a measurement of FPZ. It should be mentioned that for some of the cases there were isolated fibre bridges that originated from the static load but remain virtually unchanged

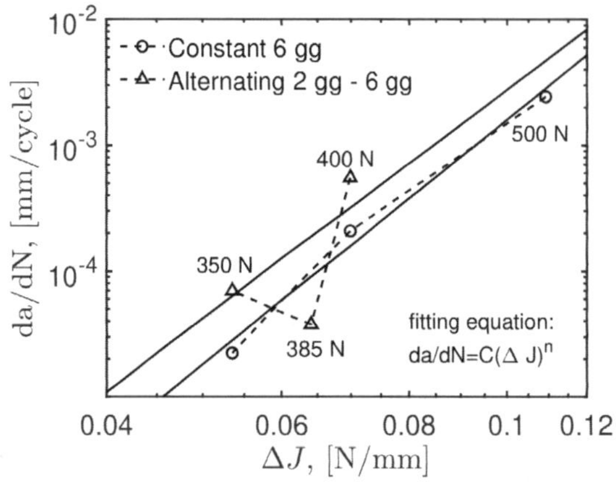

Figure 7. Crack growth rate (CGR) as a function of ΔJ for constant and alternating stitching density

during cyclic loading. This can be due to relatively small end openings. Because of the two-dimensionality of pictures, it is unclear if the observed isolated fibres are indeed isolated or there are more of these fibre across the width of the DCB. As such they are still considered as part of the FPZ.

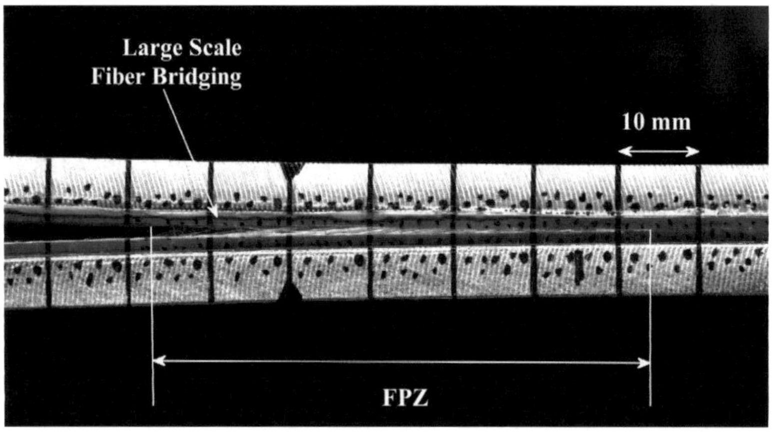

Figure 8. Illustration of large scale fibre bridging (LSB) under cyclic loading

The length of the FPZ as a function of the number of cycles was normalised with the static FPZ length. As observed from both Fig. 9a and Fig. 9b the length of the FPZ increases in a similar way as the crack-tip does. For the case of constant stitching density at $P_{max} = 500N$ an increase of up to 50% of the static FPZ is reached.

IOP Conf. Series: Materials Science and Engineering **942** (2020) 012041 doi:10.1088/1757-899X/942/1/012041

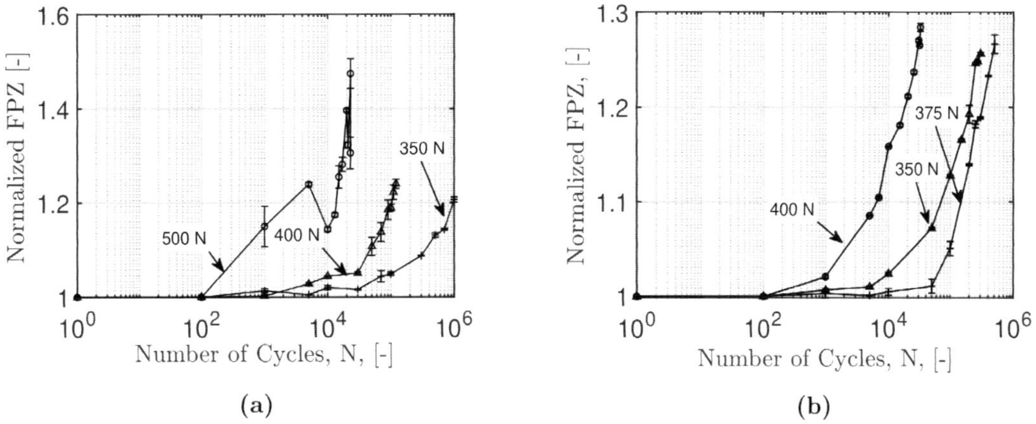

Figure 9. Normalised FPZ length a) constant stitching density, b) alternating stitching density

It can be observed that the bridging zone keeps increasing throughout the entire cycle range, which contradicts the hypothesis that the bridging zone would remain constant after reaching steady state grow [16]. However, it is important to acknowledge the difficulty of drawing conclusions from the FPZ readings without a meaningful identification of the fibre bridge zone that contributes to the cohesive traction force.

4. Conclusion

A series of DCB specimens made from constant and alternating stitching densities (6 gg and 6 gg- 2 gg respectively) were tested for static and cyclic loading with different maximum loads, but equal R-ratio. The loading was applied via constant end-moments with the same amplitude but opposite directions, i.e. pure mode I. The crack propagation is measured from pictures taken at different cycle intervals. A list of observations from the experiments is presented below.

- Contrary to previous findings [12], the stitching density was found to affect the cyclic (fatigue) fracture properties of the DCBs. More specifically, the tested DCB specimens with constant stitching density (6 gg) are more fracture-resistant under cyclic loading than those with alternating stitching density (2 gg- 6 gg).
- The stitching density is found to have an effect on the intersection of the crack growth rate curve, which is expressed by the Paris' constant C. Little to no dependency is found on the slope described by the constant n.
- A curve relating to the crack growth rate and ΔJ has been derived. This curve can be directly applied in the design and/or repair of composite structures subject to cyclic load.
- The length of the FPZ increases with the number of cycles for all of the cases.

Acknowledgments

The present article is part of a Ph.D. project funded by the Mexican National Council on Science and Technology, and the Energy Sustainability fund (CONACyT-SENER). The experimental part was funded by the European Unions Horizon 2020 Research and Innovation Programme under Grant Agreement No. 761072 (DACOMAT). Research technicians Jan Sjølin and Christian H. Madsen are acknowledged for technical assistance.

IOP Conf. Series: Materials Science and Engineering **942** (2020) 012041 doi:10.1088/1757-899X/942/1/012041

References

[1] Pagano N Jand Schoeppner G A 2000 Delamination of Polymer Matrix Composites: Problems and Assessment, *Comp. Compos. Mat.*, **2** 433–528

[2] Dugdale, D S 1960 Yielding of steel sheets containing slits. *J. Mech. Phys. Solids*, **8(2)**, 100–104.

[3] Barenblatt G I, 1962 The Mathematical Theory of Equilibrium Cracks in Brittle Fracture. *Adv. Appl. Mech.* C, **7**, 55–129.

[4] Hojo M, Tanaka K, Gustafson C and Hayashi R 1987 Effect of stress ratio on near-threshold propagation of delimination fatigue cracks in unidirectional CFRP.*Compos. Sc. Tech.* **29(4)**, 273–292.

[5] ASTM D5528-01 2014 Standard test method for mode I interlaminar fracture toughness of unidirectional fiber-reinforced polymer matrix composites. *ASTM*, **03** (Reapproved 2007), 1–12.

[6] Matsuda S, Hojo M and Ochiai S 1997 Mesoscopic Fracture Mechanism of Mode II Delamination Fatigue Crack Propagation. *JSME Int. J.*, **34(1)**, 430–433.

[7] Reeder J R, and Crews J H 1990 Mixed-mode bending method for delamination testing. *AIAA J.*, **28(7)**, 1270–1276.

[8] Sørensen B F, Jørgensen K, Jacobsen T K, Østergaard R C 2006 DCB-specimen loaded with uneven bending moments. *Int. J. Fracture*, **141(1–2)**, 163–176.

[9] Sørensen B F and Jacobsen T K 2009 Characterizing delamination of fibre composites by mixed mode cohesive laws. *Compos. Sci. Tech.*, **69(3–4)**, 445–456.

[10] Rice J R 1967 a Path Independent Integral and the Approximate Analysis of Strain Concentration by Notches and Cracks *J. Appl. Mech.*, **35**, 379-386.

[11] Bao G, Ho S, Suo Z and Fan B 1992 The role of material orthotropy in fracture specimens for composites. *Int. J. of Solids Struct.*, **29(9)**, 1105–1116.

[12] Mouritz, A P 2004 Fracture and tensile fatigue properties of stitched fibreglass composites, *J. Mat.*, **218(2)**, 87–93.

[13] Zheng J, and Powell B E 1997 A method to reduce the scatter in fatigue crack growth rate data. *Fat. Fract. Eng. Mat. Struct.*, **20(9)**, 1341–1350.

[14] Tristan Ursell 2020 Active Figure Zoom for Selecting Points, MATLAB Central File Exchange. Retrieved July 9, 2020.

[15] AAT Bioquest, Inc. (2020, July 09). Quest Graph^{TM} Linear, Logarithmic, Semi-Log Regression Calculator.". Retrieved from https://www.aatbio.com/tools/linear-logarithmic-semi-log-regression-online-calculator

[16] Sørensen B F 2016 Delamination fractures in composite materials *Comp. Sc. Eng.* **11**, 213-240

IOP Conf. Series: Materials Science and Engineering **942** (2020) 012042 doi:10.1088/1757-899X/942/1/012042

Does age matter? Impact on fire safety properties of composite materials from ageing

Anna Sandinge[12], Per Blomqvist[1] and Anne Dederichs[12]

[1] RISE Research Institutes of Sweden, Borås, Sweden
[2] DTU Technical University of Denmark, Kongens Lyngby, Denmark

anna.sandinge@ri.se

Abstract. When materials are tested and classified before entering the market, they are mainly tested as newly produced. However, it is known that material properties change with time and when exposed to temperature, humidity, wind and light. As a result, it is important to have knowledge of how material age and which parameters are affected in order to retain safety. Studies show how the mechanical properties change when the materials age. But not much can be found in literature about the ageing effect on fire properties. In the present study, accelerated ageing testing was made with a composite material of phenolic resin and basalt fibres. Selected ageing methods applied were thermal ageing at 90 °C and moisture ageing at 40 °C and 90 % Relative Humidity. Samples were collected from ageing chambers after one, two and four weeks. To investigate the ageing effect on the fire properties of the composite, fire testing was conducted using cone calorimetry according to ISO 5660-1. The test results showed that ageing does matter. There was an impact on the material and the fire properties were affected. The ignition time decreased for the aged samples and the heat release rate slightly increased. Also, the smoke production increased with ageing.

1. Introduction

When new products and materials are fire tested and classified before entering the market, they are tested as newly produced. However, it is known that material properties change with time and when exposed to harsh environments such as elevated temperature, humidity, wind and/or light. Many studies in literature show how mechanical properties change and deteriorate when the materials age. Not much can be found in the literature on the ageing effect on fire properties. An important question is what happens with fire properties when materials are exposed to temperature and humidity for a long time [1].

Composite materials are widely used in applications in ships, railway vehicles, wind turbines and construction. Fibre Reinforced Polymer (FRP) composite materials are lightweight and have good mechanical properties. However, their weakness is the fire behavior, since they are combustible they will burn, and this can harm the safety of the construction [2]. For composite materials to be more resistant to fire there is often a treatment applied with flame retardants. The flame retardants can be added to the materials surface, mixed with the material or incorporated in the material chemically. As confirmed by fire testing of newly produced materials, the criteria for fire performance are achieved. However, how these flame retardants act when aged is to a large extent unknown [1].

With respect to maintenance and the sustainable life cycle, it is essential to know the time for replacement of materials in applications such as vehicles and buildings due to degradation in the material. It is critical to know at what stage the material no longer fulfils the safety regulations and requirement regarding the fire safety. From an environmental perspective, the replacement of materials should not be done too early in a products life cycle. Altogether, the ageing effect on material properties is important.

2. Theoretical background

The only truly valid ageing method is storing materials for the corresponding time in their end use climate and then evaluate the material properties. However, this is time consuming and often not an option in development processes. Accelerated ageing can be used in order to shorten the process, which is a method to expose materials to a proper simulation of long-time use. Damage and degradation of the material can occur after a few days, weeks or months, which would normally occur after years of use in end use climate [3] [4].

2.1. Accelerated ageing methods

The selection of ageing method is important and should represent the end use climate. Important parameters to consider when selecting ageing method are the time scales, number of experimental points, choice of degradation agent and parameters to evaluate. The degradation agents are light, heat, liquids, gases, radiation, mechanical stress and electrical stress. The degradation of materials can be both physical and chemical. If the selected ageing method or climate is too tough, there can be a physical degradation or chemical reactions which would not take place when the materials are aged in real-time with end use climate. This will result in an incorrect evaluation of the material properties [3] [5]. In the following sections, thermal and moisture ageing are described more closely as they were applied in the current work.

2.1.1 Thermal ageing. Thermal ageing is a commonly used accelerated ageing method which is easy to conduct. The impact of thermal ageing with a polymer is molecular impairment within the material due to exposure of heat. As a result, the long polymer chains are broken and reactions are initiated, leading to a change in material properties [6]. Degradation mechanisms of a FRP composite material is temperature dependent and the degradation increases with increasing temperature [1]. In this paper, thermal ageing is the exposure to an elevated temperature during a discrete time.

A useful tool to correlate the time for accelerated ageing with actual ageing time in the end user application is the Arrhenius Rate Law, defining the rate of a chemical reaction as increasing exponentially with the absolute temperature. The Arrhenius Rate Law is given by equation (1).

$$\ln k(T) = -\frac{E_a}{RT} + lnA \quad (1)$$

Where k is the reaction rate of the process, E_a is the activation energy, R the universal gas constant, T the absolute temperature and lnA is the pre-exponential factor, expressing the total number of molecular collisions that may initiate a reaction [3]. High temperature and low activation energy results in a higher rate constant speeding up the degradation process [7] [5]. Thus, when a material is exposed to higher temperature the speed of chemical reactions in the material increases. This increase in kinetic energy of the molecules is described by RT in the Arrhenius rate law. If the reaction mechanism does not change with temperature the activation energy is constant. When dealing with an activation energy independent of temperature, there is a linear relationship between the logarithm of the reaction rate, ln k, and the inverse of the absolute temperature. This can be simplified with a rule of thumb for reactions at room temperature; the reaction rate doubles for every 10 K increase in temperature. If a material is stored at 30 °C for 10 weeks, the accelerated ageing climate corresponding would be 40 °C for 5 weeks. However, care should be taken with extrapolations [8] [9].

2.1.2 Moisture ageing. A high moisture content constitutes a harsh environment for many materials since it can cause damage to the material structure. The material can absorb moisture resulting in dimensional changes, swelling, reduction in glass transition temperature and/or reduction of mechanical properties [2] [10]. Accelerated ageing with moisture can be conducted in numerous ways. It can, e.g., be exposure to a specific humidity of the surrounding atmosphere or by immersion in a liquid [3] [10].

3. Experimental method

The materials used and the methods of fire testing of the aged and unaged materials are presented below.

3.1. Material

An FRP composite material consisting of Phenolic resin and Basalt fibres was used for the study. This material could, e.g., be used in wall or floor applications inside a ship or as windmill rotors. This composite also contained an expansion agent which is active during manufacturing of the composite, and its mechanism gives the possibility for different densities of the composite. The selected composite for the study had a density of 1000 kg/m^3 and a thickness of 3 mm.

3.2. Accelerated ageing

The intended end use of the Phenolic/Basalt composite material investigated in this study is an application as a wall or floor kept at 20 °C for at least 10 years. Using the rule of thumb of Arrhenius Rate law this would correspond to an accelerated ageing climate exposure of 90 °C for 4 weeks. Samples were taken from the heat chamber after one week, two weeks, as well as four weeks, in order to follow the expected degradation of the material. However, since 90 °C is a considerable higher temperature compared to the end use climate, there is a possibility that non-typical reactions can occur due to the high temperature, which would not occur when aged in real-time. This would not be possible to exclude without conducting ageing tests for longer times at lower temperatures.

For this study, moisture ageing with a specific air humidity was selected. The Phenolic/ Basalt composite material was aged at 40 °C and with a Relative Humidity (RH) of 90 %. Samples were collected from the humidity chamber after one week, two weeks and four weeks [11] [12].

3.3. Testing with TGA

To evaluate the thermal stability of the Phenolic/Basalt composite material, Thermal Gravimetric Analysis (TGA) was applied. During this test, the mass of the material was monitored over time as the temperature increased with 10 K/min. Tests were conducted with ambient air as the atmosphere [13].

3.4. Fire testing

Fire testing was conducted using cone calorimetry according to ISO 5660-1 to measure time to ignition and heat release rate (HRR). The samples were mounted horizontally and exposed to a heat flux of 50 kW/m^2. Duplicate tests were run with unaged samples and samples aged for one week, two weeks and four weeks. The test duration was 20 min [14].

When conducting cone calorimeter tests with composite materials, there is often unwanted burning originating from the specimen edges. This results in an extra contribution in heat release to the test results which would not appear in the end use of the product. To avoid this edge effect, a modified specimen holder was used. The sample specimen had the size of 136×136 mm and a modified frame with a circular opening with a diameter of 106 mm was used. This resulted in an exposed area of the specimen that was the same as that of the standard specimen holder described in ISO 5660-1. The larger distance to the specimen edges in the modified specimen holder was intended to prolong the time until start of pyrolysis of the edges. Further, between the sample and the retainer frame of the specimen holder was a 5 mm thick slab of insulation to prevent heating of the unexposed surface [15].

4. Results

Results from the tests conducted are given below. The aged samples were compared to unaged samples regarding physical appearance by visual inspection and fire safety properties by the cone calorimeter tests.

4.1. TGA test results

The results showed a relatively small gradual mass loss up to 300 °C and can therefore be seen as thermally stable at temperatures below 300 °C. Above this temperature, significant thermal degradation took place. Based on these results, using an accelerated ageing temperature of 90 °C, would be acceptable since it is below the critical temperature.

4.2. Visual inspection

A visual inspection was made of all samples after the accelerated ageing. The samples exposed to heat showed a distinct colour change. This colour change occurred already after one week. There were no shrinkage or mass change. The samples exposed to moisture and moderately increased temperature showed some colour change, not as distinct as for the thermally aged samples. There were no shrinkage or mass change.

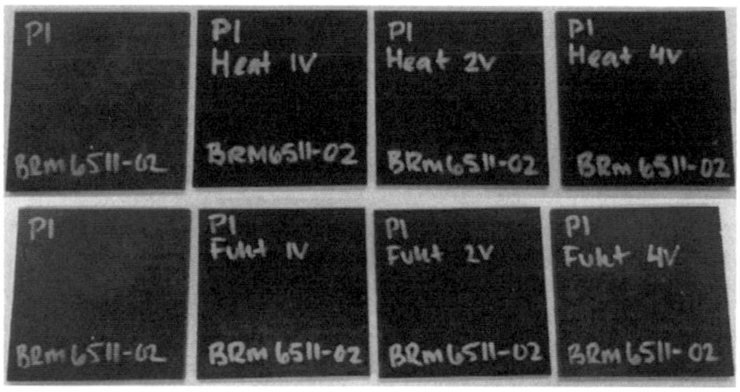

Figure 1. Samples of Phenolic/Basalt composite after thermal ageing (upper photo) and moisture ageing (lower photo). The sample to the left is unaged, then to the right, samples aged for one week, two weeks and four weeks, respectively.

4.3. Cone calorimeter testing

Table **1** shows the results of duplicate tests conducted with the cone calorimeter for unaged and thermal aged samples. Both the unaged samples and samples aged for one week show good repeatability regarding time to ignition and Total Heat Release (THR). The peak Heat Release Rate (HRR) has somewhat lower repeatability. However, the Total Smoke Production (TSP) shows very low repeatability, the value is about twice as high for one of the duplicates in both cases. The repeatability of the samples aged for two and four weeks was less good in case of time to ignition and especially peak HRR. One can note that for both the two- and four-weeks aged samples, the duplicate test with the longest time to ignition showed a higher peak HRR. The repeatability of TSP was better for the two- and four-weeks samples.

Visual observations in the tests showed the same general burning behaviour for all thermal aged samples as well as for the unaged samples. Early in test, before ignition of sample, there were a loud popping sound (from delamination) and flashing from the surface. All samples showed smoke appearing from the rear side of the specimen holder (from edge leakage of pyrolysis gases), but no ignition. The time for smoke at the rear side varied for the samples. For the unaged samples, smoke from rear side was observed at 170-200 s. This observation was noted at an earlier time for the aged samples, at 50-70 s for the one-week aged samples, at 120-170 s for two weeks aged samples and at 110 s for the four weeks aged samples.

Table 1. Cone calorimeter results of thermally aged material, duplicate tests.

Ageing time	Time to ignition (s)	Extinguished at time (s)	peak HRR (kW/m^2)	THR_{300} (MJ/m^2)	THR (MJ/m^2)	TSP_{300} (m^2/m^2)	TSP (m^2/m^2)
Un aged	93	726	110	12	35	85	87
	96	735	135	13	35	161	161
mv (rel md %)	*95 (2)*	*731 (1)*	*123 (10)*	*13 (4)*	*35 (0)*	*123 (31)*	*124 (30)*
1 week	62	835	87	11	33	41	52
	59	651	107	12	33	82	99
mv (rel md %)	*61 (2)*	*743 (12)*	*97 (10)*	*12 (4)*	*33 (0)*	*62 (33)*	*76 (31)*
2 weeks	84	813	190	16	40	206	221
	68	918	130	15	39	180	191
mv (rel md %)	*76 (11)*	*866 (6)*	*160 (19)*	*16 (3)*	*40 (1)*	*193 (7)*	*206 (7)*
4 weeks	73	699	112	14	37	120	121
	88	997	153	15	39	177	187
mv (rel md %)	*81 (9)*	*848 (18)*	*133 (15)*	*15 (3)*	*38 (3)*	*149 (19)*	*154 (21)*

Note: *mv* is the mean value of the duplicate tests and *rel md %* is the relative mean deviation in percentage from the mean value.

Despite the in some cases low repeatability of the duplicate tests, some valid conclusion can be made. The results showed clearly that the reaction-to-fire properties were affected by the ageing conditions. There was a faster ignition of all thermally aged samples compared to unaged samples. The HRR as a function of time of all samples are shown in Figure 2. The peak HRR decreased for samples aged for one week compared to the unaged samples, but the THR did not change. The THR value presented in Table 1 show the total heat release from ignition to end of test. As shown in Figure 2, the HRR reaches a steady-state production at approximately 300 s which gradually slowly decreased until the time of extinguishing. In addition, the total heat release was calculated from time for ignition to 300 s of test time, THR_{300}. This shows that even though the peak HRR for the one-week aged samples was lower than for the unaged samples, the total heat release from the samples was the same. Samples aged for two and four weeks showed an increase in the peak HRR compared to the unaged samples. The THR_{300} also showed an increase for the two- and four-weeks aged samples, as well as the THR for the complete tests.

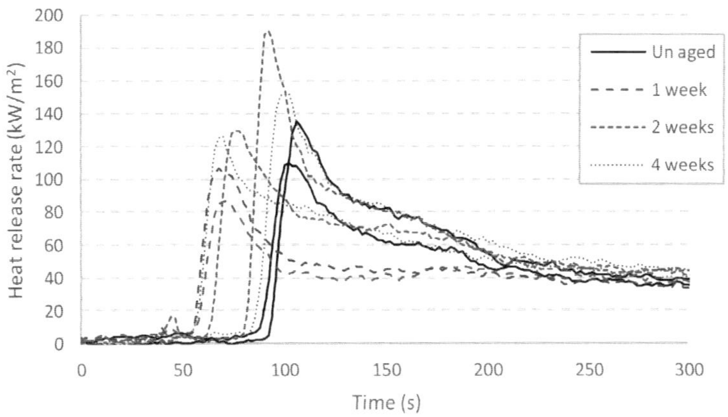

Figure 2. Heat release rate as a function of time, thermally aged material, duplicate tests.

Figure 3 shows the smoke production rate as a function of time. The smoke production increased fast at time for ignition, shown by the peak, and then decreased slowly. After 200 s, the smoke production had decreased for all samples, to stabilize close to zero. Table **1** shows the result of TSP_{300}, total smoke production from ignition to 300 s of test, and TSP, total smoke production from ignition to end of test. The repeatability of the smoke results was poor in the duplicate tests with the unaged and one-week aged samples as already discussed. The repeatability of TSP was better for the two- and four-weeks samples, and it is clear from the results that more smoke was produced from these samples compared with the unaged and one-week aged samples.

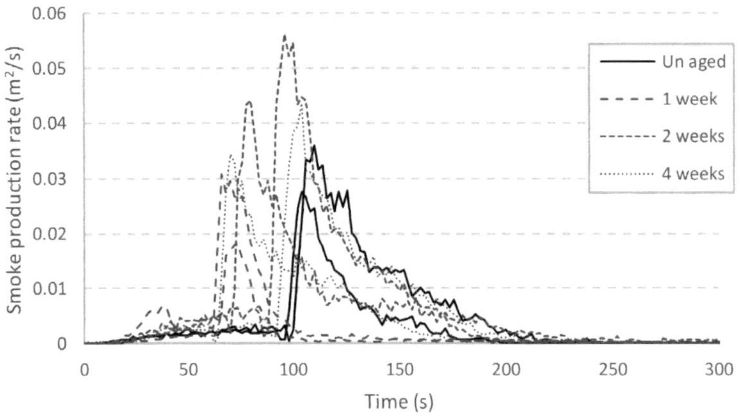

Figure 3. Smoke production rate as a function of time, thermally aged material, duplicate tests.

Table 2 shows the cone calorimeter results of the unaged samples and samples exposed to moisture and moderately increased temperature. The repeatability of the duplicate tests was acceptable for all test parameters with exception of the smoke production for the unaged samples.

Observations made during the tests showed a similar burning behaviour for both unaged samples and aged samples. Before ignition of the sample there was the loud popping sound as mentioned earlier. All samples showed smoke leakage from the rear side of the specimen holder, but no burning. Time for the smoke observation was at 170-200 s for the unaged samples as discussed above, 100-140 s for one-week moisture aged samples, 170-240 s for two weeks aged samples and 140-230 s for four weeks aged samples.

Table 2 Cone Calorimeter results of moisture aged material, duplicate tests.

Ageing time	Time to ignition (s)	Extinguished at time (s)	peak HRR (kW/m^2)	THR_{300} (MJ/m^2)	THR (MJ/m^2)	TSP_{300} (m^2/m^2)	TSP (m^2/m^2)
Un aged	93	726	110	12	35	85	87
	96	735	135	13	35	161	161
mv (rel md %)	*95 (2)*	*731 (1)*	*123 (10)*	*13 (4)*	*35 (0)*	*123 (31)*	*124 (30)*
1 week	97	930	146	14	36	198	205
	98	617	149	13	36	171	172
mv (rel md %)	*98 (1)*	*774 (20)*	*148 (1)*	*14 (4)*	*36 (0)*	*185 (7)*	*189 (9)*
2 weeks	100	721	156	15	39	234	247
	100	753	163	15	38	222	228
mv (rel md %)	*100 (0)*	*737 (2)*	*160 (2)*	*15 (0)*	*39 (1)*	*228 (3)*	*238 (4)*
4 weeks	97	784	148	15	41	234	239
	106	828	148	14	37	185	186
mv (rel md %)	*102 (4)*	*806 (3)*	*148 (0)*	*15 (3)*	*39 (5)*	*210 (12)*	*213 (12)*

Note: *mv* is the mean value of the duplicate tests and *rel md %* is the relative mean deviation in percentage from the mean value.

Comparing the test results of the unaged samples with the moisture aged samples shows that the accelerated aging had some limited effect on the reaction-to-fire properties. There was a slight increase in time to ignition for all aged samples. There was also an increase in peak HRR for all aged samples, with highest peak for the two weeks aged samples, shown in Figure 4. The heat release decreased after the initial peak at time for ignition, to reach a steady state of heat release at approximately 300 s. The THR increased somewhat with increasing ageing time for all aged samples compared to the unaged samples.

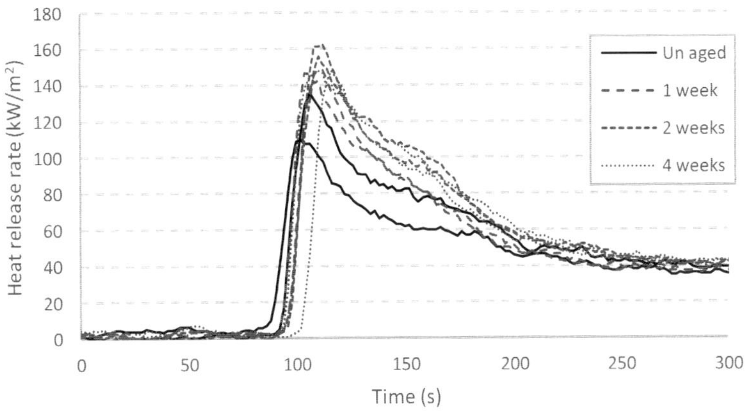

Figure 4. Heat release rate as a function of time, moisture aged material, duplicate tests.

The smoke production rate shown in Figure 5 illustrates that all aged samples had a higher smoke production compared to the unaged samples. All samples showed a peak in production at time for ignition followed by a decrease to almost no production.

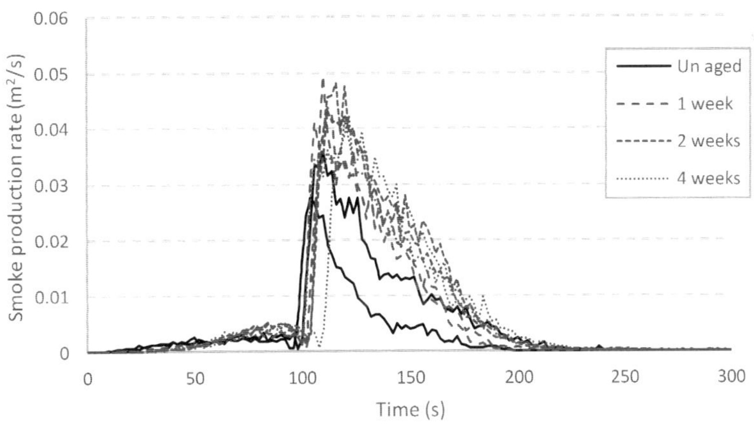

Figure 5. Smoke production rate as a function of time, moisture aged material, duplicate tests.

5. Conclusions

The test results showed that there are changes in reaction-to-fire properties when the material is aged, i.e. age matters. For the samples exposed to thermal ageing, there was a clear decrease in time to ignition, the heat release rate was increased, and the smoke production was increased. For the samples aged in moisture climate, there was a limited influence on the time to ignition, but the heat release rate and smoke production were increased compared to the unaged samples.

The main conclusion from the results is that there is a change in fire properties of the Phenolic/Basalt composite material when exposed to accelerated ageing. The results indicate that ageing deteriorates the material properties concerning fire safety. However, this study was conducted with one selected composite material only. Further testing with accelerated ageing is needed to better understand the effect of ageing on composite materials and the fire behaviour.

Acknowledgement
The research is a part of the RAMSSES project which has received funding under the European Union's Horizon 2020 research and innovation programme under the grant agreement No 723246.

References

[1] Vahabi H, Sonnier R and Ferry L. 2014. Effects of ageing in the fire behaviour of flame-retardant polymers: a review. *Journal of Polymer International* **64** 313-28.
[2] Molaba T P, Chapple S and John M J. 2018. Flame retardant treated flax fibre reinforced phenolic composites: Ageing and thermal characteristics. *Fire and Materials* 50-58.
[3] Brown R P, Kockott D, Trubiroha P, Ketola W and Shorthouse J. 1995. A review of accelerated durability tests. WAMAS Report No 18.
[4] Martins S, Borges L and D'Almeida. 2011. Effects os accelerated ageing in a PTFE matrix polymer composite. *Macromel Symp* **299/300** 92-98.
[5] Atkins P and de Paula J. 2002. Atkins physical chemistry.
[6] Ghosh S, Khastgir D, Bhowmick A K and Mukunda P G. 2000. Thermal degradation and ageing of segmented polyamides. *Polymer Degradation and Stability* **67** 427-36.
[7] Licht R R. 2005. Ageing & Environmental Exposure Properties of a Fire Protection Material. *Journal of ASTM International* **Vol 2** No 8.
[8] Emanuelsson V. Simonson M and Gevert T. 2007. The effect of accelerated aging of building wires. *Fire and Materials* **31** 311-26.
[9] Thureson P and Nilsson M. 1994. Degradation of Fire Properties of Approved Products as a Result of Ageing. SP Report 1994:61.
[10] Fiore V, Scalici T, Calabrese L, Valenza A and La Bella G. 2015. Durability of flex-basalt hybrid composites for marine applications. *International conference on composite materials, Copenhagen.*
[11] Scania TB4546. 2015. *Technical Regulation.*
[12] Volvo TR21465714. *Technical requirement.* Issue 02, Volume 01.
[13] EN ISO 11358-1. 2014. Plastics – Thermogravimetry (TG) of polymers – Part 1: General principles.
[14] ISO 5660-1. 2019. Reaction-to-fire tests – Heat release, smoke production and mass loss rate – Part 1: Heat release rate (cone calorimeter method) and smoke production rate (dynamic measurement).
[15] Sandinge A, Blomqvist P and Rahm M. 2020. A modified specimen holder for cone calorimeter testing to reduce influence from specimen edges. *Manuscript in writing process.*

DTU's blade research and demonstration platform

Philipp U Haselbach, Sergei Semenov and Peter Berring
Technical University of Denmark, DTU Wind Energy, Frederiksborgvej 399, DK-4000 Roskilde, Denmark
E-mail: phih@dtu.dk

Abstract. At DTU Wind Energy, a 12.6 m wind turbine rotor blade and the corresponding blade moulds were designed and manufactured to establish a blade research and demonstration platform. The blade research and demonstration platform shall allow manufacturing of a blade or blade segments under controlled conditions, allowing realizing theoretical concepts and testing these in full-scale for materials, manufacturing and design concepts of next generation wind turbines. For this purpose, the entire blade design with detailed airfoil and structural design as well as the manufacturing and test data of the blade will be published. Static tests, fatigue tests, post-fatigue static tests as well as a modal testing of the manufactured rotor blade are planned to characterize the blade response. In this way, the DTU 12.6 m wind turbine rotor blade and the blade moulds can serve as a reference rotor blade to benchmark numerical simulations and to test new materials, designs, manufacturing concepts or embedded sensors for future prognostics health management systems. In this article, results of a 3D scan of the blade moulds are presented as well as description of the water-based heating and cooling system and a resistance temperature detectors (RTD) measuring system.

1. Introduction

Development of new materials or structural concepts for wind turbine rotor blades requires a transformation of the initial idea via prototypes to a final product.Testing new concepts, materials or manufacturing approaches on small scale and comparing the results with reference cases is often desired to demonstrate the potential of an innovations before applying them on full-scale to large wind turbine blades. This approach reduces the time to market by simultaneously minimizing the financial risks.

Especially, for material suppliers or innovators without access to own wind turbine blade manufacturing and reference cases, it can be challenging and costly to test, benchmark and demonstrate new ideas and concepts. Currently, there exist a few public available reference wind turbine systems and wind turbine blade designs, e.g. the NREL 5 MW reference wind turbine [1] or the DTU 10 MW reference wind turbine [2]. However, these reference wind turbine concepts are mainly designed to serve as benchmark concepts for model-to-model comparison with focus on aerodynamic and aeroelastic content. The wind turbine blade structural representation is often reduced to stiffness matrix inputs [1] or only roughly described [2] without documenting structural details as e.g. the design of adhesive joints or ply drops.

At DTU Wind Energy, a blade research and demonstration platform has been designed and built to overcome the shortcomings of the aforementioned reference wind turbine concepts. DTU's new research and demonstration platform does not only provide detailed aeroelastic, aerodynamic and structural calculations of a 12.6 m long wind turbine blade but also material properties and characteristics according to DTU's curing cycles process. Moreover, DTU Wind Energy has developed an advanced blade mould, where the entire loop of numerical validation, manufacturing, testing and monitoring of new materials, manufacturing processes or structural concepts can be simulated and tested. With DTU's Large Scale Facility for structural blade testing testing of the final structural blade response can be investigated too. In the near future, all test data, simulations results as well as the entire blade and mould design will be made publicly available according to the FAIR principles, which means that the data are findable, accessible, interoperable and reusable.

The objective of DTU's research initiative is to create a blade research and demonstration platform and the necessary infrastructure to manufacture and test a wind turbine blade under controlled conditions to serve as reference wind turbine blade platform, allowing to experiment with new blade technologies and materials to create the wind turbine blades of tomorrow.

Mortensen et al. [3] demonstrated on panel level how curing process cycles and conditions can significantly affect the fatigue performance of unidirectional composite materials. By modifying the curing process, residual stresses in an epoxy-based fiber composite material could be reduced, which significantly improved the fatigue performance of these laminate by a factor of five. However, industry prefers short curing process cycles and conditions to decrease mould time per unit and at the same time to increase the production capacity. This can be achieved through the acceleration, automation and digitization of the entire composite manufacturing process compared with current production methods. Blade manufacturing processes can be improved by high precision handling methods, prefabricated inserts (precasts) and automation, where e.q. the effectiveness of the production methods can be measured by the amount of fibreglass material placed in the mould in a given amount of time (i.e. kg of fibreglass per hour). Combining modelling and real time assistance, monitoring and control of the manufacturing process through laser projection systems, augmented reality and wide range of sensors can help to optimize the manufacturing process and to reduce non-conformance in the final blade structure. Moreover, embedded sensors and heating and cooling systems in blade moulds in combination with data algorithms and control systems can improve the resin infusion process and the temperature profile of resin curing process and thus purposefully affect the residual stresses and manufacturing time. All these aforementioned steps towards faster, cheaper and improved manufacturing processes have in common that they have to be optimized and tested. DTU Wind Energy's blade mould has to offer three key features, which make the blade research and demonstration platform to an advanced testing environment. The following three aspects will be discussed more in details in this article.

- Precise scans of the mould geometry to be aware of mould geometrical deviations and imperfections as well as post-casting scans of the shells in the mould for estimation real thicknesses of the shells
- Embedded water-based heating and cooling system to control the curing process
- Dense mesh of digital resistance temperature detectors in the mould for monitoring and controlling processes

2. Methodology

DTU Wind Energy's has developed a 12.6 m long wind turbine blade. In this article, focus is paid to the wind turbine blade mould, its geometry as well as to the manufacturing process and equipment to monitor and control the blade manufacturing. The design process builds up on several iterations between aerodynamic design, aeroelasticity design and structural design. The design process itself, the applied techniques and methods as well as preliminary results are described in detail in [4].

2.1. *Mould heating*
The moulds are made of a glass fibre epoxy composition. A high temperature epoxy tooling gelcoat is used as surface coat to achieve a strong, temperature-stable mould. The epoxy tooling gelcoat allows a maximum service temperature of up to 125°C, which makes it possible to use the mould with both polyester and epoxy resins. To be in better control of the temperature in the mould on the tool side, an embedded water-based heating and cooling system has been installed to both the pressure and the suction sides of the blade moulds, which will improve the curing process fine-tune capabilities to adjust the level of residual stresses in the blade laminates. Each of the heating and cooling system consists of six independent water loops connected to a heating/cooling unit with fast-responding valves controlled by in-mould resistance temperature detectors. The maximum water temperature is designed to be around 95°C. The tubes of the heating and cooling system are arranged in that way, that the hot inflow is running parallel to the cool outflow water and thus enabling a more uniform temperature field in each of the heating regions. Heating blankets are used to control the temperature from the vacuum bag side. Initial test of the heating system have been conducted, where ca. 60°C warm water was used to heat up the blade moulds and subsequently each water loop was supplied with it.

2.2. *RTM mesh*
Both the pressure and the suction side of the blade moulds are equipped with a redundant grid mesh [5], each consisting of 110 resistance temperature detectors (RTD) measuring temperature in the mould. Only the gelcoat and two layers of biaxial fabrics separate the RTD grid mesh from the blade laminate surface. The RTM grid mesh was developed to simultaneously measure cross-sectional temperature and resin flow distribution in the mould. The sensor mesh consists of a set of 110 resistors per mould arranged in a regular grid (see Figure 1b). Platinum temperature Surface Mounted Device (SMD) resistors with a 1206 package dimensions and 1K ohm resistance were soldered to a enamelled 0.5 mm thick copper wire. The cross-sectional measurement concepts is based on electrical resistance, where a special multiplexing-excitation scheme is applied with constant voltage. The measurement system approach is based on the work from Martin Arlit et al. [5]. A first prototype of this measurement technique to determine the temperature in the mould had been successfully tested on smaller scale (see Figure 1a).

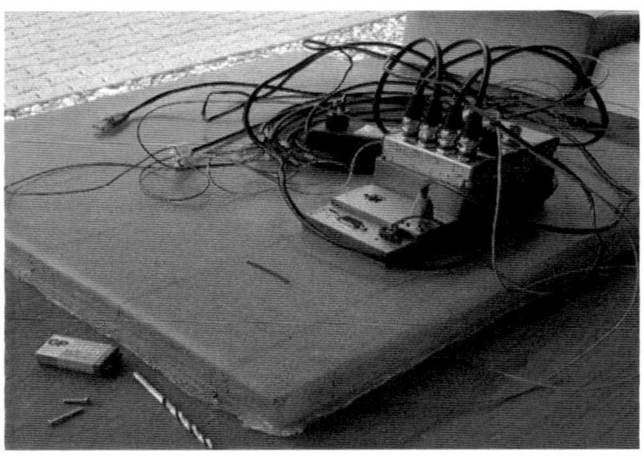
(a)

(b)

Figure 1: RTM test panel (a) and RTM grid mesh placed on the biaxial fabric layer during the wind turbine blade mould building process (b).

2.3. *Mould geometry*
In order to evaluate imperfections and manufacturing deviations of a produced wind turbine blade, it is important to know the blade mould quality and to localise and quantify mould

deviations. For this purpose, high resolution and precise three-dimensional scans of the pressure side and suction side of blade moulds were performed. Both, an ATOS III Triple Scan with 8 MP cameras and 560 mm measuring volume as well as a TRITOP photogrammetry system was used in order to perform fast and accurate 3D coordinate measurements (Figure 2). The equipment is produced by GOM GmbH, a Zeiss company and distributed by Zebicon a/s (Hedegaardsvej 11, 7190 Billund (Denmark) for region Denmark. For this project, Zebicon collaborates with DTU Wind Energy and conducted the mould scans.

The raw data consisting of 50 million measurement points was reduced to a data set with 4.2 million measurement points and post-processed with GOM Inspect Professional software [6] to evaluate the deviations of the manufactured moulds in comparison with the Computer-Aided-Designed 3D model, technical drawings and manufacturing inputs.

In addition to the three-dimensional scans, three cross-section stencils were created for locations around the highest deviations according to the measurements conducted by Zebicon a/s. Stencils at a radial blade position at 4.00 m, 4.40 m and 4.80 m were produced with a laser-cutting process and placed in the pressure side mould to conduct manually measurements. For each of the selected cross-section, both, a cross-section stencil representing the airfoil as designed and another one with a 15 mm offset were produced (Figure 3a and 3b).

Figure 2: Three-dimensional coordinate measurements and scans of the blade moulds (the figure is taken from [4]).

3. Results

3.1. *Mould heating*

The heating/cooling system allows an individual temperature regulation of the six independent regions. First tests have shown an individual heating of the separated heating/cooling loops works as planned. In the upstart phase, the water pipes on the tool side are clearly visible as shown in Figure 4, where the temperature field was measured with a FLIR A655sc FOV 25° high resolution thermal imaging and measurement camera (640 x 480 pixel microbolometer that detects temperature differences down to <30 mK) from FLIR Systems, Inc. (27700 SW Parkway Avenue, Wilsonville, OR 97070, USA). After approximately 3 min, the water pipe system leads to a much more homogeneous temperature distribution for the specific heat cycle. Figure 4 shows the temperature distribution along the blade mould. On the left hand side a heating loop has been heated up over several minutes and the temperature is more homogeneously distributed, whereas the heating cycle in the blade tip has just been activated and the heating pipes of the loop are clearly visible.

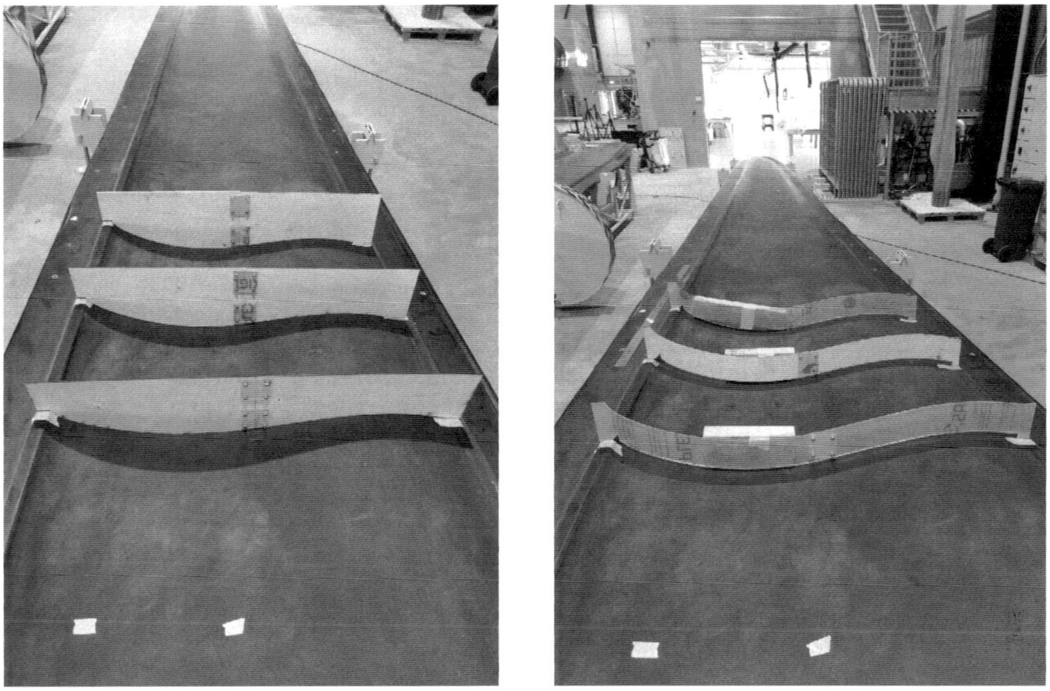

(a) (b)

Figure 3: Airfoil cross-section stencils (a) and stencils with 15 mm offset (b) to check the mould quality and to quantify the deviations in specific locations of the wind turbine blade mould.

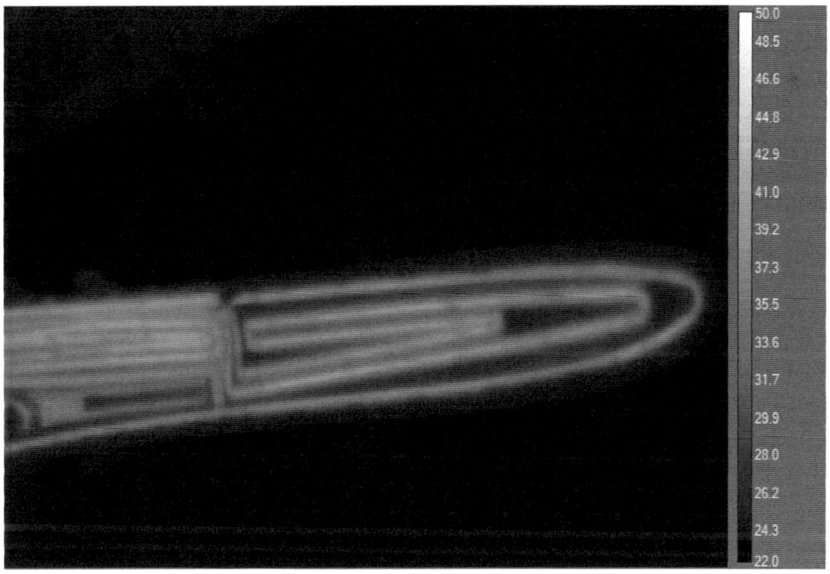

Figure 4: Temperature distribution (in °C) around the blade tip of the mould measured with a thermal camera and visualized with the FLIR ReserachIR Max software.

3.2. Mould geometry

The 3D scans conducted by Zebicon as well as manual measurement at specific cross section of the mould geometry have shown significant deviations from the mould design as visualised in Figure 5. In the figure, the best fit of the superposition between the CAD model and the 3D scan leads to a maximum deviation of 13.9 mm detected at a radial position of 4.32 m. The manually measured maximum deviations of approximately 11 mm were measured both, at the blade cross section position at 4.00 m and 4.40 m blade radial position. The distance between the manually measured maximum deviations and 3D scan is approximately 120 mm.

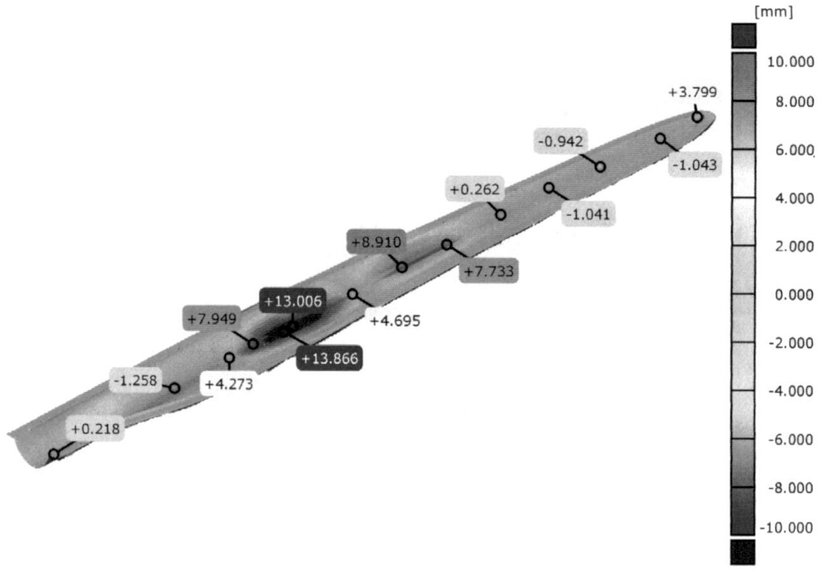

Figure 5: Best fit of 3D scan of the pressure side mould. Deviations in mm are visualized according to the colour code.

Figure 6 shows a cut-out of the blade mould around 4.20 m to 4.40 m radial position of the blade and depicts the 3D scanned mould geometry superimposed with the CAD blade model. The figure shows clearly that the pressure side mould around the trailing edge panel is significantly curved inwards towards the blade center but does it smoothly, s that no abrupt changes/grooves in the blade mould geometry can be seen. The blade mould for the suction side shows also local deviations but the discrepancy is much less pronounced compared to the pressure side mould.

3.3. RTM mesh

The dense RTM mesh grid with 110 sensors per mould has not been tested yet. However, first measurements of the test specimen were conducted successfully and accurate measurement results were acquired. In the near future, the RTM mesh grid will be tested and its performance evaluated.

4. Conclusion

In this article, the 3D scans of the blade moulds are presented as well as the water-based heating and cooling system and a resistance temperature detectors (RTD) measuring system is introduced. The water-based heating and cooling system demonstrated to heat up the blade mould effectively and also to lead to a fairly uniform heat distribution after a short time. However, further investigations have to be conducted to investigate how accurate and time-efficient the water-based heating and cooling system can be applied to tailor the curing process of the composite.

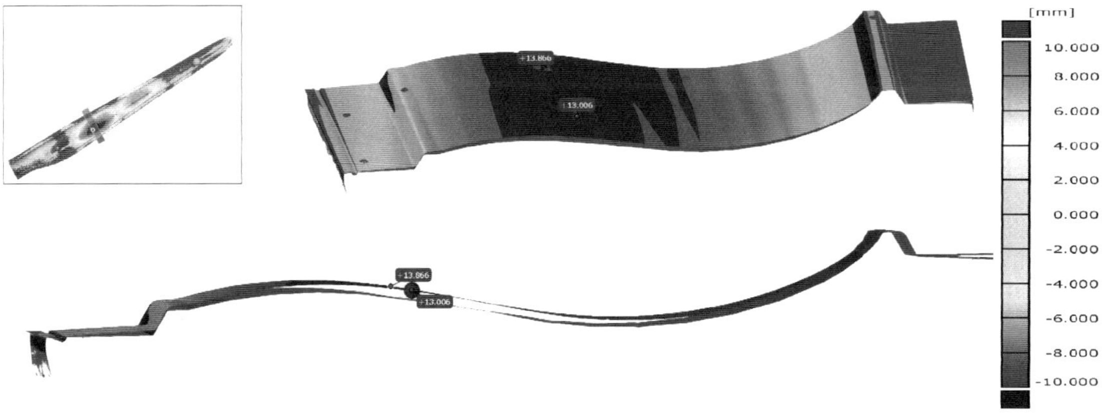

Figure 6: A slice of the best fit of 3D scan of the pressure side mould at the location of the highest deviation. Deviations in mm are visualized according to the colour code.

The installed RTD measuring grid has proven to work precise on for the test panel. The RTD mesh grid installed in the blade moulds seems to be promising but has to demonstrate in practice how precise the dense mesh can not only measure the temperature field in the mould but also detect the resin distribution/front during vacuum infusion processes.

The 3D scans of the blade moulds have detected and quantified significant deviations between the CAD model of the moulds and the manufactured blade moulds. The deviations deviate partly significantly from industrial standards, where usually $\pm$ 2 mm are the tolerable discrepancy. The deviations will not only affect the manufacturing process, where e.g. the shear web geometry has to be adapted, but also the blade structure and possibly the aerodynamic response too. However, to quantify the affect of the structural deviations/imperfections on the structural and aerodynamic response, numerical simulations have to be performed and analysed in order to evaluate and quantify how critical the significant deviations are for the structural integrity of the blade and its aerodynamic performance are. Moreover, the fitting of the shear webs has to be analysed in case modifications will be required. For this purpose, the high resolution surface scans of the blade moulds will be converted into FEM geometries and numerical simulations will be used to quantify the effect of the geometrical deviations. This deviation will allow to demonstrate one of the concepts within Digital Twin approach where feedback from latter stages should be used to update models of the earlier stages.

Acknowledgements

The work is financially supported by DTU Wind Energy's own internal CCA Blade Demonstrator project and partly supported by the RELIABLADE - Improving Blade Reliability through Application of Digital Twins over Entire Life Cycle project, a EUDP program with the grant-no. 64018-0068 from the Danish Energiforsking.dk, a Danish research, development and demonstration funding program within Energy and Climate. The support is gratefully acknowledged. The authors wish to thank and acknowledge the support from Christen Malte Markussen and Knud Stenbæk Nielsen from DTU Wind Energy. The authors also wish to thank Jeppe Hebsgaard Laursen and his team from Zebicon a/s (www.zebicon.com) for conducting and evaluating the three-dimensional coordinate measurements and scans of the blade moulds.

References

[1] Jonkman J 2007 *Technical Report NREL*

[2] Bak C, Zahle F, Bitsche R, Kim T, Yde A, Henriksen L, Natarajan A and Hansen M 2013 *DTU Wind Energy Report-I-0092*

[3] Mortensen U A, Andersen T L and Mikkelsen L P 2018 *IOP Conference Series: Materials Science and Engineering* 012–014

[4] Haselbach P U, Zahle F, Berring P, Semenov S, Voltá L and Verelst D R 2020 *Journal of Physics: Conference Series (Online), https://doi.org/10.11581/dtu:00000093*

[5] Arlit M, Schleicher E and Hampel U 2017 *Sensors* 17

[6] GOM, GOM GmbH, a Zeiss Company, Germany 2020 *https://www.gom.com/3d-software/gom-inspect-professional.html*

Preliminary multi-axial strain analysis in wind turbine blades under fatigue test loads

Oscar Castro and Kim Branner
Technical University of Denmark, DTU Wind Energy, Frederiksborgvej 399, 4000 Roskilde
E-mail: osar@dtu.dk and kibr@dtu.dk

Abstract. A preliminary experimental characterization of the longitudinal, transverse, and shear strains in wind turbine blades under fatigue loading is presented in this study. For that, strains at different cross-section regions of a 14.3 m blade under different uni-axial and biaxial fatigue tests are recorded and analyzed. The results from this analysis provide initial indications on how current testing methods used to certify and characterize the fatigue response of wind turbine blade materials can be improved to better represent what the blades experience during testing. It is shown, for example, how some of the blade regions can experience high biaxiality strain ratios, and how some of these regions can also experience broad internal loading conditions, in terms of strain levels and load orientation (i.e., compression, tension-compression, and tension), depending on the test type.

1. Introduction

Different testing methods are currently used to certify and characterize the fatigue response of wind turbine blades and their materials, which are mainly carried out at the full-length scale [1, 2] and at the coupon-length scale [3, 4]. However, there are various reasons that motivate us to question how representative these methods are. For example, are the loads applied during the coupon-scale fatigue tests representing the internal material loads of the blade during full-scale fatigue tests? In this paper, we will contribute to the discussion of this question.

At the full-length scale, only one or two blades are normally tested according to the standards [1, 2]. In these tests, the blades are commonly evaluated consecutively under flapwise and edgewise fatigue loading, although these structures are subjected to combined loads in real operational conditions [5, 6]. In addition, target bending moments in the flapwise and edgewise directions are usually used alone to design the certification tests, which may disregard the actual effects of the loads on the material response [6].

At the coupon-length scale, on the other hand, 8-10 specimens are commonly used to obtain one S-N curve for a given R-ratio (i.e., the ratio between the minimum and maximum cyclic stress, $R = \sigma_{min}/\sigma_{max}$.) and material orientation, θ, [4]. This is normally done for $\theta = 0°$ and $R = 0.1$ and, occasionally, also for $R = -1.0$ and/or $R = 10.0$. Sometimes, $\theta = 90°$ is also considered. In this way, in the best scenario, the fatigue behavior of the material in the longitudinal and transverse directions under specific R-ratios is characterized. Moreover, these tests are carried out by applying uni-axial loading to the specimens [4], which may ignore the effects of external biaxial loading conditions generally presented in wind turbine blades [5].

Some studies have attempted to assess the blade response during fatigue testing at the material level by considering strains [5] and strain-based damage targets [7] instead of only bending-moment-based damage targets. In that way, a better connection between the full-length scale and the coupon-length scale could be achieved. Nevertheless, such approaches only consider the strains in the longitudinal blade direction omitting, therefore, possible effects of transverse and shear strains that might be significant in some of the materials used in wind turbine blades, such as composite materials [8].

Little attention has been given to quantifying the multi-axial strain/stress levels and local load orientation (i.e., compression, tension-compression, and tension) in wind turbine blades under uni-axial and biaxial loading conditions, which could contribute to making a better connection between the fatigue tests at the different length scales. In [9], the estimated fatigue effects of variable wind load and non-proportional multi-axial stress states in a wind turbine blade were studied. However, the results were given by means of an equivalent multi-axial fatigue damage index, which makes difficult to quantify the multi-axial strain levels and the local load orientation in the blade.

In this sense, this study presents a preliminary quantification of the multi-axial strain distribution in wind turbine blades under full-scale testing loads. Initial measurements of longitudinal, transverse, and shear strains in different blade regions and under different loading conditions are reported and analyzed in terms of multi-axial strain levels and local load orientation. The results of this study are expected to contribute to making the fatigue coupon-length scale tests more representative respect to the fatigue full-length scale tests.

2. Methods

A commercial 14.3 m wind turbine blade [7] is used in this study. This blade type has been intensely tested at DTU Large Scale Facility, LSF, of the Wind Energy Department, Technical University of Denmark. Longitudinal, transverse, and shear strains ($\epsilon_{amp,x}$, $\epsilon_{amp,y}$, and $\epsilon_{amp,xy}$, respectively), were measured by using strain gauge rosettes installed on the outer surface of the blade, see Fig. 1. The longitudinal measurements were aligned with the global longitudinal blade direction. The strains were measured at the trailing-edge (TE), trailing-edge panel (TE-Panel), spar-cap (Cap), and leading-edge (LE1 and LE2) regions of the suction side of cross-sections located at 3.35 m and 7.8 m from the root, see Fig. 1 and Table 1. Measurements were done over 600 s for each load case, with sampling frequency of 100 Hz. Afterwards, the Gaussian method [10] was used to clean the noise from the data.

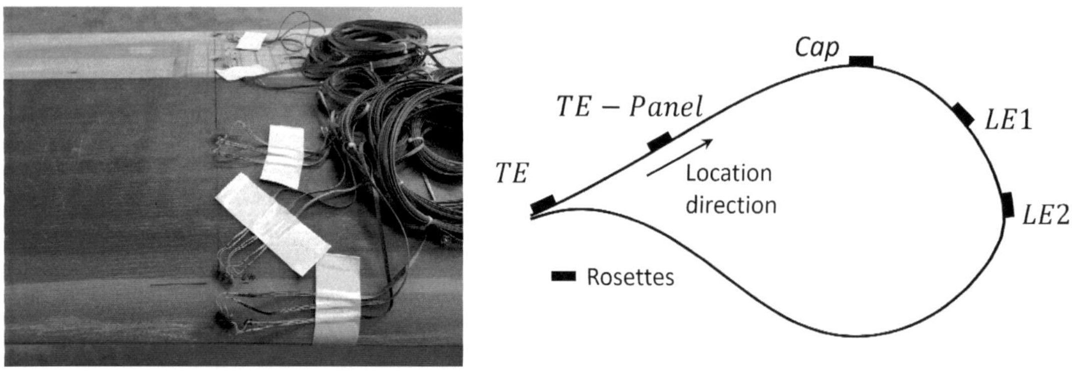

Figure 1. Strain gauge rosette locations on the blade surface.

These measurements were obtained under different loading conditions, including two flapwise loads (Flap-01 and Flap-02), two edgewise loads (Edge-01 and Edge-02), and three chaotic loads (Chao-01, Chao-02, and Chao-03). The loads were applied to the blade by using a ground-based resonance exciter located at 5.2 m from the root. More details about the test setup and loading

Table 1. Strain gauge rosette locations on the suction-side surface, measured from the trailing-edge towards the leading-edge. Values in mm.

Cross-section location [m]	TE	TE-Panel	Cap	LE1	LE2
3.35	20	323	782	1141	1281
7.80	20	165	441	640	684

conditions are shown in Table 2. In addition, an explanation of chaotic loading can be found in [5, 7].

Table 2. Forces, F, applied to the blade in the flapwise and edgewise directions for the different evaluated load cases. The flapwise frequency, f_{flap}, and edgewise frequency, f_{edge}, for all cases were 2.315Hz and 4.595Hz, respectively.

Load case	F_{flap} [kN]	F_{edge} [kN]	Load case	F_{flap} [kN]	F_{edge} [kN]
Flap-01	2.00	-	Flap-02	3.00	-
Edge-01	-	0.50	Edge-02	-	0.75
Chao-01	2.00	0.50	Chao-02	3.00	0.50
Chao-03	2.00	0.75	-	-	-

The different strain time series at the cross-section regions were reduced into a number of constant-amplitude strain-cycle blocks by using the rainflow counting method [11]. Each of these blocks were characterized using the Q-ratio (i.e., $Q = \epsilon_{mean}/(\epsilon_{amp} - \mid \epsilon_{mean} \mid)$, where ϵ_{mean} is the mean strain and ϵ_{amp} is the strain amplitude [12]), as an alternative of the R-ratio. With the Q-ratio, all possible load conditions can be included continuously into a range from -1 to 1. This means that, for compression-compression (C-C), $-1.0 \leq Q < -0.5$; for compression-tension (C-T), $-0.5 \leq Q < 0$; for tension-compression (T-C), $0 \leq Q < 0.5$; and for tension-tension (T-T), $0.5 \leq Q < 1.0$.

3. Results and Discussion

In this section, the results and an analysis on the characterization of the multi-axial strain cycles at the different regions of the blade, under different types of load, are presented. The results of the LE1 region are not graphically shown in the paper as they were found to be similar than the ones obtained from the LE2 region. These similarities are due to the fact that both regions were made of the same material and layup, and that the measuring points were relatively close to each other.

The percentage of number of cycles, N/N_{total}, for a given Q-ratio and strain level, $\epsilon_{amp,i}/\epsilon_{amp,x,max}$, for different cross-section regions and under flapwise, edgewise, and chaotic loading conditions are presented in Figs. 2, 3, and 5, respectively. Where, $\epsilon_{amp,x,max}$ is the maximum strain amplitude in the longitudinal direction, N is the actual number of cycles under a given Q-ratio and $\epsilon_{amp,i}$; N_{total} is the total number of cycles within the time series; and $i = x$, y, or xy, denoting the longitudinal, transverse and shear directions, respectively.

As seen in Fig. 2, when a flapwise test is carried out, all cycles from the different strain components and at the different cross-sections regions are under C-T or T-C. In fact, most of them tend to a Q-ratio equal to zero, which would be equivalent to $R = -1.0$. This is because,

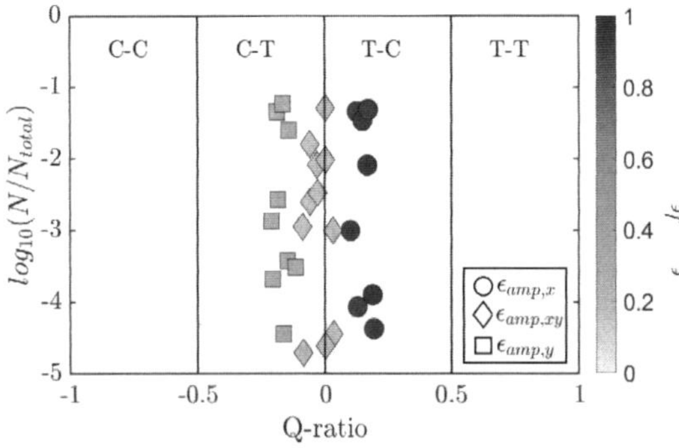

(a) TE, $\epsilon_{amp,x,max} = 70.2\mu m/m$

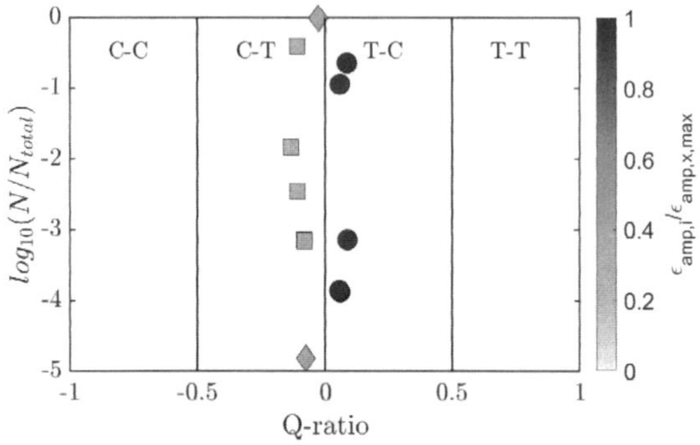

(b) TE-Panel, $\epsilon_{amp,x,max} = 100.0\mu m/m$

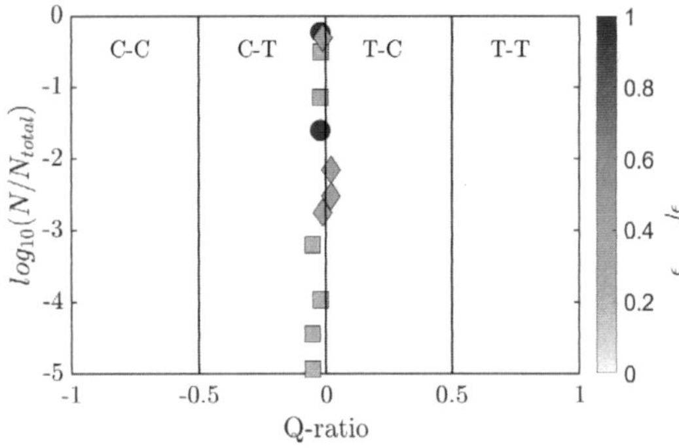

(c) Cap, $\epsilon_{amp,x,max} = 354.1\mu m/m$

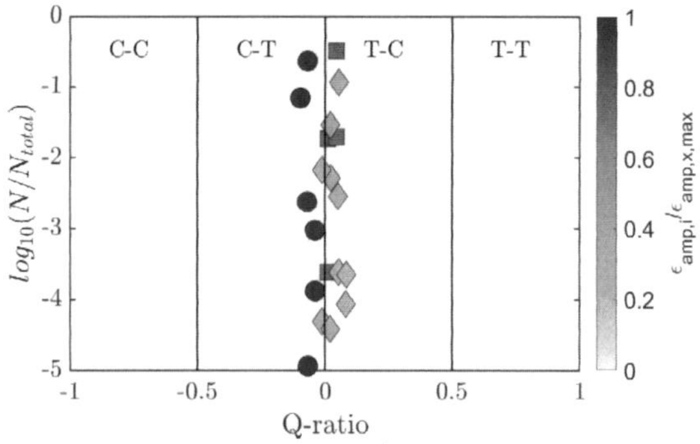

(d) LE2, $\epsilon_{amp,x,max} = 122.8\mu m/m$

Figure 2. $N/N_{total} - Q - \epsilon_{amp,i}/\epsilon_{amp,x,max}$ matrix of different cross-section regions: (a) TE, (b) TE-Panel, (c) Cap, and (d) LE2, at 3.35m span. Flap-02 test case.

during the test, the load applied by the exciter also has a $Q = 0$, meaning that the maximum and minimum applied load have the same value but with opposite signs. Moreover, although the load is applied in the global flapwise direction (i.e., with respect to the root-coordinate system), a small load component in the edgewise direction is also presented in the local cross-section-coordinates due to the orientation of the local principal axes. This can be seen as the TE and LE experience some deformations as well, see Figs. 2-a and 2-d, respectively.

Similarly, when an edgewise test is carried out, most of cycles from the different strain components and at the different cross-sections regions are under C-T or T-C loading conditions, see Fig. 3. This except for some transverse and shear strain cycles at the Cap region, which were under T-T and C-C, see Fig. 3-c. These cycles can be related to noise in the signal as the strain amplitudes in these two directions were low and noisy, see Fig. 4, and the data was not properly cleaned. Omitting this noise, it can be seen that the strain amplitude in the transverse direction was almost zero (i.e., $Q = -1.0$), whereas the main strain components in the shear direction had a Q-ratio close to 0.35.

It is worth to note that, for the flapwise and edgewise tests, the strain time series for the different strain components and at the different cross-section regions can be simplified as a few constant-amplitude blocks with similar strain level and Q-ratio, see Figs. 2 and 3. This is because the load is applied in only one direction and the frequency with which it is applied remains, in theory, constant during the test.

However, when a chaotic test is carried out, several block cycles with a bigger difference in the strain level and Q-ratio define the different strain time series for the different strain

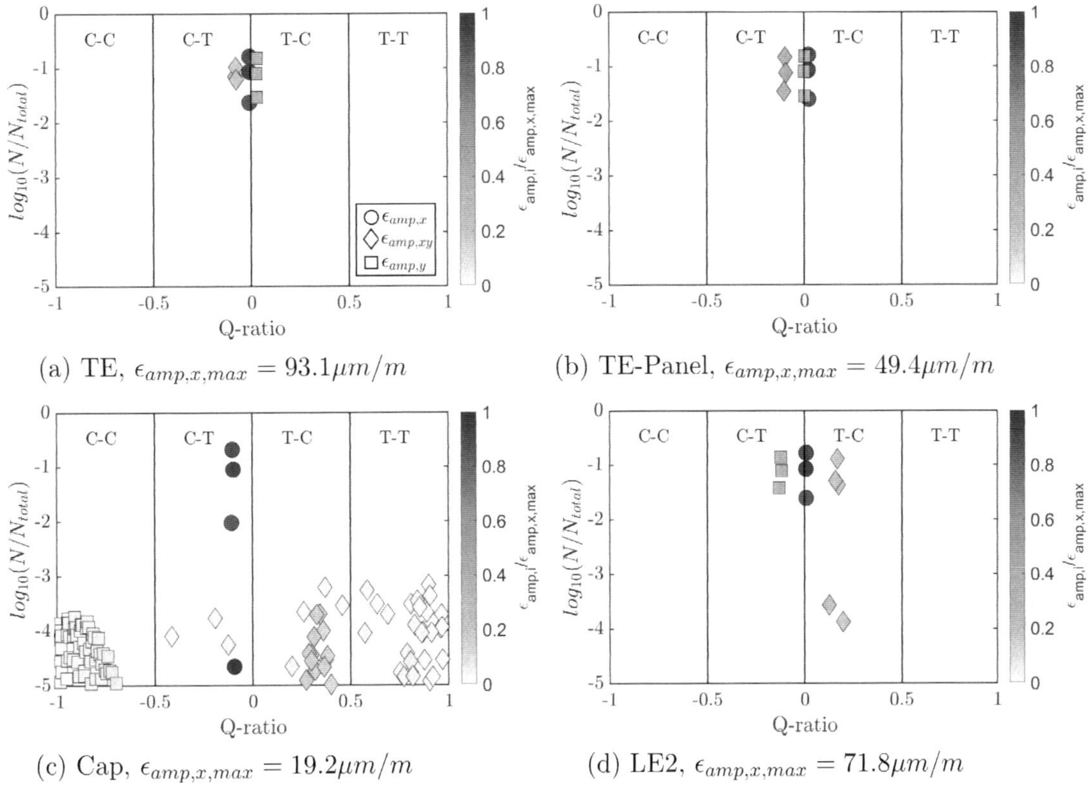

(a) TE, $\epsilon_{amp,x,max} = 93.1\mu m/m$

(b) TE-Panel, $\epsilon_{amp,x,max} = 49.4\mu m/m$

(c) Cap, $\epsilon_{amp,x,max} = 19.2\mu m/m$

(d) LE2, $\epsilon_{amp,x,max} = 71.8\mu m/m$

Figure 3. $N/N_{total} - Q - \epsilon_{amp,i}/\epsilon_{amp,x,max}$ matrix of different cross-section regions: (a) TE, (b) TE-Panel, (c) Cap, and (d) LE2, at 3.35m span. Edge-01 test case.

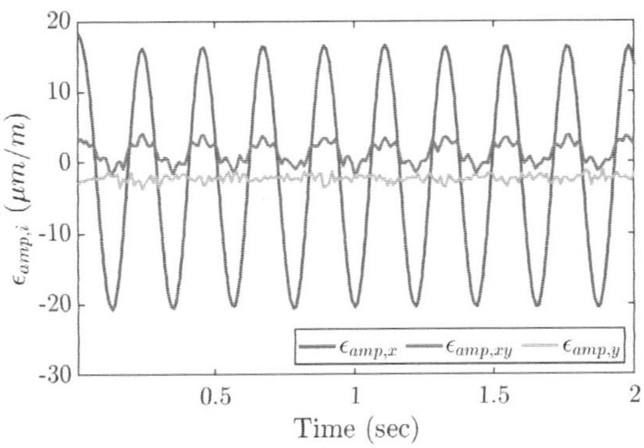

Figure 4. Strain time series at the Cap region for the Edge-01 test case. Note the noisy signals for $\epsilon_{amp,xy}$ and $\epsilon_{amp,y}$.

components, see Fig. 5. This is because, during the chaotic tests, the flapwise and edgewise forces have different frequencies, and both the magnitude of each of them and the phase angle between them change with time, which causes strains to change significantly over time, see Fig. 6. Moreover, as seen in Fig. 5, most of the cycles from the different strain components and at the different cross-section regions are under C-T and T-C, except for some cycles with low strain level that are under C-C and T-T, see Fig. 5-d. In this case, these cycles are not necessarily related to noise in the signal as in the edgewise-test case. In fact, they do exist as a consequence of the superposition of the two external forces, see e.g. in Fig. 6.

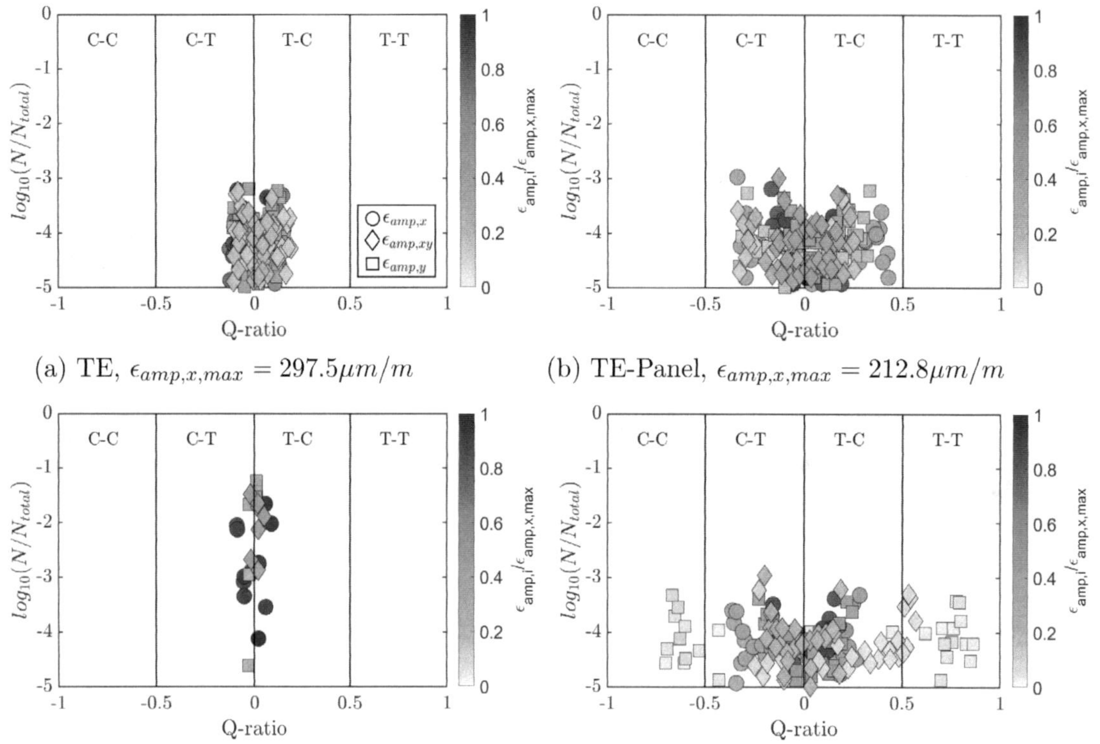

(a) TE, $\epsilon_{amp,x,max} = 297.5\mu m/m$ (b) TE-Panel, $\epsilon_{amp,x,max} = 212.8\mu m/m$

(c) Cap, $\epsilon_{amp,x,max} = 312.8\mu m/m$ (d) LE2, $\epsilon_{amp,x,max} = 287.6\mu m/m$

Figure 5. $N/N_{total} - Q - \epsilon_{amp,i}/\epsilon_{amp,x,max}$ matrix of different cross-section regions: (a) TE, (b) TE-Panel, (c) Cap, and (d) LE2, at 3.35m span. Chao-02 test case.

On the other hand, it was found that, although the longitudinal strains are the dominant strain component in the blade, the transverse and shear strain components can also have significant magnitudes in some of the cross-section regions, see e.g. in Fig. 2, 3, and 5. In general, for the shear strains, significant magnitudes (i.e., $\epsilon_{amp,xy}/\epsilon_{amp,x,max} \geq 0.4$) were found at the TE-Panel for all test cases, and at the Cap for the flapwise and chaotic test cases. Whereas, for the transverse strains, significant magnitudes (i.e., $\epsilon_{amp,y}/\epsilon_{amp,x,max} \geq 0.4$) were found at the TE, TE-Panel, and LE2 regions, for all test cases.

Moreover, for each cross-section, there was not a considerable qualitatively difference between the strain behavior at the different cross-section regions with the variation of the load level, see e.g. in Figs. 7-a and 7-b, and 7-c and 7-d, respectively. The variation was mainly quantitatively, as the strains increase with the increase of the load. In addition, for a given load level applied to the blade, some qualitatively and quantitatively differences can occur between the different cross-sections, see e.g. in Figs. 7-a and 7-c, and 7-b and 7-d, respectively. This is because the cross-section geometry, the cross-section properties, and the loads, among other factors, change along the blade span.

All this suggests that, a more comprehensive fatigue characterization at the coupon-length scale level might be needed to better predict the material response due to the loads that the blades experience when subjected to more realistic loading conditions, such as chaotic loads. In addition, the consideration of the transverse and shear strain components might be relevant for most of the cross-section regions, specially at the TE-Panel.

A similar study to this one, in which the actual operational loads that the blades experience

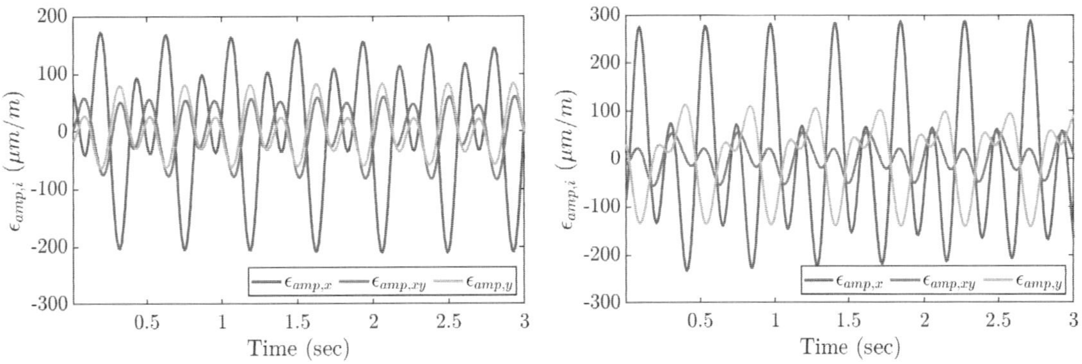

(a) TE-Panel (b) LE2

Figure 6. Strain time series at different cross-section regions: (a) TE-Panel; and (b) LE2, at 3.35m span. Chao-02 test case.

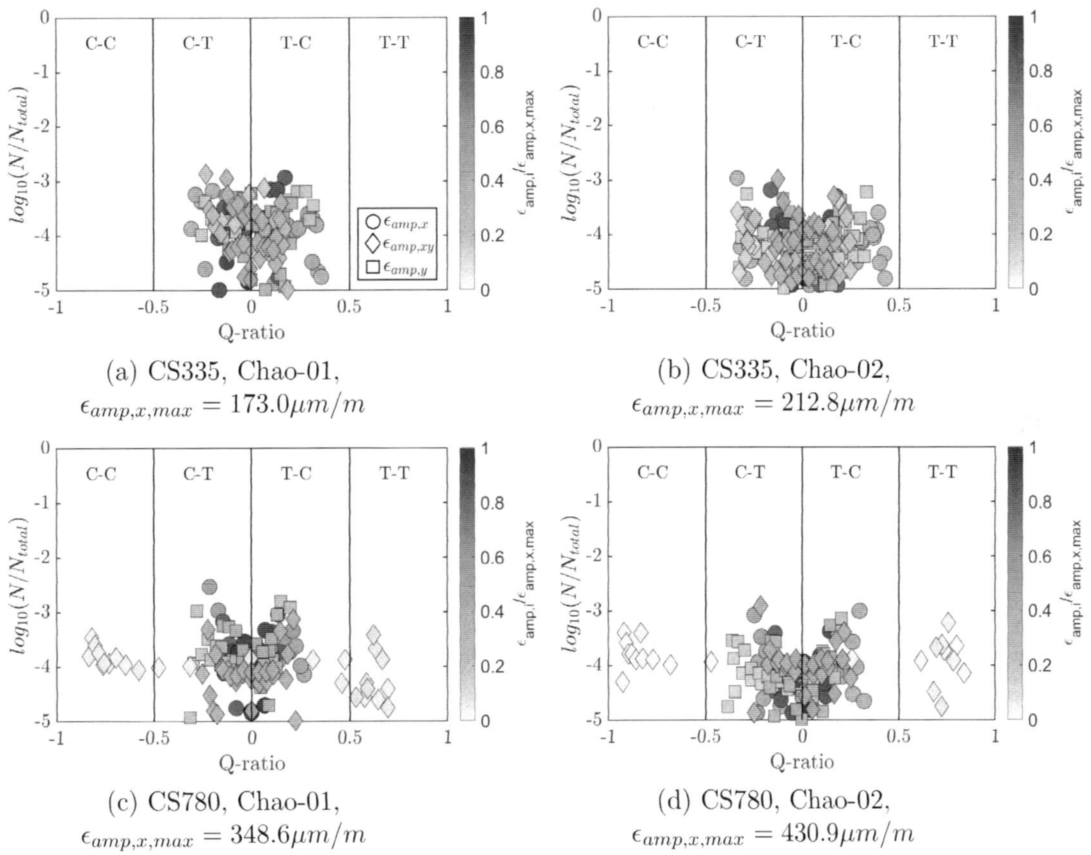

(a) CS335, Chao-01, $\epsilon_{amp,x,max} = 173.0\mu m/m$

(b) CS335, Chao-02, $\epsilon_{amp,x,max} = 212.8\mu m/m$

(c) CS780, Chao-01, $\epsilon_{amp,x,max} = 348.6\mu m/m$

(d) CS780, Chao-02, $\epsilon_{amp,x,max} = 430.9\mu m/m$

Figure 7. $N/N_{total} - R - \epsilon_{amp,i}/\epsilon_{amp,x,max}$ matrix of the TE-Panel under chaotic loading. At 3.35m span, (a) Chao-01 and (b) Chao-02. At 7.8m span, (c) Chao-01 and (d) Chao-02.

during their lifetime are considered, would provide a better indication of how well current fatigue tests methods at the different length scales represent reality. Furthermore, it would be also good to quantify the local strain behaviour at each of the layers within the different cross-section regions (i.e., not only at the surface), as damages can develop at any layer. Moreover, it might be also necessary to quantify the interaction between the different strain components.

4. Conclusion

A preliminary experimental characterization of the longitudinal, transverse, and shear strain cycles at different regions of a 14.3 m blade under different uni-axial and biaxial fatigue test loading conditions, was presented in this paper. Under uni-axial loading, the time series from all strain components and at all evaluated cross-section regions could be simplified as a few blocks of cycles with similar Q-ratio and strain levels, being most of them under C-T and T-C. Whereas, under chaotic loading, the time series from all strain components were composed by a larger number of block cycles under C-T and T-C and with broader Q-ratio and strain levels. Some of the strain block cycles at the leading-edge were also under T-T and C-C. Furthermore, it was also found that the transverse and shear strain components can have a significant magnitude in most of the cross-sections regions, specially at the TE-Panel, if compared with the respective longitudinal strains (i.e., more than 40% of the longitudinal strains). Hence, this work provides initial indications about how current certification fatigue test methods for wind turbine blade materials can be improved in order to account for the biaxiality ratio between the different strain components and the broad internal loading conditions (i.e., T-T, C-C, T-C, and C-T) that the different blade cross-section regions can undergo, specially under external biaxial loading.

Acknowledgments

The work is supported by the Danish Energy Agency through the Energy Technology Development and Demonstration Program (EUDP), Grant No. 64016-0023. The supported project is named "BLATIGUE: Fast and efficient fatigue test of large wind turbine blade." The financial support is greatly appreciated. The authors would also like to thank the technicians and engineers from the LSF, who contributed to setting up the blade and test.

References

[1] IEC-61400-23 2014 Wind turbines – part 23: Full-scale structural testing of rotor blades Standard International Electrotechnical Commission
[2] DNVGL-ST-0376 2015 Rotor blades for wind turbines Standard DNV-GL
[3] ISO-527-5 2009 Determination of tensile properties - part 5: Test conditions for unidirectional fibre-reinforced plastic composites Standard International Organisation for Standardization (ISO)
[4] ISO-13003 2003 Fibre reinforced plastics – determination of fatigue properties under cyclic loading conditions Standard International Organisation for Standardization (ISO)
[5] Greaves P 2013 *Fatigue analysis and testing of wind turbine blades* Ph.D. thesis Durham University Durham (Eng)
[6] Hughes S D, Musial W D and Stensland T 1999 Implementation of a two-axis servo-hydraulic system for full-scale fatigue testing of wind turbine blades Technical Report NREL/CP-500-26896 National Renewable Energy Laboratory Golden (CO)
[7] Castro O, Belloni F, Stolpe M, Yeniceli S C, Berring P and Branner K 2020 *Optimized method for multi-axial fatigue testing of wind turbine blades. Compos Struct (submitted for publication)* xx–xx
[8] Quaresimin M, Susmel L and Talreja R 2010 *Int J Fatigue* **32** 2 – 16
[9] Hu W, Zhupanska O, Choi K and Buchholz J 2015 *Proceedings of the American Society for Composites 30th Technical Conference, 2015* (American Society for Composites)
[10] Haddad R A and Akansu A N 1991 *IEEE Transactions on Signal Processing* **39** 723–727
[11] ASTM-E1049-85 2017 Standard practices for cycle counting in fatigue analysis Standard ASTM International
[12] Westphal T 2012-10-08 Personal communication

Strength analysis of a 5-m composite wind turbine blade under static and fatigue loading conditions

C Muyan[1,2] and D Coker[1,2*]

[1]Department of Aerospace Engineering, Middle East Technical University, Ankara, 06800, Turkey
[2]Structures and Materials Laboratories, RUZGEM (METUWIND) Center for Wind Energy Research, Middle East Technical University, Ankara, 06800, Turkey

*Correspondence to: Demirkan Coker (coker@metu.edu.tr)

Abstract. This study focuses on the static and fatigue strength analysis of the existing 5-m RUZGEM composite wind turbine blade with and without defect in the form of debonding. Puck progressive damage analysis is used to investigate the effect of defect introduced to the spar/pressure side interface on the ultimate strength under extreme flap-wise (min) by single-loading point application. In addition, fatigue life according to the Germanischer Lyod (GL) guidelines is determined. For the original blade, collapse is observed after 85% loading, whereas due to defect, failure region changes and the blade collapses after 69% loading. Blade is found to exhibit sufficient resistance against fatigue. However, the inclusion of defect decreases the fatigue life of the blade significantly.

1. Introduction

Wind turbines are environmentally-friendly renewable energy resources, which are designed to operate over a lifespan of 20 years. According to Holmes et al. [1], long-term structural reliability of wind turbine components is vital when high cost of manufacturing, inspection and repair, especially for turbines located in remote regions are considered. Composite blades are important components of a wind turbine, that are subjected to complex loading conditions. During a service time of 20 years, several damage modes caused by extreme gusts in addition to cyclic loading can occur in rotor blades [2, 3].

Static and fatigue strength analysis are needed for validating the design of wind turbine blades. In the literature, there are many studies regarding the static or fatigue strength of blades. However, there are a smaller number of papers, where both static and fatigue strength analysis of wind turbine blades are presented.

For investigating the failure behavior of large-scale composite turbine blades researchers conducted full-scale tests under static loading. Jensen et al. [4] tested a 34-m blade and found that Brazier effect and further delamination buckling is the main reason for the blade collapse. In contrast, according to Overgaard et al. [5], the main failure mechanism which leads to blade collapse is argued to be delamination and its interaction with local buckling. Haselbach and Branner [6] highlight the influence of buckling on the damage onset in the trailing edge and sandwich panel failure during the full-scale test of a 34-m wind turbine blade. In another study, conducted by Chen et al. [7], delamination failure at the root transition were found to be the main failure mechanism for 52.3-m blade collapse. Muyan and Coker [8], investigated the strength characteristics of the 5-m wind turbine blade without any defects utilizing progressive damage analysis under extreme flap-wise, edgewise and combined loading conditions.

Regarding the fatigue strength of blades, Marin et al. [9] estimated the fatigue life of a 300 kW wind turbine blade using a simplified procedure proposed by GL standard. They found that an increase in the stress state due to the combined effect of an abrupt change of laminate thickness and existence of a superficial crack leads to a crack generation in the laminate that results in total failure. Castelos and Balzani [10] studied the effect of geometric nonlinearities on the fatigue analysis of the trailing edge bonding in wind turbine blades. They point out that the superposition of stresses for the fatigue may be misleading for modern, flexible rotor blades where geometric nonlinearities must be considered. In addition to this, they propose a novel methodology for calculating stresses with a new load application method that reduces geometric nonlinear behavior of the blade.

Recently, using white light imaging technique and ex situ X-ray computed tomography technique, Mikkelsen [11] investigated the fatigue damage mechanisms in quasi-directional load-carrying laminates of wind turbine blades. He was able to observe the three-stage stiffness degradation during tension/tension fatigue. Moreover, he found that for R=0.1 tension/tension fatigue and R=-1 tension/compression fatigue, damage mechanisms were similar, whereas for R=10 compression/compression case, damage mechanism was different.

For both static and fatigue strength analysis, Montesano et al. [12] introduced a physically based multi-scale damage model, which can account for failures of a 33.25 m rotor blade. The simulation results show the capability of the model to predict the evolution of sub-critical ply cracks between spar webs located near the blade root at maximum chord length. In the follow up work by Zu et al. [13], the multi-scale model is further expanded to include cohesive zone elements to predict structural debonding failure at the spar/skin interface located near the blade root at maximum chord. Structural analysis of large-scale vertical axis wind turbine components is reported by Lin et al. [14]. Strain history is used to evaluate the fatigue damage according to GL Guidelines [15]. Ultimate strength is determined based on Tsai-Wu failure criterion. The study reveals that compressive loads lead to larger fatigue damage than tensile cyclic loads. Highest damage is observed at the blade mid-span.

This study focuses on the static and fatigue strength analysis of the existing 5-m RUZGEM composite wind turbine blade with and without defect in the form of debonding. The main blade components are depicted in Figure 1. Previous study by Muyan and Coker [8], investigated the strength characteristics of the 5-m wind turbine blade without any defects utilizing progressive damage analysis under distributed pressure loading. In this study the investigation is limited to the redistribution of the stresses due to the debonding defect and its influence on the blade static and fatigue strength. The modeling method is not extended to investigate the debonding crack growth. Puck progressive damage analysis is used to investigate the ultimate strength under extreme flap-wise (min) single-loading point application. In addition, blade fatigue life according to GL guidelines is investigated.

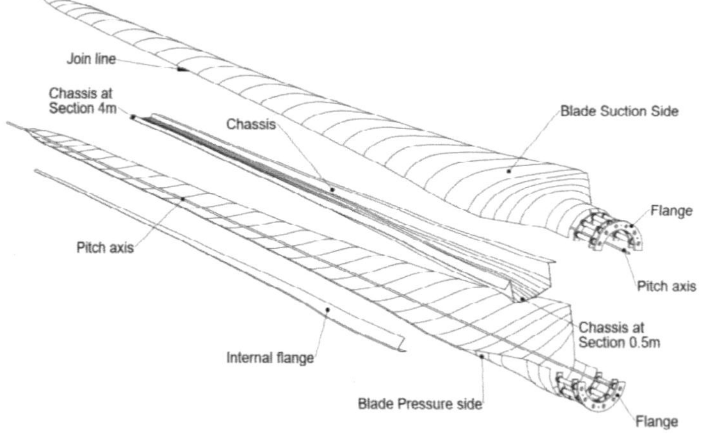

Figure 1. Blade assembly for the 5-meter RUZGEM turbine blade [16].

2. Methods

2.1. Static strength analysis method

For this static analysis Puck's physically based phenomenological model is implemented in the 5-m composite RUZGEM blade FE Model using ANSYS Parametric Design Language (APDL) [17]. The blade is modelled with plane stress elements. According to Puck's failure criteria [18] the degradation rules are applied to the elements as explained in detail in the previous work of the authors [8]. In Table 1, the material properties and design allowables of the blade materials for static analysis is listed [16]. Referring to GL Guidelines [15], design allowables are obtained from the knockdown of the experimental strength values by the material safety factor 2.406.

Table 1. Material properties and design allowables of RUZGEM blade for static analysis

Material Property		Unidirectional Laminate	Steel	Gel Coat	CSM 300	Divinycell H45
Density, ρ	[kg/mm^3]	1896	7850	1200	1896	200
Thickness, h	[mm]	0.716	5.3	0.9	0.358	5 or 10
E_1	[GPa]	24.84	210	3.98	9.14	55x10^{-3}
E_2	[GPa]	9.14				55x10^{-3}
v_{12}		0.29	0.3	0.34	0.29	0.4
G_{12}	[GPa]	2.83				15 x10^{-3}
X_T	[MPa]	191.73	581.8	35.29	16.86	1.4
X_C	[MPa]	101.16				0.6
Y_T	[MPa]	16.86				1.4
Y_C	[MPa]	50.41				0.6
S	[MPa]	11.29				0.56

For the design of the RUZGEM blade, the turbine specifications are obtained from the meteorological data in Ankara, Turkey [19]. Based on this information, the turbine specifications were selected according to the IEC 61400-2 standard [20]. Loads for the structural design were selected as the worst load case scenario for the complete set of IEC extreme loads.

Extreme external loads are given over 28 stations of the blade suction and pressure sides for flap-wise(min) loading. Figure 2 shows the calculated values of the flap- bending moment in sections along the blade span length. These flap-wise moment values are computed from flap-wise loads. In order to simulate testing condition using single load application method, the load which resembles the bending moment over the blade segment 0-3.25 m range is applied at station 3.25-m as shown in Figure 3. The load at this station is distributed among the nodes along the spar width on the suction and pressure sides equally.

As the boundary condition all rotational and translational degrees of freedom at the blade root are fixed. A total number of approximately 35,000 nodes are used in the finite element model, which is determined based on the mesh convergence study. SHELL 181 elements with size 15x15 mm are used for the mesh. This element size correlates to the typical element size for small scale wind turbine blades. Adhesive materials used for the blade connections are modeled using bonded contact with multi point constraint (MPC) algorithm in the FE Model. Geometric nonlinearity is included in the static strength analysis.

FE model of the existing 5-m METUWIND blade with debonding defect is depicted in Figure 4. Debonding defect is a common failure type observed in wind turbine blades [21]. Size of the debonding region is approximately 450x186 mm^2 as highlighted in blue in Figure 4. The defect is modeled between the spar-pressure side interface and is located approximately 1490 mm away from the blade root. This debonding defect is created by deleting the bonded contact elements (CONTACT174 and TARGET170) in the defect region.

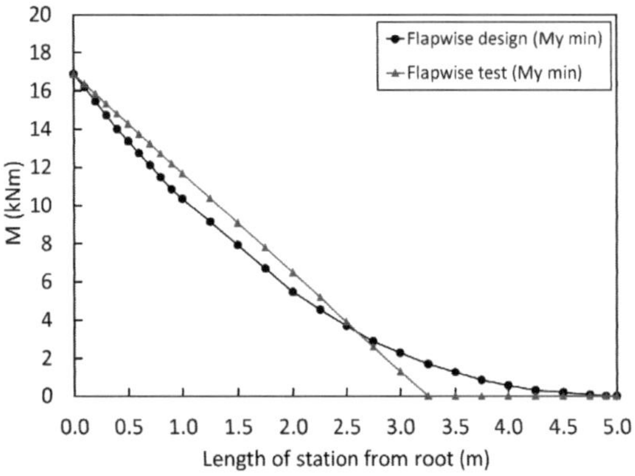

Figure 2. Extreme moment distribution along span length.

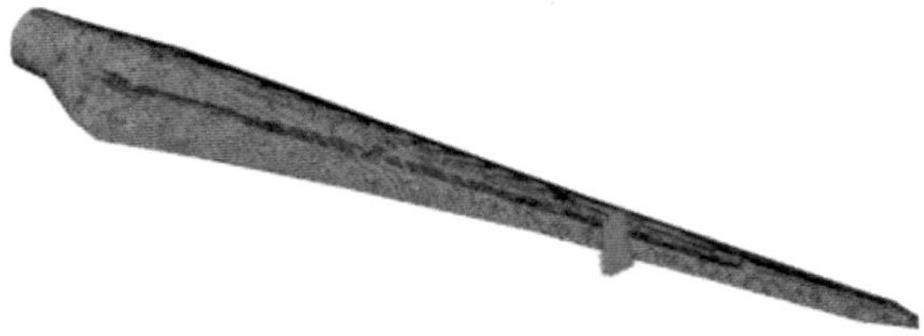

Figure 3. Extreme flap-wise(min) load application to the FE Model.

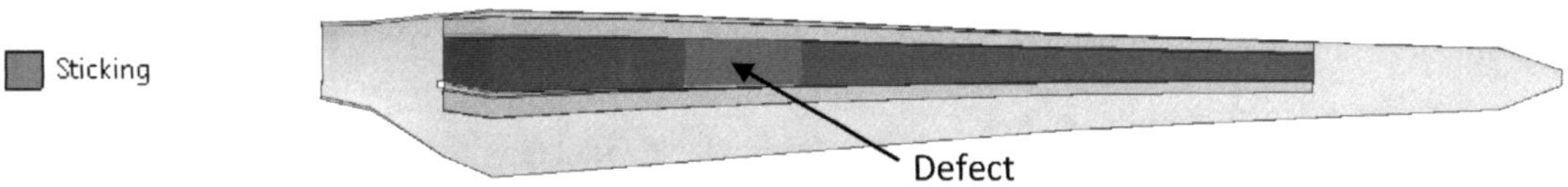

Figure 4. Status of the bonded contact interface between spar-pressure side of the RUZGEM blade with defect.

2.2. *Fatigue strength analysis method*

Fatigue calculation is carried out based on the approach suggested by Germanischer Lyod (GL) Guidelines [15]. This approach suggests simplified and conservative rules by utilizing tensile and compressive strengths of the composite material in the Goodman diagram. Fatigue loading is chosen to be 75% of the extreme flap-wise (min) loading and as stated in [11], the stress ratio is assumed to be R=0.1. According to the recommendations of IEC 61400-23:2002 standard [22], a partial safety factor of 1.2 is included in the fatigue loads Afterwards, cycle mean stress $\sigma_{k,m}$ and cyclic stress amplitude $\sigma_{k,A}$ are generated from FE element simulation. For the computation of stresses geometric nonlinearity is included. Depending on this information, permissible number of cycles, N_f for each element is calculated as follows:

$$N_f = \left[\frac{R_{k,t} + |R_{k,c}| - \left|2 \cdot \gamma_{Ma} \cdot \sigma_{k,m} - R_{k,t} + |R_{k,c}|\right|}{2(\gamma_{Mb}/C_{1b})\sigma_{k,A}}\right]^m \quad (1)$$

where $R_{k,t}$ and $R_{k,c}$ are the characteristic tensile and compressive strength of the material, respectively. $\sigma_{k,m}$ and $\sigma_{k,A}$ are the mean value and amplitude of the stresses obtained from the blade FE Model under cyclic loading. For this calculation, σ_k are calculated as the average resultant axial stresses on laminate level. In expression (1), the values of the material partial safety factors γ_{Ma}, γ_{Mb} constant C_{1b} and slope m are selected according to GL Wind Guidelines [15], m is selected as 10 for glass reinforced epoxy composites. γ_{Ma} and γ_{Mb} are taken as 2.406 and 1.96 for UD and Triax materials, respectively. Characteristic strength values for the materials of the RUZGEM blade fatigue analysis are obtained from [23] and are listed in Table 2.

Table 2. Characteristic strength values of the RUZGEM blade materials for fatigue analysis

	Unidirectional Laminate	Triax $[0/\pm45]_T$
Characteristic static tensile strength ($R_{k,t}$) [MPa]	461	301
Characteristic static compressive strength ($R_{k,c}$) [MPa]	243	287

3. Results and discussion

In this section static and fatigue strength of existing 5-m RUZGEM blade is investigated using FE Models with and without defect. First the static strength of the blade with and without defect is presented followed by the fatigue life of the blade with and without defect.

3.1. Static strength analysis

Using progressive damage analysis method, for the extreme flap-wise (min) loading case, the load-displacement curves of the blade with and without defect are obtained as displayed in Figure 5. The load is application and the displacement, which is measured at the load application point, is in the flap-wise (min) direction. As seen from the figure, RUZGEM blade without defect and with defect collapses after 85% and 69% of the extreme flap-wise (min) loading, respectively.

Based on the information element failure progression in the damaged blade components (pressure side, internal flange, spar and suction side) are displayed for model without defect and with defect in Figure 6 (a) and Figure 6 (b), respectively. According to the methodology [8], an element fails if fiber-failure (FF) or inter-fiber-failure mode C (IFF(C)) in at least one third of the plies of a laminate is detected. The stiffness of the failed elements is set to zero and they do not contribute to the blade strength anymore. Failed elements are colored in red in the figures. For the model without defect depicted in Figure 6 (a), root transition region is the main damage region which leads to blade collapse. This failure region shows similarity with the findings in [24], where FEA strain were found to be high at the root transition region for the 9-m blade. Chen et al. [7] reports that the root transition was the main failure mechanism region of the 52.3-m blade collapse. The failure region for the blade with defect differs considerably from the blade model without defect as the laminate failure region before collapse is concentrated in defect region as shown in Figure 6 (b).

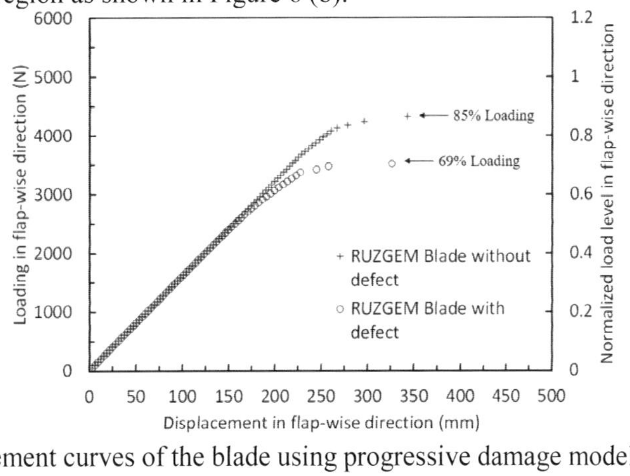

Figure 5. Load displacement curves of the blade using progressive damage model (Puck) under extreme flap-wise (min) loading.

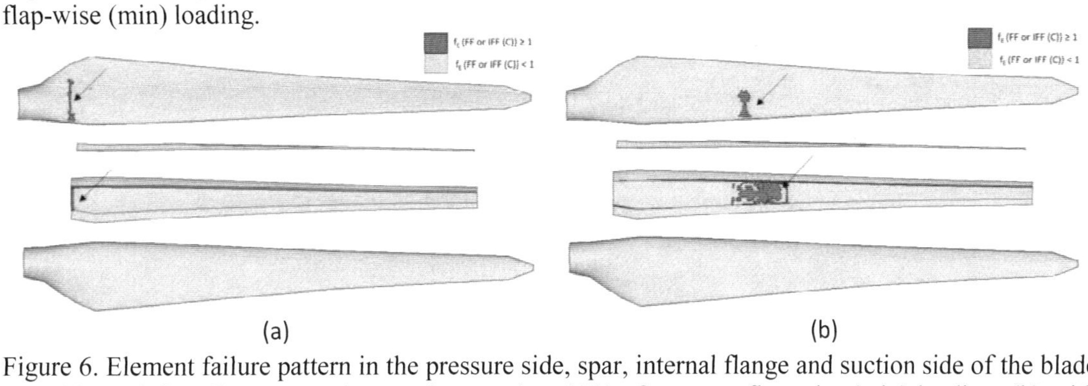

Figure 6. Element failure pattern in the pressure side, spar, internal flange and suction side of the blade (a) without defect (from top to bottom in a row) at 85% of extreme flap-wise (min) loading, (b) with defect (from top to bottom in a row) at 69% of extreme flap-wise (min) loading.

3.2. *Fatigue strength analysis*

Under fatigue flap-wise (min) loading, average axial stress distribution of the RUZGEM blade without and with defect is depicted in Figure 7 (a) and (b), respectively. Suction side and pressure side are the blade components with the highest stress levels. As seen from the figures, suction side is subjected to tension stresses, whereas pressure side is under compressive stresses. On the suction side, stress concentration location is observed between the spar caps near the maximum chord. This variation of the stress concentration shows similarity with the stress distribution of the 33.25-m blade studied by Montesano et al. [12]. When the stress levels of the blade without defect is compared with the original blade, approximately 16% average compressive axial stress increase is noted in the vicinity of the debonding defect as seen in Figure 7. The increase in the stress level is caused by the local buckling observed at the debonding defect region. In addition, in the model with defect a slight average tensile axial stress increase is observed.

As explained in the methods section, fatigue analysis is conducted according to Germanischer Loyd (GL) 2010 Guidelines [15]. Since the pressure side, the suction side and the spar are the primarily damaged blade components, fatigue analysis results are presented for three components. Fatigue damage locations and the corresponding permissible number of cycles (N_f) are shown for the blade without defect and with defect in Figure 8 and Figure 9, respectively. In Figure 8, the first 60-100 elements with the lowest number of permissible fatigue cycles are marked in black and considered to be damaged due to fatigue. For comparison, in Figure 9, the elements which fall into the same range of permissible cycles are marked in black, as well. Due to flap-wise (min) loading, suction side is loaded under compression-compression cyclic loads. Since the compressive strength of the blade is less than its tensile stress a smaller number of permissible fatigue cycles are found for the pressure side of the blade compared to suction side. The result that compressive loads demonstrated larger fatigue damage than tensile fatigue loads was also reported by Lin et al. [14]. In Figure 8, for the blade without defect, fatigue damage in the pressure side is located at the pressure side along the spar width near maximum chord. Fatigue damage location of the suction side is between the spar webs towards the mid span and accumulates at the ply drop. In Figure 9, for the blade with defect, fatigue damage in the pressure side is located in the vicinity of the debonding defect. Moreover, it is observed that there is a significant increase in the number of damaged elements in the pressure side and spar of the blade compared to model without defect. Similar to the blade without defect, in for the blade with defect fatigue damage location of the suction side is between the spar webs towards the mid-span and accumulates at the ply drop and compared to the original blade a slight increase in the number of damage elements is seen. For the blade with defect, a significant decrease in blade fatigue life of approximately 1.56E+09 cycles are observed.

According to Spera [24] under stochastic loading wind turbine blades undergo up between 10^8-10^9 million load cycles during their service life. Wingerde et al. [26] states that under constant amplitude bending moment fatigue tests for wind turbine blades are conducted for 1-5 million. As the fatigue simulations are performed under constant amplitude bending moment, results show that RUZGEM blade without damage satisfies the desired number of load cycles and is not critical in terms of fatigue strength.

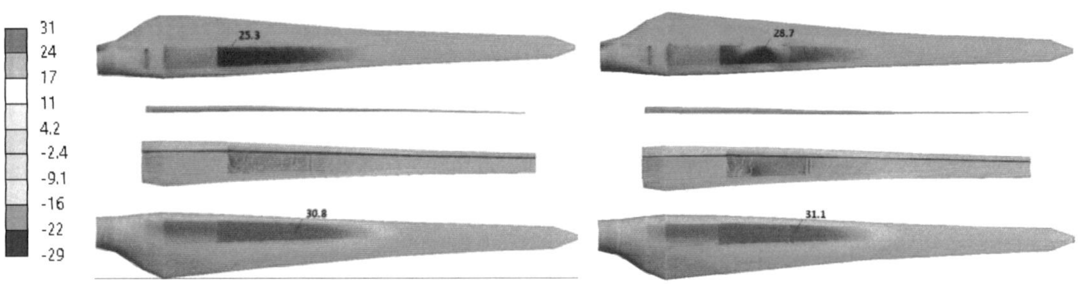

Figure 7. Average axial stress in the pressure side, internal flange, spar and suction side (from top to bottom in a row) of the blade (a) without defect (b) with defect under fatigue flap-wise (min) loading.

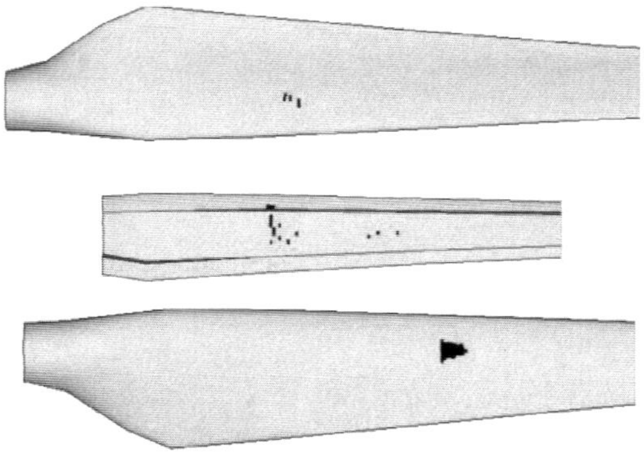

Figure 8. Fatigue damage location on the (a) pressure side (1.69E+09 <Nf <2.01E+09) (b) spar (2.21E+10<Nf<5.20E+10) (c) suction side (3.66E+10<Nf<5.87E+10) of the blade with defect under constant amplitude flap-wise (min) cyclic loading with R=0.1.

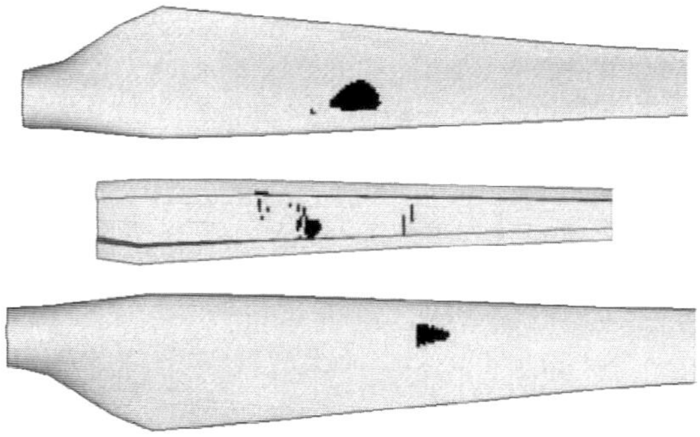

Figure 9. Fatigue damage location on the (a) pressure side (1.31E+08 <Nf <1.99E+09) (b) spar (5.87E+09<Nf<5.20E+10) (c) suction side (3.33E+10<Nf<5.87E+10) of the blade without defect under constant amplitude flap-wise (min) cyclic loading with R=0.1.

4. Conclusions

In this study, static and fatigue strength analysis of the existing 5-m RUZGEM composite wind turbine blade with and without defect is conducted. The main conclusions are as follows:

- Under extreme flap-wise (min) loading, the blade without defect collapses after 85% loading at the root transition region. When the debonding defect is introduced at the pressure side-spar interface, blade collapses after 69% loading and the laminate failure region changes to the defect region, where debonding is modeled.
- The fatigue analysis results show that the blade without defect satisfies the desired number of load cycles and is not critical in terms of fatigue strength. In the blade FE model with defect, stress redistribution due to the debonding defect is investigated and the debonding crack growth is not simulated. Local buckling in the vicinity of the defect leads to a stress redistribution and an increase of 16% in the average stress level, which further causes a drop in the fatigue life of the blade by approximately 1.56E+09 load cycles.
- As the compressive strength of the blade is less than its tensile stress, compared to the suction side a smaller number of permissible fatigue cycles for the pressure side of the blade are determined.

As a future study it may be useful to investigate the effect of defects by analyzing the blade on ply level using more advanced fatigue analysis models, such as physically based pheonomelogical fatigue damage model as explained in [27].

Acknowledgement

This work was partly supported by RUZGEM, METU Center for Wind Energy at the Middle East Technical University.

References

[1] Holmes, J, Sørensen, B, and Brøndsted P 2007 Reliability of wind turbine blades An overview of materials testing *Proceedings of the Wind Power Shanghai 2007* pp. 310–315 Chinese Renewable Energy Industry Association

[2] Chen C P, Kam T Y 2011 Failure analysis of small composite sandwich turbine blade subjected to extreme wind load Procedia Eng. 14 1973-1981

[3] P Brøndsted, H Lilholt A Lystrup 2005 Composite materials for wind power turbine blades Annu. Rev. Mater. Res. 35 505-538

[4] Jensen, F, Falzon, B, Ankersen, J, and Stang, H 2006 Structural testing and numerical simulation

of a 34m composite wind turbine blade Compos. Struc. 76 52 – 61
[5] Overgaard, L C, Lund, E, & Thomsen, O T 2010 Structural collapse of a wind turbine blade. Part A: Static test and equivalent single layered models Compos. Part A Applied Science and Manufacturing 41(2)
[6] Haselbach, P U and Branner, K 2016 Initiation of trailing edge failure in full-scale wind turbine blade test Engineering Fracture Mechanics 162 136-154
[7] Chen, X, Zhao, W, Zhao, X L, and Xu, J Z 2014 Failure test and finite element simulation of a large wind turbine composite blade under static loading Energies 7(4), 2274-2297
[8] Muyan, C, and Coker, D 2020 Virtual full-scale testing for investigating strength characteristics of a composite wind turbine blade Wind Energy Science Discussions 1-28
[9] Marin, J C, Barroso, A., Paris, F., and Canas, J. 2009 Study of fatigue damage in wind turbine blades Engineering failure analysis 16(2), 656-668
[10] Noever Castelos, P, and Balzani, C 2016 The impact of geometric non-linearities on the fatigue analysis of trailing edge bond lines in wind turbine rotor blades Journal of Physics: Conference Series Bristol: IOP Publishing Ltd
[11] Mikkelsen, Lars P 2020 "The fatigue damage evolution in the load-carrying composite laminates of wind turbine blades" Fatigue Life Prediction of Composites and Composite Structures Woodhead Publishing 569-603
[12] Montesano, J, Chu, H, and Singh, C V 2016 Development of a physics-based multi-scale progressive damage model for assessing the durability of wind turbine blades Compos. Struc. 141, 50-62
[13] Zuo, Y, Montesano, J, and Singh, C V 2018 Assessing progressive failure in long wind turbine blades under quasi-static and cyclic loads Renew. Energy 119, 754-766
[14] Lin, J, Xu, Y L, and Xia, Y 2019 Structural Analysis of Large-Scale Vertical Axis Wind Turbines Part II Fatigue and Ultimate Strength Analyses Energies 12(13), 2584
[15] Germanischer Lloyd Guidelines 2010 Rules and Guidelines Industrial Services, Guideline for the Certification of Wind Turbines
[16] Philippidis, T and Roukis, G 2013 Structure design report of METUWIND small rotor Blade, Confidential Interim Report
[17] ANSYS Inc. http://www.ansys.com, Release17.2, 2017
[18] Puck, A and Schürmann, H 1998 Failure analysis of FRP laminates by means of physically based phenomenological models Compos. Sci. and Technol. 58 1045-1067
[19] Weinzierl G and Pechlivanoglou G 2013 Blade Design Loads METU-5.0, Rev. 1.5, SMART BLADE GmbH Internal Technical Specification
[20] IEC 61400-2: Wind Turbines, Part 2 2016 Design requirements for small wind turbines.
[21] Sørensen, B F, Jørgensen, E, Debel, C P, Jensen, F M, Jensen, H M, Jacobsen, T K, and Halling, K 2004 Improved design of large wind turbine blade of fibre composites based on studies of scale effects (Phase 1). Summary Report. Risø National Laboratory, Denmark
[22] IEC 61400-23:2002 Part 23 2015 Full-Scale Structural Testing of Rotor Blades
[23] Philippidis, T, Assimakopoulou, T, and Roukis, G 2013 Static tests on GFRP composites made of METYX Glass NCF, Confidential Interim Report
[24] Paquette, J A, and Veers, P S 2007 Increased Strength in Wind Turbine Blades through Innovative Structural Design (No. SAND2007-2632C). Sandia National Lab.(SNL-NM), Albuquerque, NM.
[25] Wilson, R E, and Spera, D A 2009 Wind Turbine Technology: Fundamental Concepts in Wind Turbine Engineering. ASME, New York
[26] Wingerde, A V. Sayer, F, and Putnam, E 2015 Testing for Certification of Rotor Blades – Today and in the Future, Brazil WindPower 2015 Conference and Exhibition, Rio De Janeiro
[27] Passipoularidis, V A, Philippidis, T P, and Brondsted, P 2011 Fatigue life prediction in composites using progressive damage modelling under block and spectrum loading International Journal of Fatigue 33, 132–144, 2011

AUTHOR INDEX